DESCRIPTIVE GEOMETRY

Prentice-Hall Books by Louis Gary Lamit

Piping Systems: Drafting and Design (1981)
Piping Systems: Drafting and Design Workbook (1981)
Industrial Model Building (1981)
Descriptive Geometry (1983)
Descriptive Geometry Workbook (1983)

DESCRIPTIVE GEOMETRY

LOUIS GARY LAMIT
CADD Department
De Anza College
Cupertino, CA

Illustrations By

John James Higgins
Patrick Joseph Scheetz

Technical Assistance By

Walter N. Brown

PRENTICE-HALL, INC., *Englewood Cliffs, NJ 07632*

Library of Congress Cataloging in Publication Data

Lamit, Louis Gary. (date)
   Descriptive geometry.

   Includes index.
   1. Geometry, Descriptive. I. Title.
QA501.L36 1983   516.2   82-18532
ISBN 0-13-199802-1

Editorial/production supervision: **Karen Skrable**
Manufacturing buyer: **Anthony Caruso**
Cover design: **Ben Santora**
Cover art: *Equinox* by **Jeff Gunion**

© 1983 by Prentice-Hall, Inc., Englewood Cliffs, NJ 07632

*All rights reserved. No part of this book
may be reproduced in any form or
by any means without permission in writing
from the publisher.*

Printed in the United States of America

10 9 8 7 6

ISBN 0-13-199802-1

Prentice-Hall International, Inc., *London*
Prentice-Hall of Australia Pty. Limited, *Sydney*
Editora Prentice-Hall do Brasil, Ltda., *Rio de Janeiro*
Prentice-Hall Canada Inc., *Toronto*
Prentice-Hall of India Private Limited, *New Delhi*
Prentice-Hall of Japan, Inc., *Tokyo*
Prentice-Hall of Southeast Asia Pte. Ltd., *Singapore*
Whitehall Books Limited, *Wellington, New Zealand*

Dedication

To my teacher, **Bill Kwong**
Om Mani Padme Hum

Acknowledgements

The author would like to thank the following for contributions to the text. John J. Higgins and Pat J. Scheetz for the illustrations, which included inking and paste-up of each page. My wife Margaret K. Lamit for illustration paste-up and checking the manuscript.

Walt N. Brown of Santa Rosa J.C. wrote the vector chapter and contributed to the geometric construction, lettering, mining and geology, and problem sections.

NASA photographs were provided through B. Michael Donahoe, Educational Programs Officer, Ames Research Center, Mountain View, California 94035.

CONTENTS

PREFACE		*XIII*
INTRODUCTION		*1*
1	**LETTERING, NOTATION, AND LINEWORK**	*2*
1.1	Descriptive Geometry	*3*
1.2	Notation and Labeling	*4*
1.3	Line and Symbol Key	*5*
1.4	Notation Key	*6*
1.5	Notation	*6*
1.6	Linework	*7*
1.7	Pencil and Lead Selection	*7*
1.8	Linework and Technique	*7*
1.9	Instrument Drawings	*8*
1.10	Line Types	*8*
1.11	Erasing/Keeping the Drawing Clean	*8*
1.12	Lettering	*9*
1.13	Pencil Technique	*9*
1.14	Lettering Forms	*9*
1.15	Scales	*13*
1.16	The Architect's Scale	*13*
1.17	The Engineer's Scale	*13*
1.18	The Metric Scale	*13*
1.19	Scales and Descriptive Geometry	*13*
1.20	Problem Page Setup	*14*
1.21	Problem Specifications	*14*
2	**GEOMETRIC CONSTRUCTION**	*15*
2.1	Geometric Constructions	*16*
2.2	Curved Lines	*16*
2.3	The Bow Compass and Dividers	*17*
2.4	The French Curve	*18*
2.5	Lines	*19*
2.6	Construction of Parallel and Perpendicular Lines	*19*
2.7	Angles and Circles	*20*
2.8	Dividing a Line into Equal Parts	*21*
2.9	Vertical Method	*22*
2.10	Proportional Division of a Line	*22*
2.11	Bisectors for Lines and Angles	*23*
2.12	Polygons	*24*
2.13	Regular Polygons	*25*
2.14	Solving for the Center of a Known Circle	*26*
2.15	Construction of a Circle Through Three Given Points	*26*
2.16	Tangency of a Line and Arc	*27*
2.17	A Line Tangent to a Circle Through a Point	*27*
2.18	Tangent Arcs	*28*
2.19	The Inscribed Circle of a Triangle	*29*
2.20	The Circumscribing Circle of a Triangle	*29*
2.21	To Rectify the Circumference of a Circle	*30*
2.22	The Approximate Rectification of an Arc	*30*
2.23	Ellipse Construction	*31*
2.24	Geometric Forms, Polyhedra	*33*
2.25	Curved Surfaces	*34*
2.26	Warped Surfaces	*35*
3	**PROJECTION**	*36*
3.1	Projection	*37*
3.2	Multiview Projection	*38*
3.3	Orthographic Projection	*39*

3.4	The Six Principal Views	40
3.5	First and Third Angle Projection	41
3.6	The Glass Box and Fold Lines	42
3.7	Line of Sight	43
3.8	Auxiliary Views	44
3.9	Primary Auxiliary Views	45
3.10	Frontal Auxiliary Views	46
3.11	Horizontal Auxiliary Views	47
3.12	Profile Auxiliary Views	48
3.13	Auxiliary View Applications	49
3.14	Secondary Auxiliary Views	50
3.15	Related/Adjacent Views	51

4 POINTS AND LINES 57

4.1	Points and Lines	58
4.2	Views of Points	59
4.3	Rectangular Coordinates of a Point	60
4.4	Primary Auxiliary Views of a Point	61
4.5	Secondary Auxiliary Views of a Point	62
4.6	Lines	64
4.7	Orthographic Projection of a Line	65
4.8	Auxiliary Views of Lines	66
4.9	Principal Lines	67
4.10	Definitions	67
4.11	Level and Vertical Lines	68
4.12	Inclined Lines	69
4.13	Frontal Lines	71
4.14	Horizontal Lines	72
4.15	Profile Lines	73
4.16	Oblique Lines	74
4.17	True Length of Oblique Lines	75
4.18	True Length of a Line	76
4.19	True Length Diagrams	77
4.20	Point View of a Line	78
4.21	Points on Lines	80
4.22	Point on Line by Spatial Description and Coordinate Dimension	81
4.23	Typical Characteristics of Two Lines	82
4.24	Visibility of Lines	83
4.25	Visibility of Lines and Surfaces	84
4.26	Skew Lines	85
4.27	Intersecting Lines	86
4.28	Nonintersecting Lines	87
4.29	Parallelism of Lines	89
4.30	Parallel and Nonparallel Lines (Special Cases)	90
4.31	Construction of a Line Parallel to a Given Line	91
4.32	Perpendicularity of Lines	93
4.33	Intersecting Perpendicular Lines	94
4.34	Nonintersecting Perpendicular Lines	95
4.35	Construction of a Line Perpendicular to a Given Oblique Line	96
4.36	Line Drawn Perpendicular to a Given Line at a Specific Point	97
4.37	Shortest Distance between a Point and a Line (Line Method)	99
4.38	Shortest Distance between a Point and a Line (Plane Method)	100
4.39	Shortest Distance between Two Skew Lines	101
4.40	Shortest Distance between Two Skew Lines (Plane Method)	102
4.41	Shortest Horizontal (Level) Distance between Two Skew Lines	104
4.42	Steepest Connection between a Point and a Line	105
4.43	Shortest connector between Two Lines and through a Given Point	106
4.44	Connector between a Point and a Line at a Specified Angle	107
4.45	Angle between Two Intersecting Lines	109
4.46	Angle between Two Skew Lines	110
4.47	Angle between a Line and a Principal Plane	111
4.48	Bearing of a Line	113
4.49	Azimuth of a Line	114
4.50	Slope of a Line	115
4.51	Grade of a Line	116
4.52	Slope Designations	117
4.53	To Draw a Line Given the True Length, Bearing, and Slope (Grade)	118
4.54	Shortest Line of a Given Slope (or Grade) between Two Skew Lines	119
4.55	Revolution/Rotation	121
4.56	Revolution of a Point	122
4.57	Revolution of a Point about an Oblique Axis	123
4.58	Revolution of a Line	124
4.59	True Length of Line in Horizontal View by Revolution	125
4.60	True Length of a Line by Revolution	126
4.61	Slope of a Line by Revolution	127
4.62	Revolution of a Line about an Oblique Axis	128
4.63	Revolution of a Line about a Horizontal Line (Axis)	129
4.64	Cone Locus of a Line	131
4.65	Construction of a Line at Given Angles to Two Principal Planes	132
4.66	Locus of a Line at Given Angles with Two Principal Planes	133

5 PLANES 136

5.1	Planes	137
5.2	Representation of Planes	138
5.3	Principal Planes	139
5.4	Vertical Planes	140
5.5	Oblique and Inclined Planes	141
5.6	Auxiliary Views of a Plane	143
5.7	Points on Planes	144
5.8	Lines on Planes	145
5.9	True Length Lines on Planes	146
5.10	Locating Parallel and Nonparallel Lines on Planes	147
5.11	Edge View of a Plane	149
5.12	Edge View of Plane by Primary Auxiliary View	150
5.13	True Size (Shape) of Oblique Plane	151
5.14	Views of Circular Planes	152
5.15	Plane Figures on a Given Plane	153
5.16	Circles on Planes	154

5.17	Largest Possible Circle on a Given Plane	*155*
5.18	Strike of a Plane	*157*
5.19	Slope of a Plane	*158*
5.20	Visibility	*160*
5.21	Parallelism of Lines and Planes	*161*
5.22	Parallelism of Planes	*162*
5.23	Plane through a Point Parallel to a Given Plane	*163*
5.24	Perpendicularity of Planes	*165*
5.25	Line Perpendicular to a Plane (Edge View Method)	*166*
5.26	Line Perpendicular to a Plane (Two View Method)	*167*
5.27	Plane Perpendicular to a Line and through a Given Point	*168*
5.28	Plane through a Given Line Perpendicular to a Plane	*169*
5.29	Plane through a Point and Perpendicular to Two Given Planes	*170*
5.30	Shortest Distance between a Point and a Plane	*172*
5.31	Shortest Grade or Slope Line between a Point and a Plane	*173*
5.32	Angle between a Line and a Plane (Plane Method)	*174*
5.33	Angle between a Line and a Plane (Line Method)	*175*
5.34	Angle between a Line and a Plane (Complementary Angle Method)	*176*
5.35	Angle between a Plane and a Principal Projection Plane	*177*
5.36	Angle between Two Planes	*179*
5.37	Angle between Planes (Dihedral Angle)	*180*
5.38	Angle between Two Limited Planes without a Common Intersection Line	*181*
5.39	Revolution of Planes	*183*
5.40	Edge View of Plane Using Revolution	*184*
5.41	True Size of a Plane by Revolution	*185*
5.42	True Shape of a Plane Using Revolution	*186*
5.43	Double Revolution of a Plane	*187*
5.44	True Size of a Plane by Double Revolution	*188*
5.45	Angle between a Line and a Plane by Revolution	*190*
5.46	Angle between Two Planes by Revolution	*191*
5.47	Dihedral Angle by Revolution	*192*
5.48	Revolution of Planes at Specified Angles	*194*
5.49	Revolution of a Plane about a Given Line	*195*
5.50	Restricted Revolution and Clearance	*196*
5.51	Revolution of a Solid	*197*
5.52	Double Revolution of a Solid	*198*
5.53	Piercing Points (Edge View Method)	*200*
5.54	Piercing Point of a Line and a Plane (Edge View Method)	*201*
5.55	Piercing Points (Cutting Plane Method)	*202*
5.56	Piercing Point of a Line and a Plane (Individual Line Method)	*203*
5.57	Piercing Point by Line Extension (Edge View Method)	*204*
5.58	Projection of a Point on a Plane	*206*
5.59	Projection of a Line on a Plane (Cutting Plane Method)	*207*
5.60	Projection of a Line on a Plane (Edge View Method)	*208*

6	**INTERSECTIONS**	**211**
6.1	Intersections	*212*
6.2	Intersection of Two Planes (Edge View Method)	*213*
6.3	Intersection of Two Oblique Planes (Edge View Method)	*214*
6.4	Intersection of Two Planes (Cutting Plane Method)	*215*
6.5	Intersection of Planes (Cutting Plane Method)	*216*
6.6	Intersection of Two Infinite Planes (Cutting Plane Method)	*217*
6.7	Intersection of a Line and a Prism	*219*
6.8	Intersection of a Plane and a Right Prism (Edge View Method)	*220*
6.9	Intersection of a Plane and an Oblique Prism (Edge View Method)	*221*
6.10	Intersection of an Oblique Plane and an Oblique Prism (Edge View Method)	*222*
6.11	Intersection of a Plane and a Right Prism (Cutting Plane Method)	*223*
6.12	Intersection of an Oblique Plane and an Oblique Prism (Cutting View Method)	*224*
6.13	Intersection of a Plane and a Pyramid (Edge View Method)	*226*
6.14	Intersection of a Plane and a Pyramid (Cutting Plane Method)	*227*
6.15	Cylinders	*228*
6.16	Oblique Cylinders	*229*
6.17	Intersection of a Line and a Cylinder	*230*
6.18	Intersection of a Plane and a Cylinder	*231*
6.19	Intersection of an Oblique Plane and an Oblique Cylinder (Edge View Method)	*232*
6.20	Intersection of an Oblique Plane and an Oblique Cylinder (Cutting Plane Method)	*233*
6.21	Cones	*235*
6.22	Intersection of a Line and a Cone	*236*
6.23	Intersection of a Line and an Oblique Cone	*237*
6.24	Conic Sections (Intersection of a Plane and a Cone)	*238*
6.25	Intersection of a Plane and a Cone	*239*
6.26	Intersection of an Oblique Plane and a Cone (Cutting Plane Method)	*240*
6.27	Spheres	*242*
6.28	Intersection of a Line and a Sphere	*243*
6.29	Intersection of a Plane and a Sphere	*244*
6.30	Intersection of a Plane and a Torus	*245*
6.31	Intersection of Prisms	*247*
6.32	Intersection of Two Prisms (Edge View Method)	*248*
6.33	Intersection of Two Prisms (Cutting Plane Method)	*249*

6.34	Intersection of a Prism and a Pyramid	*251*
6.35	Intersection of a Prism and a Pyramid (Edge View Method)	*252*
6.36	Intersection of a Prism and a Pyramid (Cutting Plane Method)	*253*
6.37	Intersection of Cylinders	*255*
6.38	Intersection of Two Cylinders (Not at Right Angles)	*256*
6.39	Intersection of Two Oblique Cylinders	*257*
6.40	Intersection of a Prism and a Cylinder	*258*
6.41	Intersection of a Cylinder and a Prism (Edge View Method)	*259*
6.42	Intersection of Cones	*261*
6.43	Intersection of Two Right Circular Cones	*262*
6.44	Intersection of Two Oblique Cones	*263*
6.45	Intersection of a Cone and a Horizontal Cylinder	*264*
6.46	Intersection of a Cone and a Cylinder	*265*
6.47	Intersection of a Cone and a Vertical Cylinder	*267*
6.48	Intersection of an Oblique Cone and a Cylinder	*268*
6.49	Intersection of an Oblique Cone and an Oblique Cylinder	*269*
6.50	Intersection of a Cone and a Prism	*270*
6.51	Intersection of a Sphere and a Prism	*272*
6.52	Intersection of a Sphere and a Cylinder	*273*
6.53	Pictorial Intersections	*275*
6.54	Pictorial Intersection of a Line and a Plane	*276*
6.55	Pictorial Intersection of Two Planes	*277*
6.56	Pictorial Intersection of Planes	*279*
6.57	Pictorial Intersection of a Plane and a Solid	*280*
6.58	Pictorial Intersection of a Plane and a Solid Object	*281*
6.59	One-Point Perspective (Pictorial Intersection)	*284*
6.60	Fair Surfaces	*286*
6.61	Fairing	*287*

7 DEVELOPMENTS *291*

7.1	Developments	*292*
7.2	Basic Developments	*293*
7.3	Types of Developments	*294*
7.4	Sheet Metal Developments	*295*
7.5	Development of Models	*297*
7.6	Development of a Right Prism	*298*
7.7	Development of a Truncated Right Prism	*299*
7.8	Development of a Truncated and Cutout Right Prism	*300*
7.9	Development of an Oblique Prism	*302*
7.10	Development of a Prism (Top Face and Lower Base Included)	*303*
7.11	Development of an Intersected Prism	*304*
7.12	Development of a Right Pyramid	*306*
7.13	Development of a Truncated Right Pyramid	*307*
7.14	Development of a Truncated Right Pyramid (Top Face Included)	*308*
7.15	Development of an Oblique Pyramid	*309*
7.16	True Length Diagrams	*312*
7.17	Development of a Truncated Oblique Pyramid	*313*
7.18	Curved Surfaces	*315*
7.19	Development of Single Curved Surfaces	*316*
7.20	Development of a Right Circular Cylinder	*317*
7.21	Development of a Truncated Right Circular Cylinder	*318*
7.22	Development of an Oblique Cylinder	*320*
7.23	Development of Intersecting Cylinders	*322*
7.24	Development of an Elbow Joint	*323*
7.25	Development of Cones	*325*
7.26	Development of a Right Circular Cone	*326*
7.27	Development of a Truncated Right Circular Cone	*327*
7.28	Development of an Oblique Cone	*329*
7.29	Development of a Truncated Oblique Cone	*330*
7.30	Development of a Conical Offset	*331*
7.31	Development of an Oblique Cone without Using the Vertex	*332*
7.32	Transition Pieces	*334*
7.33	Development of Transition Pieces	*335*
7.34	Triangulation	*336*
7.35	Development of a Transition Piece by Triangulation	*337*
7.36	Development of a Transition Piece: Circular to Rectangular	*338*
7.37	Transition Piece Development	*339*
7.38	Development of a Convolute Transition Piece	*340*
7.39	Development of a Warped Transition Piece	*341*
7.40	Double-Curved Surfaces	*343*
7.41	Spheres	*344*
7.42	Development of a Sphere (Gore Method)	*345*
7.43	Development of a Sphere (Zone Method)	*346*

8 MINING AND GEOLOGY *353*

8.1	Mining and Topographic Applications	*354*
8.2	Contour Maps and Plan-Profiles	*355*
8.3	Plan-Profiles and Vertical Sections	*356*
8.4	Bearing, Slope, and Grade of a Line	*357*
8.5	Plan-Profile of a Pipeline	*359*
8.6	Cut and Fill for a Level Road	*361*
8.7	Cut and Fill for a Grade Road	*362*
8.8	Cut and Fill for a Roadway and a Dam	*363*
8.9	Cut and Fill for a Curved Roadway and a Dam	*364*
8.10	Mining Applications	*367*
8.11	Mining and Geology Terms	*368*
8.12	Strike, Dip, and Thickness of an Ore Vein	*369*
8.13	Strike, Dip, and Outcrop of an Ore Vein	*370*
8.14	Outcrop and Distances to Ore Vein	*371*

8.15	Intersection of Unlimited Planes (Mineral Strata)	*373*
8.16	Line of Intersection between Two Ore Veins	*374*
8.17	Line of Intersection between Two Overlapping Ore Veins	*375*

9   WARPED SURFACES *377*

9.1	Warped Surfaces	*378*
9.2	Cow's Horn and Warped Cone	*379*
9.3	The Conoid	*380*
9.4	The Cylindroid	*382*
9.5	Hyperbolic Paraboloids	*383*
9.6	Hyperboloid of Revolution	*384*

10   THE HELIX *386*

10.1	Helices	*387*
10.2	The Right Helicoid	*388*
10.3	The Oblique Helicoid	*389*

11   TANGENCIES *391*

11.1	Tangencies	*392*
11.2	Plane Tangent to a Cone	*393*
11.3	Plane Tangent to a Sphere	*394*
11.4	Plane Tangent to a Convolute	*395*

12   SHADES AND SHADOWS *397*

12.1	Shades and Shadows	*398*
12.2	Shades and Shadows: Cylinders	*399*
12.3	Shades and Shadows: Cones and Pyramids	*400*
12.4	Oblique Pictorial Shade and Shadow	*401*
12.5	Isometric Pictorial: Shade and Shadow	*402*

13   VECTORS AND GRAPHICAL STATICS *403*

I	Vector Representations of Forces for Graphical Solutions and Statics Problems	*405*
13.1	Definition of Terms	*406*
13.2	Definitions of Force Systems	*406*
13.3	Definitions of Diagrams	*407*
13.4	Necessary Conditions for Expected Results	*407*
13.5	Coplanar Vector Summation	*408*
II	Coplanar Force Systems	*411*
13.6	Two Forces Acting on an Object	*411*
13.7	Three Forces at a Pin Collection	*413*
13.8	Three Forces Acting on a Beam	*416*
13.9	Three Parallel Forces Acting on a Beam	*419*
13.10	Four or More Forces Acting on a Beam	*421*
13.11	Trusses and the Maxwell Diagram	*424*
III	Noncoplanar Force Systems	*430*
13.12	The Special Case	*432*
13.13	The General Case	*436*

INDEX *441*

PREFACE

This text has been written to provide both the prospective engineer and drafter/designer with a lucid coverage of the field of Descriptive Geometry. It can be used for a one-term technical school class or as a comprehensive text for a two-term college course. The sections on vectors and on mining and geology could be used to supplement offerings in other courses. Those teaching sheet metal layout might find the text applicable to their area as well.

Individuals in industry will find the text useful as a reference book for the many problems encountered out in the field.

A workbook keyed to the text with problems and assignments is also available. Using a workbook will avoid much setup time and possible mistakes in project construction. The text itself has over 80 pages of problems and many alternative assignments for instructors' convenience.

The basics of drafting, lettering, geometric construction, and projection are covered in the first three chapters of the text. These chapters are not meant to be exhaustive presentations on their particular subjects, which are covered in detail in every drafting/drawing text on the market, but are meant as an essential "bare bones" coverage of the mechanics of drawing and lettering. In most cases, courses in this subject have as a prerequisite either high school drafting or one term of engineering drafting. My own experience is that at least 25% of every descriptive geometry class is composed of students who have never even been in a drafting room, let alone to have drawn anything of a technical nature. The beginning chapters will provide these students with the theory and practice of drafting, along with being a review for others who may need to brush up on their lettering or projection skills. The projection section covers in detail the essential elements of orthographic projection, which is the foundation of descriptive geometry.

Chapters 4, 5, 6, and 7, are the basic procedures and practices of descriptive geometry: Points and Lines, Planes, Intersections, and Developments. The last part of the text deals with specific applications: Mining and Geology, Warped Surfaces, The Helix, Tangencies, Shades and Shadows, and Vectors.

The instructor need not cover the material in the same sequence as presented in the text. Many teachers introduce developments before intersections or prefer to present mining and geology after planes. Each page is inclusive, covering a concept or application without stretching to the next page. The instructor could map out a page by page or section by section individualized format to be used to cover just what is necessary for the course. Obviously it would be impossible to cover every page and problem in the text in only one term, so some selectivity is necessary.

A step by step method for many of the basic concepts has been provided in order to facilitate understanding of the most important principals. Space restrictions prevented this procedure from being used throughout.

Problems are introduced every two to six pages at specific intervals determined by concept coverage and problem page restrictions.

The four main chapters—Points and Lines, Planes, Intersections, and Developments—have a quiz at the end of the chapter and a page of test problems. All problems on the last page of these chapters are test problems.

The text has been written so as to be used by those using either SI or English units. Although metrics is a very important aspect of American industry, it is by no means universally advocated in all fields. This author would rather give the instructor the choice as to the unit type and not impose a particular approach as do some texts. Many areas in industry such as piping, structural, and most building trades still use English units almost exclusively. Dimensions in the text are shown in both English and SI units. Problems are not assigned units of measurement but are laid out on grid sheets or are shown large enough to be scaled from the text by dividers. The instructor can assign the unit type and the enlargement.

As has been mentioned, each page covers a specific topic and the text material does not stretch from page to page, but each page is inclusive unto itself. The illustration, explanation, and in many cases a photo are provided on each page and do not extend beyond that page. The large $8 \frac{1}{2}'' \times 11''$ drafting book format reflects the author's belief that much of the misunderstanding and student frustration with this subject matter stems from the size of the current texts on descriptive geometry and not from the subject itself. The text has a drafting book format with full-scale illustrations and large photographs so the student can see the concepts better and can even scale the text and follow along with the steps using dividers.

The illustrations in this book were drawn full scale and then inked; they were not reduced. The notation, labeling, and other descriptive material on the illustrations were phototypeset in Univers Medium and then pasted up on each page. Hand lettering was not used because of possible clarity problems. But in size and layout each illustration still represents what the end result would look like if a student was to complete such an assignment.

Present texts in this field are numerous and in most cases very well done. But the format size, small illustrations, little or no photographs, and complicated, wordy explanations reflect the style and perspective of education in the previsual media age. Many of the existing texts were originally written before 1950 and have been revised as time required.

We have tried to present descriptive geometry so that the present day student who has been saturated with the visual side of life can more readily comprehend this interesting but many times frustrating subject. Over the years I have taught descriptive geometry to four-year college, technical school, and junior/community college students and have found without exception that reading the text is not the primary means of learning this subject, or for that matter drafting in general. The person raised on TV, computer games, movies, and calculators many times is not willing to, or sadly in some cases, is not capable of learning from the written word, as opposed to some combination of media, photographs, illustrations, and personal teacher contact. This is not to say that all who take descriptive geometry are the same, but any instructor can relate to the frustrating situation where a student asks a simple question (and one that is explained very clearly in the text) and the instructor looks at the student and says "It's in the book. Did you read the book?"

"Uh well, not yet, but I guess that I should, I usually just get what I need from the picture."

This book was written with that student in mind. In many cases the "picture" will tell the story and the words are there to help. Not that the wording is unimportant but there has been a concerted effort to simplify the text discussion and relate it to the given page illustration and photograph. Each illustration is drawn full scale and has as many notes and notations as possible, in order to describe the problem and process of solution.

The author will give careful consideration to feedback from the student or instructor using the text. Suggestions, ideas, criticisms, and corrections are encouraged.

Louis Gary Lamit

DESCRIPTIVE GEOMETRY

INTRODUCTION

Descriptive Geometry is the use of orthographic projection in a manner that will solve three dimensional problems by a graphical two dimensional procedure. The user is allowed to subtract from a specific problem everything but the essential points, lines, or planes and therefore, to more simply find a solution by a graphical method. Most solutions to descriptive geometry problems require the use of specific methods of projection in order to establish the true length or point view of a line, angle between lines or surfaces, true shape of a surface, clearances, etc. All of descriptive geometry is based on orthographic projection and its underlying principles.

A developed sense of visualization is essential for the mastering of this subject. This skill can be cultivated by a step by step presentation of the theory and practice of projection as is provided in the text. The designer encounters many situations where visualization and a mastery of the principles of projection allows for the solution of complex engineering and technical problems. The ability to analyze a specific problem, visualize its spacial considerations, and translate the problem into a viable graphical projection is essential for the designer.

The text is heavily illustrated and exhibits a variety of photographs which introduce applications, actual industry uses, and end results of the use of descriptive geometry. Practical industrial applications include sheet metal layout, piping clearances, intersection situations in mechanical and structural engineering, topographical, mining and geology problems, and vector analysis. Every applied field of engineering: aerospace, automotive, power, petrochemical, etc., makes use of the concepts and applications of descriptive geometry. There are few lines drawn in engineering work that do not apply some aspect of projection and the basic geometry of engineering.

Traditionally descriptive geometry has been the most difficult of engineering drawing courses. With perseverance and effort it is possible to view this subject as the most satisfying and useful of all drafting courses. The text slowly leads the user through each new concept, method, application, and problem, building on the previous section and providing many illustrations and explanations. Some who have had previous experience or are quick learners might feel that the first four or five chapters dwell on certain aspects too much. For these fortunate ones it is obviously possible to cover some aspects lightly and move on. For the rest of us who are not so blessed, this approach will hopefully fill the understanding gaps that are frequently experienced by descriptive geometry students.

Many problems are provided at the ends and throughout each chapter. The instructor should choose problems as required by the course requirements and perspective. Since the problems are printed on a grid paper students can sketch many of the simpler ones right in the text.

Each of the four main chapters ends with a quiz and a problem page. Students are encouraged to use shading and a variety of colored pencils to differentiate between the types of lines and surfaces encountered. Instructor's permission should be requested before using this method.

The text consists of 13 chapters, an index, and a comprehensive contents page for easy location of a particular method or procedure. Chapter sequence is by no means the only possible ordering of the subject matter. Instructors are encouraged to experiment with different possibilities. It is suggested that the first five chapters be presented in their given order, since they are the building blocks of the subject.

Please note that many of the point subscripts have been eliminated or reduced in the later chapters so as not to clutter the more complicated illustrations or where they should be obvious by that stage of the subject material.

1
LETTERING, NOTATION, AND LINEWORK

1.1 DESCRIPTIVE GEOMETRY

Descriptive geometry problems are drawings that are used to solve for specific information using a graphical medium. Linework, lettering, and drawing standards are no less important here than for mechanical, architectual, piping, or other forms of drafting. Lettering and notation are the primary means of communication on drawings. No matter how accurate and precise the drawing, if it is poorly lettered and inadequately labeled it cannot communicate its solution or present its ideas properly. Therefore, concise well-formed lettering, properly positioned notes, and sufficient labeling are essential to the solution of a descriptive geometry work sheet or drawing.

Many students feel that, since this subject is not a formal drafting/design field, light lines, poor lettering, and a sketchlike appearance are adequate. This could not be further from the truth. In reality, because of the inherent accuracy problems associated with graphical solutions to engineering problems, all linework, measuring, and lettering must be of extremely high quality. Descriptive geometry is not only a means to communicate a particular aspect of a technical problem, it is the actual solution in a graphical form. The descriptive geometry worksheet/drawing is equivalent to the final answer in numerals when using a mathematical method.

This chapter will introduce the reader to the form that will be used to construct descriptive geometry solutions to typical engineering problems. Page 4, Notation and Labeling, exhibits a typical descriptive geometry drawing using the special language and notation that has been developed for this subject. It is not necessary for the student to understand this page completely at this stage but it *is* essential that the format, symbols, and notation become part of the students' technical vocabulary as they progress through the text. Frequent referrals to this page as you read the text will reinforce this new language.

Page 5 presents a Line and Symbol Key that will define the types and thickness of the lines and symbols used in descriptive geometry. Many of the line weights and line types are similar to those found in mechanical/engineering drafting. Only the fold line (reference line) and development element (bend) line should be unfamiliar to the student. The fold line is used to divide each view and to establish a reference from which to take dimensions when projecting from view to view. This line is probably the most important and often used and will be explained in detail later in the text. The development element is used extensively when developing curved surfaces and for triangulation of surfaces and will be explained in Chapter 7. All line types, symbols, abbreviations, and notation are explained in detail as one progresses through the text.

1.2 NOTATION AND LABELING

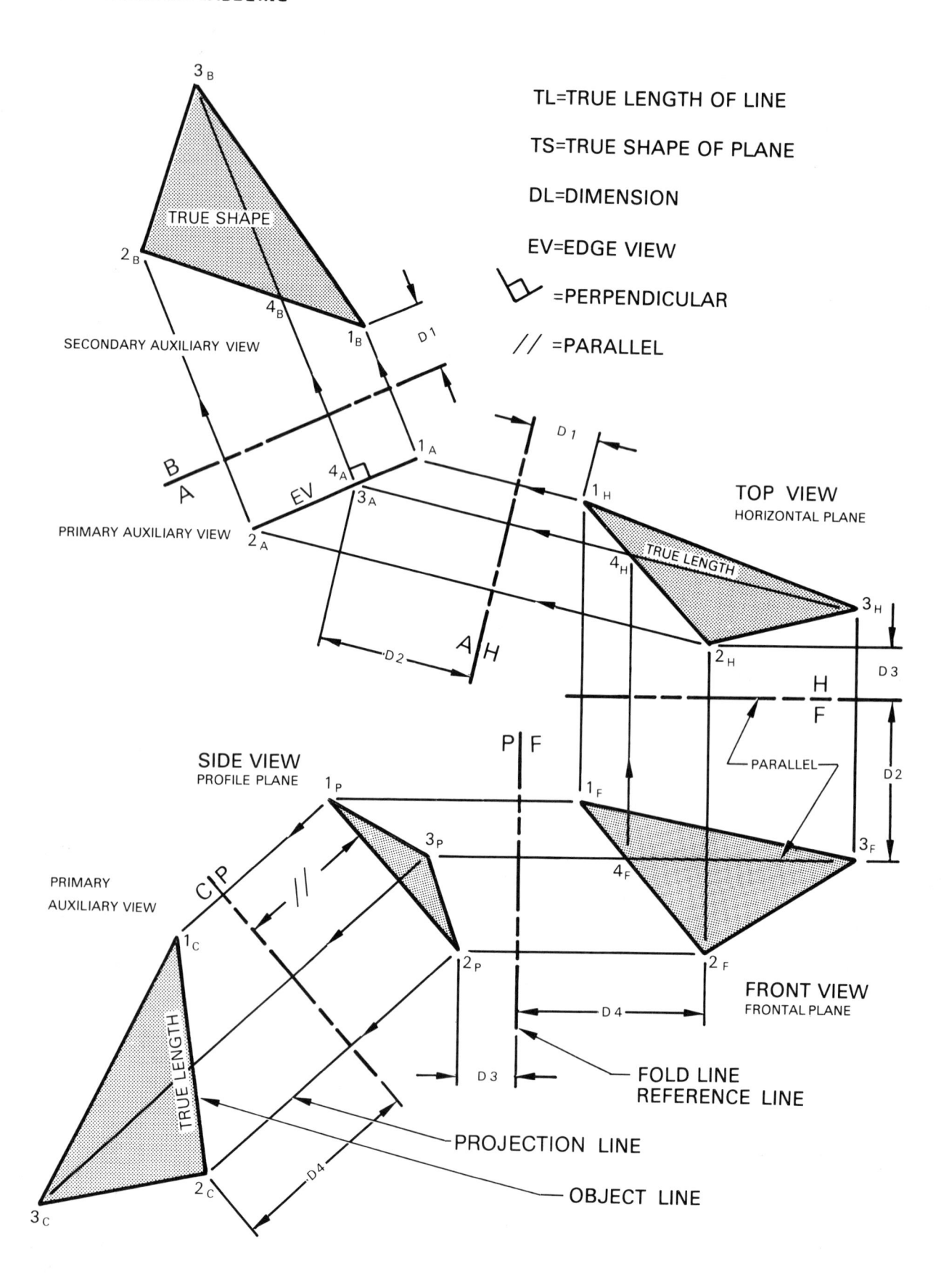

TL = TRUE LENGTH OF LINE
TS = TRUE SHAPE OF PLANE
DL = DIMENSION
EV = EDGE VIEW
⦜ = PERPENDICULAR
// = PARALLEL

1.3   LINE AND SYMBOL KEY

Border Line (extra heavy)

Cutting Plane Line (heavy)

Visible Object Line (medium)

Hidden Object Line (medium)

Fold Line/Reference Line (medium)

Break Line (medium)

Dimension Line (thin)

Extension Line/Projector Line (very thin)

Development Element (very thin)

Parallel Lines

Perpendicular Lines

Piercing/Intersection Point

Edge View

True Length Line

True Size/True Shape

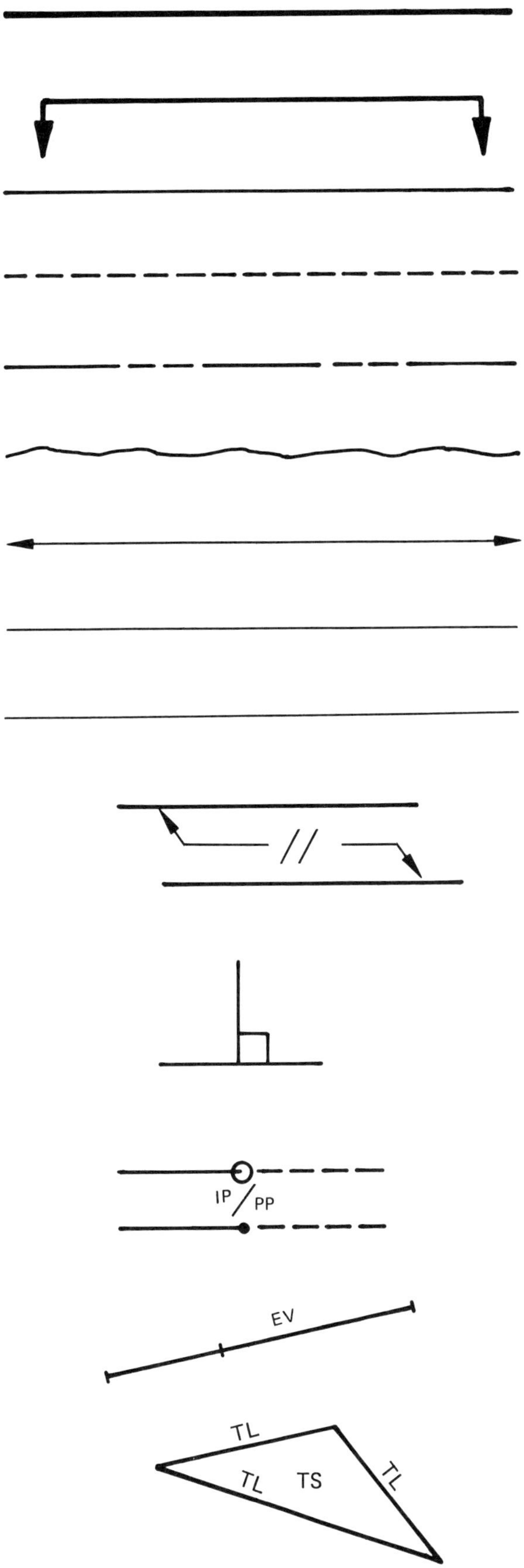

KEY

EV = Edge View

IP = Intersection Point

PP = Piercing Point

PV = Point View

TL = True Length

TS = True Shape

TS = True Size

D = Dimension

H = Horizontal View

F = Frontal View

A,B = Auxiliary Views

P = Profile View

1, 2, 3, 4, etc. = Points

H, F, P, A, B, etc. = View identification

$1_H, 2_F, 3_P, 4_A$, etc. = View subscript

$1^1, 2^1, 3^2, 4^2$, etc. = Superscript

$1_R, 2_R, 3_R, 4_R$, etc. = Revolved points

$1_S, 2_S, 3_S, 4_S$, etc. = Shade/shadow points

1.5 NOTATION

The notation key at the left gives the abbreviations and notational elements used throughout the text. EV is the edge view of a plane. IP has been used to mean the intersection of a line and surface, whereas PP is the piercing point of a line (that is, part of a plane) and another surface; theoretically, IP and PP are the same. PV is the point view of a line. True shape and true size mean the same thing and are abbreviated as TS. TL is the true length of a line. D has been used to note a dimension.

H, F, and P are used to identify the three primary views in orthographic projection: horizontal, frontal, and profile. A will always be the first auxiliary view on a problem, followed by B, C, D, etc.

Whole numbers 1, 2, 3, 4, 5, 6, etc. establish points in space. They can be points, or determine the extent of lines, planes, or solids. In a few cases capital letters are used as points for clarity.

Subscripts establish the view that a point is in, such as 2_H, which means point 2 in the H (horizontal) view.

Superscripts are used where an aspect of a point appears more than one place in a view, as when a line of a prism is called $3\text{-}3^1$, or where for clarity the piercing point of a line is noted as an aspect of the original point, e.g., 2^1.

A point number followed by an R establishes a point in its revolved position, e.g., 3_R. Points followed by an S are the extended points of the intersection of a light ray and a point, used to define a shade or shadow of an object, e.g., 6_S.

After reading the text and completing a few of the problems these notations will become second nature and enable the user readily to label, notate, and communicate using descriptive geometry and its specialized language.

1.6 LINEWORK

Drafting skills associated with all forms of engineering and technical work are essential when using the method of descriptive geometry to solve problems. An engineer's, designer's, or drafter's graphical calculations must be neat, clean, and accurate in order to convey the proper message. In many cases the drawing will need to be reproduced by blueprinting, photocopying, or other methods. Lettering and linework must be dark and of high quality. The following is a short overview of the basics of drafting and will provide the reader with a set of guidelines, procedures, and techniques for developing high quality drawing skills. All assigned problems are to be completed using the proper line weights, notation, labeling, and projection techniques provided in this chapter and throughout the text.

1.7 PENCIL AND LEAD SELECTION

Drafting pencils are graded according to the hardness of their lead. The hardness of the lead determines the quality and shade of the line that can be drawn. A hard lead can make a very sharp, accurate, and thin line, which is quite beneficial when doing descriptive geometry calculations that will not need to be reproduced. Hard leads do not reproduce well in most cases. A soft lead will make dark lines but they are difficult to keep sharp. The "H" grades are, from hardest to softest: 9H, 8H, 7H, 6H, 5H, 4H, 3H, 2H, H, F, and HB. The recommended hardness of lead for descriptive geometry drawings, using a good grade of drafting paper, are 4H for layout and construction of linework, 2H for reproduceable lines, and H for lettering. These three hardnesses of lead combined with proper drafting techniques will produce excellent drawings having sharp, dense lines and dark, readable lettering that will make good prints. For descriptive geometry calculations in industry where only the solution is important, not the drawing itself, a hard lead like 3H or 4H will be sufficient for the whole drawing.

1.8 LINEWORK AND TECHNIQUE

All drawings consist of a combination of lines that are intended to convey an idea from the drafter to the user. The control drafters have over their pencils and the techniques determine the quality of the drawings produced. Serious drafters/designers/engineers constantly strive to improve technique through practice and attention to detail for as long as they use graphical methods of communication.

A properly drawn line will be uniform for its entire length. This can be accomplished by the application of two techniques.

1. Incline the pencil so that it makes an angle of about 60° with the surface of the paper and then pull in the direction that it is leaning.
2. Rotating the pencil slowly as the line is drawn and maintaining a semisharp point enables the drawer to control the thickness and quality of the line.

These techniques take practice but will soon become automatic and lines will be uniform from end to end and from one line to another.

The printability of these lines is determined by their density; that is, how dark they are. Density is controlled first by the hardness of the lead and second by the pressure applied while drawing the line. The width and sharpness of the line is determined by the size of the point touching the paper. A pencil point must always be smoothed and rounded on scratch paper after being repointed. It can also be resharpened on scratch paper. Fine line pencils are also widely used; these excellent lead holders are available in different lead thicknesses. A .5mm or .7mm H or 2H lead works well for lettering and object lines; .4mm can be used for construction. The beauty of these instruments is that they require no sharpening and help maintain a high quality, consistently uniform line. Drawbacks include: A different pencil must be purchased for each line thickness required and the thinner leads tend to break often when using soft grades.

Using a dust pad or erasing powder will help keep linework and the paper clean and unsmudged. Frequent brushing and using an extra piece of paper to rest the hand on while drawing also helps.

1.9 INSTRUMENT DRAWINGS

All problems for descriptive geometry are to be drawn using straight, sharp lines with the aid of triangles and some form of straightedge (T-square, drafting machine, parallel bar). As with all other forms of drafting, descriptive geometry drawings are to be considered instrument drawings and are not to be constructed using anything except high quality drafting instruments. All lines are to be drawn using a proper technique and utilizing the correct instrument. Vertical lines are to be constructed using a straightedge and triangle, horizontal lines are to be constructed using a straightedge that will give consistent parallel lines. Curved lines are to be plotted and then drawn with the aid of a compass, template, or French curve. No lines in this subject are to be formed without instruments, or freehand, except lettering.

The drawing of an instrument line is a two-step process. First the position and length of the line is determined using dividers or a scale, and then the line is drawn with correct width and density. This process requires that two completely different lines be drawn. The first line is for positioning and is thin and gray, using 4H lead. It is suggested that each descriptive geometry problem be completed in total using this type of line. This will insure that all dimensions and measurements will be taken from thin, sharp, accurate points and lines, therefore creating a more precise and correct drawing. After the problem is complete, a second type of line is required. This second line is drawn uniform, thicker, and denser using 2H lead. It will be necessary to darken only object lines, points, fold lines, dimension, and extension lines. Lines used for construction purposes can in most cases be eliminated or left light. The student should study each example of a problem as it is shown in the text in order to establish which will be construction lines and which will need darkening so as to be printable.

1.10 LINE TYPES

Printable lines in descriptive geometry are drawn with different widths to provide specific information. In reality each line type is a symbol and will express an idea or communicate a special situation. On page 5 a line key is given. These lines are divided by thickness.

1. Fine lines: Thin black lines are used to provide information about the drawing or used in its construction. These include dimension lines, construction lines, extension lines, and center lines.

2. Medium lines: Intermediate black lines that are used to outline and represent planes, lines, solids, etc. Medium lines are also used to represent hidden object lines, fold lines, and break lines.

3. Heavy lines: Solid, thick black lines are used for the border, and to represent cutting planes where they will section an object.

The one thing that all of these lines will have in common is that they are all *black*, with the exception of temporary construction lines.

1.11 ERASING/KEEPING THE DRAWING CLEAN

Erasing is a necessary part of drafting and, when done properly, improvements to drawings are easily made. The eraser should have good "pick up" power without smudging, such as an Eberhard Faber Pink Pearl. To protect adjacent areas that are to remain, all erasing is done through the perforations of a stainless steel erasing shield. The erasing shield is held firmly in place on the drawing with one hand while erasing through a particular slot or hole with the other. Care must be taken not to erase other areas through adjacent openings. After each erasing, the drawing should be brushed so as not to grind erasing particles into the drawing. Electric erasers are excellent for nonplastic papers. All drawings attract dirt and the amount is in direct proportion to the habits of the drafter. Cleanliness does not just happen; it must be consistently cultivated over the years as correct habits. Procedures that will help to keep drawings clean include the following:

1. Clean hands: A periodic washing of the hands is necessary to remove accumulations of graphite, dirt, perspiration and body oils.

2. Equipment: All tools that come in contact with the drawing require a periodic washing with soap and water. Tools that contain wood should be cleaned with a damp sponge. Drawing boards must be scrubbed down when they become soiled.

3. Graphite: Most "dirt" on a drawing is actually graphite from the chalked dust resulting from line drawing and from the lines themselves. Constant use of the desk brush to remove this graphite dust, before other tools smear it around, will contribute significantly to clean drawings.

4. Pencil pointer: The pencil pointer will leave dust clinging to the lead after sharpening. It is suggested that after sharpening the lead that it is wiped with tissue or poked into an erasing dust pad. Of course do not use the same dust pad to clean the drawing!

5. Equipment use: Proper use of the straight edge and triangles will always place these instruments between the hands and the paper, except when lettering. While lettering, use a clean sheet of scratch paper on which to rest the hands. Even clean hands will put body oils onto the paper and this has a magnetic effect on dirt.

1.12 LETTERING

Line drawings are never complete until additional information is given in the form of "notes," dimensions, notation, and titles. This information is carefully hand lettered using a freehand single-stroke gothic alphabet. This gothic alphabet does not have short bars or serifs at the ends of strokes, as does the roman alphabet. Legibility is the first requirement for any lettering, followed by ease of printing and speed. All drafters, designers, and engineers must master the art of freehand lettering through studying the form of letters, the direction of strokes, and practice. The importance of good lettering cannot be overemphasized. The lettering can "make or break" an otherwise excellent drawing. In descriptive geometry sloppy or misplaced lettering can completely change the final calculation or answer.

Unlike typical drafting projects, descriptive geometry problems could not, in most cases, be constructed without notation. For many problems there are so many points, lines, and planes that without the labeling of most or every point it would be quite impossible to complete the project. From the beginning it is very important to establish the habit of clearly identifying each point in every view, labeling every fold line, and providing view subscripts for each point number. Noting every situation where a line appears as a point view (PV), or is true length (TL), a plane as an edge view (EV) or as true shape (TS), will prevent mistakes and expedite the answer.

For all problems in the text the following lettering guidelines should be used.

1. Label all points using numbers 1/8" (3.17mm) in height.
2. Use 3/32" (2.3mm) subscripts for view identification of each point.
3. Label all situations where part of the problem will appear as EV, TS, PV, PP, IP, TL, etc., using 1/8" (3.17mm) lettering.
4. Use problem examples in the text for information as to what other notes, labels, or dimensions may be necessary; use 1/8" (3.17mm) lettering in most cases.
5. Use 3/32" (2.3mm) lettering for R (revolved point identification), S (shade or shadow point notation), superscripts, and subscripts.
6. Fold line identification can be slightly larger than that used for points and general notes. Always use capital letters as shown throughout the text.

1.13 PENCIL TECHNIQUE

Freehand lettering places a requirement on linework that is different from what is possible with instrument lines. Instrument lines are drawn by going over the line to increase its density. It is impossible consistently to retrace freehand lines; therefore, they must be put down with the proper density using only one stroke. To help get the proper density, one grade softer lead is used, the H lead. This softer lead also contributes to the "dirt" on the drawing, as it chalks more easily. Frequent use of the desk brush is necessary. Due to its tendency to smear, lettering is usually the last step in the completion of a drawing, but since in descriptive geometry problems the lettering is essential to the construction it must be added as the drawing is in progress. Therefore 2H lead can be used for the necessary problem notation.

The preparation of the pencil point is the same as would be used to draw a medium line. The strokes of letters need to be kept consistent in both width and density. The density of the lettering should approach the density of the linework. Obvious variation in the densities between lettering and linework should be avoided.

Freehand lettering for words and notes is to be completed with the aid of guide lines at the top and bottom of the letters (except fold line notation, subscripts, superscripts, and other point identification that will usually be only one character long). Guidelines are the same as construction lines —very thin, sharp, and gray and made with 4H lead. All lettering is done with capital letters and in most cases whole numbers. Therefore only two lines placed $\frac{1}{8}''$ (3.17mm) apart are necessary.

1.14 LETTERING FORMS

On page 10, Fig. 1-1, an alphabet of vertical and inclined letters and numerals is shown, along with space for practicing each character. Guidelines are shown thicker and darker than is necessary for actual descriptive geometry problems. After completing this short assignment compare it to Figure 1-2 on pages 11 and 12. Figure 1-2 gives a set of common mistakes, comments, and notes concerning typical lettering formation problems. It is very important to catch any bad habits early, in order not to ingrain them in your lettering style. Please note that in the beginning it is important to eliminate any individualized style until your lettering becomes clear, concise, dark, and well formed. Later, through use and practice a new, more attractive personal style will emerge and become obviously "yours."

FIG. 1-1 Lettering forms.

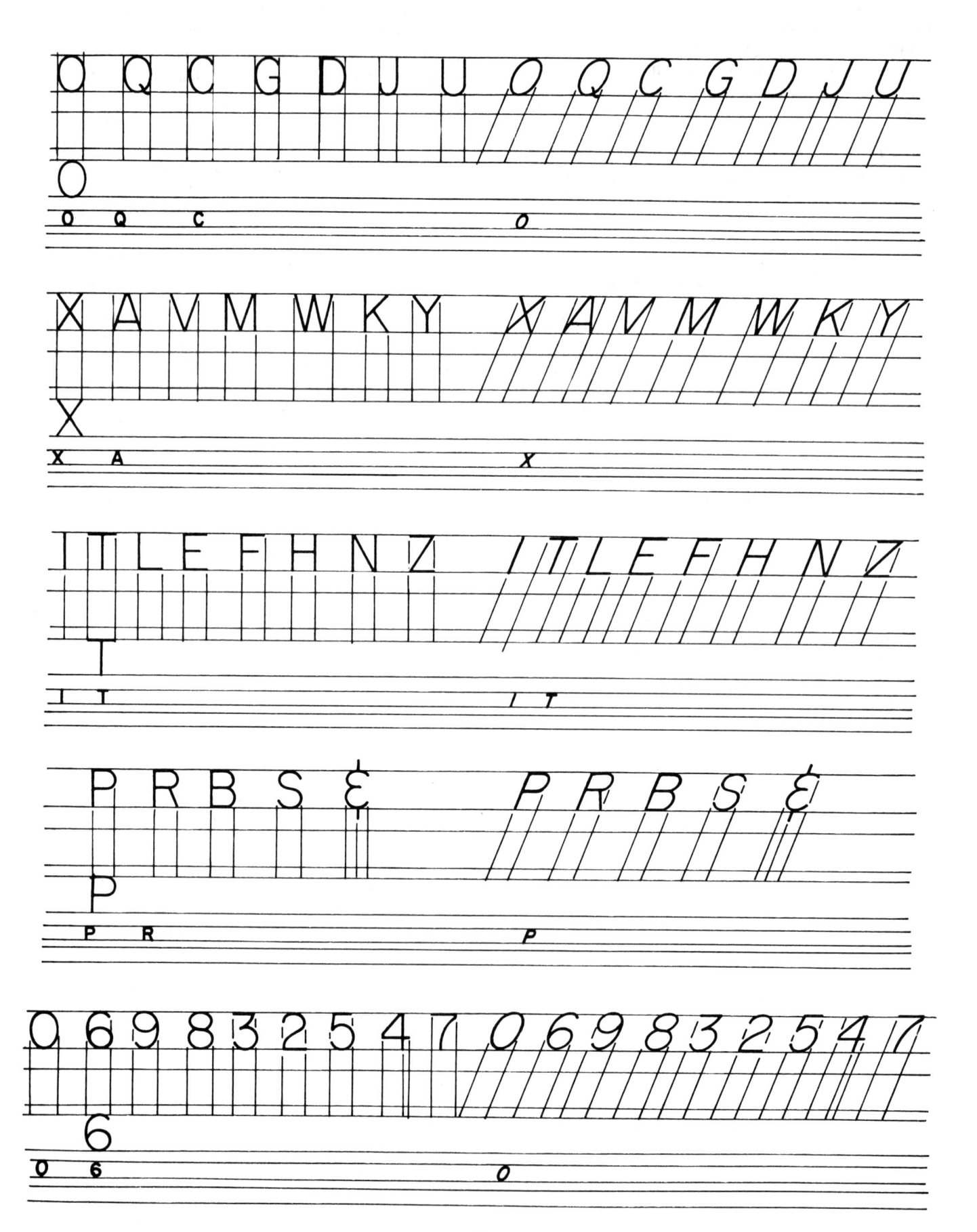

FIG. 1-2 Lettering problems.

Number or Letter		Notations and Comments	Wrong	Possible Mistakes
A	A	Make upper part larger than bottom.	A	
B	B	Lower part slightly larger than upper part.	B	8
C	C	Full open area, elliptical letter body.	C	O
D	D	Horizontal bars and straight line back.	D	O
E	E	Short line bar slightly above center.	E E	L
F	F	Short bar slightly above centerline.	F F	T E
G	G	Based on true ellipse, short horizontal line above centerline.	G G G	C O 6
H	H	Bar slightly above centerline.	H H	
I	I	No serifs, except when next to Number 1.	I 1	
J	J	Wide full hook with no serifs.	J J	
K	K	Extend lower branch from upper branch.	K K	R
L	L	Make both lines straight.	L	
M	M	Not as wide as W: center part extends to bottom of letter.	M M	
N	N	Do not cram lines together.	N N	V U
O	O	Full true ellipse.	0 O	Q
P	P	Middle bar intersects at letter's middle.	P P	K T D
Q	Q	Based on true wide ellipse.	Q	O
R	R	Make upper portion large.	R R	K
S	S	Based on number 8; keep ends open.	S	8

FIG. 1-2 Lettering problems (continued).

Number or Letter		Notations and Comments	Wrong	Possible Mistakes
T	T	Draw full width of letter E.	T T	7
U	U	Lower portion elliptical, vertical bars parallel.	U	V
V	V	Bring bottom to points.	U	U
W	W	Widest letter; center extends to top of letter.	W W	N
X	X	Cross lines above centerlines.	X	
Y	Y	Upper part meets below center.	Y Y	V T
Z	Z	Horizontal lines parallel.	Z	2
O	O	Same as letter O.	O	Q
I	I	Same as letter I.	I I	7
2	2	Based on number 8; open hook.	2 2	Z
3	3	Based on number 8; upper part smaller than lower.	3 3	8 5
4	4	Horizontal bar below center of figure.	4 4 4	7 9 H
5	5	Based on ellipse; keep wide.	5 5	6 3 S
6	6	Based on ellipse; open.	6	8
7	7	Keep as wide as letter E.	7 7 7	1
8	8	Based on ellipse; keep wide.	8 8 8	B
9	9	Composed of two ellipses; keep full.	9	8

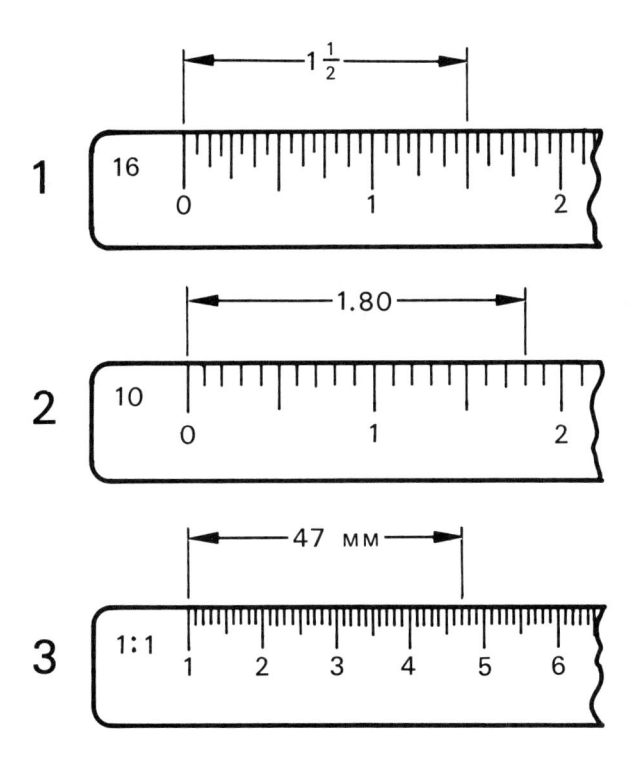

1.15 SCALES

All instrument drawings are drawn accurately in size so that each aspect of the object or other graphical form is shown in proper proportion. The size of the drawing may be the full size of the shape or it may be a reduced size. A drawing that is accurately drawn in this manner is said to be drawn "to scale." To make measurements that are accurate and correct it is necessary to use a high quality scale. Scales are instruments that are designed and made to represent accurately specific units of measurement. The scale is a precision instrument and, with proper use, will help produce consistent drawings.

There are four basic scales available: architect's, engineer's, metric, and mechanical engineer's. The engineer's scale, Fig. 1-3(2), is used to draw very large projects: earth works, roads, and surveys of property. Chapter 9, Mining and Geology, will require the use of this type of scale. The architect's scale, Fig. 1-3(1) is used to make drawings of buildings and structures. Metric scales, Fig. 1-3(3), are used for all types of projects and are easily adaptable to all forms of technical and engineering work. The basic shape and size of these three scales is triangular and about 12 inches long. The triangular shape makes six surfaces available for the different sized scales. Specialty flat scales are also available, such as the mechanical engineer's scale, which is two-sided. One side is the full inch scale divided into decimal units of .10 inches or as many as 50 divisions (every .2 inches). The opposite side is at half scale.

The markings on the scales are arranged in two ways, fully divided and open divided. The fully divided scale has each main unit throughout its length completely divided into specific units, as in the engineer's and mechanical engineer's scales. The open divided scale has each main unit of the scale undivided and an extra main unit is fully divided at the "0" end. The architect's scale is open divided with the exception of the full scale side, Fig. 1-3(1).

1.16 THE ARCHITECT'S SCALE

The *architect's* scale has a "foot ruler" full-size scale, Fig. 1-3(1), and 10 different reduced size open divided scales. The open divided scales are used to permit these 10 different scales to be placed on the remaining five sides. All scales on a architects scale are divided in feet and inches. The open divided scale has only full one foot units reading in one direction from the "0" and has a fully divided one foot unit reading in the opposite direction. Therefore, the number of feet is read along the length of the scale and the number of inches is read in the fully divided unit at the "0" end of that same scale, both numbers becoming larger as the distance from the "0" becomes greater. Each scale is identified by the number or fraction at the "0" end, and indicates the unit of length in inches that represents one foot of real size.

1.17 THE ENGINEER'S SCALE

The *engineer's* scale, Fig. 1-3(2), has six scales that are equally divided. These scales are 10, 20, 30, 40, 50, and 60 divisions per inch and are numbered at each 10th division along the length of each scale. The number of divisions per inch is marked at the "0" end of each scale. The user can assign any units needed to the divisions of these scales, although the usual usage is that each division equals one foot.

1.18 THE METRIC SCALE

Standard International (SI) units are measured with a *metric* scale. Figure 1-3(3), shows a 1:1 metric scale. This full-size scale is fully divided into major units of centimeters and smaller units of millimeters. There are 10 millimeters in each centimeter. Metric units in full scale, reductions, and enlargements are used in all forms of engineering work.

1.19 SCALES AND DESCRIPTIVE GEOMETRY

This text has been designed to be used with all types of scales. In most cases where units of mea-

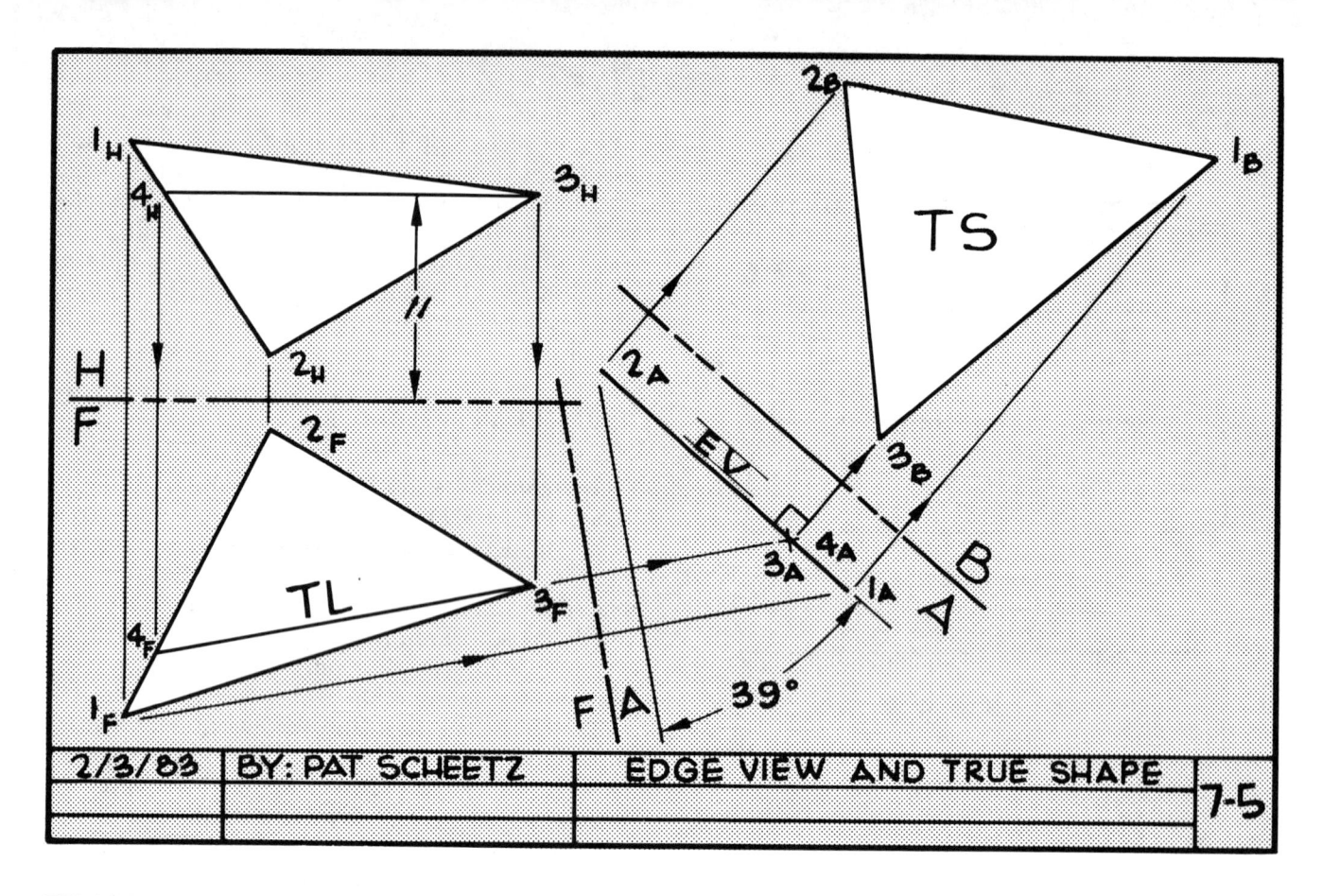

FIG. 1-4 Problem page setup.

surement are given they are shown in both English and SI units. Most problems in the text are undimensioned, allowing the instructor to choose the unit of measurement. Use of all forms of scales is suggested, alternating among the four types. Descriptive geometry drawings can be in any existing unit of measurement.

1.20 PROBLEM PAGE SETUP

Problems are presented on grid systems that represent an area of 8″ × 10″ with a $\frac{1}{4}$″ grid. The student is to transfer these to $8\frac{1}{2}$″ × 11″ drafting vellum with a $\frac{1}{4}$″ grid system. Establish a $\frac{1}{4}$″ border on all four sides, and a $\frac{1}{2}$″ title block as shown in Figure 1-4. Note the date, your name, the name of the problem, and the problem number. Follow all given instructions as presented in the text; do not deviate unless given alternative instructions by your instructor.

Problem pages are provided throughout the text at frequent intervals. At the end of the four main chapters a page of test problems is provided along with a short quiz to test retention of the subject matter.

1.21 PROBLEM SPECIFICATIONS

The following will provide a series of specifications for the completion of the texts problems:

1. Draw all lines black and sharp. Lines representing points, lines, and planes must be dense and black in order to print well.

2. Letter with $\frac{1}{8}$″ (3.17mm) high letters, using guide lines except for isolated numbers and letters.

3. Complete all views used in the problem solution. Project solutions solved for in auxiliary views back into preceding views.

4. Label all fold lines and points. Show view subscripts for all points unless otherwise instructed.

5. Label revolved points (R) and shade or shadow points (S).

6. Label all true lengths (TL), true shapes (TS), edge views (EV), point views (PV), piercing points (PP), intersection points (IP), true angles (TA), etc.

7. Note with the appropriate symbol parallel and perpendicular situations.

8. Show all important extension lines, dimension lines, and centerlines. Do not show arrow heads on extension/projector lines unless instructed to do so.

9. Complete all of the problem lightly before darkening lines, and label points as the problem is being constructed.

10. Space problems for optimum clarity and use the given space efficiently by choosing the proper view from which to project auxiliary views.

11. When acceptable, use colored pencils for clarity on complicated problems.

12. Shade planes and intersections with instructor's permission.

2

GEOMETRIC CONSTRUCTION

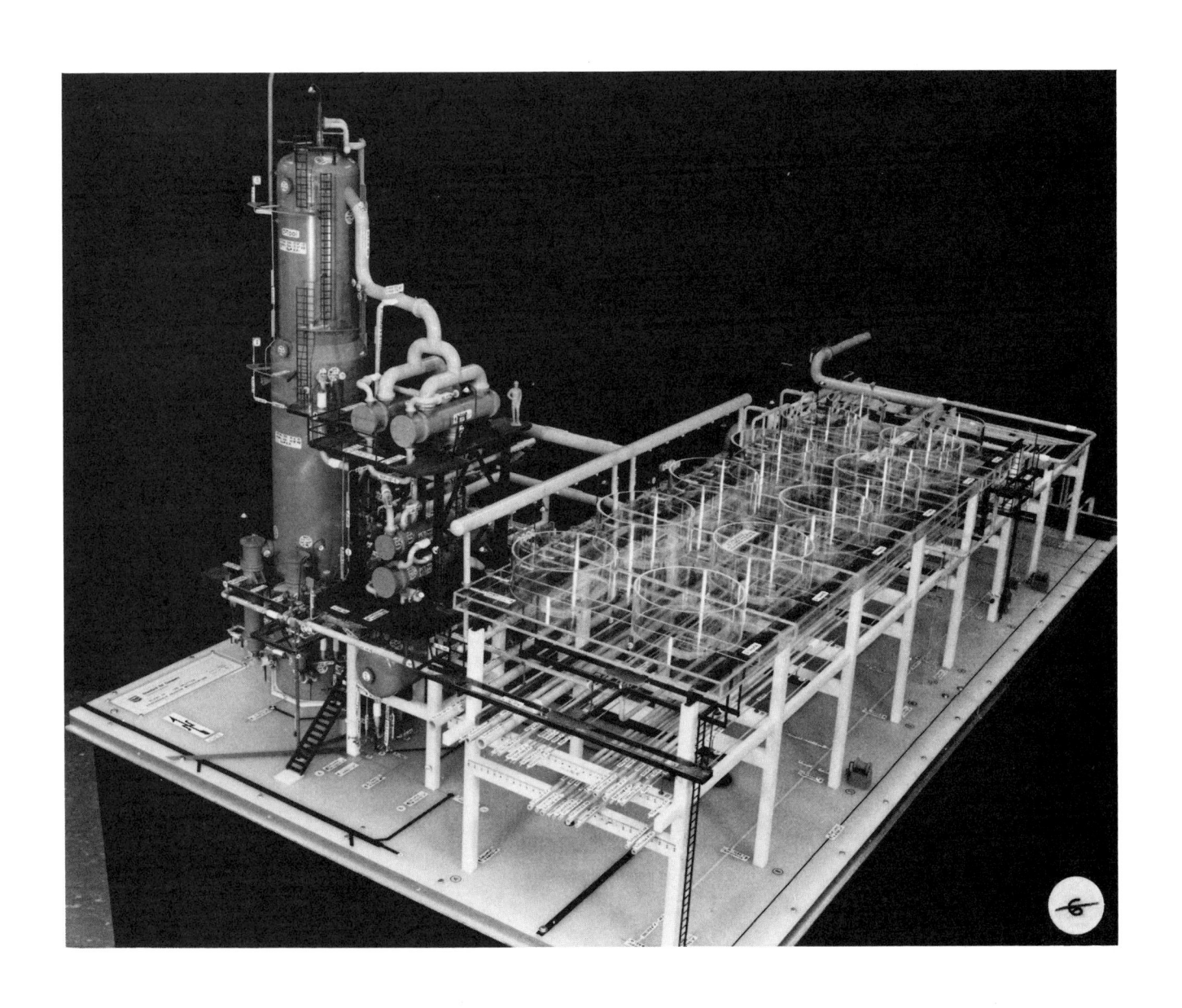

2.1 GEOMETRIC CONSTRUCTIONS

A variety of standard geometrical constructions are used in the solving of descriptive geometry problems. Many of these will already be familiar to the reader, such as bisecting an angle or line, ellipse construction, perpendicular and parallel constructions, etc. Throughout the text geometric constructions will be called for within assigned descriptive geometry problems. The student is not to use a template for situations where geometric construction would be better suited and more accurate, as for inscribing and circumscribing planes with circles.

The geometry of drafting is a simple systematic procedure for drawing figures and shapes in the most efficient manner. Efficient drafting occurs when the same figure is consistently drawn, using a minimum of effort to obtain a maximum result. While the geometry of drafting is based on the geometry of mathematics, a knowledge of mathematical geometry is not required for the mastery of geometric construction. Of more value is a simple understanding of the mechanics and shapes of geometric figures and the ability visually to reason out a problem.

Drawings consist of straight lines, angles, circles, arcs, and curves assembled in a multitude of ways to create a view of an object. The geometry of drafting uses the most straightforward approach to draw these views so that they are consistently accurate in size and shape. Scale, uniformity of linework, and the smooth joining of lines and curves are all equally important to produce useful drawings. A systematic approach to drawing figures contributes to the efficiency of drafting and to the usefulness of the drawings produced.

2.2 CURVED LINES

Geometric constructions introduce the drawing of arcs, circles, and other curved lines that require new linework techniques. The compass lead is fixed in the compass and cannot be rotated in the same manner as the pencil while drawing. Noncircular curves are drawn using a French curve as a "straightedge" but it only fits the curve for a short distance. Moreover, these lines must be drawn equal in width and density to the straight lines in order to produce a uniform drawing. Figure 2-1 shows a student using a set of French curves to construct a curved line.

As with straight lines, a few basic principles need to be established and then good results will come with practice. The use of the compass and the French curve to create dark, consistent linework is typically one of the most frustrating parts of learning to draft. Circle, ellipse, and other curved templates are available in most standard sizes. These excellent tools can be applied to many of the constructions encountered in drafting. Unfortunately they are limited in sizes and shapes. It is suggested that no construction of curved surfaces be completed with the use of templates until the use and application of the compass and French curve are mastered.

FIG. 2-1 Student using French curve.

2.3 THE BOW COMPASS AND DIVIDERS

A good bow compass and dividers are essential to the accurate construction of descriptive geometry drawings. A bow compass has a center-wheel that is used to set and hold the spacing between the center point and the lead. Dividers do not have a center-wheel and are used to set off measurements quickly from one view to another, which is extremely useful in the construction of descriptive geometry problems.

The centering point for the compass is either a tapered point as shown in Fig. 2-3, or a short needle point projecting from a wider shaft to create a shoulder. This second type is somewhat better for beginners because it provides a stop to limit the point's penetration into the paper. Dividers have two identical tapered metal points, one of which can be exchanged for a piece of 4H lead and used to set off dimensions instead of using two metal points.

The compass lead should be a piece broken off of the drafting pencil lead. By using a piece from the pencil, straight and curved lines are drawn using the same lead, resulting in greater uniformity. The lead is secured in the compass with about $\frac{3}{8}''$ (9.52mm) exposed, as shown in Fig. 2-3, and is sharpened using a sandpaper block. Use care while sharpening to keep the line through the point and the lead perpendicular to the sandpaper, and make a flat cut that gives an oval surface. This surface should be about three times as long as the diameter of the lead. The resulting "point" is chisel shaped and should have about the same taper, when viewed from the side, as the cone shaped taper of the drafting pencil. Do not adjust the lead in the compass after it is sharpened; it is almost impossible to properly reposition the chisel shape. Adjust the centering point so that the needle point end is even with the lead end. The beveled end can be on either side of the lead as shown in Fig. 2-3. The compass can now be adjusted to the required radius and used to draw a circle or arc. Locate the center of the required circle or arc and draw a horizontal construction line. Set a distance equal to the radius of the circle or arc to be drawn. Set the compass to this distance and draw a construction circle.

Measure the diameter of the circle. The reading should be twice the given radius, as in Fig. 2-2. To get an accurate diameter reading, be sure the measurement is taken along a line that passes through the center point of the circle. Any difference between the measured diameter and twice the given radius is twice the error of the compass setting. The width of the line to be drawn is determined entirely by the width of the lead (thickness of the bevel) at the moment. As a circle is drawn, the point begins to shorten and the line to widen.

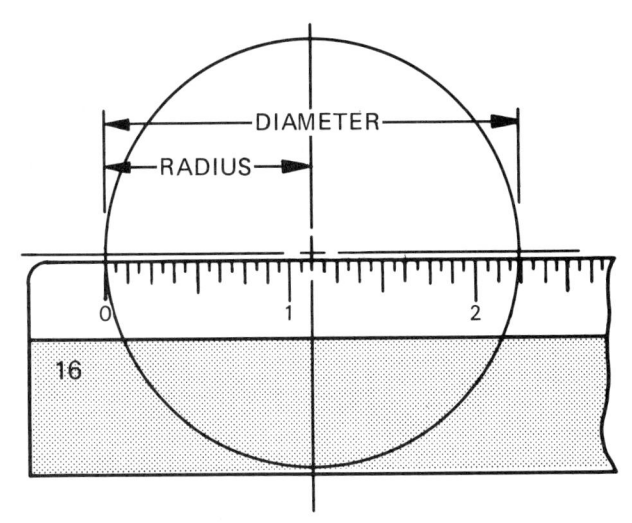

FIG. 2-2 Measurements for circle construction.

This will tend to make the line thicker, but may also create a fuzzy line. To get a crisp, clean, dark line keep the bevel very sharp and draw the line thin and dark by rotating the compass a couple of times. Then resharpen the lead and draw another line touching the first line but slightly larger or smaller depending on the required dimension and the size of the first line. This procedure will always give a sharp, clear line. A longer taper on the lead holds a line width longer.

When using dividers to set off dimensions, always measure from the center of a line, not the edge.

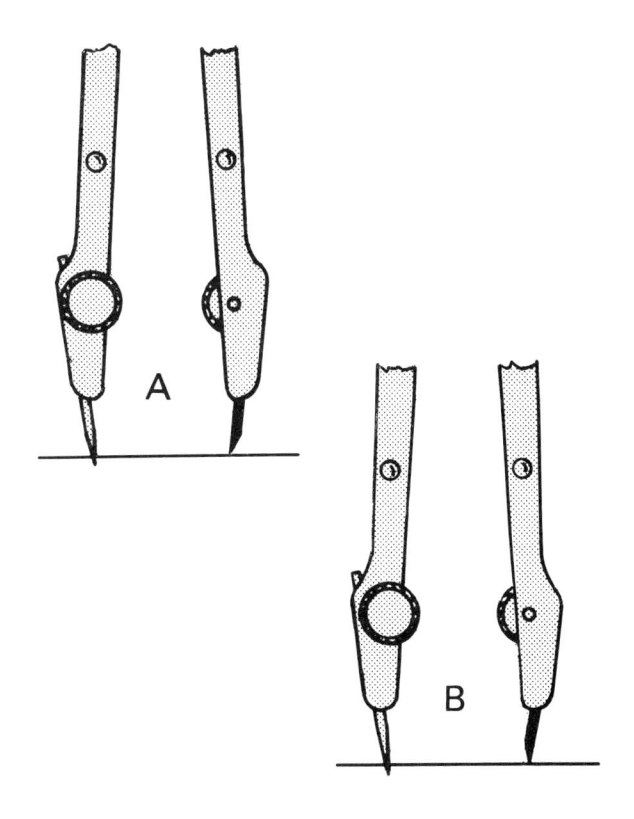

FIG. 2-3 Compass lead points.

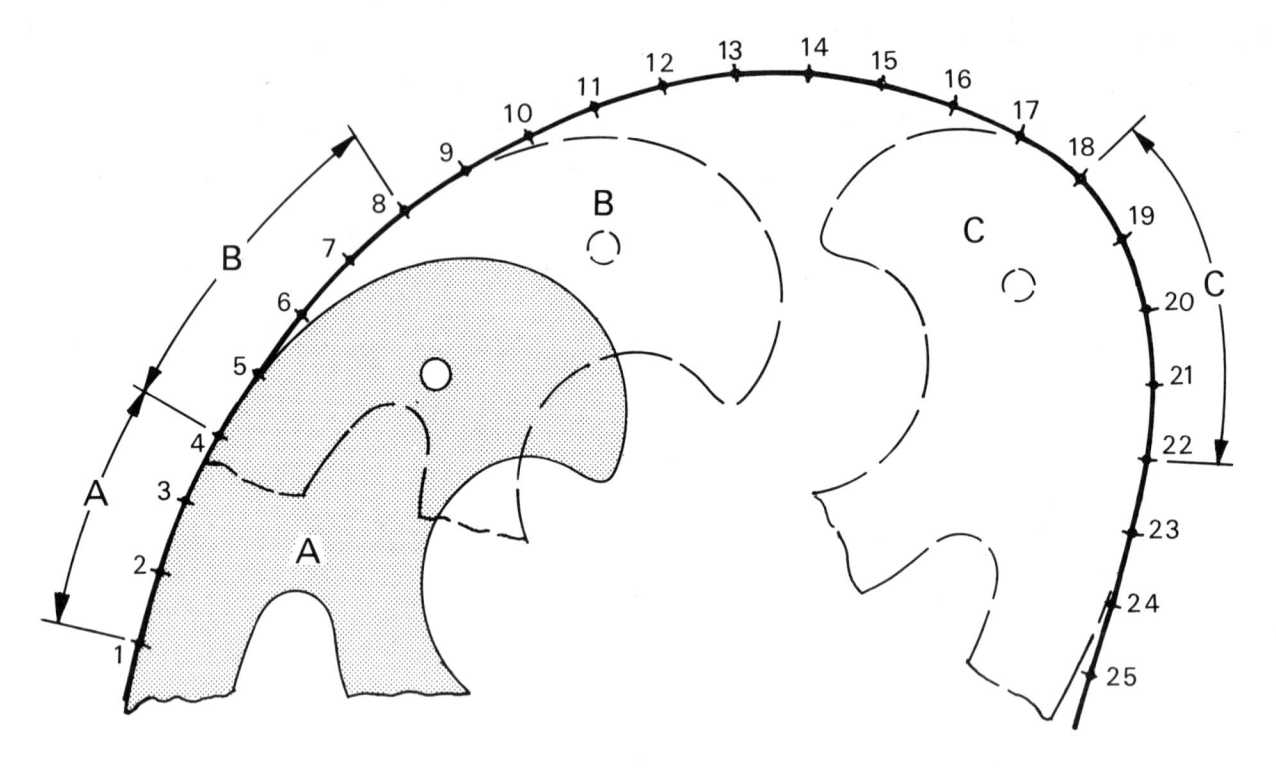

FIG. 2-4 Irregular curve construction using a French curve.

2.4 THE FRENCH CURVE

Noncircular curves require the use of the **French curve** to make smooth printable lines. Examples of such curves are the ellipse (an angular view of a circle), the helix, etc. French curves are manufactured in a great many shapes and sizes, and it is a good idea to have a variety of types from which to choose.

The curves that are drawn using French curves are usually determined by first plotting a series of points that are known to lie on the curve, as in Fig. 2-4. Then a curve is drawn that includes all of these points. Good results can be obtained if the following steps are taken.

1. Sketch lightly by eye a smooth line that includes the plotted points. It is easier to set the French curve to a line than to a series of points.

2. Set the French curve so that it fits to a part of a line, Fig. 2-4, usually a minimum of four points.

3. Draw the line that fits the curve but stop one point short, before the end of the fit. In Fig. 2-4, **A** fits from point 1 to point 5, but is drawn from 1 to 4. **B** fits from point 3 to point 9, and is drawn from 4 to 8. **C** fits from point 17 to point 23, and is drawn from 18 to 22.

4. Reset the French curve to fit each next part of the curve and include the last portion of the already drawn line. This overlapping will give a smoother curve.

Neater work will result if the sketched curve and the first series of fitting the French curve are done on a tracing paper overlay. When doing this, mark the ends of each segment of the line as it fits the French curve so that the same "fits" can be used in the next step. When all fits are made, place the tracing paper overlay under the drawing and carefully align the curve under the plotted points. Trace the curve onto the drawing using the French curve fits marked on the overlay. This technique has two advantages. First, all the fitting is made on a throwaway paper so that erasures can easily be made without erasing the plotted points. Second, the accuracy of the fit can be seen before the final drawing of the curve, when the overlay is positioned under the drawing.

The overlay technique is particularly valuable when the curve is symmetrical. For example, an ellipse has four identical curves; two are mirror images of the other two. All are symmetrical about the major and minor axes. It is only necessary to fit one of these curves and then to duplicate this fit on the other three.

The plotting of the points of an irregular curve is particularly important if a smooth curve is to result. A small error in the position of a point can easily cause false irregularities in the curve. The number of plotted points should be maximum where the curve is sharpest and minimum where the curve is the straightest.

2.5 LINES

Lines and their relationships are the most important part of descriptive geometry and geometric construction. The student must become familiar with the terms that are used to name the relationships between lines, and the definitions of the terms as they apply to lines.

A line is considered to have length but no width. A straight line is the shortest distance between two points and is inferred when speaking of a "line." A line that bends is called a "curve." In Fig. 2-5 a **horizontal** line (1), **vertical** line (2), and **curved** line (3) are shown.

Parallel lines are equally spaced along their entire length, becoming neither closer together nor further apart. A symbol for parallel lines is //. Numbers (1) and (2) in Fig. 2-6 are examples of parallel lines.

Perpendicular lines are at an angle of 90 degrees. A symbol for perpendicular lines is shown in Fig. 2-6(4); the small box indicates perpendicularity. (3) and (4) are perpendicular lines.

2.6 CONSTRUCTION OF PARALLEL AND PERPENDICULAR LINES

Using a straightedge and a triangle it is easy to construct parallel and perpendicular lines. In Fig. 2-7 position 1 is the first line drawn. Slide the triangle along the T-square to position 2 and draw a parallel line. By rotating the triangle to position 3 a perpendicular line can be drawn. This last line will be perpendicular to both of the other lines. Remember to draw on the same edge of the triangle as before the triangle was rotated.

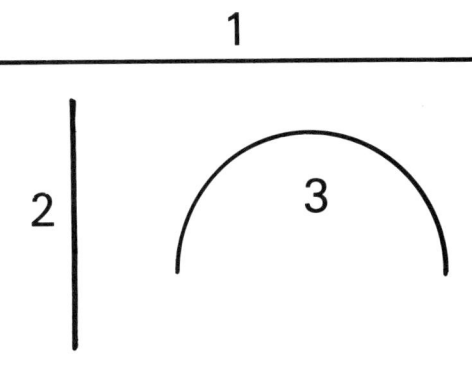

FIG. 2-5 Lines, 1 (horizontal), 2 (vertical), 3 (curved).

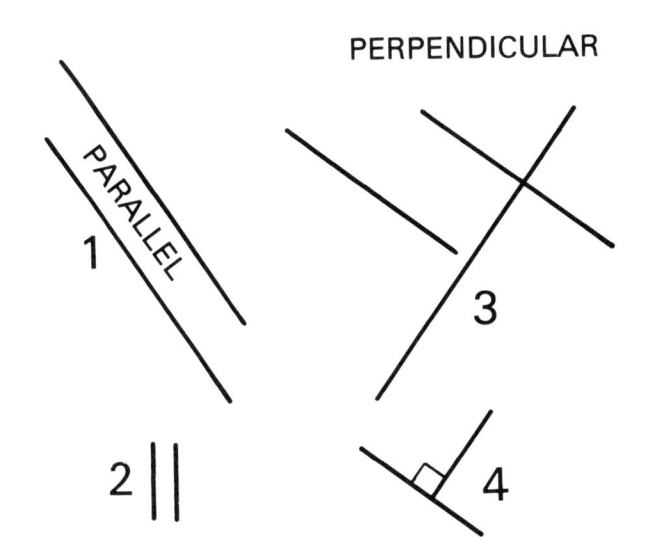

FIG. 2-6 Parallel and perpendicular lines.

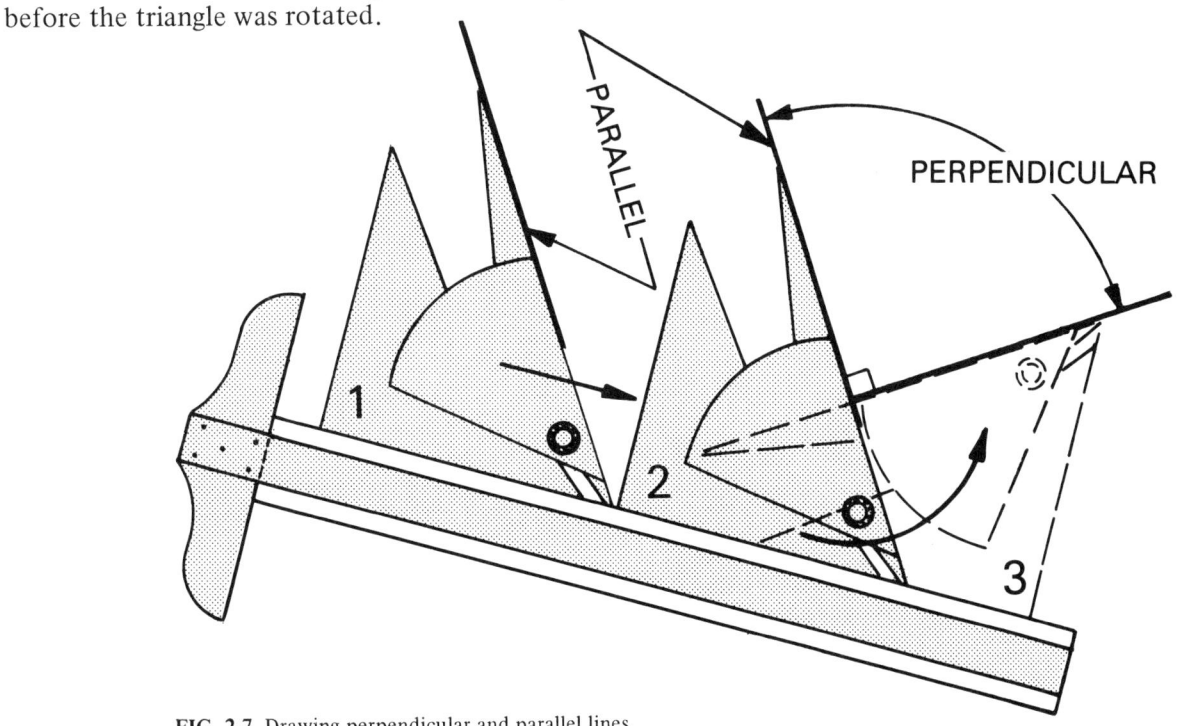

FIG. 2-7 Drawing perpendicular and parallel lines.

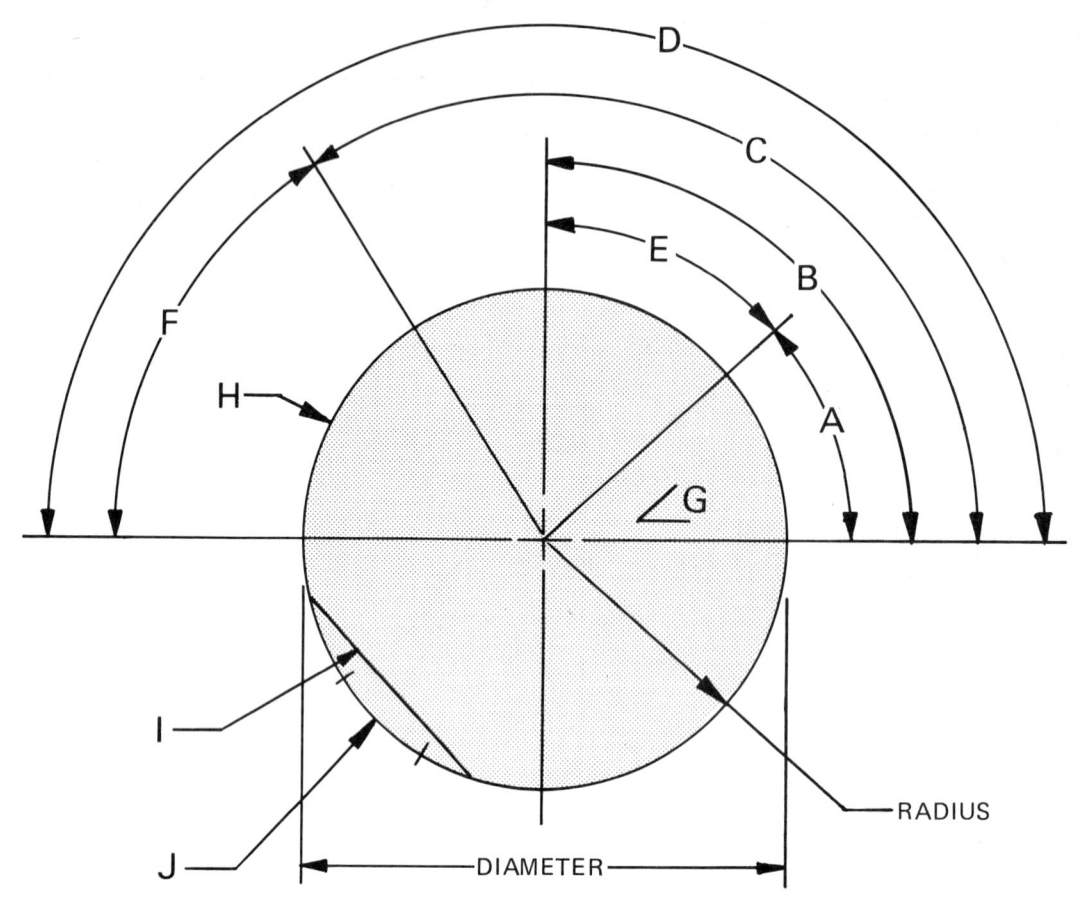

FIG. 2-8 Angles and circles.

2.7 ANGLES AND CIRCLES

Angles are formed by two intersecting lines. Angles are named for the space between the intersecting lines and for the relationship between adjacent angles. The amount of space between the intersecting lines is expressed in degrees. There are 360° in a full circle. A symbol for representing an angle is shown in Fig. 2-8(G).

The parts of a circle that are used in the construction of geometric shapes are: the **circumference** (distance around the circle), the **diameter** (distance measured from edge to edge and through the center of the circle), the **radius** (one half the diameter, from the center of the circle to one edge), the **chord**, and the **arc**. A chord is a straight distance from any two points on the circumference. An **arc** is a portion of the circumference from one fixed point to another. **Concentric circles** are of different radius and share the same center point. **Eccentric circles** overlap but have different center points.

The following describes the parts of a circle and angles in relationship to circles as illustrated in Fig. 2-8.

A = **Acute angle** (less than 90°)

B = **Right angle** (90°, two perpendicular lines)

C = **Obtuse angle** (more than 90° but less than 180 degrees)

D = **180°** (a straight line)

E = **Complementary angles** (two angles that together equal 90°); A plus E are complementary angles

F = **Supplementary angles** (two angles that together equal 180°); F plus C equal 180° and are therefore supplementary

G = **Angle symbol**

H = **Circle** (360°)

I = **Chord** (a straight line connecting two points on a circle or curve)

J = **Arc** (a continuous portion of a circle or curve)

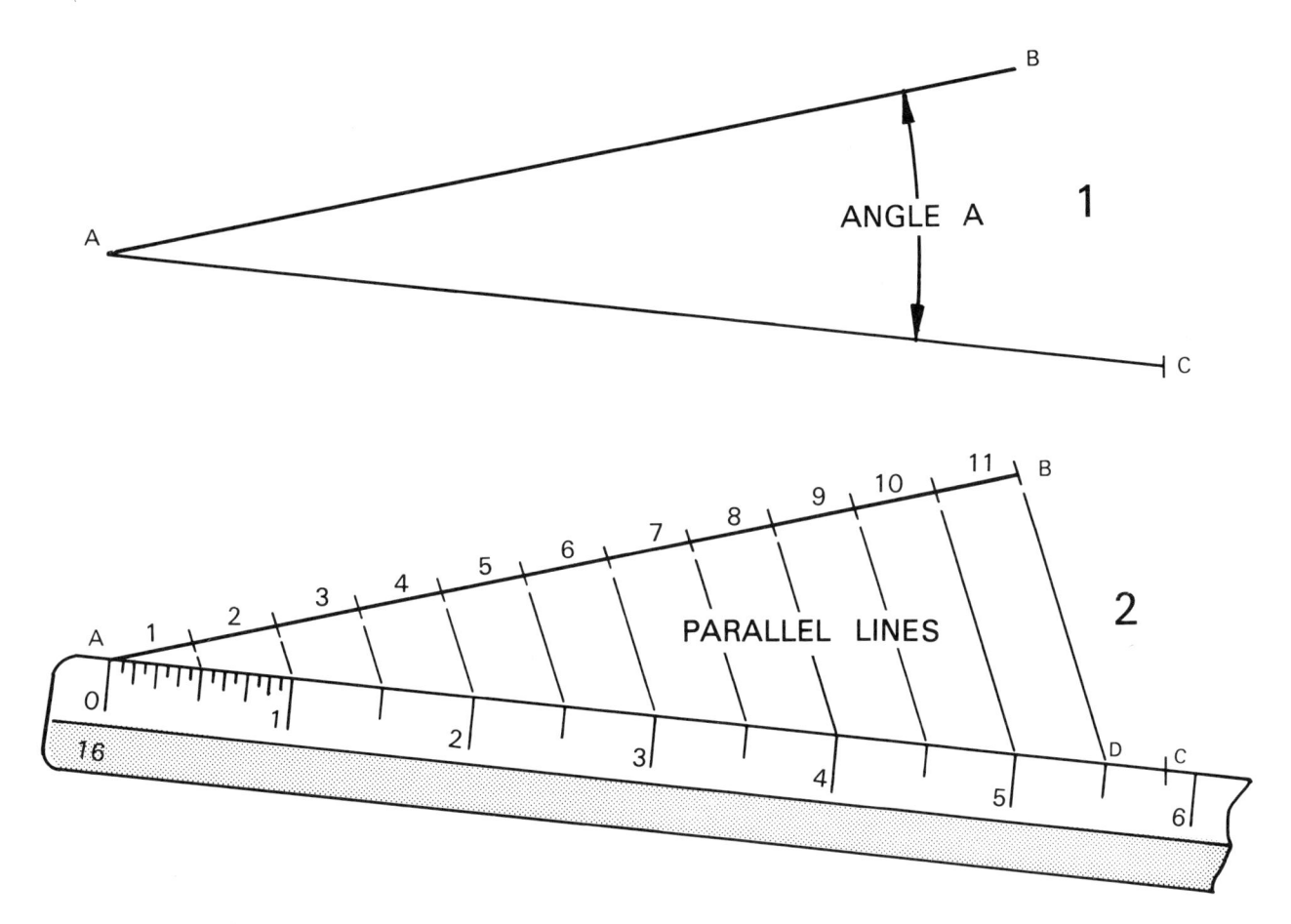

FIG. 2-9 Dividing a line.

2.8 DIVIDING A LINE INTO EQUAL PARTS

Descriptive geometry problems have many situations where it is necessary to divide a line equally. As when constructing a development of a cylinder, the elements along its lateral surface are evenly divided so as to expedite its rollout (development). It is possible to calculate the length of a line, divide by the number of required parts, and then use this measurement to set off each division using a scale or dividers. This method results in a large accumulated error in the overall dimension of the sum of all of the segments.

A much better technique is to determine the overall dimension of all of the spaces and then to divide this one dimension into equal spaces using the **parallel (adjacent line) line method** described below. This technique does not even require making an accurate real measurement of any of the spaces; only easily read measurements are used.

Figure 2-9 illustrates the parallel line method of dividing a line. The example shown is to divide a line into 11 equal parts. Line AB is given. Any type of scale and unit of measurement can be used in this method. The idea here is to simplify the construction and measuring process; therefore, the unit type and scale that is the most convenient is the best. The following steps describe the process.

1. Draw a construction line AC that starts at end A of given line AB. This new line is longer than the given line and makes an angle of not more than 30 degrees with it. Angle A should not be less than 20 degrees because it will be hard to project the divisions from the construction line AC to the original line AB.

2. Find a scale that will approximately divide line AB into the number of parts needed (in this case, 11) and mark these 11 divisions on line AC. Here the full-size inch scale was used, using $\frac{1}{2}''$ for each division. There are now 11 equal divisions from A to D that lie on line AC.

3. Set the adjustable triangle to draw a construction line from point D to point B. Then draw construction lines through each of the remaining 10 divisions parallel to the first line BD by sliding the triangle along the straightedge. The original line AB will now be accurately divided.

It is also possible to use dividers for step 2 and divide the construction lines into the required number of equal parts. This method is less accurate than using a scale.

2.9 VERTICAL METHOD

Figure 2-10 illustrates the **vertical line** method of dividing a line into equal parts. This is the preferred method because all of the projector lines are vertical lines and are easily drawn using a straightedge and any triangle. This example is to divide line AB into seven equal parts.

1. Draw a vertical construction line BC through point B of line AB.
2. Using point A as the pivot point, position a scale that will give the proper number of divisions so that it equally divides a distance from point A to some point on line BC. Here full-scale, English units were used, each $\frac{1}{2}''$ unit of the scale corresponding to each division to mark points 1 through 7. Note that it was necessary to use a scale that gives an overall length of seven units that is slightly longer than the line AB.
3. Using the vertical side of a triangle, draw construction line projectors from points 1 through 7 down to line AB. This establishes seven equally spaced segments along line AB.

For the vertical and parallel methods it was not necessary to make any measurements that were less than one whole easily measured unit, regardless of the mathematical value of the resulting divisions. Also note that the scale is used only to measure equal units without regard to what those units may be. Therefore, any scale that will measure equal units may be used: architect's, engineer's, mechanical, or metric.

Remember, the best results are obtained when the angle between the given line and the measured line is kept to a maximum of 30 degrees.

2.10 PROPORTIONAL DIVISION OF A LINE

It may be required that a line be divided into two or more spaces that are not equal, but rather are at some specified proportion, or ratio, to each other. Suppose that a line were to be divided so that the first part is three times as long as the second part. This is written as 3:1 and is read as "three parts to one part" or "three to one." The construction used is almost the same as was used to divide a line into equal parts.

Figure 2-11 illustrates dividing a line into four parts that have proportions of 4:3:5:9.

1. Mathematically total the proportions of the four parts: 4 + 3 + 5 + 9 = 21. This is the number of equal parts that are to be measured on the scale.
2. Draw a construction line at an angle to, and longer than, the given line. Set the scale to make 21 equal divisions. Make the first mark at 4 units, add three units and make the second mark at 7 units, add five more units and make the third mark at 12 units. Adding nine units brings the total to the 21 unit setting on the given scale.
3. Project these marks to position points 2, 3, 4, and 5 onto the given line using the parallel line method. This will create line segments that are in proportions of 4:3:5:9.

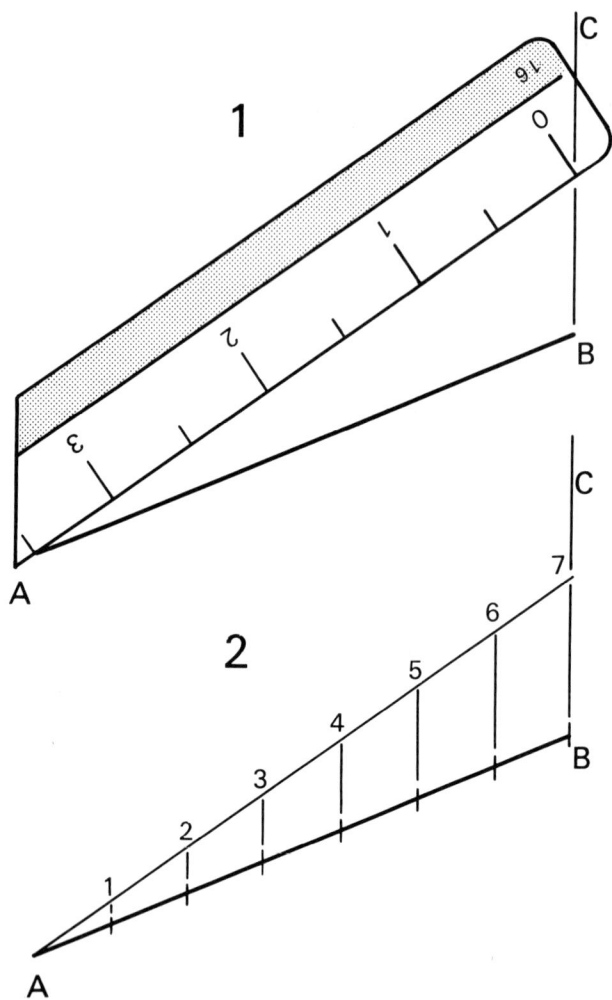

FIG. 2-10 Vertical method of dividing a line.

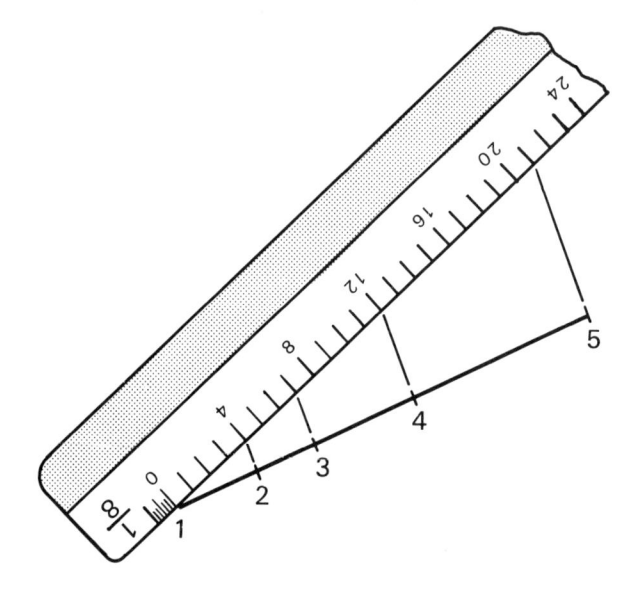

FIG. 2-11 Proportional division of a line.

1. Lines 1-2 and 2-3 intersect and form angle 1-2-3.
2. Set the compass at any convenient radius that allows accurate construction. For small angles and short lines it is necessary to extend the lines that form the angle. With point 2 as the center, swing an arc that will establish points 4 and 5 on their respective lines. The length of 2-4 will be equal to 2-5.
3. Draw the perpendicular bisector of the distance 4-5. The line 4-5 is shown with a broken line but it need not be drawn. An extension of bisector 6-7 passes through point 2 and will be the bisector of angle 1-2-3.
4. Because the perpendicular bisector of distance 4-5, the line 6-7, always passes through the point 2, it is actually only necessary to find one point of the bisector, point 6 in this example. The line 2-6 is then the bisector of angle 1-2-3.

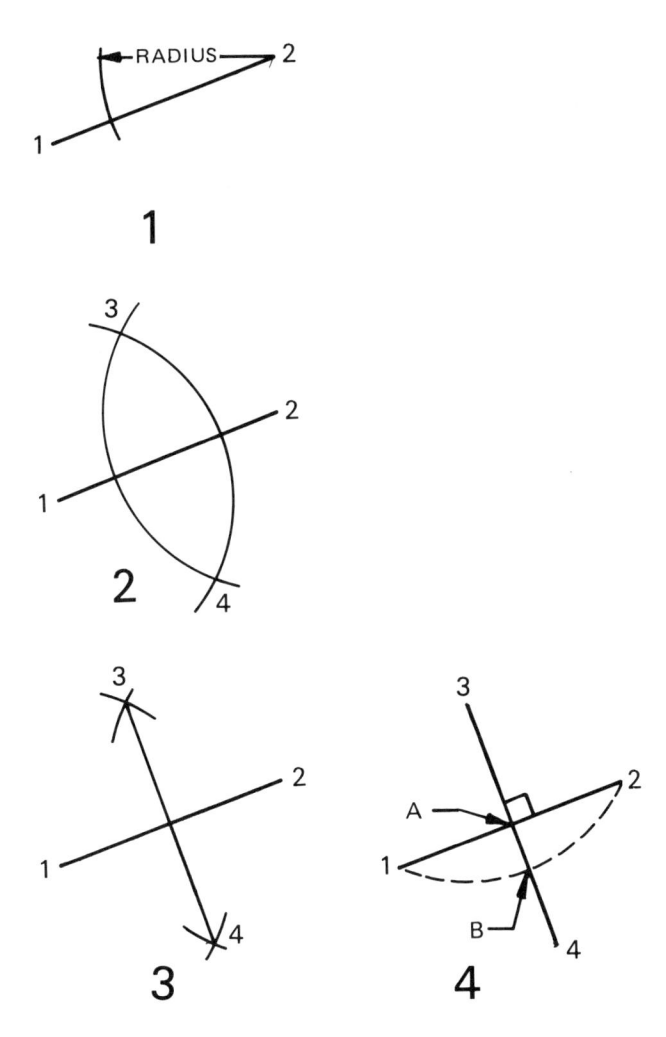

FIG. 2-12 Perpendicular bisector.

2.11 BISECTORS FOR LINES AND ANGLES

A **bisector** of a line is another line that divides the given line into two equal parts. It is also perpendicular to the given line. It is useful in determining either or both of these conditions. A perpendicular bisector can be constructed using only compass and straightedge as illustrated in Fig. 2-12 and explained in the following steps.

1. Given line 1-2, set the compass at a radius equal to about three-fourths of the length of 1-2.
2. Using points 1 and 2 as centers, draw arcs to establish points 3 and 4.
3. Draw line 3-4 by connecting the two new points.
4. The line 3-4 intersects the line 1-2 at its midpoint (point A) and is perpendicular to it. If points 3 and 4 were the ends of an arc, shown with the broken line, line 3-4 would bisect the arc 1-2. Also, the center of arc 1-2 is on the bisector line 3-4 at point B.

A bisector is used to divide an angle into two equal parts as illustrated in Fig. 2-13. The following describes the process. The given intersecting lines 1-2 and 2-3 form the angle 1-2-3.

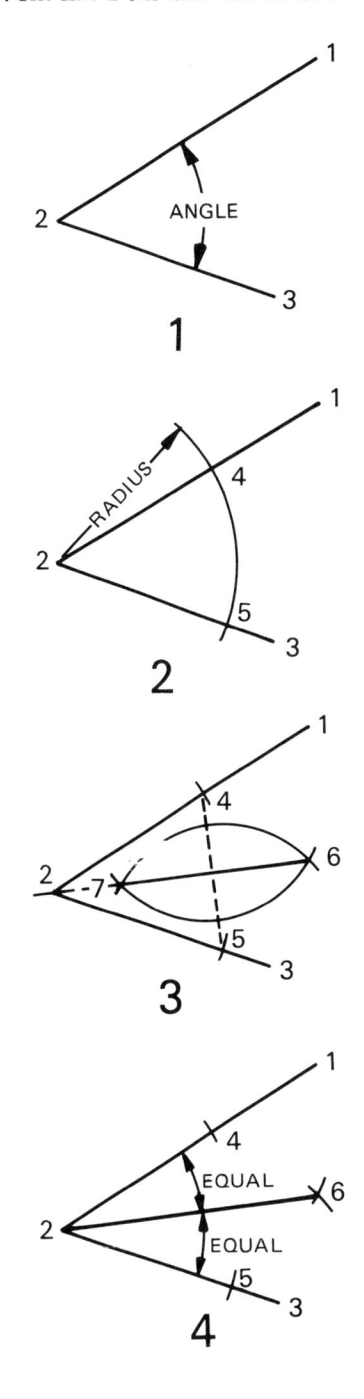

FIG. 2-13 Bisecting an angle.

FIG. 2-14 Triangles.

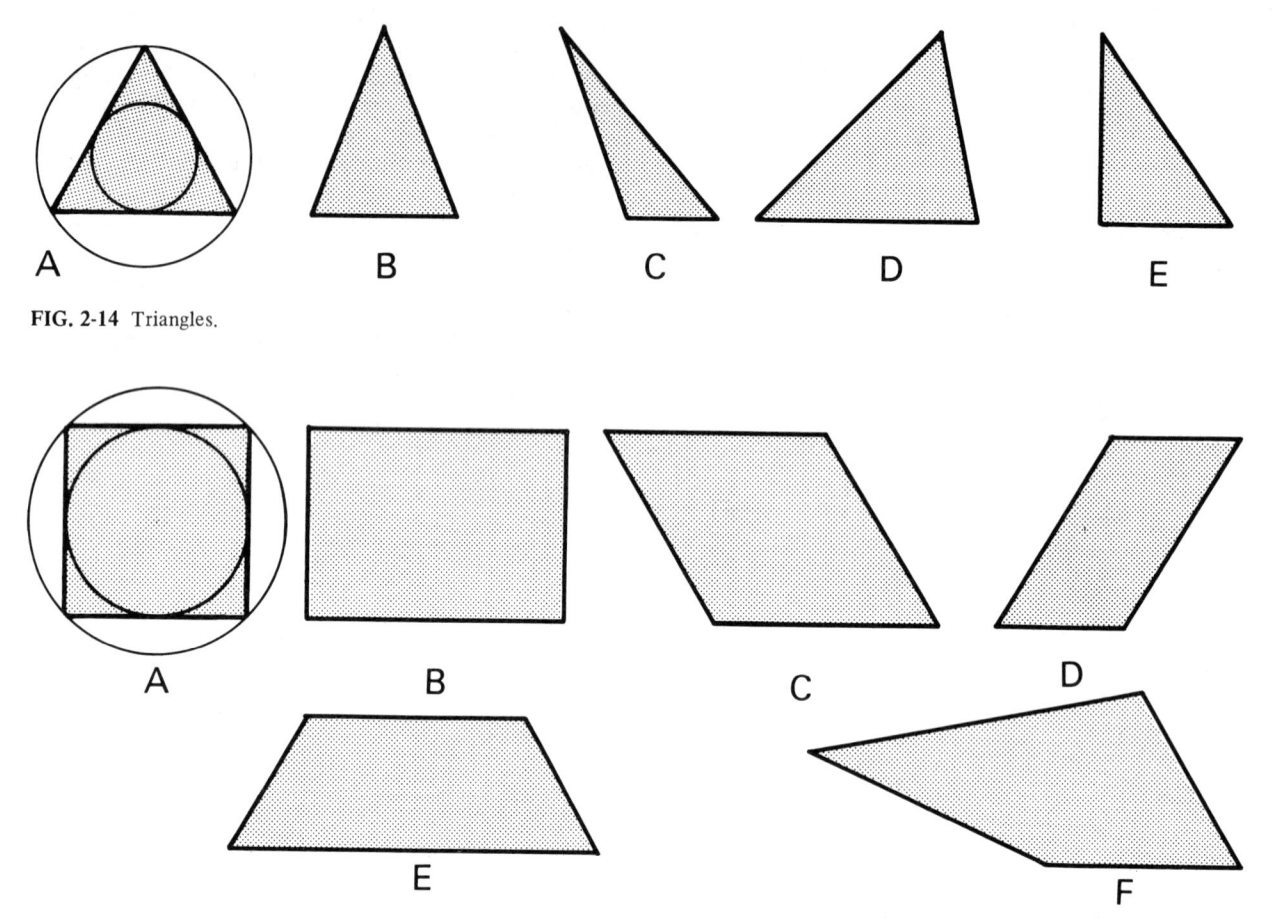

FIG. 2-15 Quadrilaterals.

2.12 POLYGONS

A *polygon* is any closed figure that has three or more straight sides. A *regular polygon* has all sides of equal length and angles of equal size. A regular polygon can be inscribed within a circle with points touching the circle and can be circumscribed about a circle (not the same circle) with sides touching the circle.

For a **triangle**, the sum of its interior angles always equals 180°. In Fig. 2-14, triangle A has equal sides and equal angles and is called an *equilateral* triangle; it is a regular triangle and is shown inscribed and circumscribed. B is an *isosceles* triangle, two equal sides and two equal angles; the unequal side is the base and the corner opposite is the apex or vertex. A line through the apex and perpendicular to the base divides an isosceles triangle into two equal right triangles. C and D have no equal angles or sides and are called *scalene* triangles. E is a right triangle because one angle equals 90 degrees.

A **quadrilateral** has four sides and the sum of its interior angles is 360°. In Fig. 2-15 the six types of quadrilaterals are shown; A through D have opposite sides equal in length, and are therefore also *parallelograms*. The first shape (A) is a *square*, which can be inscribed and circumscribed and therefore is a regular quadrilateral. B is a *rectangle*; opposite sides are equal and adjacent. C has four equal sides and is called a *rhombus*; D has opposite sides parallel and is called a *rhomboid*. E is a *trapezoid*, two sides parallel. When a quadrilateral has no equal sides it is called a *trapezium* (F).

Figure 2-16 exhibits three other regular polygons that are commonly found in the geometry of drafting: A is a *pentagon* (five sides), B is a *hexagon* (six sides), C is an *octagon* (eight sides). All three are regular polygons and can be inscribed and circumscribed as shown.

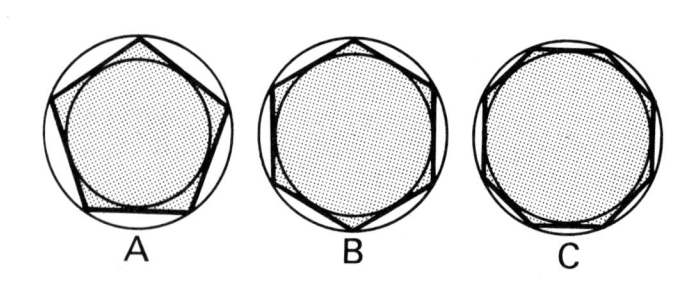

FIG. 2-16 Inscribed and circumscribed.

FIG. 2-17 Hexagon, given across corners.

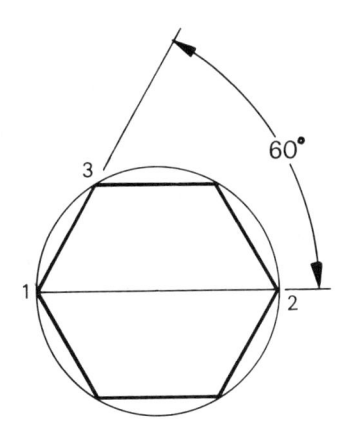

2.13 REGULAR POLYGONS

Polygons are closed figures having three or more sides. *Regular polygons* have equal length sides and equal angles at the corners. All regular polygons will fit inside a circle with the corners touching the circle. The term for the polygon in a circle like this is "inscribed within a circle." Both Fig. 2-16 and Fig. 2-17 provide examples of this condition. All regular polygons will also fit around the outside of a circle with the sides tangent to the circle at the midpoint of each side. The term for a polygon in this condition is "circumscribed about a circle." Figures 2-16 and 2-18 are examples of this situation. It can also be said that the circle drawn tangent to the polygon's sides is "inscribed in the polygon."

Because of the relationship of circles to regular polygons, each polygon shares a common centerpoint with its inscribing and with its circumscribing circle. This centerpoint is the usual positioning point for these figures. Horizontal and vertical axes through the centerpoint are usually drawn to establish the position of the centerpoint. Measurements are made from the axes when drawing these figures.

To draw a particular regular polygon, it is necessary to have at least one specific dimension for that figure. The two dimensions that are used for the even-number sided figures (square, hexagon, octagon, etc.) are "across corners" and "across flats." "Across corners" is the maximum measurable straight-line distance across the figure and is equal to the diameter of its circumscribing circle, as shown in Fig. 2-17 where line 1-2 is the distance across the corners and the diameter of the circumscribing circle. "Across flats" is the minimum measurable straight-line distance across the figure and is equal to the diameter of its inscribed circle. In Fig. 2-18 the distance across the flats is equal to the diameter of the inside circle. These inscribing and circumscribing circles are used to construct easily and draw consistently regular polygons. The measurement across flats is the size of a wrench used on a hexagonal or square-headed bolt or nut.

To draw a hexagon where the distance across the corners is known, establish the diameter of the circle that will inscribe the hexagon (as in Fig. 2-17 where line 1-2 is drawn first). Draw the circle. In this example a template was used; therefore, it was unnecessary to find the center of the circle. At each end of line 1-2 construct a chord using a 60° angle (line 1-3). Draw chords above and below line 1-2; where they cross the circle connect each point.

In Fig. 2-18 the distance across the flats was known. Establish the center of the required hexagon. Draw a circle equal to the distance across the flats (A), then use a 30° angle to construct tangents to the circle.

To construct a regular polygon with a specific number of sides, divide the given diameter using the parallel line method as shown in Fig. 2-19. In this example seven sides were required. Construct an equilateral triangle (0-7-8) with the diameter (0-7) as one of its sides. Draw a line from the apex (point 8) through the second point on the line (point 2). Extend line 8-2 until it intersects the circle at point 9. Radius 0-9 will be the size of each side of the figure. Using radius 0-9 step off the corners of the seven sided polygon and connect the points.

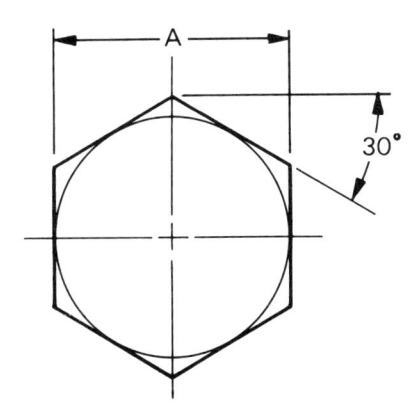

FIG. 2-18 Hexagon, given across flats.

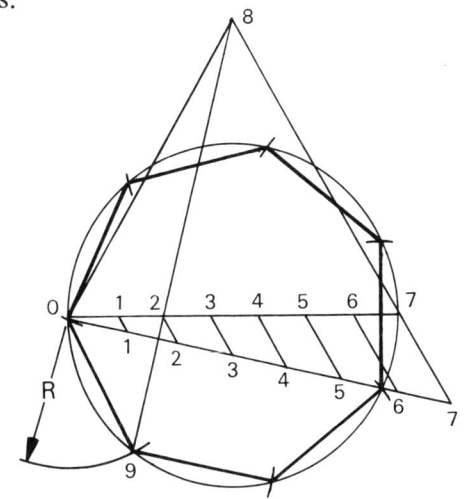

FIG. 2-19 Construction of a polygon in a circle.

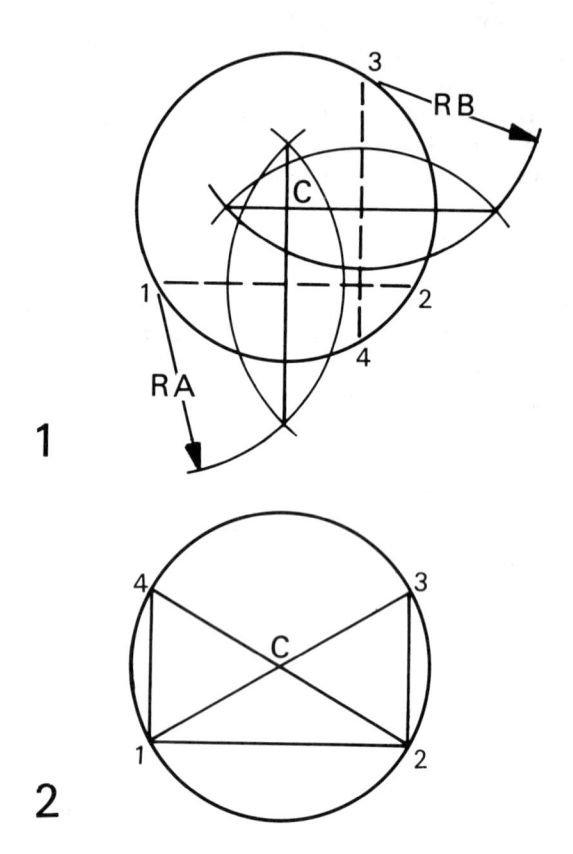

FIG. 2-20 Finding the center of a known circle.

2.14 SOLVING FOR THE CENTER OF A KNOWN CIRCLE

If a significant portion of a circle is known, its center can be located by establishing two perpendicular bisectors of any two chords of the circle, Fig. 2-20(1). Lines constructed between any two points on the edge of a circle are known as chords. The perpendicular bisector of any chord will pass through the center of the circle. Two bisectors will establish the center of the circle. To get the best accuracy when reading the intersection of the two perpendicular bisectors, select chords that are at right angles to each other (90 degrees), such as one horizontal (line 1-2) and one vertical (line 3-4) as in Fig. 2-20(1). The two bisectors will cross at the center of the circle (point C).

Part (2) of Fig. 2-20 gives an alternate method: First, a horizontal chord 1-2 is drawn. Vertical chords 1-4 and 2-3 are projected from its ends. Diagonals are drawn from points 2 to 4 and 1 to 3. These diagonals cross at the center of the circle (C). Note that both diagonals are diameters of the circle.

This illustrates that the three corners of any right triangle lie on a circle that has a diameter equal to the longest side of the triangle.

2.15 CONSTRUCTION OF A CIRCLE THROUGH THREE GIVEN POINTS

In Fig. 2-21(1) points 1, 2, and 3 are given. In order to draw a circle that passes through all three points, perpendicular bisectors are drawn for the distances 1-2 and 1-3. These distances are chords of the circle to be found. The perpendicular bisectors of two chords intersect at the center of the required circle (C). The circle is drawn using the distance from C to any of the three points as the radius. To check the solution, pass another perpendicular bisector through the remaining chord 2-3, Fig. 2-21(2).

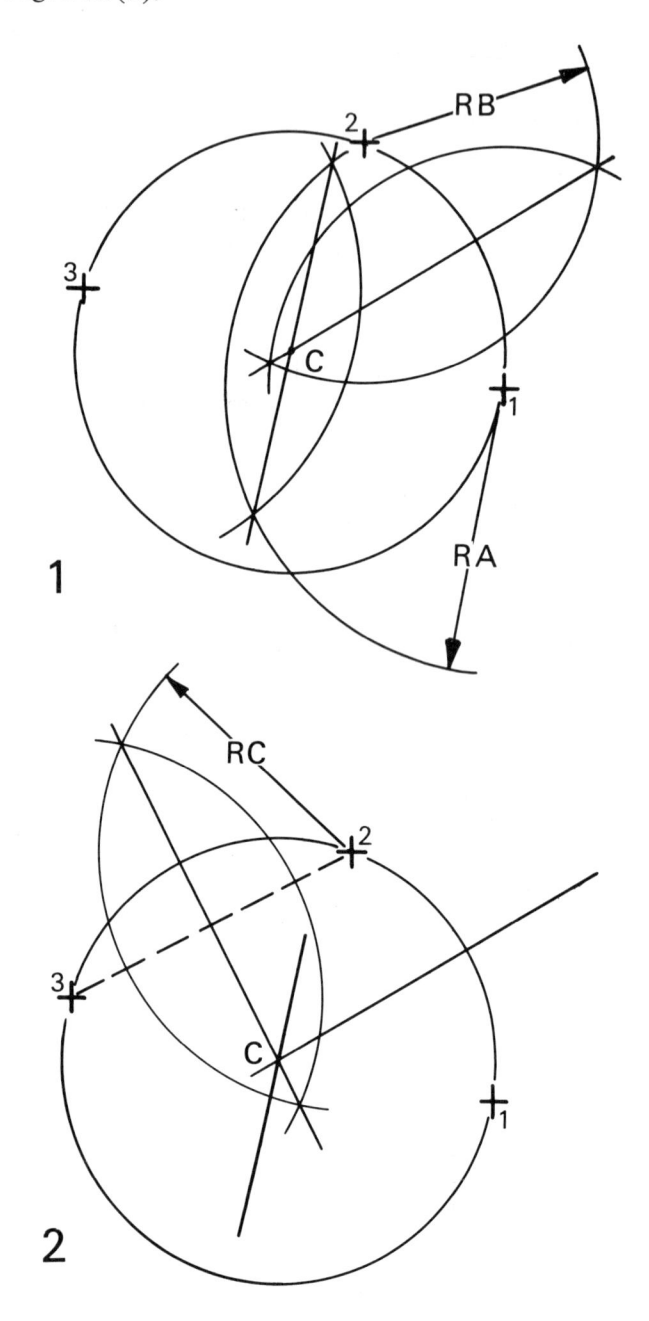

FIG. 2-21 Drawing a circle through three known points.

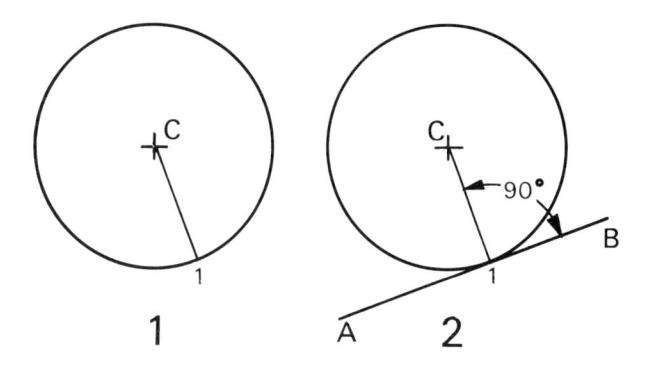

FIG. 2-22 Construction of a line tangent to a circle.

2.16 TANGENCY OF A LINE AND ARC

An arc that touches a line at only one point is **tangent** to that line and the line is **tangent** to the arc. Likewise two curves can be tangent. Note that the line and arc will touch at only one place even if extended. If a line and an arc are tangent the following are true:

1. The tangent line is perpendicular to the radius of the arc at the point of tangency (touching).
2. The center of the arc is on a line that is perpendicular to the tangent line and extends from the point of tangency.

Figure 2-22 illustrates the first principle. To draw a line tangent to the circle at point 1, first draw radius C-1. Then construct line AB perpendicular to the radius line C-1 passing through point 1. Line AB is then tangent to the circle at point 1.

Figure 2-23 illustrates principle 2. To draw a circle tangent to a given line, first project a line perpendicular to the given line AB from point T (tangent point). The center of the desired circle will lie on this line. Locate the center point by striking an arc from point T using the radius of the required circle. Using the same radius (line CT), draw the tangent circle.

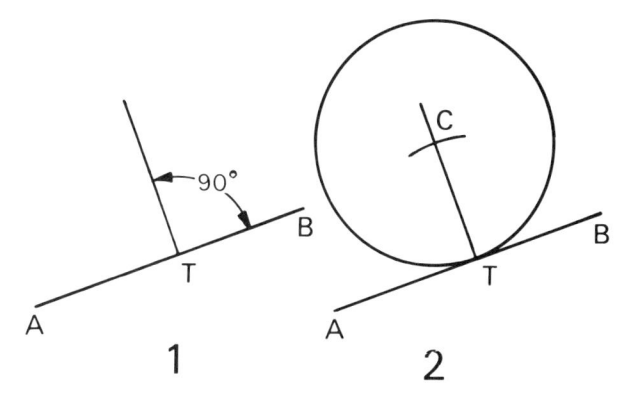

FIG. 2-23 Construction of a circle tangent to a line.

2.17 A LINE TANGENT TO A CIRCLE THROUGH A POINT

Figure 2-24 illustrates the procedure for finding the points of tangency of two lines originating at a given point that are tangent to a circle.

1. Point A and the circle are given.
2. Draw a construction line from point A to C and then bisect it with a perpendicular bisector to locate its midpoint M.
3. Swing RA to establish points 4 and 5, which are the points of tangency.
4. Draw lines A-4 and A-5. Lines A-4 and A-5 are tangent to the circle and at 90 degrees to lines C-4 and C-5 respectively.

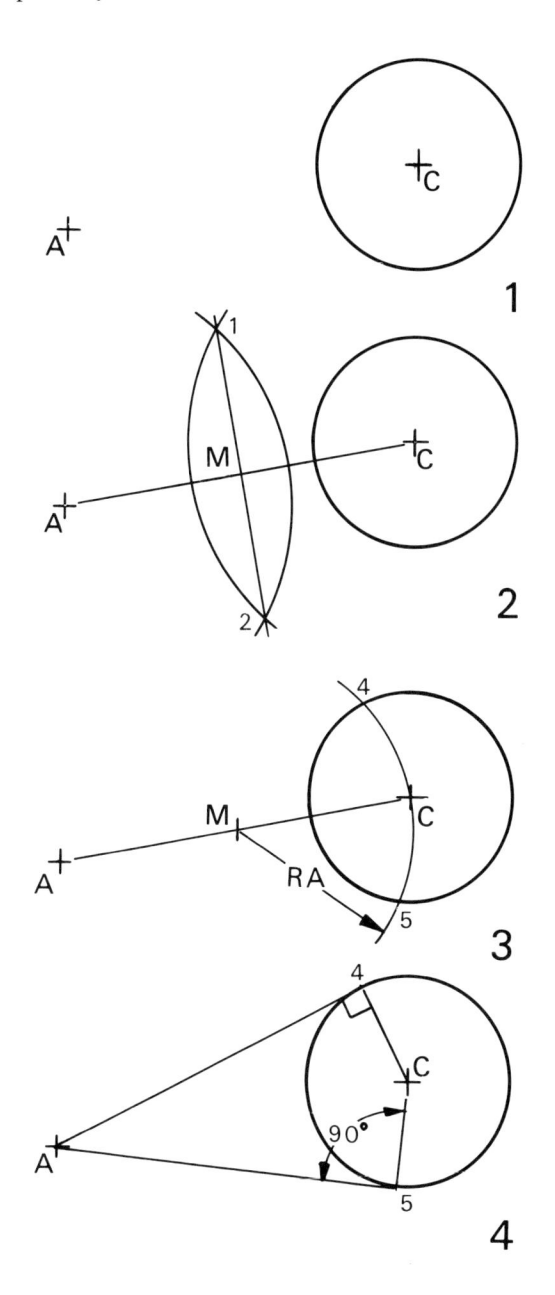

FIG. 2-24 Line tangent to a circle through a given point.

27

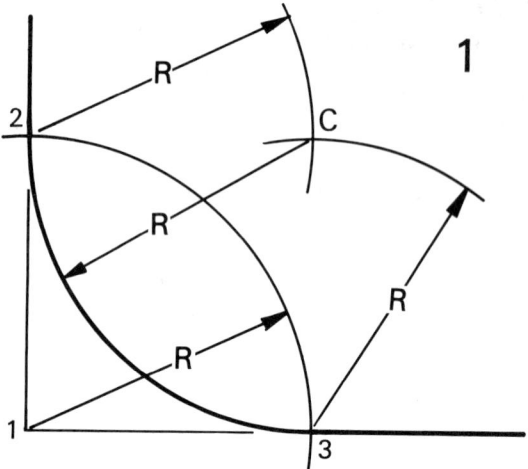

FIG. 2-25 Arc tangent to perpendicular lines.

2.18 TANGENT ARCS

To draw an arc between two perpendicular lines there are two possible methods. Figure 2-25(1) illustrates the construction of the arc 2-3 using only a compass.

1. The two given perpendicular lines can be extended to meet at point 1.
2. From point 1 strike a radius equal to the required radius of the tangent arc. Where this radius crosses the given lines it will establish tangent points 2 and 3.
3. Use the same radius, strike construction arcs from points 2 and 3. Where these two arcs cross at point C, the center of the required tangent arc is established.
4. From point C swing arc 2-3 tangent to both perpendicular lines.

The easiest method is shown in Fig. 2-25(2).

1. Extend the given perpendicular lines so that they meet at point 1.
2. Draw a line parallel and equal distance from each of the given lines using the required tangent arc radius for dimension R. Point C is at the intersection of these two lines.
3. Locate tangent points 2 and 3 by extending construction lines from C perpendicular to the given lines.
4. From center point C swing the required tangent arc from point 2 to 3.

For arcs that are tangent to nonperpendicular lines the same procedure can be used as the second method described above. For both examples in Fig. 2-26 the given lines have been extended to meet at point 1. Draw a line parallel to and equidistant from the given lines using the required tangent arc radius for dimension R. Where these two lines intersect (point C), draw construction lines perpendicular to the given lines to establish points 2 and 3 as the points of tangency. Finally, swing radius R from point 2 to 3 in order to establish an arc tangent to both given lines. This procedure can be used for lines at acute or obtuse angles. These same principles can be applied to the construction of arcs tangent to given curved lines.

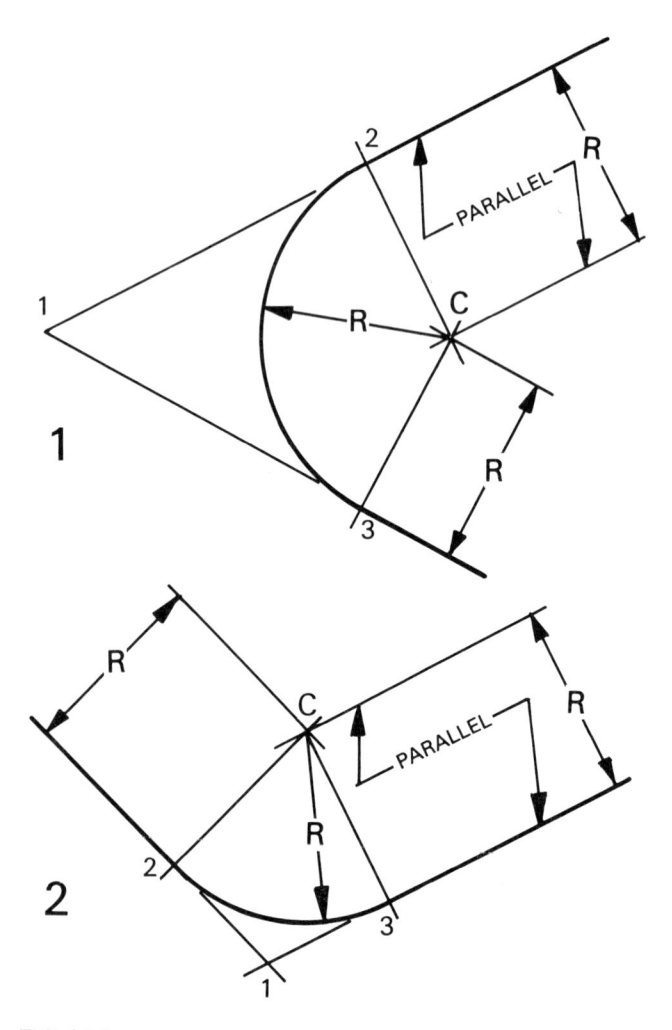

FIG. 2-26 Arcs tangent to nonperpendicular lines.

2.19 THE INSCRIBED CIRCLE OF A TRIANGLE

The procedure for constructing the inscribed circle of a triangle is illustrated in Fig. 2-27. In this example the given triangle is represented by points 1-2-3. An inscribed circle is tangent to each side of the triangle. First it is necessary to solve for the center of the circle by bisecting at least two of the triangle's angles.

In Fig. 2-27 the angle at vertex point 1 is bisected by swinging arc RA to establish points 4 and 5 (RA is any convenient length). From points 4 and 5 swing equal arcs (RB). Point 6 is the intersection of the two arcs. Draw line 1-6 to establish the bisector of the angle and extend this line beyond point 6. Bisect the angle at vertex point 2 by swinging arc RC to locate points 7 and 8. Establish point 9 by drawing equal arcs (RD) from points 7 and 8. A line drawn from point 2 through point 9 and extended until it intersects the first bisector will meet at the center of the required circle (C). To check this point a third bisector can be constructed that will also meet at point C. Draw a line from point C perpendicular to one of the triangle's sides (side 1-3 is used here) to solve for radius CR. To complete the solution draw the inscribed circle using C as the center and CR for the radius.

FIG. 2-27 The inscribed circle of a triangle.

FIG. 2-28 The circumscribing circle of a triangle.

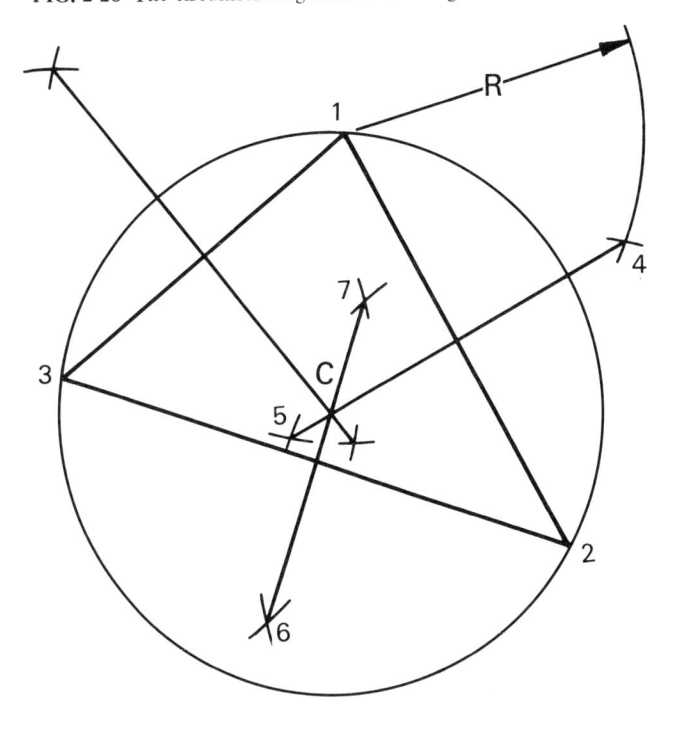

2.20 THE CIRCUMSCRIBING CIRCLE OF A TRIANGLE

The same procedure for drawing a circle through three points is used to construct the circumscribed circle of a triangle. In Fig. 2-28 sides 1-2 and 2-3 have been bisected. The intersection of bisectors 4-5 and 6-7 establish the center of the required circle (C). A third bisector (of line 1-3) can be drawn to check for accuracy. The radius of the required circle is the distance from C to any of the three points on the triangle (C-1, C-2, or C-3).

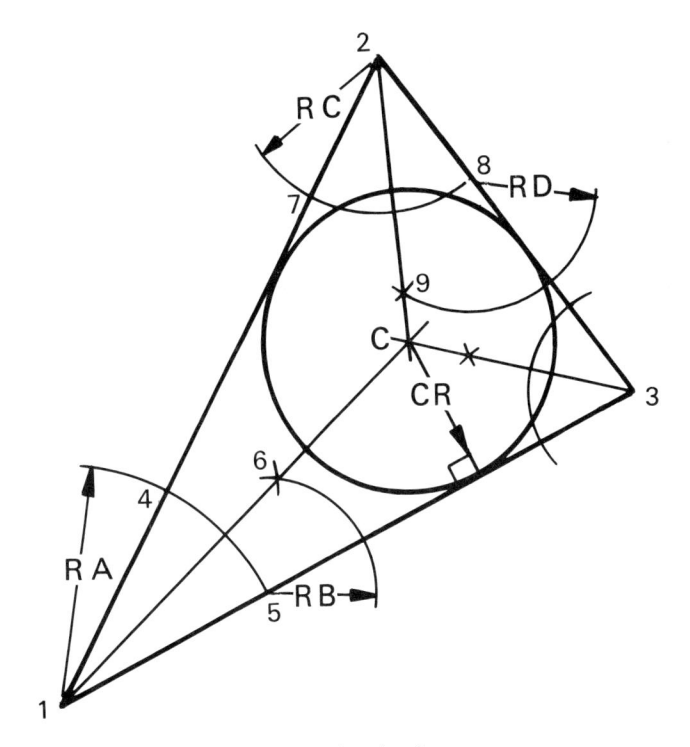

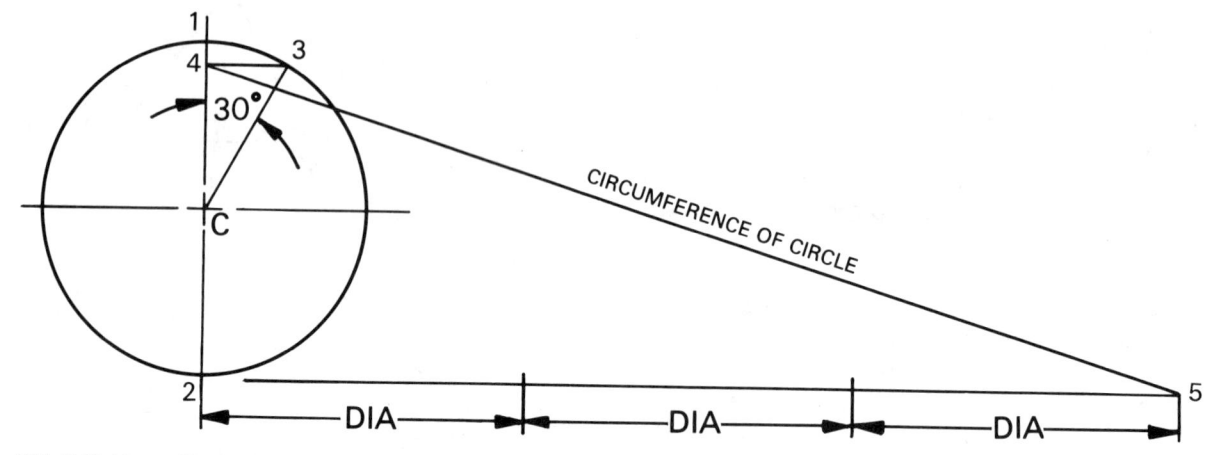

FIG. 2-29 To rectify the circumference of a circle.

2.21 TO RECTIFY THE CIRCUMFERENCE OF A CIRCLE

To rectify the circumference of a circle means to solve graphically for the circumference, "to make it straight." In Fig. 2-29 the circumference of the given circle has been established by rectification. Line 2-5 is drawn tangent to the bottom of the circle and exactly three times its diameter. Draw radius C-3 at an angle of 30 degrees as shown. Line 3-4 is then drawn perpendicular to the vertical centerline (line 1-2) of the circle. Connect point 4 to point 5. Line 4-5 will be equal to the circumference of the given circle. Note that the line 4-5 is 1/22,000 longer than the circumference, not exactly a measurable difference.

2.22 THE APPROXIMATE RECTIFICATION OF AN ARC

To rectify any given arc or curved line it is necessary to draw a line tangent to one end. In Fig. 2-30 the line was drawn tangent to the curved line at point 1. Note it is not necessary to have point A, though it does help to establish the exact tangent points. Use dividers to step off very small equal distances. The smaller the more accurate, because each distance will be the chord measurement of its corresponding arc segment and therefore will be somewhat shorter. Starting at the opposite end of the arc, away from the side with the tangent line, step off equal chords, point 14 to point 13, 13 to 12, 12 to 11, 11 to 10, etc. Continue stepping off each division until less than one full space remains, which is at point 2 in the given example. Note that in the example it works out perfectly, which will seldom be the case. Without lifting the dividers, start dividing the tangent line into the same number of segments, 2 to 3, 3 to 4, 4 to 5, etc. The tangent line 1-14 will approximately equal the length of the given arc.

FIG. 2-30 The approximate rectification of an arc.

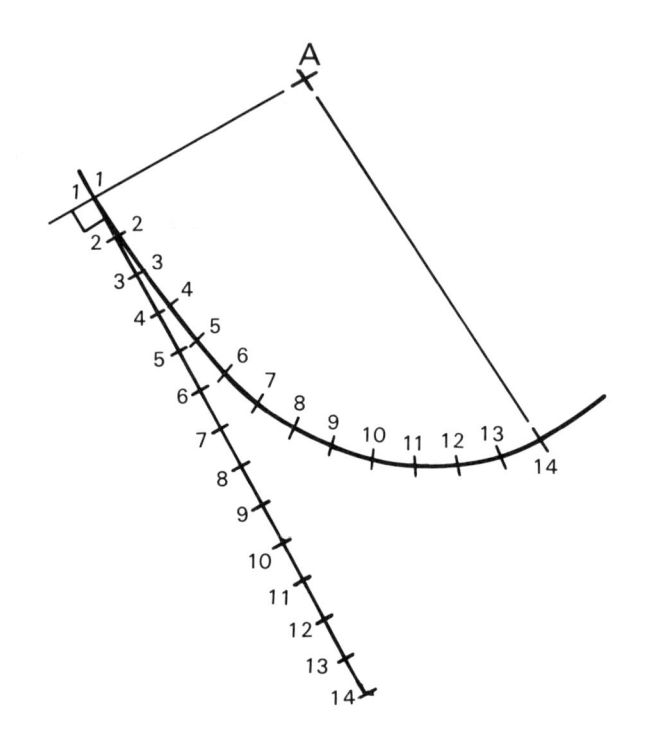

30

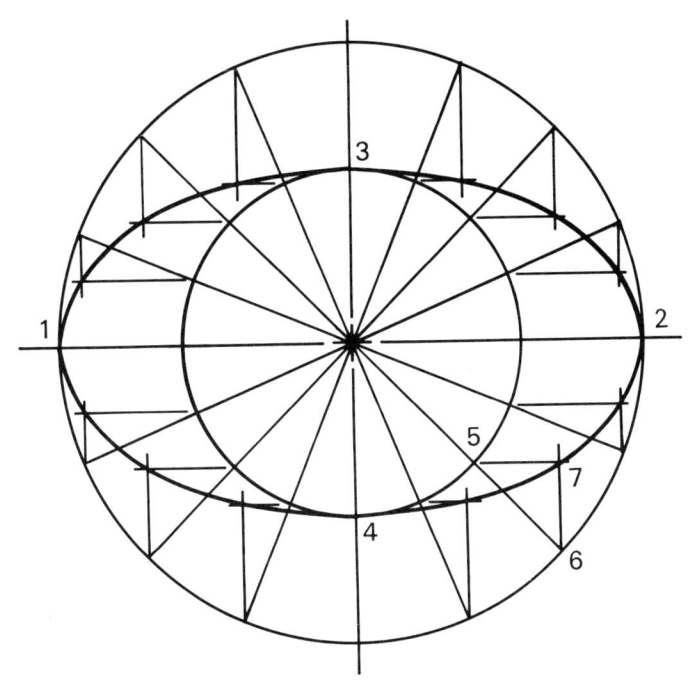

FIG. 2-31 Ellipse construction (the concentric circle method).

2.23 ELLIPSE CONSTRUCTION

There are many methods available for the construction of an ellipse. Whenever possible use a template. For odd sizes or for a very large ellipse one of the two methods given here will be sufficient. Both methods approximate the shape of a true ellipse.

In Fig. 2-31 the *concentric method* is shown. Given the major axis 1-2 and the minor axis 3-4, draw concentric circles (circles of a different size with the same center point) using the axes as diameters. Divide the circles into an equal number of sections; in the example, 16 equal divisions were used. Where each line crosses the inner circle (point 5) draw a line parallel to the major axis, and where the same line crosses the outer circle (point 6) draw a line parallel to the minor axis. The point of intersection of these two lines (point 7) will be on the required ellipse. Repeat this process for each division of the circles. Use a French curve to connect the points smoothly. Note that the ellipse will be easier to draw and more accurate in direct proportion to the number of divisions used and points located.

In Fig. 2-32 the *approximate method*, or *four-center method* as it sometimes called, was used to construct the ellipse. With the major axis (1-2) and minor axis (3-4) drawn, connect points 1 and 3. Using the distance from the center of the ellipse to point 1 as the radius, strike arc RA. Point 5 is the intersection of RA and the extended minor axis. The distance from point 3 to point 5 is used to establish RB. Swing arc RB so that it intersects line 1-3 at point 6. Bisect the line 1-6, and extend the bisector 7-8 so that it crosses the minor axis at point 10. Point 10 will be the center point for radius RD. Where bisector 7-8 crosses the major axis (point 9) swing radius RC to establish the sides of the ellipse at point 1 and point 2. RD is the radius for the upper and lower arc at points 3 and 4 of the required ellipse.

These methods work best when the minor axis is at least 75% of the major axis. The example in Fig. 2-31 and Fig. 2-32 shows what happens when the minor axis is too small in comparison to the major axis. The top and bottom of the ellipse are flattened.

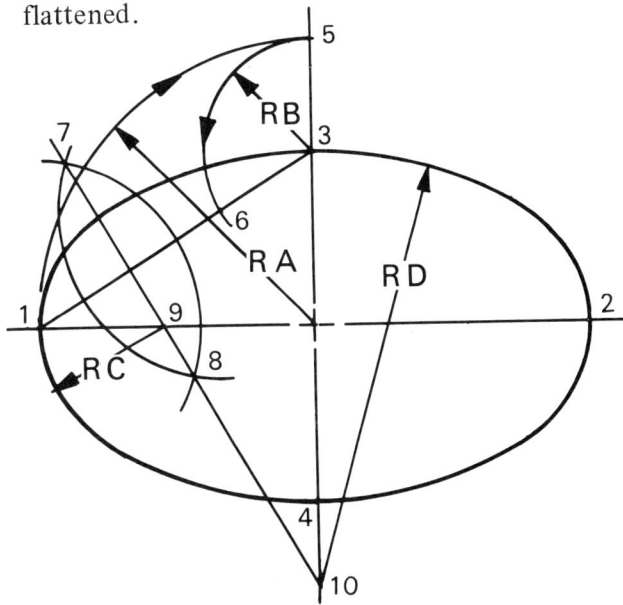

FIG. 2-32 The approximate ellipse.

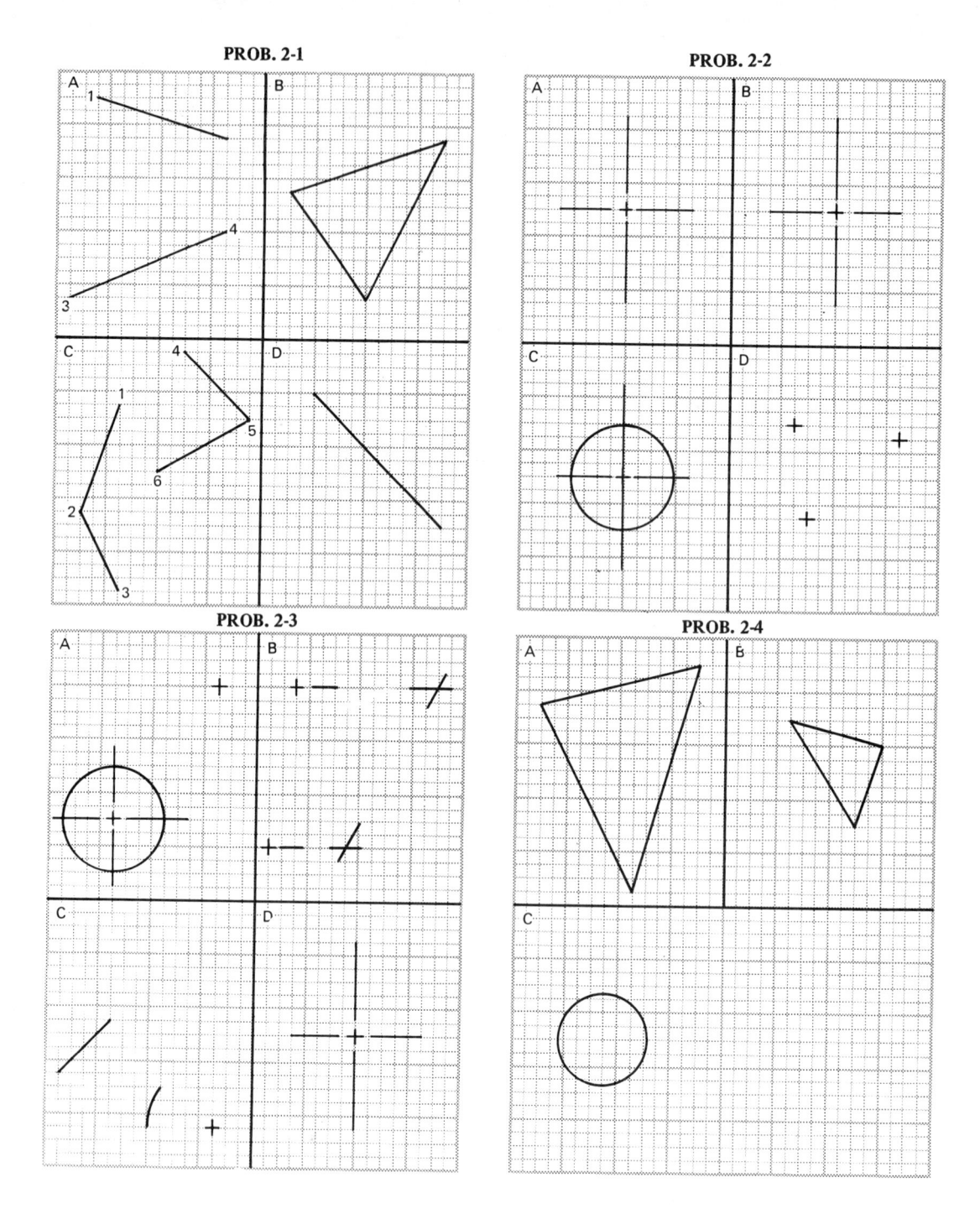

PROB. 2-1A. Divide line 1-2 into five equal parts. Starting from point 3 divide line 3-4 into proportional parts having ratios of 3:2:6.
PROB. 2-1B. Construct bisectors of the angles of the triangle.
PROB. 2-1C. Bisect both angles.
PROB. 2-1D. Draw a triangle having sides with the proportions 3:4:5. The given line is five units long.
PROB. 2-2A. Draw a hexagon that is 3″ (76.2mm) across the flats.
PROB. 2-2B. Draw a hexagon that is 3″ (76.2mm) across the corners.
PROB. 2-2C. Construct a seven-sided regular polygon in the given circle.
PROB. 2-2D. Draw the circle that passes through the three given points.

PROB. 2-3A. Draw two lines that pass through point 1 and that are tangent to the circle. Indicate the angle between the lines.
PROB. 2-3B. Connect the given lines with 1″ (25.4mm) radius arcs.
PROB. 2-3C. Connect the given lines with a 1¼″ (31.75mm) radius arc. Extend the lines if necessary.
PROB. 2-3D. Draw an ellipse having axes of 3″ (76.2mm) and 2″ (50.8mm).
PROB. 2-4A. Construct the inscribed circle of a triangle.
PROB. 2-4B. Draw the circumscribed circle of the given triangle.
PROB. 2-4C. Find the center of the circle by perpendicular bisectors. Rectify the circle.

FIG. 2-33 Prisms.

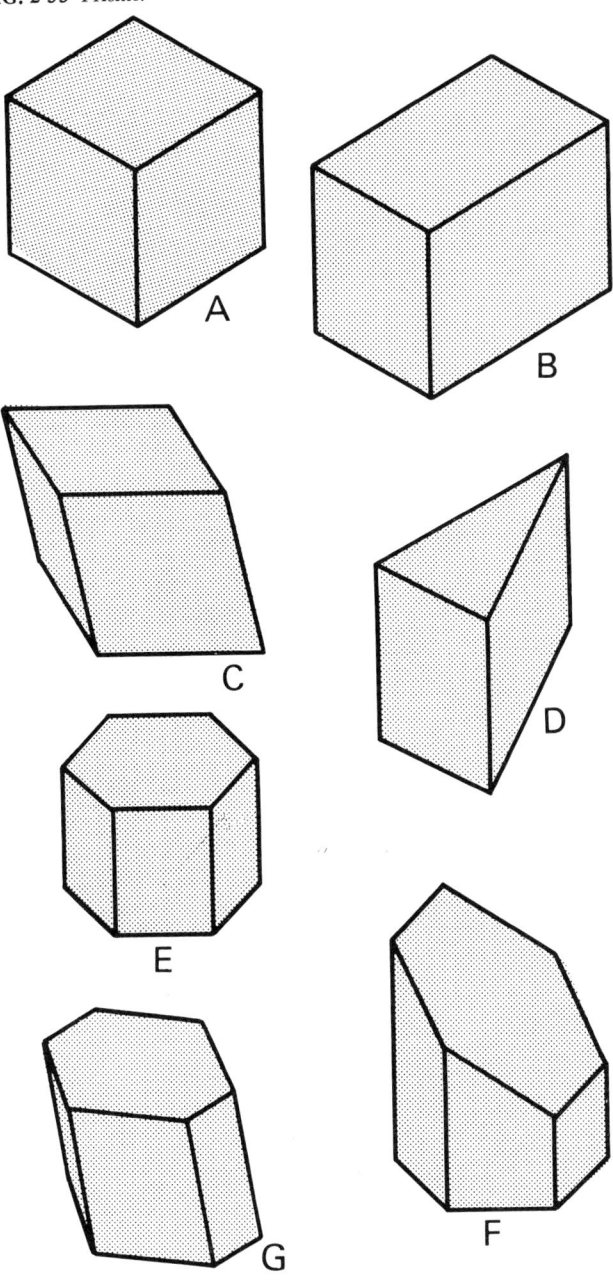

(C) is an *oblique prism*, (D) is a *right triangular prism*, (E) is a *right hexagonal prism*, (F) is a *truncated prism*, and (G) is an *oblique hexagonal prism*.

Figure 2-34 shows different types of **pyramids**. A *pyramid* is a *polyhedron* that has a polygon for a base and triangles with a common vertex for faces. (A) has four equal sides and is called a *tetrahedron*. (B) is a *right square pyramid*, (C) and (E) are *oblique pyramids*. (D) is a *right hexagonal pyramid*, (F) is a *frustrum of a pyramid*, and (G) is a *truncated pyramid*.

These examples are just a few of the possible variations of polyhedra that one may encounter doing engineering work. The text defines and gives examples for many different types of prisms and pyramids and their relationship with other solids, planes, lines, and points.

FIG. 2-34 Pyramids.

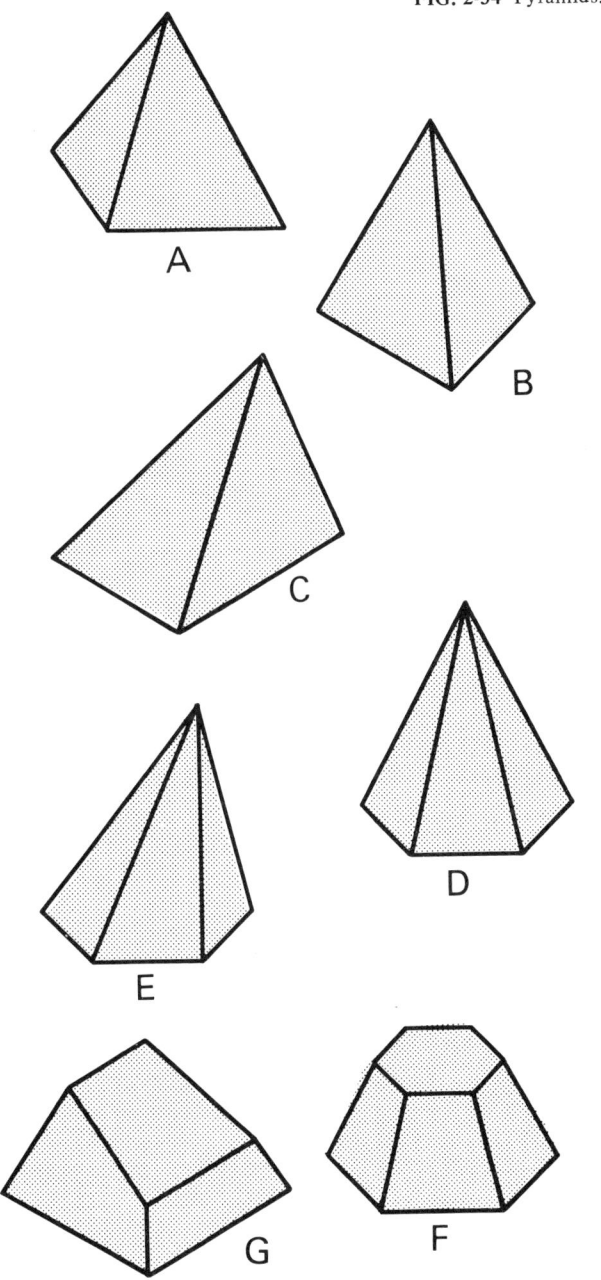

2.24 GEOMETRIC FORMS, POLYHEDRA

Geometric forms include a wide range of shapes and plane figures such as squares, triangles, and circles, and solids created by plane, single-curved, double-curved, and warped surfaces. All of these forms will be explained in the text as they are used to present various types of descriptive geometry applications.

The geometrical forms shown in Fig. 2-33 and Fig. 2-34 are **polyhedra**, solids formed by plane surfaces. Every surface (face) of each solid is a polygon. **Prisms** are *polyhedra* that have two parallel polygon shaped ends and sides that are composed of parallelograms. In Fig. 2-33, (A) is a *cube* (six equal sides), (B) is a right *rectangular prism*,

33

(C), a *frustrum of a cone* (D), *oblique cone* (E), and *truncated cone* (F).

Double-curved surfaces are *surfaces of revolution* generated by curved lines. Double-curved surfaces have no straight line elements and are generated by a curved line that moves according to a particular mathematical law; therefore, they can only be approximately developed. Figure 2-36 illustrates seven types of double-curved surfaces: *sphere* (A), *half sphere* (hemisphere) (B), *ellipsoid* (C), *paraboloid* (D), *hyperboloid* (E), *serpentine* (F), and *torus* (G).

Single-curved and double-curved surfaces are described in detail later in the text. Descriptive geometry procedures and applications are used throughout industry when working with these types of solid shapes or hollow forms. Intersection and development applications involving ruled, curved, and warped surfaces are presented in Chapters 6 and 7.

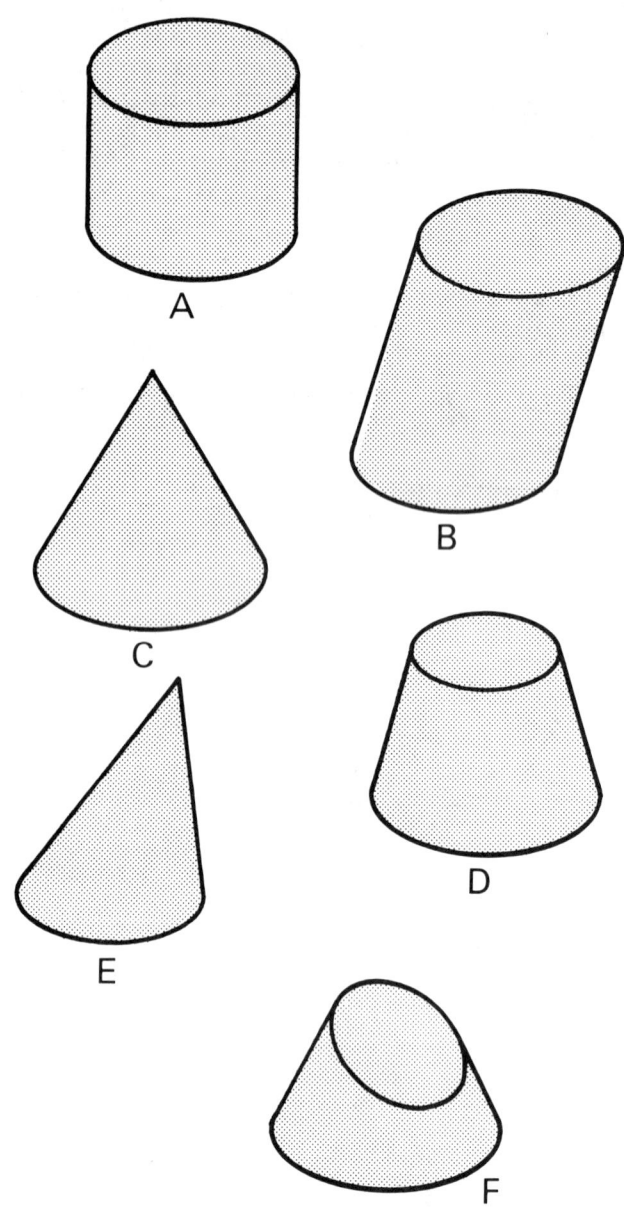

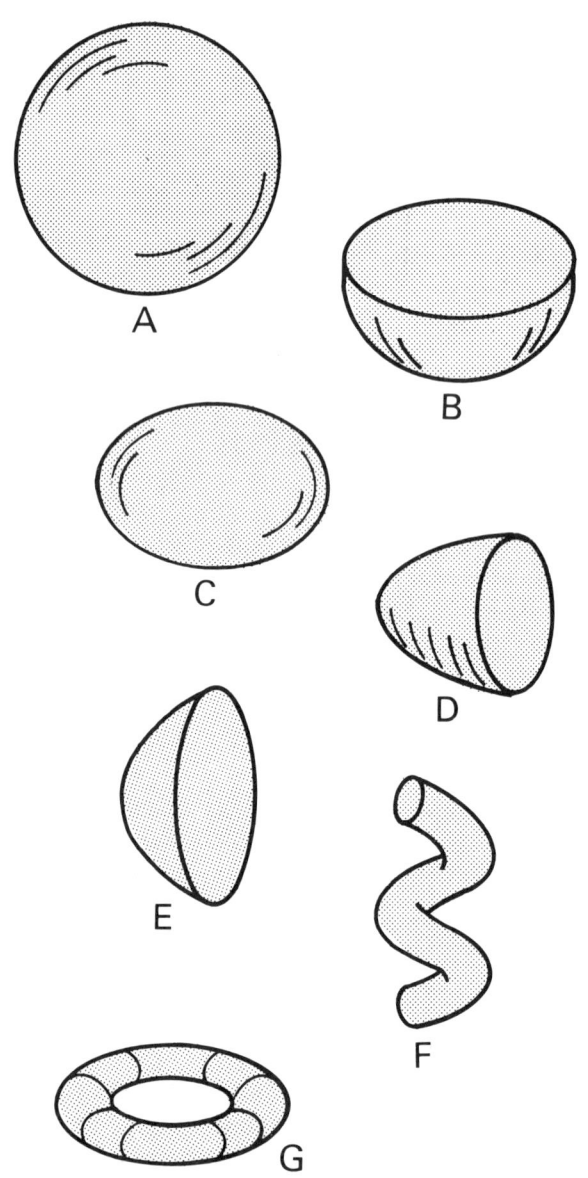

FIG. 2-35 Single-curved surfaces.

2.25 CURVED SURFACES

Curved surfaces are divided into two categories, *single-curved* and *double-curved*. **Ruled surfaces** like those in Figs. 2-33, 2-34, and 2-35 are generated by moving a straight line. *Double-curved* surfaces are generated by the movement of a curved line. The **generatix** is any position of the generating line (element of the surface).

Single-curved surfaces, Fig. 2-35, are generated by moving a straight line so that in any two of its closest positions it will lie in the same plane. Forms that are bounded by single-curved surfaces include *cones* and *cylinders* (both are generated by moving a straight line). In Fig. 2-35, (A) is a *right cylinder* and (B) is an *oblique cylinder*. Variations of cones are also singled-curved surfaces, *right cone*

FIG. 2-36 Double-curved surfaces.

2.26 WARPED SURFACES

When a straight line is moved so that it does not lie in the same plane with any of its two closest positions, it will create a warped surface. Chapter 9 describes warped surfaces in greater detail. The generatix of a warped surface is a straight line and each of its positions will lie in different planes; in other words they are not parallel and do not intersect. Therefore a warped surface can only be approximately developed.

Figure 2-37 shows only five variations of **warped surfaces**: *hyperboloid* (A), *hyperbolic paraboloid* (B), *helicoid* (C), *oblique helicoid* (D), and *conoid* (E). Other types of warped surfaces that are not shown here are the *cylinoid* and *warped cone*.

There are many industrial uses for these specialized forms, including screw threads (oblique helicoid), screw conveyors (right helicoid), and cooling towers (hyperboloid of revolution). The aerospace industry applies single-curved surfaces, double-curved surfaces, and warped surfaces alone and in combination, along with a variety of plane surfaces in the design of spacecraft, airplanes, missiles, etc. Figure 2-38 shows an example of the application of these shapes and forms. The full-scale model of one of the space shuttle concepts is shown here in a wind tunnel, where feasibility tests are conducted concerning its shape, the craft's reaction to reentry, and its flying characteristics. Notice the different types and combinations of shapes and contours involved in this example. There are many examples of aircraft and other applications of these types of surfaces throughout the text.

FIG. 2-37 Warped surfaces.

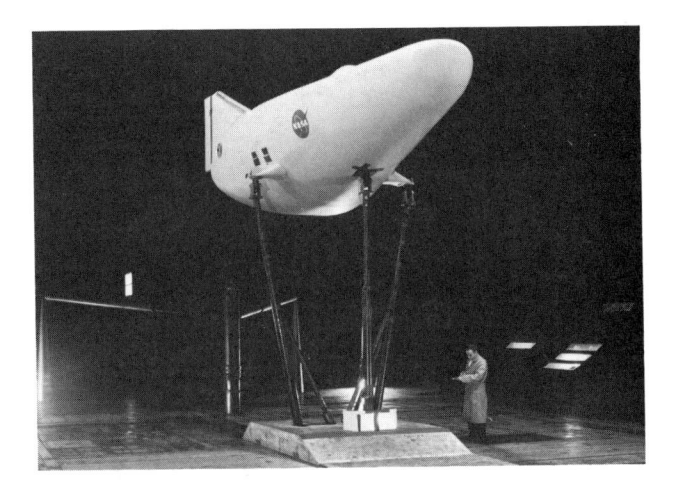

FIG. 2-38 One of the original space shuttle concepts. *(Courtesy NASA.)*

3
PROJECTION

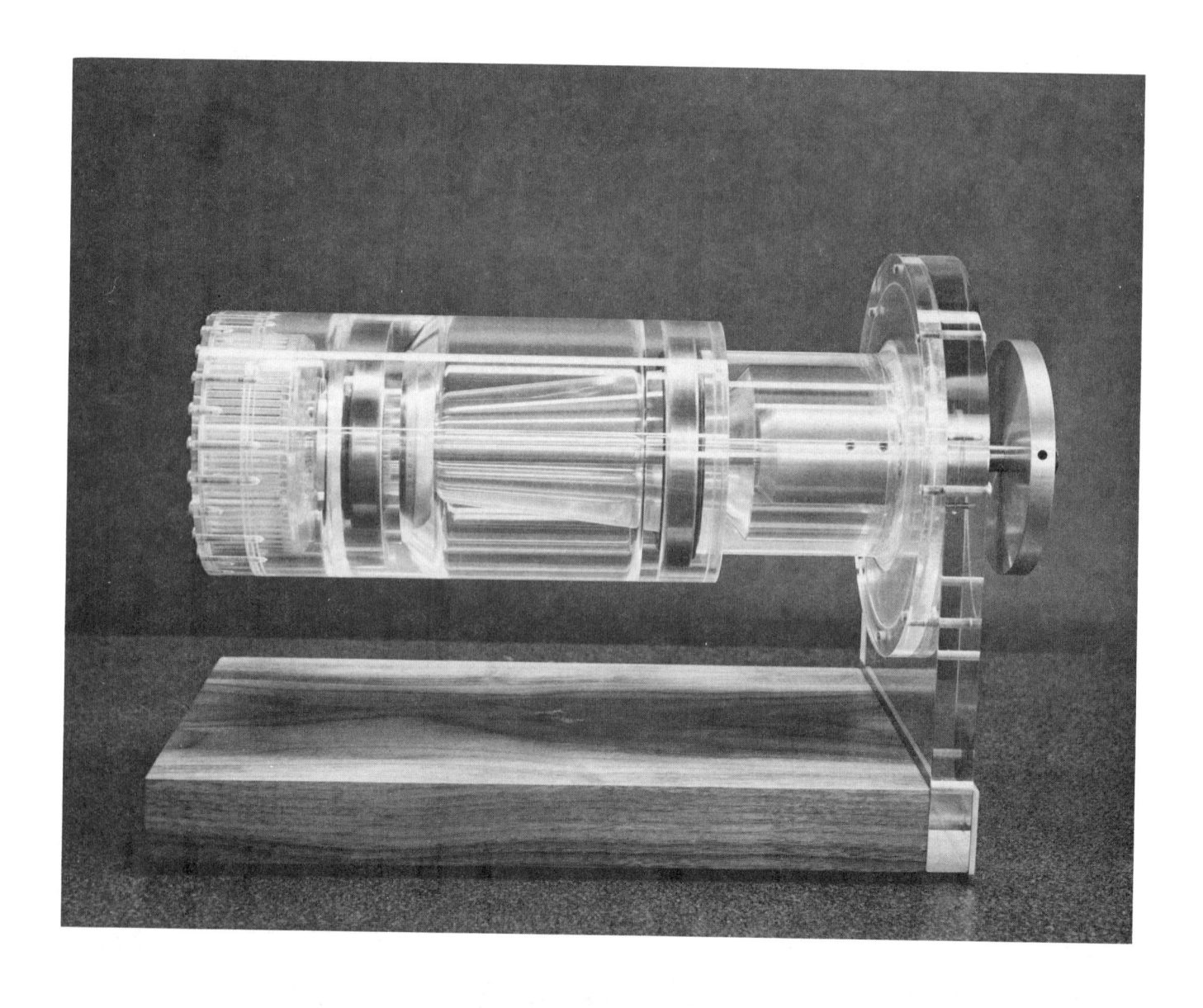

PICTORIAL

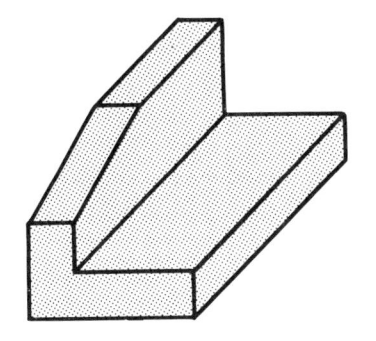

OBLIQUE PROJECTION

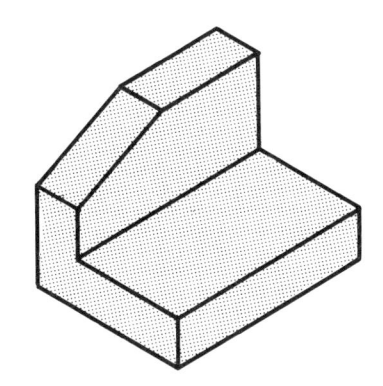

ISOMETRIC PROJECTION

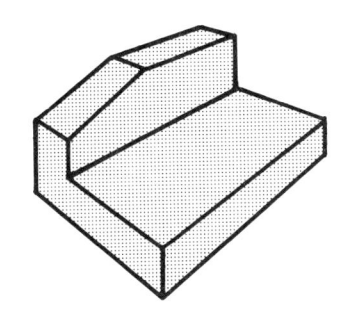

PERSPECTIVE PROJECTION

MULTIVIEW

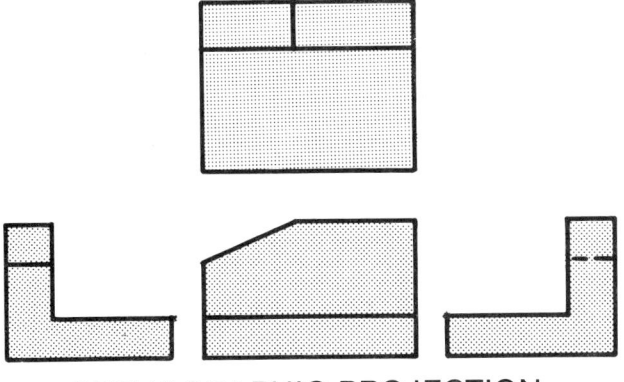

ORTHOGRAPHIC PROJECTION

3.1 PROJECTION

All forms of engineering and technical work require that a two-dimensional surface (paper) be used to communicate ideas and the physical description of a variety of shapes. Here projections have been divided into two basic categories: pictorial and multiview. This simple division separates single-view projections (**oblique, perspective,** and **isometric**) from multiview projections (**orthographic**). Theoretically, projections can be classified as convergent and parallel, or divided into three systems of projection: perspective, oblique, and orthographic. Division of types based on whether the drawing is a one or multiview projection sufficiently separates projection types into those used for engineering working drawings (orthographic) and those used for display (architectural rendering, technical illustration, etc.).

Descriptive geometry uses multiview orthographic projection almost exclusively, except where a pictorial is used as a tool in understanding aspects of intersections and shading.

In Fig. 3-1 the angle block was drawn using each of four projection types and the same scale. This figure illustrates the difference between the types of projections and their shortcomings.

Pictorial projections are single-view drawings that do not lend themselves to the communication of engineering data except as rough sketches for the presentation of preliminary ideas. **Perspective** projections are constructed using projecting lines that converge at a point, and thus will not show the true dimensions of the object. The **oblique** method distorts when the object's depth becomes too great. Foreshortening may eliminate some distortion but then the drawing's dimensions cannot be scaled. The **isometric** method uses full-size dimensions for all lines that are vertical or are parallel to the axes (recede at 30 degrees) and is therefore more useful for engineering sketching. Isometric angles distort as do lines not vertical or parallel to the axes, and cannot be scaled from the drawing.

Multiview drawings, because they show the object in more than one view, are not lifelike. This is their primary defect. This method of projection presents the object's top, front, and side simultaneously. All dimensions are drawn to scale and the three basic views can be used to project any number of needed auxiliary views in order to solve for specific engineering data.

FIG. 3-1 Types of projection.

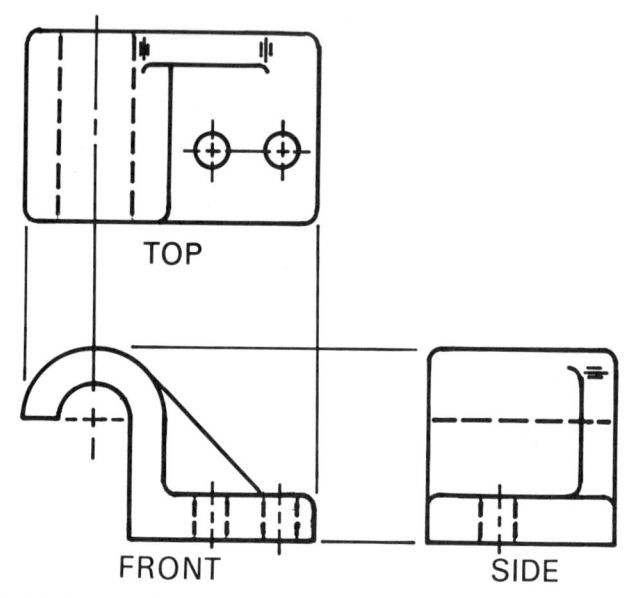

FIG. 3-2 Orthographic Drawing.

3.2 MULTIVIEW PROJECTION

Multiview/orthographic projection is the primary means of graphic communication used in engineering work. Drawings like Fig. 3-2 are used to convey ideas, dimensions, shapes, and procedures for the manufacture of an object or construction of a system. Orthographic projection is the basis of all descriptive geometry procedures. Multiview projection is a procedure that can be used to completely describe an object's shape and dimensions using two or more views that are normally projected at 90 degrees to each other, or at specified angles for auxiliary views. In general, engineering work is completed using this method of projection. The finished drawing is then reproduced and sent to the shop or to the job site.

Another method that has gained increasing acceptance is shown in Fig. 3-3. The engineering model provides a three dimensional scale representation of a particular project. The model method works well for many types of projects, especially those too complicated for graphic description, but is limited because of the impossibility of accurately reproducing the model's views with a photograph. This figure shows the front view of a reactor model. Distortion and the inability to scale the photo reduce effectiveness. Models do, however, reduce the total number of necessary drawings by 20% to 50%.

With the advent of widespread use of computer-aided drafting and design (CADD) computers are now being used to design and illustrate many of the projects formerly hand drawn. This new medium eliminates the need for hand-completed linework and lettering but still requires that the operator be well versed in the practice and theory of multiview projection. Knowledge and understanding of engineering drawing and descriptive geometry, which are based on orthographic projection, is essential for the aspiring engineer, designer, drafter, or CADD operator.

Descriptive geometry is basically the use of orthographic projection in order to solve for advanced technical data involving the spatial relationship of points, lines, planes, and solid shapes. There are two primary means of understanding these types of orthographic projection: the **normal/natural method** or the **glass box method**. The *normal* or *natural method* is illustrated in Fig. 3-2 and is typical of mechanical and other engineering fields. In this method the front view will show the height and width; the top view, the depth and width; and the side view, the depth and height. Therefore the width dimension will vertically align the top and front views, and the height dimension will horizontally align the front and side views. This method requires that the object be viewed perpendicular to each of its three primary surfaces. The observer changes position for each view.

The **glass box method**, used primarily for descriptive geometry problems, requires that the user imagine that the object, points, lines, planes, etc., are enclosed in a transparent "box." Each view of the object is established on its corresponding glass box surface by means of perpendicular projectors originating at each point of the object and extending to the related box surface. The box is hinged so that it can be unfolded onto one flat plane (the paper). The following pages discuss this method in detail.

FIG. 3-3 Reactor containment design verification model.

FIG. 3-4 Orthographic projection.

ORTHOGRAPHIC PROJECTION

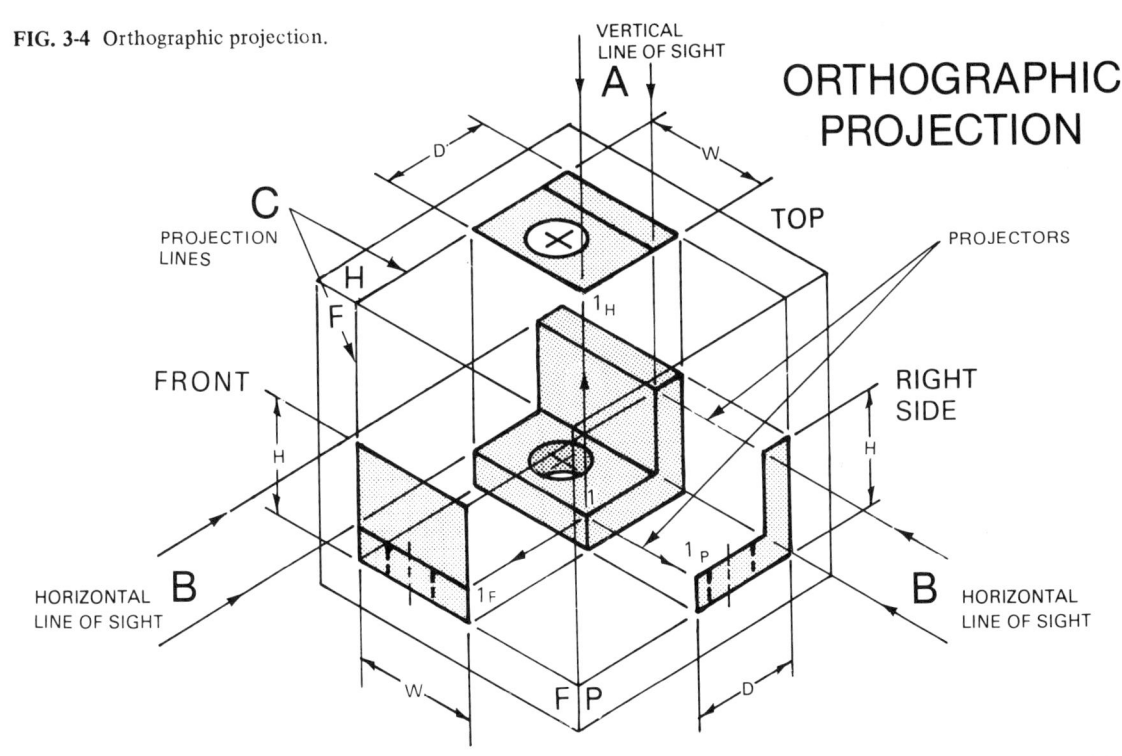

3.3 ORTHOGRAPHIC PROJECTION

The glass box method of projection introduced in the last section is illustrated in its closed (folded) position in Fig. 3-4 and open (unfolded) position in Fig. 3-5. The given object has been theoretically enclosed in the transparent box. The concepts and definitions presented here will be used throughout the text.

LINES

A = Vertical lines of sight
B = Horizontal lines of sight
C = Projection lines

DIMENSIONS

D = Depth, H = Height, W = Width

IMAGE PLANES

F = Front (*frontal plane*)
H = Top (*horizontal plane*)
P = Side (*profile plane*)

The lines of sight represent the direction from which the object is viewed. The vertical lines of sight (A) and horizontal lines of sight (B) are assumed to originate at infinity. The line of sight is always perpendicular to the image plane, represented by the surfaces of the glass box (top, front, and right side). Projection lines (C) connect the same point on the image plane from view to view, always at right angles. Remember that "the object" could be any graphical form: point, line, plane, etc.

A point is projected upon the image plane where its projector, or line of sight, pierces that image plane. In these figures point 1, which represents a corner of the given object, has been projected onto the three primary image planes. Where it intersects the horizontal plane (top image plane) it is identified as 1_H, when it intersects the frontal plane (front image plane) it is identified as 1_F, and where it intersects the profile plane (right side image plane) it is labeled 1_P.

Figure 3-5 shows the position of the unfolded image planes which now lie in the same plane as the paper. Note that the position and labeling of point 1 is the same here as in Fig. 3-4.

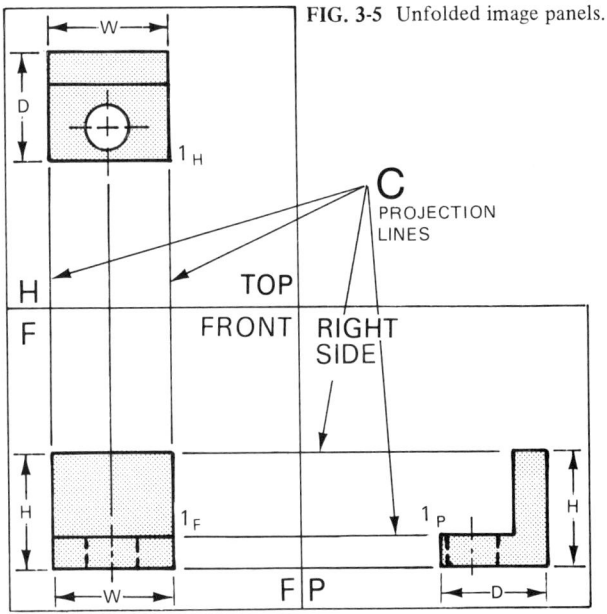

FIG. 3-5 Unfolded image panels.

3.4 THE SIX PRINCIPAL VIEWS

When the glass box is opened, its six sides are revolved outward so that they lie in the plane of the paper. With the exception of the back plane, all are hinged to the front plane. The back plane is normally revolved from the left side view but can also be hinged to the right side view as shown in Fig. 3-6. Each image plane is perpendicular to its adjacent image plane and parallel to the image plane across from it before it is revolved around its hinged fold line (reference line). *A fold line is the line of intersection between any hinged (adjacent) image plane. This includes principal and auxiliary views.*

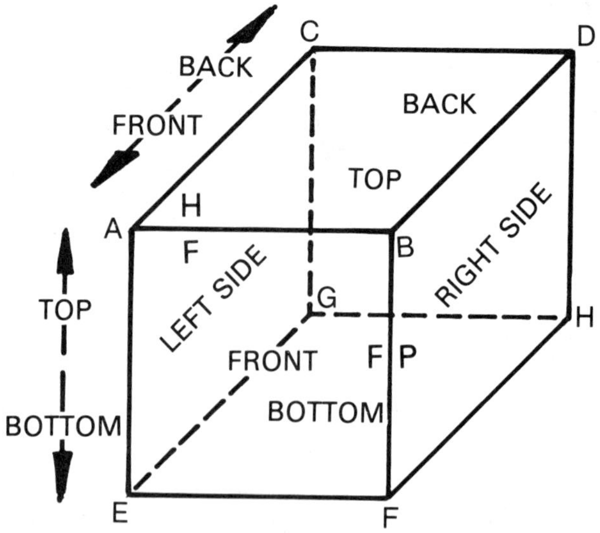

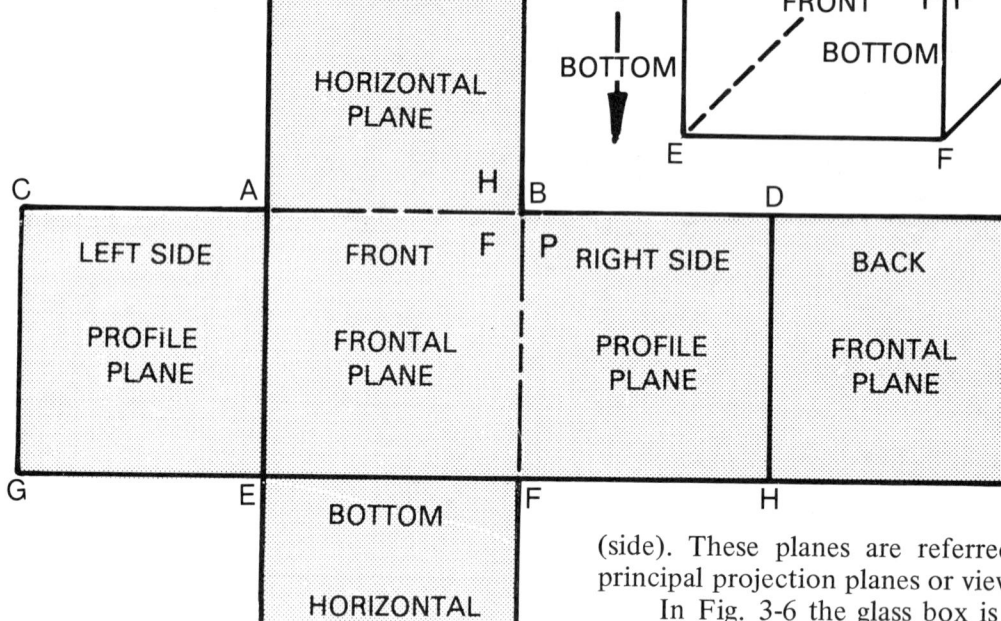

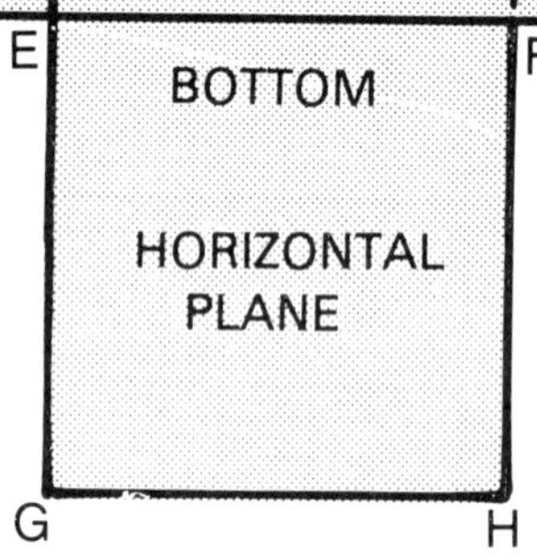

The left side, front, right side, and back are all elevation views. Each is vertical. In these views the height dimension, elevation, and top and bottom of the view can be determined and dimensioned.

The top and bottom planes are in the horizontal plane. The depth dimension, width dimension, and front and back can be established in these two horizontal planes.

In most cases the top, front, and right sides are required. These are consistently labeled throughout descriptive geometry as the **horizontal** plane, H (top), **frontal** plane, F (front), and **profile** plane, P

FIG. 3-6 The six principal views or image planes.

(side). These planes are referred to as the three principal projection planes or views.

In Fig. 3-6 the glass box is shown in oblique projection before it is revolved. Its revolved position is also provided. Notice that the top, front, and bottom are in line vertically, and the left side, front, right side, and back are aligned horizontally. An exception to this alignment happens when the glass box is revolved around the top (horizontal) view, which is advantageous when the object has great depth and width in comparison to the height.

Auxiliary views of an object are not parallel to any of the six principal views and will be discussed later. This type of view may be used to establish and dimension a portion of the object tha does not lie in the plane of a principal view. Note that point, line, plane, solid, or a combination of these forms can be substituted for the word "object."

When using directions to establish the location of a point, object, etc., the *top* and *bottom* are shown in the frontal plane; the terms *above* and *below* are also used to describe directions in this plane. The horizontal view can be used to determine if a point is "in *front* of" or "in *back* of" a particular starting point or fold line. To locate a point to the *right* or *left* of a particular fold line or established point the frontal or horizontal plane can be used. In the profile plane the *top, bottom, front,* and *back* can be determined.

FIG. 3-7 First and third angle projection.

1 FIRST ANGLE PROJECTION

2 THIRD ANGLE PROJECTION

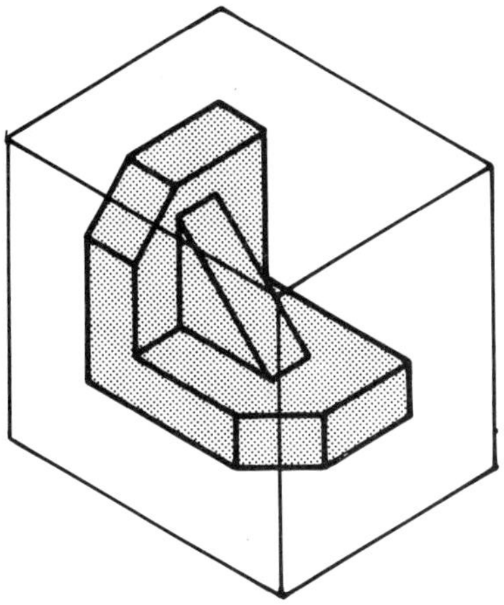

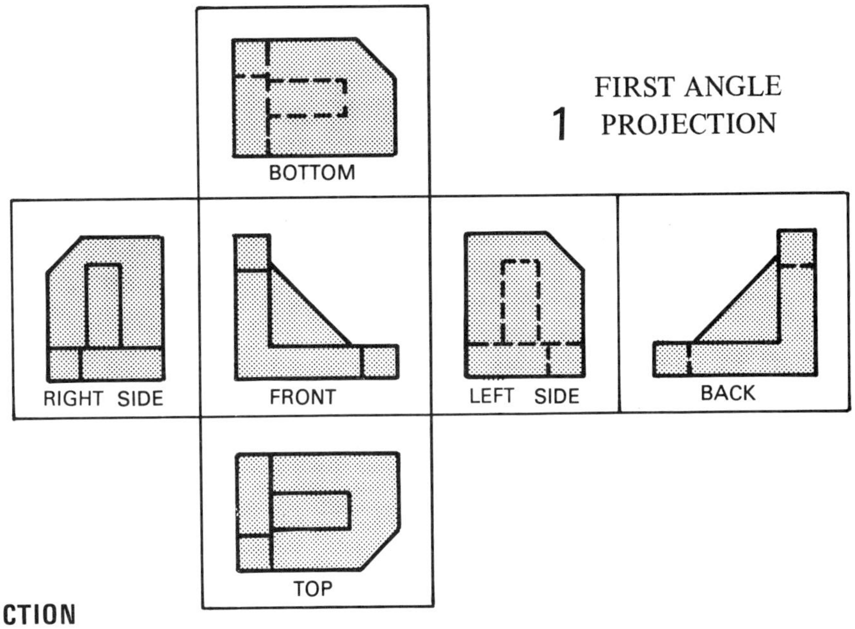

3.5 FIRST AND THIRD ANGLE PROJECTION

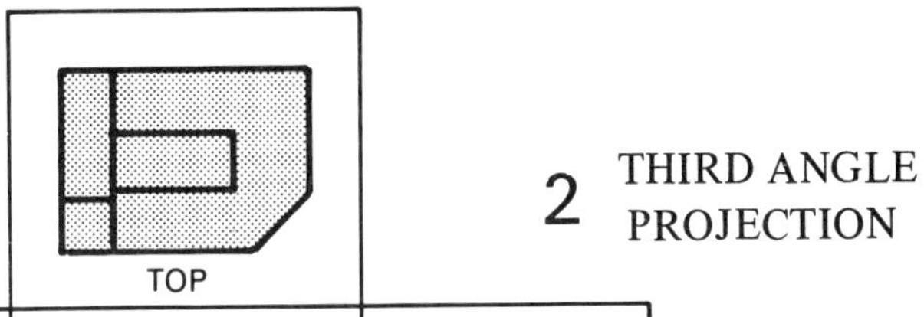

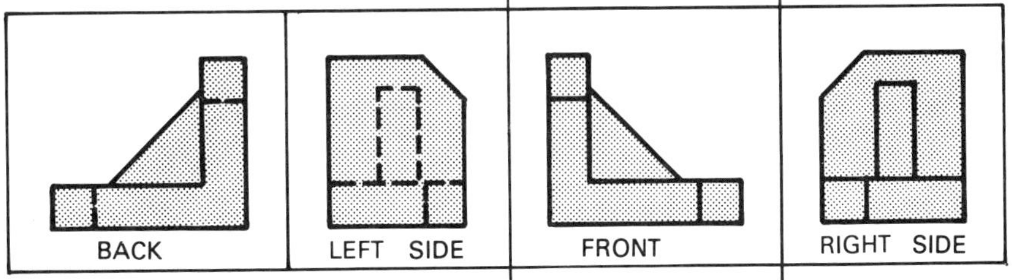

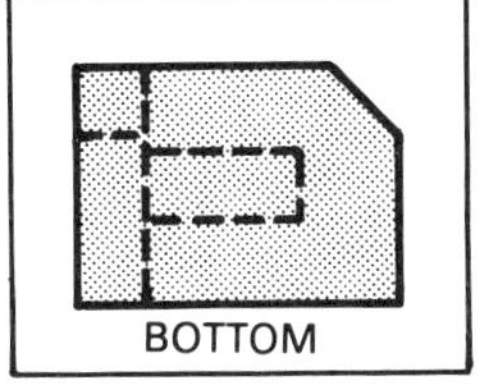

The six principal views of an object or the glass box have previously been presented in the type of orthographic projection known as *third angle orthographic projection*. This form of projection is used throughout descriptive geometry and is the primary form of projection found in all of American industry with the exception of some special cases in the architectural and structural fields. In third angle projection the line of sight goes through the image plane to the object, as shown in Fig. 3-4. To obtain each view of the object it is then necessary to assume that the object is projected back (along the lines of sight) to the image plane. **Projectors** are used to illustrate this projection from the object to where they intersect the image plane. Figure 3-7(2) illustrates third angle projection and the normal procedure for unfolding the glass box. Notice how each view is hinged to the front with the exception of the back view which is normally rotated from the left side.

The type of projection used in most foreign countries and on many American structural and architectural drawings is called *first angle orthographic projection*. In this form of projection the object is assumed to be in front of the image plane. Each view is formed by projecting through the object and onto the image plane. Figure 3-7(1) shows how each view looks when using first angle projection. Notice that the left side view is placed on the right of the front view, the right side view is shown to the left of the front view and the top view is placed below the front view, exactly the opposite as in third angle projection. Each view in third angle projection is hinged to the front view except the back, which is normally rotated from the left side view.

Regardless of which form is used, orthographic projection is still right angle projection, where each view is projected at right angles to the previous image plane (view). All projectors for a given view are parallel and are at right angles to the image plane (of a given view).

Throughout the text only third angle projection is used.

41

3.6 THE GLASS BOX AND FOLD LINES

Each image plane, surface of the glass box, is connected at right angles to an adjacent view. The top view is "hinged" to the front view as is the right side view. These hinge lines are the intersection of the perpendicular image planes and are called **fold lines** (reference). Normally fold lines are not shown on technical drawings, but in descriptive geometry all dimensions are taken from fold lines as will be seen.

In Fig. 3-8(2) the object is enclosed in a glass box. The top image plane (horizontal plane) is shown being rotated about the line of intersection, H/F, fold line, between itself and the front image plane (frontal plane). This fold line is labeled **H/F**. The side image plane (profile plane) is rotated about the line of intersection between it and the front image plane. This fold line is labeled **F/P**.

Figure 3-8(3) shows the glass box opened into the plane of the paper. Remember it is assumed that the front view is stationary and all other views are hinged to it. Each required view is then rotated so as to be in the same plane as the front view.

Measurements are taken from fold lines to locate the object and its dimensions. Dimension A is the distance from the H/F fold line to the top of the object. Dimension B is the distance from the H/F fold line to the front of the object, and dimension C is the distance that the right edge of the object is from the F/P fold line. The width (W) of the object is found in the top and front views, the height (H) in the front and side views, and the depth (D) in the top and side views.

Point 1 is located on the front right corner of the object and is shown projected onto each of the three image planes. The subscripts next to the point establish what projection plane it is in: H for horizontal, F for frontal and P for the profile plane.

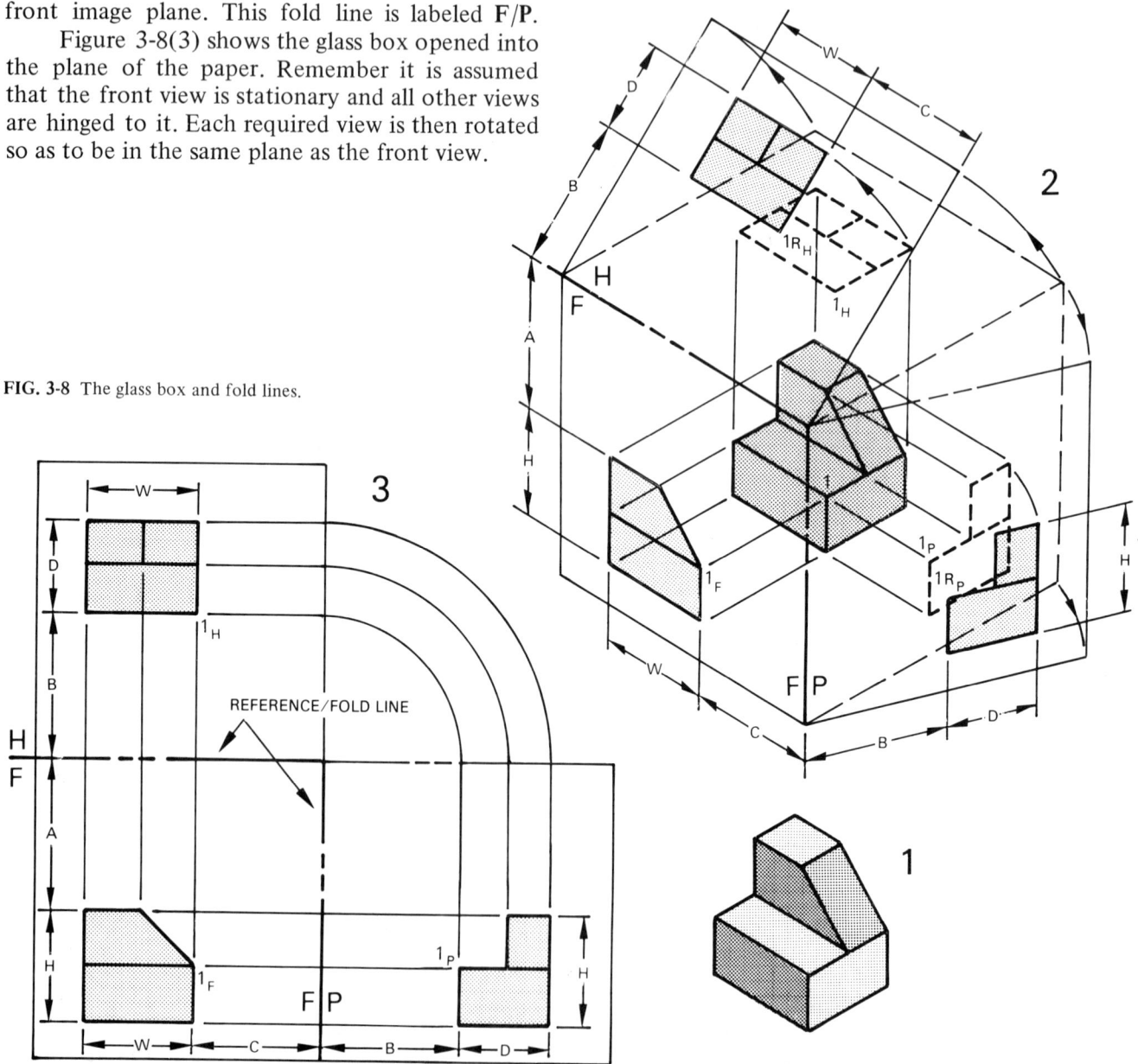

FIG. 3-8 The glass box and fold lines.

MULTIVIEW ORTHOGRAPHIC PROJECTION

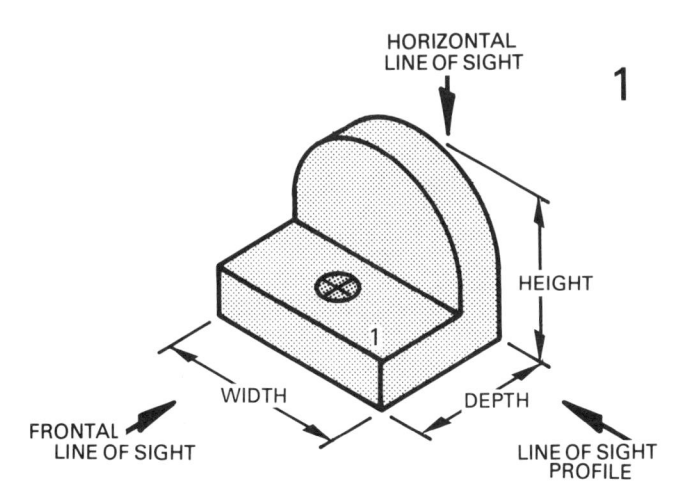

3.7 LINE OF SIGHT

When an object is projected onto an image plane it creates a "view" of that object. In Fig. 3-9(1) the **line of sight** for each view is shown. These lines of sight establish the direction of viewing that the observer will take when completing each view. In Fig. (3), (4), and (5), the top, front, and side views are broken apart and analyzed separately. Notice that all points on each surface of the object are projected onto their corresponding image plane (view). Where these projectors pierce the image plane they create the view of the object. Figure 3-9(2) shows the three views properly aligned so as to form an orthographic drawing.

The line of sight is always at right angles to the projection plane. In order to properly visualize this, one must imagine that they are standing in front of the object with the image plane between them and the object. The position of the observer will change for every view of the object so that the observer's line of sight is at right angles to each image plane.

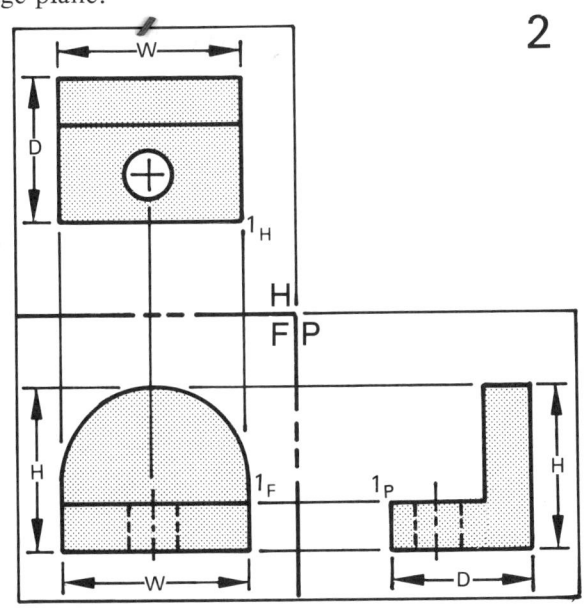

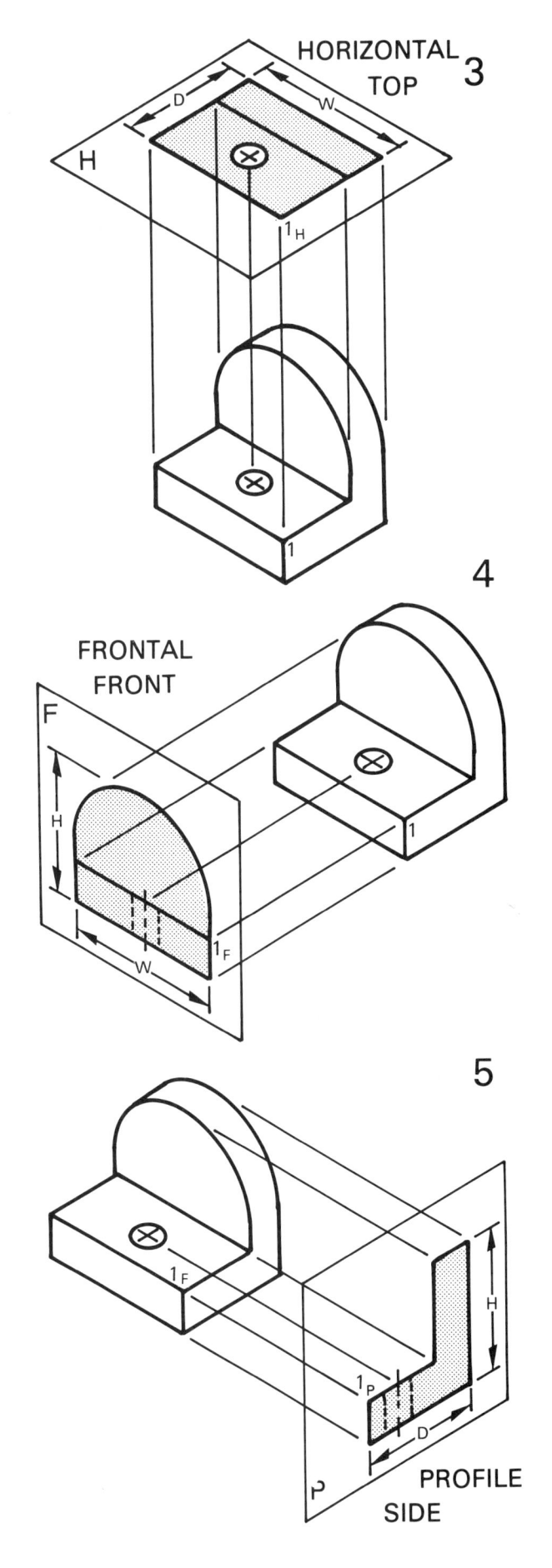

FIG. 3-9 Line of sight.

3.8 AUXILIARY VIEWS

Any view that lies in a projection plane other than the horizontal, frontal, or profile plane is considered an auxiliary view. This type of projection is essential where the object to be drawn is complex and has a variety of lines or planes that are not parallel to one of the three principal planes. In Figs. 3-10, 3-11, and 3-12 the adequate representation of each piece of machinery, structure, or object using orthographic projection would require the construction of many primary, secondary, and successive auxiliary views.

Primary auxiliary views are projected from one of the principal views. A primary auxiliary view will be perpendicular to one of the three principal planes and inclined to the other two. *Secondary auxiliary views* are taken off of a primary auxiliary view and are inclined to all three principal planes of projection. *Successive auxiliary views* are projected from secondary auxiliary views.

In industry, auxiliary views are used to describe the true configuration of an object and to dimension it in views that will show inclined lines' or planes' true size. In most cases only partial auxiliary views are constructed. For descriptive geometry problems, primary, secondary, and successive auxiliary views are normally drawn in their

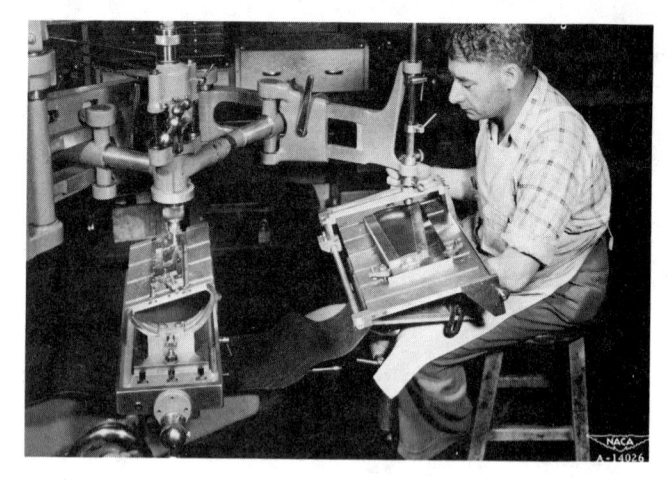

FIG. 3-11 A three dimensional duplicator being used on an airfoil section. *(Courtesy NASA.)*

entirety in order to solve for aspects of points, lines, planes, or three-dimensional shapes.

In Fig. 3-10 the rotational device and its supporting structure were designed by applying the principles of orthographic projection and the use of auxiliary views in order to solve for and dimension the true length of each structural element and the angles between elements. The Groton three-dimensional duplicator in Fig. 3-11 required the use of descriptive geometry principles in its design. Angles, rotational clearances, and a variety of other technical data were used in its design and manufacturing stages.

FIG. 3-10 Rotational device for maneuvering practice. *(Courtesy NASA.)*

FIG. 3-12 Construction crane model. *(Courtesy Engineering Model Associates, Inc.)*

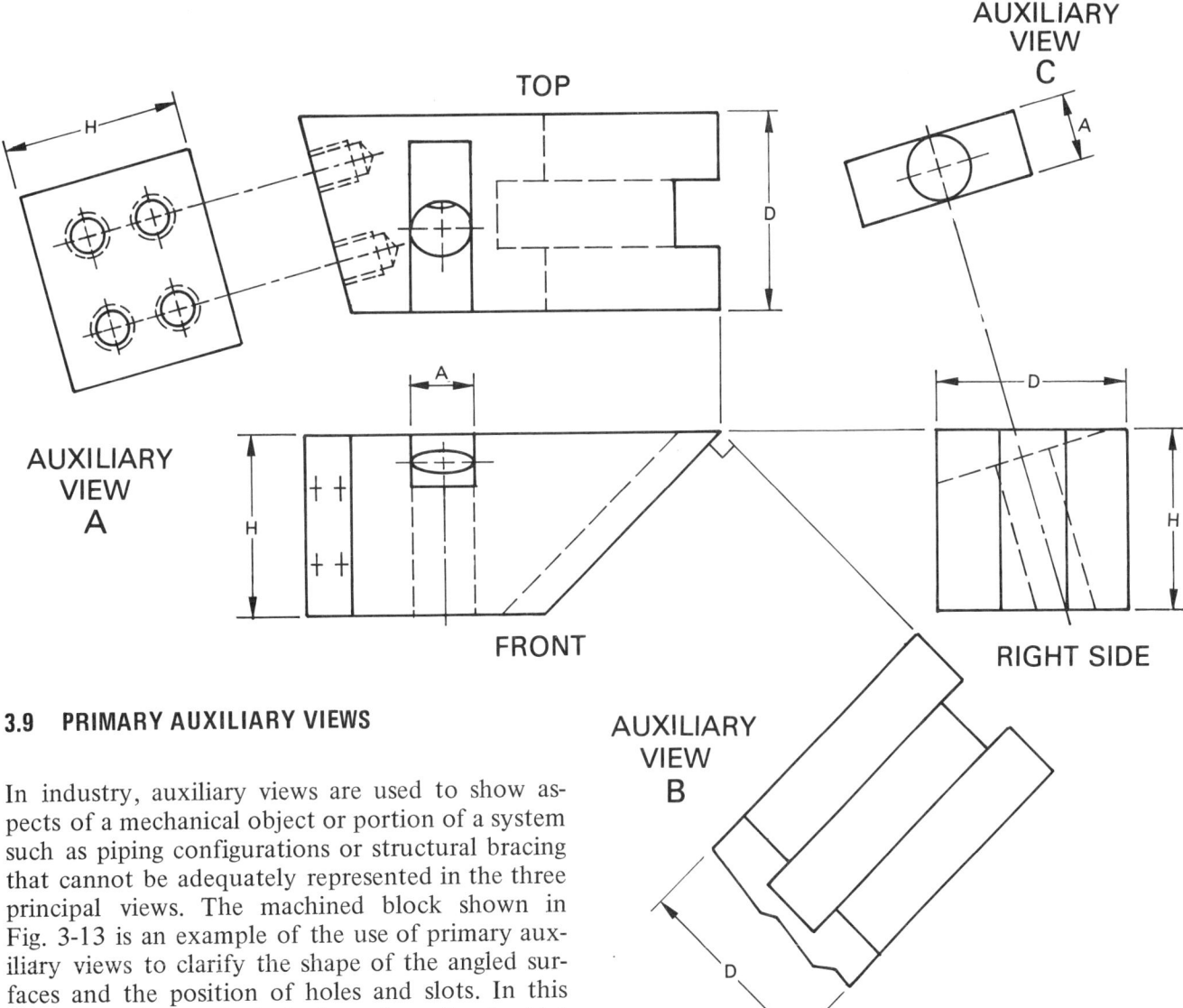

3.9 PRIMARY AUXILIARY VIEWS

In industry, auxiliary views are used to show aspects of a mechanical object or portion of a system such as piping configurations or structural bracing that cannot be adequately represented in the three principal views. The machined block shown in Fig. 3-13 is an example of the use of primary auxiliary views to clarify the shape of the angled surfaces and the position of holes and slots. In this figure the three principal views (top, front, side) do not provide true shape/size views of each surface. It is necessary to project three primary auxiliary views in order to describe the angled surfaces in detail.

Primary auxiliary views are divided into three types depending on which principal view they are projected from.

1. Primary auxiliary views projected from the top (horizontal) view. This can be termed a *horizontal auxiliary* view.
2. Primary auxiliary views projected from the front (frontal) view. This can be termed a *frontal auxiliary* view.
3. Primary auxiliary views projected from the side (profile) view. This can be termed a *profile auxiliary* view.

These three categories are represented in Fig. 3-13 where **auxiliary view A** is projected from the top (horizontal) view, **auxiliary view B** is projected from the front (frontal) view, and **auxiliary view C** is projected from the side (profile) view. The auxiliary projections in this figure are only partial views, showing only the inclined surfaces as true shape. This is normal industry practice, since the projection of the total object would not only add little to the understanding of the object's configuration, but may actually confuse the view. Hidden lines that fall behind the true shape surface in an auxiliary view can normally be eliminated for the same reason.

Each primary auxiliary view, besides being projected from one of the three principal views, will have one common dimension with at least one other principal view. The height (H) dimension in the front view is used to establish the limits of auxiliary view A by using dimension H. The depth (D) of the object can be found in the top view (and side view) and is used to establish the D dimension in auxiliary view B. Dimension A in auxiliary view C is taken from the front view where the width of the slot is drawn true size.

FIG. 3-14 Frontal auxiliary view.

FRONTAL
PRIMARY AUXILLIARY VIEW

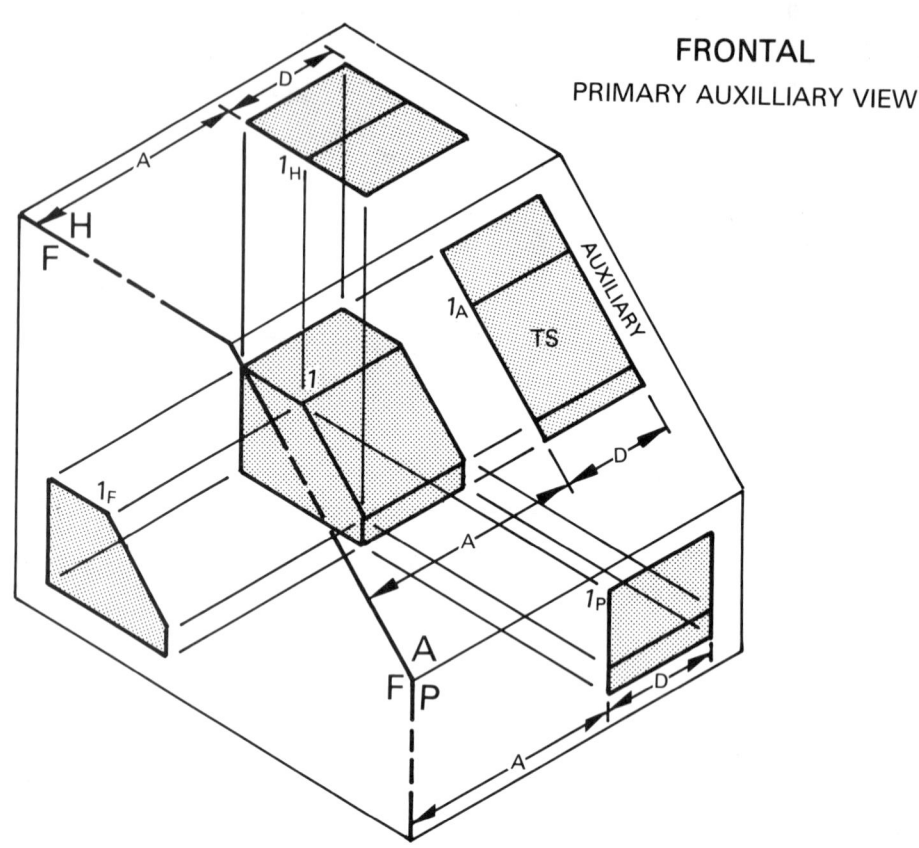

3.10 FRONTAL AUXILIARY VIEWS

The true shape of an inclined plane that appears as an edge in the front view must be projected from that view. This type of auxiliary view is called a *frontal auxiliary view*.

The glass box method is applied here to illustrate the procedure for drawing this type of auxiliary view. The following overview of steps describes the projection of the frontal auxiliary view shown in Fig. 3-15 and represented pictorially in Fig. 3-14.

1. **Line of sight:** The line of sight for a frontal auxiliary view is perpendicular to the inclined surface which appears as an edge in the front view.
2. **Fold line:** Fold line F/A is established perpendicular to the line of sight and parallel to the inclined surface (edge view).
3. **Projectors:** Projectors are drawn from all points on the front view, parallel to the line of sight and perpendicular to the fold line. Note that hidden lines were omitted in this example.
4. **Measurements:** Measurements are taken (using dividers for speed and accuracy) from fold line H/F or F/P to establish the front face of the object in the auxiliary view. Dimension A is transferred from the top or side view to establish the distance from the F/A fold line to the front face of the object in the auxiliary view. The depth dimension (D) of the object is then transferred as shown.

FIG. 3-15 Orthographic projection of a frontal auxiliary view.

FIG. 3-16 Horizontal auxiliary view (pictorial).

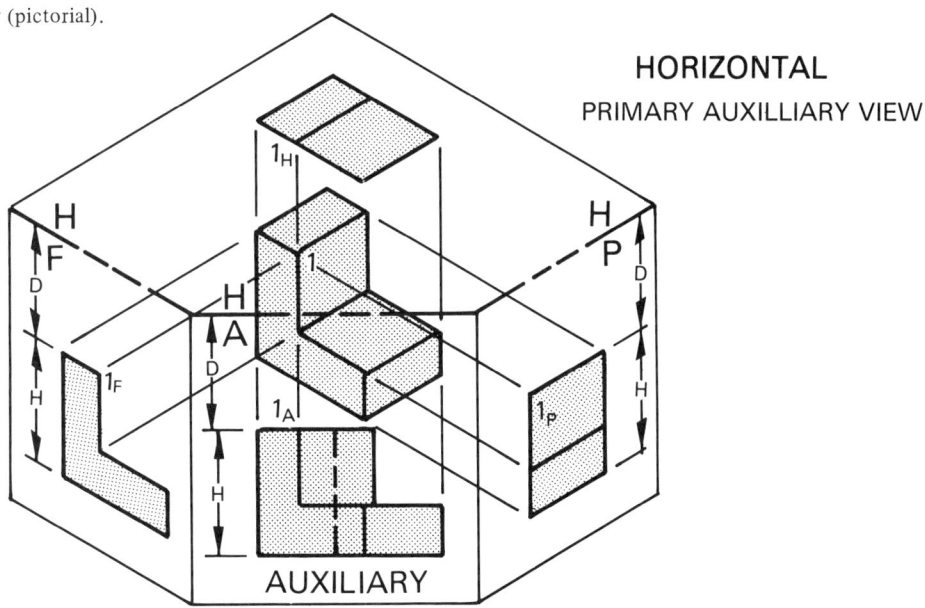

3.11 HORIZONTAL AUXILIARY VIEWS

The second type of primary auxiliary view is the *horizontal auxiliary view*. In this case the auxiliary view is taken perpendicular to the horizontal plane and is inclined to the other two principal planes. The glass box method is shown in Fig. 3-17 and represented pictorially in Fig. 3-16. In this example the auxiliary view is projected at a required viewing angle, and is not being used to solve for the true shape of an inclined surface as in Fig. 3-15. Because of this, the object will not show the true shape of a surface but instead will provide a view that has no horizontal true length lines. Auxiliary views are projected to provide different perspectives (viewing angles) as well as to solve for specific information such as the true shape of a surface or true view of a hole.

In situations where an object has a surface that is inclined to the front and side views but is an edge view in the top view, a horizontal auxiliary view can be projected with a line of sight perpendicular to the edge view to solve for the true shape of the surface.

Horizontal auxiliary views are projected from a principal view using the same basic steps as when projecting a frontal auxiliary view. The following steps provide an overview of the process used to complete the orthographic projection in Fig. 3-17. Figure 3-16 is a pictorial representation of the same example.

1. Line of sight: Establish a line of sight at a required angle of viewing; 45 degrees was used here.

2. Fold line: Fold line H/A is drawn perpendicular to the line of sight.

3. Projectors: From each point in the top (horizontal) view, extend a projector parallel to the line of sight and perpendicular to the fold line. In this example even lines that appear hidden are shown.

4. Measurements: Dimension D is transferred from the side or front view to establish the distance from the H/A fold line to the top of the object. The height (H) dimension is then transferred to locate the bottom of the object. Visibility is determined and the view completed with the required lines.

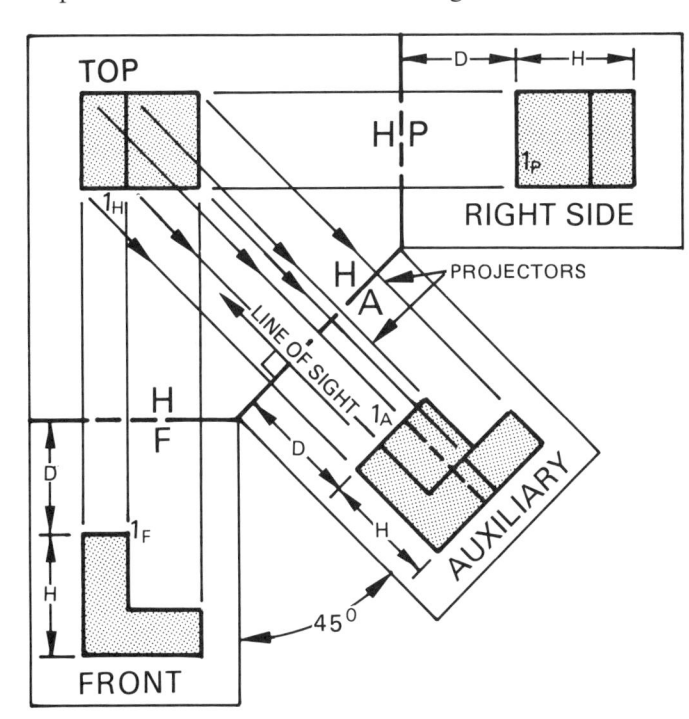

FIG. 3-17 Orthographic projection of a horizontal auxiliary view.

FIG. 3-18 Profile auxiliary view (pictorial).

**PROFILE
PRIMARY AUXILLIARY VIEW**

3.12 PROFILE AUXILIARY VIEWS

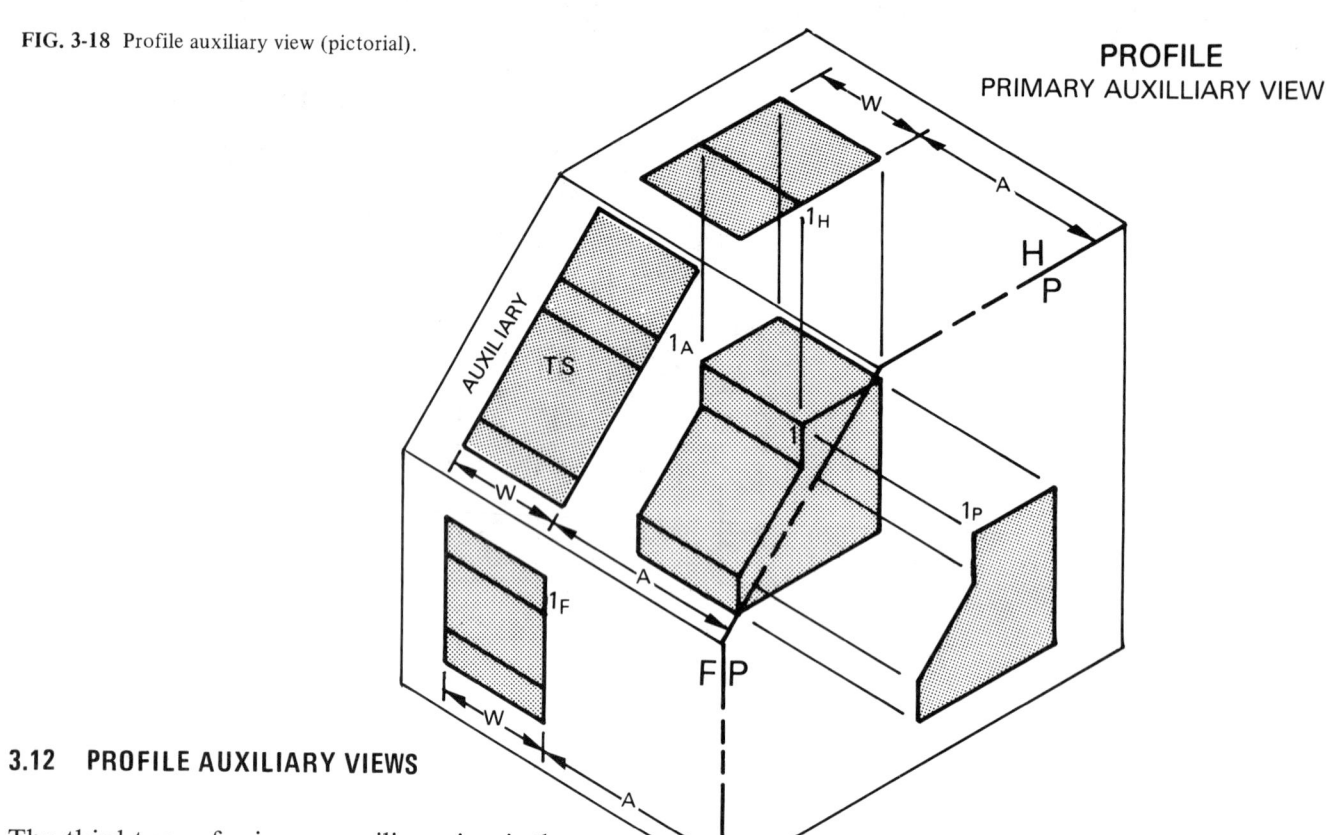

The third type of primary auxiliary view is the *profile auxiliary view*. It will always be projected perpendicular to the side view (profile plane). In Fig. 3-19 the orthographic projection of a profile auxiliary view is shown. This projection is represented pictorially in Fig. 3-18. In this example one of the surfaces of the object is inclined to the front and top view and appears as an edge in the side view. By projecting an auxiliary view perpendicular to the side view it appears true shape in the profile auxiliary view as shown. The following steps describe an overview of this process.

1. Line of sight: Establish a line of sight perpendicular to the edge view of the inclined surface in the profile view.
2. Fold line: Draw the fold line A/P perpendicular to the line of sight and parallel to the inclined surface (edge view).
3. Projectors: Projector lines are drawn parallel to the line of sight and perpendicular to fold line A/P from all points in the profile view. Hidden lines were omitted.
4. Measurements: Transfer dimension A and the width dimension (W) from the top or front view to the auxiliary view; determine the proper visibility and complete the view. Note that only the surface that appears as an edge shows true shape in the auxiliary view; the rest of the object is inclined to the auxiliary image plane.

FIG. 3-19 Orthographic projection of a profile auxiliary view.

FIG. 3-20 Housing cover.

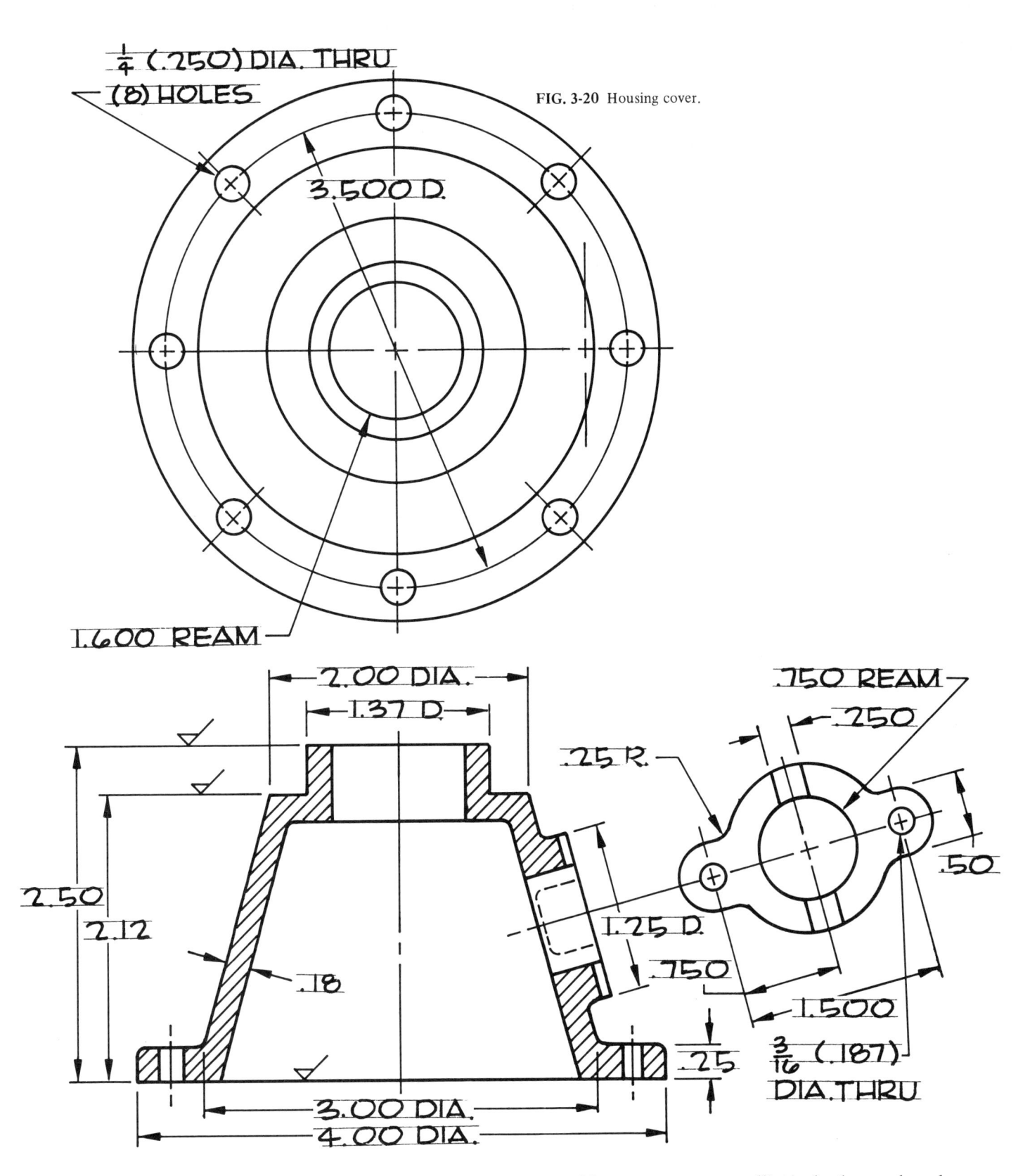

3.13 AUXILIARY VIEW APPLICATIONS

In industry, auxiliary views are frequently needed to describe and dimension a variety of mechanical objects or aspects of architectural, structural, or piping configurations. The housing cover detailed in Fig. 3-20 is an example of this type of drawing. Only true shape parts of the housing are drawn in each view. In the top view the inclined boss, which would appear as an elliptical shape, has been omitted since it would add little to the clarification of the object.

The auxiliary view shows the true shape and dimensions of the castings boss, the true size of the .750″ reamed hole, $\frac{3}{16}$″ diameter holes, and the .25″ slot. The projection of the entire object would in all probability detract from the original purpose of the view. The cost and time of projecting the entire view would also be considerable.

49

FIG. 3-21 Secondary auxiliary view.

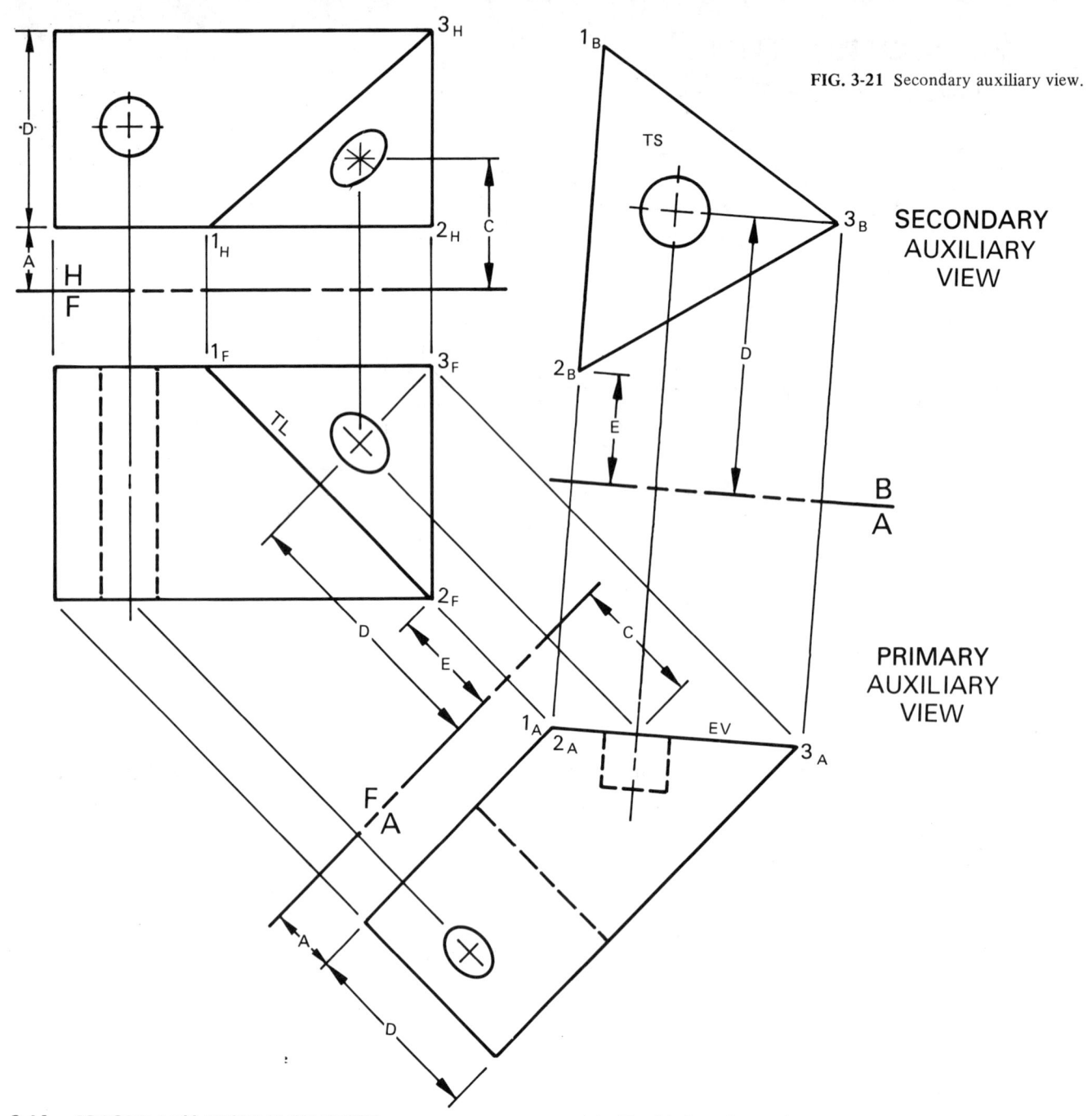

3.14 SECONDARY AUXILIARY VIEWS

Primary auxiliary views are projected from one of the six principal views. *Secondary auxiliary views* are projected from primary auxiliary views. In Fig. 3-21 the mechanical object has one surface that is not inclined to any of the principal planes of projection. Therefore it is not possible to solve for the true shape of the surface in a primary auxiliary view. This type of surface is normally referred to as an oblique surface (not to be confused with oblique projection). Secondary auxiliary views are perpendicular to primary auxiliary views, since all consecutive views of an object are at right angles. Note that any view projected from a secondary auxiliary view is called a *successive auxiliary view.* The following steps were used to draw the object in Fig. 3-21:

1. Establish a line of sight parallel to the true length (TL) line 1-2 in the front view.

2. Draw fold line F/A perpendicular to the line of sight and a convenient distance from the front view.

3. Complete the primary auxiliary view by transferring dimensions A, C, and D from the top view and draw the object.

4. Establish a line of sight perpendicular to the edge view (EV) of surface 1_A-2_A-3_A in the primary auxiliary view.

5. Draw fold line A/B perpendicular to the line of sight and at a convenient distance from auxiliary view A.

6. Complete the secondary auxiliary view by transferring dimensions from the front view. Draw only plane 1-2-3, which will show true shape. Dimensions D and E will establish points 2 and 3.

50

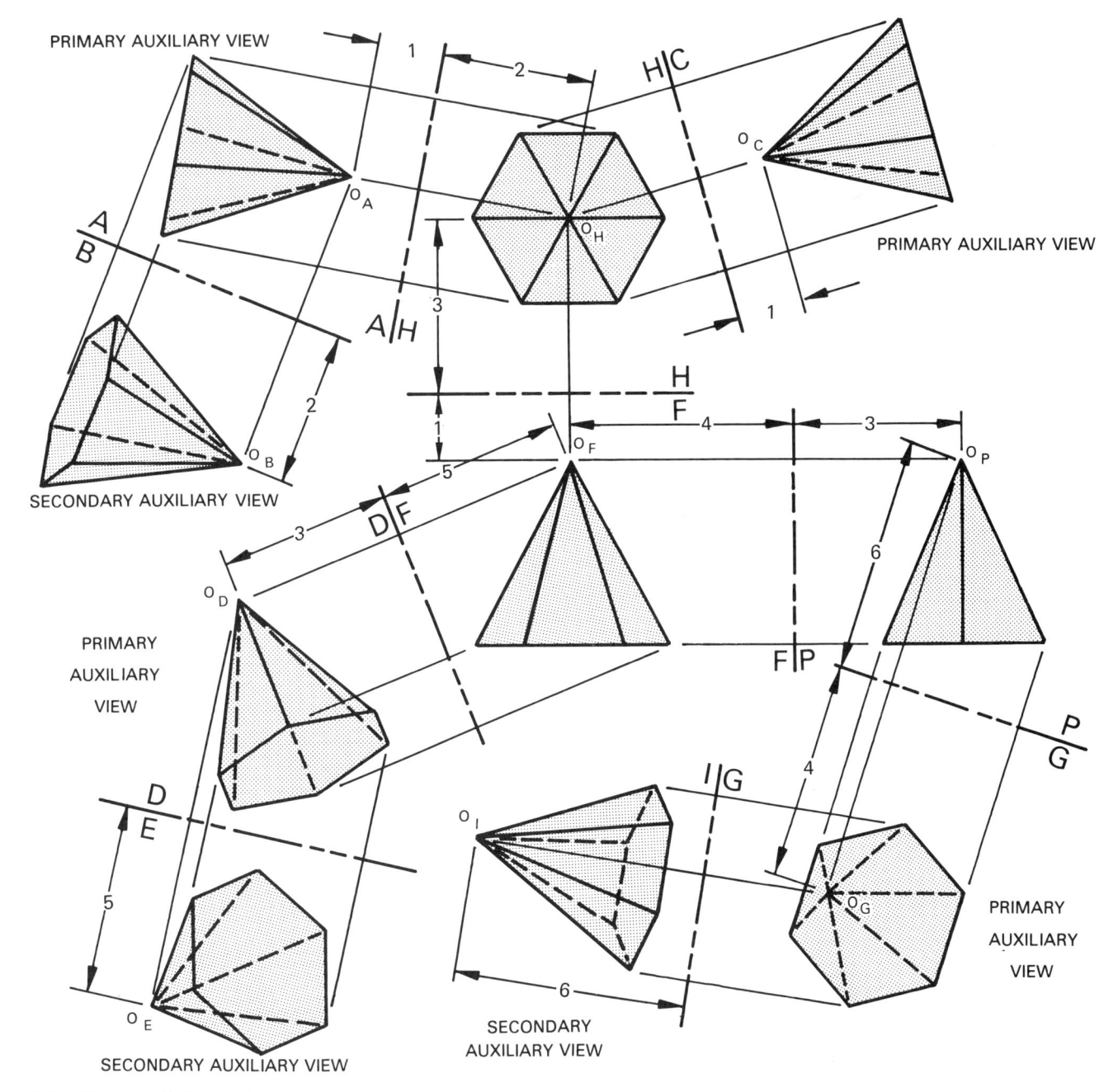

FIG. 3-22 Related/adjacent views.

3.15 RELATED/ADJACENT VIEWS

All projections of consecutive views are at right angles. Each view and its preceding and following view are considered *related views*. A related or adjacent view is therefore any view that is connected to another view by *projecting lines* and, in the case of descriptive geometry procedures and the glass box theory of projection, are separated by a *mutual fold line*. Each of the primary auxiliary views (A, C, D, and G) of the pyramid in Fig. 3-22 are projected from their related principal views. Secondary auxiliary views (B, E, and I) are then taken from their related primary auxiliary view. Notice that the H view has two auxiliary views projected from it, view A and view C.

It is important to understand that principal views can also be related/adjacent views. The top (H) view is related (adjacent) to the front (F), the front (F) to both the side (P) and top (H), and the side (P) to the front (F). Principal views are also related (adjacent) to their primary auxiliary views.

In Fig. 3-22 each dimension is used to establish apex point O from view to view. These dimensions are used on alternate views, not consecutive (related) views. Therefore dimension 1 is the distance from fold line H/F to point 1 in the front view, and is used to locate point O in primary auxiliary views A and C.

PROB. 3-1A through 3-1I

A

B

C

D

E

F

G

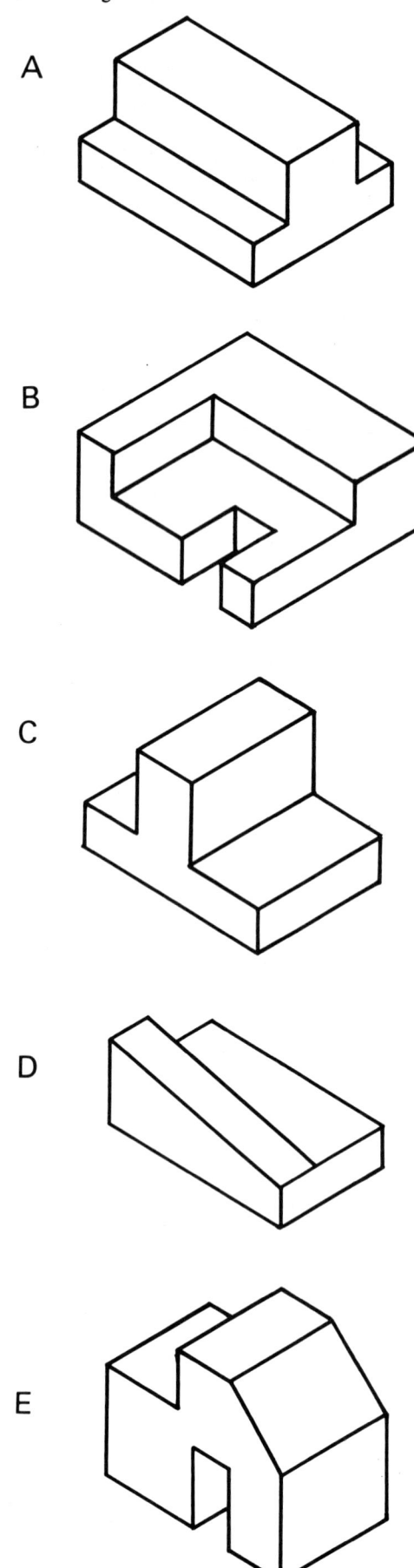

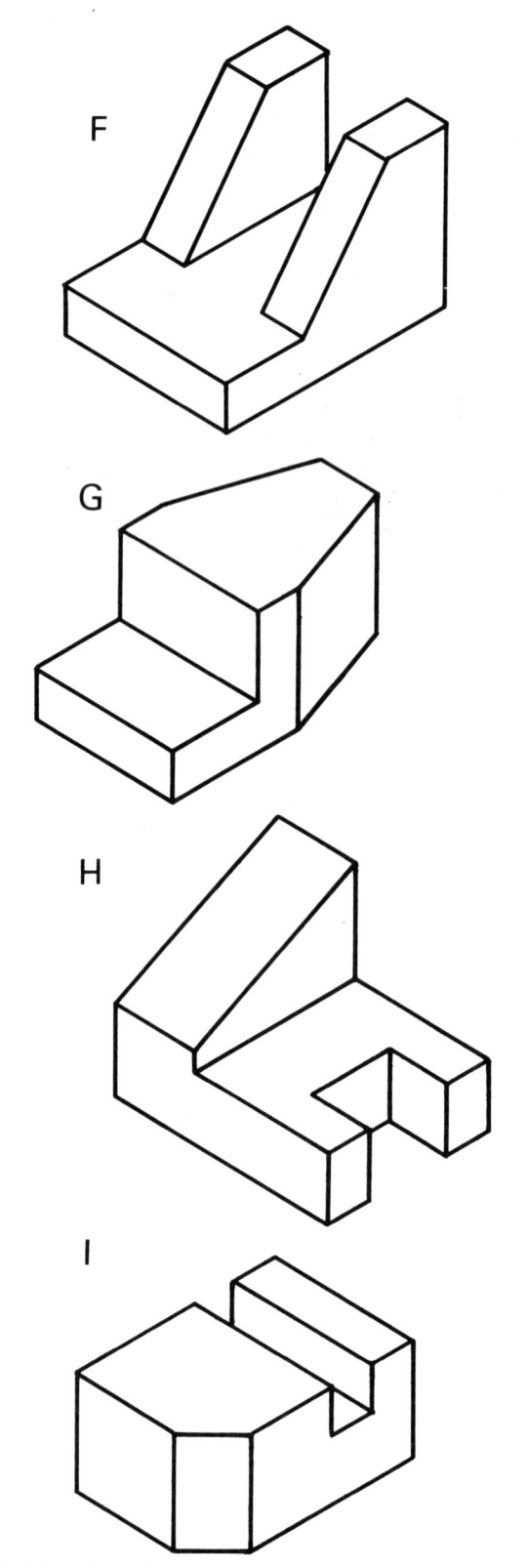

H

I

PROBS. 3-1A through 3-1I. These problems can be assigned as three-view orthographic drawings completed as freehand sketches or instrument drawings. The use of fold lines can be assigned by the instructor. Each of the figures is shown drawn to full scale, all are in $\frac{1}{8}$ inch increments, and can be scaled from the book. The instructor can assign alternative scales or metric units. This set of problems can also be used for oblique or perspective projects.

PROB. 3-2A through 3-2J

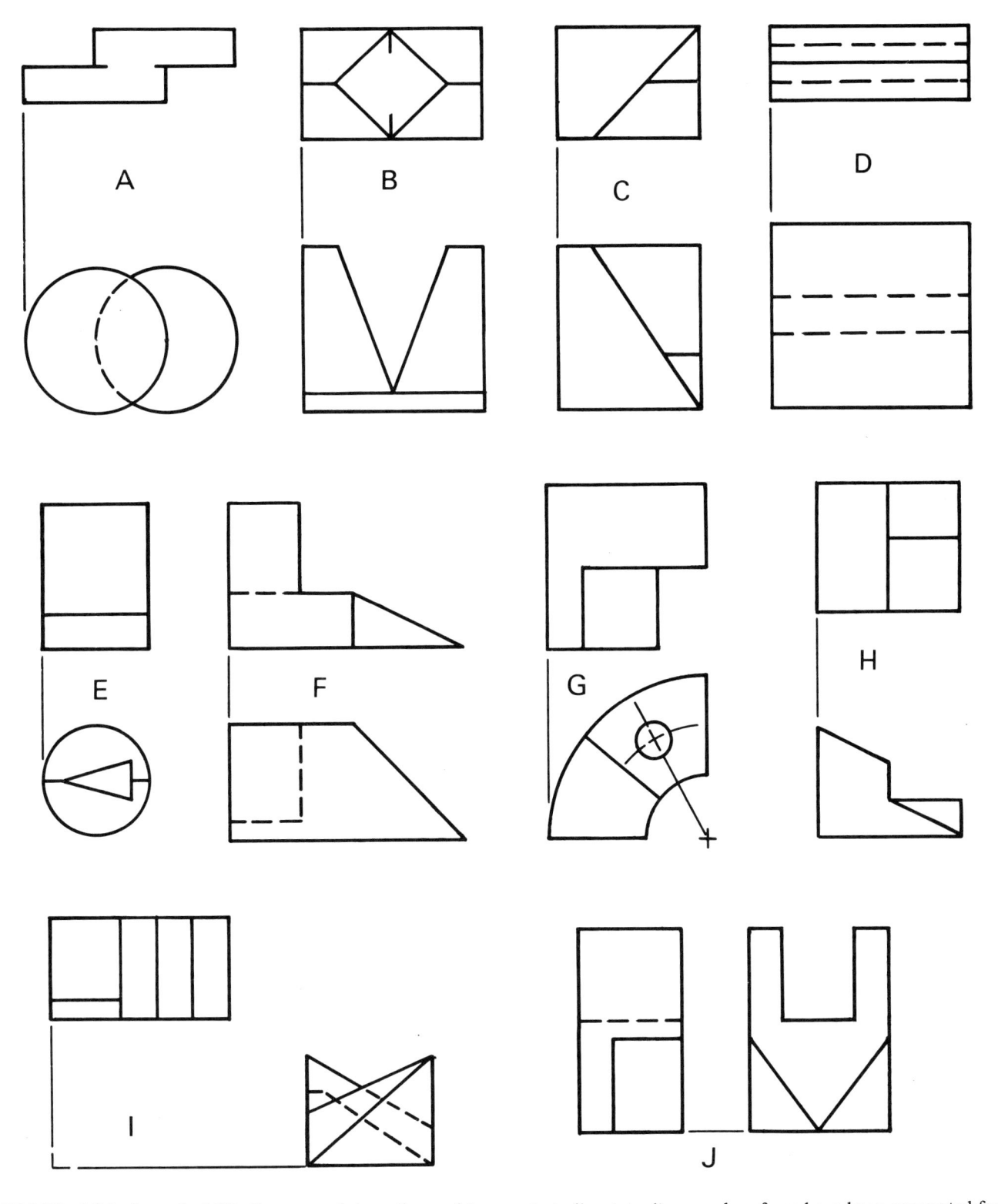

PROBS. 3-2A through 3-2J. Two complete orthographic views of an object can often give all the information necessary to describe the object. From these two views, it is possible to project a third view as well as to draw an isometric projection. Problems C, D, F, H, and I are exercises in making these projections.

When making three-view drawings, it is necessary to verify that all points, lines, and surfaces have been accounted for in each of the views. Problems A, B, E, G, and J, are exercises in completing three views when lines have been omitted. The objects do not extend beyond the silhouettes given.

All figures are shown full scale in $\frac{1}{8}$ inch increments and can be scaled from the text. Alternative scale reductions or enlargements in English or metric units can be assigned.

53

PROB. 3-3A through 3-3I

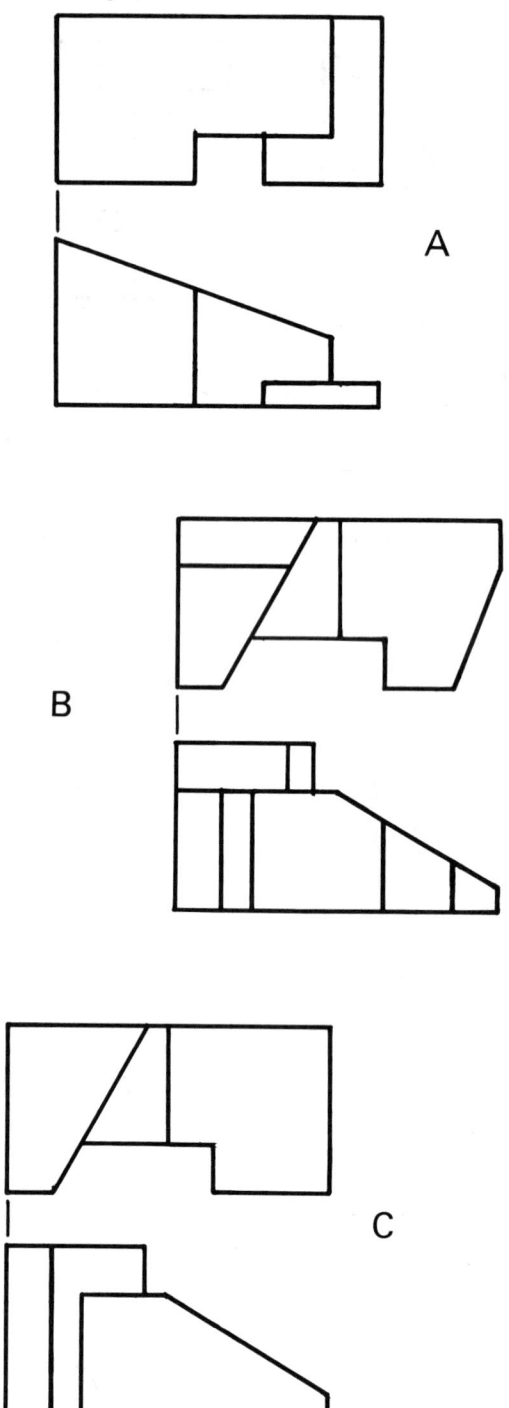

PROBS. 3-3A through 3-3I. For problems A, B, and C draw the missing right side view and an auxiliary view showing the inclined surface true shape. Problem B will need two auxiliary views.

Problems D through I all have the same front view. For each assigned problem, draw the given front and top views and then project an auxiliary view that will show the true shape of the inclined surface that is shown EV in the front view.

All problems can be scaled from the text. Problems D through I should be drawn 2X scale. Use fold lines for all auxiliary views.

Instructor can assign alternative scales, isometric or oblique projections, or other variations.

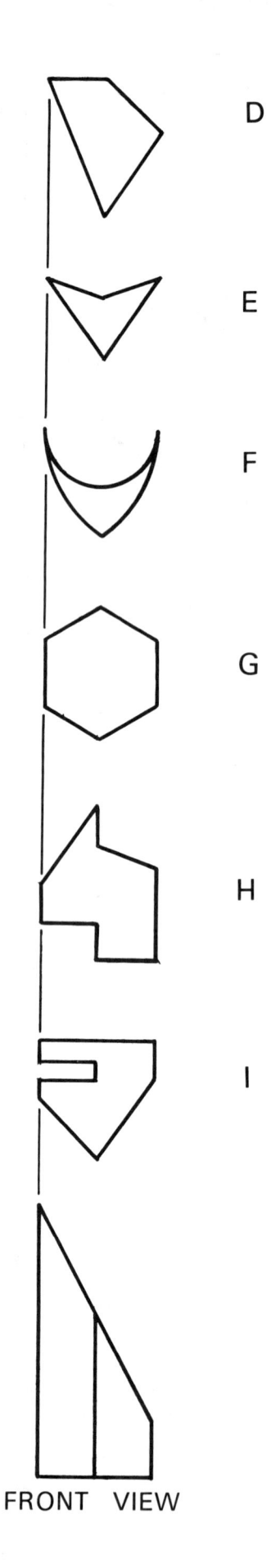

54

PROB. 3-4A through 3-4D

A

B

C

D

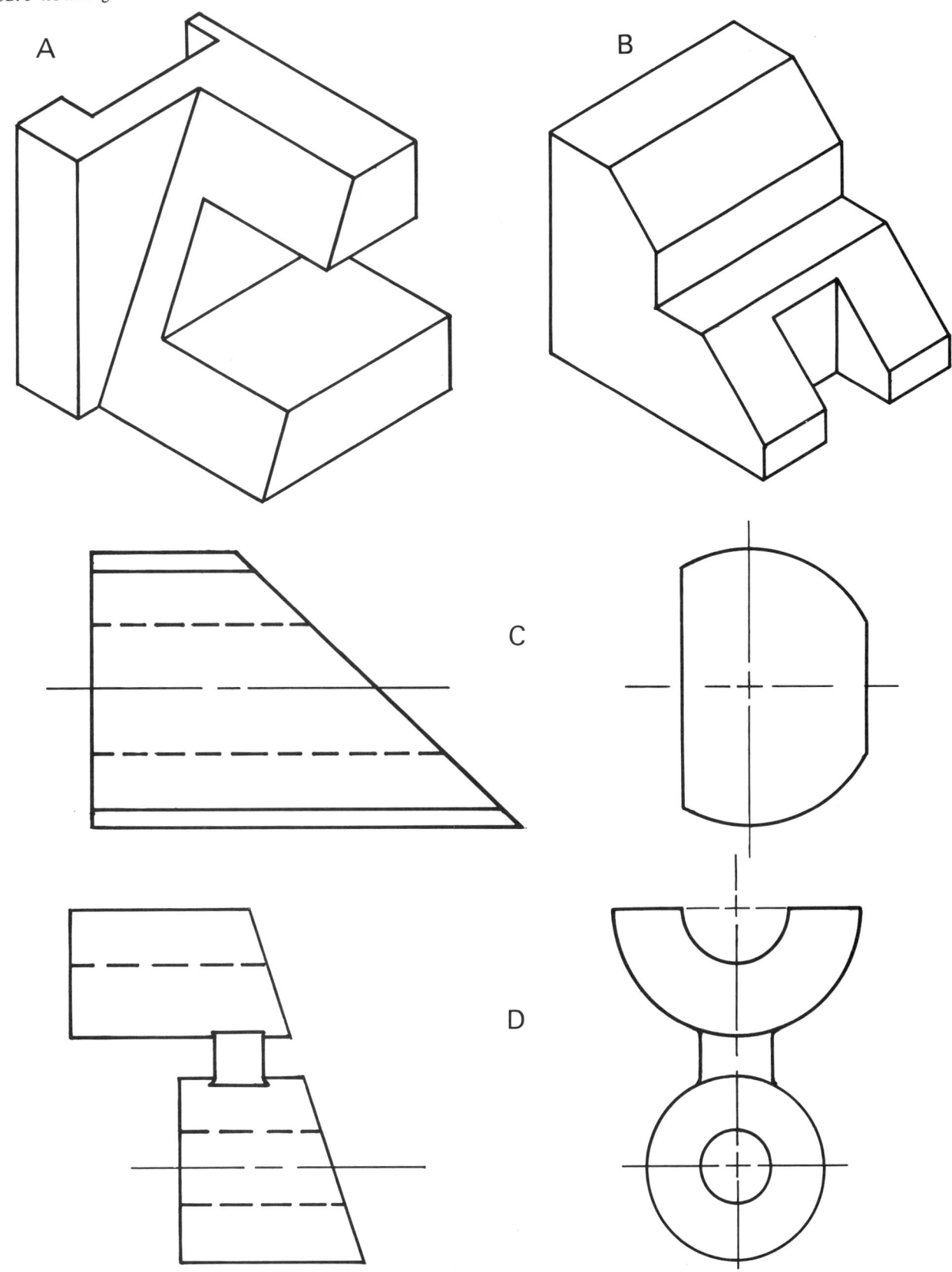

PROBS. 3-4A through 3-4D. For problems A and B draw three views of the object and the auxiliary view that will show the inclined surface in true shape. For problems C and D, draw the given two views and the auxiliary view needed to exhibit the inclined surfaces in true shape.

Draw only the inclined surface in the auxiliary view. Instructor can assign a variety of scales. Projects are shown full size and can be scaled from the text.

PROB. 3-5

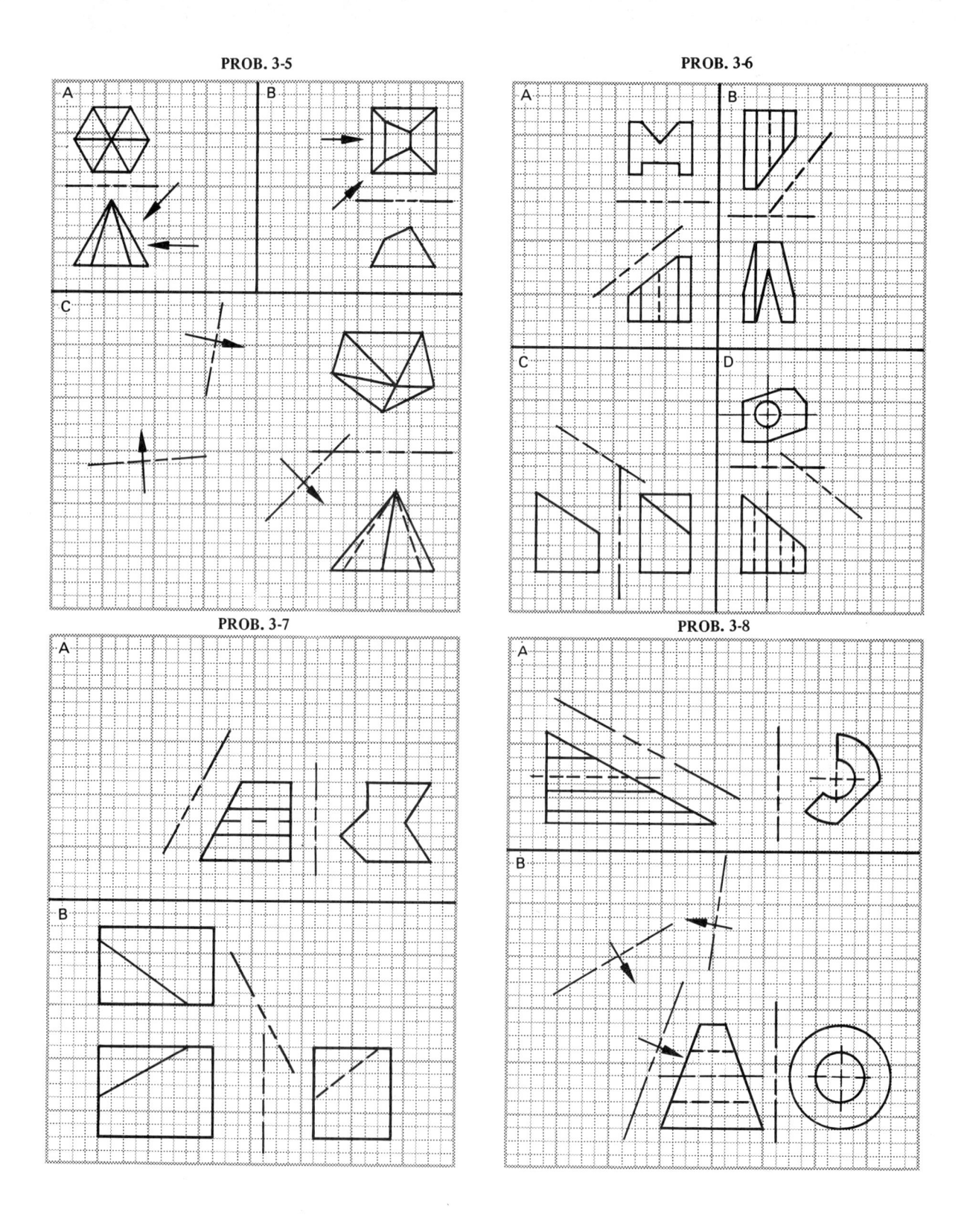

PROB. 3-6

PROB. 3-7

PROB. 3-8

PROBS. 3-5A through 3-5C. Draw the required auxiliary views as indicated by the arrows. Problem C will require successive auxiliary views.

PROBS. 3-6A through 3-6D. Using measurements taken from the given fold lines, construct the missing auxiliary views.

For problem C place a 3/8" (9.5mm) dia. through hole in the exact middle of and perpendicular to the triangular auxiliary face of the object and project back into all views.

PROBS. 3-7A and 3-7B. Draw the required auxiliary views to show the inclined surfaces in true shape. Problem B will require both a primary and secondary auxiliary view. Establish a $\frac{1}{2}''$ (12.7mm) dia. through hole in the center of the inclined surface and show in all views.

PROBS. 3-8A and 3-8B. Draw the required auxiliary views as indicated by the arrows and fold lines.

4

POINTS AND LINES

FIG. 4-1 Flight simulator for advanced aircraft. *(Courtesy NASA.)*

4.1 POINTS AND LINES

All geometric shapes are composed of points and their connectors, lines. In descriptive geometry points are the most important geometric element and the primary building block for any graphical projection of a form. All projections of lines, planes, or solids can be physically located and manipulated by identifying a series of points that represent the object.

A point can be located in space and therefore illustrated by establishing it in two or more projections. Throughout the text, points are identified by a thin cross (+) or dot. Numbers are used to label most points and are established in a specific projection by subscripts which denote the principal plane of projection (H, F, P) or auxiliary view (A, B, C, etc.). Two points that are connected are considered a line. Points can also be used to describe a plane or solid, or can be located in space by themselves, though they have no real physical dimension. All of descriptive geometry is based on the orthographic projection of points in space.

In actual industry practice points are used to locate and identify aspects of practical and theoretical problems concerning a variety of engineering applications. The flight simulator in Fig. 4-1 and the petrochemical installation in Fig. 4-2 are just two examples of the applied use of descriptive geometry and graphical methods. The design and manufacture of industrial products, the construction of industrial and architectural structures, and engineering of processing facilities all require the use of descriptive geometry methods and techniques. Points and lines can be used to establish a variety of engineering data, including the angle between structural elements, shortest distance from structural beams to pipe supports, clearance between piping runs, intersection and development of heating and ventilation ducting, and the revolution of forms in mechanical design.

FIG. 4-2 Petrochemical facility.

FIG. 4-3 Pictorial view of a point.

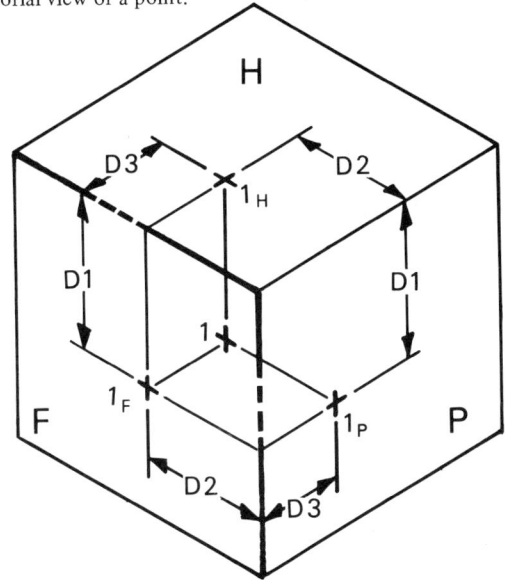

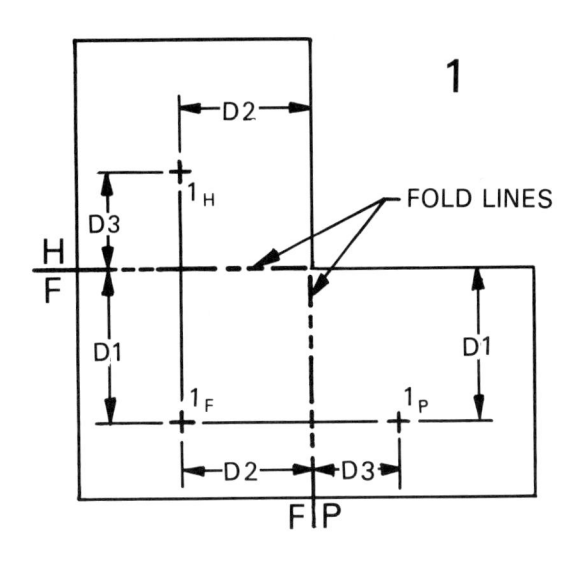

4.2 VIEWS OF POINTS

Since a point is a location in space and not a dimensional form, it must be located by measurements taken from an established reference line as is used in the glass box method of orthographic projection illustrated pictorially in Fig. 4-3 and orthographically in Fig. 4-4. These two figures represent the projection of point 1 in the three principal planes, frontal (1_F), horizontal (1_H), and profile (1_P). In the glass box method it is assumed that each mutually perpendicular plane (Fig. 4-3) is hinged so as to be revolved into the plane of the paper as an orthographic projection, Fig. 4-4(1). *The intersection line of two successive (perpendicular) image planes is called a **fold line/reference line***. All measurements are taken from fold lines to locate a point (line, plane, or solid) in space. A fold line/reference line can be visualized as the edge view of a reference plane.

Point 1 is shown in Fig. 4-4 as it would look with the principal planes defined (1), with its locating dimensions and subscripts (2) and as it would appear in a typical descriptive geometry illustration (3). In all three forms extension/projection lines connect the point from view to view. 1_H and 1_F are vertically in line, 1_F and 1_P are horizontally in line, and 1_P is the same perpendicular distance behind the frontal plane as 1_H (dimension D3).

A point can be located by means of verbal description by giving dimensions from fold/reference lines. Point 1 is below the horizontal plane (D1), to the left of the profile plane (D2), and behind the frontal plane (D3). D1 establishes the elevation or height of the point in front and side view, D2 the right-left location or width in the front and top view, and D3 the distance behind (depth) the frontal plane in the top and side view.

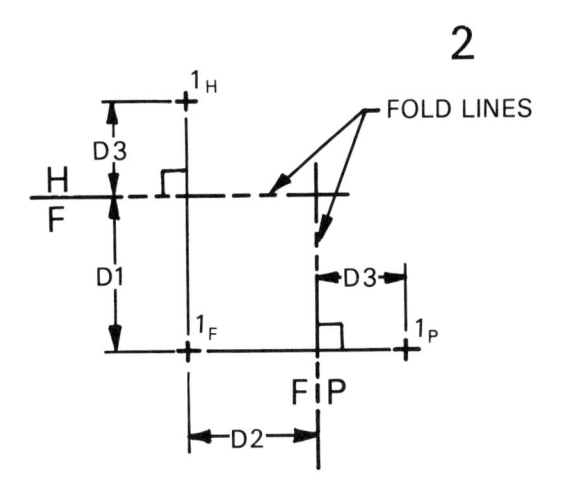

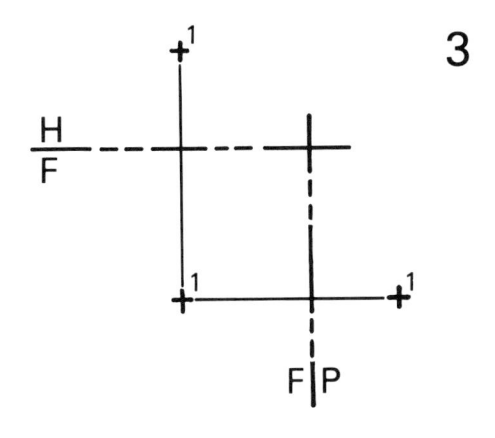

FIG. 4-4 Orthographic projection of a point.

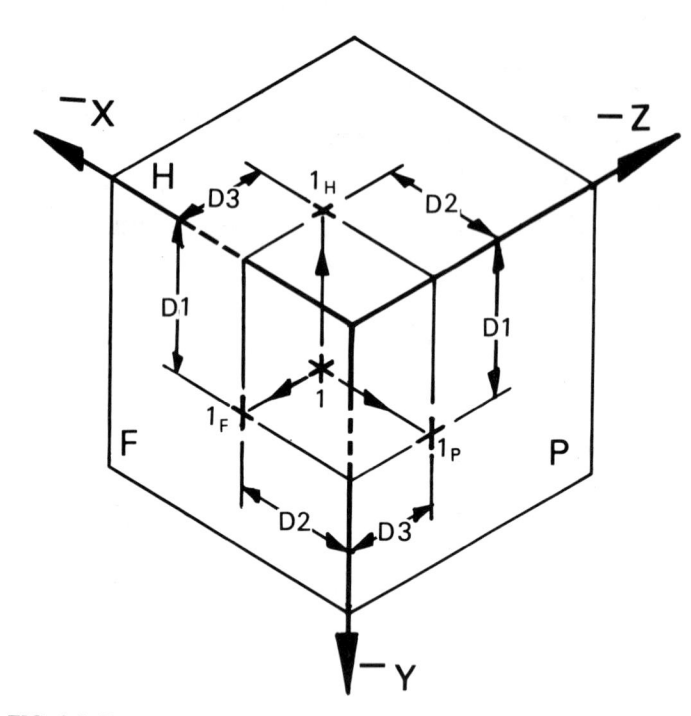

FIG. 4-5 Pictorial view of a point located by rectangular coordinates.

4.3 RECTANGULAR COORDINATES OF A POINT

Another method of locating a point in space uses *rectangular coordinates*. A variety of different origins can be used. The pictorial view of point 1 in Fig. 4-5 and shown orthographically in Fig. 4-6 has been established in the third quadrant using rectangular coordinates. In this example the −X and −Z axes form the horizontal plane, the −X and −Y axes the frontal plane, and the −Y and −Z axes the profile plane. Point 1 has X, Y, Z coordinates of −5, −7, −4. In other words point 1 is five units (−5 X) to the left of the profile plane, seven units (−7 Y) below the horizontal plane, and four units (−4 Z) behind the frontal plane.

Note that when the glass box is rotated into the plane of the paper the −Z axis appears both to the right of the F/P fold line and above the H/F fold line. This happens because both the profile and the horizontal plane are rotated from the frontal plane.

Normally in descriptive geometry problems the coordinate system is not used. Dimensions are used to locate all points from fold lines, as in Fig. 4-5 and Fig. 4-6 where D1 locates point 1 vertically (in elevation, the distance below the horizontal plane) from the H/F fold line, D2 horizontally (to the left of the profile plane) from the F/P fold line, and D3 from the F/P or H/P fold line to establish its depth (distance behind the frontal plane) dimension. Rectangular coordinates can be used to establish objects in space on CAD/CAM systems (see Fig. 4-12).

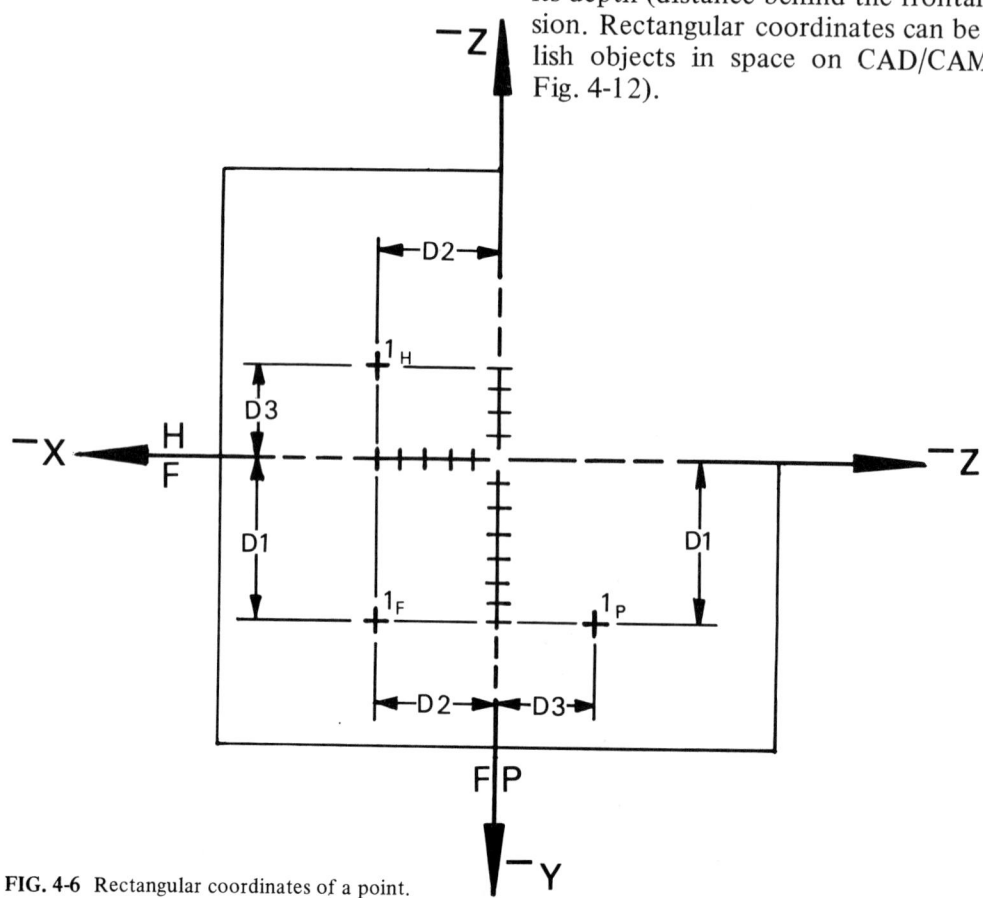

FIG. 4-6 Rectangular coordinates of a point.

60

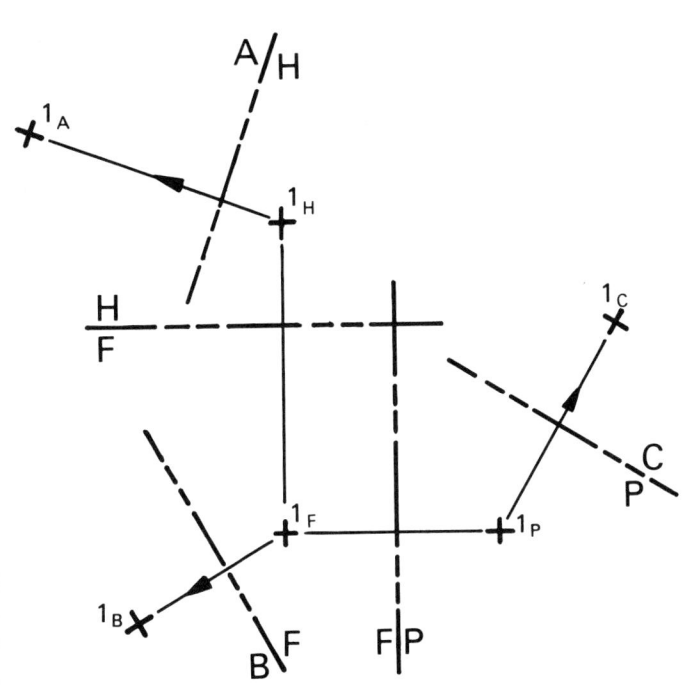

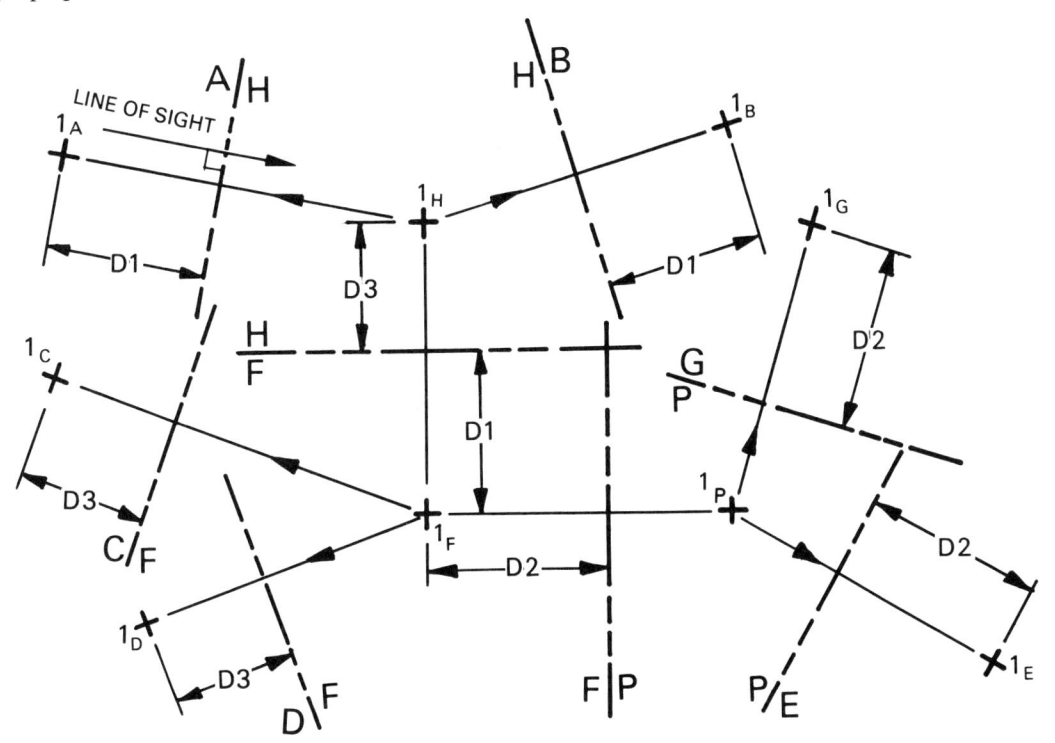

4.4 PRIMARY AUXILIARY VIEWS OF A POINT

Auxiliary views taken from one of the three principal views are *primary auxiliary views*. A primary auxiliary view of a point will be perpendicular to one of the principal planes and inclined to the other two. Another name for this type of view is a *first auxiliary view*, being the first view off of a principal plane. Figure 4-7 shows a primary auxiliary view taken from each of the three principal planes. Primary auxiliary view A is taken perpendicular to the horizontal plane, primary auxiliary view B is drawn perpendicular to the frontal plane, and primary auxiliary view C is perpendicular to the profile plane.

In Fig. 4-8 each principal view has two primary auxiliary views projected from it. The following steps describe this construction, assuming that the three principal views of point 1 (1_H, 1_F, and 1_P) already exist. Use dividers and follow along with the steps.

FIG. 4-7 Primary auxiliary views of a point.

1. Determine the required line of sight for auxiliary view A (there are an infinite number of possibilities).
2. Draw fold line H/A perpendicular to the line of sight and a convenient distance from point 1_H.
3. Draw a projection line from 1_H perpendicular the fold line H/A and parallel to the line of sight.
4. Using dividers transfer D1 from the front view to locate 1_A along the projection line.
5. Repeat this process for primary auxiliary view B.
6. Determine line of sight for auxiliary C.
7. Draw F/C perpendicular to the line of sight.
8. Draw a projection line from 1_F perpendicular to the fold line.
9. Transfer D3 from the horizontal (or profile) view to locate 1_C along the projection line.
10. Repeat for auxiliary D.
11. Determine line of sight for auxiliary E.
12. Draw P/E perpendicular to line of sight.
13. Draw a projection line from 1_P perpendicular to P/E.
14. Transfer D2 from front view to locate 1_E along the projection line.
15. Repeat for auxiliary view G.

FIG. 4-8 Auxiliary views of a point.

FIG. 4-9 Secondary auxiliary views of a point.

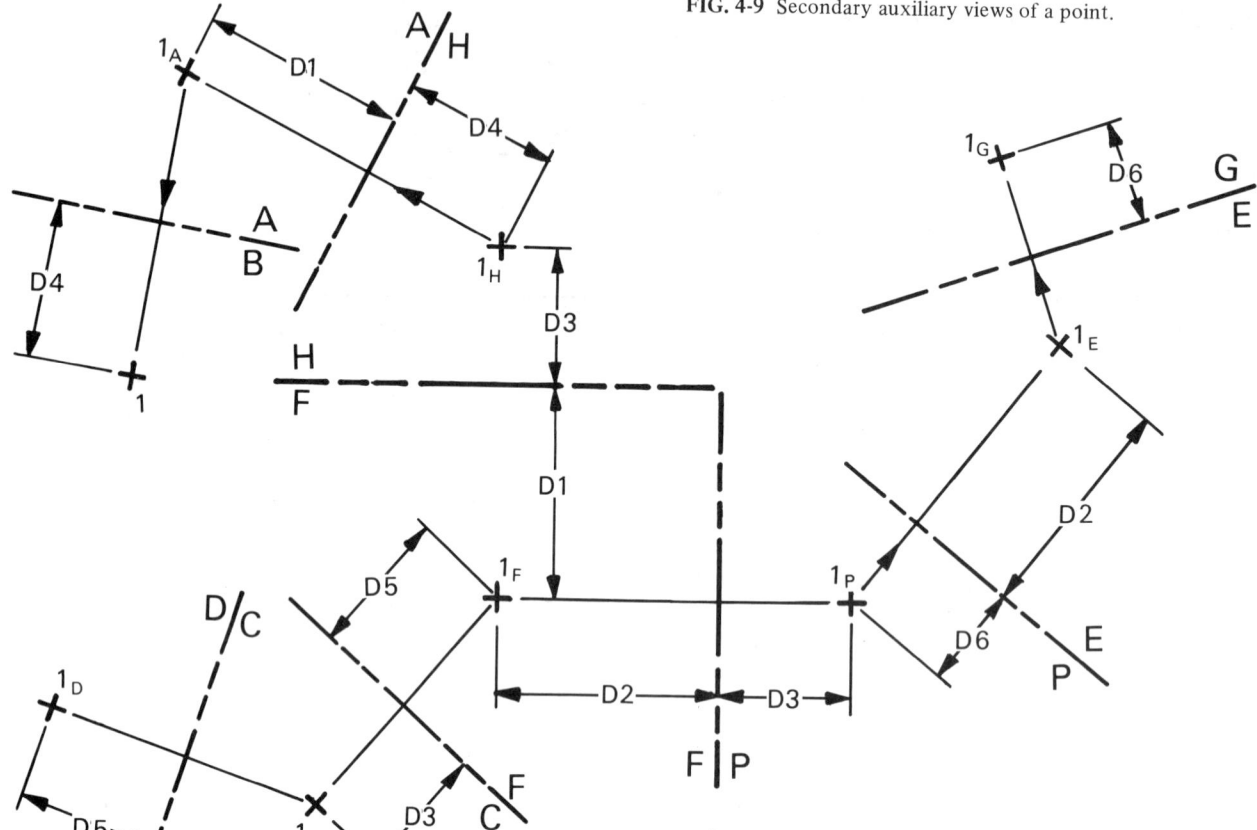

4.5 SECONDARY AUXILIARY VIEWS OF A POINT

Auxiliary views projected from a primary auxiliary view are called *secondary auxiliary views*. Secondary auxiliary views are drawn perpendicular to one primary auxiliary view and will therefore be oblique projections, since they will be inclined to all three principal views (Fig. 4-9). All views projected from a secondary auxiliary view are called *successive auxiliary views*, as are all views thereafter. In most cases, solutions to descriptive geometry problems require only secondary auxiliary projections. Auxiliary views were used to design the landing legs of the lunar vehicle shown in Fig. 4-10.

In Fig. 4-9, primary and secondary auxiliary views have been projected from each principal view. Given the frontal, horizontal, and profile views of point 1 the following steps describe this process. Follow along with dividers.

1. Determine line of sight for primary auxiliary view A.
2. Draw fold line H/A perpendicular to the line of sight, at a convenient distance from point 1_H.
3. Extend a projection line from point 1_H parallel to the line of sight and perpendicular to fold line H/A.
4. Using dividers, transfer dimension D1 from the frontal view to locate 1_A along the projection line in auxiliary view A.
5. Determine the line of sight for secondary auxiliary view B.
6. Draw fold line A/B perpendicular to the line of sight and at a convenient distance from point 1_A.
7. Draw a projection line from point 1_A perpendicular to fold line A/B and parallel to the line of sight.
8. Transfer dimension D4 from the horizontal view to locate point 1_B in secondary auxiliary view B.
9. Repeat the above steps to establish secondary auxiliary views D and G.

Note that each auxiliary projection of a point (as 1_B) is extended from its preceding view (from 1_A) along a projection line. Its position in this new view is determined by skipping its preceding view (A) and transferring the distance (D4) from the fold line (A/H) to the point (1_H) in the previous view (horizontal view here).

FIG. 4-10 Small lunar transportation vehicle. *(Courtesy NASA.)*

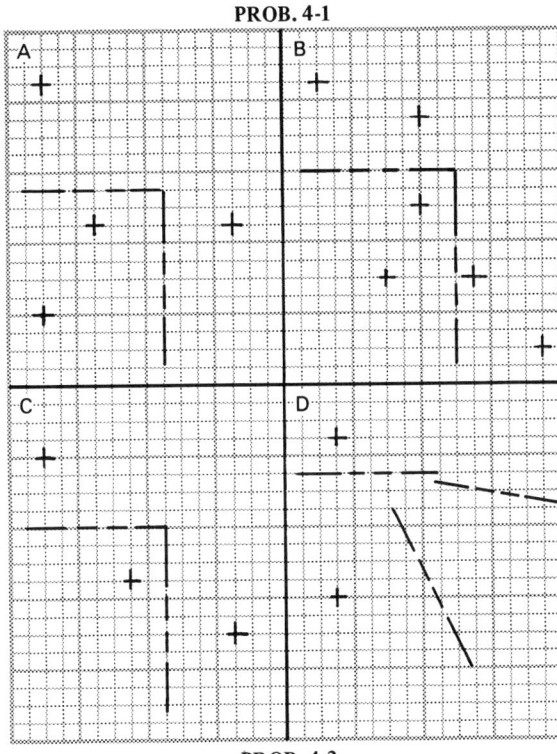

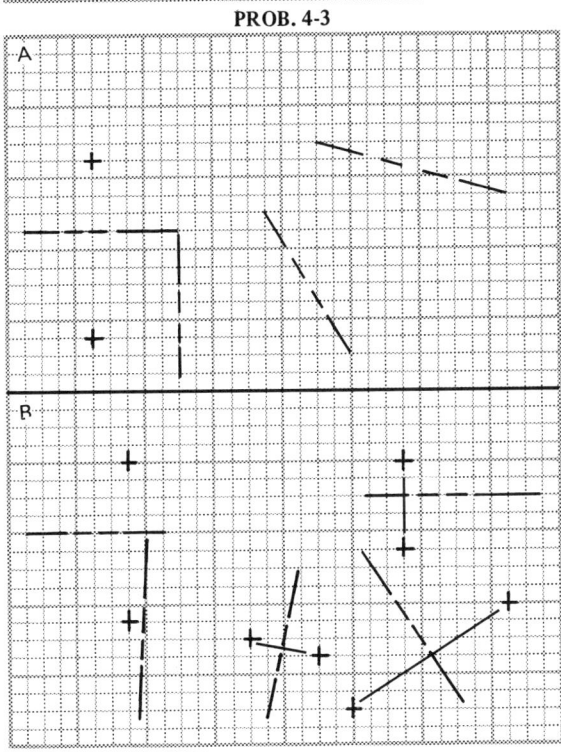

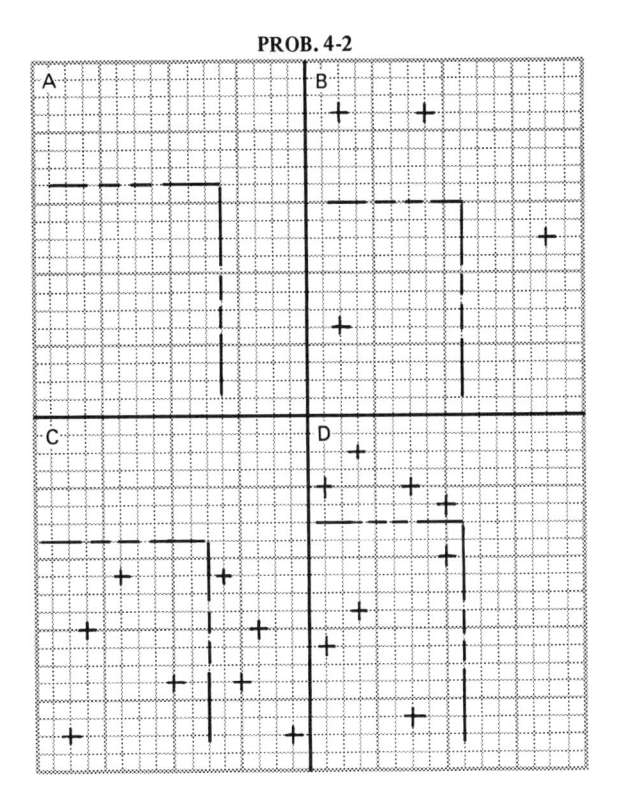

PROB. 4-1A. Locate the two points in all views. Label completely and show the projection lines between points.
PROB. 4-1B. Locate the three points in all views.
PROB. 4-1C. Locate the following three points in the given views. Point 1 is seven units below point 2. Point 2 is two units behind point 1. Point 3 is three units to the left of point 2. Point 1 is given in the H view, point 2 is given in the F view, and point 3 is given in the P view.
PROB. 4-1D. Locate points 1 and 2 in all four views. Point 1 is given. Point 2 is $\frac{1}{4}''$ (6.35mm) in front of, $\frac{3}{4}''$ (19mm) to the right of, and $1\frac{1}{4}''$ (31.7mm) below point 1.
PROB. 4-2A. Locate the following points. Point 1 is four units behind the frontal plane, nine units to the left of the profile plane, and twelve units below the horizontal plane. Point 2 is three units behind the frontal plane, seven units below the horizontal plane, and seven units to the right of point 1. What is the distance between the two points in the front view?
PROB. 4-2B. Locate the points in the three views. What is the true distance between the points?
PROB. 4-2C. Show the points in the top view. Label each point.
PROB. 4-2D. Project the points into the profile view.
PROB. 4-3A. Locate points 1 and 2 in the primary and secondary auxiliary views. Point 1 is given, point 2 is three units to the right, two units in front, and four units above point 1. Does line 1-2 show up as true length in any view?
PROB. 4-3B. Locate the four given points in each view.

FIG. 4-11 Projection of a line using a glass box.

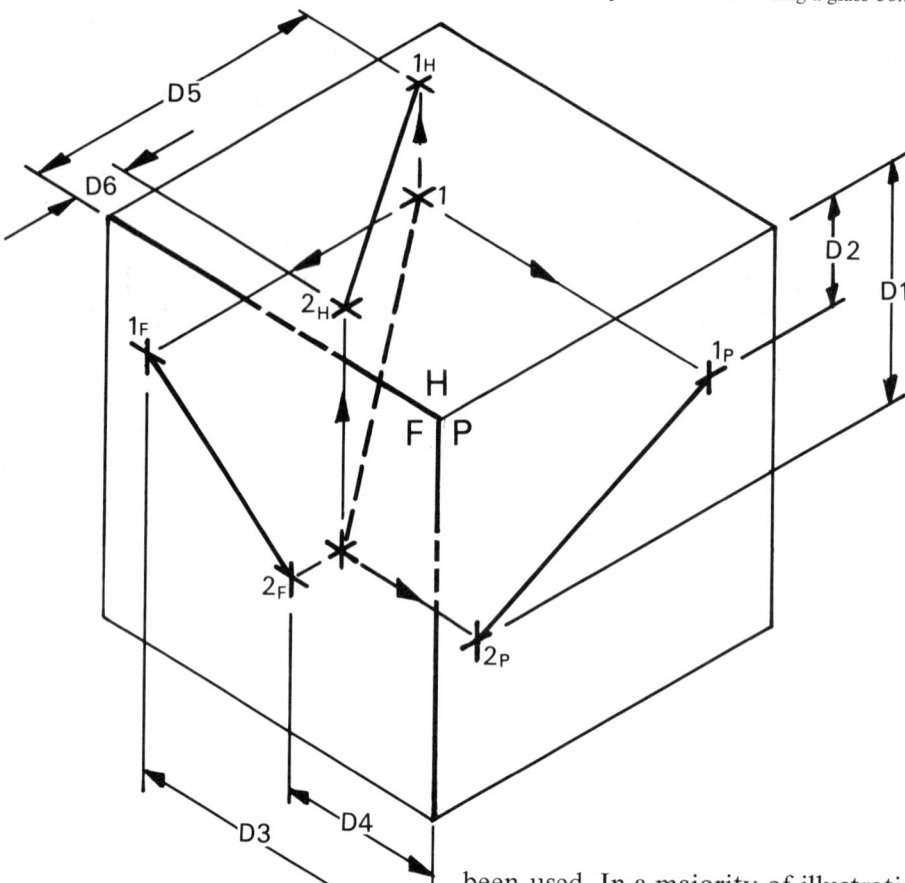

4.6 LINES

Lines can be thought of as a series of points in space, having magnitude (length) but not width. It will be assumed that a line is straight unless otherwise stated. Every line in order to be drawn must have a thickness. The line key in Chapter 1 should be consulted as to thickness of each "kind" of line used in this text. Though a line may be located by establishing its end points and may be of a definite specified length, all lines can be extended in order to solve a problem. Therefore a purely theoretical definition of a line could be: *Lines are straight elements that have no width, but are infinite in length (magnitude); they can be located by two points which are not on the same spot but fall along the line.* When two lines lie in the same plane they will either be parallel or intersect. Lines can be used to establish surfaces, or solid shapes, as in Fig. 4-12 where the experimental Delta Wing aircraft is composed of a series of straight and curved surfaces. The Delta Wing is being designed on a CRT using a light pen. The IBM terminal is part of a CAD system. Note the coordinate arrows in the upper left portion of the monitor.

Throughout the text, numbers have been used to designate the end points of a line, except in special cases or where the total quantity of points (and therefore numbers) would create a lack of clarity. In these circumstances capital letters have been used. In a majority of illustrations the view of a line and its locating points are labeled with a subscript corresponding to the plane of projection, as in Fig. 4-11 where the end points of line 1-2 are notated 1_H and 2_H in the horizontal view, 1_F and 2_F in the frontal view, and 1_P and 2_P in the profile plane. Note that in some examples subscripts are eliminated where they would clutter the illustration. Subscripts are also not shown on illustrations where the view is obvious or only one point may be labeled per view.

FIG. 4-12 Computational Fluid Dynamics at IBM terminal showing Delta Wing phase. *(Courtesy NASA.)*

4.7 ORTHOGRAPHIC PROJECTION OF A LINE

Lines are classified according to their orientation to the three principal planes of projection or how they appear in a projection plane. They can also be described by their relationship to other lines in the same view. All lines are located from fold lines/reference lines, which represent the intersection of two perpendicular planes of projection.

In Fig. 4-11, line 1-2 is pictorially represented. Line 1-2 (dashed) is projected onto each principal projection plane and located by dimensions taken from fold lines. This same example is shown with the glass box unfolded into the plane of the paper (in orthographic projection) in Fig. 4-13.

Each end point of line 1-2 is located from two fold lines in each view, using dimensions or projection lines that originate in a previous view. Dimension D1 and D2 establish the elevation of the end points in the profile and frontal view, since these points are horizontally in line in these two views. D3 and D4 locate the end points in relation to the F/P fold line (to the left of the profile plane), in both the frontal and horizontal views, since these points are aligned vertically. D5 and D6 locate each end point in relation to the H/F and the F/P fold line since these dimensions are the distance behind the frontal plane and will show in both the horizontal and profile views. Subscripts are provided for each point to establish the plane of projection.

Lines are used throughout descriptive geometry to represent a variety of industrial elements or shapes. In Fig. 4-14 the design of the clean room structure alone required the use of lines to establish structural beams, angle and true length of bracing, location of floors and scaffolding, not to mention the graphical solutions required for the development and production of the Apollo spacecraft.

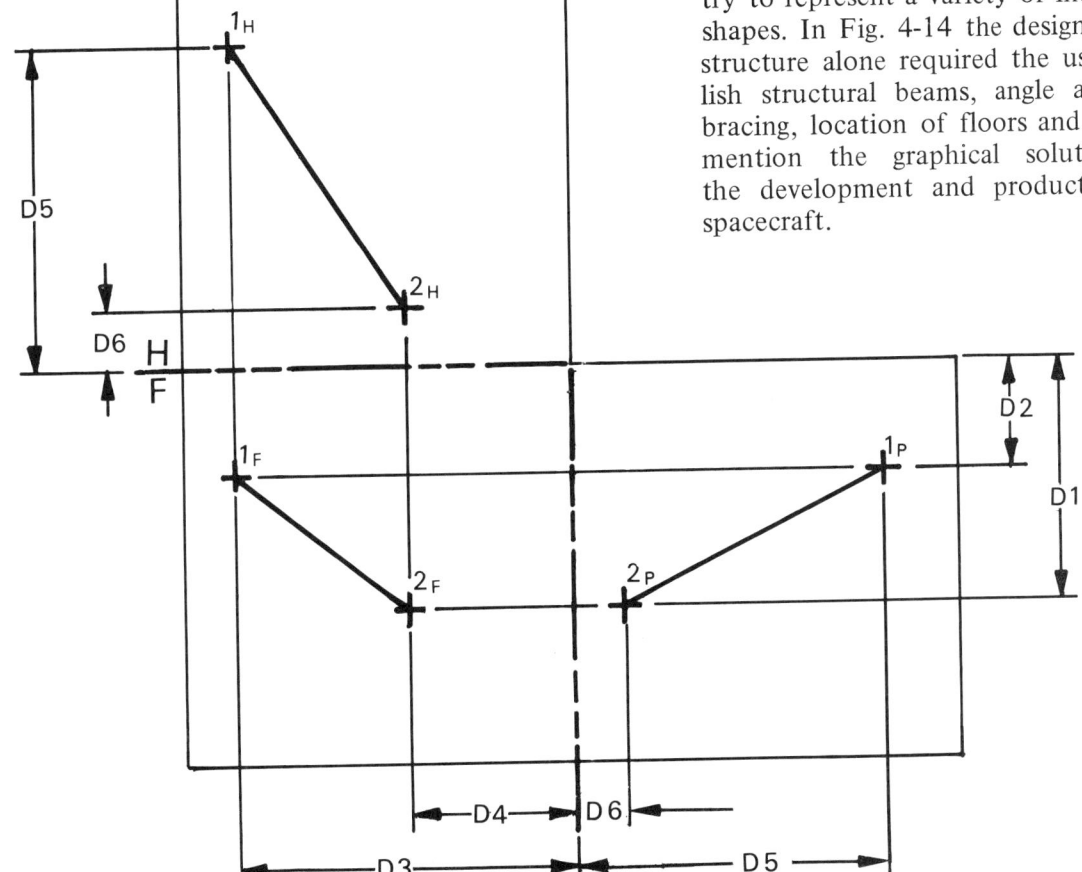

FIG. 4-14 Apollo clean room. *(Courtesy NASA.)*

FIG. 4-13 Orthographic projection of a line.

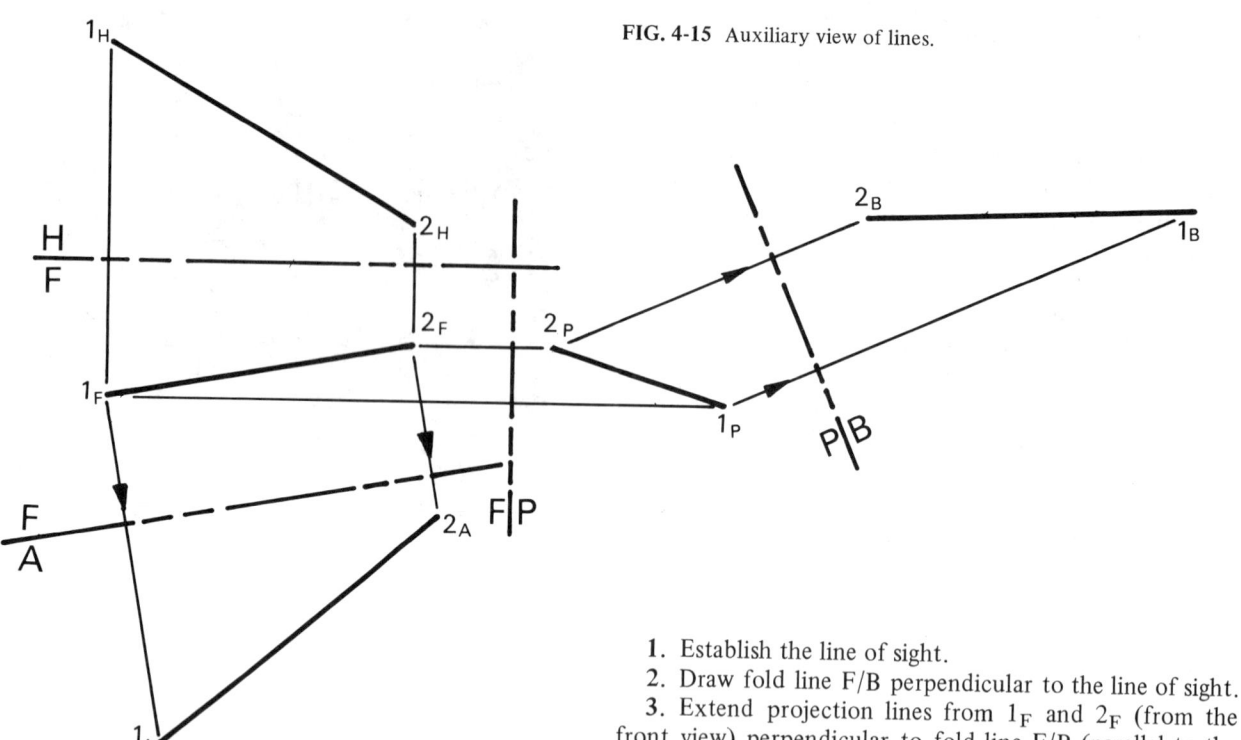

FIG. 4-15 Auxiliary view of lines.

1. Establish the line of sight.
2. Draw fold line F/B perpendicular to the line of sight.
3. Extend projection lines from 1_F and 2_F (from the front view) perpendicular to fold line F/B (parallel to the line of sight).
4. Transfer dimension D1 to locate point 1_B along its projection line in view B.
5. Transfer dimension D2 to locate point 2_B along its projection line in view B.
6. Connect 1_B and 2_B.

4.8 AUXILIARY VIEWS OF LINES

Lines can be projected onto an infinite number of successive projection planes. As with points, the first auxiliary view from one of three principal planes is called a *primary auxiliary view*. Any auxiliary view projected from a primary auxiliary is a *secondary auxiliary view*, and all auxiliary views projected from these are called *successive auxiliary views*. A line will appear as a point, true length, or foreshortened in orthographic projections.

In Fig. 4-15, line 1-2 is shown in the frontal, horizontal, and profile views. Primary auxiliary view A is projected perpendicular to the frontal view (and is inclined to the other two principal planes). Primary auxiliary view B is perpendicular to the profile view (and inclined to the other two principal views). The line of sight for an auxiliary view is determined by the requirements of the problem. In this example, the line of sight for view A is perpendicular to line 1-2 in the frontal view. View B is a random projection. An infinite number of auxiliary projections can be taken from any view.

In Fig. 4-16 only two principal views of line 1-2 are shown: horizontal and frontal. The following steps describe the construction of primary auxiliary view B.

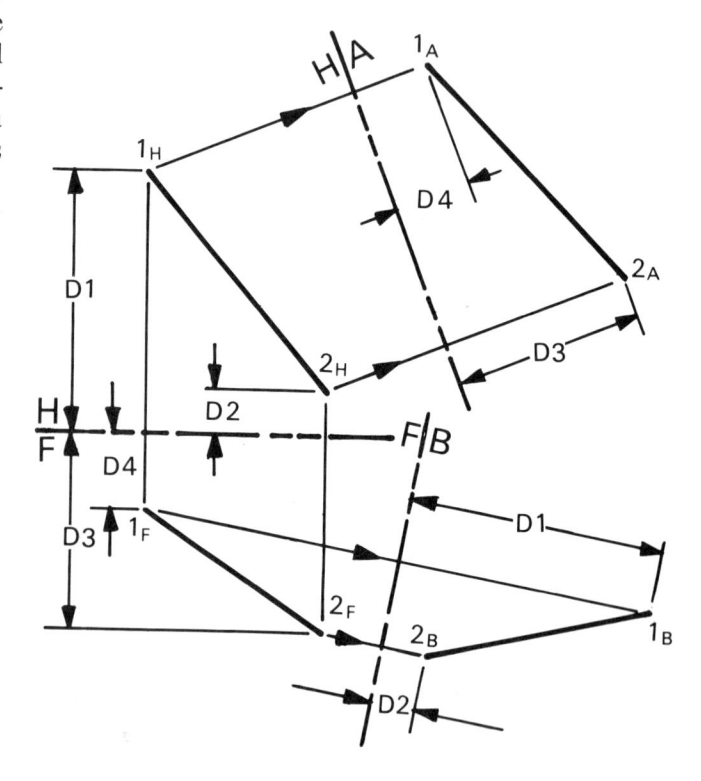

FIG. 4-16 Primary auxiliary view of lines.

66

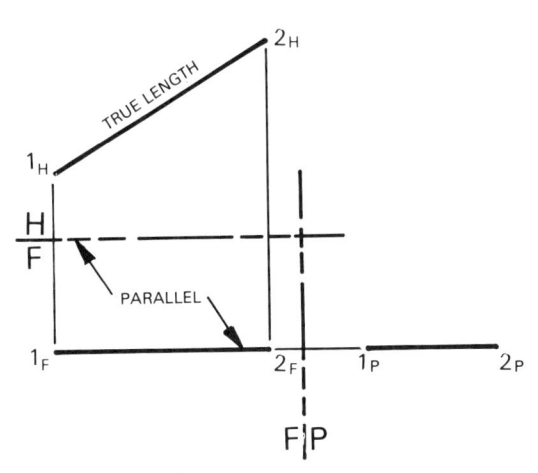

FIG. 4-17 Horizontal line.

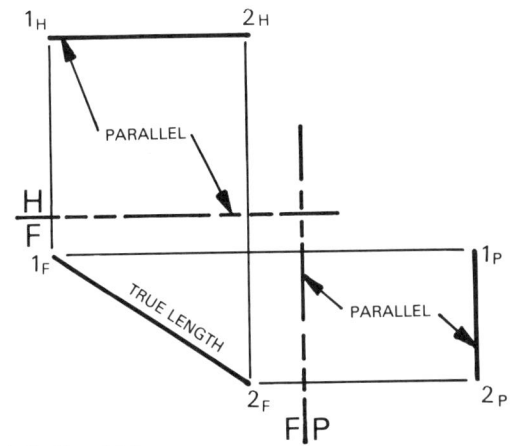

FIG. 4-18 Frontal line.

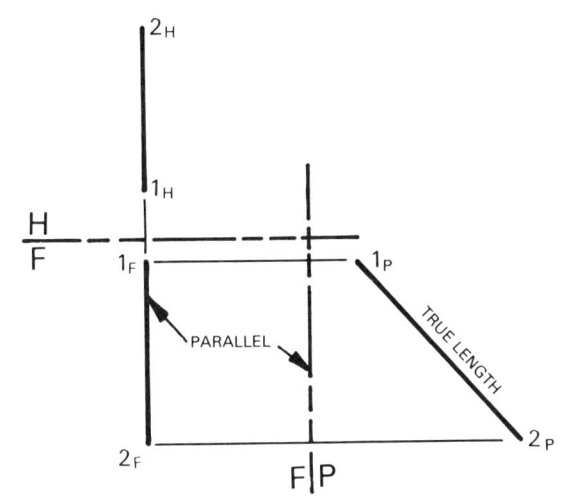

FIG. 4-19 Profile line.

4.9 PRINCIPAL LINES

A line that is parallel to a principal plane is called a ***principal line***, and is *true length* in the principal plane to which it is parallel. Since there are three principal planes of projection, there are three principal lines: ***horizontal, frontal*** and ***profile:***

1. A ***horizontal line***, Fig. 4-17, is a principal line. It is parallel to the horizontal plane and appears true length in the horizontal view.
2. A ***frontal line***, Fig. 4-18, is a principal line. It is parallel to the frontal plane and appears true length in the frontal view.
3. A ***profile line***, Fig. 4-19, is a principal line. It is parallel to the profile plane and appears true length in the profile view.

4.10 DEFINITIONS

The following terms are used to describe lines according to *direction* and *type* (in a particular view).

DIRECTION

Vertical line: Vertical lines are perpendicular to the horizontal plane and appear true length in the frontal and profile planes (consequently will be *both* frontal and profile principal lines). Vertical lines appear as a point (point view) in the horizontal view.

Level line: Any line that is parallel to the horizontal plane is a level line. Level lines are horizontal lines.

Inclined line: Inclined lines will be parallel to the frontal or profile planes (and will therefore be a profile *or* frontal principal line) and at an angle to the horizontal plane. An inclined line appears true length in the view where it is parallel to a principal plane.

TYPE

Oblique: Oblique lines are inclined to all three principal planes and therefore will not appear true length in a principal view.

Foreshortened: Lines that are not true length in a specific view appear shorter (foreshortened) than their true length measurement.

Point view: Where a view is projected perpendicular to a true length line, that line appears as a point view; the end points are therefore coincident. A point view is a view of a line in which the line is perpendicular to the viewing plane (the line of sight is parallel to the line). A point view is an "end view" of a line.

True length: A view in which a line can be measured true distance between its end points shows the line as true length (*true length line*). A line appears true length in any view where it is parallel to the plane of projection.

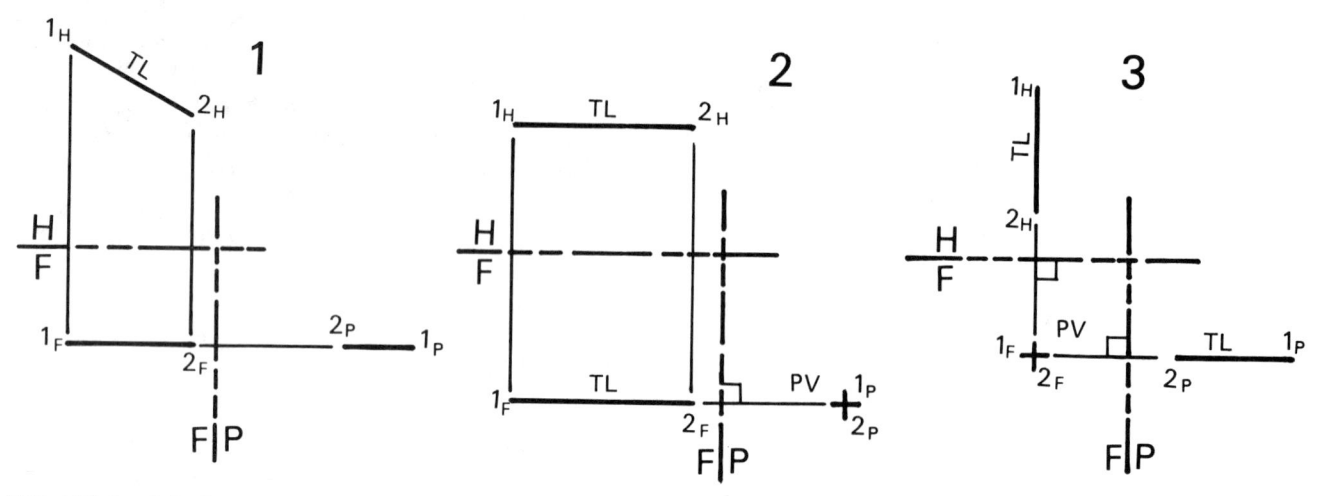

FIG. 4-20 Level (horizontal) lines.

4.11 LEVEL AND VERTICAL LINES

A line that is parallel to the horizontal projection plane is a *level* line and appears true length in the horizontal view. A level line is a horizontal line (principal line) since it is parallel to the horizontal projection plane. In Fig. 4-20, three variations of level (horizontal) lines are illustrated. In example (1) line 1-2 is parallel to the horizontal plane and inclined to the frontal and profile planes. It appears true length in the horizontal view and foreshortened in the other two views.

Example (2) shows level line 1-2 parallel to both the frontal and horizontal projection planes and is a point view in the profile view. Line 1-2 is true length in two principal views and is therefore both a horizontal line and a frontal line.

Another case where a level line appears true length in two principal views is illustrated in example (3), where line 1-2 is parallel to both the profile and horizontal projection planes. Here the line shows as a point view in the frontal view, and true length in both other principal views. Line 1-2 is a profile line, and a horizontal line.

Note that a line that appears true length in a view and perpendicular to the adjacent projection plane projects as a point view. The point view of a line is parallel to each successive projection plane. An example of this condition is shown in Fig. 4-20(3), where the frontal view of line 1-2 shows as a point view (1_F, 2_F) and is parallel to the horizontal and profile planes. Line 1-2 appears as true length in both the horizontal and profile view where it is shown perpendicular to the frontal projection plane.

Vertical lines are perpendicular to the horizontal plane (view) and appear true length in both the frontal and profile views and as a point in the horizontal view, Fig. 4-22. A vertical line is parallel to the profile and frontal projection planes.

Piping systems as that shown in Fig. 4-21 are composed of a series of vertical, horizontal, and inclined lines arranged as banks or runs on pipe racks. Descriptive geometry is used to solve various problems associated with the design and drafting of piping projects.

FIG. 4-21 Pipe rack.

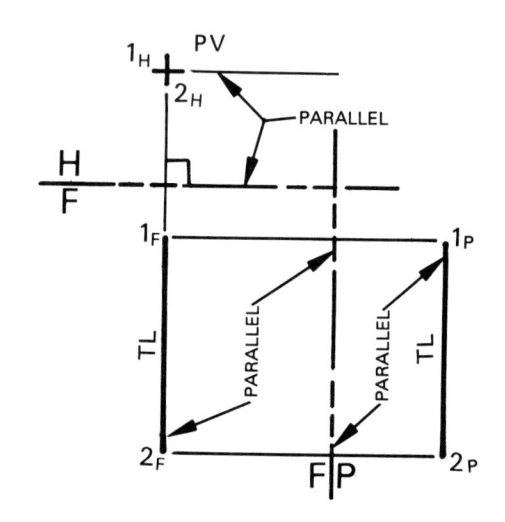

FIG. 4-22 Vertical line.

FIG. 4-23 Inclined lines.

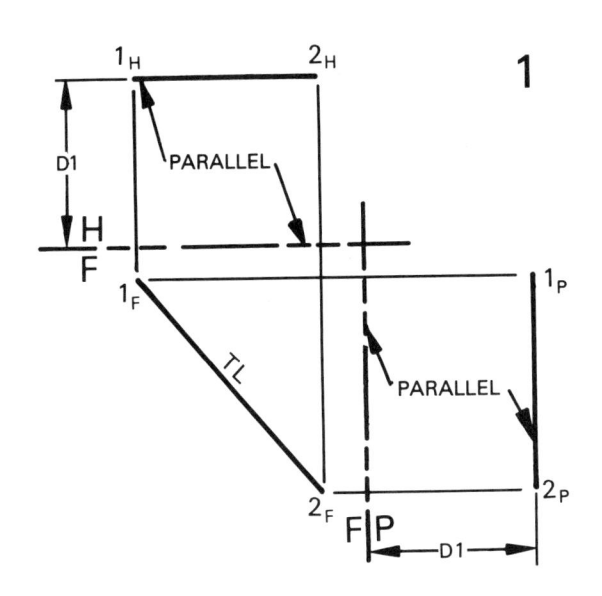

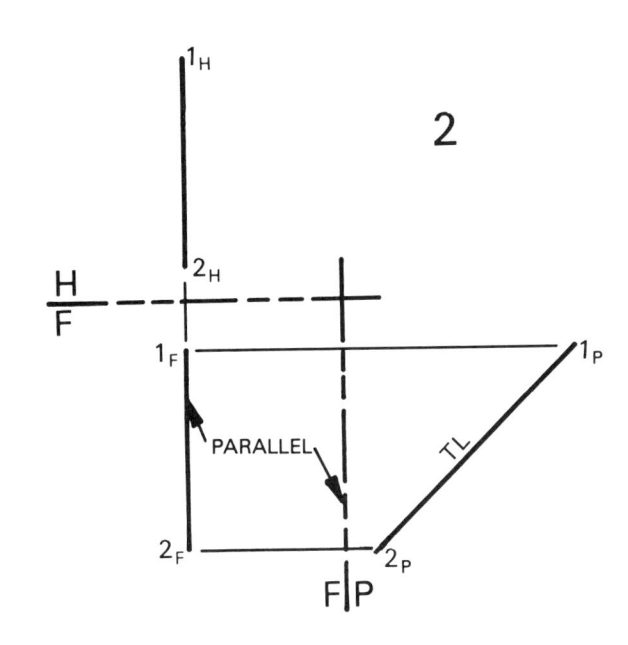

4.12 INCLINED LINES

Lines that appear true length in the frontal or profile view (but not both) are inclined lines. Inclined lines, Fig. 4-23, will be parallel to the frontal plane (1), or parallel to the profile plane (2). In example (1) line 1-2 is true length in the frontal view and parallel to the frontal projection plane in the profile and horizontal view. In example (2) line 1-2 is true length in the profile view and parallel to the profile projection plane in the frontal view. Horizontal lines (that are not point views in the frontal or profile views) could be considered *"inclined"* to the frontal and profile projection planes; however, this is not a universally accepted definition.

Inclined lines will appear foreshortened in two principal views and true length in the other principal view. In Fig. 4-23 (1), line 1-2 is foreshortened in the H and P views, and true length in F. In example (2) line 1-2 is foreshortened in the H and F views, and true length in the profile view. Note that line 1-2 in example (1) is the same distance behind the frontal plane in both the horizontal and profile view, dimension D1.

Complex processing facilities, Fig. 4-24, can be reduced to points and lines to establish essential engineering and design data.

FIG. 4-24 Petrochemical facility.

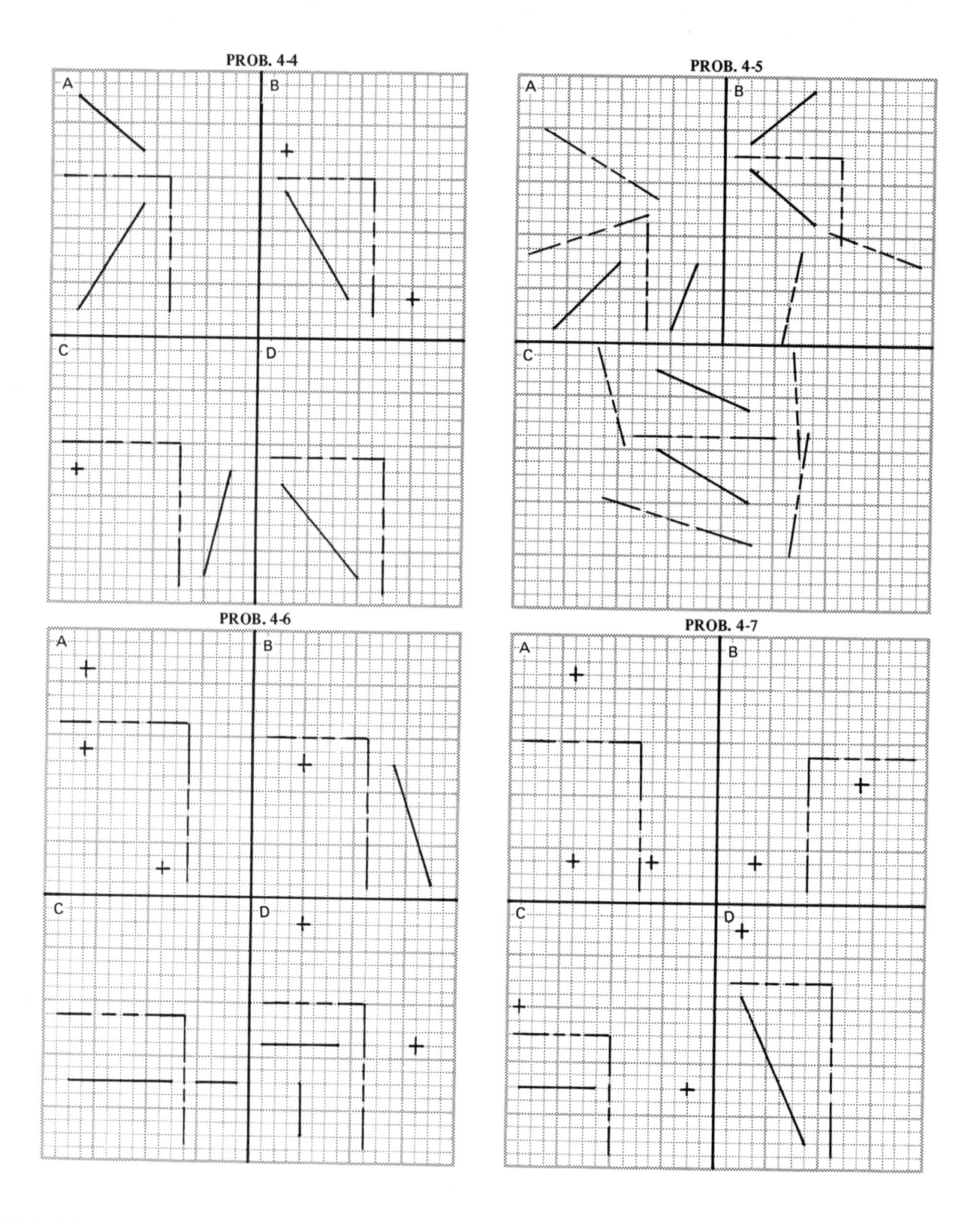

PROB. 4-4A. Draw the profile view of the line.
PROB. 4-4B. Complete the three views of the line.
PROB. 4-4C. Draw line 1-2 in all views. Point 1 is given in both views. Point 2 is six units to the right of point 1.
PROB. 4-4D. Line 1-2 slants downward. Point 1 is five units behind the frontal plane and point 2 is two units behind the frontal plane.
PROBS. 4-5A, 4-5B, 4-5C. Locate the given line in the required auxiliary views.
PROB. 4-6A. Draw a frontal line using the given points.
PROB. 4-6B. Complete the three views of the profile line.
PROB. 4-6C. Draw the horizontal line in all three views.
PROB. 4-6D. Complete the views of the two lines. Label the principal lines.
PROB. 4-7A. Complete the views of the profile (and horizontal) line.
PROB. 4-7B. Draw the three views of the vertical line.
PROB. 4-7C. Draw a level line using the given information. Is it a principal line?
PROB. 4-7D. Complete the three views of the inclined line.

70

FIG. 4-25 Frontal lines.

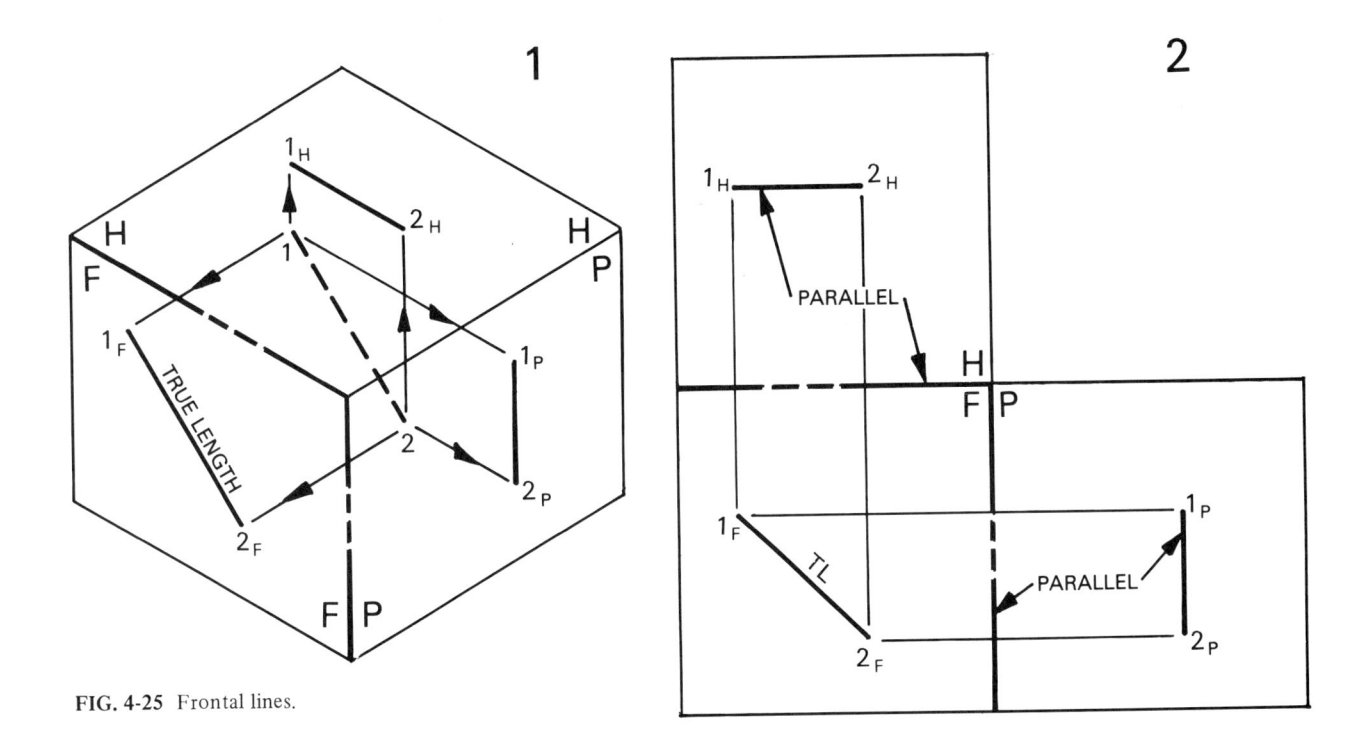

4.13 FRONTAL LINES

*A line that is parallel to the frontal projection plane is a **frontal line**.* Frontal lines are principal lines and always appear true length in the front view. A frontal line can be inclined, level, or vertical, but must be true length. A vertical frontal line appears as a point view in the horizontal view and true length in both the frontal and profile views, therefore a vertical line is both a frontal line and a profile line since it is true length in both of these views. If a level frontal line appeared as a point in the profile plane it would be true length in the frontal and horizontal views and is therefore a horizontal as well as a frontal line.

In Fig. 4-25(1) a pictorial illustration is shown of frontal line 1-2. An orthographic projection of this line appears in (2), where the glass projection box has been revolved into the plane of the paper. Line 1-2 is an inclined frontal line, it is foreshortened in the profile and horizontal views, and true length in the frontal view. Line 1-2 is parallel to and equidistant from the frontal plane in both the horizontal and profile views.

In Fig. 4-26 line 1-2 lies behind the frontal plane the same distance in the horizontal and profile views, dimension D1. Note that it is not possible to tell if a line is a frontal line given the front view alone. Only in the profile and horizontal views can it be established that a line is a frontal line, since in these two views the line is parallel to the fold lines (and therefore the frontal plane).

The angle formed by a frontal line and the F/H fold line is the angle the line makes with the horizontal plane, Fig. 4-26 angle H. The angle formed by a frontal line and the F/P fold line is the angle that the line makes with the profile plane, Fig. 4-26 angle P. The true angle between a plane and a line requires that the line be true length and the plane show as an edge view.

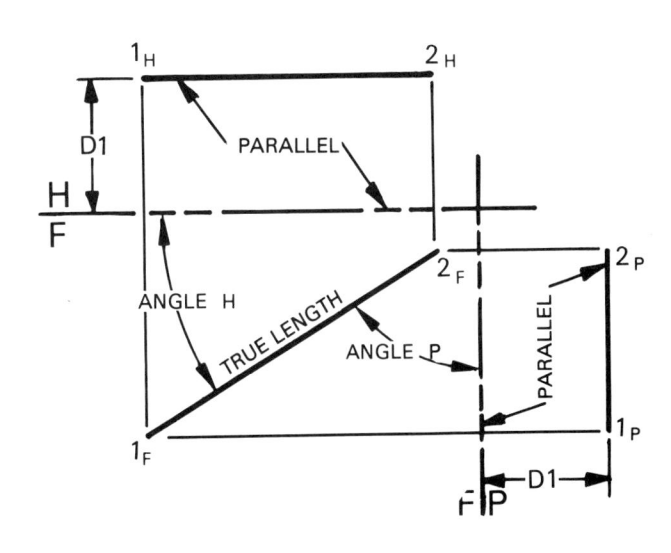

FIG. 4-26 Orthographic projection of a frontal line.

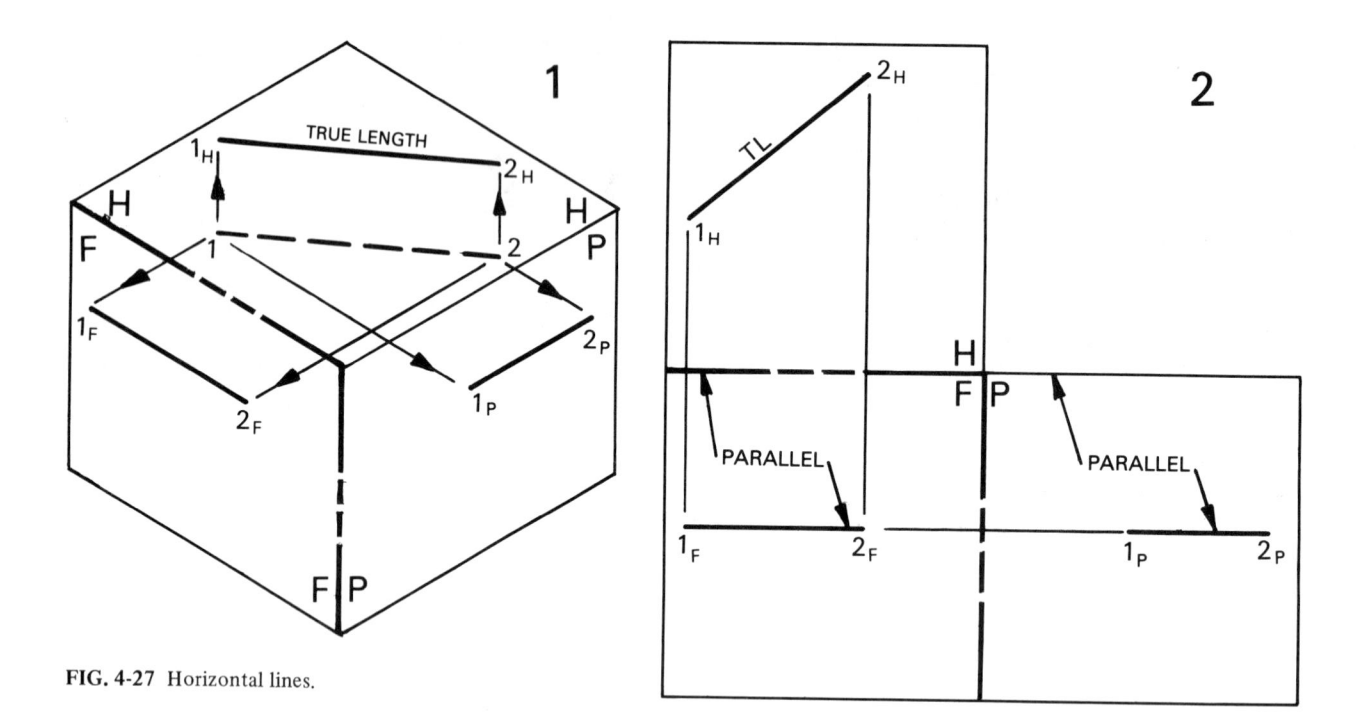

FIG. 4-27 Horizontal lines.

4.14 HORIZONTAL LINES

*A line that is parallel to the horizontal projection plane is a **horizontal line***. Horizontal lines appear true length in the horizontal view and are therefore principal lines. A horizontal line will always be level in the frontal and profile views. If a horizontal line is perpendicular to the frontal plane (and parallel to the profile view) it appears as a point view in the front view and true length in the profile view and is a combination horizontal and profile line. Likewise, if a horizontal line is parallel to the frontal plane it appears as a point view in the profile view and true length in the frontal view and thus is a combination horizontal and frontal line.

In Fig. 4-27, horizontal line 1-2 is shown pictorially (1), and unfolded into the plane of the paper as an orthographic projection (2). The glass box method has been used here to illustrate the position of a horizontal line in the three principal planes of projection. Line 1-2 is a level line (horizontal) since it is parallel to the horizontal plane. It appears true length in the horizontal view and is at an angle to the other two principal planes, therefore it is foreshortened in the frontal and profile views.

In order to tell if a line is in or parallel to the horizontal plane, it is necessary to have either the frontal or profile views of the line. Only the frontal and profile views show the line as a point view or parallel to the horizontal plane.

In Fig. 4-28, line 1-2 is parallel to the H/F fold line and therefore true length in the horizontal view. This horizontal line is not parallel to any other principal plane and therefore shows as foreshortened in the frontal and profile views. Dimensions D1 and D2 locate the end points of the line behind the frontal plane. The angle that line 1_H-2_H forms with the H/F fold line (angle F) is the angle between the line and the frontal plane. Angle P is the angle between line 1_H-2_H and the profile plane.

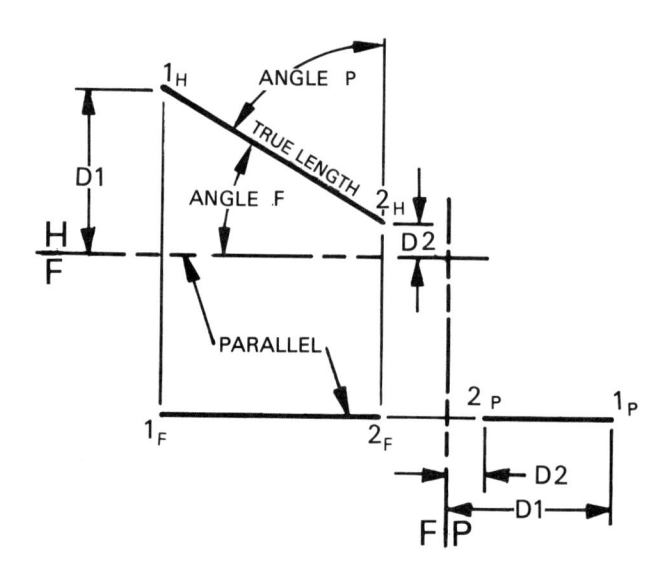

FIG. 4-28 Orthographic projection of a horizontal line.

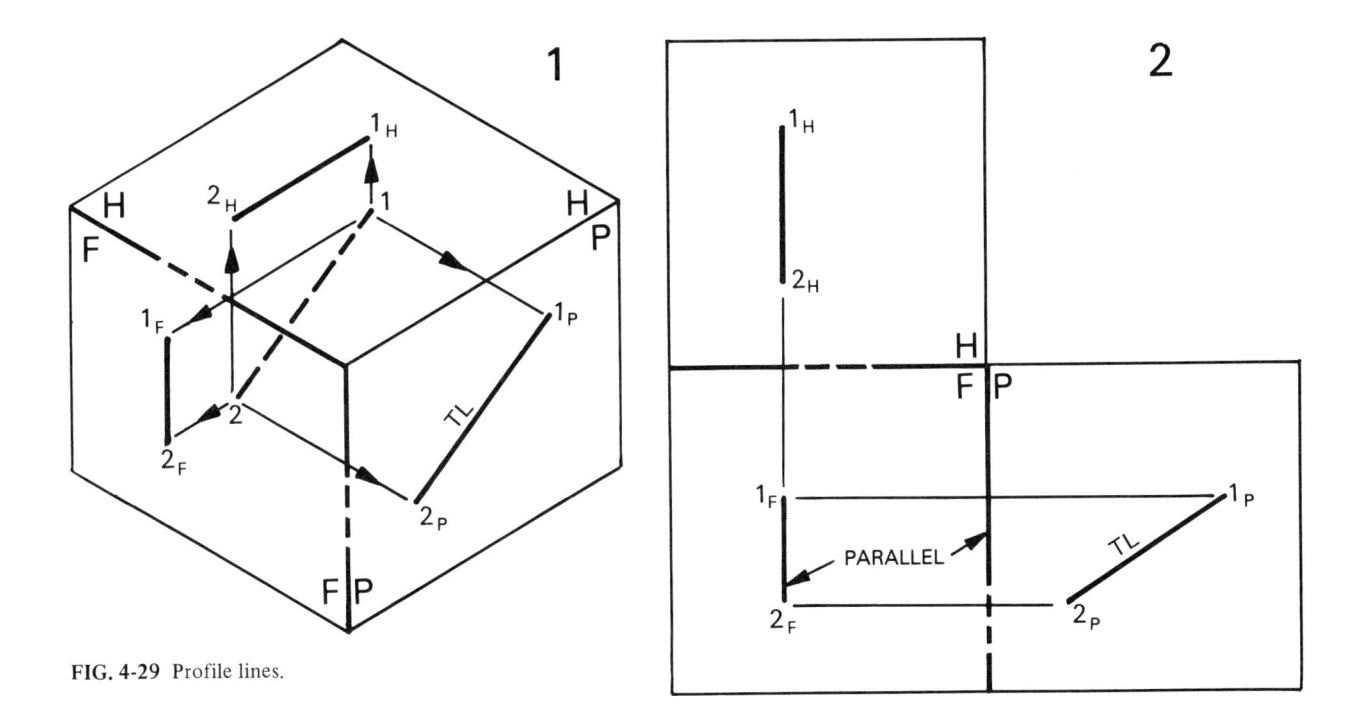

FIG. 4-29 Profile lines.

4.15 PROFILE LINES

A *profile line* is parallel to the profile plane and shows as true length in the profile view. The frontal and horizontal view of a profile line always shows the line as a point view or foreshortened, in either case the line is parallel to the profile plane.

Vertical lines are both profile and frontal lines since they appear true length in the frontal and profile views and as a point in the horizontal view. Where a profile line appears as a point in the front view, the line is both a frontal and profile line.

In Fig. 4-29(1) line 1-2 is shown projected onto the three principal planes represented pictorially by the glass box. Example (2) shows the glass box rotated into the plane of the paper with line 1-2 in orthographic projection. Notice that profile lines are parallel to the profile plane and look vertical in the frontal view, though they are normally foreshortened. In this example, line 1-2, being a profile line, is true length in the profile view and foreshortened in the frontal and horizontal views.

Line 1-2 in Fig. 4-30 is true length in the profile view and appears foreshortened in the other two principal views. Dimension D1 and D2 locate the end points of the line from the F/P fold line and the H/F fold line. These measurements represent the distance point 1 and point 2 are behind the frontal plane. The angle that profile line 1$_P$-2$_P$ makes with the F/P fold line is the angle between the line and the frontal plane (angle F). Angle H is the angle that line 1$_P$-2$_P$ makes with the horizontal plane.

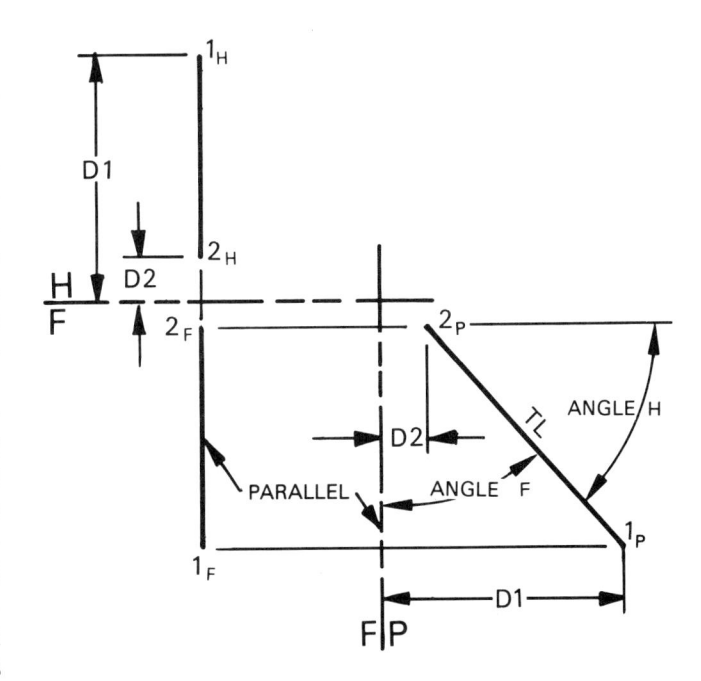

FIG. 4-30 Orthographic projection of a profile line.

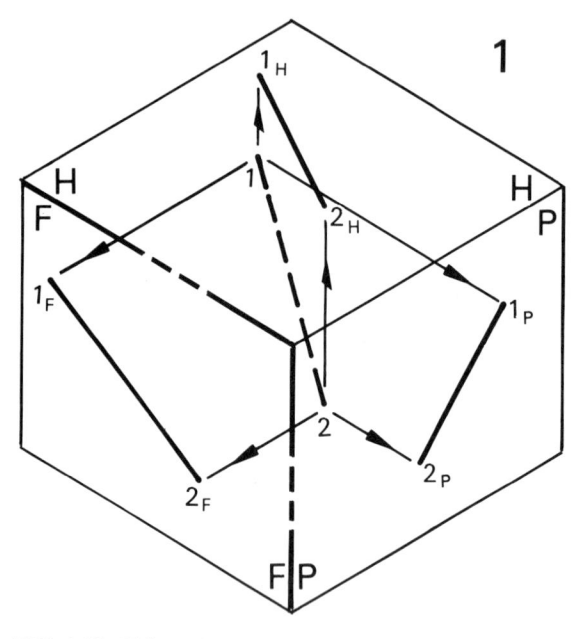

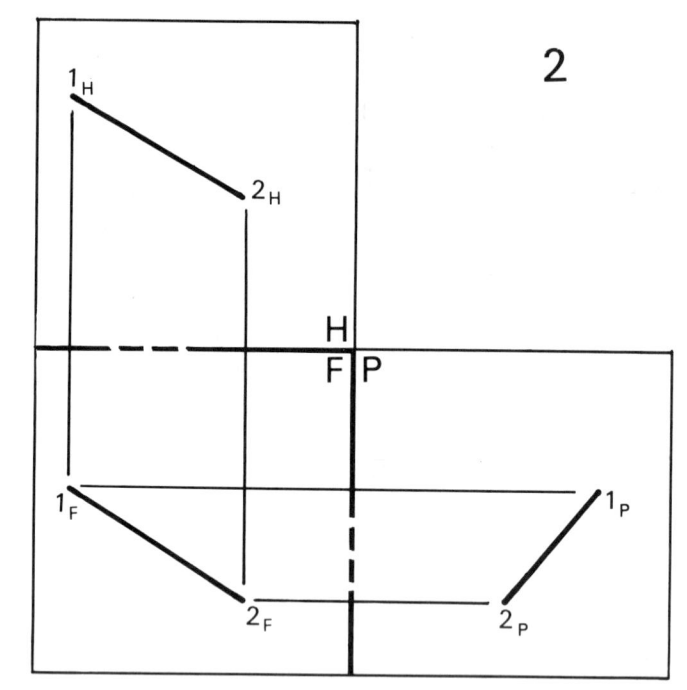

FIG. 4-31 Oblique lines.

4.16 OBLIQUE LINES

Oblique lines are inclined to all three principal planes: horizontal, frontal, and profile. An oblique line is not vertical, parallel or perpendicular to any of the three principal planes, and therefore does not appear true length in the frontal, horizontal, or profile views. All three principal views of an oblique line appear foreshortened. In Fig. 4-31 an oblique line is shown projected onto the three principal planes of projection represented pictorially (1) by the glass box. In each principal view, line 1-2 appears foreshortened, as shown in example (2) where the glass box is rotated into the plane of the paper to form an orthographic projection.

In order to solve for the true length of an oblique line, an auxiliary view with a line of sight perpendicular to a view of the oblique line must be projected from any existing view. The fold line between these two views will be parallel to the oblique line. This procedure is covered in detail in the next section.

Line 1-2 in Fig. 4-32 is an oblique line since it is not parallel to any principal plane of projection and appears foreshortened in every view. In order to locate an oblique line in space, dimensions must be taken from fold lines and projection lines extended from an existing view. Dimension D1 and D2 locate the end points of line 1-2 from the H/F and F/P fold line and represent the distance line 1-2 is behind the frontal plane.

Note that the angle formed between an oblique line and any fold line is not a true angle between the line and a principal plane since the line is not true length.

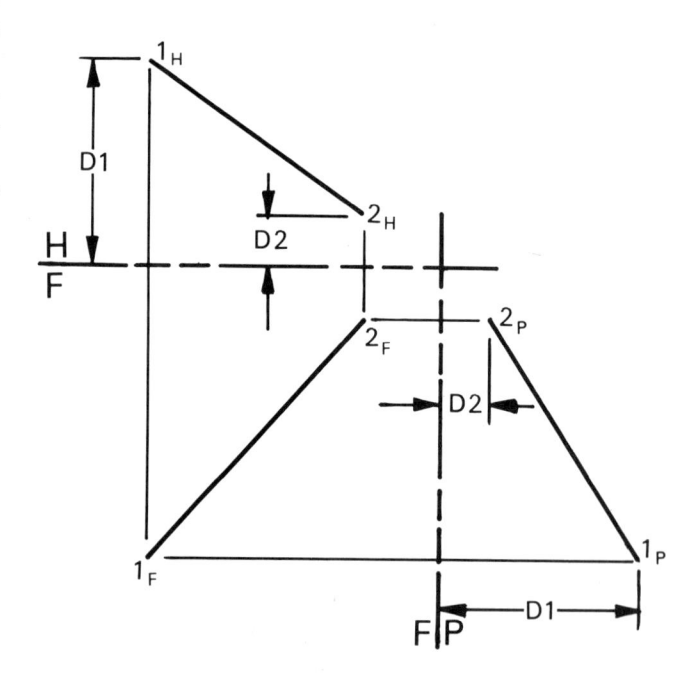

FIG. 4-32 Orthographic projection of an oblique line.

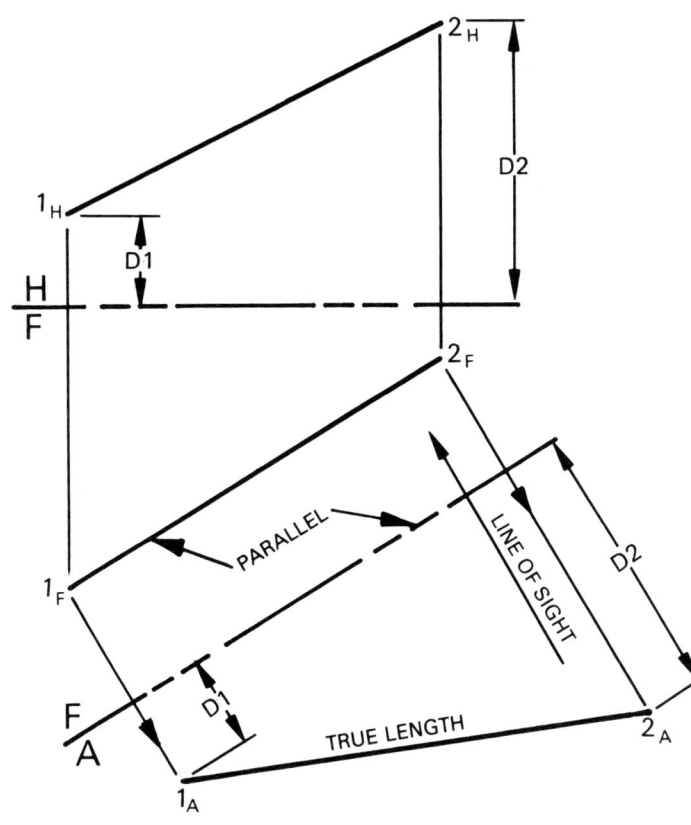

FIG. 4-33 True length of oblique line using a frontal auxiliary view.

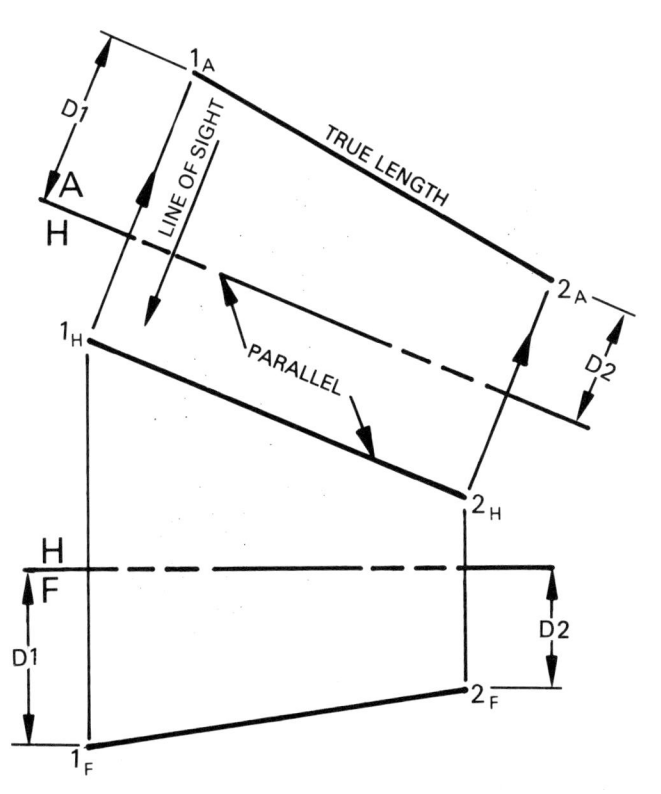

FIG. 4-34 True length of oblique line using a horizontal auxiliary view.

4.17 TRUE LENGTH OF OBLIQUE LINES

Since oblique lines are not true length in a principal plane, it is necessary to project an auxiliary view in which the given line appears true length. An oblique line appears foreshortened in all three principal views: frontal, horizontal, and profile. In order to have a true length of a line, the line must be parallel to the projection plane in the adjacent view. Since oblique lines are not parallel to any of the principal planes of projection they do not appear true length in a principal view.

A true length view of an oblique line can be projected from any existing view by establishing a line of sight perpendicular to a view of the line and drawing a fold line parallel to the line (perpendicular to the line of sight).

The following steps describe the procedure for drawing a true length projection of an oblique line from the frontal view, Fig. 4-33.

1. Establish a line of sight perpendicular to oblique line 1-2 in the frontal view.
2. Draw fold line F/A perpendicular to the line of sight and parallel to the oblique line 1_F-2_F.
3. Extend projection lines from point 1_F and 2_F perpendicular to the fold line and parallel to the line of sight.
4. Transfer dimension D1 and D2 from the horizontal view to locate point 1_A and 2_A along the projection lines in auxiliary view A.
5. Connect points 1_A and 2_A. This is the true length of line 1-2.

The true length of an oblique line can also be projected from the horizontal view (Fig. 4-34) using the same basic steps as described above.

In Fig. 4-35 the wing of the supersonic transport is oblique when in its rotated position.

FIG. 4-35 Oblique wing aircraft. *(Courtesy NASA.)*

FIG. 4-36 True length of a line by auxiliary view.

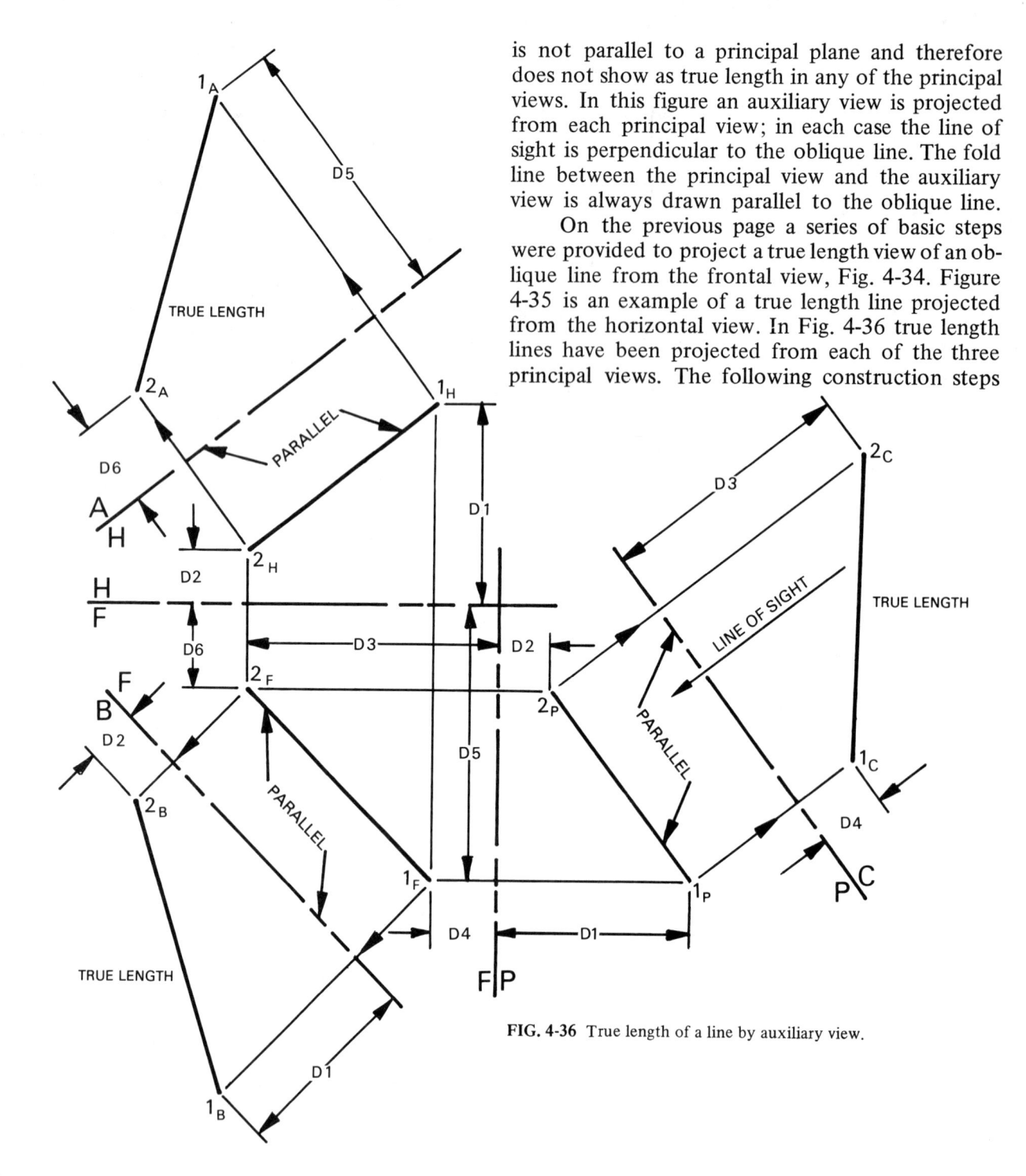

4.18 TRUE LENGTH OF A LINE

*The **true length of a line** can be measured in a view where the line is parallel to the projection plane of that view.* If the line of sight for a view is not perpendicular to the line, the line will be foreshortened, as in the three principal views of line 1-2 in Fig. 4-36. Line 1-2 is an oblique line. Note that the true length of an oblique line can be found in an auxiliary projection that is parallel to the oblique view of the line. This new view can be projected from the frontal, horizontal, or profile views.

Line 1-2 in Fig. 4-36 is an oblique line since it is not parallel to a principal plane and therefore does not show as true length in any of the principal views. In this figure an auxiliary view is projected from each principal view; in each case the line of sight is perpendicular to the oblique line. The fold line between the principal view and the auxiliary view is always drawn parallel to the oblique line.

On the previous page a series of basic steps were provided to project a true length view of an oblique line from the frontal view, Fig. 4-34. Figure 4-35 is an example of a true length line projected from the horizontal view. In Fig. 4-36 true length lines have been projected from each of the three principal views. The following construction steps describe the projection of a true length view from the profile view. Note that the same basic steps are used here as on page 75, only the principal plane that the view is projected from is different.

1. Establish a line of sight perpendicular to line 1_P-2_P.
2. Draw fold line P/C parallel to line 1_P-2_P and perpendicular to the line of sight.
3. Extend projection lines from point 1_P and 2_P, perpendicular to fold line P/C and parallel to the line of sight.
4. Transfer dimensions D3 and D4 from the frontal view to locate the end points of line 1-2 in auxiliary view C.
5. Connect point 1_C and point 2_C to complete the true length line.

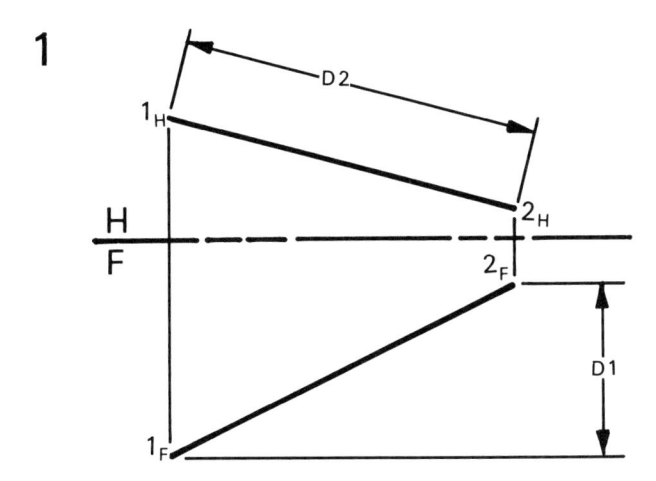

FIG. 4-37 True length of a line by true length diagram.

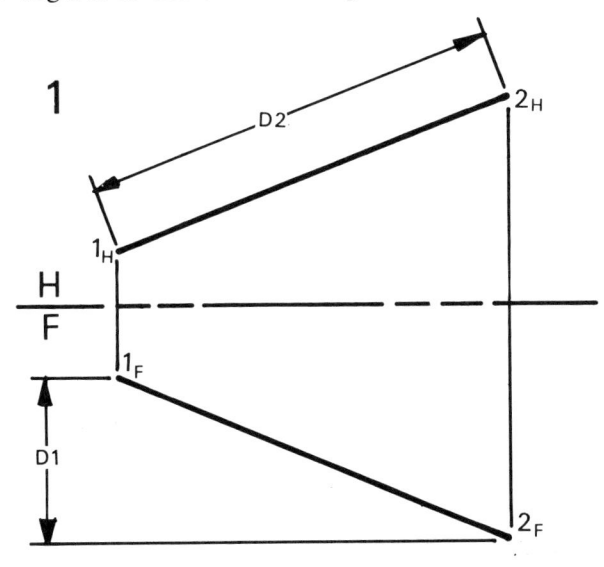

4.19 TRUE LENGTH DIAGRAMS

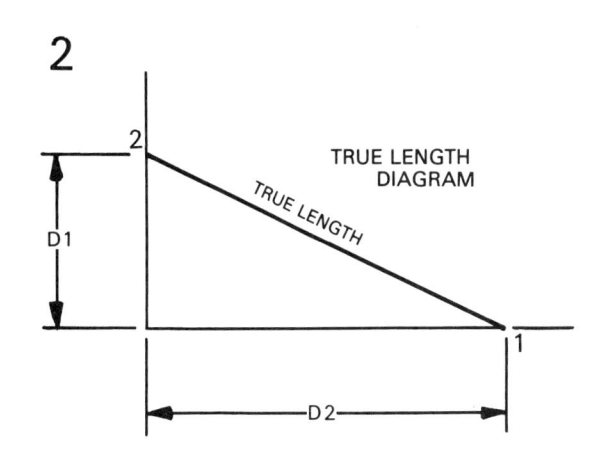

An alternative to the auxiliary view method of solving for the true length of a line is the *true length diagram*. This method is used extensively when developing a variety of complicated shapes such as transition pieces and other developments where there may be a large number of elements in one view that are oblique and not parallel to one another. In this type of situation it would be impossible to project auxiliary views of every line; see Chapter 7. A true length diagram can be used to establish the true length measurement of an oblique line using any two adjacent (successive) views of that line, thereby eliminating the necessity of projecting an auxiliary view.

In Fig. 4-37(1) oblique line 1-2 is shown in the frontal and horizontal views. Instead of projecting an auxiliary view to establish its true length, a true length diagram (2) has been used. To construct the diagram, draw two construction lines at 90 degrees to the side of the given views. Transfer the vertical dimension D1 from the frontal view to the vertical leg of the construction line, to locate point 2. Dimension D2 can then be transferred from the horizontal view to the horizontal construction line to locate point 1. This newly formed right triangle has a hypotenuse equal to the true length of line 1-2, and can be measured from the drawing. The true length can also be mathematically calculated using the Pythagorean theorem to solve for the hypotenuse, $C = \sqrt{A^2 + B^2}$. C is the hypotenuse (true length of line 1-2), A is the altitude/height (D1), and B is the base (D2),

$$\text{Hyp} = \sqrt{D1^2 + D2^2}.$$

Note that placing the right triangle in line with the front view so that the height dimension (D1) could be projected horizontally would simplify construction of the true length diagram. This method is applied in the development chapter, but is not shown here because it is important to realize that a true length diagram can be constructed from any two adjacent views and need not be taken from the horizontal and frontal principal views.

Figure 4-38 is another example of the use and construction of a true length diagram; the same steps as given above describe this example.

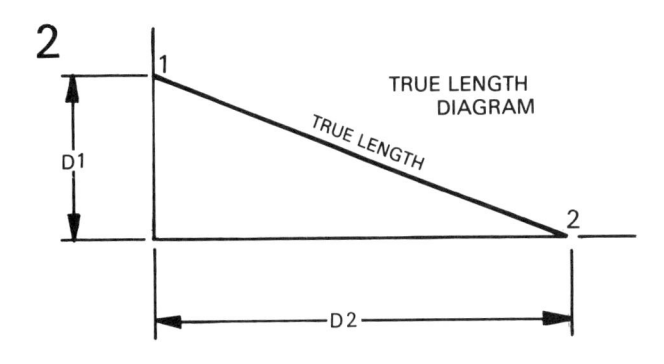

FIG. 4-38 True length diagrams.

FIG. 4-39 Point view of a principal line.

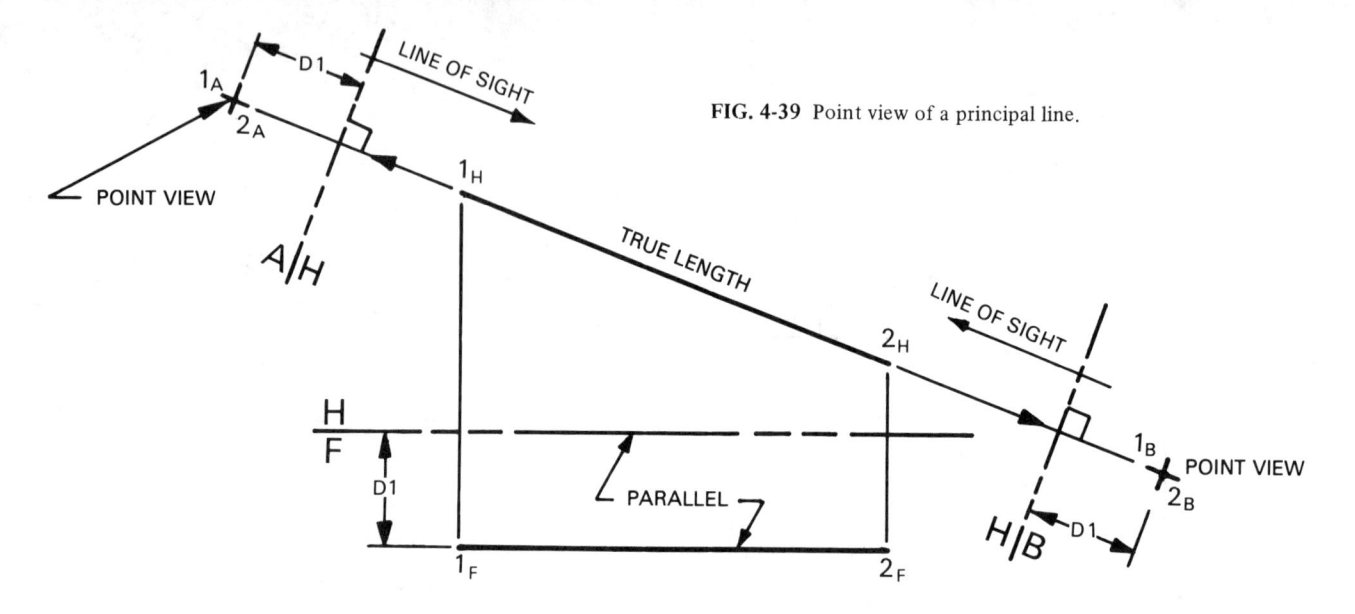

4.20 POINT VIEW OF A LINE

*A line will project as a **point view** when the line of sight is parallel to a true length view of the line; in other words the point view is projected on a projection plane that is perpendicular to the true length line.* Finding the true length and the point view of a line will be required for many situations involving the application of descriptive geometry to engineering problems. The first requirement for a point view is that the line be projected as true length. This procedure has been discussed in detail in previous sections.

The point view of a principal line is established by the use of a primary auxiliary view as shown in Fig. 4-39. The following steps describe this process.

1. Establish a line of sight parallel to the true length line 1_H-2_H (note that two possibilities exist).
2. Draw the fold line perpendicular to the line of sight (H/A or H/B). Note that the fold line is also perpendicular to the true length line.
3. Draw a projection line from line 1_H-2_H perpendicular to the fold line and therefore parallel to the line of sight.
4. Transfer dimension D1 from the front view to locate both points along the projection line, in either auxiliary view A or B.

The point view of an oblique line can be drawn only after the line is projected as true length in an auxiliary view. In Fig. 4-40, line 1-2 is projected as true length in two separate primary auxiliary views, A and C. To establish the point view, a secondary auxiliary view (B or D) is projected perpendicular to the true length line. For auxiliary view B transfer the distance between H/A and line 1_H-2_H along the projection line from fold line A/B to locate point view 1_B-2_B. For auxiliary view D, transfer the distance between F/C and line 1_F-2_F along the projection line from fold line C/D to locate the point view of line 1_D-2_D.

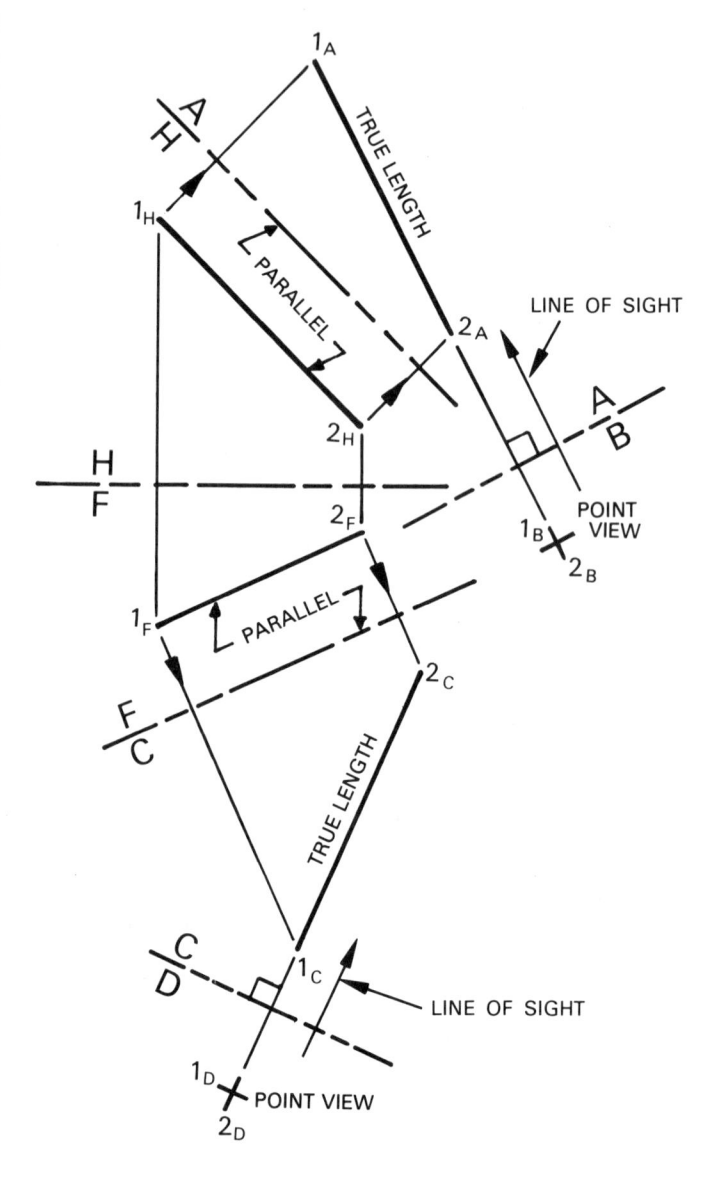

FIG. 4-40 Point view of an oblique view.

PROB. 4-8

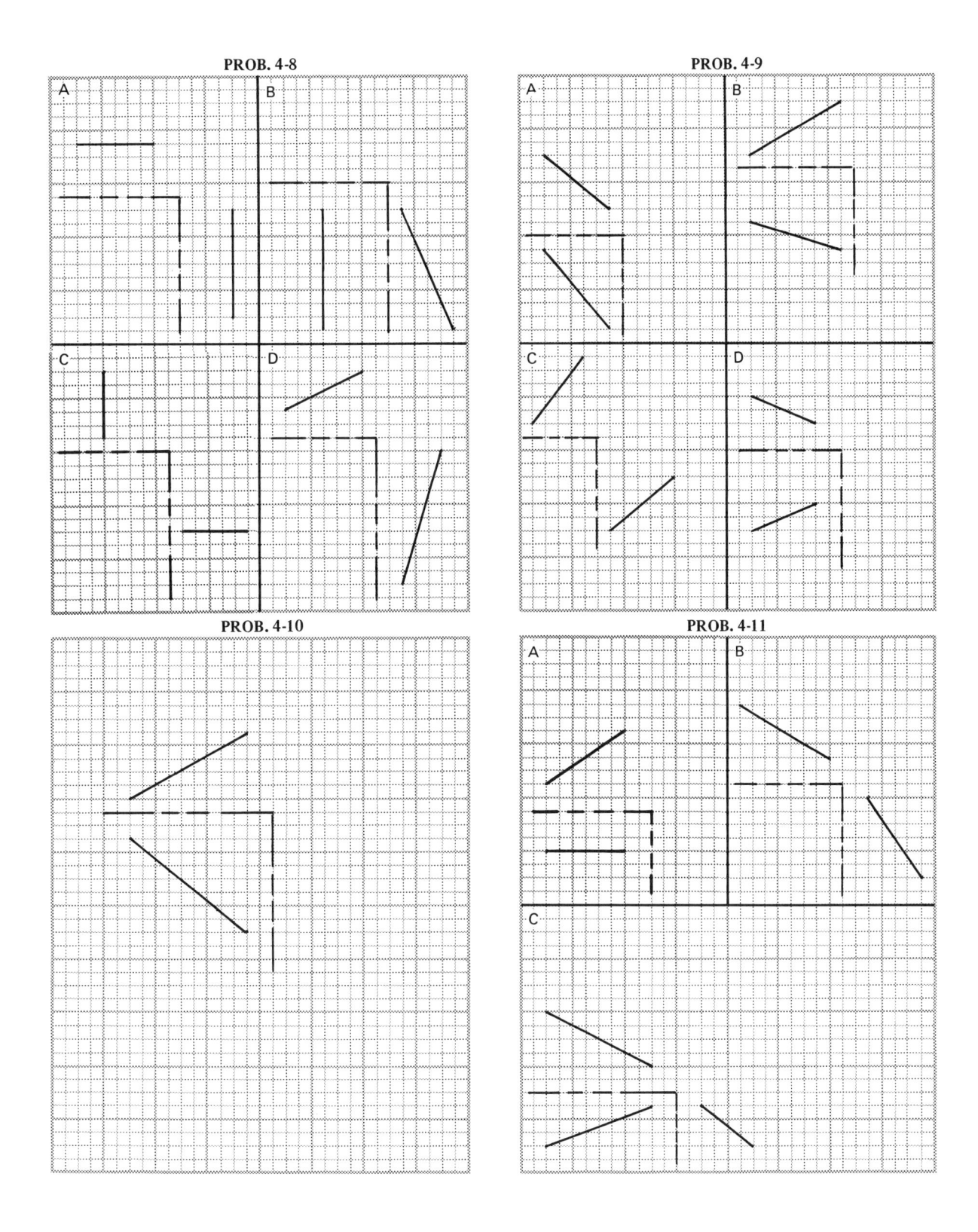

PROB. 4-9

PROB. 4-10

PROB. 4-11

PROBS. 4-8A, 4-8B, 4-8C, 4-8D. Complete the three views of the given line in each of the problems for this page. Label all lines where they appear as principal lines, and note if a line is oblique, inclined, true length, or parallel with a projection plane.

PROBS. 4-9A, 4-9B, 4-9C, 4-9D. Complete the three views of each oblique line and then draw an appropriate auxiliary view to establish its true length. In problem D solve for the true length in two separate views.

PROB. 4-10. Draw the three primary auxiliary views where the given line will appear true length. Show all notation and label each view completely. On the bottom portion of the page construct a true length diagram of the line.

PROBS. 4-11A, 4-11B, 4-11C. In each part solve for the true length of line, and the point view. In problem C establish the true length and point view twice, using auxiliary view projections from the profile and horizontal views.

FIG. 4-41 Point on a line.

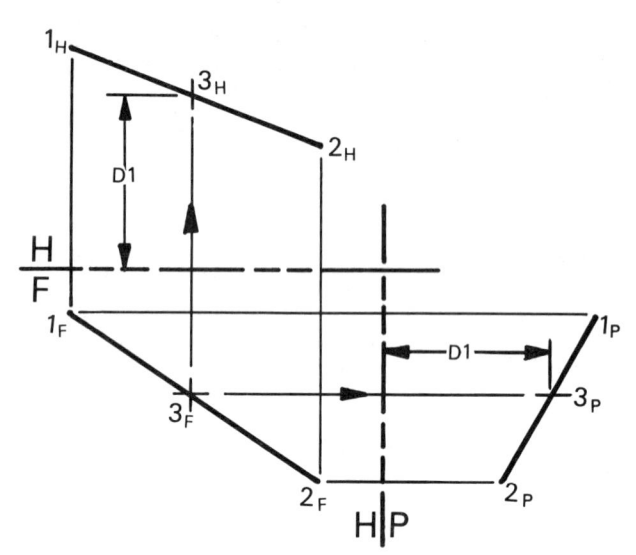

4.21 POINT ON LINES

In most cases, successive views (principal or auxiliary) of a point on a line may be projected to all adjacent views by extending a projection line from the point, perpendicular to the fold line, until it crosses the line in the next view. This typical case is illustrated in Fig. 4-41 where point 3 is located on line 1-3 in the front view. Projection lines are extended from 3_F perpendicular to the fold line for each adjacent view. In the profile and horizontal views point 3 is the same distance behind the frontal plane (D1) and on the line. Point 3 is aligned horizontally in the frontal and profile views and vertically in the frontal and horizontal views.

When a line is perpendicular or appears as if it is perpendicular to a fold line this procedure is not satisfactory. In Fig. 4-42 point 3 is on line 1-2 but cannot be projected from view to view since line 1-2 is parallel to fold line H/P and appears perpendicular, in its foreshortened views, to the H/F fold line. Point 3 is given on the line in the frontal view (1) and projected to the profile view as shown (2). Since foreshortened line 1_F-2_F appears as if perpendicular to H/F it is necessary to transfer dimension D1 from the profile view to the horizontal view to locate point 3_H (3).

In Fig. 4-43 point 3 and 4 appear to be on line 1-2 in the horizontal view, but their frontal and auxiliary views show that only point 3 is on the line, whereas point 4 lies directly above the line (horizontal view). This example shows that two views of a point are necessary to establish its position in space and whether or not it is on a line. If a point is centered on a line, then it is centered on the line (true length or oblique) in all views.

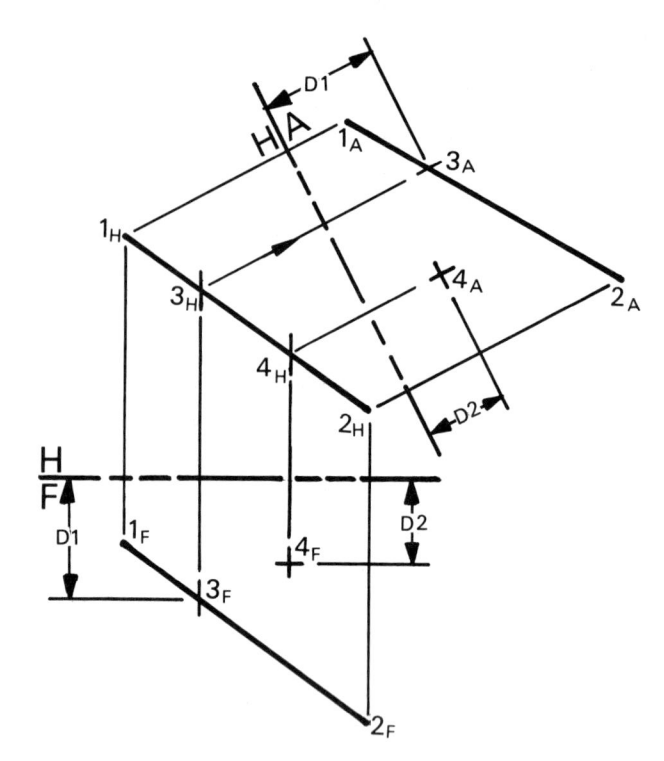

FIG. 4-43 Points on and off line.

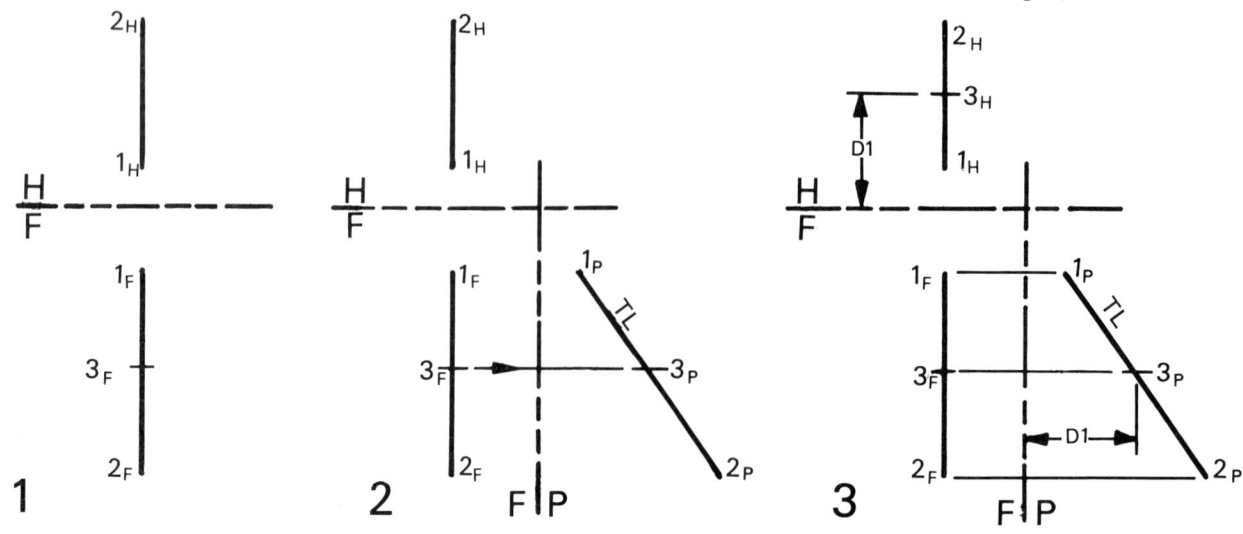

FIG. 4-42 Point on profile line.

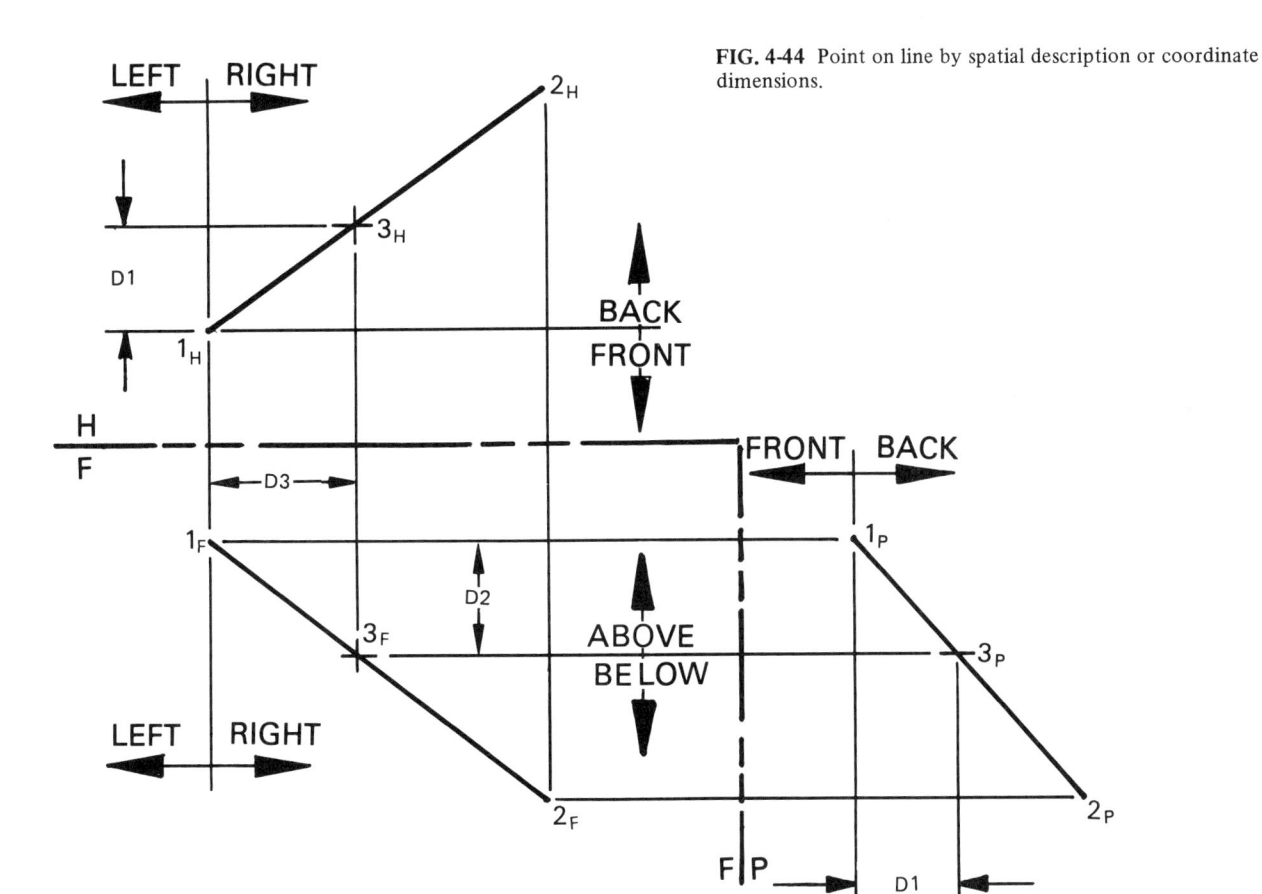

FIG. 4-44 Point on line by spatial description or coordinate dimensions.

4.22 POINT ON LINE BY SPATIAL DESCRIPTION AND COORDINATE DIMENSION

The location of a point on a line can be determined by spatial description and one coordinate dimension.

Points can be located by describing their relationship to another point, Fig. 4-44. The frontal view locates a point above or below and to the right or left of a given point. The horizontal view locates a point to the front or back and to the right or left of a given point, and the profile view locates a point above or below and to the front or back of a given point. Notice that each view has one location direction in common with an adjacent view: in the frontal and horizontal views, the left/right distance; in the frontal and profile views, the above/below distance; and in the horizontal and profile views, the front/back distance. In Fig. 4-44, point 3 is on line 1-2. To locate the point it is necessary to know only one coordinate distance and its spatial description. Point 3 can be said to lie on line 1-2, distance D1 behind point 1. This would fix the point in all views by measurement or projection. Another way of describing the location of point 3 would be to say point 3 is on line 1-2, and distance D2 below point 1; or point 3 is on line 1-2, distance D3 to the right of point 1. Of course point 3 could also be located in respect to point 2.

Note that the distance dimension would be given in specific units of measurements, English or metric.

If point 3 were to lie midpoint of line 1-2 then it would only be necessary to state that fact, since a point on the midpoint of a line is at its midpoint in every view.

Location of a point using spatial description and coordinate dimensions is useful when locating aspects of a pipeline, Fig. 4-45 (pipe supports, valves, flanges), on hand-drawn or computer generated isometric spools.

FIG. 4-45 Piping control station.
(Courtesy Grove Valve and Regulator.)

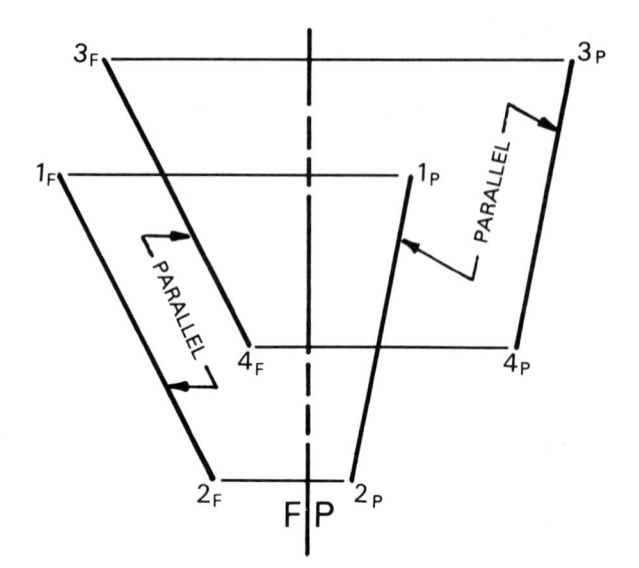

FIG. 4-46 Parallel lines.

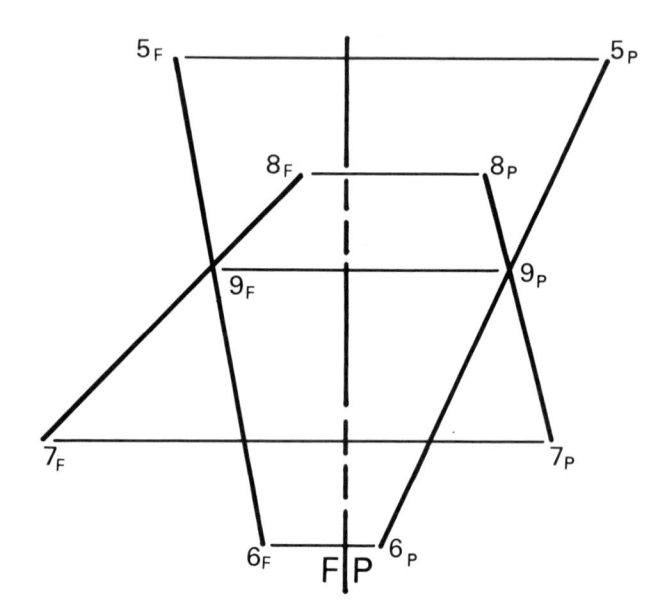

FIG. 4-47 Intersecting lines.

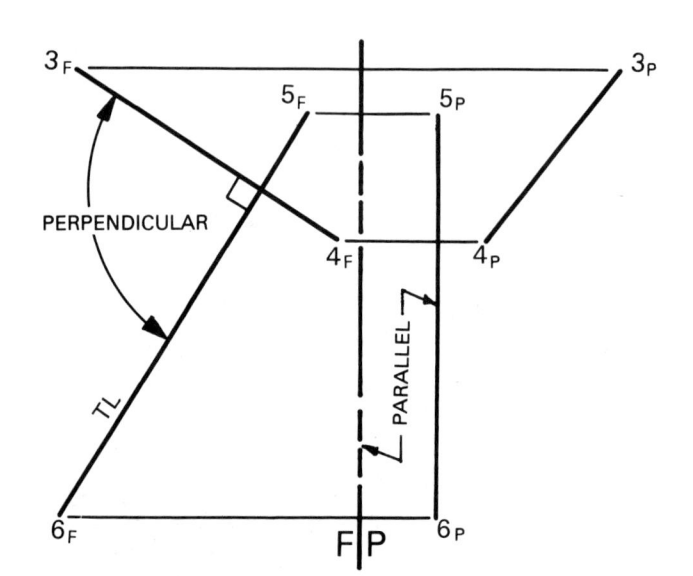

FIG. 4-48 Perpendicular lines.

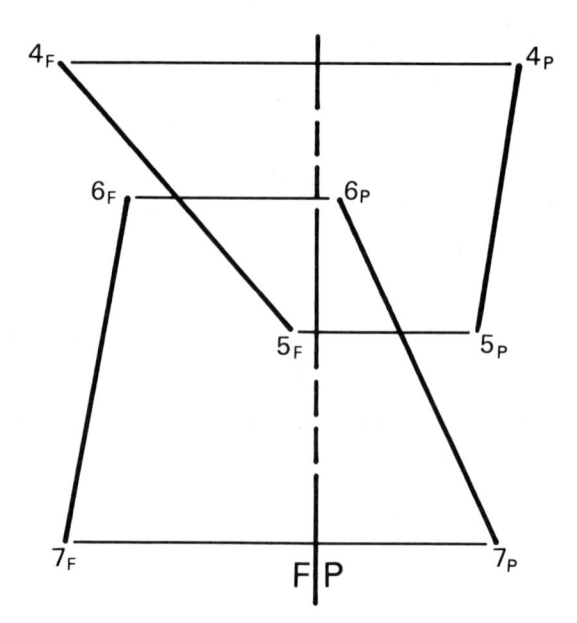

FIG. 4-49 Skew lines.

4.23 TYPICAL CHARACTERISTICS OF TWO LINES

In this section the characteristics of two lines are discussed. There are four basic categories or situations that can define the relationship of two lines in space. Lines can be ***parallel, intersecting, perpendicular,*** or ***skew.***

Lines are *parallel,* Fig. 4-46, when the distance between them remains the same. They are parallel in all projections. Lines that are parallel to a reference plane may appear parallel in two principal views, but only the third view will establish if they are in fact parallel or not. Two parallel lines form a plane.

When two lines lie in the same plane and share a common point, they are *intersecting lines.* Intersecting lines form a plane. The common point (intersecting point) of two lines is aligned in all adjacent projections, as is point 9 in Fig. 4-47.

Perpendicular lines are at right angles (90 degrees) in a view where one or both of the lines show true length, Fig. 4-48. Perpendicular lines can be intersecting or nonintersecting. In Fig. 4-48 line 3-4 and line 5-6 are nonintersecting, perpendicular lines. Intersecting perpendicular lines form a plane.

Skew lines do not intersect and are not parallel or perpendicular, Fig. 4-49. Skew lines do not form a plane.

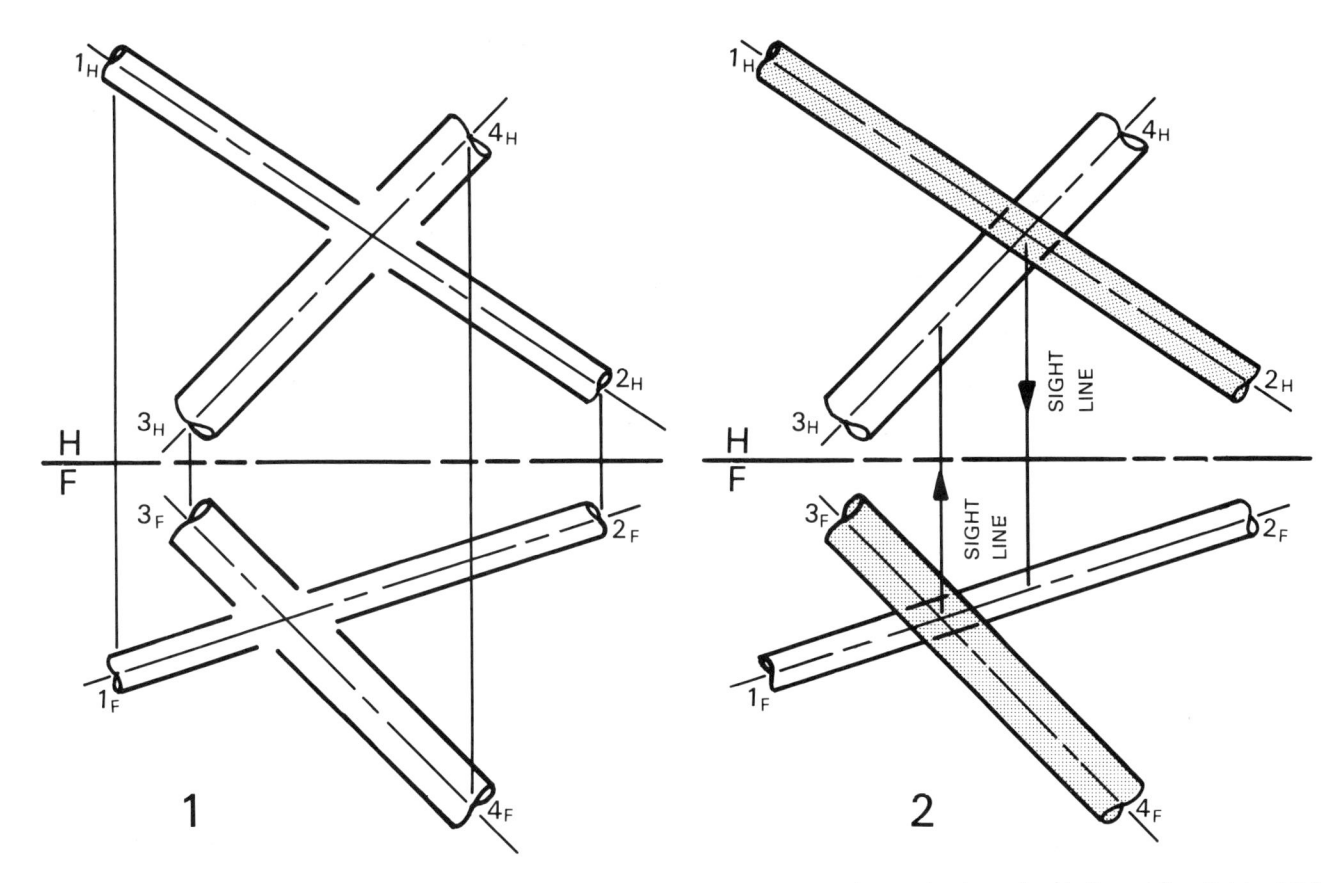

FIG. 4-50 Visibility of lines.

4.24 VISIBILITY OF LINES

When two lines cross in space they may intersect or one is visible and the other hidden at the crossing point. A visibility check will determine the proper relationship of the lines. Note that the visibility of two lines can change in every view; first one line may be visible then in the next view the other line, or the same line may be visible in adjacent views.

In Fig. 4-50(1) lines 1-2 and 3-4 cross. It must be determined which line lies in front of the other in the frontal view and which line is on top in the horizontal view. A visibility check must be made. The following steps describe this process:

1. Where line 1_H-2_H crosses line 3_H-4_H in the horizontal view, extend a **sight line** perpendicular to H/F until it meets one of the lines in the frontal view (2). Here, line 1_F-2_F is the first line to be encountered, therefore line 1_F-2_F is the visible line in the *horizontal* view.

2. Extend a **sight line** from the crossing point of line 1_F-2_F and 3_F-4_F in the frontal view until it meets the first line in its path in the horizontal view. Since line 3_H-4_H was encountered first, it will be the visible line in the *frontal* view.

3. Complete the visibility of lines by showing the proper solid (visible) and dashed (hidden) lines in both views. In Fig. 4-50 the visible line has been shaded for clarity, though this is not standard practice. Note that line 1-2 is visible in the horizontal view (is above line 3-4), and line 3-4 is visible in the frontal view (is in front of line 1-2).

If two lines have the same crossing point in adjacent projections, they are intersecting lines and both are visible in all views. This possibility should be determined first.

In Fig. 4-51 the visibility check shows that line 1-2 is visible in both views. Note that the visibility of crossing lines in one view can only be determined in its adjacent view.

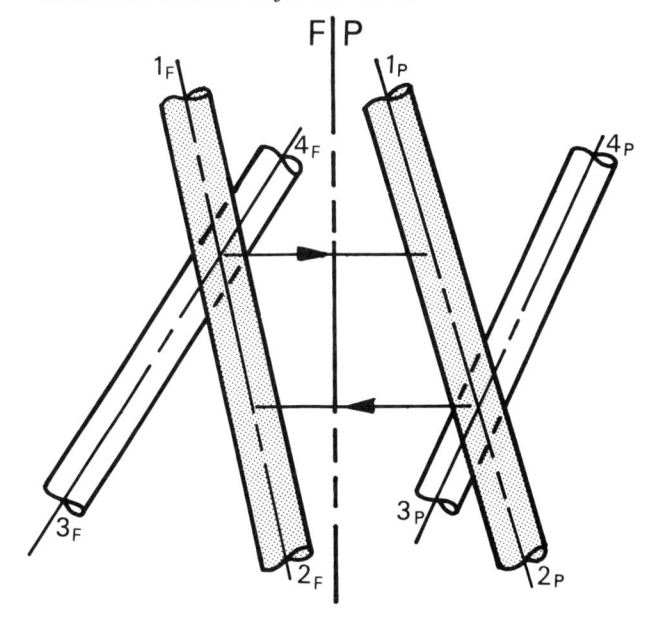

FIG. 4-51 Visibility.

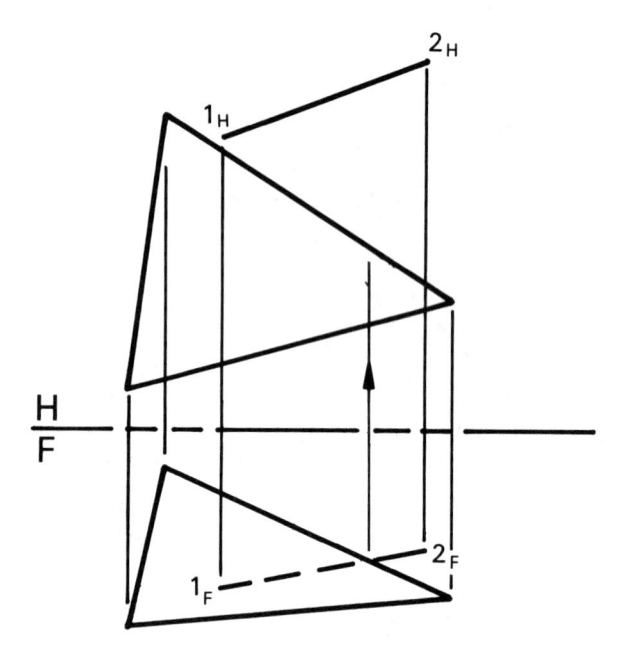

FIG. 4-52 Visibility of line and plane.

4.25 VISIBILITY OF LINES AND SURFACES

The visibility of a line and a surface or two lines of an object can be determined using the visibility check as explained in the last section.

In Fig. 4-52, line 1-2 and the given plane cross in the front view. In the horizontal view, line 1_H-2_H does not cross the plane, consequently a visibility check for the front view is all that is required. Extend a sight line (along the line of sight, perpendicular to H/F) from the crossing points in the frontal view until it encounters either the line or part of the plane in the horizontal view. Here it is shown that the plane is met first and will therefore be solid (visible) in the *frontal view*.

The *inspection method* could be applied here since the line and plane only cross in the frontal view. For this method it is only necessary to find out which line or object is closer to the fold line in the adjacent view. As can be seen, the plane is closer to the H/F fold line in the horizontal view and is therefore visible (solid) in the *frontal* view. This procedure should be used only in uncomplicated situations.

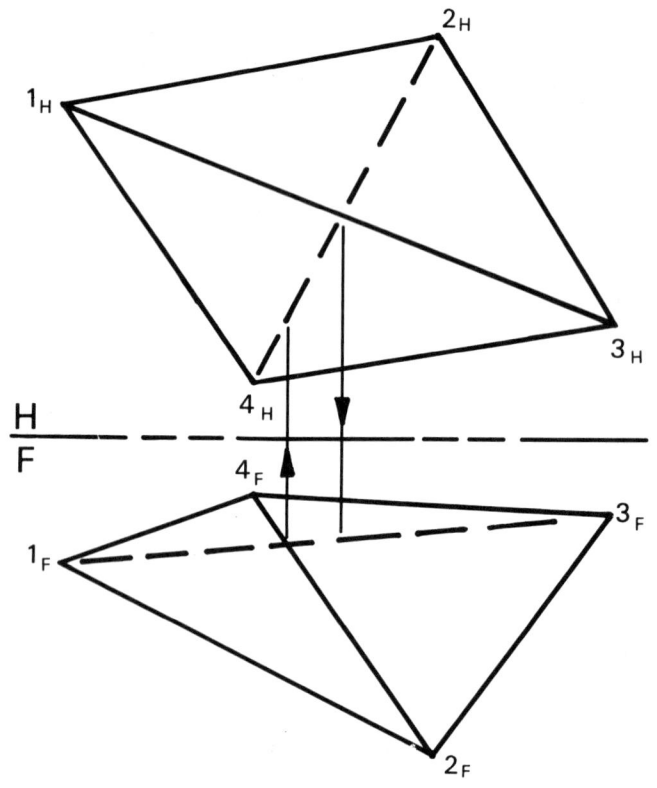

FIG. 4-54 Visibility of lines and surfaces.

When two lines that are part of the same object cross, the visibility check can be used to solve for which line, and therefore which surface, is visible. Note that the outer boundaries are always solid. In Fig. 4-53 line 1-4 and 2-3 of the tetrahedron cross. A sight line is extended from their crossing point in the frontal view until it meets either line in the adjacent view. In this case, line 1_H-4_H is encountered first in the horizontal view and will therefore be visible in the *frontal* view. A sight line is then extended from the crossing point in the horizontal view until it hits one of the two crossing lines in the frontal view. Line 1_F-4_F is encountered first and so will be visible in the *horizontal* view. This same procedure was used to solve for the visibility of the crossing lines in Fig. 4-54.

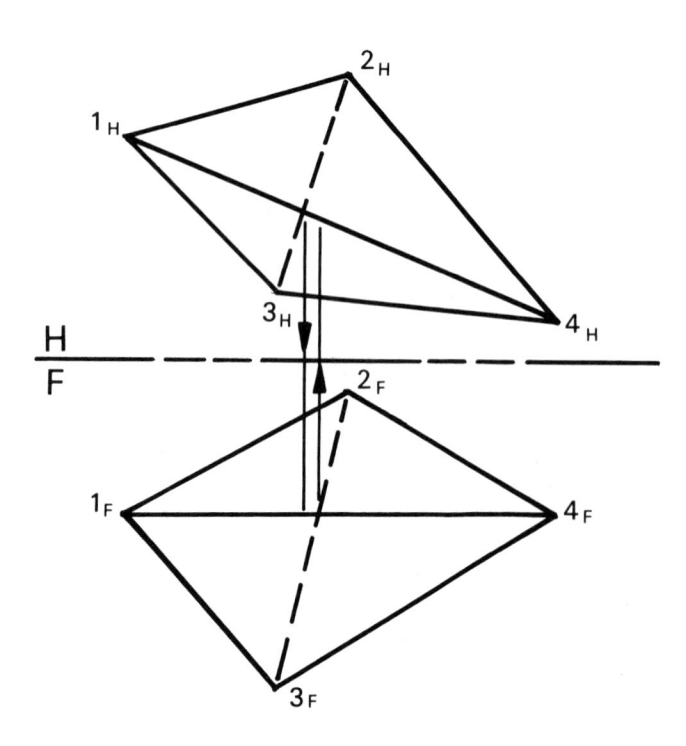

FIG. 4-53 Visibility of crossing lines.

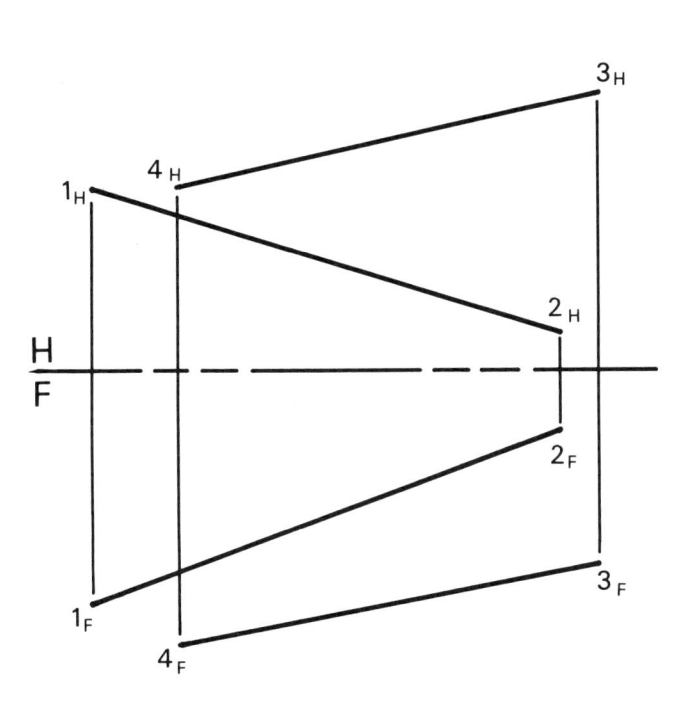

FIG. 4-55 Skew lines.

4.26 SKEW LINES

Skew lines do not intersect and are not parallel. In some cases skew lines will not cross, as in Fig. 4-55, where line 1-2 and 3-4 do not cross in either view. A visibility check is unnecessary for this situation.

In Fig. 4-57 the two nonparallel, nonintersecting pipelines cross in both views. In this case the visibility of the skew lines must be determined in order to correctly draw the two views. A sight line is extended from the crossing point of 7_F-8_F and 9_F-10_F in the frontal view, parallel to the line of sight (perpendicular to the fold line H/F) until it

FIG. 4-56 Piperack in petrochemical plant.

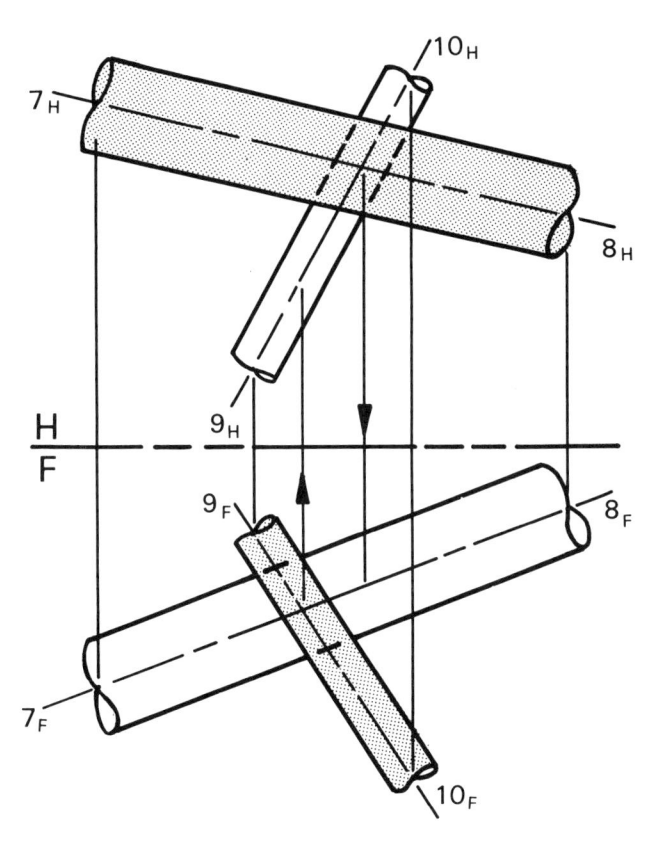

FIG. 4-57 Visibility of skew lines.

meets one of the two lines in the horizontal view. In this example, pipe 9_H-10_H is encountered first, and so will be drawn solid (visible) in the *frontal* view. Pipe 7_F-8_F is hidden (behind) in the *frontal* view. A sight line is then drawn from where the lines cross in the horizontal view until it meets one of the two lines in the adjacent view. Pipe 7_F-8_F is encountered first and is therefore solid (visible) in the *horizontal* view. Pipeline 9_H-10_H is hidden since it is below pipe 7_H-8_H, in the *horizontal* view.

Note that the visibility test can originate in either view and need not be taken from the frontal view first. *The most important thing to remember when doing a visibility test is that the adjacent view must be examined in order to find the visibility for a particular view. Visibility cannot be determined in one view.*

The use of skew lines and their relationships are a major aspect of descriptive geometry and its application to a variety of engineering problems. The solution of piping problems such as the shortest connector between two skewed pipelines, clearances between piping and equipment, or structural elements are just two such applications. The piping system shown in Fig. 4-56 is an excellent example of parallel, perpendicular, intersecting, and skew lines in space and their interrelationship.

FIG. 4-58 Intersecting lines.

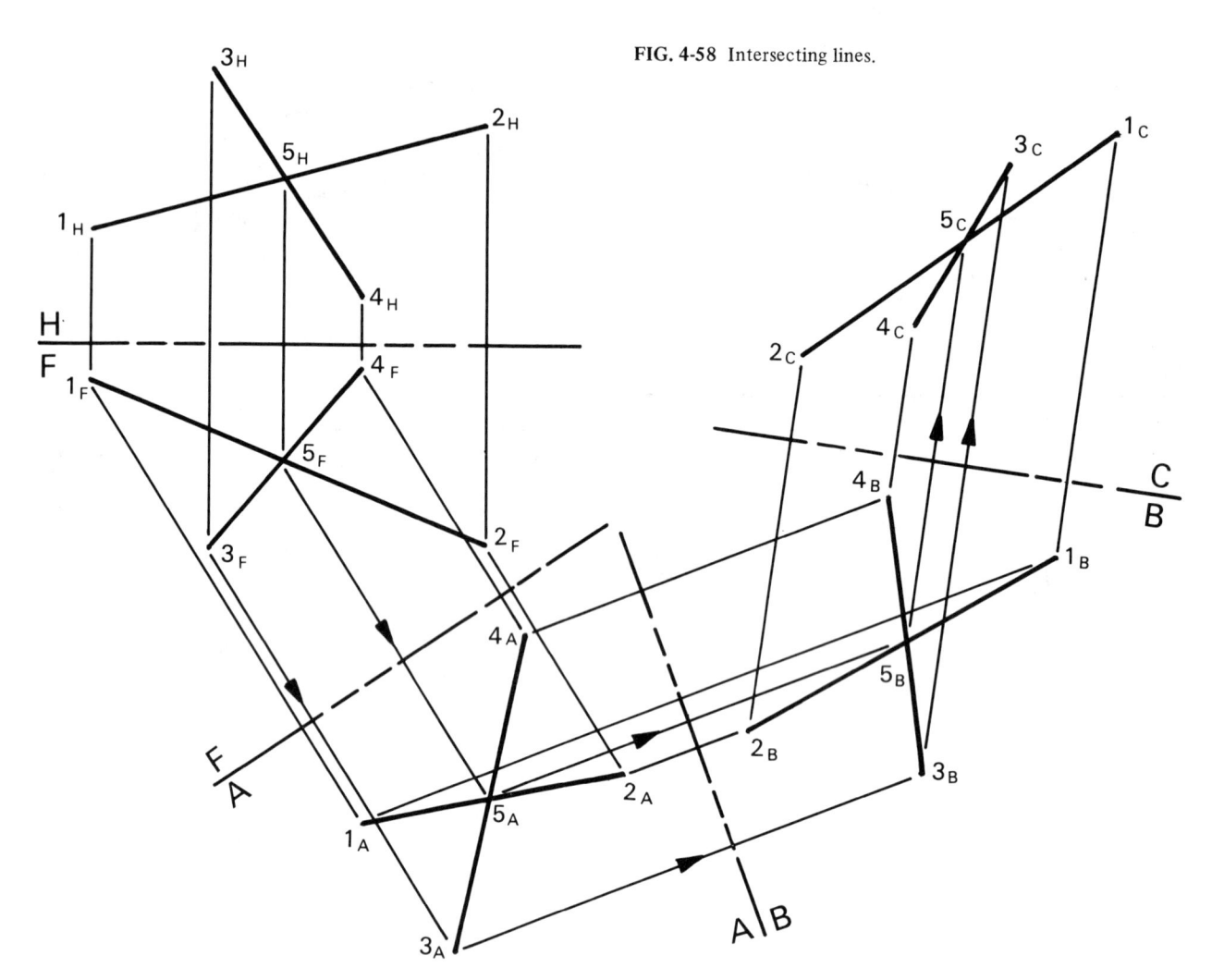

4.27 INTERSECTING LINES

Intersecting lines *must have a common point, one where the two lines meet and therefore "intersect."* Parallel lines and skew lines do not intersect. Perpendicular lines can be intersecting or nonintersecting.

The two lines in Fig. 4-58 are intersecting lines since they have a common point, point 5, which remains the same in all projections. In this example two principal views and three auxiliary views are shown of lines 1-2 and 3-4. Notice that the intersecting (common) point is aligned vertically between the horizontal and frontal views and that all projection lines between the views are parallel. If the apparent common point projected between two lines is not parallel to the other projection lines (and therefore to the line of sight), the two lines do not intersect.

For all auxiliary projections of intersecting lines the common point must be projected parallel to the other projection lines and perpendicular to the fold line, as in Fig. 4-58 where point 5 projects perpendicular to fold line F/A, A/B, and B/C for each auxiliary view.

The launcher in Fig. 4-59 was designed using intersecting structural elements.

FIG. 4-59 Launcher. *(Courtesy NASA.)*

FIG. 4-60 Nonintersecting lines.

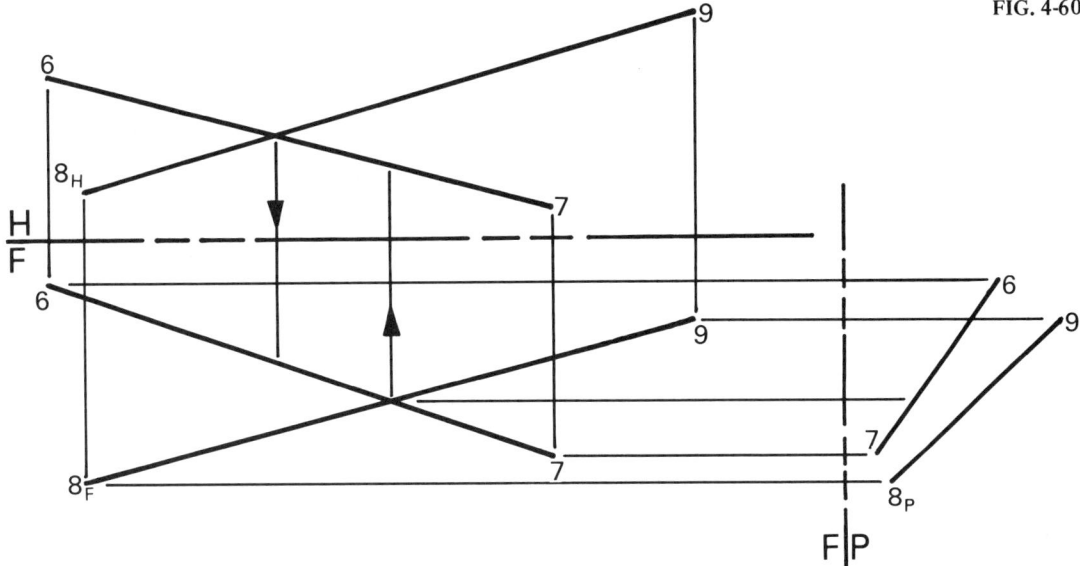

4.28 NONINTERSECTING LINES

Intersecting lines have a common point that will project parallel to the line of sight for each adjacent view. The projection/extension line of a common point will be perpendicular to the fold line between the views. In the case of *nonintersecting lines*, the crossing point of two lines is different in each adjacent view. In Fig. 4-60 line 6-7 crosses line 8-9 in the horizontal view. By projecting this point to the frontal view it can be seen that the crossing point is not the same for both views. In this example a profile view was drawn to illustrate that the two lines are nonintersecting. In general only two views are necessary to solve for intersecting lines.

Piping systems are normally designed so that pipelines run in only two directions (perpendicular) within each plane (horizontal and vertical). The manifold shown in Fig. 4-61 is composed of a series of perpendicular intersecting and nonintersecting lines and parallel lines.

In Fig. 4-62 lines 5-6 and 7-8 appear to intersect. By extending a projection line from the apparent common point (point 9) it can be seen that the crossing point in each projection is aligned between views and that the lines intersect. Remember, two lines may cross in all projections but intersect only if they share a common point. The common point is on a single projection line parallel to all other projection lines between adjacent views and perpendicular to the fold line between the views. In problems where the lines are too close together or the crossing point is not clearly different in adjacent views, a third projection should be taken.

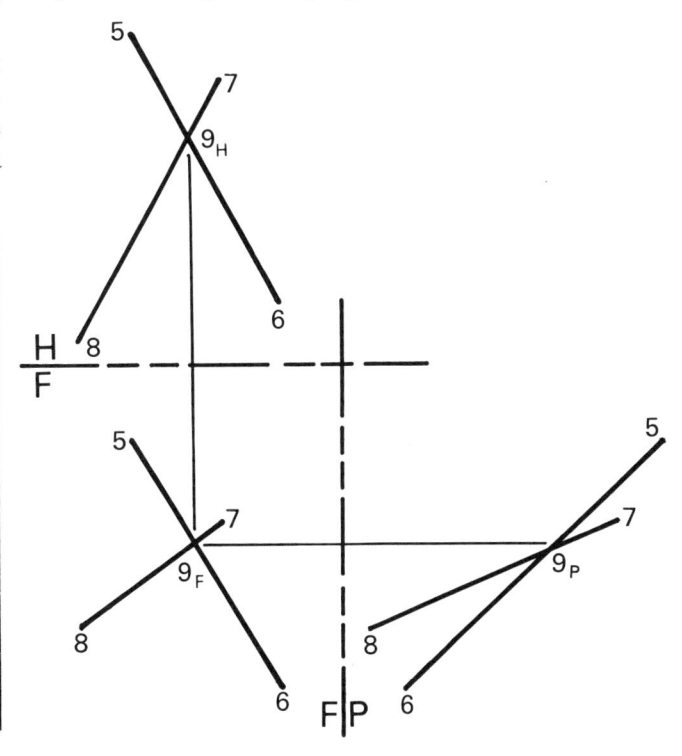

FIG. 4-61 Pipe manifold. *(Courtesy Grove Valve and Regulator.)*

FIG. 4-62 Intersecting lines.

PROB. 4-12

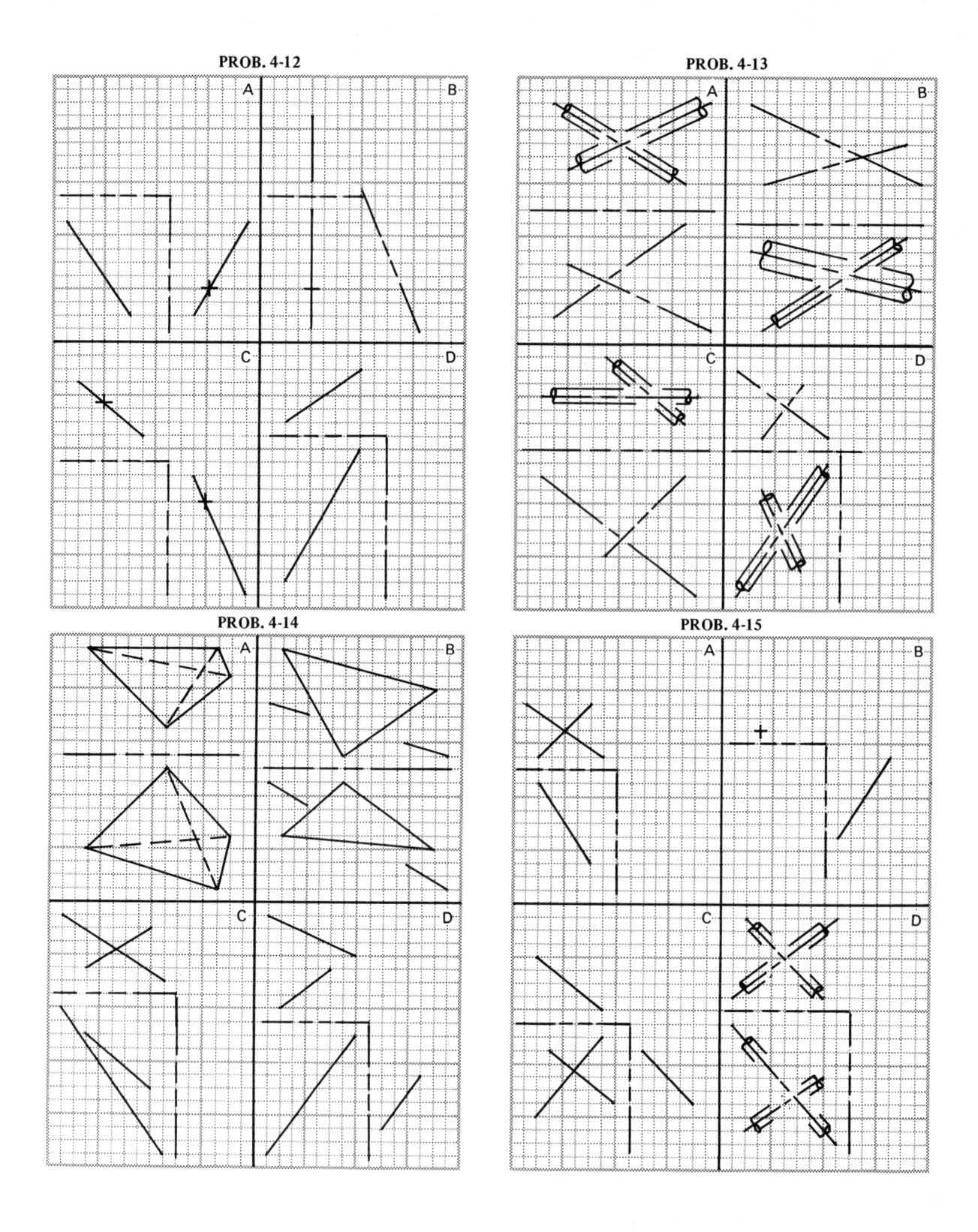

PROB. 4-13

PROB. 4-14

PROB. 4-15

PROBS. 4-12A, 4-12B, 4-12C. Draw the given line in each view and locate the point on the line in each projection.

PROB. 4-12D. Point 3 is three units to the right of point 1 and lies on the line.

PROBS. 4-13A, 4-13B, 4-13C, 4-13D. Complete the views of the pipes and solve for visibility. Shade the pipe which is visible in each view.

PROBS. 4-14A, 4-14B, 4-14C, 4-14D. Solve for the correct visibility in each problem.

PROB. 4-15A. Project the frontal and profile views of the two lines so that they intersect. One line is a horizontal line. Label the intersection point in each view.

PROB. 4-15B. Project three views of profile line 1-2, given in the P view. Then draw three views of line 3-4 so that the lines intersect at their midpoints and line 3-4 appears as a point in the profile view.

PROB. 4-15C. Complete the three views of the lines. Do these lines intersect?

PROB. 4-15D. Project the profile view of the two lines. Show proper visibility.

4.29 PARALLELISM OF LINES

Two lines in space will be *intersecting, skew, parallel,* or *perpendicular*. **Parallel lines project parallel in all views.** Figure 4-63 is an example of parallel lines. Note that in each view, lines 6-7 and 8-9 are parallel though they are oblique. Parallel lines may appear as points (in the same view) or their projections may coincide.

Two oblique lines that project parallel in two or more views will always be parallel. Two lines that are parallel to a principal plane and appear parallel to each other may not be parallel lines, and a third view will be needed to establish their relationship.

The true distance between two parallel lines is shown in a view where the lines appear as points. In Fig. 4-64 oblique lines 1-2 and 3-4 are parallel. Auxiliary view A is projected parallel to both oblique lines from the profile view (fold line P/A is drawn parallel to 1_P-2_P and 3_P-4_P). View A shows both lines as true length. Note that parallel lines both show true length in the same view. Auxiliary view B is then projected perpendicular to the true length lines (fold line A/B is drawn perpendicular to 1_A-2_A and 3_A-4_A). In auxiliary B both lines appear as point views. The true distance between the lines can be measured in this view.

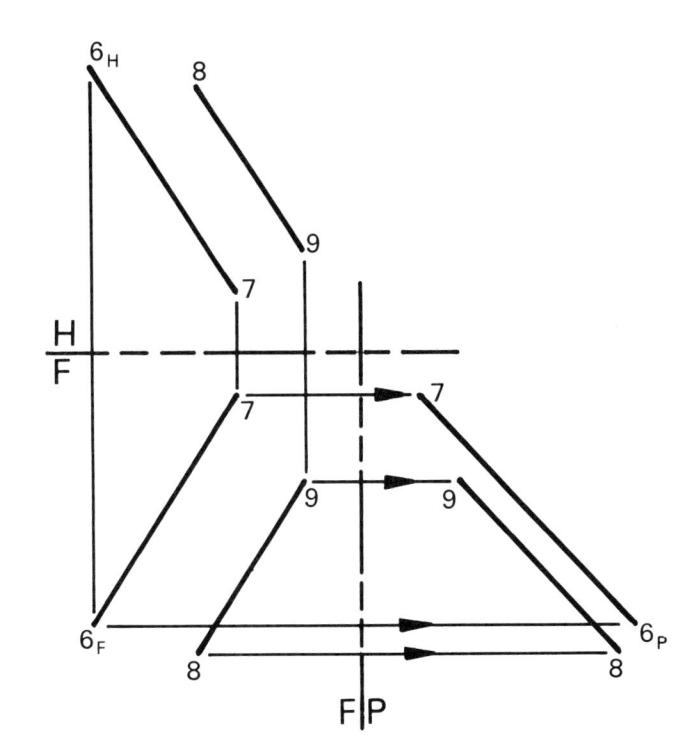

FIG. 4-63 Parallel lines.

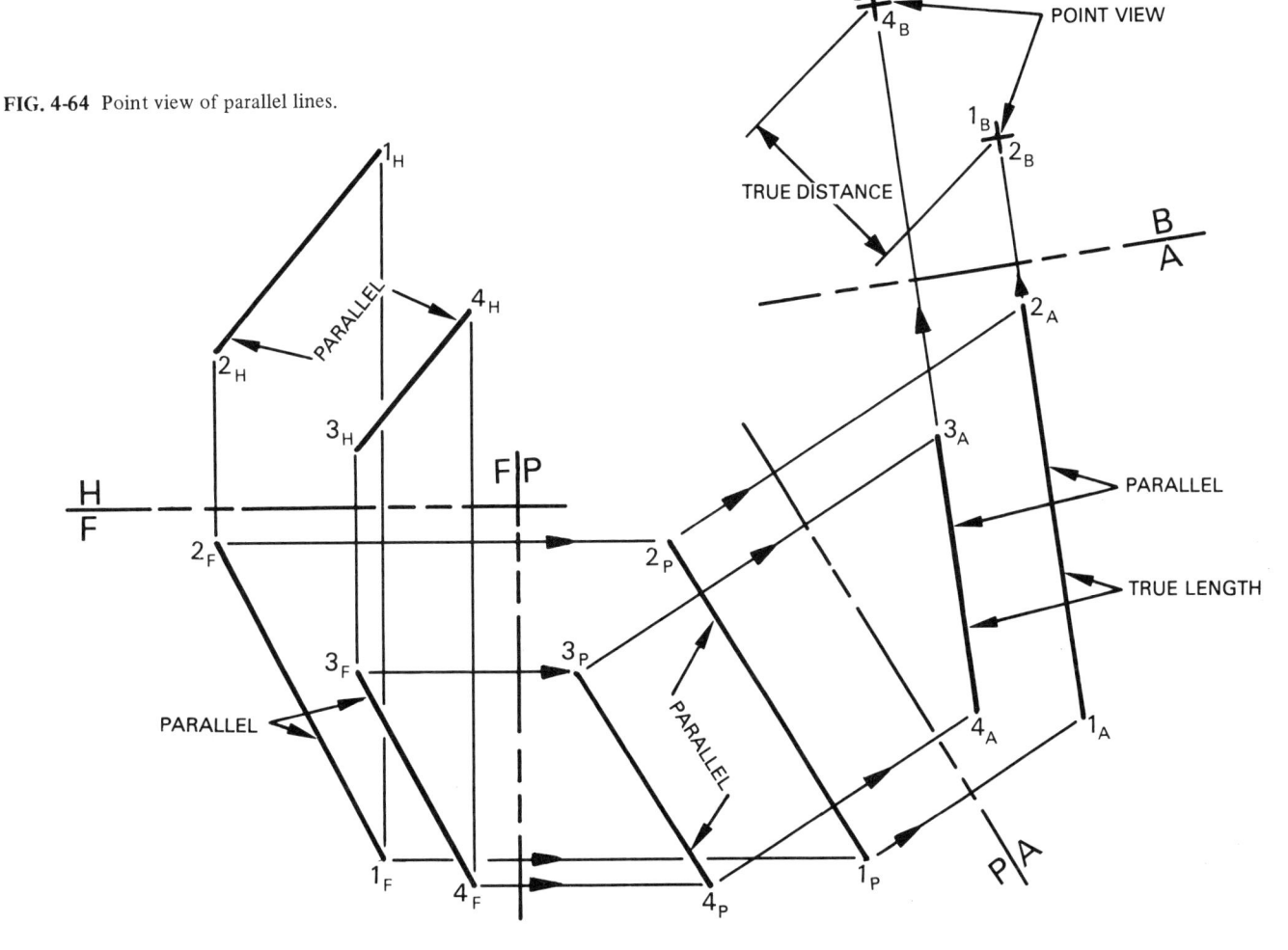

FIG. 4-64 Point view of parallel lines.

FIG. 4-65 Parallel profile lines.

FIG. 4-66 Nonparallel profile lines.

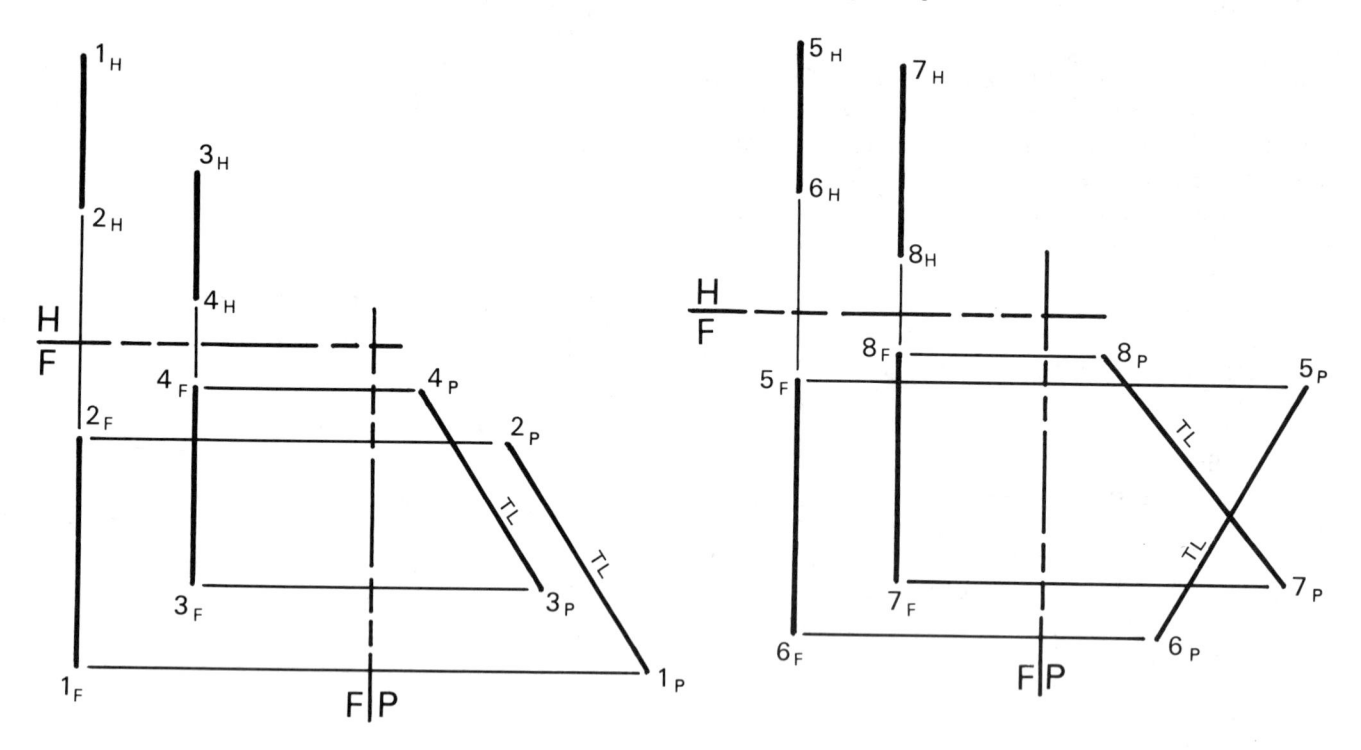

4.30 PARALLEL AND NONPARALLEL LINES (SPECIAL CASES)

When two lines are parallel to a principal plane and appear parallel to each other, a third view where they show in true length is necessary to determine if they are in fact parallel. In this situation the lines could project parallel, perpendicular, or skew.

In Fig. 4-65, lines 1-2 and 3-4 are parallel to the profile plane (and will therefore be profile lines). They appear parallel to each other in the frontal and horizontal views and the profile projection shows them to be parallel lines. In Fig. 4-66, lines 5-6 and 7-8 are parallel to the profile plane, but the profile projection shows that they are not parallel or perpendicular. Since both lines show true length in the profile view they are profile lines.

Two lines that are parallel to the horizontal plane and appear parallel to each other in the profile and frontal views also require a third view to determine parallelism. In Fig. 4-67 lines 7-8 and 9-10 are parallel to the horizontal plane and appear parallel in the frontal and profile views. The horizontal projection shows that they are not parallel. Both lines show true length in the horizontal view and are therefore horizontal lines.

In Fig. 4-66 and Fig. 4-67 the given lines are not parallel, perpendicular, or intersecting, so they are skew lines.

Understanding the theory and practice of using parallelism is extremely helpful in solving many descriptive geometry problems. Parallelism is used by the drafter, designer, and engineer to solve and draw a variety of technical problems and is utilized throughout this text along with perpendicularity to complete many of the assignments.

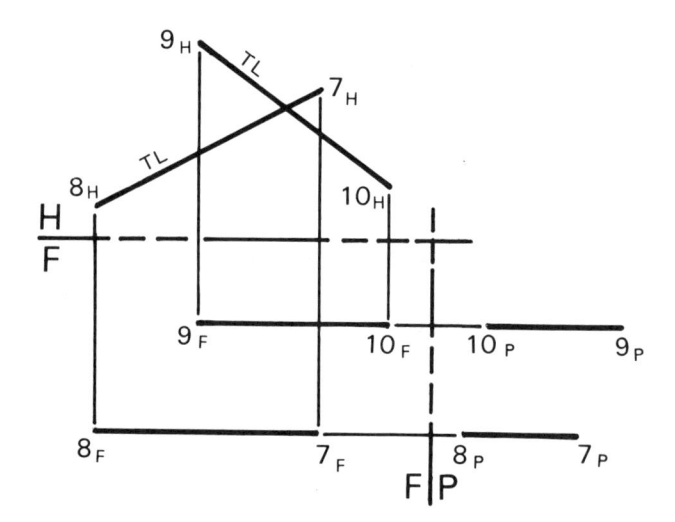

FIG. 4-67 Nonparallel horizontal lines.

4.31 CONSTRUCTION OF A LINE PARALLEL TO A GIVEN LINE

A commonly required construction in descriptive geometry is drawing a line parallel to a given line and through an established point. This procedure helps solve many problems encountered when doing intersections of surfaces and solids and for other assignments in the text. Since parallel lines are parallel in all views it is simply necessary to draw a line through a point parallel to another line.

Normally only two views are required for oblique lines. When the given line is parallel to the horizontal or profile planes it is necessary to draw three views of the lines.

In Fig. 4-68(1), line 1-2 and point 3 are given. A line is to be drawn parallel to line 1-2 with its midpoint at point 3. Since line 1-2 is oblique, only two views are necessary. The new line is drawn through point 3 and parallel to line 1-2 in both views (2). A specific length was not required, only that the new line be parallel and have point 3 at its midpoint. The end points of the new line must be aligned so that the line is equal length in both views.

When the given line is parallel to the horizontal or profile plane (is a horizontal or profile line) it is necessary to project a third view. In Fig. 4-69 a line parallel to line 1-2 and originating at point 3 is required. Line 1-2 is parallel to the profile plane, therefore it is necessary to draw the profile view. Line 3-4 is first located in the profile view since it is here that both lines show true length. Line 3_P-4_P is drawn parallel to line 1_P-2_P and originating from point 3_P. Note that a specific length was not required. After drawing line 3-4 in the profile view and establishing its length, a projection line is drawn to locate point 4 in the frontal view. The horizontal view of line 3-4 is drawn parallel to line 1-2, but point 4_H must be established by transferring dimension D1 from the profile view.

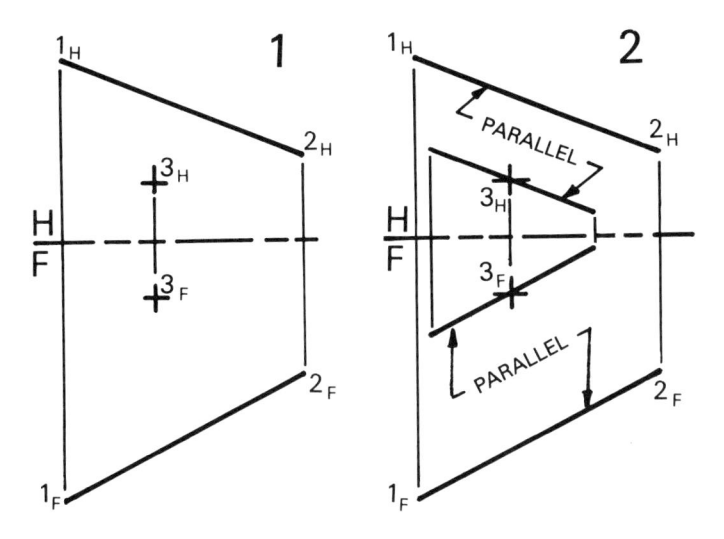

FIG. 4-68 Line parallel to a given line.

FIG. 4-69 Line through a point parallel to given profile line.

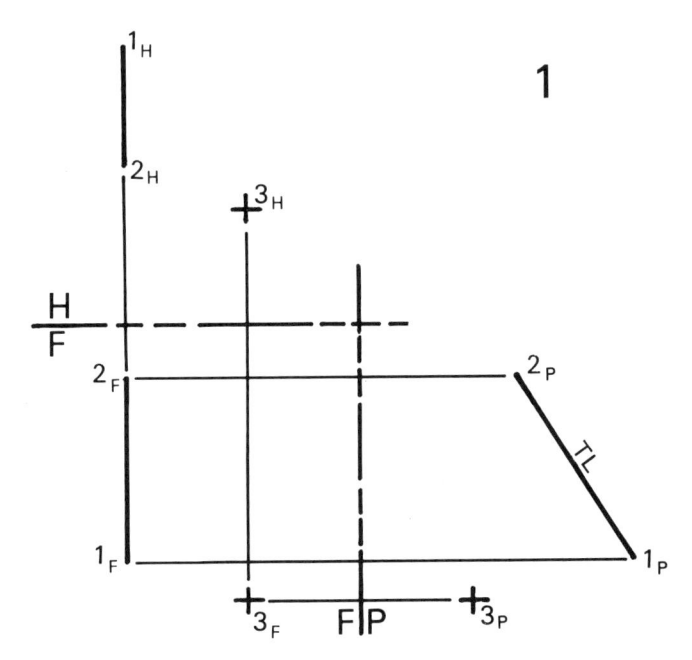

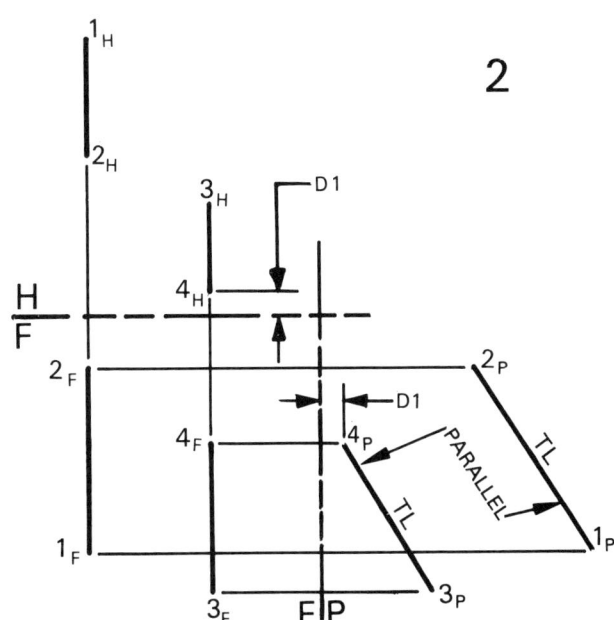

91

PROB. 4-16

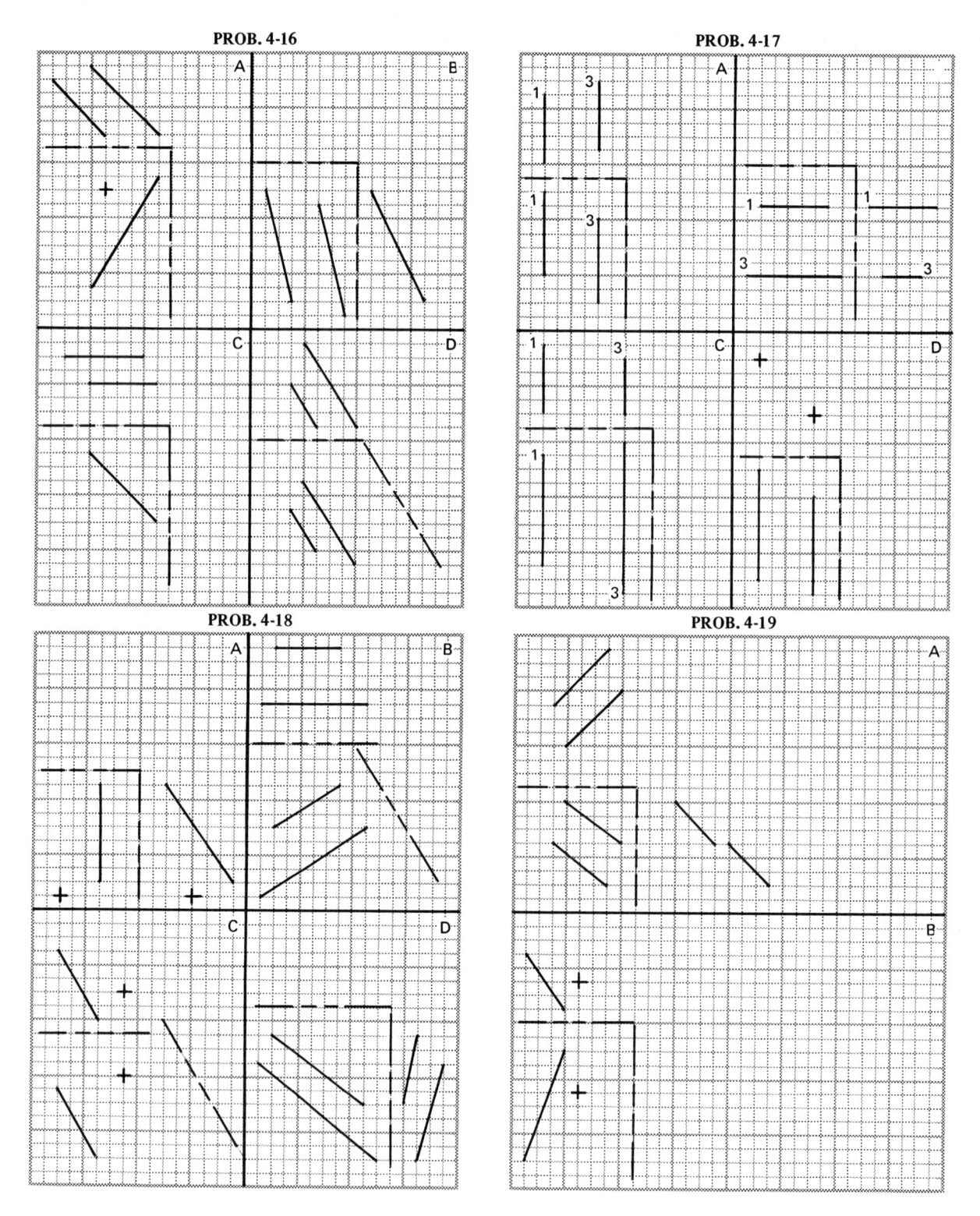

PROB. 4-17

PROB. 4-18

PROB. 4-19

PROBS. 4-16A, 4-16B, 4-16C, 4-16D. Complete the required views of the parallel lines. In 4-16B, the missing profile view of one of the lines is 3 units behind the other line. In 4-16C, the missing frontal line is 5 units below the other line.

PROBS. 4-17A, 4-17B, 4-17C, 4-17D. Complete the problems by projecting the lines into the required views. Label each view where the lines show as true length and parallel.

PROB. 4-18A. From the given point, draw a 1.25" (31.7mm) line parallel to the profile line.

PROB. 4-18B. Solve for the point view and true distance between the lines.

PROB. 4-18C. Complete the three views of the parallel lines. Line 1-2 is given. The given point is the midpoint of line 3-4 which is 1" (2.54cm) long.

PROB. 4-18D. Complete the views of the lines. Are they parallel?

PROB. 4-19A. Solve for the point view of the given lines. What is the true distance between the lines?

PROB. 4-19B. Line 1-2 is given. The given point is the midpoint of line 3-4, which is 2" (50.8mm) long. What is the true distance between the two lines? The lines are parallel.

92

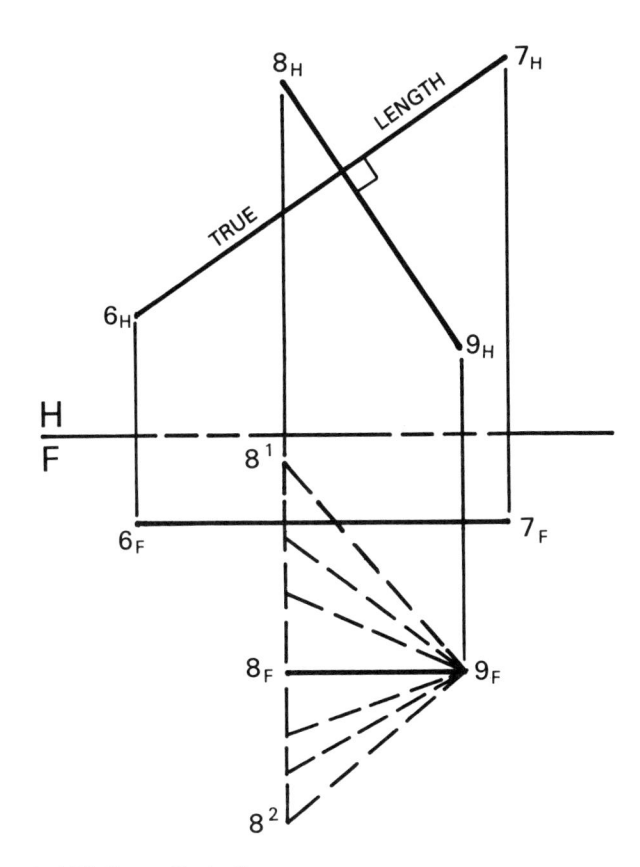

FIG. 4-70 Perpendicular lines.

4.32 PERPENDICULARITY OF LINES

Perpendicularity, along with parallelism, is used throughout descriptive geometry to solve a wide range of graphical problems. *Lines that are perpendicular will show perpendicularity in any view in which one or both of the lines is true length.* Since two lines may be oblique in their given views, it is necessary to project a view that shows one or both of the lines as true length in order to check for perpendicularity. If two lines appear perpendicular in a given view and neither one is true length, then the lines are not perpendicular. Perpendicular lines can be intersecting or nonintersecting lines.

In Fig. 4-70 line 6-7 is parallel to the horizontal plane and therefore is a horizontal line and shows true length in the horizontal view. Line 8-9 has been drawn perpendicular in the horizontal view. The frontal view of line 8-9 can be an infinite number of possibilities from vertical to horizontal. Since the horizontal view of line 8-9 remains the same, variations shown in the frontal view do not affect perpendicularity.

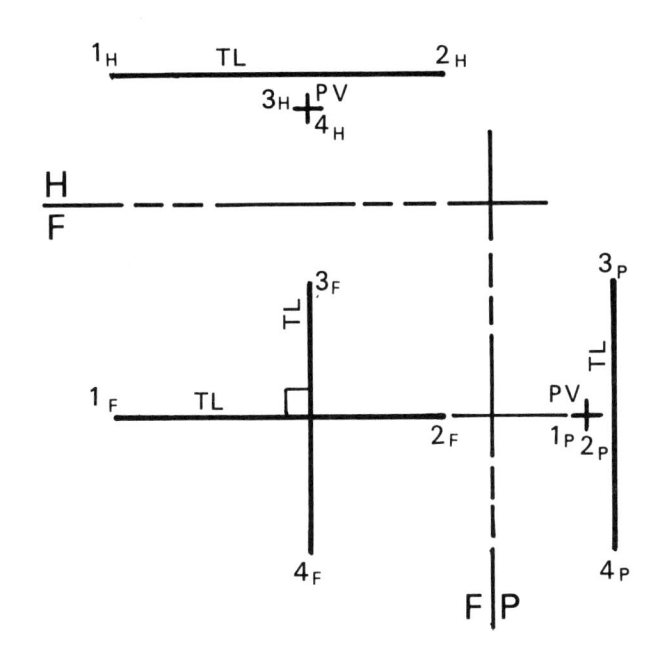

FIG. 4-72 Perpendicular lines (special case).

Frontal perpendicular lines appear parallel in the horizontal and profile views and perpendicular in the frontal view, Fig. 4-71. Both lines show true length in the frontal view since they are frontal lines.

In a view where one line is a point view and the other line is true length, the lines are perpendicular, Fig. 4-72. In this example, line 1_H-2_H is true length in the horizontal view and line 3_H-4_H shows as a point view since it is a vertical line. The reverse happens in the profile view where line 1_P-2_P is a horizontal line and shows as a point view. Line 3_P-4_P is true length. The frontal projection of both lines shows lines 1_F-2_F and 3_F-4_F true length and perpendicular.

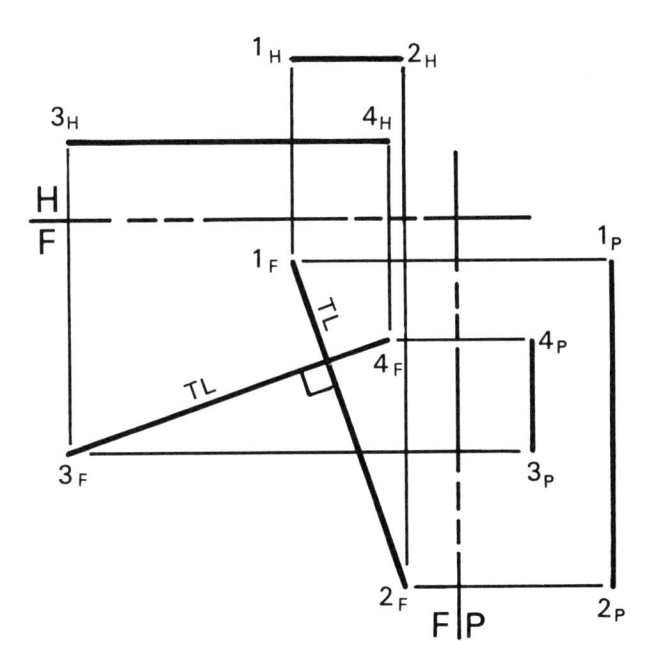

FIG. 4-71 Perpendicular frontal lines.

FIG. 4-73 Intersecting perpendicular lines.

the lines is true length. Thus, two lines that intersect at a common point, and form 90 degrees with each other where one or both lines appears as true length, are *intersecting perpendicular lines*.

When two intersecting lines are oblique in the frontal and horizontal views, it is necessary to project a view where one or both of the lines is true length to check for perpendicularity. Lines 6-7 and 8-9 in Fig. 4-73 are intersecting lines since they have a common point that is aligned in adjacent views. In this example fold line H/A is drawn parallel to oblique line 6_H-7_H (auxiliary view A is parallel to line 6_H-7_H). Both lines are then projected into the auxiliary view. Line 6_A-7_A is true length and forms a 90-degree angle (is perpendicular) with line 8_A-9_A.

In Fig. 4-74, three special cases of intersecting perpendicular lines are shown. In case (1), both lines are true length in the horizontal view (are horizontal lines). In the frontal view line 1-2 is true length (frontal line) and line 3-4 appears as a point view. Point 2 lies on line 3-4 and is the common point. In this case perpendicularity can be proven in either view.

In case (2), point 3 is the common point and lies on line 1-2. Line 3-4 is a horizontal line and line 1-2 is a frontal line. Perpendicularity can be proven in either view since two lines are perpendicular if they form a 90-degree angle where one (or both) is shown as true length and the other is oblique or a point view.

Because two intersecting lines form a plane, the true angle (and therefore perpendicularity) between the lines is shown in a view where both lines show true length. In case (3), lines 1-2 and 3-4 intersect at point 5 and are parallel to the frontal plane. The true angle between the lines shows in the frontal view where both lines are true length and measure 90 degrees (perpendicular).

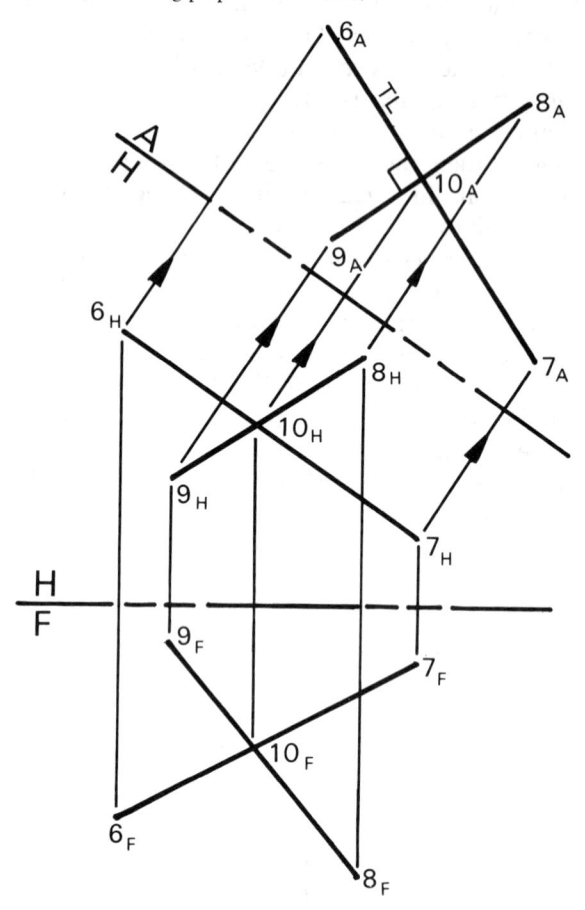

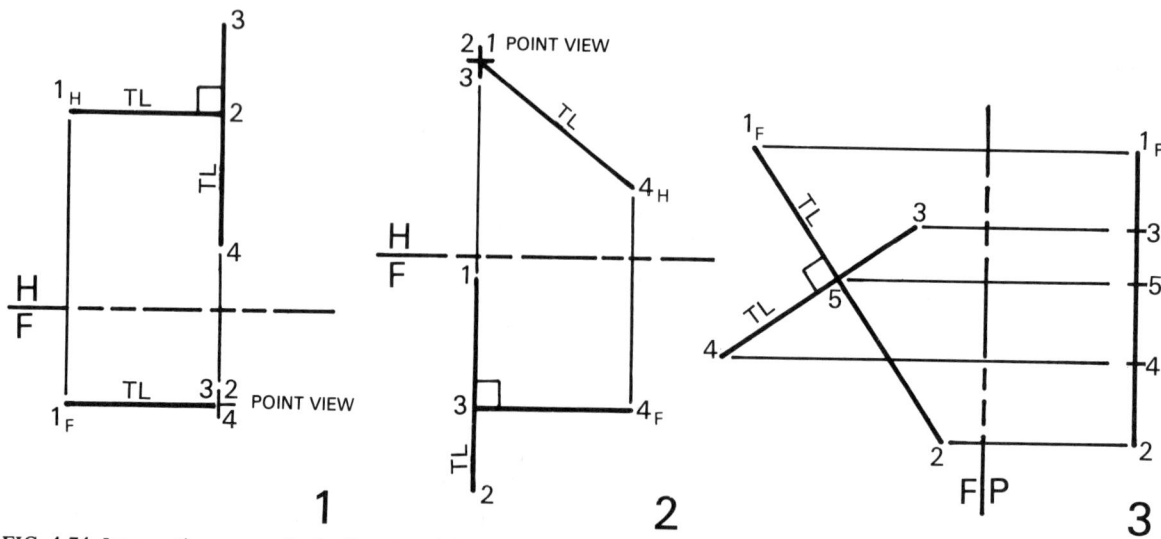

4.33 INTERSECTING PERPENDICULAR LINES

Intersecting lines have a common point which lies on a single projection line, parallel to all other projection lines between adjacent views. Perpendicular lines make right angles with one another, and appear perpendicular in a view where one or both of

FIG. 4-74 Intersecting perpendicular lines (special cases).

FIG. 4-75 Nonintersecting perpendicular lines.

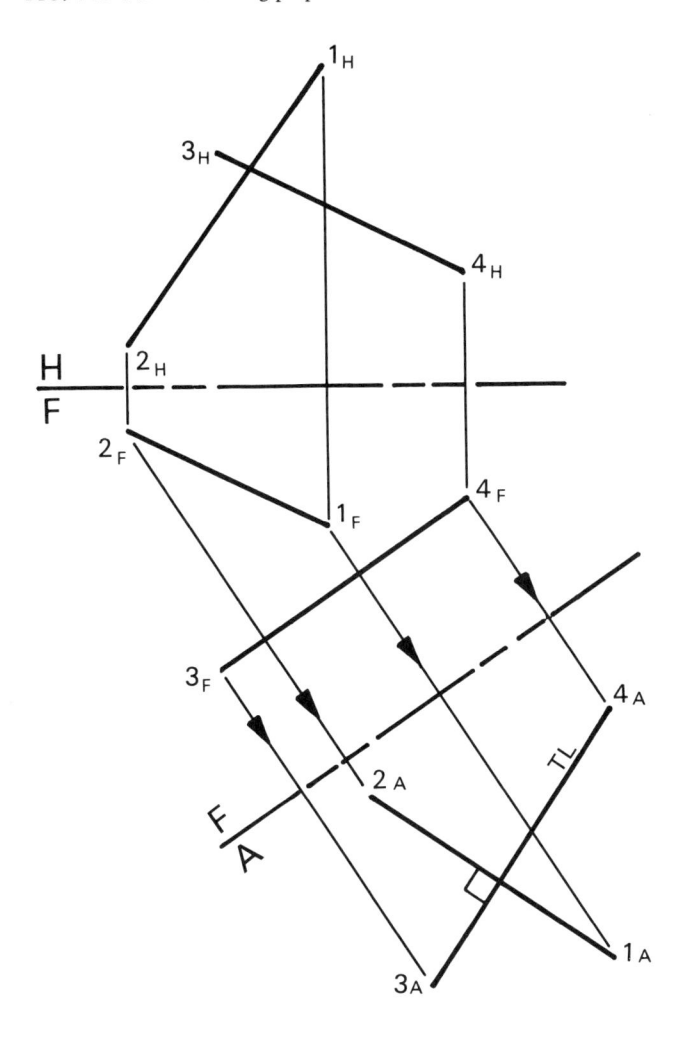

FIG. 4-76 Perpendicular lines.

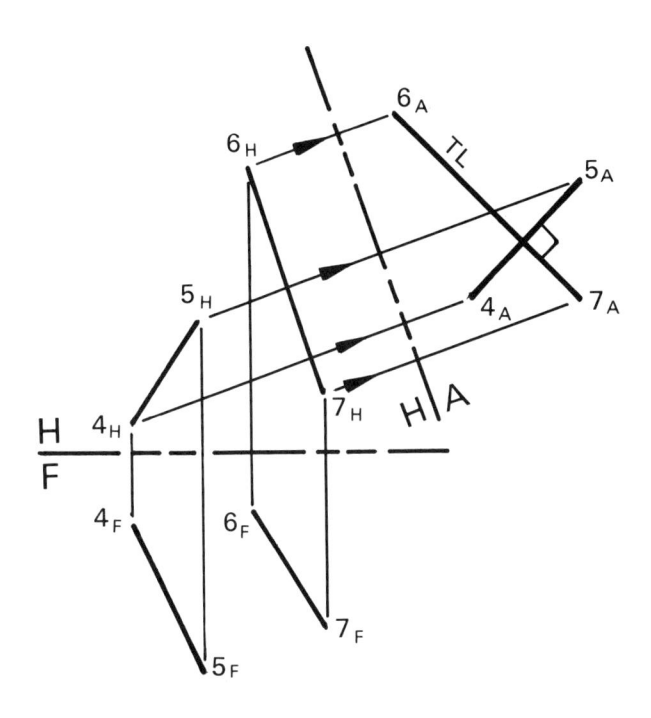

4.34 NONINTERSECTING PERPENDICULAR LINES

Two nonintersecting lines are perpendicular lines if they form right angles in a view where one or both are shown true length. For oblique lines it is necessary to project an auxiliary view where at least one of the lines is true length and measure the angle between the lines in that new view.

In Fig. 4-75 the principal views of the two lines establish that they are nonparallel, nonintersecting, and oblique. Auxiliary view A is projected parallel to line 3-4 by drawing fold line F/A parallel to line 3_F-4_F. Projection lines are then drawn perpendicular to the fold line from all points in the frontal view. Measurements to locate each point are transferred from the horizontal view to establish the points along the projection lines in auxiliary view A. Line 3_A-4_A is true length and line 1_A-2_A is oblique. The lines are perpendicular since they appear at right angles in auxiliary view A.

Lines 4-5 and 6-7 in Fig. 4-76 are nonparallel and nonintersecting oblique lines. To determine if they are perpendicular, a view is projected that shows one of the lines as true length. In this example, H/A is drawn parallel to line 6_H-7_H. Line 6_A-7_A shows true length and line 4_A-5_A oblique. Since the two lines are at right angles, they are perpendicular lines.

The fossil fuel power plant in Fig. 4-77 was designed and modeled with the aid of descriptive geometry principals and practical applications for construction of its equipment, structural and piping configuration. The structural steel column bents and bracing and the piping runs are a maze of intersecting and nonintersecting perpendicular lines and elements.

FIG. 4-77 Power plant model.

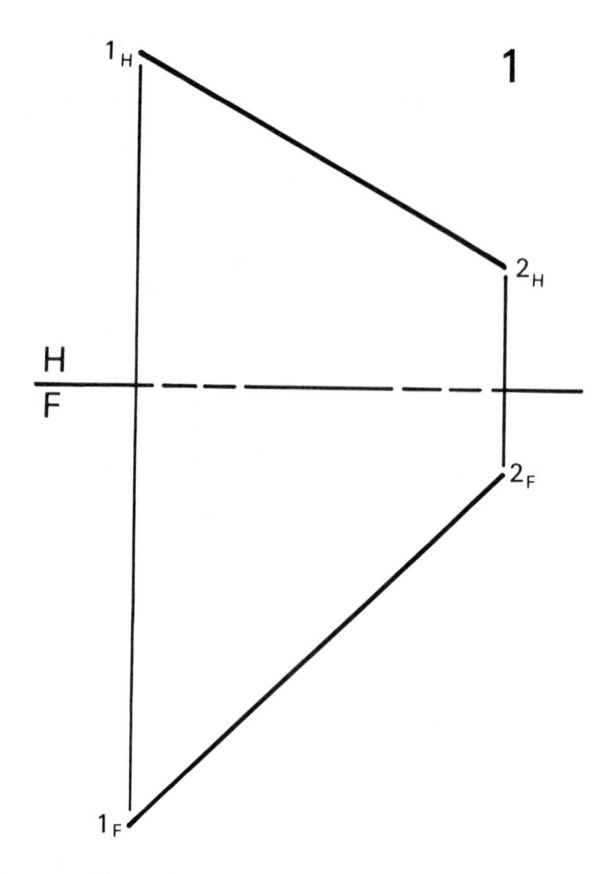

FIG. 4-78 Construction of a line perpendicular to a given oblique line.

4.35 CONSTRUCTION OF A LINE PERPENDICULAR TO A GIVEN OBLIQUE LINE

To draw a line perpendicular to a given oblique line it is necessary to show the new line as true length in one of the views and parallel to the fold line in the adjacent view. In Fig. 4-78(1), oblique line 1-2 is given. A line perpendicular to and at the midpoint of line 1-2 is required. The following steps describe this procedure.

1. Locate point 5 at the midpoint of line 1-2 in both views.
2. Draw line 3_H-4_H through point 5_H and parallel to the frontal plane (parallel to fold line H/F).
3. Draw a line through point 5, perpendicular to line 1_F-2_F in the frontal view.
4. Extend projection lines from point 3_H and 4_H perpendicular to fold line F/H to establish the end points of line 3-4 in the frontal view. Line 3-4 is a frontal line (true length in the frontal view).

In Fig. 4-78(2), the dashed lines represent other frontal lines that have been drawn perpendicular to line 1-2 (not at its midpoint). An infinite number of possibilities exist. Note that horizontal lines could also have been used, which would appear perpendicular in the horizontal view.

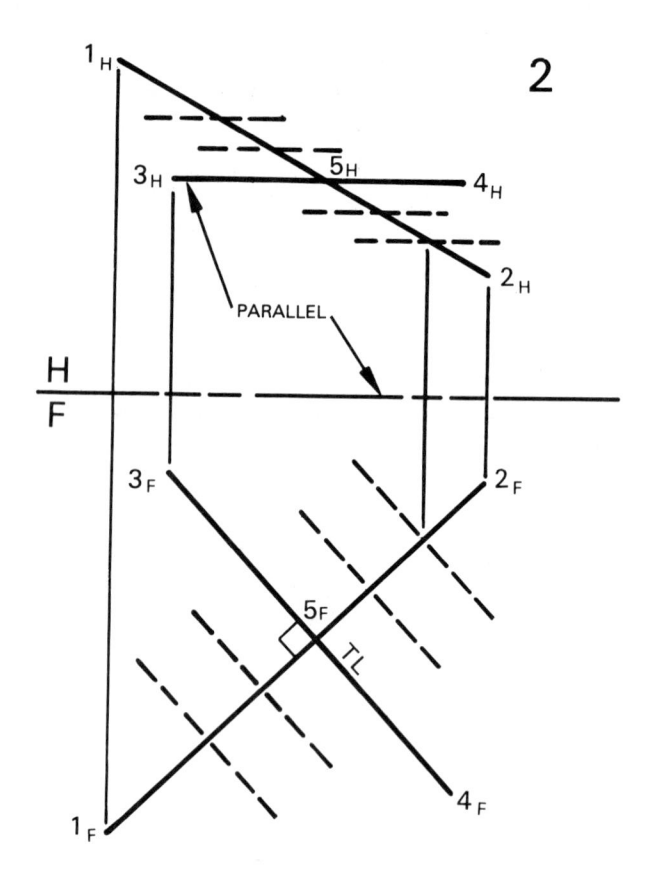

A line can be drawn nonintersecting and perpendicular to a given oblique line by constructing it parallel to one projection plane and perpendicular to the given line in the adjacent view. In Fig. 4-79, line 3_F-4_F is drawn parallel to the horizontal plane. The horizontal view shows line 3_H-4_H true length and perpendicular to line 1_H-2_H. Again, there are an infinite number of possibilities.

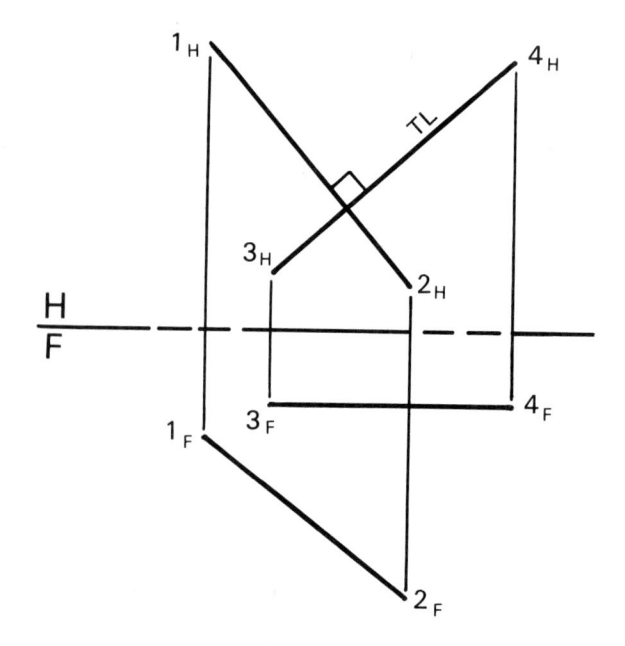

FIG. 4-79 Line perpendicular to a given oblique line (nonintersecting).

4.36 LINE DRAWN PERPENDICULAR TO A GIVEN LINE AT A SPECIFIC POINT

Structural bracing for the scaffolding shown in Fig. 4-80 is composed of perpendicular and angled elements welded into one solid unit. This is but one application of perpendicularity in industry.

From the previous examples it can be seen that an infinite number of lines can be drawn perpendicular to a given line and through a point. A different situation arises when the required perpendicular line is given in one view. Only one position for its adjacent projections will be possible.

In Fig. 4-81, oblique line 1-2 is given in both views. Line 2-3 is fixed in the horizontal view but missing in the frontal view. The frontal view is required. The following steps describe the process:

FIG. 4-80 Aircraft being secured for wind tunnel testing.

FIG. 4-81 Line perpendicular to a given line at a specified point.

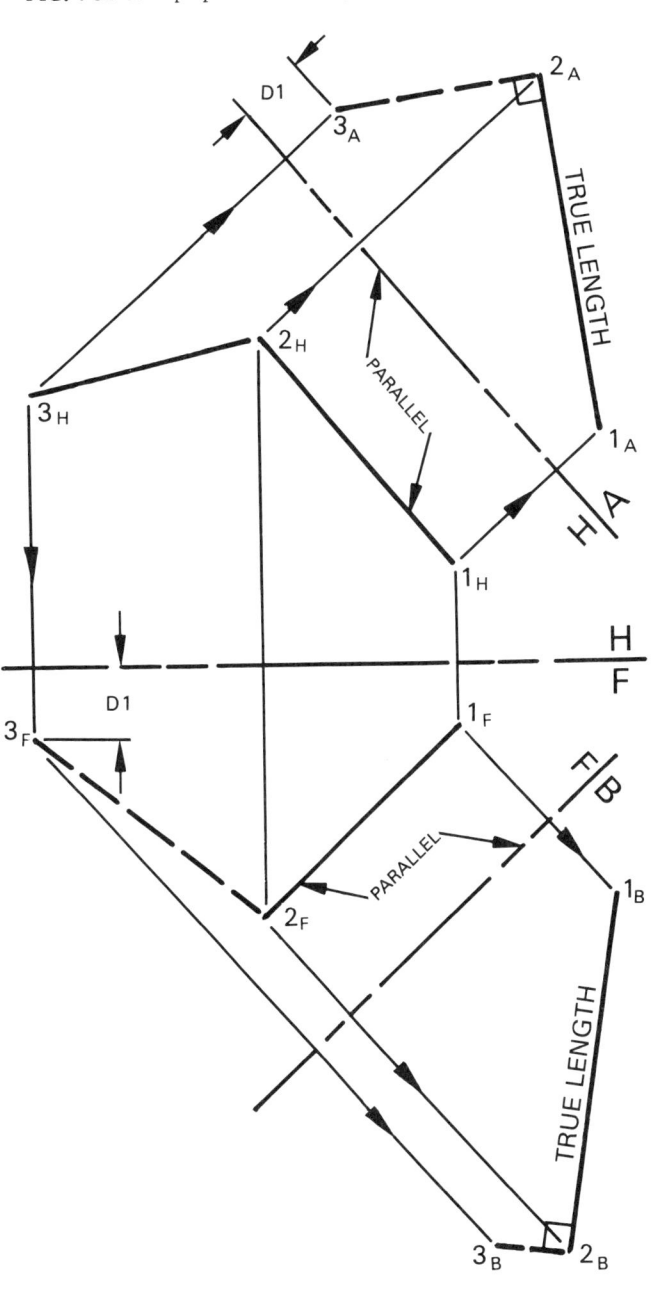

1. Draw fold line H/A parallel to line 1_H-2_H. Project auxiliary view A parallel to line 1_H-2_H.
2. Draw projection lines from all points in the horizontal view, perpendicular to H/A. Line 1_A-2_A shows true length in auxiliary view A.
3. Draw a line from point 2_A, perpendicular to line 1_A-2_A until it intersects the projection line extended from point 3_H. Line 2_A-3_A is oblique in this view and perpendicular to line 1_A-2_A which is true length.
4. Extend a projection line from point 3_H perpendicular to H/F.
5. Transfer dimension D1 from auxiliary view A to locate point 3_F in the frontal view.
6. Auxiliary view B is projected parallel to line 1_F-2_F (F/B is drawn parallel to line 1_F-2_F). Note this view was not required.
7. Line 2_B-3_B is perpendicular to true length line 1_B-2_B in auxiliary view B.

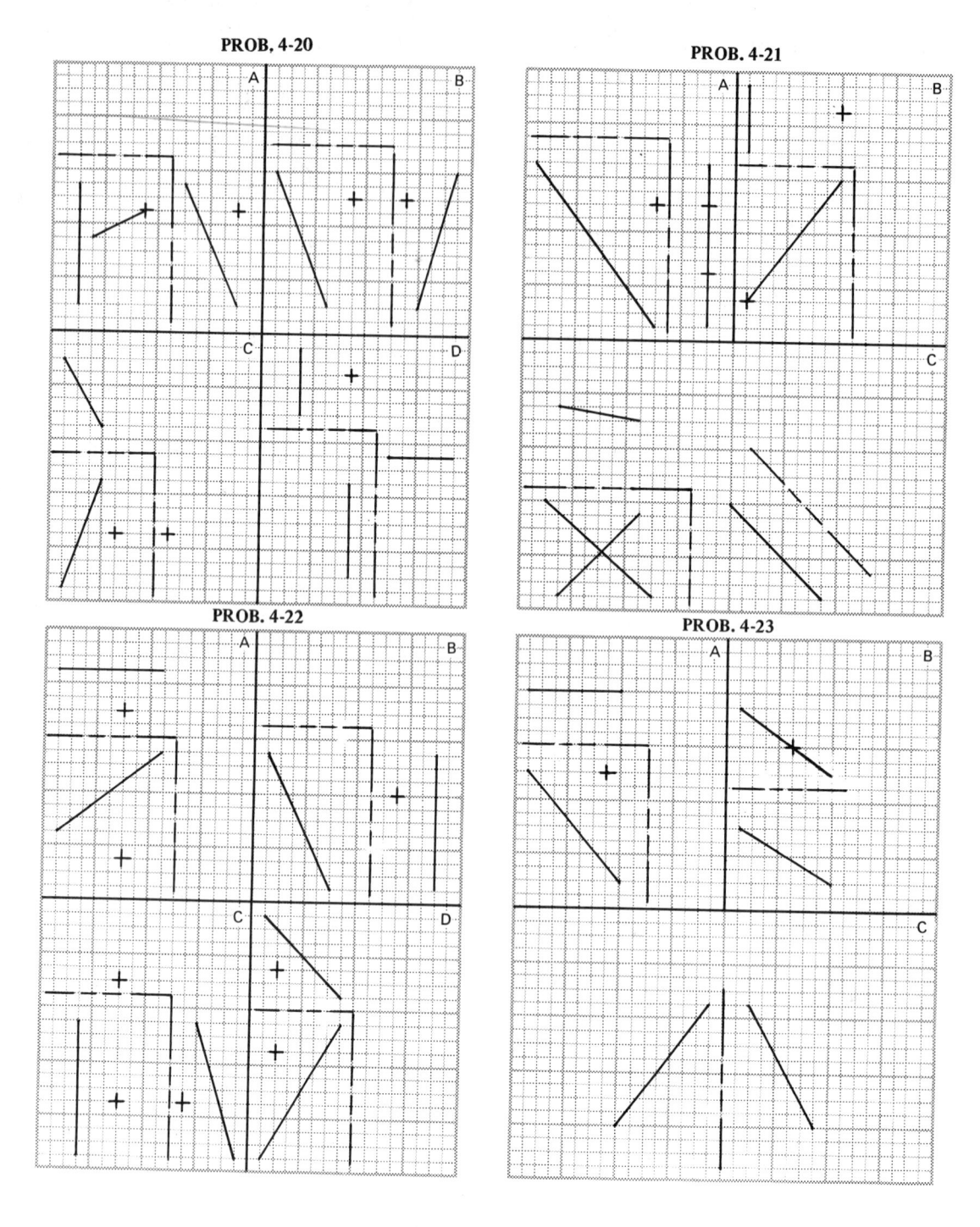

PROB. 4-20A. Complete the three views of the perpendicular lines.

PROB. 4-20B. The two lines are perpendicular. Line 3-4 is 1.625" (41.27mm) long. Point 3 is given.

PROB. 4-20C. Line 1-2 is given. Line 3-4 is 1" (25.4mm) long and shows as a point view in the front view. Are the two lines perpendicular?

PROB. 4-20D. Complete the views of the two lines. Line 3-4 shows as a point view in the horizontal view. Are they perpendicular?

PROB. 4-21A. Project the three views of the intersecting perpendicular lines.

PROB. 4-21B. Finish the views of the intersecting perpendicular lines. Line 1-2 projects as a point view in the frontal plane.

PROB. 4-21C. Complete all required views of the two lines. Are they perpendicular?

PROB. 4-22A. Construct a line through the point, perpendicular to and on the given line.

PROB. 4-22B. Construct line 3-4 perpendicular to line 1-2 and through the given point.

PROB. 4-22C. Through the given point, draw a horizontal line which ends three units behind the frontal plane. The new line is to be perpendicular to the given line.

PROB. 4-22D. Draw a line through the point and perpendicular to the other line.

PROB. 4-23A, 4-23B. Construct perpendicular intersecting lines using the given lines and points.

PROB. 4-23C. Draw perpendicular lines through the end points of the given line.

FIG. 4-82 Shortest distance between a point and line (line method).

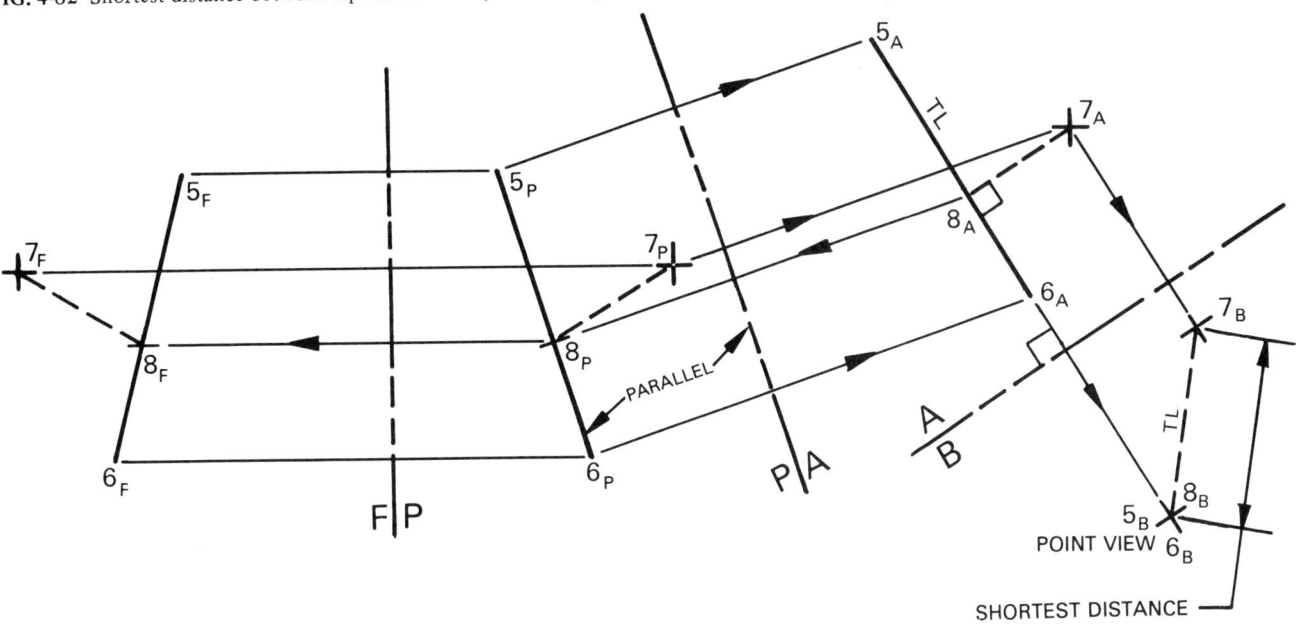

4.37 SHORTEST DISTANCE BETWEEN A POINT AND A LINE (LINE METHOD)

A perpendicular line between a given point and line is its shortest connection (distance). The **shortest distance** between a point and a line is measured along a perpendicular connector in a view where the line appears as a point view. To obtain a view where a perpendicular connector can be established it is first necessary to project the line as true length.

In Fig. 4-82 oblique line 5-6 and point 7 are given. The shortest connector between the line and point is required. This connector needs to be shown in all views. The following steps describe the procedure for finding the shortest distance between a point and line using the "**line method**."

1. Draw auxiliary view A parallel to oblique line 5_P-6_P, start by drawing fold line P/A parallel to the line.

2. Project line 5-6 and point 7 into auxiliary view A. Line 5_A-6_A shows true length.

3. Draw a perpendicular connector between point 7_A and true length line 5_A-6_A, and label this new point 8_A.

4. Project auxiliary view B parallel to line 7_A-8_A (and perpendicular to true length 5_A-6_A). Note that fold line A/B is parallel to line 7_A-8_A and perpendicular to line 5_A-6_A.

5. Auxiliary view B shows line 5_B-6_B as a point view and 7_B-8_B as true length. This true length dimension is the *shortest distance* between the point and the line.

6. Line 7-8 can now be projected back into the profile and frontal view. Draw a projection line from point 8_A until it intersects line 5_P-6_P. Connect 7_P to 8_P. From point 8_P draw a projection line until it intersects line 5_F-6_F. Connect point 7_F to 8_F.

In Fig. 4-83 the shortest distance (clearance) between the pipeline and point 0 is required. Line 1_H-2_H is true length in the horizontal view. Draw fold line H/A perpendicular to true length line 1_H-2_H. Auxiliary view A shows line 1_A-2_A as a point view. The diameter of the pipe is drawn here so that the clearance dimension can be measured between the fixed point and the pipe wall.

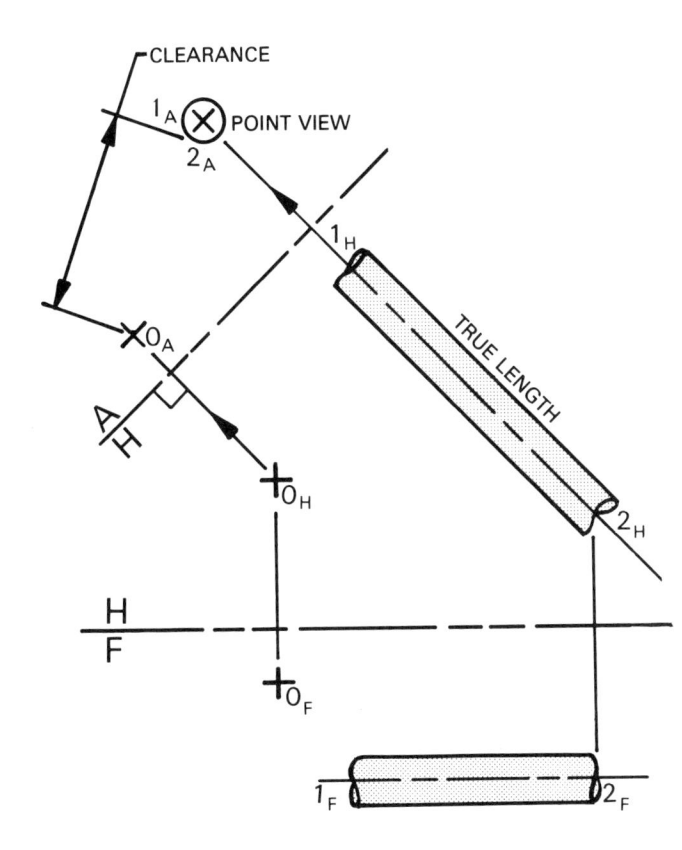

FIG. 4-83 Clearance between a fixed point and a pipe.

FIG. 4-84 Shortest distance between point and line (plane method).

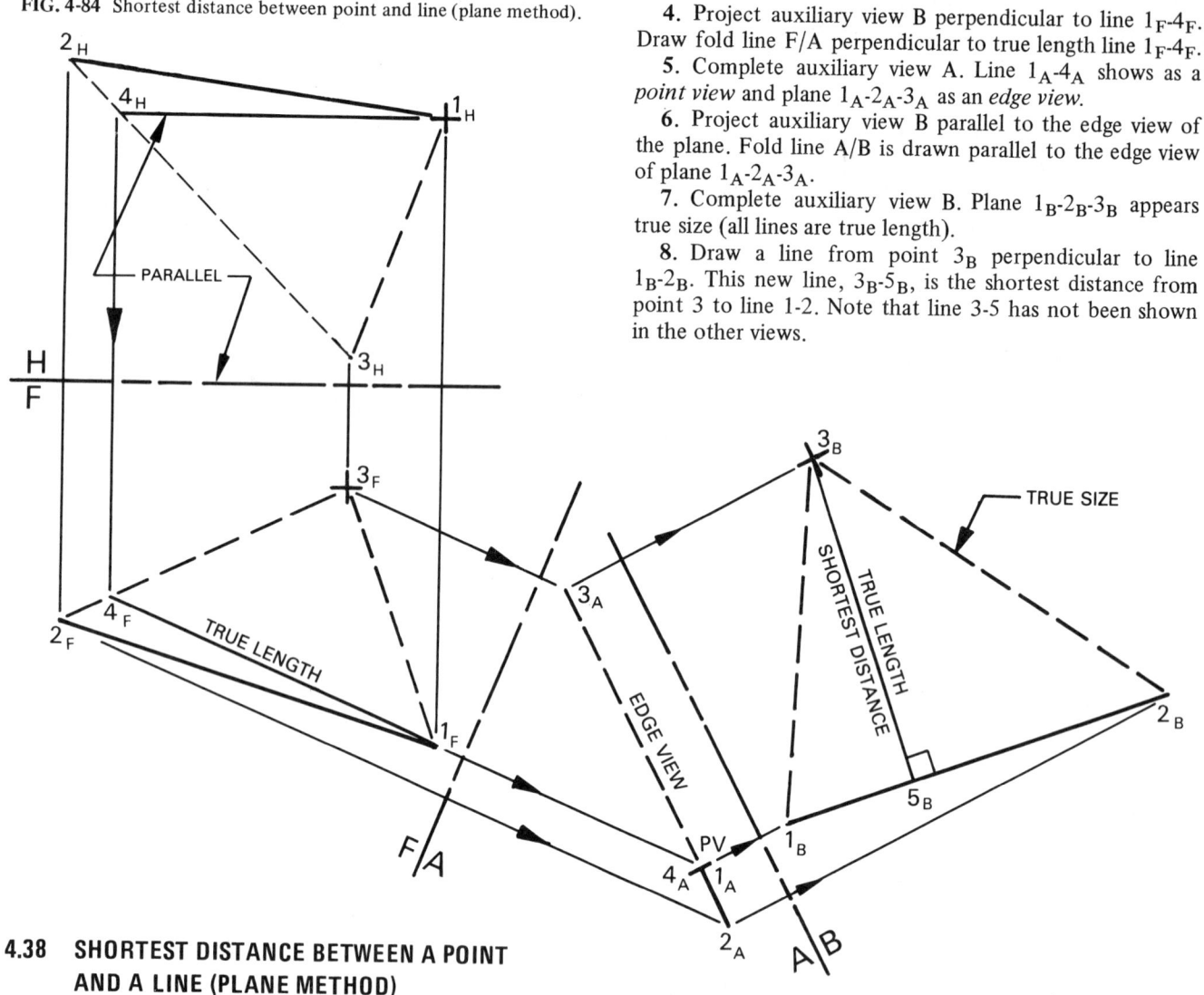

4. Project auxiliary view B perpendicular to line 1_F-4_F. Draw fold line F/A perpendicular to true length line 1_F-4_F.

5. Complete auxiliary view A. Line 1_A-4_A shows as a *point view* and plane 1_A-2_A-3_A as an *edge view*.

6. Project auxiliary view B parallel to the edge view of the plane. Fold line A/B is drawn parallel to the edge view of plane 1_A-2_A-3_A.

7. Complete auxiliary view B. Plane 1_B-2_B-3_B appears true size (all lines are true length).

8. Draw a line from point 3_B perpendicular to line 1_B-2_B. This new line, 3_B-5_B, is the shortest distance from point 3 to line 1-2. Note that line 3-5 has not been shown in the other views.

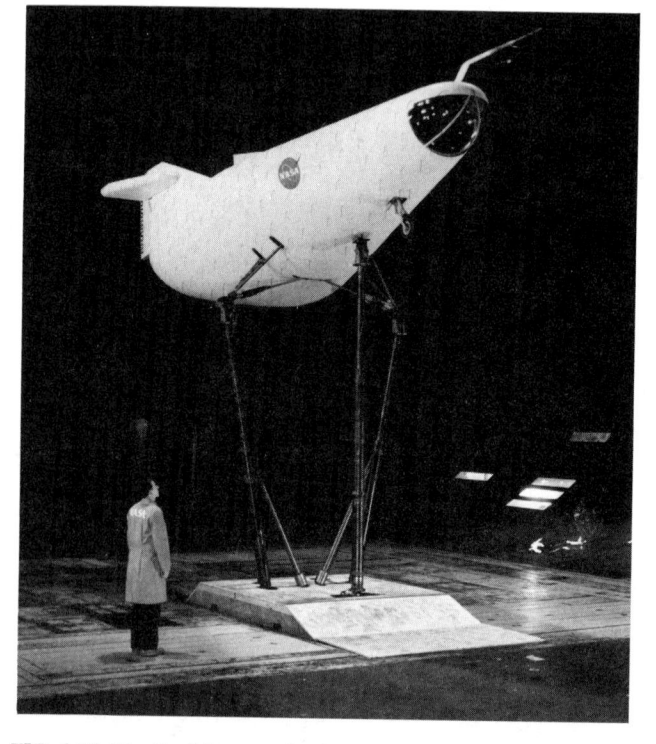

4.38 SHORTEST DISTANCE BETWEEN A POINT AND A LINE (PLANE METHOD)

An alternative procedure for finding the shortest distance between a point and a line is the "**plane method.**" The shortest distance between a point and line can be measured where the plane formed by the point and line shows true size. An edge view of the plane is necessary before a true shape (size) view can be obtained. An edge view can be constructed by projecting a point view of a true length line that lies in the plane.

In the Fig. 4-84 only the shortest distance is required, not its projection in every view. The point and line are given in the horizontal and frontal views. The following steps describe the construction of this example.

1. Connect point 3 and line 1-2 to form plane 1-2-3.
2. A true length line must be constructed so that it lies in plane 1-2-3. Draw line 1_H-4_H parallel to the H/F fold line.
3. Project point 4 into the frontal view by extending a projection line from 4_H perpendicular to H/F until it intersects line 2_F-3_F. Line 1_F-4_F is a frontal line and shows true length.

FIG. 4-85 The flexible mounting braces for the spacecraft were designed by finding the shortest distance between a point and a line.

FIG. 4-86 Shortest distance between two skew lines (line method).

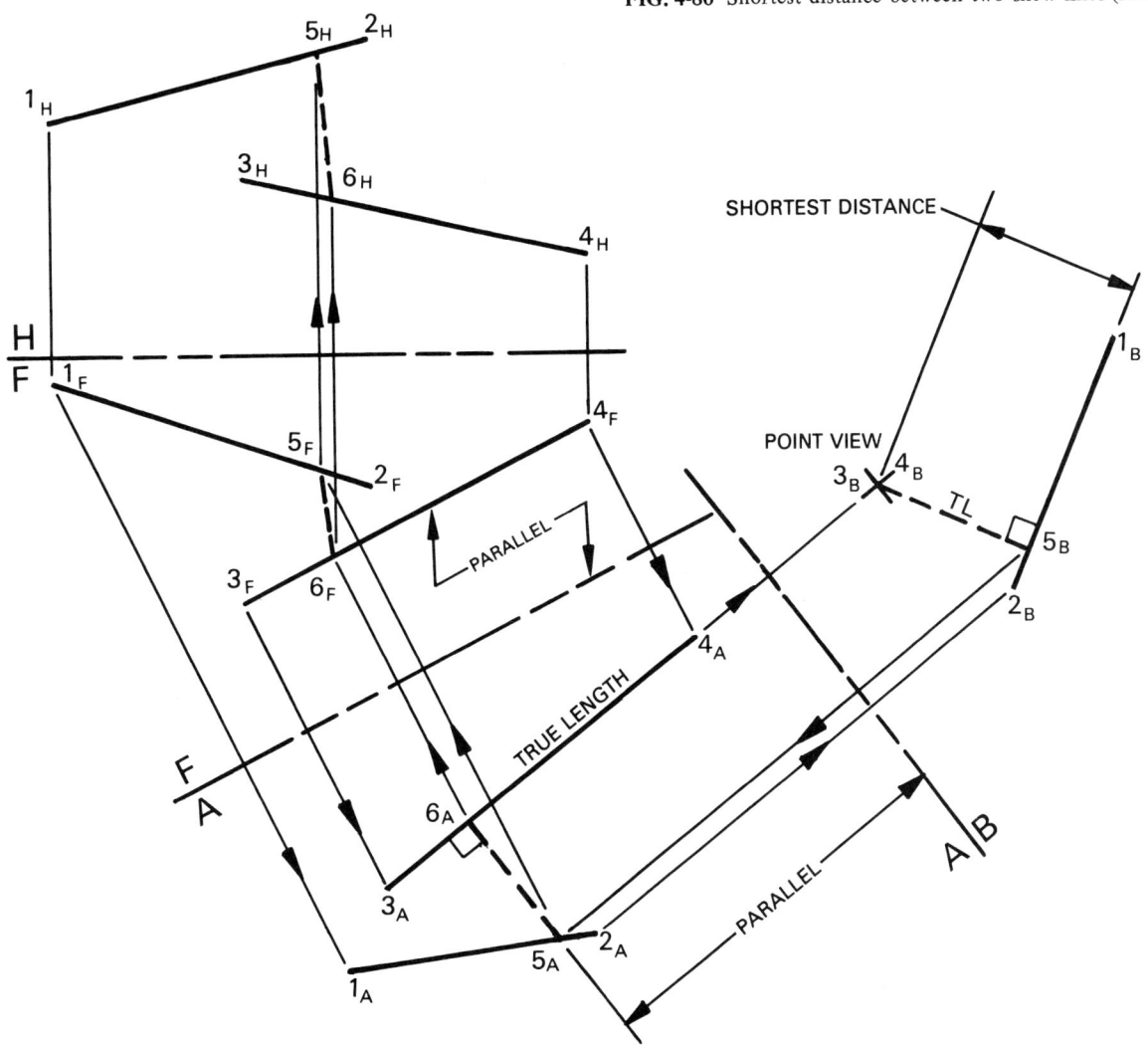

4.39 SHORTEST DISTANCE BETWEEN TWO SKEW LINES (LINE METHOD)

The shortest distance between two skew lines is required in a variety of industrial situations. In Fig. 4-85 the support structure for holding the spacecraft during wind tunnel testing can be reduced to lines and points in space. The procedure for finding the shortest distance between a point and line or between two lines could have been used to design these moveable supporting elements.

As with finding the shortest distance between a point and a line, there are two basic methods for finding the shortest distance between two skew lines: the "**line method**" and the "**plane method.**" On this page the line method is reviewed.

Two nonparallel, nonintersecting lines are called *skew lines*. The shortest distance between two skew lines is a line that is perpendicular to both lines. Only one solution is possible. This common perpendicular is shown as true length in a view where one line appears as a point view and the other oblique.

Given lines 1-2 and 3-4 in the horizontal and frontal views, the following steps describe the "**line method**" for drawing the shortest distance between skew lines, Fig. 4-86.

1. Draw fold line F/A parallel to line 3_F-4_F, and project auxiliary view A. Line 3_A-4_A is true length and line 1_A-2_A is oblique.

2. Draw fold line A/B perpendicular to true length line 3_A-4_A, and complete auxiliary view B. Line 3_B-4_B is a point view and line 1_B-2_B is oblique.

3. Draw a line from point view 3_B-4_B perpendicular to line 1_B-2_B. This is the shortest distance between the two skew lines. Note that this shortest distance line is perpendicular to *both* skew lines.

4. The shortest line can now be drawn in auxiliary view A. Project point 5_B from auxiliary view B to where it intersects line 1_A-2_A in auxiliary A. Draw line 5_A-6_A parallel to A/B and perpendicular to line 3_A-4_A. Point 6_A is on line 3_A-4_A. Since the shortest connector is perpendicular to both skew lines, it follows that perpendicularity shows in any view where the line or the shortest distance appears as true length.

5. Project points 5_A and 6_A into the frontal and horizontal views. Remember that point 5 remains on line 1-2 and point 6 on line 3-4 in all views.

FIG. 4-87 Shortest distance between two skew lines (plane method).

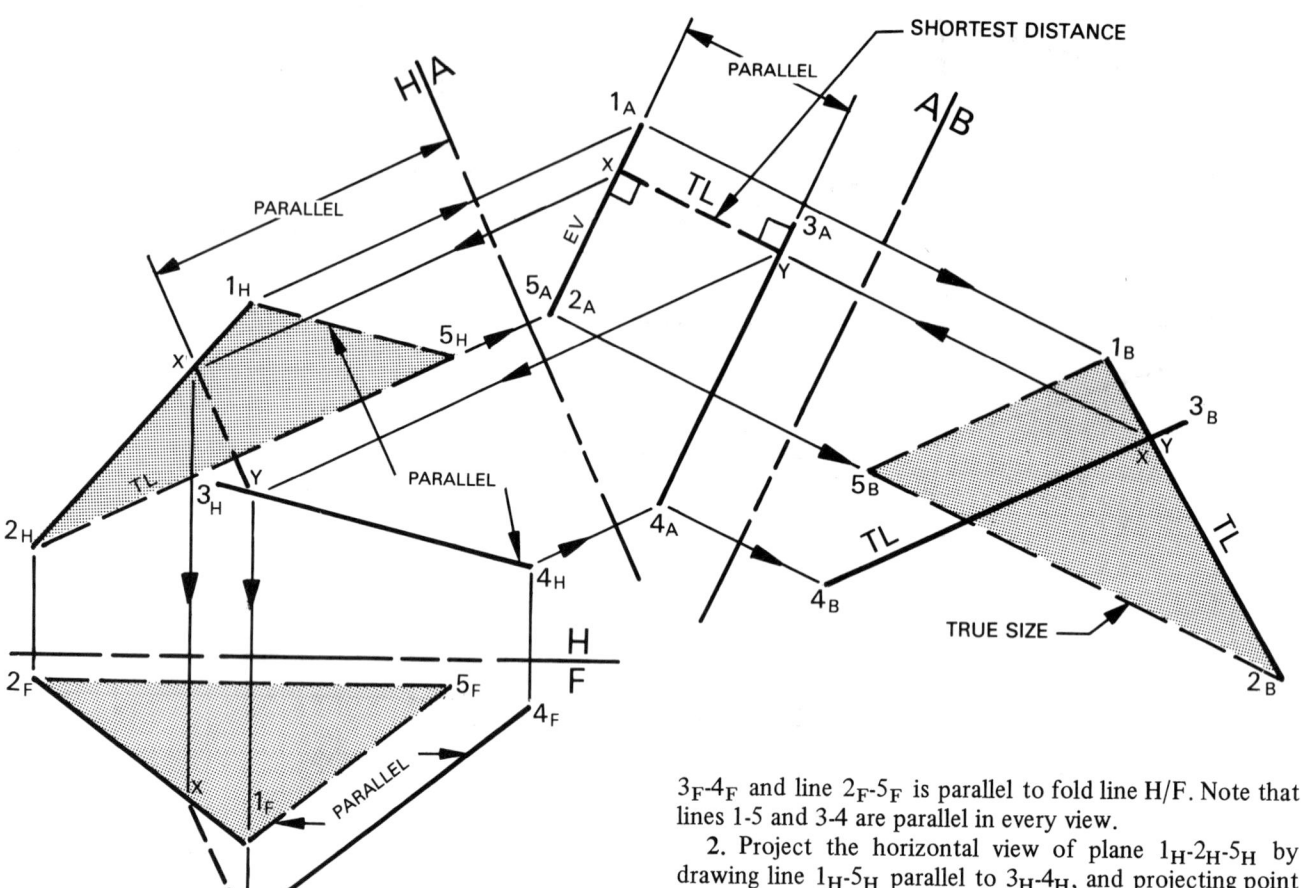

4.40 SHORTEST DISTANCE BETWEEN TWO SKEW LINES (PLANE METHOD)

The shortest distance (shortest perpendicular connector) between two skew lines can also be solved by the "**plane method.**" To find this connector, a plane is constructed containing one of the lines and parallel to the other. Where this newly formed plane appears as an edge view, the shortest distance can be measured. This primary auxiliary view shows the lines parallel to one another, and the perpendicular distance between them is the shortest distance. The exact location of the shortest connector requires an auxiliary view where the plane appears as true size. In this secondary auxiliary view the shortest connector appears as a point view.

In Fig. 4-87 skew lines 1-2 and 3-4 are given. The shortest connector between them is required. In this example the plane was formed in the frontal view using line 1-2. The following steps describe this procedure:

1. Form a plane containing line 1_F-2_F and parallel to line 3_F-4_F in the frontal view. Line 1_F-5_F is parallel to line 3_F-4_F and line 2_F-5_F is parallel to fold line H/F. Note that lines 1-5 and 3-4 are parallel in every view.

2. Project the horizontal view of plane 1_H-2_H-5_H by drawing line 1_H-5_H parallel to 3_H-4_H, and projecting point 5_H from the frontal view. Note that line 2_H-5_H is true length, a horizontal line, since it was drawn parallel to the H/F fold line in the frontal view.

3. The *edge view* of a plane shows in a view where a line in that plane projects as a point view. Since line 2_H-5_H is true length in the horizontal view a point view can be obtained by projecting auxiliary view A perpendicular to it. Draw fold line H/A perpendicular to true length line 2_H-5_H.

4. Complete auxiliary view A. Plane 1-2-5 contains line 2-5 which is parallel to line 3-4, therefore they remain parallel in this view. Line 2_A-5_A appears as a point view, consequentially plane 1_A-2_A-5_A shows as an edge view. The shortest distance can be measured but not located in this view by measuring the perpendicular distance between the plane and line.

5. In order to fix the connector in space, it is necessary to project a view where the plane appears true size and the connector shows as a point view. Auxiliary view B is constructed by drawing A/B parallel to the edge view of plane 1-2-5. Note that A/B is also parallel to oblique line 3_A-4_A.

6. Complete auxiliary view B. Plane 1-2-5 shows as true size. Both line 3_B-4_B and line 1_B-2_B project as true length. The shortest connector, X-Y, appears as a point view where the lines cross.

7. Connector X-Y can now be located in all views by projection. Auxiliary view A shows the true length of X-Y since it appears as a point view in auxiliary B. In the horizontal view, X-Y can be located by projection, and is parallel to H/A, since it is true length in auxiliary A. Finally, X-Y is projected into the frontal view where it is oblique. Remember that X is always on line 1-2, and Y remains on line 3-4, in all views.

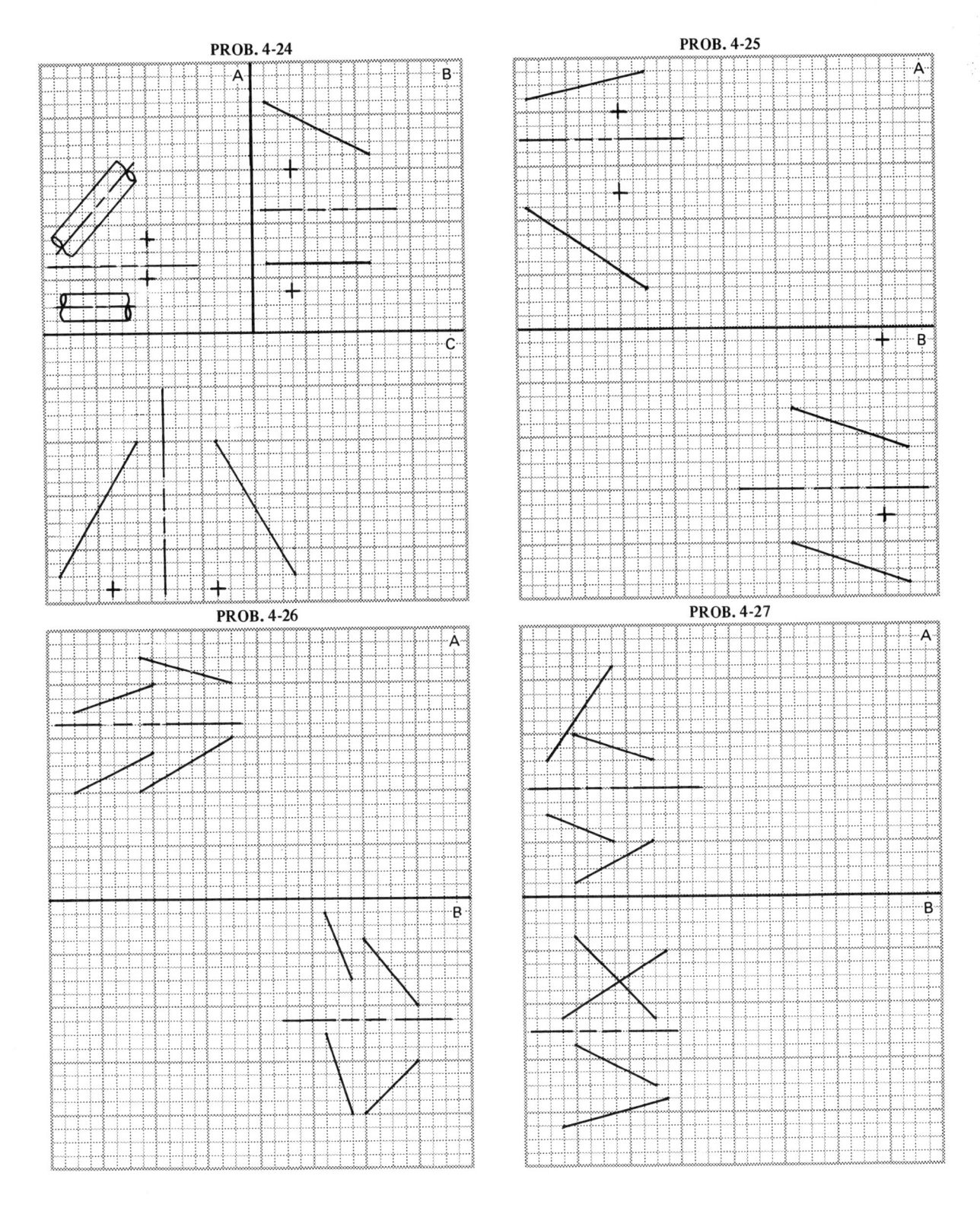

PROB. 4-24A. What is the closest distance between the pipe wall and the point?

PROBS. 4-24B, 4-24C. Using the line method solve for the true distance between the lines, show the connector in each view.

PROBS. 4-25A, 4-25B. Using the plane method solve for the shortest distance between the line and point. Show the new line in each view.

PROBS. 4-26A, 4-26B. Find the shortest (perpendicular) distance between the two lines, using the line method. Project the line back into all views. Note that the line may need to be extended in 4-26A.

PROB. 4-27A. The two skew lines represent two pipelines. A cross connecting pipe is to be installed so that it is of minimum length. What is this minimum length, using a scale of $1'' = 50$ ft? Use the plane method.

PROB. 4-27B. The skew lines represent the centerlines of two mine tunnels. A cross connecting ventilating shaft is to be dug so that it is of minimum length. What is the minimum length? Use the plane method and a scale of $1'' = 50$ ft.

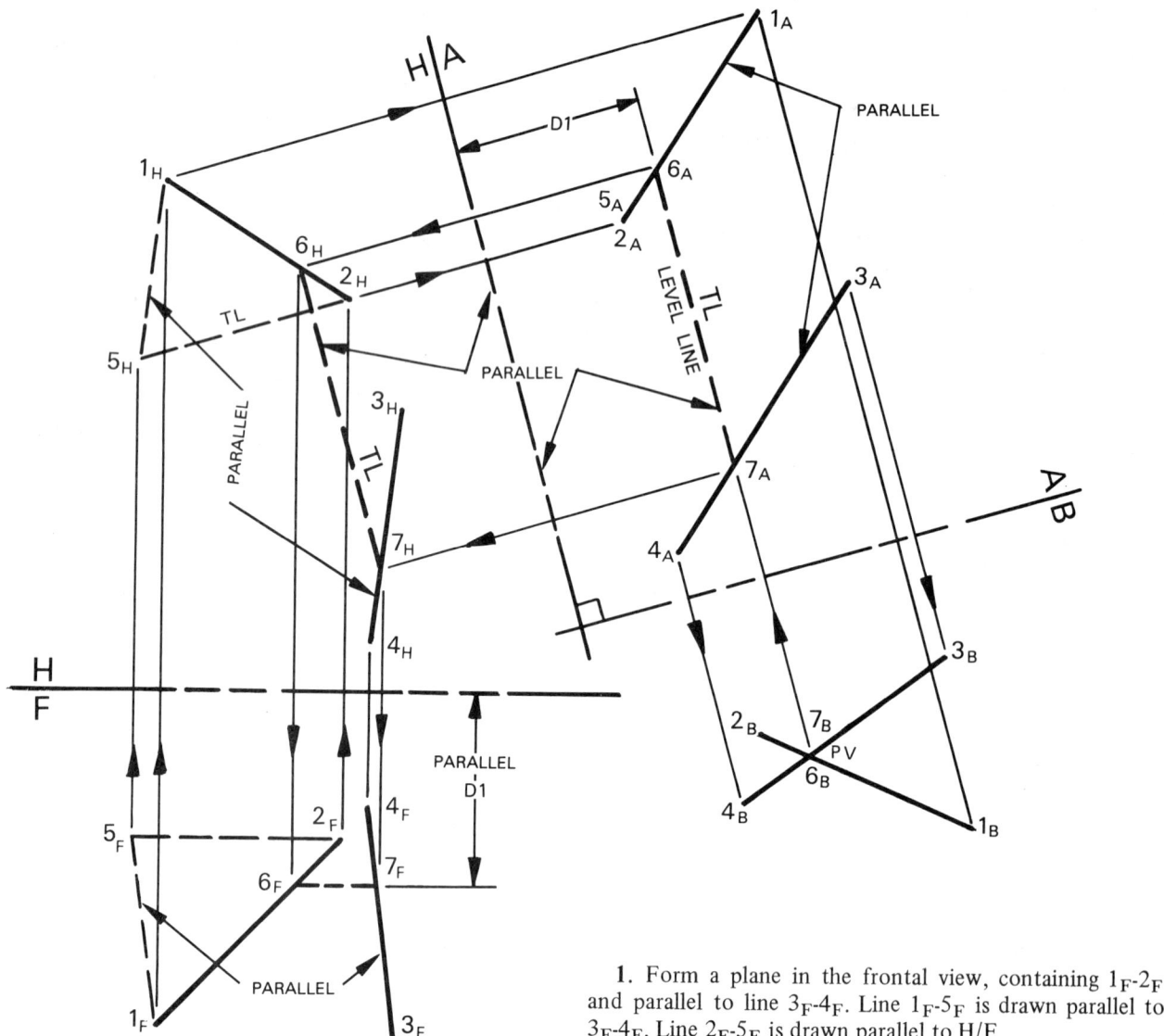

FIG. 4-88 Shortest horizontal distance between two skew lines.

4.41 SHORTEST HORIZONTAL (LEVEL) DISTANCE BETWEEN TWO SKEW LINES

The **shortest horizontal distance** between two skew lines is a level line, found by using the same general method as described for the shortest (perpendicular) distance. Here, the first auxiliary view must be taken from the horizontal view in order to show the horizontal plane as an edge. The shortest horizontal distance can only be measured in a view where the horizontal plane appears as an edge and the connector shows as a level (horizontal) line and therefore true length. The shortest level line will project as true length in the horizontal view. Unlike the previous problem, the second auxiliary view is projected perpendicular to the first auxiliary view. The following steps describe this process. In Fig. 4-88 lines 1-2 and 3-4 are given and the shortest horizontal (level) connector is required.

1. Form a plane in the frontal view, containing 1_F-2_F and parallel to line 3_F-4_F. Line 1_F-5_F is drawn parallel to 3_F-4_F. Line 2_F-5_F is drawn parallel to H/F.
2. In the horizontal view, form plane 1-2-5 by drawing line 1_H-5_H parallel to line 3_H-4_H and projecting point 5 from the frontal view. Line 2_H-5_H is true length.
3. Draw H/A perpendicular to true length line 2_H-5_H and project auxiliary view A. Line 2_A-5_A appears as a point view, therefore plane 1_A-2_A-5_A shows an edge view and is parallel to line 3_A-4_A.
4. To locate the shortest horizontal connector between the two skew lines, a view where the connector shows as a point view is required. Draw fold line A/B *perpendicular* to fold line H/A. Auxiliary view B is at a right angle to auxiliary view A.
5. Complete auxiliary view B. Lines 1_B-2_B and 3_B-4_B show oblique. The plane need not be shown. Where the two lines cross establishes the point view of the shortest horizontal connector, line 6_B-7_B.
6. Project connector 6_B-7_B into all views. Auxiliary view shows line 6_A-7_A as true length and parallel to H/A. Line 6_A-7_A can be projected into the horizontal view where it appears true length (a horizontal line) and parallel to H/A. In the frontal view, line 6_F-7_F is parallel to H/F. In cases where the two skew lines are almost vertical, dimension D1 can be used to check the accuracy of the projection.

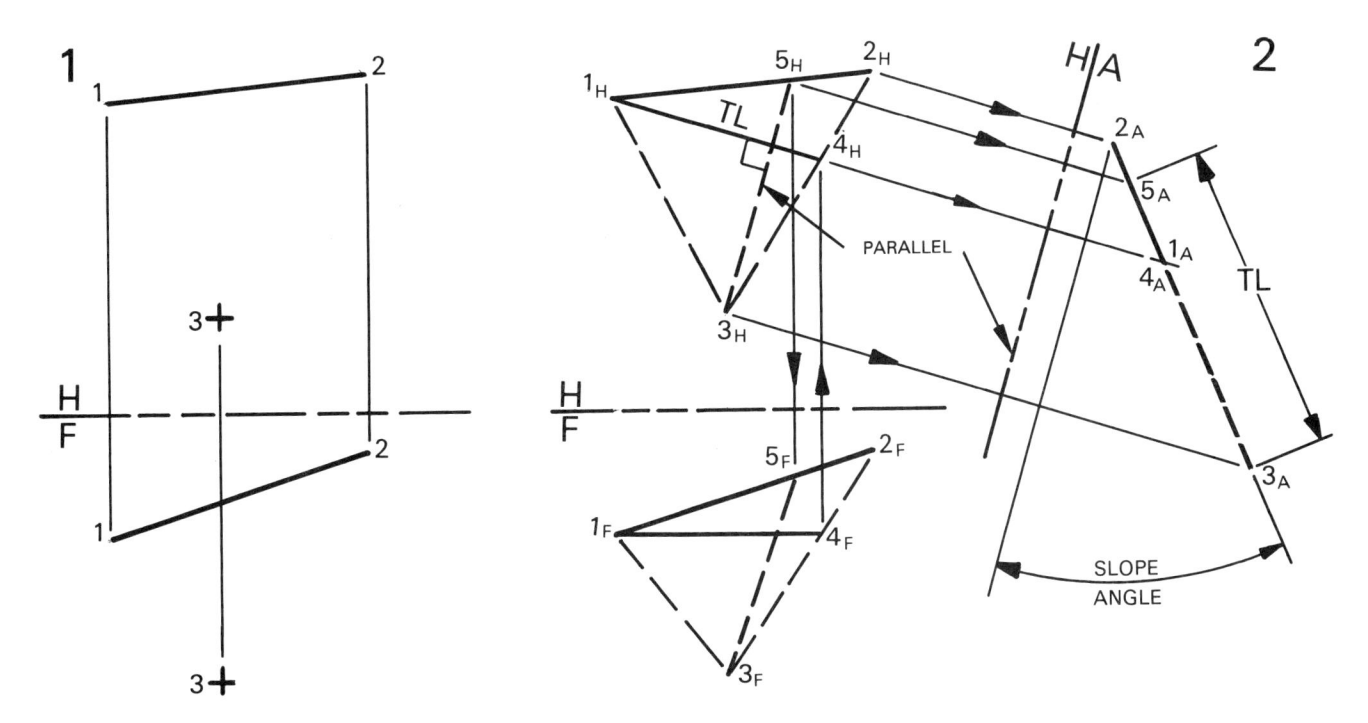

FIG. 4-89 Steepest connection between a point and a line.

4.42 STEEPEST CONNECTION BETWEEN A POINT AND A LINE

The design of industrial chutes as shown in Fig. 4-90 requires the solution of a variety of problems using the graphical method. The **steepest connection** between a point and a line is a typical example. The steepest line on a plane formed by the point and the line is the steepest connection between them. The steepest connector appears true length (measurable) in an elevation view where the plane formed by the point and line show as an edge. This first auxiliary view must be projected from the *horizontal view* in order to show the edge in an elevation view. Note that the angle formed by the edge view of the plane and the horizontal plane shows the slope of both the plane and the steepest connection. The slope of a line or plane is the true angle that they make with the horizontal plane and can only be measured in a view where the line is true length (or the plane is an edge) and the horizontal plane appears as an edge.

In Fig. 4-89(1), the point and line are given and the steepest connector is required.

1. Form a plane between the point and line (2).
2. In the frontal view draw line 1_F-4_F parallel to H/F. Horizontal line 1_H-4_H is true length.
3. The steepest connector, line 1_H-5_H, is drawn perpendicular to 1_H-4_H, and can be projected back into the frontal view. Line 1-5 is oblique in both views.
4. Draw fold line H/A perpendicular to 1_H-4_H (and therefore parallel to 3_H-5_H) and complete auxiliary view A.

Line 1_A-4_A shows as a point view and plane 1_A-2_A-3_A appears as an edge view. Since the steepest connector, line 3_A-5_A, is on plane 1-2-3 and was drawn parallel to H/A in the horizontal view, it shows as a true length in auxiliary A and can be measured in this view only.

FIG. 4-90 The steepest connector between a point and a line is a frequently required situation in the design of an industrial chute.

105

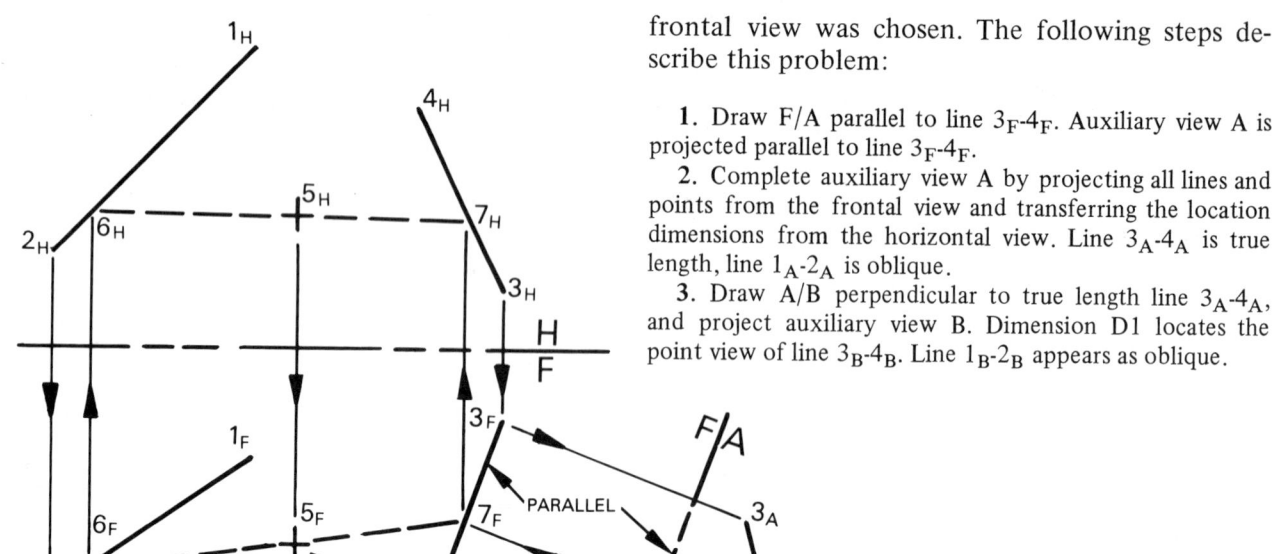

FIG. 4-91 Shortest connector between two lines through a given point.

4.43 SHORTEST CONNECTOR BETWEEN TWO LINES AND THROUGH A GIVEN POINT

In this section, a connector between two skew lines passes through a given point. This procedure is similar to finding the shortest (perpendicular) connector between two lines. The difference is that the required connector is the shortest connector through a specific fixed point. The connector is not perpendicular to either given line and projects as oblique in most views (a true length projection could be taken parallel to the connector if required). In Fig. 4-91, the true length of the connector is not solved for, only the views necessary for the construction of the connector have been drawn, which are normally all that is required for this type of problem.

Lines 1-2 and 3-4 represent two pipes or structural elements that require a connecting pipe or brace that passes through a fixed point. Lines 1-2 and 3-4 and point 5 are given. Note that four possible beginning projections could be used depending upon space availability; auxiliary view A could be drawn parallel to any of the oblique views of lines 1-2 or 3-4. In this example 3-4 in the frontal view was chosen. The following steps describe this problem:

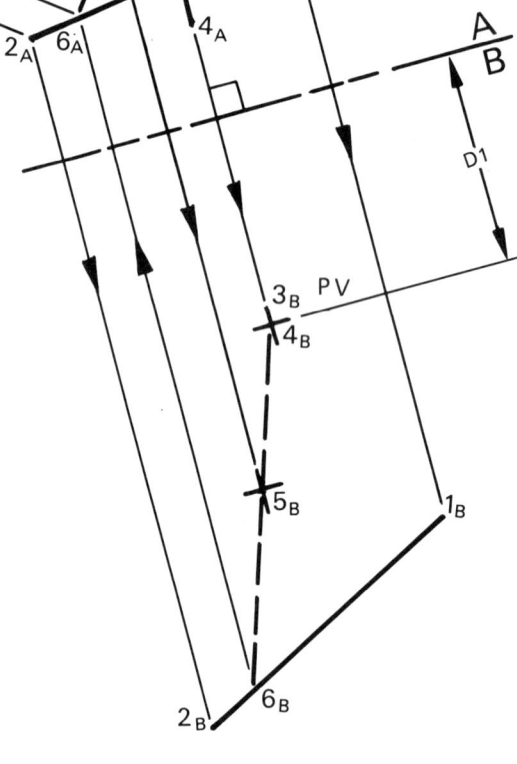

1. Draw F/A parallel to line 3_F-4_F. Auxiliary view A is projected parallel to line 3_F-4_F.
2. Complete auxiliary view A by projecting all lines and points from the frontal view and transferring the location dimensions from the horizontal view. Line 3_A-4_A is true length, line 1_A-2_A is oblique.
3. Draw A/B perpendicular to true length line 3_A-4_A, and project auxiliary view B. Dimension D1 locates the point view of line 3_B-4_B. Line 1_B-2_B appears as oblique.
4. Draw a line from the point view of line 3_B-4_B, through point 5_B to line 1_B-2_B, point 6_B. This new line is the connector.
5. Point 6 is located in auxiliary A by projection. 6_A is on line 1_A-2_A. Draw a line from 6_A through 5_A to locate point 7_A on line 3_A-4_A.
6. Locate connecting line 6-7 in the frontal and horizontal views by projection. Remember that it passes through point 5 in all views, and that point 6 is on line 1-2 and point 7 is on line 3-4 in all projections.

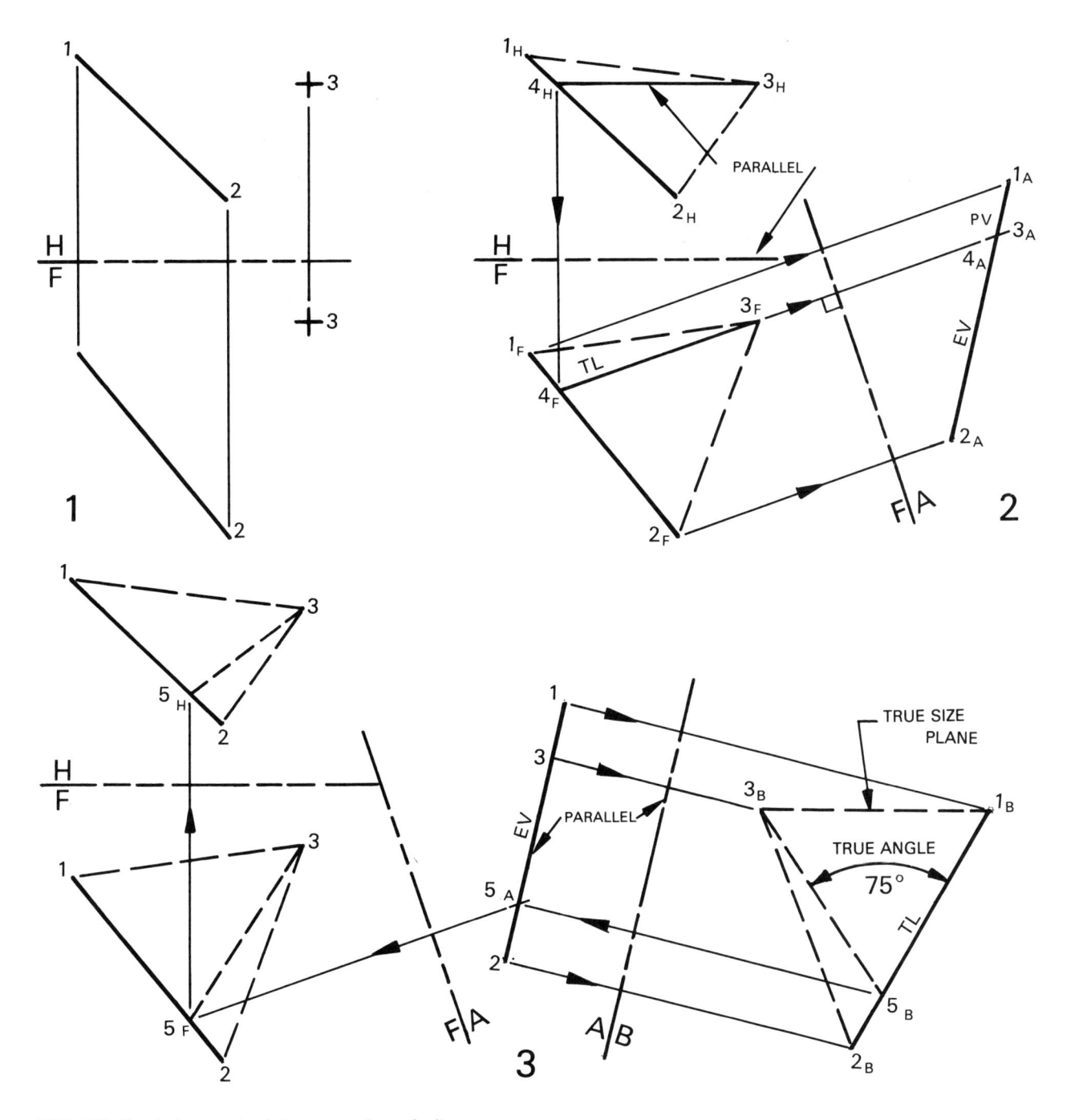

FIG. 4-92 Shortest connector between a point and a line at a specified angle.

4.44 CONNECTOR BETWEEN A POINT AND A LINE AT A SPECIFIED ANGLE

A typical engineering design problem would be to solve for a connector between a point and a line at a given angle. In Fig. 4-92(1), line 1-2 and point 3 are given and a connector sloping downward at 75° between them is required:

1. Form a plane with the given line 1-2 and point 3 (2).
2. Draw frontal line 3_H-4_H parallel to H/F and show its true length projection in the frontal view.

3. Draw F/A perpendicular to true length line 3_F-4_F and project auxiliary view A. Line 3_A-4_A appears as a point view, therefore plane 1_A-2_A-3_A shows as an edge in auxiliary A.
4. Draw A/B parallel to the edge view of plane 1_A-2_A-3_A and project auxiliary view B. Plane 1_B-2_B-3_B shows true size in this view (3).
5. From point 3_B draw a connector downward at 75° to line 1_B-2_B. Note that point 2_F is the low point of line 1_F-2_F in the frontal (elevation) view, so connector 3_B-5_B slants towards point 2_B (in all views).
6. Project point 5_B back to auxiliary A, frontal and horizontal views. Point 5 is on line 1-2 in all views. Draw connector 3-5 in each view.

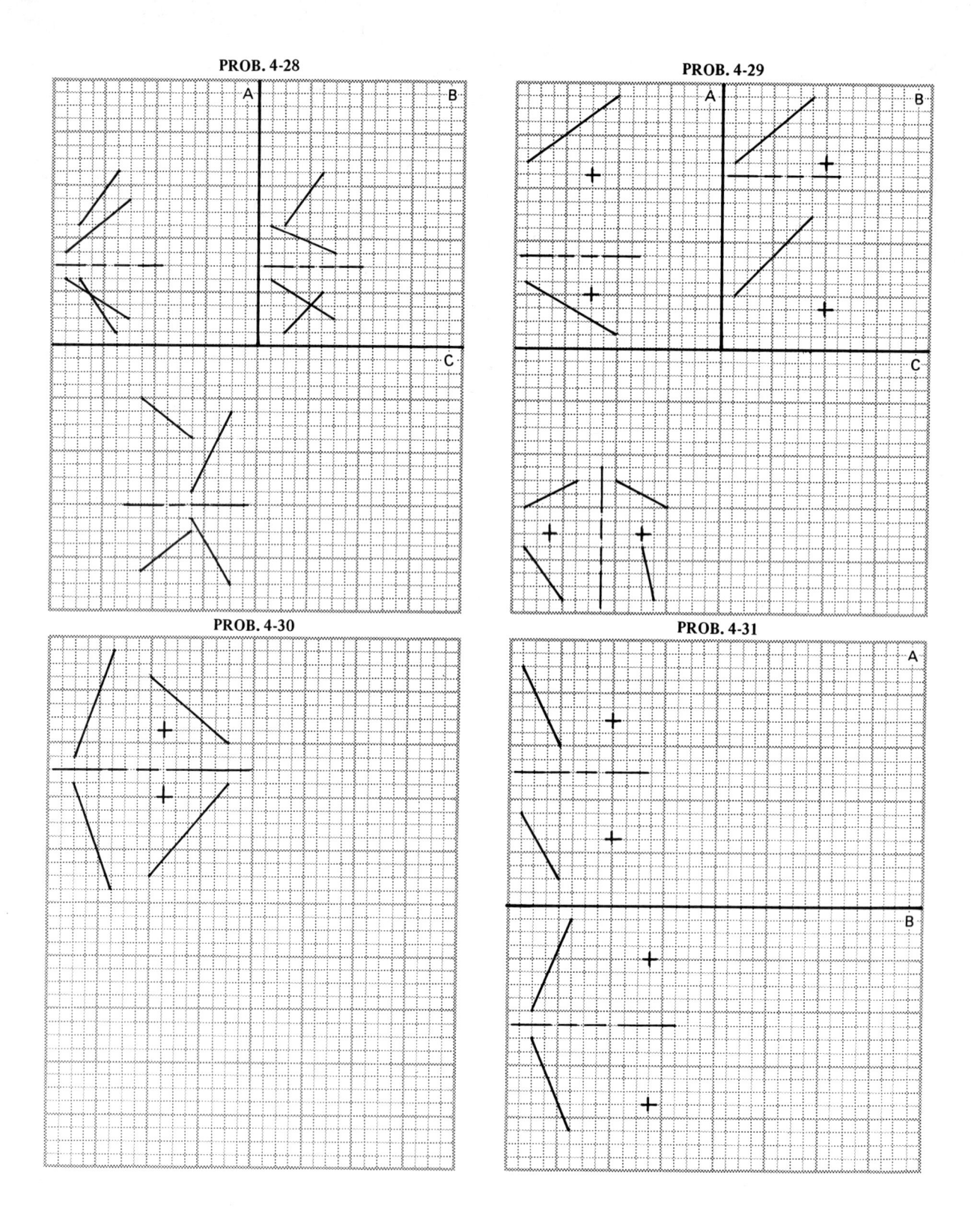

PROBS. 4-28A, 4-28B, 4-28C. Project the shortest horizontal line between the skewed lines. Show in the H and F views.

PROBS. 4-29A, 4-29B. Establish the steepest connector between the point and line.

PROB. 4-29C. Establish the shortest connector between the two lines and passing through the given point. Start with an auxiliary view off the lower line in the profile view. Show the connector in all views.

PROB. 4-30. Construct the shortest connector between the skew lines and through the given point. Show the line in all views.

PROB. 4-31A. Construct a connector between the point and line, upward at a 35° angle. Show in all views.

PROB. 4-31B. Same as *A* but construct the connector upward at 28°.

108

FIG. 4-93 Angle between two intersecting lines.

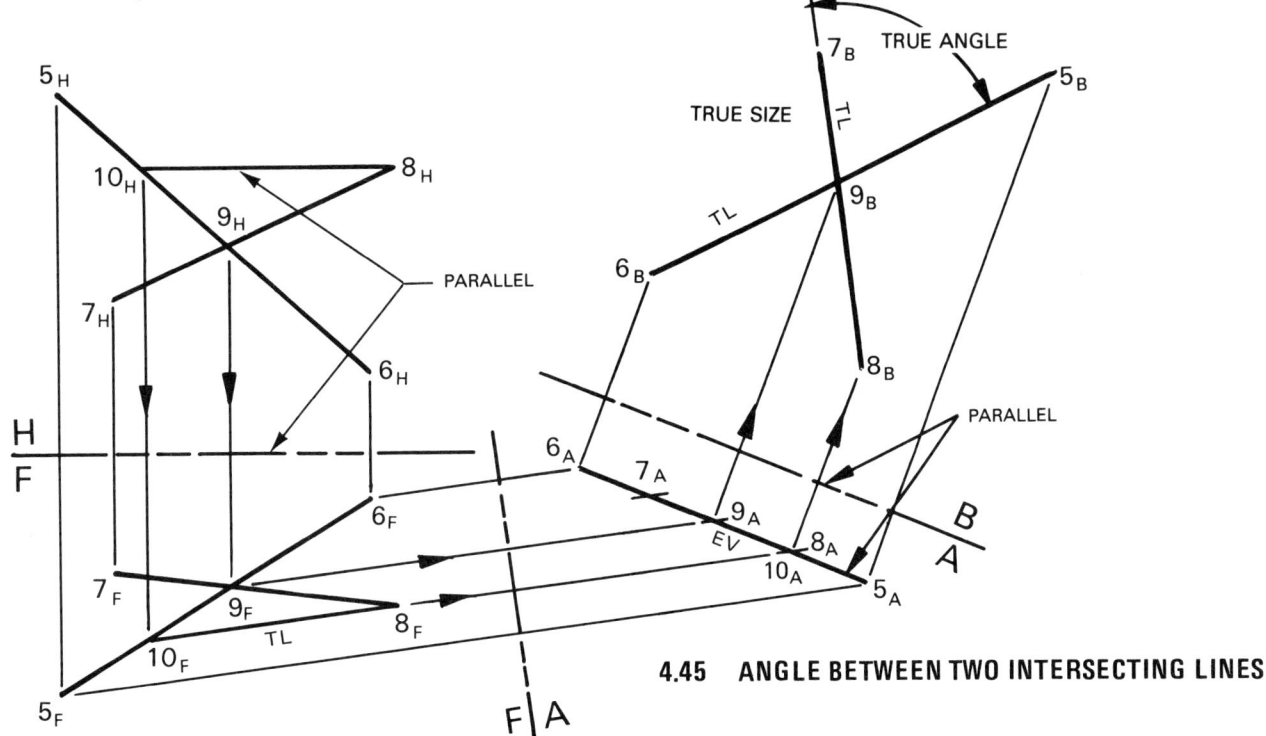

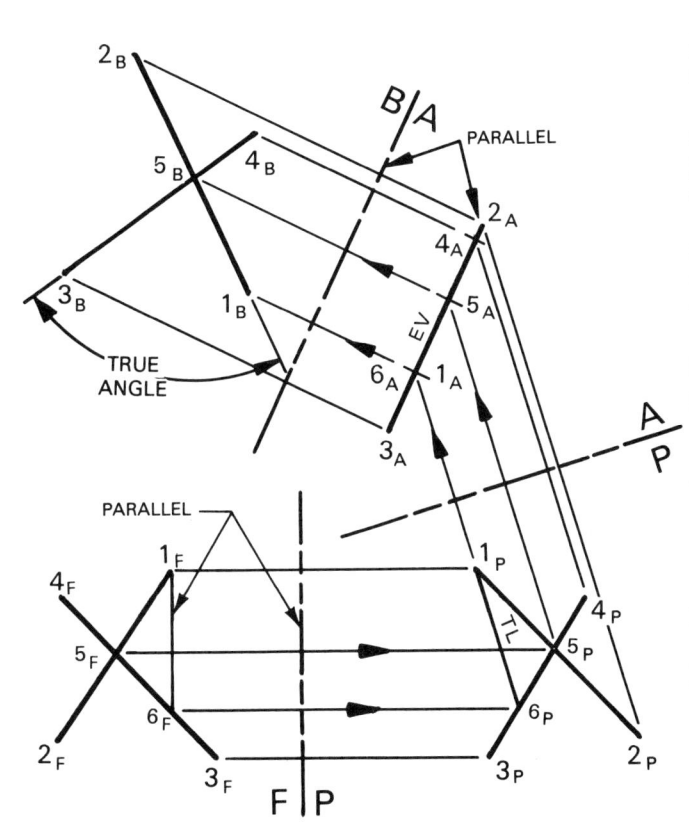

4.45 ANGLE BETWEEN TWO INTERSECTING LINES

Since two intersecting lines form a plane, the true angle between the lines is seen in a view where the plane appears as true shape. In Fig. 4-93, lines 5-6 and 7-8 are intersecting lines; the true angle between them is required.

1. Assuming that lines 5-6 and 7-8 are a plane, draw frontal line 8_H-10_H parallel to H/F, and project to the frontal view where it appears true length.
2. Draw F/A perpendicular to true length line 8_F-10_F, and project auxiliary view A. Line 8_A-10_A appears as a point view, therefore "plane" 5_A-6_A-7_A-8_A shows as an edge.
3. Draw A/B parallel to the edge view of "plane" 5_A-6_A-7_A-8_A, and project auxiliary B. Intersecting lines 5_B-6_B and 7_B-8_B are both true length and thus determine the true size of plane 5_B-6_B-7_B-8_B.
4. The true angle formed by lines 5_B-6_B and 7_B-8_B can be measured in this view. Note that the acute angle is measured.

In Fig. 4-94, lines 1-2 and 3-4 intersect. The F and P views are given. The true angle between the lines is required:

1. Draw profile line 1_F-6_F parallel to F/P, and true length in the profile view.
2. Fold line P/A is drawn perpendicular to true length line 1_P-6_P. Complete auxiliary A. Line 1_A-2_A appears as a point view, therefore "plane" 1_A-2_A-3_A-4_A shows as an edge.
3. Draw A/B parallel to edge view 1_A-2_A-3_A-4_A, and complete auxiliary B. "Plane" 1_B-2_B-3_B-4_B shows true size since lines 1_B-2_B and 3_B-4_B appear true length.
4. Measure the acute angle formed by intersecting lines 1_B-2_B and 3_B-4_B.

FIG. 4-94 True angle between two intersecting lines.

FIG. 4-95 Angle between two skew lines

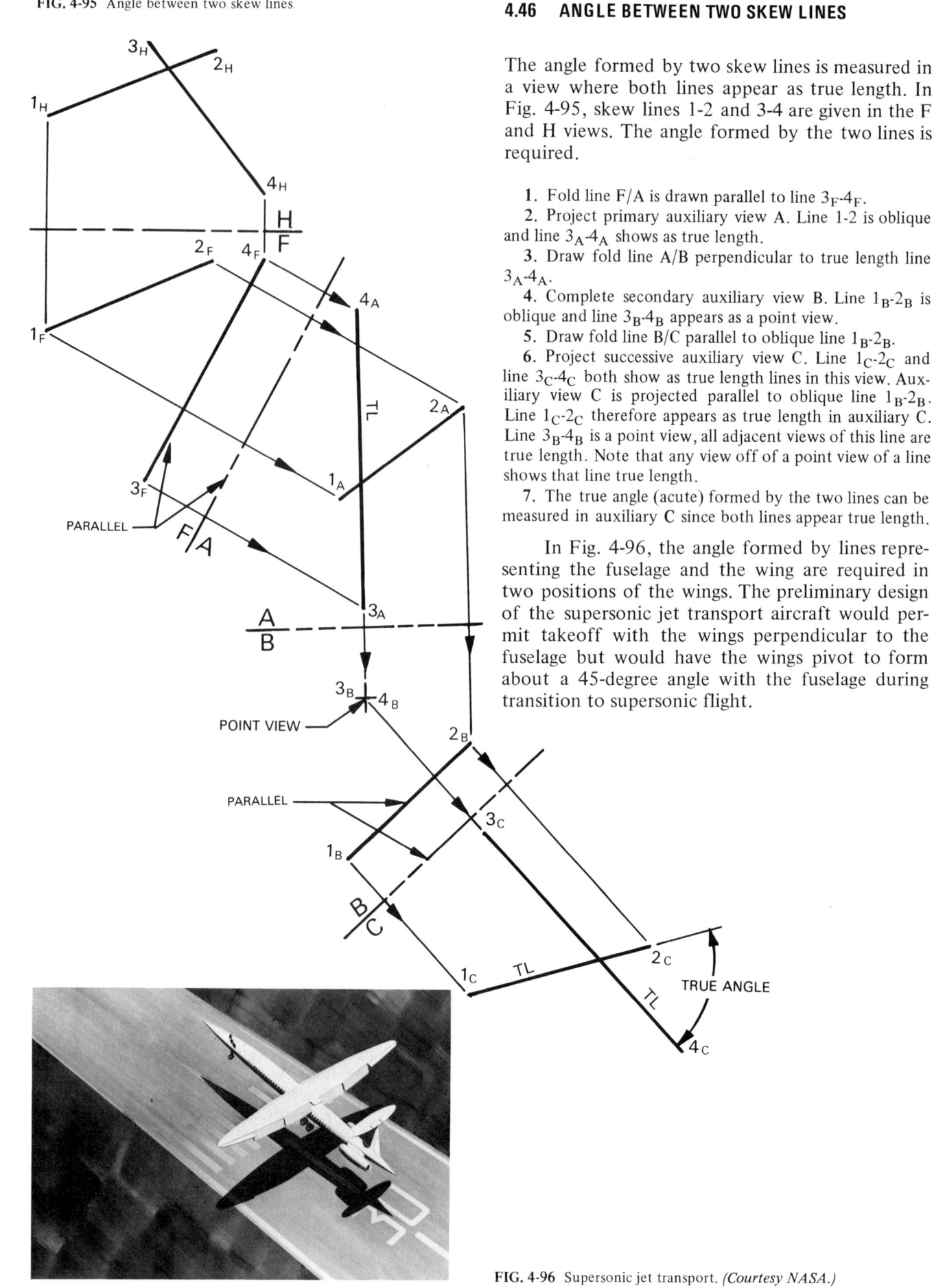

4.46 ANGLE BETWEEN TWO SKEW LINES

The angle formed by two skew lines is measured in a view where both lines appear as true length. In Fig. 4-95, skew lines 1-2 and 3-4 are given in the F and H views. The angle formed by the two lines is required.

1. Fold line F/A is drawn parallel to line 3_F-4_F.
2. Project primary auxiliary view A. Line 1-2 is oblique and line 3_A-4_A shows as true length.
3. Draw fold line A/B perpendicular to true length line 3_A-4_A.
4. Complete secondary auxiliary view B. Line 1_B-2_B is oblique and line 3_B-4_B appears as a point view.
5. Draw fold line B/C parallel to oblique line 1_B-2_B.
6. Project successive auxiliary view C. Line 1_C-2_C and line 3_C-4_C both show as true length lines in this view. Auxiliary view C is projected parallel to oblique line 1_B-2_B. Line 1_C-2_C therefore appears as true length in auxiliary C. Line 3_B-4_B is a point view, all adjacent views of this line are true length. Note that any view off of a point view of a line shows that line true length.
7. The true angle (acute) formed by the two lines can be measured in auxiliary C since both lines appear true length.

In Fig. 4-96, the angle formed by lines representing the fuselage and the wing are required in two positions of the wings. The preliminary design of the supersonic jet transport aircraft would permit takeoff with the wings perpendicular to the fuselage but would have the wings pivot to form about a 45-degree angle with the fuselage during transition to supersonic flight.

FIG. 4-96 Supersonic jet transport. *(Courtesy NASA.)*

4.47 ANGLE BETWEEN A LINE AND A PRINCIPAL PLANE

The true angle between a line and a principal plane shows in a view where the line is true length and the principal plane appears as an edge. It follows that principal lines form a true angle with the edge view of the adjacent principal plane. The angle formed by a horizontal line and the H/F fold line is the true angle between the line and frontal plane. The angle formed by a frontal line and the H/F fold line is the true angle between the line and the horizontal plane, and the angle it makes with F/P is the true angle between it and the profile plane. The angle formed by a profile line and the F/P fold line is the angle that the line makes with the frontal plane.

When a given line is oblique, it is necessary to project a primary auxiliary view where the line is true length and the principal plane shows as an edge. In Fig. 4-97, oblique line 5-6 is given and the angle between it and the horizontal plane is required. Primary auxiliary A is projected parallel to line 5_H-6_H. Draw H/A parallel to line 5_H-6_H, and complete auxiliary A. In this new view line 5_A-6_A

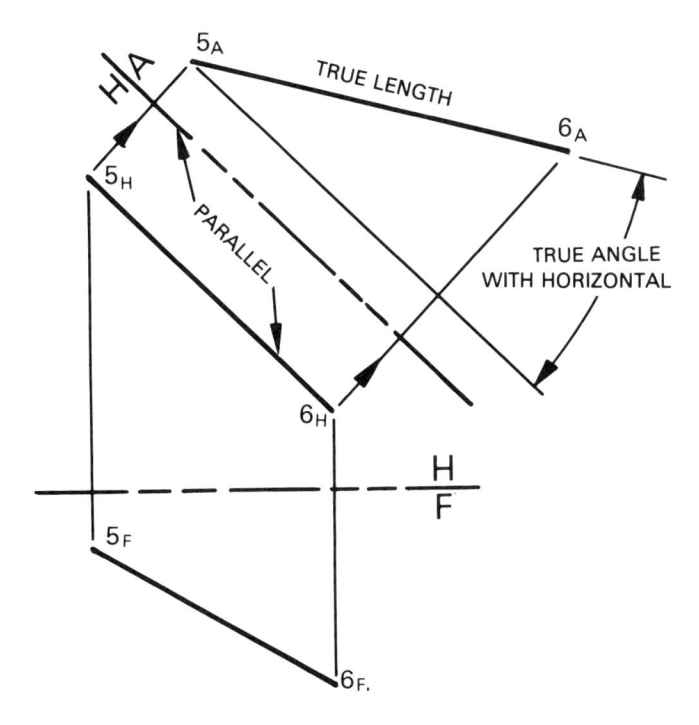

FIG. 4-97 Angle a line makes with the horizontal plane.

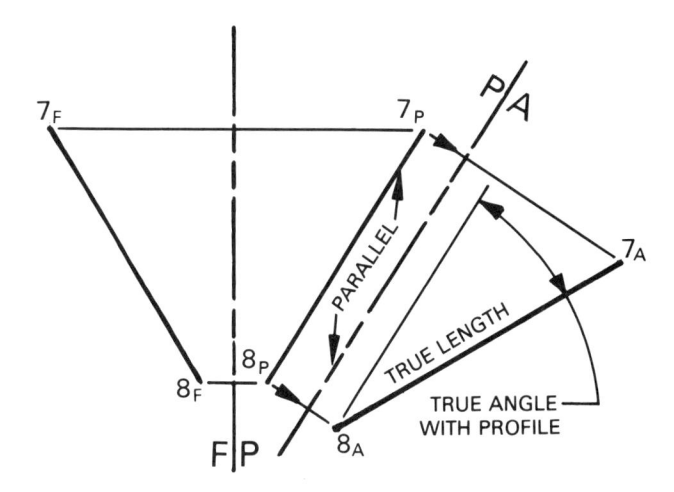

FIG. 4-99 Angle a line makes with the profile plane.

appears true length and the horizontal plane shows as an edge. The true angle between line 5_A-6_A and fold line H/A is the true angle between the line and the horizontal plane.

The true angle formed by oblique line 1-2 and the frontal plane in Fig. 4-98 can be measured in a primary auxiliary view that shows the line as true length and the frontal plane as an edge. In this example F/A is drawn parallel to line 1_F-2_F. Auxiliary view A shows line 1_A-2_A as true length and the frontal plane as an edge. Therefore the true angle between them is measured between line 1_A-2_A and the F/A fold line.

In Fig. 4-99, line 7-8 is oblique and the angle it makes with the profile plane is required. Auxiliary view A is projected parallel to line 7_P-8_P. The true angle can be measured in this view since line 7_A-8_A is true length and the profile plane appears as an edge.

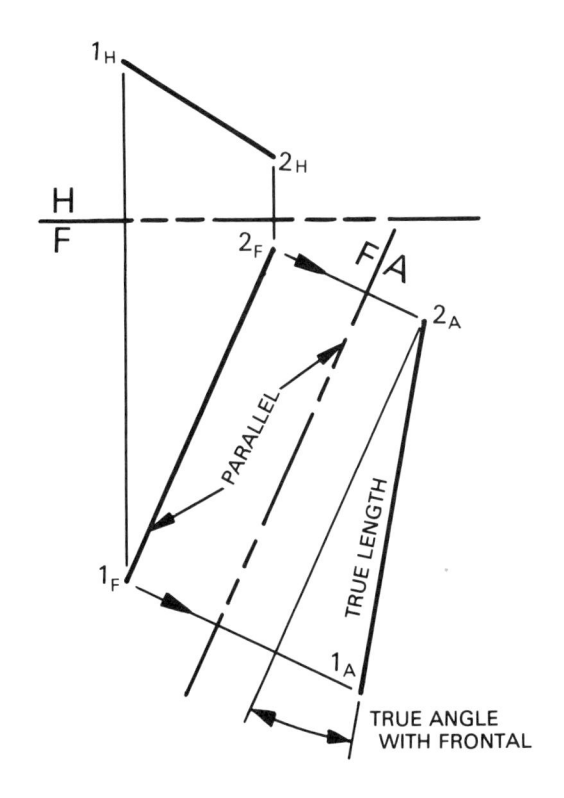

FIG. 4-98 Angle a line makes with the frontal plane.

PROB. 4-32

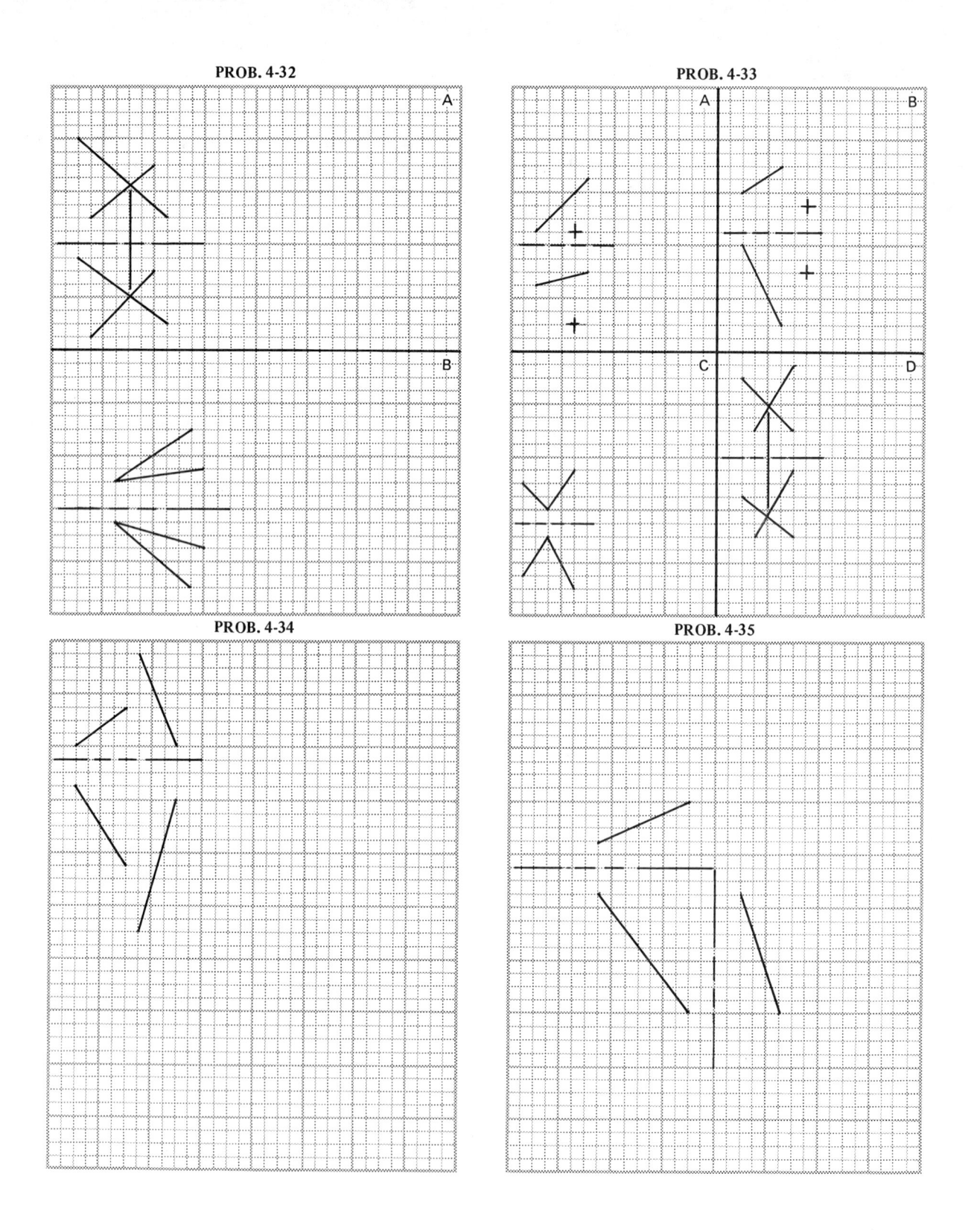

PROB. 4-33

PROB. 4-34

PROB. 4-35

PROBS. 4-32A, 4-32B. Find the true angle between the two intersecting lines.

PROB. 4-33A. Project the H and F views of a line that is through the given point and at a 60-degree angle with the given line. Project from the H view.

PROB. 4-33B. Project the H and F views of an equilateral triangle. One leg of the triangle lies on the given line. The triangle's vertex is at the given point.

PROBS. 4-33C, 4-33D. Project and measure the true angle formed by the two lines. Do the lines intersect in 4-33D?

PROB. 4-34. Solve for the angle between the nonintersecting lines.

PROB 4-35. Solve for the angle that the oblique line makes with each principal plane, and the true length of the line.

FIG. 4-100 Bearing of a line.

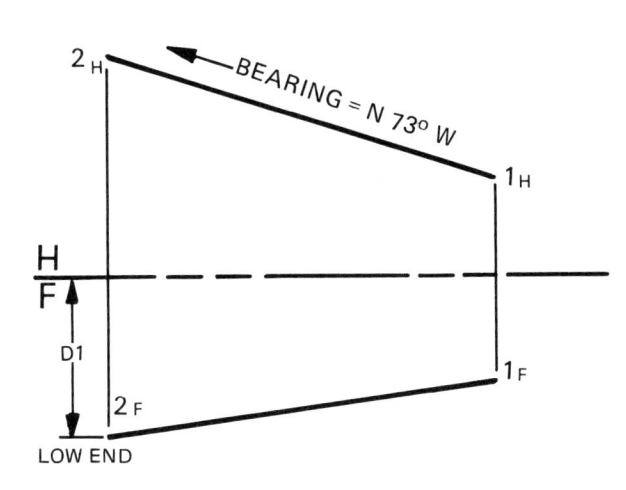

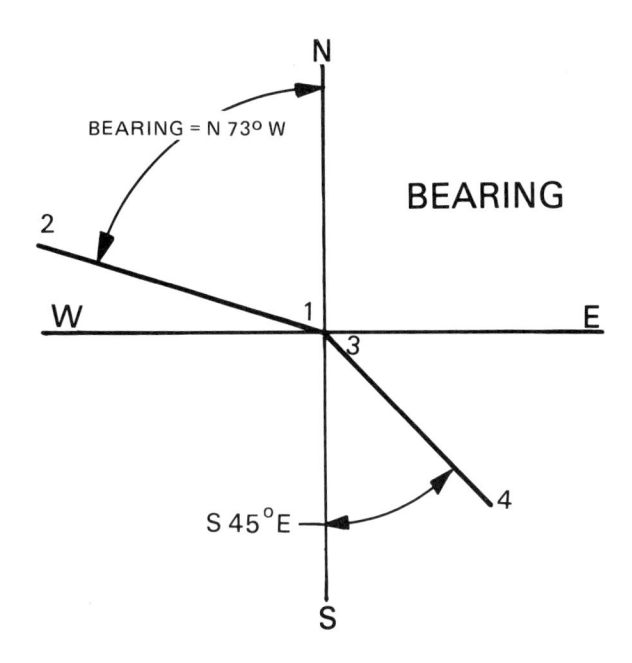

FIG. 4-102 Bearing measurements.

4.48 BEARING OF A LINE

*The angle that a line makes with a north-south line in the horizontal view is the **bearing** of that line.* The ***bearing*** can only be measured in the horizontal view. Since the bearing of a line is the angle that the line makes with the north-south meridian, it is measured from the north or south towards the east or west. *The bearing is the map direction of a line and measured in degrees with a compass from the north or south.* The bearing indicates the quadrant that the line lies in and is always measured from the north or south.

Normally the originating point is the lowest numerical value or sequential letter, such as line 1-2 (A-B), which will start at point 1 (A). The low end is the lowest point on a line as seen in a frontal or elevation view. In some cases, the bearing is measured from the high end of the line toward the low end as for a sloping cross-country pipeline.

Lines that fall on a meridian have a bearing of due north, due south, due east, or due west.

The horizontal angle that a line makes with the north-south line is the bearing and is always an acute angle measured from the north or south. In Fig. 4-100, line 1-2 has a bearing of N 73°W, measured from the north, 73° toward the west. The bearing is measured from the north towards the west, from point 1 towards point 2.

Figure 4-102 shows the horizontal view of line 1-2, located in relation to the compass meridians. Line 1-2 lies in the second quadrant. Therefore it is measured from the north towards the west.

In Fig. 4-101, the bearing of the pipeline, line 3-4, is S 45°E. This means that line 3_H-4_H forms a 45-degree angle with the north-south meridian and is measured from the south toward east. Here the concept of low end has been applied. The low end is always determined in the frontal view where the elevation of the line is shown. In Fig. 4-102 line 3-4 is located in relation to the meridians and lies in the fourth quadrant since it measured from the south towards the east.

The bearing of a line is used in engineering work to locate lines by compass directions. The bearing of a road, etc., would be measured on a map, normally from the north. Note that, in surveying, the concept of low end is useless, since the elevation may not be known or needed in regard to the bearing.

Chapter 8, Mining and Geology, discusses specific applications of the bearing of a line.

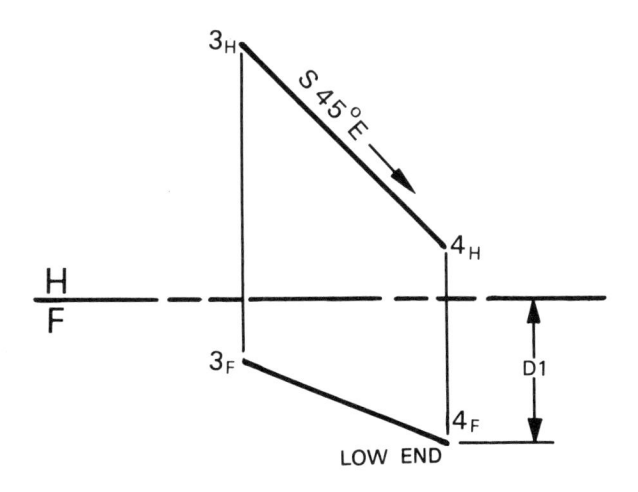

FIG. 4-101 Bearing.

FIG. 4-103 Azimuth readings.

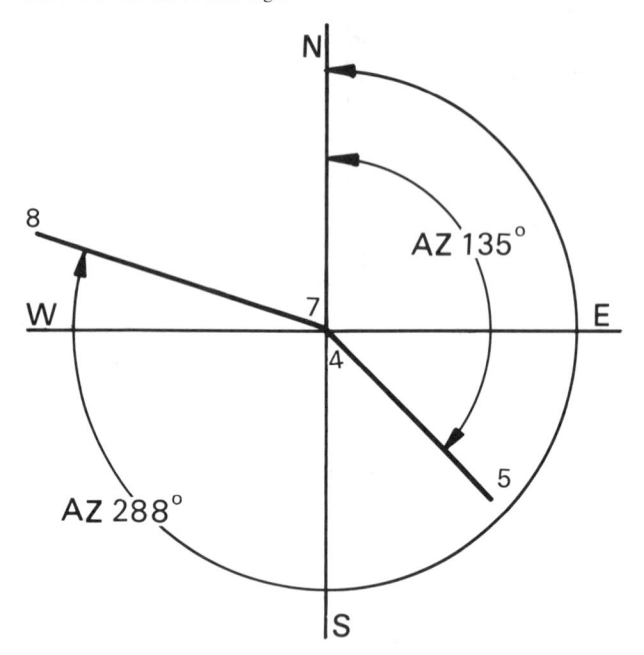

4.49 AZIMUTH OF A LINE

*The **azimuth bearing** of a line is the angle the line makes with the north-south meridian and is always measured from the north in a clockwise direction.* In Fig. 4-103, line 4-5 has an azimuth reading of 135° and line 7-8 has an azimuth of 288°. Note that the azimuth is always measured from the north and that the directions of the compass are not required. Both the azimuth bearing and the compass bearing are used in engineering and mapping work. The bearing for line 7-8 is N 72°W, and S 45°E for line 4-5. Measurements of azimuth or bearing are always taken in the horizontal view, since a compass direction will only show in the plan view and north can only be determined looking down on a map as in Fig. 4-104.

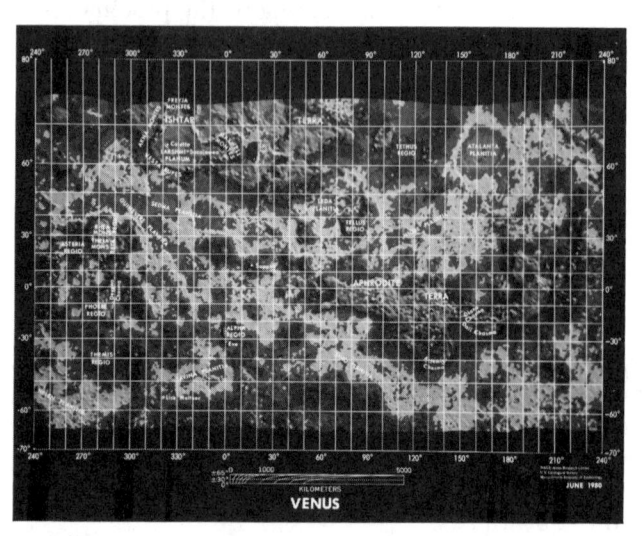

FIG. 4-104 Venus contour map. *(Courtesy NASA.)*

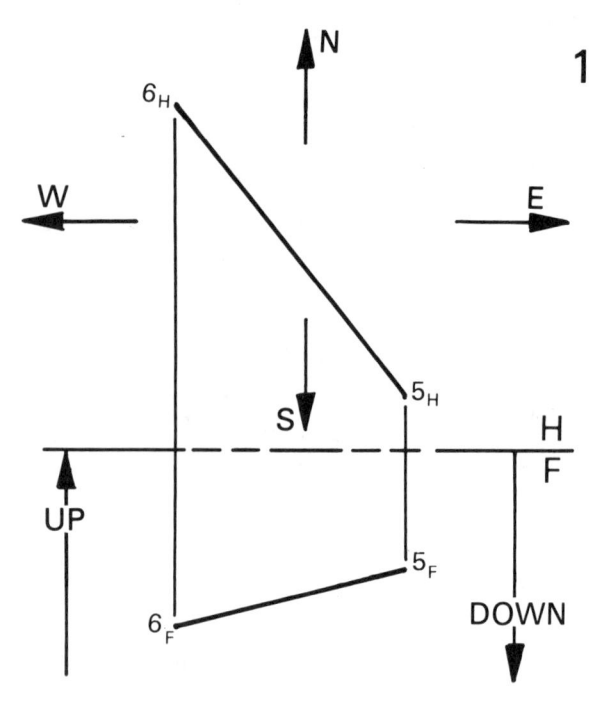

FIG. 4-105 Bearing N 38° W, Azimuth 322°.

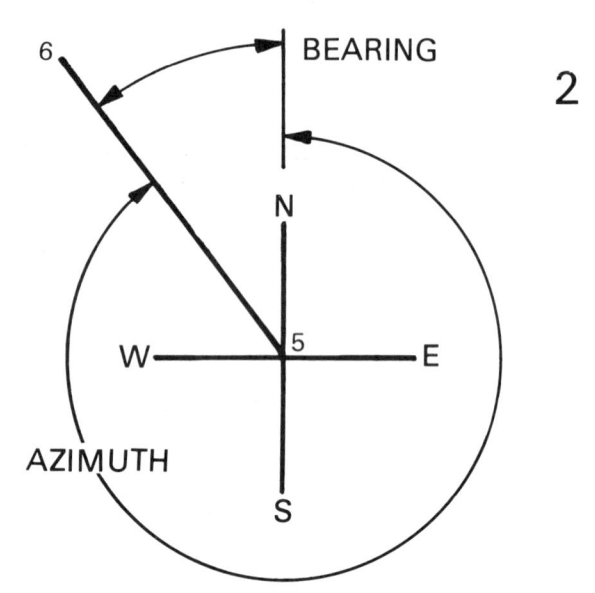

In Fig. 4-105, line 5-6 has a bearing of N 38°W and a corresponding azimuth of 322°. Normally a line's bearing is not affected by its elevation view, though for some applications such as the slope of a tunnel or angle of slope for a pipeline the low point determines the direction of bearing. In this figure the bearing of line 5-6 could be given as N 38°W or S 38°E, but since the low end or down side of the line is at point 6, the bearing was given from the north towards the west. This method of determining the direction of bearing is not accepted in all engineering fields but is used for portions of this text. Note the first point listed for a line could be assumed to be the starting point for the direction of the line instead of using the low side concept.

FIG. 4-106 Slope of a frontal line.

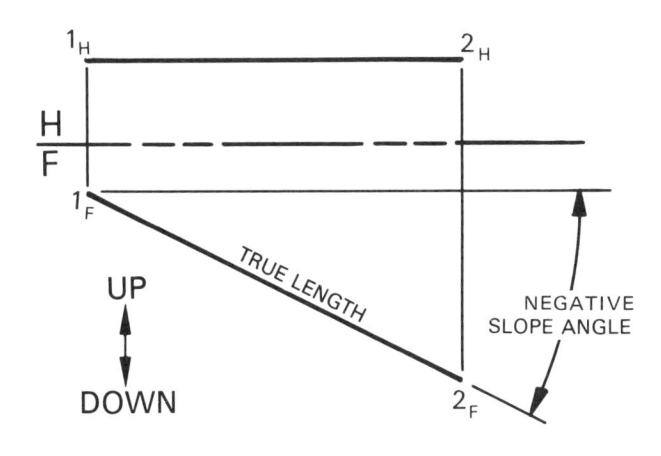

4.50 SLOPE OF A LINE

*The angle that a line (true length) makes with the horizontal plane is called the **slope** of a line.* Normally the slope of a line is given in degrees as a slope angle. The slope can only be measured in a view where the line is true length and the horizontal plane appears as an edge. Thus the slope is seen in a elevation view where the line is true length.

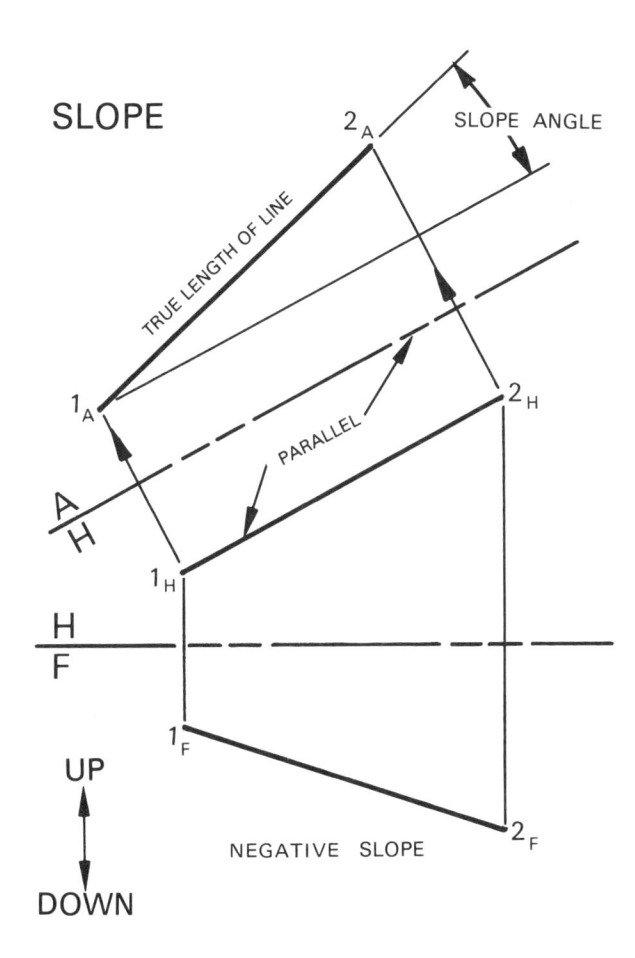

FIG. 4-107 Slope of an oblique line.

The slope cannot be determined in the horizontal view.

The slope of a frontal line is measured in the frontal (elevation) view since it is parallel to the frontal plane in the horizontal view and therefore shows as true length in the frontal view, Fig. 4-106. In this figure line 1-2 is a frontal line (true length in the frontal view). The slope angle is the angle formed by true length line 1_F-2_F and the H/F fold line. Since point 1_F is above point 2_F, the line slopes down; in other words it has a *negative slope* (-26 degrees). The bearing of line 1-2 is due east if the low end method is used. The bearing would also be due east if the first numerical value procedure was followed since the line slants down from point 1_F to 2_F. Note that the slope would be positive if the line originated at point 2_F and consequently sloped upwards.

The slope of a profile line is measured in the profile view. A horizontal line is not a slope since it is a level line and is parallel to the horizontal plane. To establish the slope of an oblique line, a primary auxiliary view must be projected from the horizontal view, parallel to the oblique line.

In Fig. 4-107, line 1-2 is oblique. To measure its slope, auxiliary view A is projected parallel to line 1_H-2_H. Draw fold line H/A parallel to line 1_H-2_H. Line 1_A-2_A appears true length in auxiliary view A, and the slope angle ($-16°$) is measured between the line and fold line H/A. Line 1-2 has a negative slope since it slants from point 1 downward toward point 2.

In Fig. 4-108 many of the sloping structural braces and angled elements were designed using concepts and practices from basic descriptive geometry.

FIG. 4-108 Model of five-degree-of-freedom simulator. *(Courtesy NASA.)*

FIG. 4-109 Grade of a line.

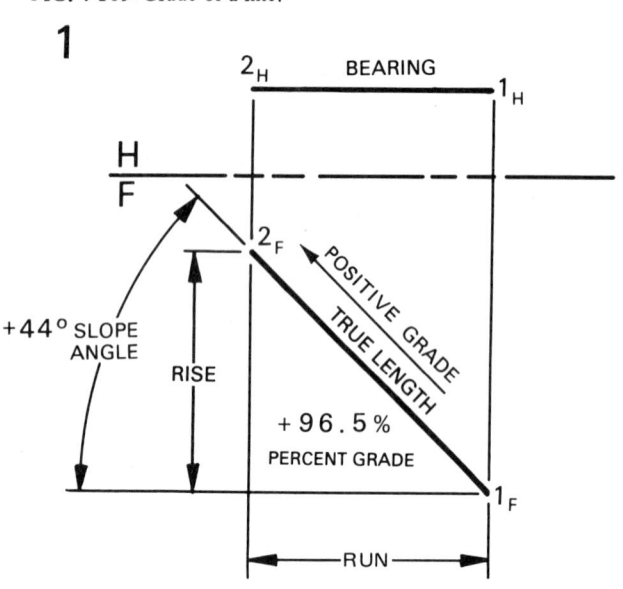

4.51 GRADE OF A LINE

Another way of stating a lines slope is to give the **grade of the line**. The **grade** or **percent grade** *is the ratio of its rise (vertical height) to its run (horizontal distance)*. The percent grade is calculated in a view where the line appears as true length and the horizontal plane is an edge.

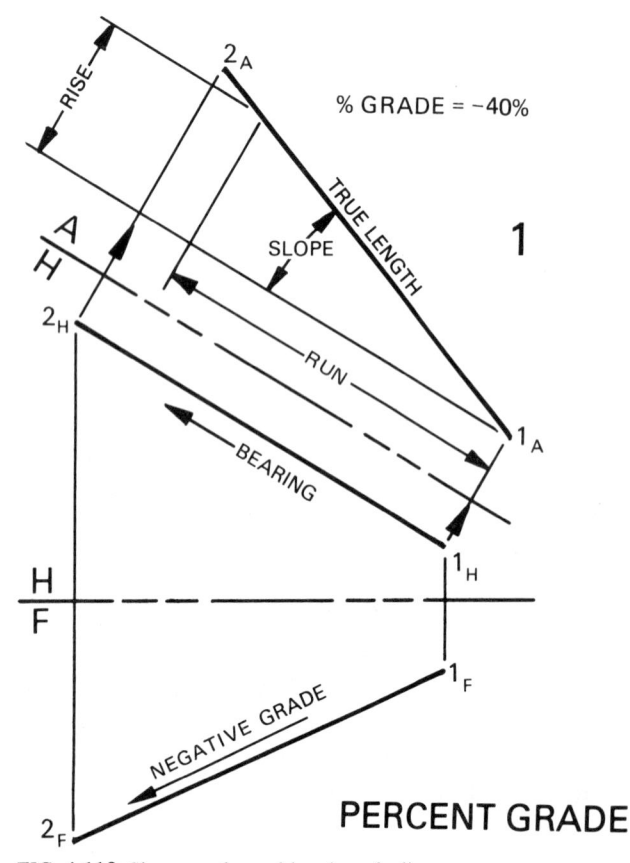

FIG. 4-110 Slope, grade, and bearing of a line.

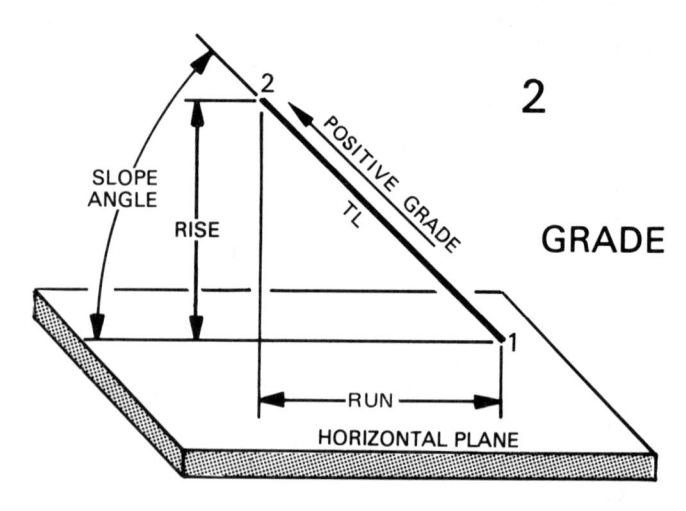

In Fig. 4-109, line 1-2 is a frontal line. The slope angle and grade can be calculated in the F view since the line is true length and the horizontal plane shows as an edge. Note that the percent grade can also be calculated by changing the tangent of the slope angle into a percent. In this figure, line 1-2 has a slope angle of 44 degrees.

The tangent of 44 degrees equals .9656. Multiply the tangent .9656 by 100 in order to convert it to a percent: .9656 × 100 = 96.56%. Line 1-2 has a +96.56% grade since it slopes upward from point 1. The bearing of line 1-2 would be due west if taken from point 1.

When calculating the percent grade using the ratio of rise to run, always use 100 units for the run and measure the rise with the same type of units. This method will yield the percent grade. In Fig. 4-110 line 1-2 is oblique. Auxiliary view A is projected parallel to line 1_H-2_H (1). Line 1_A-2_A is true length and the grade can be calculated in this view. In (2), line 1-2 has been drawn so as to illustrate this procedure better. Note that a true length diagram could have been used.

One hundred units are set off along the run and the rise has been measured at 40 units (the type of units is irrelevant). The percent grade equals 40 divided by 100 multiplied by 100 (40%). The grade of line 1-2 is −40% since it slopes downward from point 1. The tangent of the slope angle is equal to the percent grade divided by 100; −40% divided by 100 equals −.4. Converting tangent −.40 to an angle gives the slope angle of −21°48′.

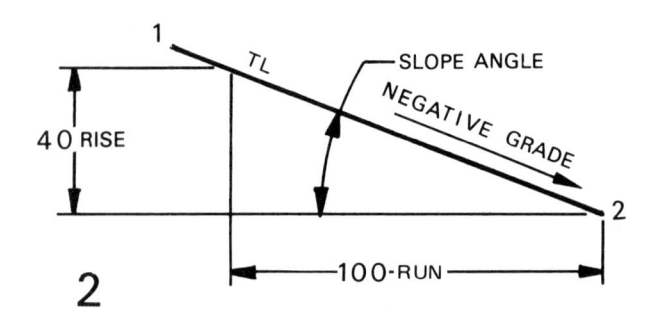

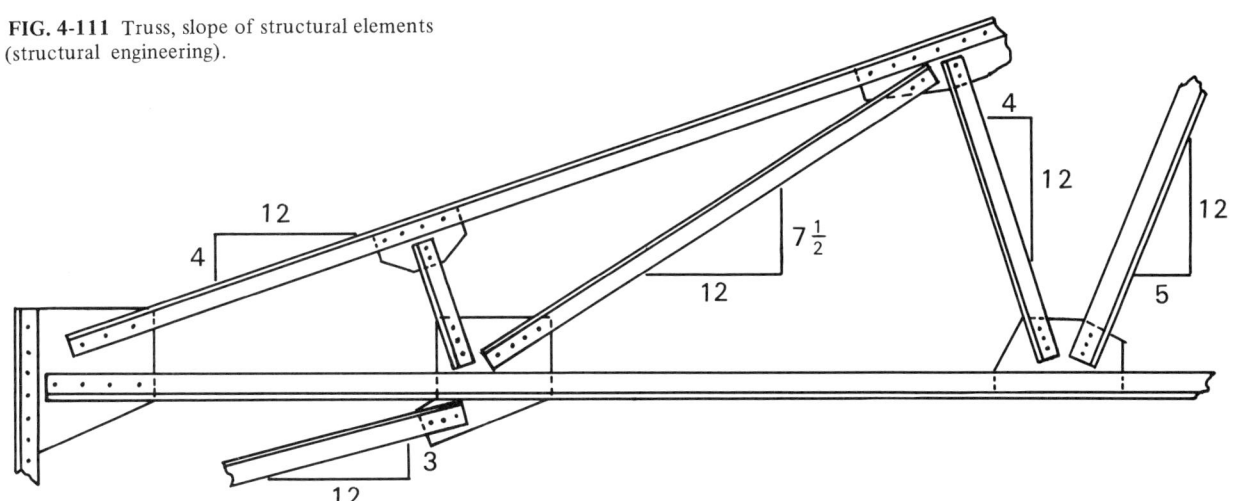

FIG. 4-111 Truss, slope of structural elements (structural engineering).

4.52 SLOPE DESIGNATIONS

The slope of a line can be noted in a variety of ways. The **slope ratio** (vertical rise over horizontal run) can be expressed as *percent grade*, a *fraction*, a *decimal*, or as a *slope angle*. Each engineering field has developed a specific procedure and name to designate the slope of a line as it pertains to a given aspect of their work. In structural engineering the angle of slope is called the *slope* or *bevel* of a structural member (beam, truss element) and is designated by a slope triangle as shown in Fig. 4-111. The longest leg of the slope triangle is always 12 units and the shorter one is measured in the same units and designated as in Fig. 4-111. For architectural projects the slope is designated as the ratio of rise to span (*run*), as in Fig. 4-112 where the roof pitch = rise/span (10/12 = 5/6, 5/10 = 1/2).

The slope angle of a highway, road, cross-country pipeline in civil engineering is normally indicated as the *grade*. The percent grade = rise/

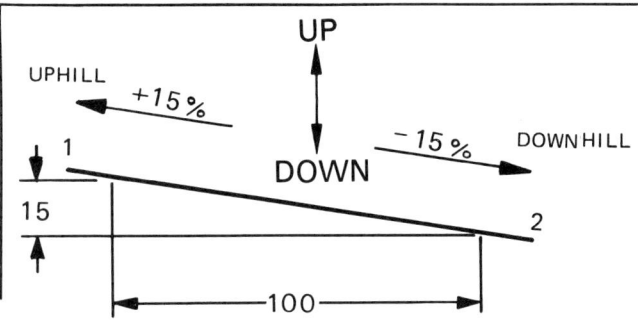

FIG. 4-113 Grade (slope) of roadway (civil engineering).

run × 100. In Fig. 4-113, the grade of the roadway is calculated by setting off 100 horizontal units and measuring the rise with the same units. Note that 100 is always used for the run (horizontal) units. In this figure the percent grade equals: grade = 15/100 × 100 = 15%. If the road sloped from point 1, downhill towards point 2, it would be −15% grade. If it sloped uphill from point 2 towards point 1, it would be +15% grade. The direction of slope is usually determined by which number (end) is designated first and which second. Chapter 8, Mining and Geology, explains slope designations for cut and fill of roads and provides examples of applications for slope angle and grade in geological, civil, and mining work.

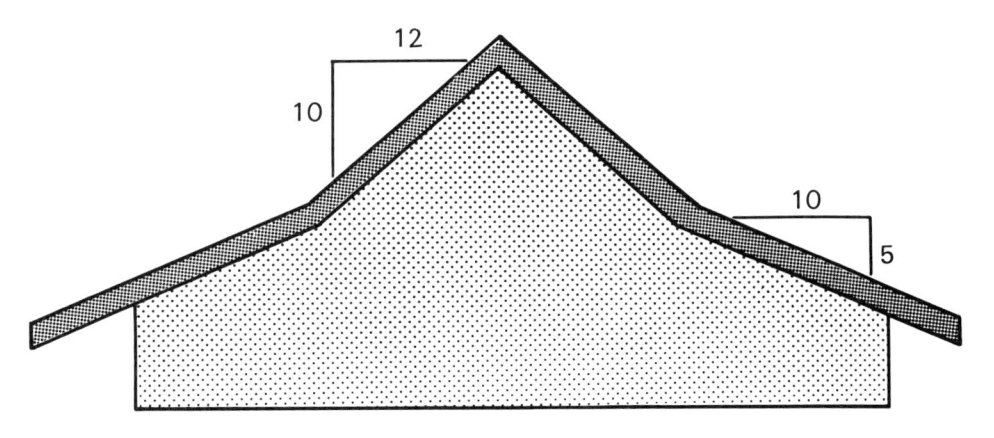

FIG. 4-112 Pitch (slope) of roof (architectural).

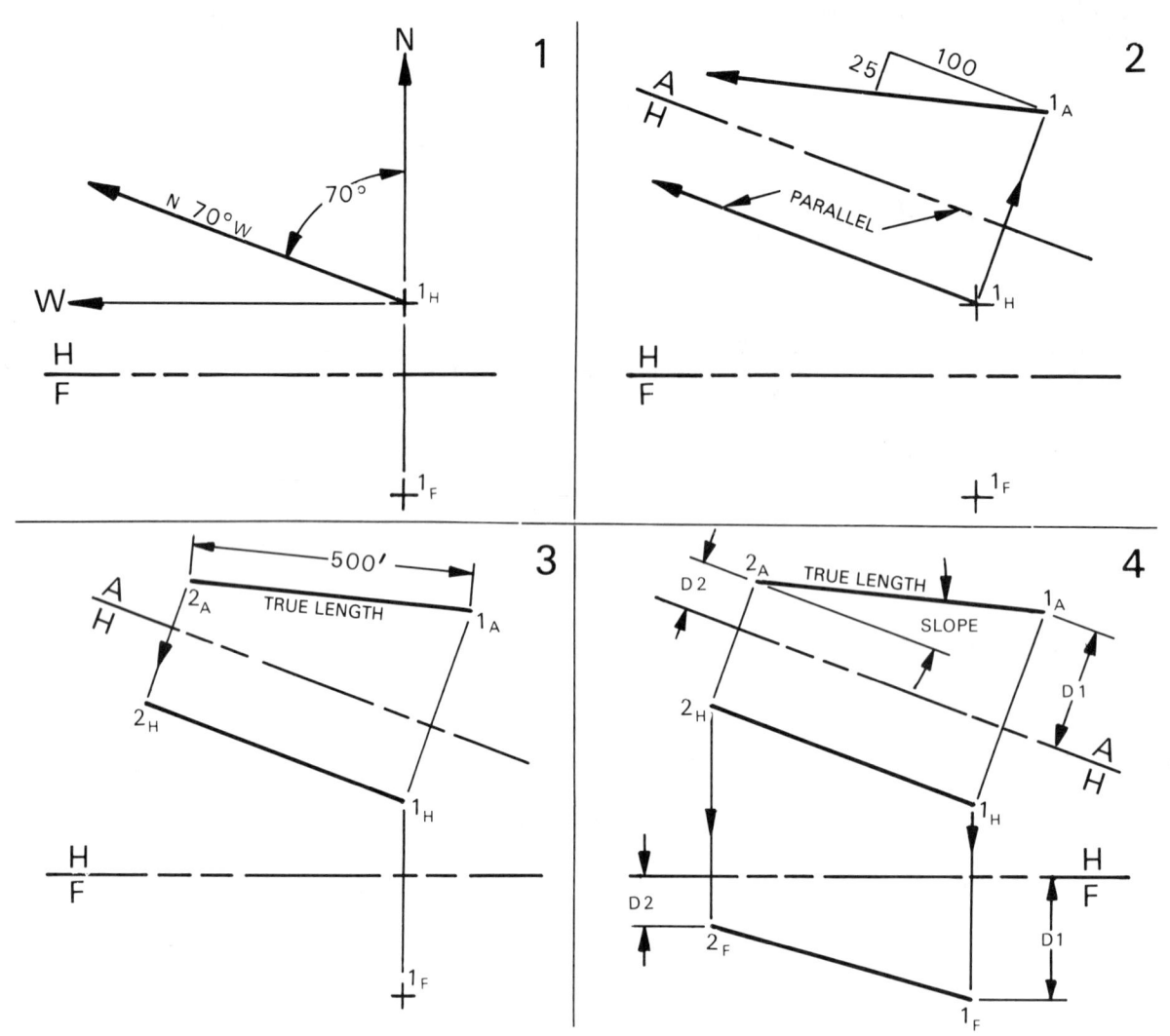

FIG. 4-114 Locating a line given the bearing, slope (grade), and true length.

4.53 TO DRAW A LINE GIVEN THE TRUE LENGTH, BEARING, AND SLOPE (GRADE)

A line can be located in space if its length, bearing, and slope (or grade) are known. The bearing of a line will fix the line's position in the horizontal view which can be drawn without regard to its true length. Since the slope and the grade of a line shows in a view where the line is true length and the horizontal plane appears as an edge, a primary auxiliary view projected from the horizontal view will fix the line in space. This auxiliary view must be projected parallel to the line which is established in the horizontal view by its bearing only. In the auxiliary view, the slope or grade of the line can be used to draw the line an indefinite length and the true length can then be established by measurement along the slope line. With both ends of the line fixed it can be projected back to the horizontal and frontal views.

The following steps describe the construction of line 1-2 in Fig. 4-114. Line 1-2 is 500 ft long, has a bearing of N 70°W, and an upward grade of +25% from point 1 to point 2. Note that the bearing in this problem is not orientated toward the low end of the line.

1. Establish and label point 1 in the frontal and horizontal view (1). Draw a line from point 1 having a bearing of N 70°W. Draw this line a convenient length.

2. Draw A/H parallel to the bearing line and project point 1 in auxiliary view A (2). Draw a construction line from point 1_A parallel to H/A and lay off 100 units for the run and 25 units for the rise as shown. The rise is perpendicular to the run and extends toward the H/A fold line since the line has a positive grade. Draw the line from point 1_A an indefinite length and touching the 25 unit rise. This fixes the grade and slope angle of the line.

3. Measure off 500 ft along the line from point 1_A and label the other end point 2_A (3). Locate point 2_H in the horizontal view by projection.

4. Locate point 2_F in the frontal view by projection and transferring dimension D2 from auxiliary A (4). Connect the two points to complete the frontal view.

118

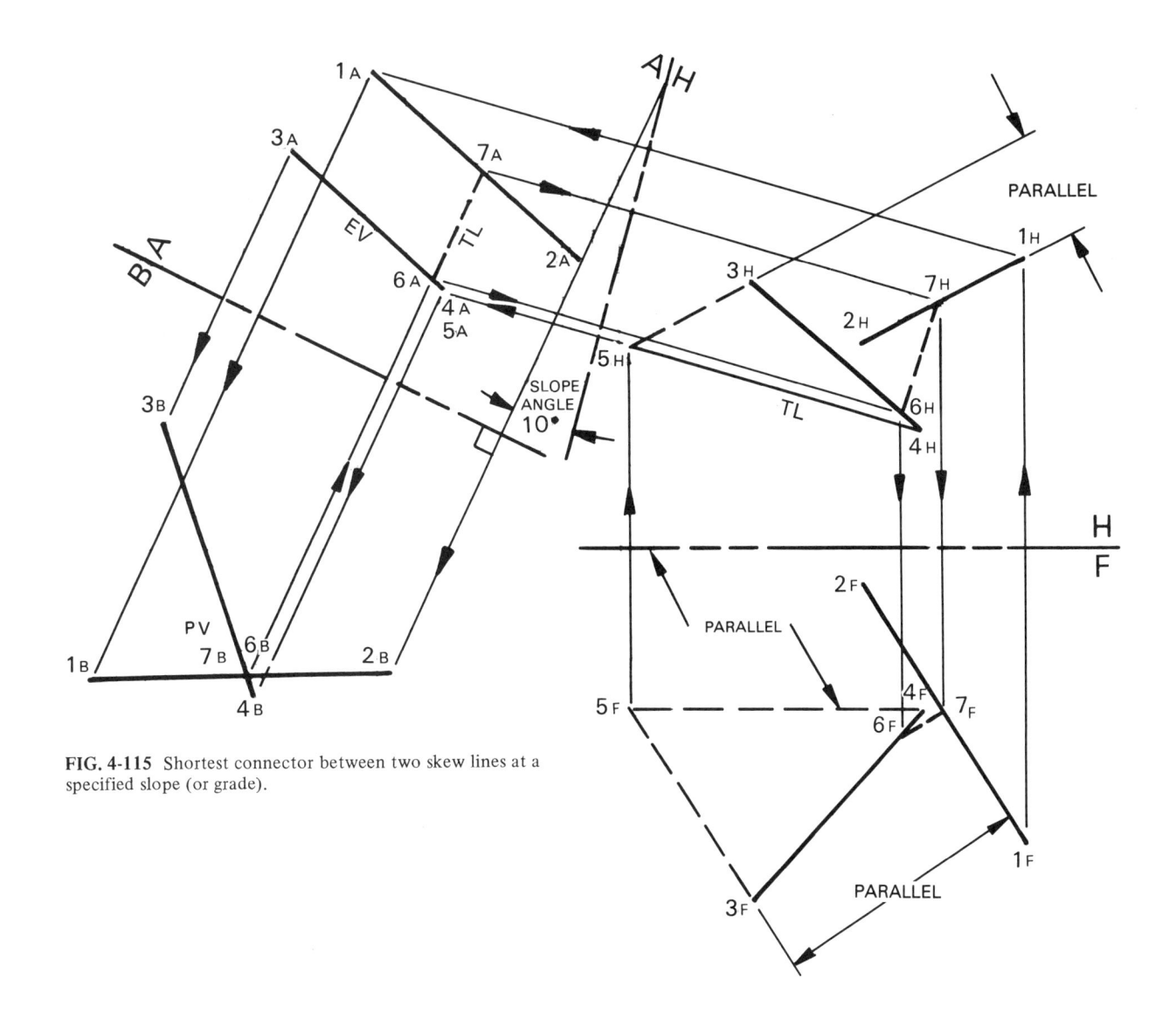

FIG. 4-115 Shortest connector between two skew lines at a specified slope (or grade).

4.54 SHORTEST LINE OF A GIVEN SLOPE (OR GRADE) BETWEEN TWO SKEW LINES

The shortest connector of a given slope (or grade) between two lines is a typical problem encountered in industry when trying to connect two pipes or tunnels at a specific angle. To solve for this connector it is necessary to use the "plane method" as explained in section 4.40 where the procedure for finding the shortest distance between two skew lines is explained. Note that the "line method" will not work, and that the primary auxiliary must be projected from the *horizontal view* not the frontal.

In Fig. 4-115, lines 1-2 and 3-4 are given. The shortest connector with a slope of 10° is required. The connector will slope upward (+10°) from line 3-4 to 1-2. The direction of the slope must be indicated and its originating line given, since the connector could slope at 10 degrees up or down from line 1-2 or from 3-4. Note that the grade could be used instead of the slope.

1. Form a plane with one line and parallel to the other. Line 4_F-5_F is drawn parallel to H/F, and line 3_F-5_F is parallel to line 1_F-2_F. The horizontal projection of this plane shows line 3_H-5_H parallel to line 1_H-2_H and line 4_H-5_H true length.

2. Auxiliary view A is projected perpendicular to true length line 4_H-5_H. H/A is drawn perpendicular to line 4_H-5_H and the view completed. Line 1_A-2_A is parallel to the edge view of plane 3_A-4_A-5_A in this view.

3. The slope angle makes an angle of 10° with the H/A fold line, slanting upward from line 3_A-4_A towards 1_A-2_A. Note that the rise and run of the grade could also be used. The run would be set off along the H/A fold line with rise perpendicular to the run (and fold line). The true length of the shortest connector with a given slope can be measured in this view, but another auxiliary projection is necessary to locate its exact position.

4. Fold line A/B is drawn perpendicular to the slope direction and auxiliary B projected perpendicular to the slope angle line. The point where line 3_B-4_B crosses line 1_B-2_B is a point view of the required connector (6_B-7_B).

5. Connector 6-7 is located in all other views by projection. In auxiliary A it shows as true length.

PROB. 4-36

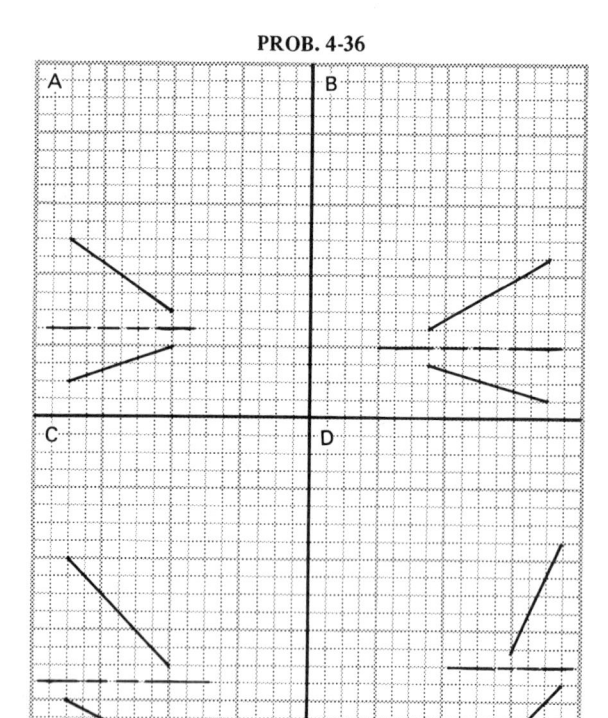

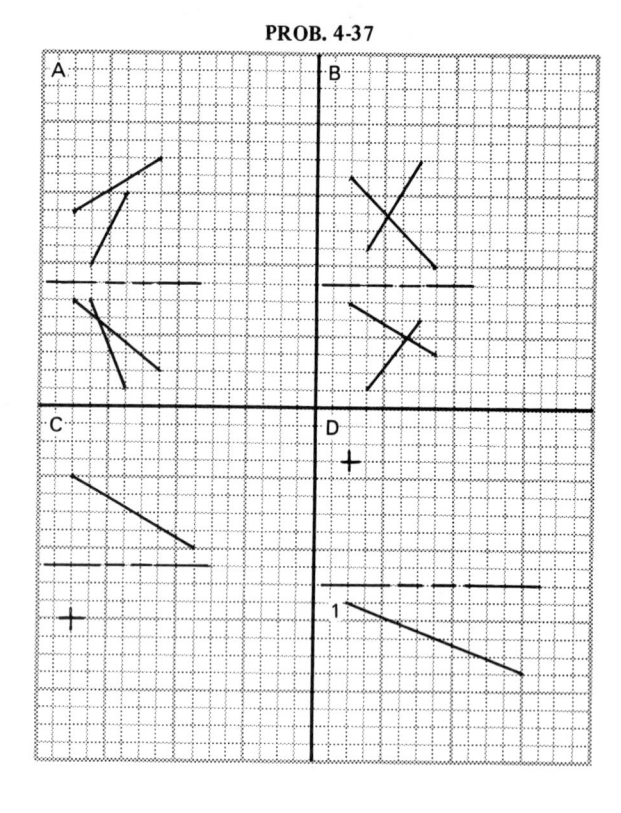

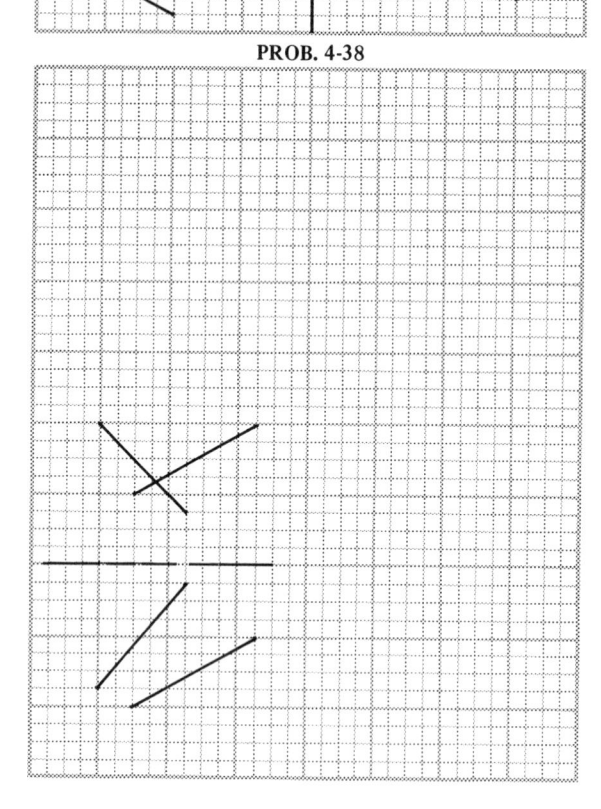

PROB. 4-37

PROB. 4-38

PROBS. 4-36A, 4-36B, 4-36C, 4-36D. For each of the problems solve for the bearing, azimuth, slope angle, true length, and percent grade (positive or negative).

PROB. 4-37A. Project the shortest line between the skew lines that has a grade of −25%. Show in all views.

PROB. 4-37B. Project the shortest line that connects the two skew lines and has a grade of +15%.

PROB. 4-37C. Draw the frontal view of the pipeline that has a grade of −30%. What is the length (scale to be assigned by instructor) and the bearing of the pipeline?

PROB. 4-37D. The Red Baron, limping home on a wing and a prayer, approaches the runway along the glide path 1-2 that is shown in the frontal view (point 1 is shown in the H view). The azimuth of the glide path is 120°. What is the slope angle and the length of the glide path? Scale: 1″ = 5000 ft. Metric units can be substituted.

PROB. 4-38. Determine the shortest line between the skewed lines that has a slope of 35°. Show in all views. What is the length and bearing of this line? Scale 1″ = 50 ft. Instructor can assign metric units.

EXTRA ASSIGNMENTS

The instructor can assign extra problems or a quiz assignment by requiring the construction of a line given its true length, slope (or grade), and bearing. The possibilities are unlimited for this type of problem.

FIG. 4-116 Rotating cylinder flap. *(Courtesy NASA.)*

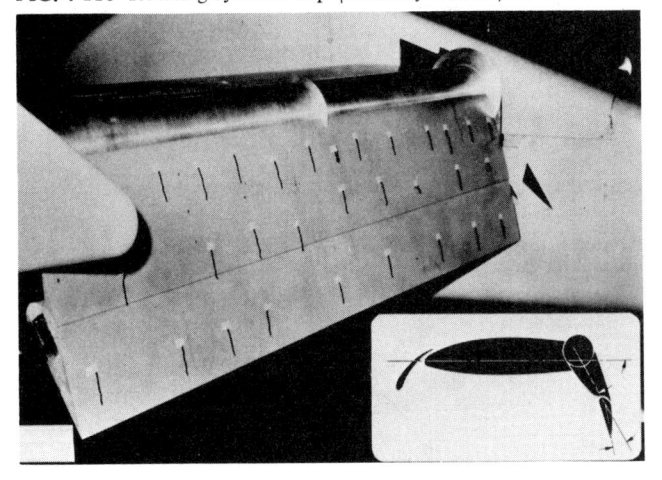

FIG. 4-118 Telescope satellite. *(Courtesy NASA.)*

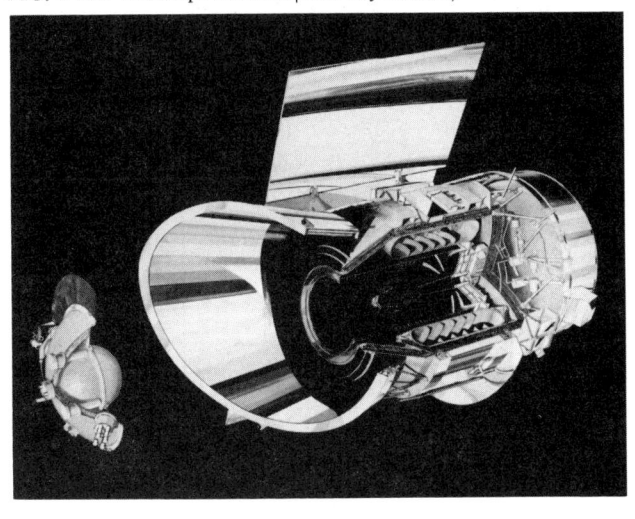

4.55 REVOLUTION/ROTATION

Descriptive geometry and orthographic projection are normally associated with the projection of an object or form onto a projection plane. The view of each projection plane assumes that the object is stationary and that the person drawing the object observe the object from a different vantage point for each view. In other words, the viewer changes position and the object is stationary. *Revolution/Rotation requires that the object be revolved or rotated and the observer remain stationary.* This procedure allows for fewer views to accomplish a specific task, though it may clutter or crowd the given views.

Another use for revolution is to show the rotated position of a mechanical device as in all figures on this page. The clearance between a fixed point and a rotating form can also be solved for by using revolution. Whatever the specific use of revolution, the object, line, or point to be revolved must be rotated about an established axis line. The following pages cover the basic theory of revolution and its application to points and lines. In Chapter 5, Planes, revolution is presented as it pertains to planes, and in Chapter 7, Developments, revolution is applied to specific developmental applications.

FIG. 4-117 Rotating flight hardware for satellite. *(Courtesy NASA.)*

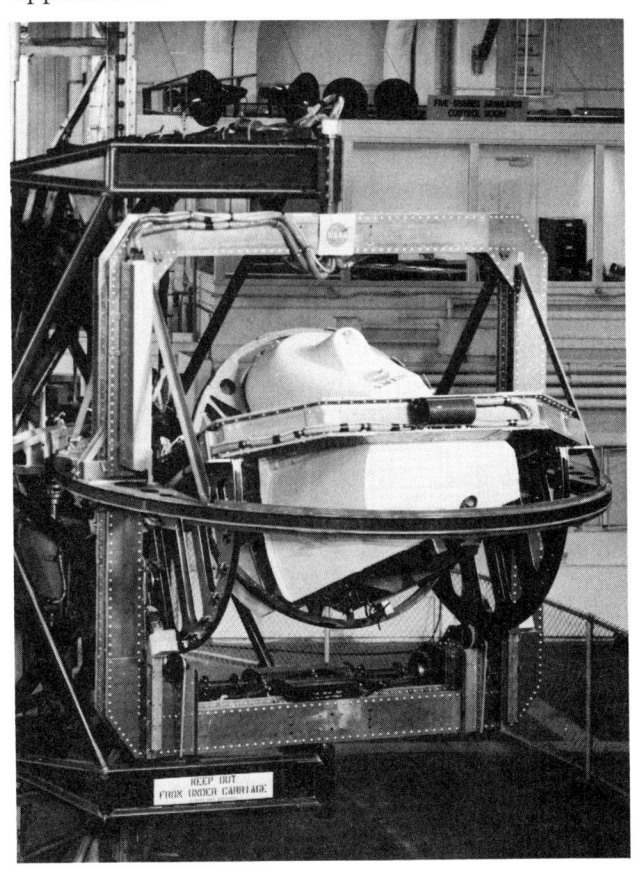

FIG. 4-119 Five-axis reentry simulator. *(Courtesy NASA.)*

FIG. 4-120 Revolution of a point around a vertical axis.

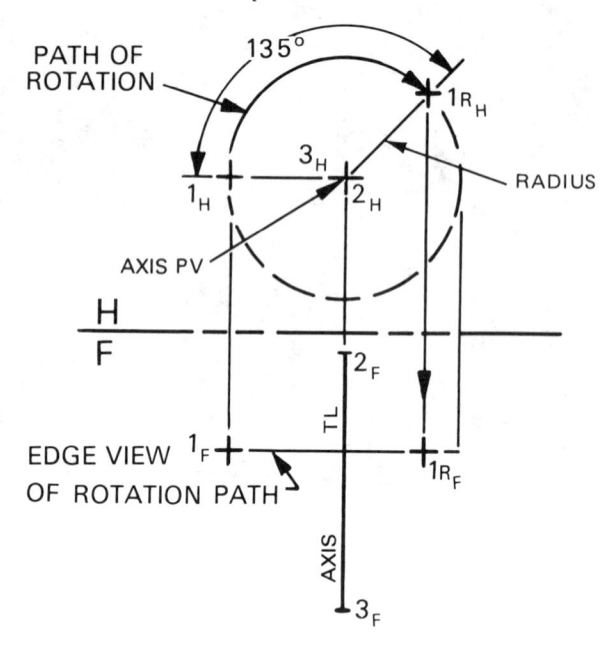

FIG. 4-121 Revolution of a point around a horizontal axis.

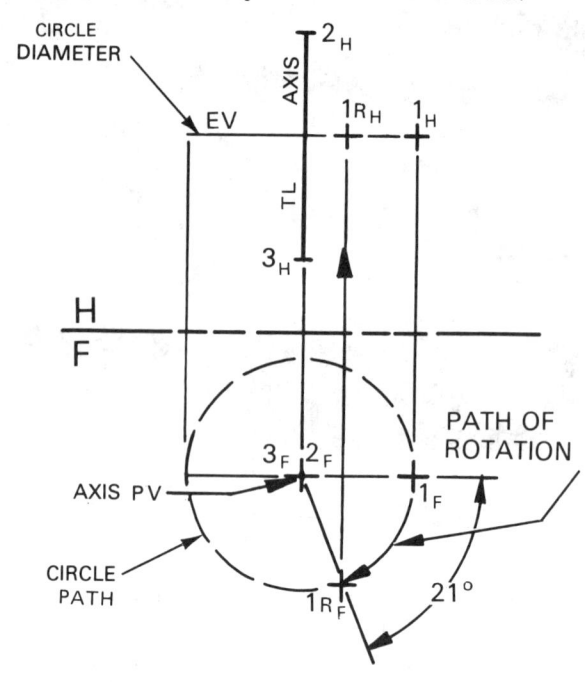

4.56 REVOLUTION OF A POINT

For all problems using revolution the observer remains stationary and the object (point) is rotated (revolved) about a straight line axis. Each revolved point moves in a circular path of rotation perpendicular to the axis line. Revolution and regular orthographic projection can be combined to solve a variety of engineering problems such as clearances between moving machine parts, or for quickly solving numerous design problems using as few views as possible.

The following principles apply to the revolution of a point, and since all objects, lines, and planes are composed of points these principles form the theoretical foundation for all revolution problems.

1. The *axis of revolution (rotation)* is always a straight line and must be established before a point can be revolved. The axis is a *point view* where the path of rotation is a circle, and appears *true length* where the path of rotation is an edge.
2. The *revolution of a point* is always *perpendicular* to the axis and moves in a circular path around the point view of the axis line. This circular path forms a plane perpendicular to the axis, which appears as a circle (or portion of a circle) when the axis is a point view and as an edge where the axis is true length.
3. The *path of rotation* is formed by the revolving point and a line connected from it to the axis. This line is the radius of the circle (or arc) formed by the revolution. When the axis shows as true length, the path of rotation appears as an edge with a length equal to the diameter of the circle.

In Fig. 4-120, point 1 is revolved around vertical axis line 2_H-3_H. Axis 2-3 is true length in the frontal view and is a point view in the horizontal view. Point 1_H is revolved clockwise 135° to position $1R_H$. The path of rotation is an edge view in the frontal view where the axis is true length and is a circular path in the horizontal view.

In Fig. 4-121, point 1 is revolved clockwise 21° about horizontal axis line 2_F-3_F. The axis is true length in the horizontal view and is a point view in the frontal view. The path of rotation created by rotating point 1_F to its revolved position $1R_F$ forms a circular path in the frontal view and appears as an edge in the horizontal view.

In Fig. 4-122, the helicopter rotor test shows a stress break; the rotor blade revolves in one plane about a vertical axis.

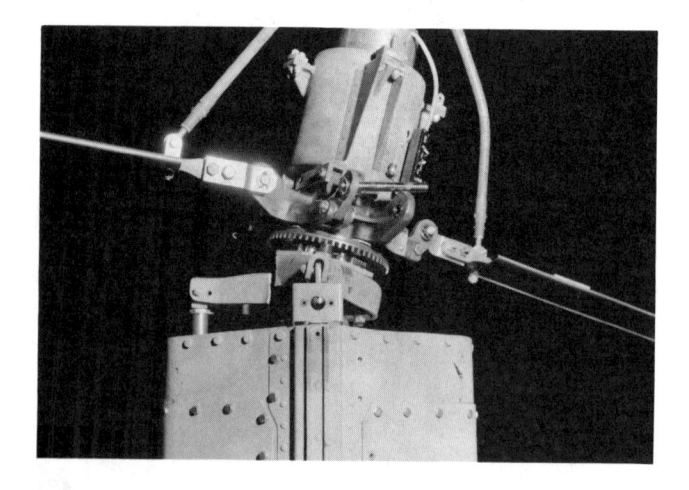

FIG. 4-122 Rotor stress test. *(Courtesy NASA.)*

FIG. 4-124 Revolution of a point about an oblique axis.

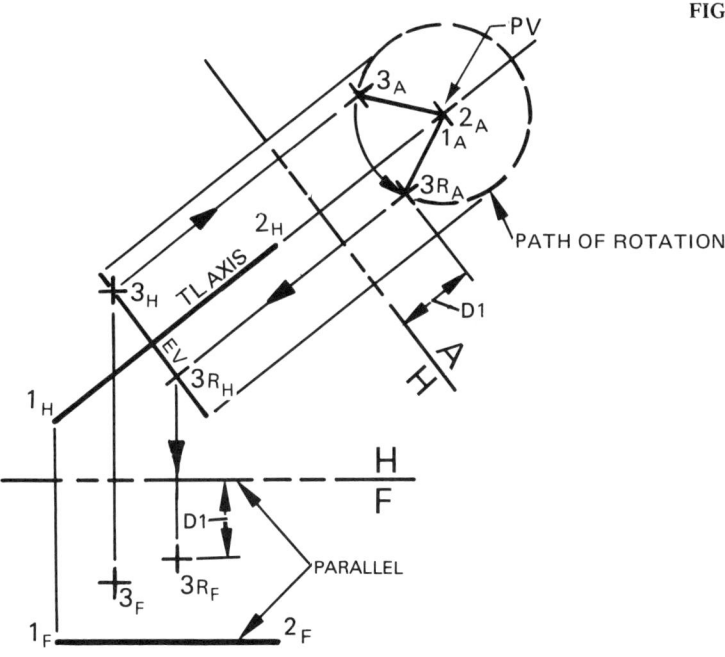

FIG. 4-123 Revolution of a point about a horizontal line.

4.57 REVOLUTION OF A POINT ABOUT AN OBLIQUE AXIS

When a point is revolved about an axis that does not appear as a point in the frontal or horizontal views, an auxiliary projection is required where the axis appears as a point view. In Fig. 4-123, point 3 is to be revolved about line 1-2, which does not project as a point view in the frontal or horizontal views; therefore the path of rotation would appear as an ellipse in the frontal projection. In order to revolve point 3 about horizontal line 1-2, an auxiliary view is projected perpendicular to the true length of axis 1_H-2_H. Fold line H/A is drawn perpendicular to line 1_H-2_H, and axis 1_A-2_A shows as a point view in this primary auxiliary view. Point 3_A is revolved to position $3R_A$ in this view. The path of rotation generated by moving the point creates a circular plane in this view. Point $3R_A$ can be located in the horizontal plane by simple projection since it falls on the edge view of the path of rotation. The frontal position of point 3R is located by transferring D1 from auxiliary A to the frontal view along its projection line.

In Fig. 4-124, point 3 is to be revolved 180 degrees about oblique line 1-2.

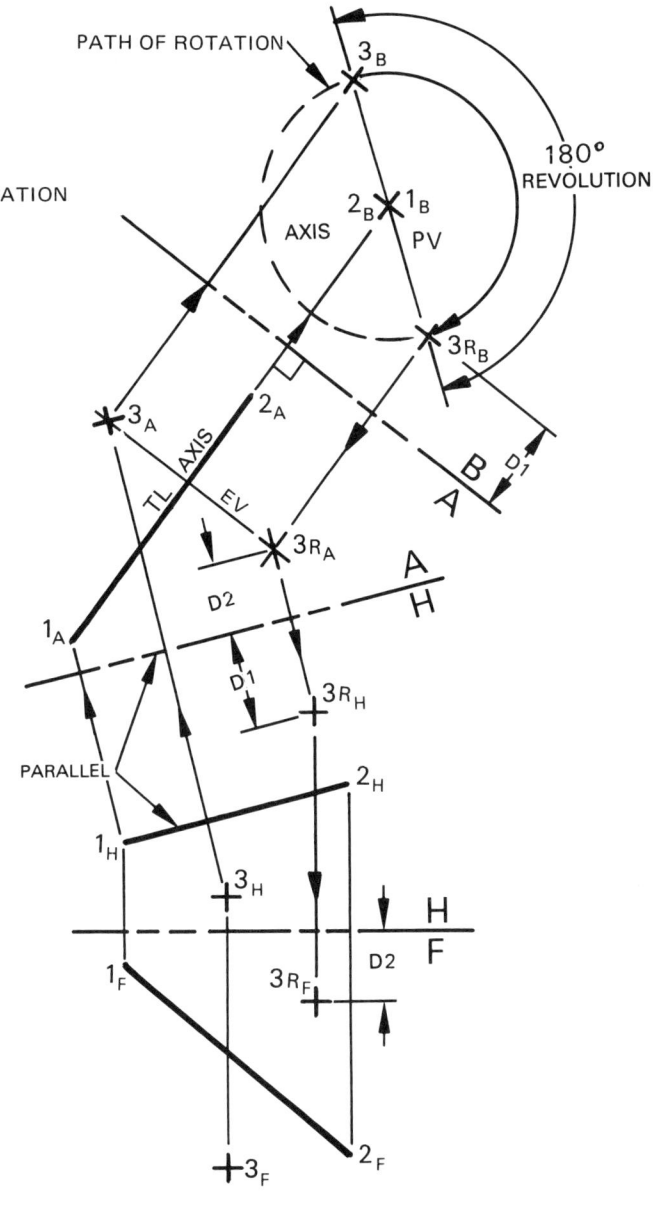

1. Draw H/A parallel to line 1_H-2_H and project auxiliary view A. Axis line 1_A-2_A is true length.
2. Draw A/B perpendicular to the true length axis line 1_A-2_A and complete auxiliary view B. Axis line 1_B-2_B is a point view.
3. In auxiliary view B revolve point 3_B 180° about axis 1_B-2_B to position $3R_B$.
4. Locate point $3R_A$ in auxiliary A by projection, where it falls on the edge view of the path of rotation. The horizontal view of point $3R_H$ is found by transferring D1 from auxiliary B along its projection line. The location of point $3R_F$ is established by drawing its projection line and transferring D2 from auxiliary A to the frontal view.

The highest and lowest location of point 3 as it is revolved about the axis can be located in the frontal view at the extreme points of the ellipse formed by the path of rotation.

FIG. 4-125 True length of line by auxiliary view (1) and by revolution (2).

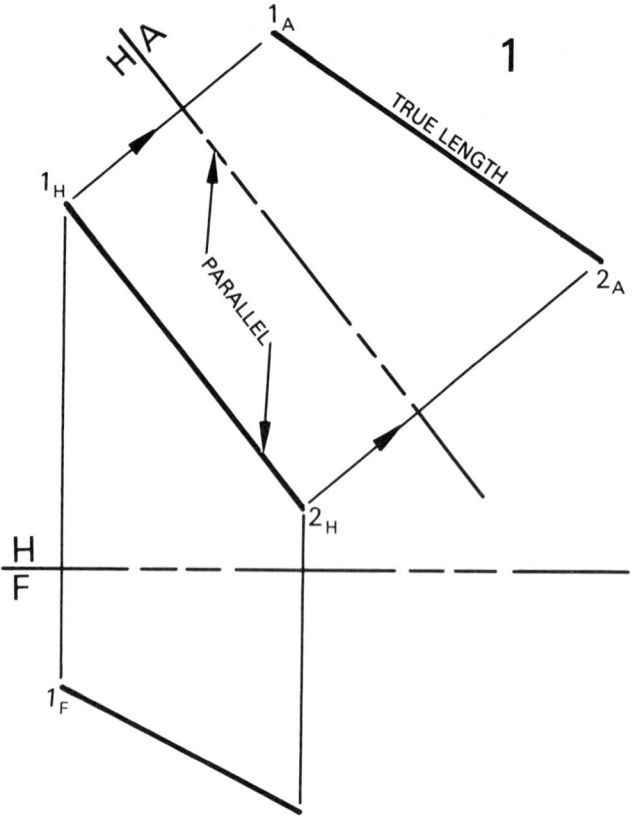

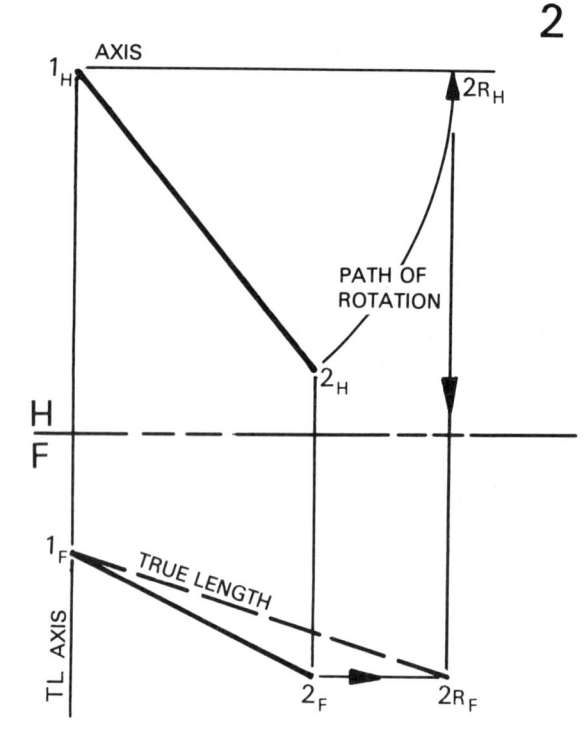

4.58 REVOLUTION OF A LINE

A line can be revolved in the same manner as a point. The axis must be established before a line can be revolved. The axis can be through the end point of a line, a point on the line, or a point off the line. In the first case the line revolves about a single end point and generates a cone, as in Fig. 4-126 where the reentry recovery rotor revolves about one end and generates a 60-degree cone. In the second case both ends revolve (generating two cones). Where the axis is independent of the line, the whole line revolves and changes position. Each end point remains in the same plane created by their paths of rotation.

One of the most common uses of revolution is to find the true length of an oblique line without the use of an extra view. Since *revolution changes the position of a line but not its length* it is possible to revolve an oblique line so that it is parallel to the adjacent projection plane. The line projects as true length in the adjacent view. The axis can be located through an end point or on the line and revolved in either given view. The axis line is a point view in one view and assumed true length in the other, though it need not have a specific length.

Figure 4-125 compares the auxiliary view method (1) with the revolution method (2). The auxiliary view method (1) requires the projection of a new view. In (2) point 2_H is revolved about a vertical axis line located through point 1_H until it is parallel to H/F in position $2R_H$. Point 2 is located in the frontal view by simple projection since it falls on the edge view of the path of rotation. In other words, a projection line can be drawn from $2R_H$ into the frontal view. A construction line (which is really a portion of the edge view of the path of rotation) is drawn parallel to H/F until it intersects the projection line and locates $2R_F$. Note that the revolution of point 2 changes its position in the horizontal and frontal view but does not alter its elevation since it must remain in a plane perpendicular to the true length axis. Line 1_F-$2R_F$ is true length.

FIG. 4-126 Rotor testing with 60 coning. *(Courtesy NASA.)*

FIG. 4-127 True length in horizontal view by revolution.

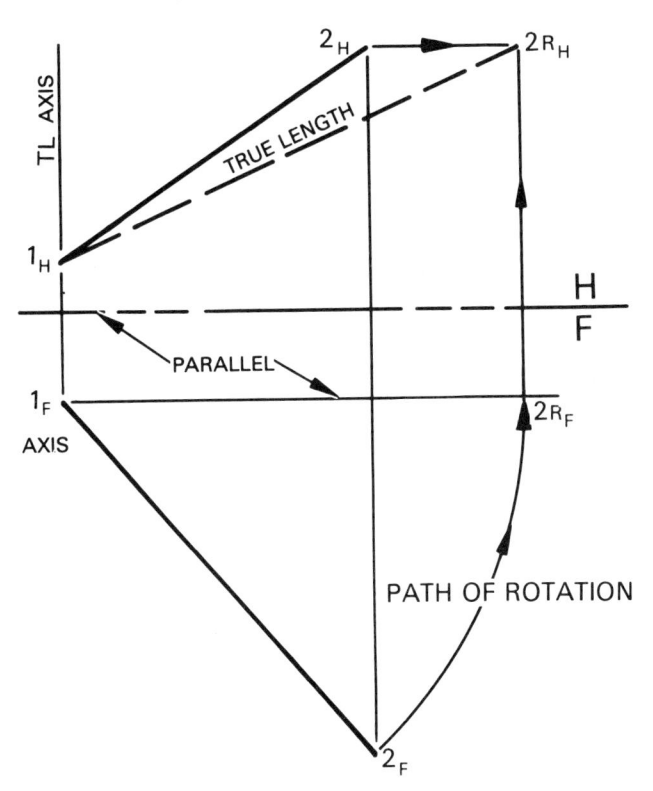

FIG. 4-128 Electrical substation. *(Courtesy NASA.)*

4.59 TRUE LENGTH OF A LINE IN THE HORIZONTAL VIEW BY REVOLUTION

To establish the true length of an oblique line in the horizontal view, the axis of revolution must be parallel to the horizontal plane. In Fig. 4-127, point 2_F is rotated about point 1_F. The axis has been located parallel to the horizontal plane and is a point view in the frontal view where it passes through point 1_F. Point 2_F is rotated about the axis to position $2R_F$ until line 1_F-$2R_F$ lies parallel to the horizontal projection plane (and therefore H/F). Point $2R_F$ is projected to the horizontal view and located by drawing a construction line from 2_H perpendicular to the true length axis. Point $2R_H$ is on the edge view of the path of rotation.

The true lengths of the structural members that form the double bus structure of the electrical substation in Fig. 4-128 could have been found by revolution of a line.

In Fig. 4-129, line 1-2 is revolved so that it is true length in the horizontal view. Note that it is revolved around point 1_F and again about point 2_F. Both axes A and B are parallel to the horizontal plane and are point views in the frontal view. Line 1_F-$2R_F$ and 2_F-$1R_F$ are rotated about their axis points until parallel to H/F. The horizontal view shows the line true length in two places.

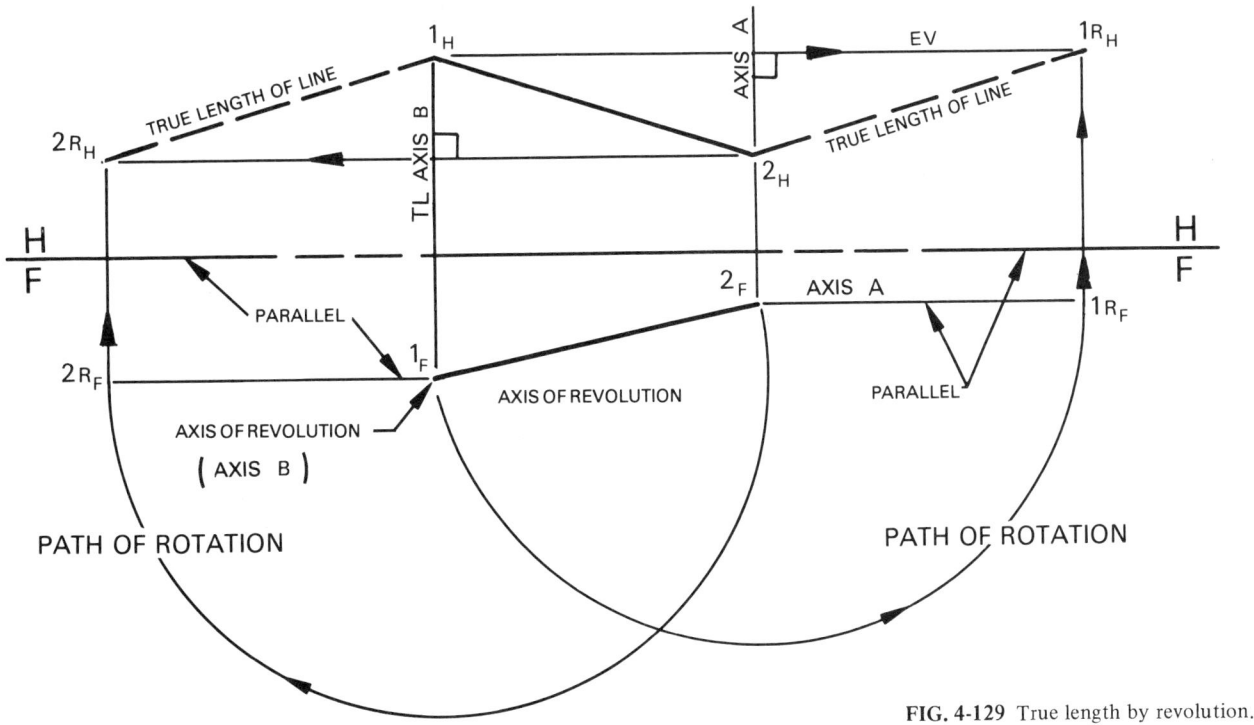

FIG. 4-129 True length by revolution.

125

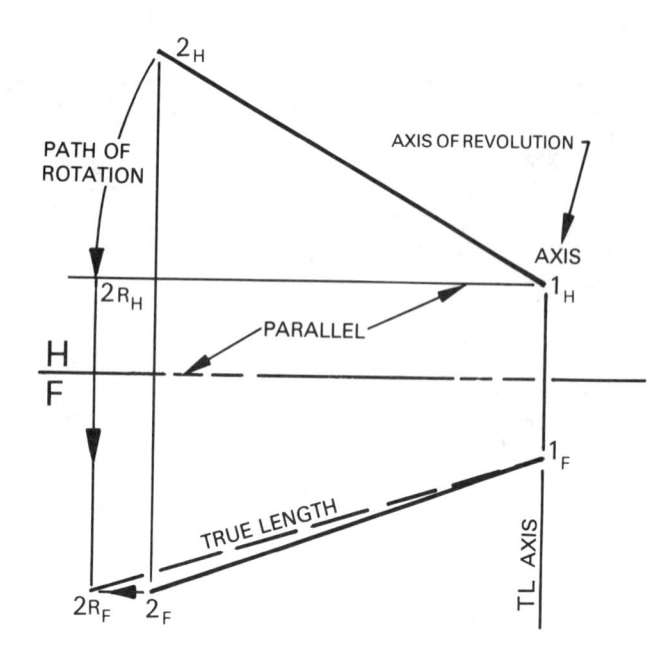

FIG. 4-130 True length of line in frontal view by revolution.

4.60 TRUE LENGTH OF A LINE BY REVOLUTION

The true length of a line can be found by revolution using any two adjacent views of the line. The axis about which a line rotates determines the view in which the true length is measured. When the axis is parallel to the frontal plane (appears as a point view in the horizontal view) and the line is revolved parallel to the frontal plane the true length appears in the frontal view as in Fig. 4-130 and Fig. 4-131. In this situation the axis line is true length in the frontal view and is drawn perpendicular to the H/F fold line. In both figures, point 2_H is revolved about an axis line that passes through point 1_H until line 1_H-$2R_H$ is parallel to the frontal plane (H/F). In the frontal view, point $2R_F$ lies on the edge view of the path of rotation. A projection line is drawn from point $2R_H$ to the frontal view and point 2_F is extended perpendicular to the true length axis until it intersects the projection line and locates $2R_F$. Line 1_F-$2R_F$ is true length.

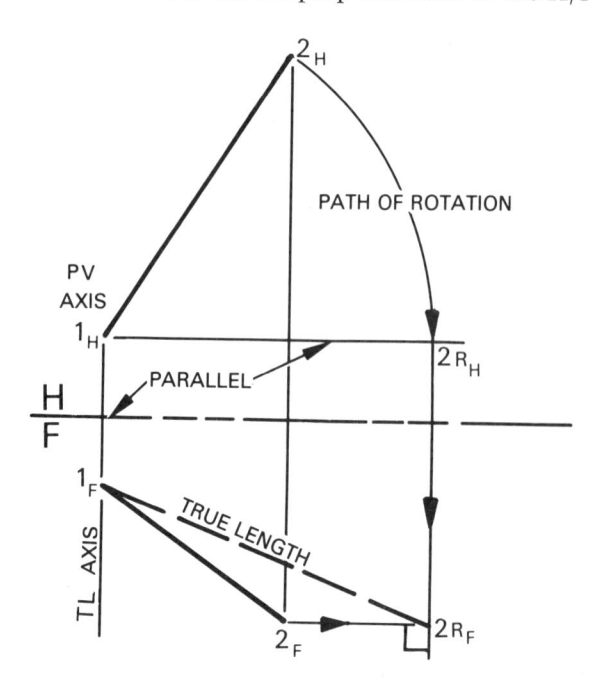

In Fig. 4-132, the axis of revolution is drawn parallel to the profile plane and passes through point 1_F. Point 2_F is revolved about the axis at point 1_F until line 1_F-$2R_F$ is parallel to the profile plane (F/P). The revolved position of line 1_P-$2R_P$ is true length in the profile view. Note that a line can be revolved in any view to solve for the true length in the adjacent projection plane. The axis of revolution can pass through any point on the line, though the ends points are the most convenient since only one end will need to be revolved.

The propeller blades in Fig. 4-133 revolve about a horizontal axis. The wing tips rotate 90 degrees in the frontal view of the aircraft.

FIG. 4-132 True length of line in profile view by revolution.

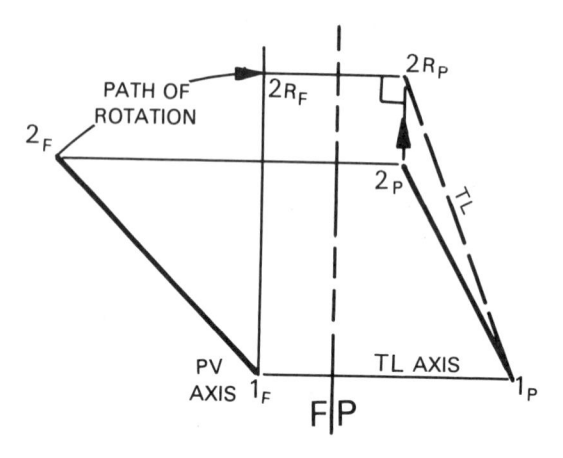

FIG. 4-131 True length of line in frontal view.

FIG. 4-133 Aircraft with rotating wing tips. *(Courtesy NASA.)*

to find the true length in an elevation view. In Fig. 4-134 the slope of line 1-2 is found by revolving the line about a vertical axis that passes through point 2_H. Point 1_H is revolved about the point view of the axis at 2_H to position $1R_H$. Line $1R_H$-2_H is parallel to H/F therefore its frontal projection shows as a true length line (frontal line). The slope angle is measured between true length line $1R_F$-2_F and a *level* line (or the edge view of the horizontal plane as represented by fold line H/F).

Revolution can be used to locate a line in space given its bearing, slope, and true length. Only two views are necessary when revolution is used. In Fig. 4-135, the bearing is given as N 37°W, with a −39° slope from point 1 towards point 2, and a true length of 200 ft.

1. Locate point 1 in the frontal and horizontal views. Draw a line from point 1_H an indefinite length with a bearing of N 37°W (1).
2. Draw line 1_F-2_F at −39° from point 1_F downward towards point 2_F, using 200 ft as the true length. Line 1_H-$2R_H$ is parallel in the horizontal plane (H/F). Line 1-2R represents the revolved position of the required line (1-2) (2).
3. Pass a vertical axis through point 1_H. The frontal view shows the axis in true length. Revolve point $2R_H$ about the axis at point 1_H until it intersects the bearing line. Point $2R_H$ is thus revolved to position 2_H, and line 1_H-2_H is the horizontal view of the required line. Point 2_F is located in the frontal view by projection; 2_F is located on the edge view of the path of rotation by extending point $2R_F$ perpendicular to the true length axis until it intersects the projection line drawn from point 2_H (3).

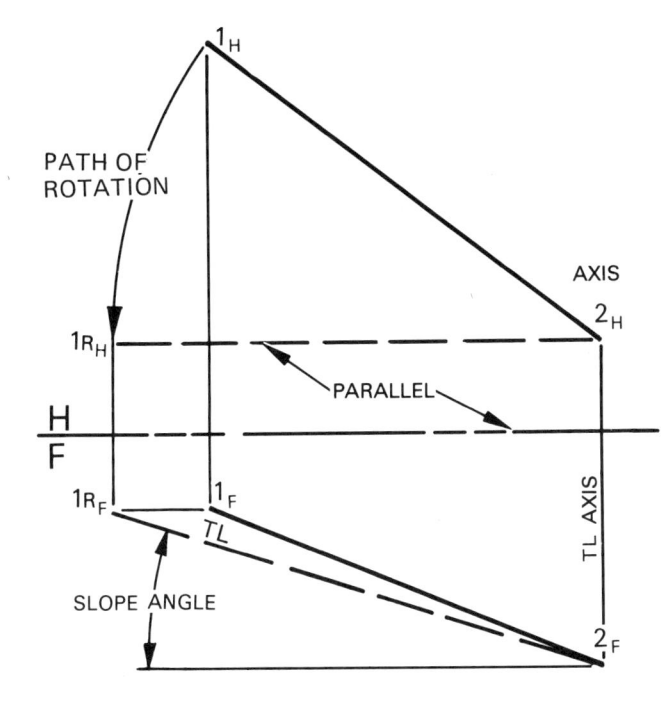

FIG. 4-134 Slope angle by revolution.

4.61 SLOPE OF A LINE BY REVOLUTION

The slope of a line can be measured in an elevation view where the line appears true length and the horizontal plane shows as an edge. To measure the slope of a line that is not a frontal or profile principal line, a primary auxiliary view projected parallel to the line in the horizontal view is normally used

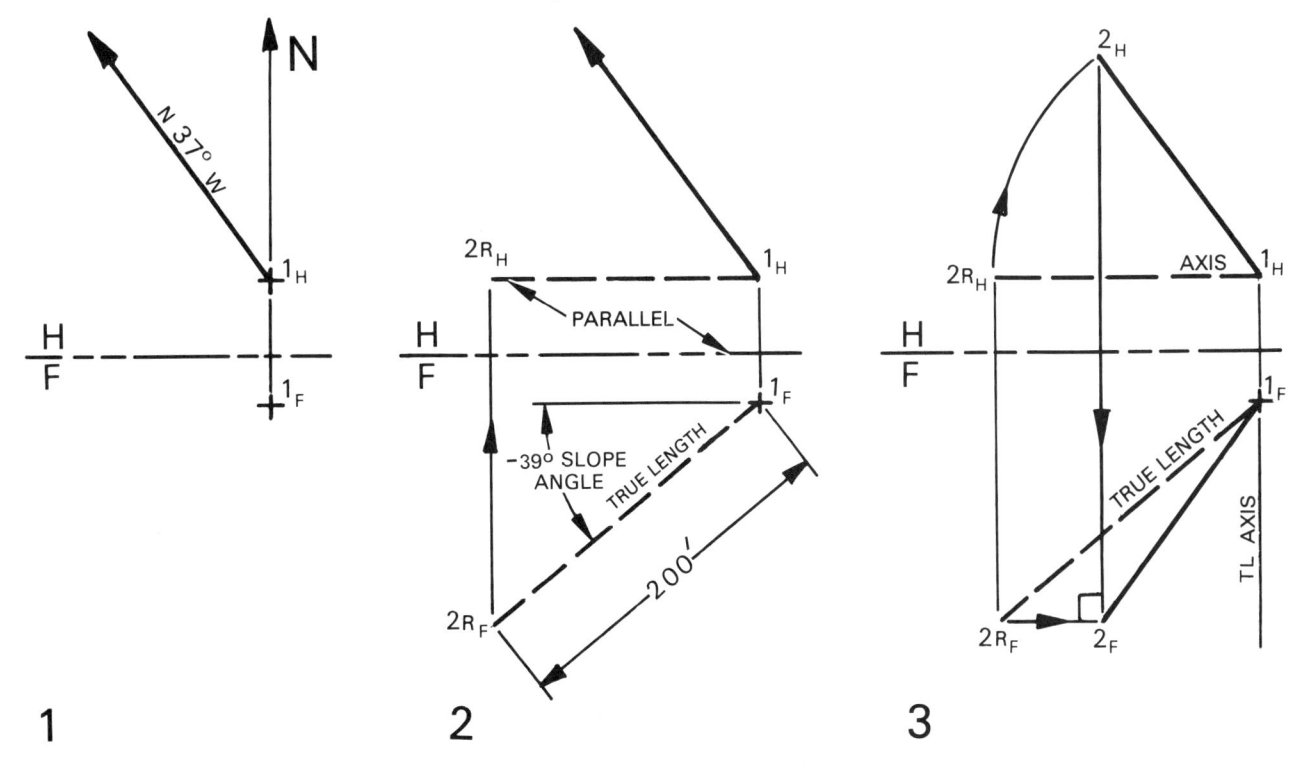

FIG. 4-135 Construction of a line by revolution given its bearing, slope, and true length.

4.62 REVOLUTION OF A LINE ABOUT AN OBLIQUE AXIS

When a line is revolved about another line, it generates a *cylinder* if the lines are parallel, a *circle* if they are perpendicular, a *cone* if intersecting, or a *hyperboloid* of revolution if oblique. In Fig. 4-136, oblique line 1-2 is revolved clockwise 110° about vertical line 3-4. Line 3_H-4_H is used as the axis of revolution and appears as a point in the horizontal view and true length in the frontal view. Line 1_H-2_H is revolved about the point view of axis line 3_H-4_H in the horizontal view by rotating both end points 110°. Because the whole line is revolved it does not change its oblique shape in the horizontal view; only its position ($1R_H$-$2R_H$) is altered. The frontal projection of the revolved line is located by moving point 1_F and 2_F perpendicular to the true length axis 3_F-4_F until it intersects the projection line drawn from each revolved point in the horizontal view.

When both lines are oblique lines a primary auxiliary view is projected parallel to the oblique view of the axis. The axis line shows true length in this view. The point view of the axis line is then found in a secondary auxiliary view projected perpendicular to the true length axis. Each point is revolved in this view and then projected back into all preceding views. In Fig. 4-137, line 3-4 is re-

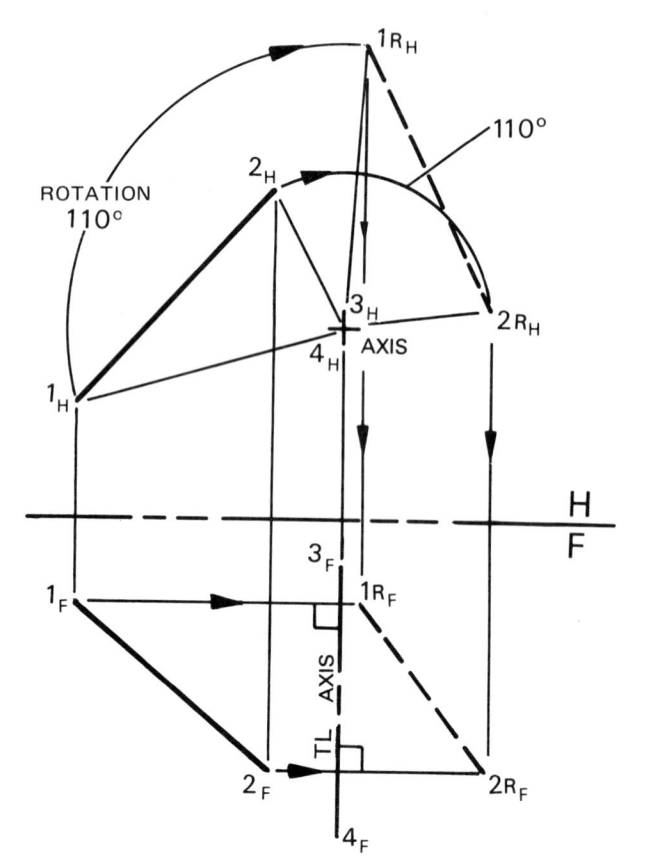

FIG. 4-136 Revolution of a line about a vertical line (axis).

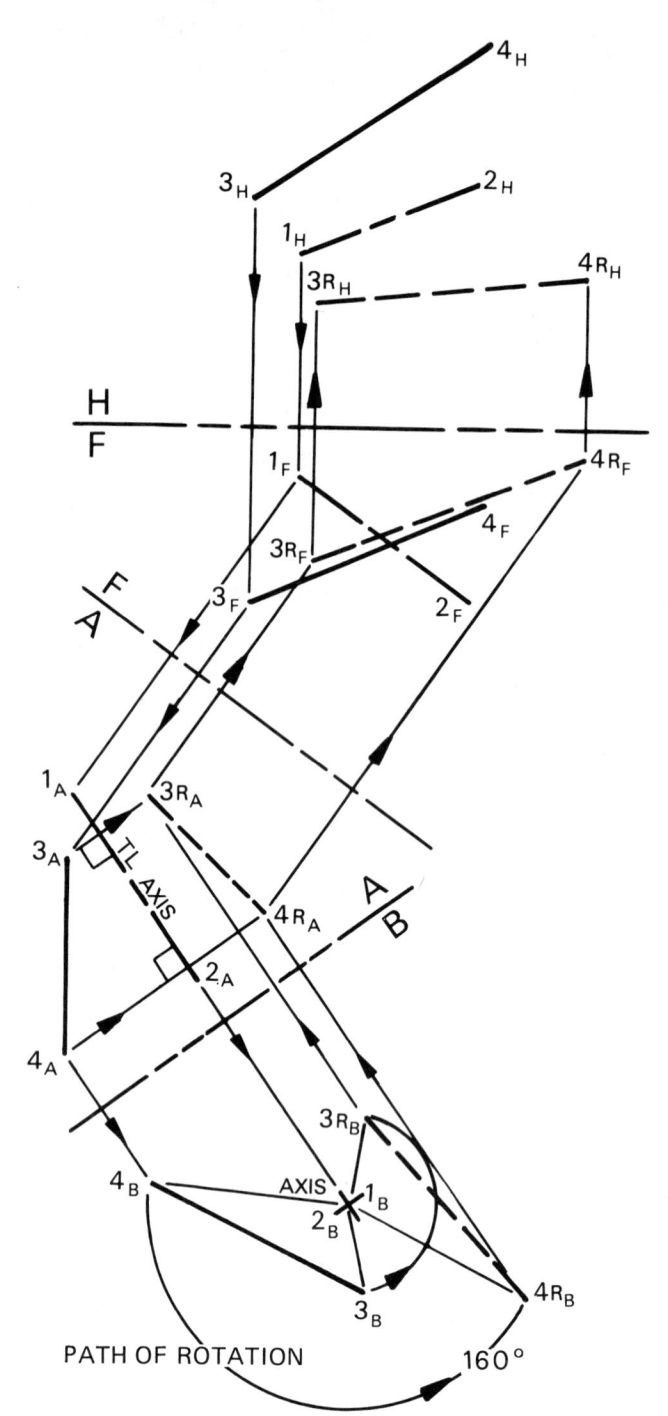

FIG. 4-137 Revolution of a line about an oblique axis.

volved 160° counterclockwise (when viewed from point 2 towards point 1) about line 1-2.

1. Draw F/A parallel to axis line 1_F-2_F.
2. Project auxiliary A. The axis line 1_A-2_A is true length.
3. Draw A/B perpendicular to true length axis line 1_A-2_A.
4. Project auxiliary B. Axis 1_B-2_B appears as a point view.
5. Revolve both points 160° counterclockwise to position $3R_B$-$4R_B$.
6. Project revolved line $3R_B$-$4R_B$ into auxiliary A. Points 3_A and 4_A are extended perpendicular to the axis along the edge view of their corresponding paths of rotation to revolved position $3R_A$-$4R_A$.
7. The frontal and horizontal views of the revolved line are established by projection.

FIG. 4-138 Revolution of a line about a horizontal line (axis).

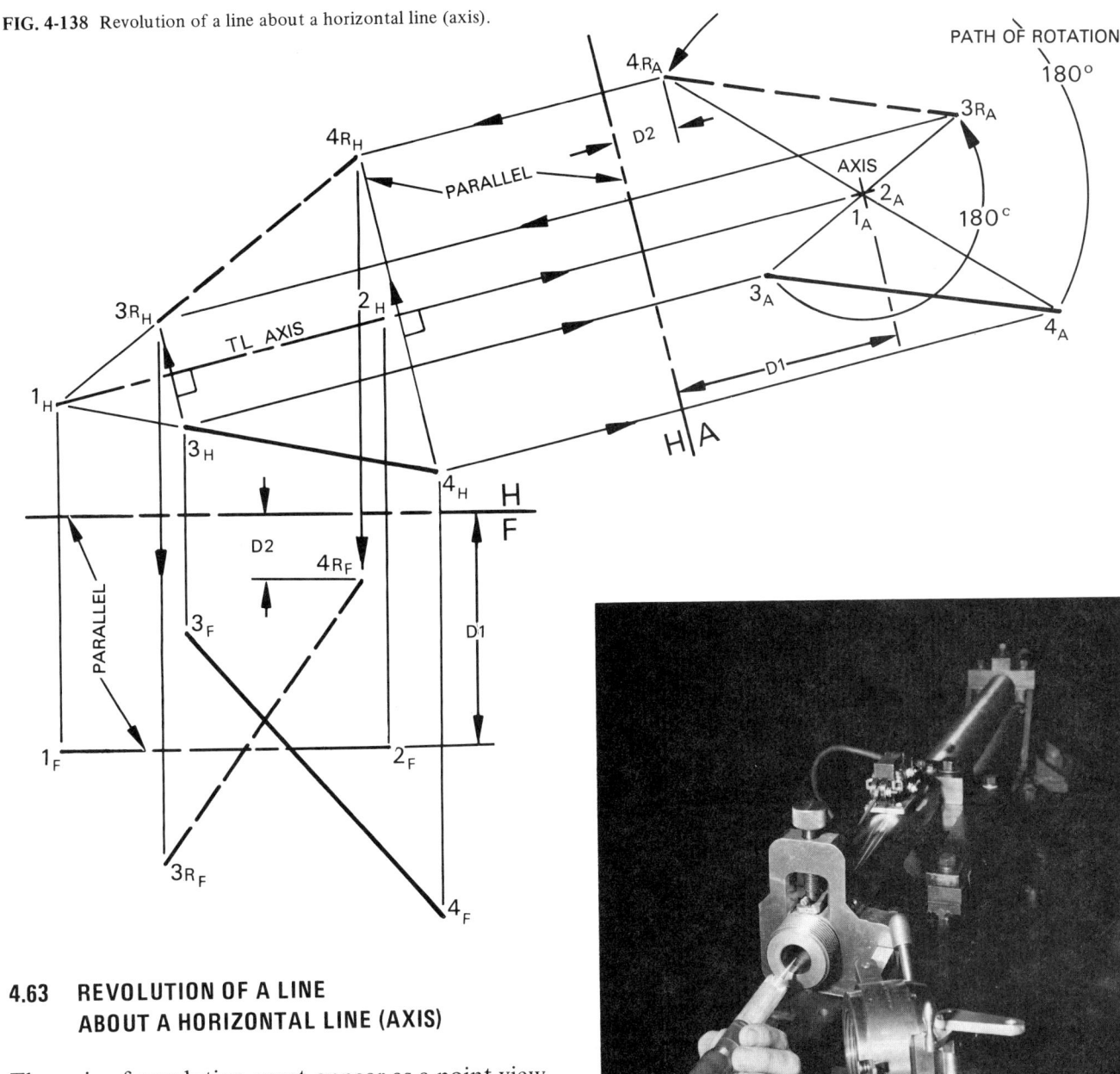

4.63 REVOLUTION OF A LINE ABOUT A HORIZONTAL LINE (AXIS)

The axis of revolution must appear as a point view before rotating a given point, line, or object. When the axis is a horizontal line, which does not appear as a point in the frontal view, a primary auxiliary view is projected perpendicular to the true length view of the axis to establish its point view. In Fig. 4-138, oblique line 3-4 is rotated 180° counterclockwise (when viewed from point 2 toward point 1) about horizontal line 1-2.

1. Draw H/A perpendicular to the true length view of line 1_H-2_H.
2. Project auxiliary view A. Axis line 1_A-2_A appears as a point, D1.
3. Revolve point 3_A 180° counterclockwise to position $3R_A$, and point 4_A 180° counterclockwise to position $4R_A$.
4. Draw projection lines from points $3R_A$ and $4R_A$ perpendicular to H/A. Project points 3_H and 4_H perpendicular to the true length axis to locate the revolved position of line $3R_H$-$4R_H$. Points $3R_H$ and $4R_H$ are on the edge view of their corresponding paths of rotation.

FIG. 4-139 The hatch of the missile launching gun revolves in the horizontal plane. *(Courtesy NASA.)*

5. The frontal view of revolved line $3R_F$-$4R_F$ is established by projection and transferring distances from the auxiliary view (D2).

An industrial application of revolution around an axis is shown in Fig. 4-139, where the barrel hatch revolves about a horizontal axis.

PROB. 4-39

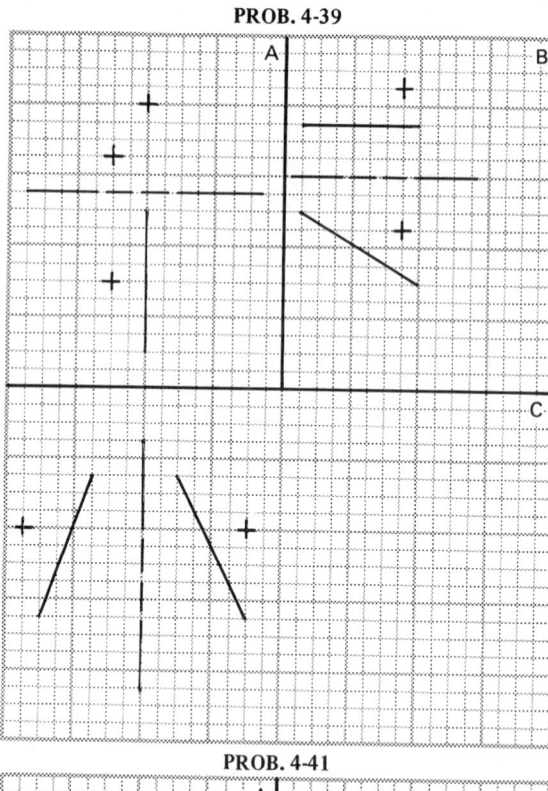

PROB. 4-40

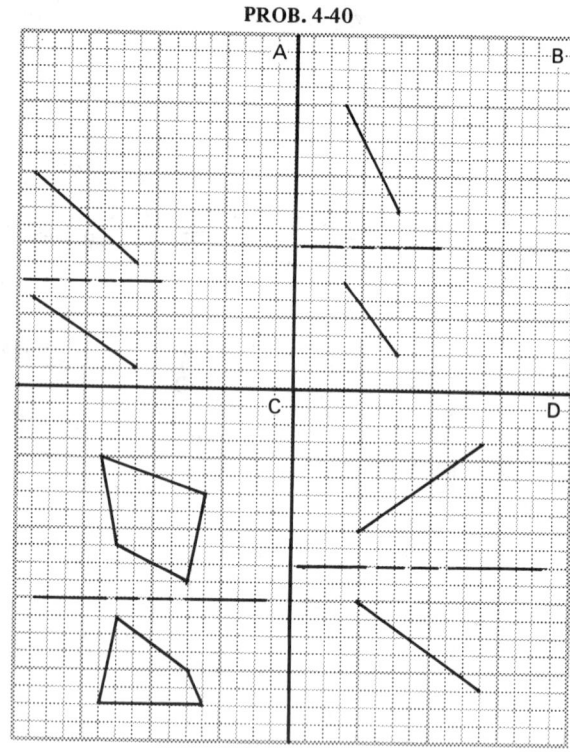

PROB. 4-41

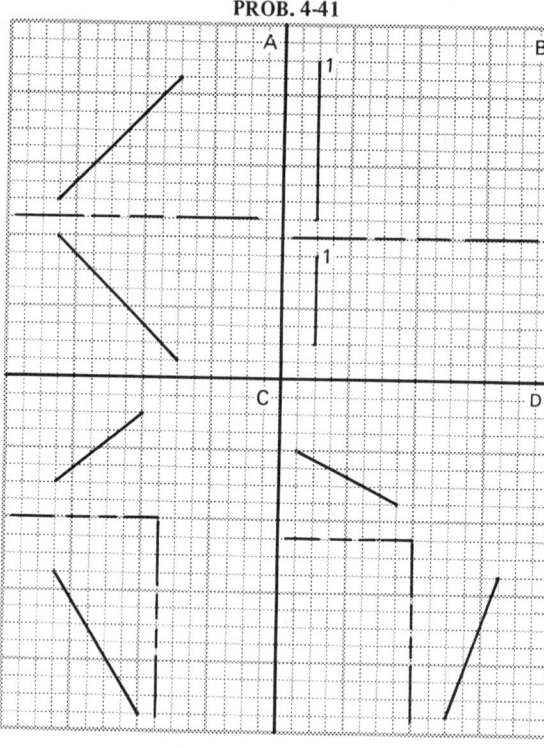

PROB. 4-42

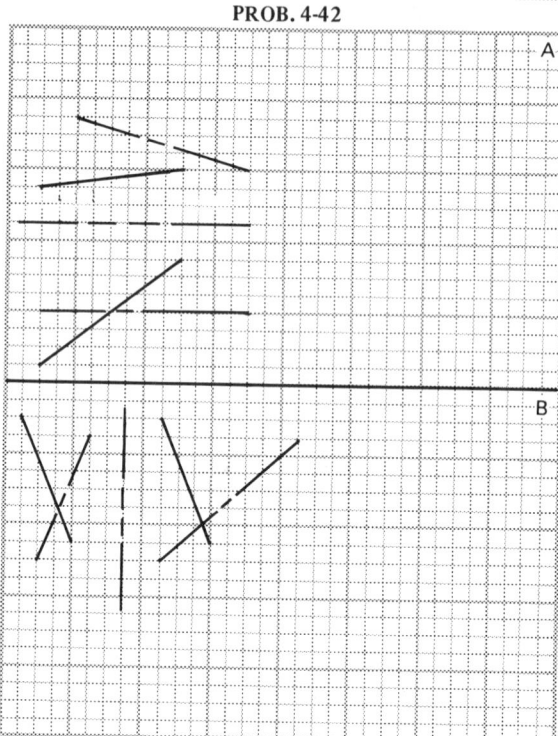

PROB. 4-39A. Rotate the point 200° clockwise around the frontal line. Show in all views.

PROB. 4-39B. Rotate the point around the line 100° counterclockwise.

PROB. 4-39C. Project an auxiliary view to establish a true length of the given line. Then solve for a point view of the line and rotate the given point 180°. Show in all views.

PROB. 4-40A. Using rotation, determine the true length of the line and the angle the line makes with the horizontal plane. Verify by projecting an auxiliary view.

PROB. 4-40B. Determine the angle between the line and the profile plane.

PROB. 4-40C. Using rotation, project and measure the true lengths of the sides of the given figure.

PROB. 4-40D. Solve for the angle that the line makes with the F view, the slope, bearing, azimuth, and grade.

PROBS. 4-41A, 4-41B. What is the true length, bearing, and slope of the line? What is the angle between the line and the F view?

PROBS. 4-41C, 4-41D. Using rotation, what is the angle that the line makes with each of the principal planes?

PROB. 4-42A. Revolve the line 180° around the given centerline using the true length view to establish a primary auxiliary view and point view of the centerline.

PROB. 4-42B. Rotate the line 90° clockwise around the centerline.

FIG. 4-140 The locus of a line.

4.64 CONE LOCUS OF A LINE

FIG. 4-141 Each of the three aircraft has rotating elements, props, flaps, etc. *(Courtesy NASA.)*

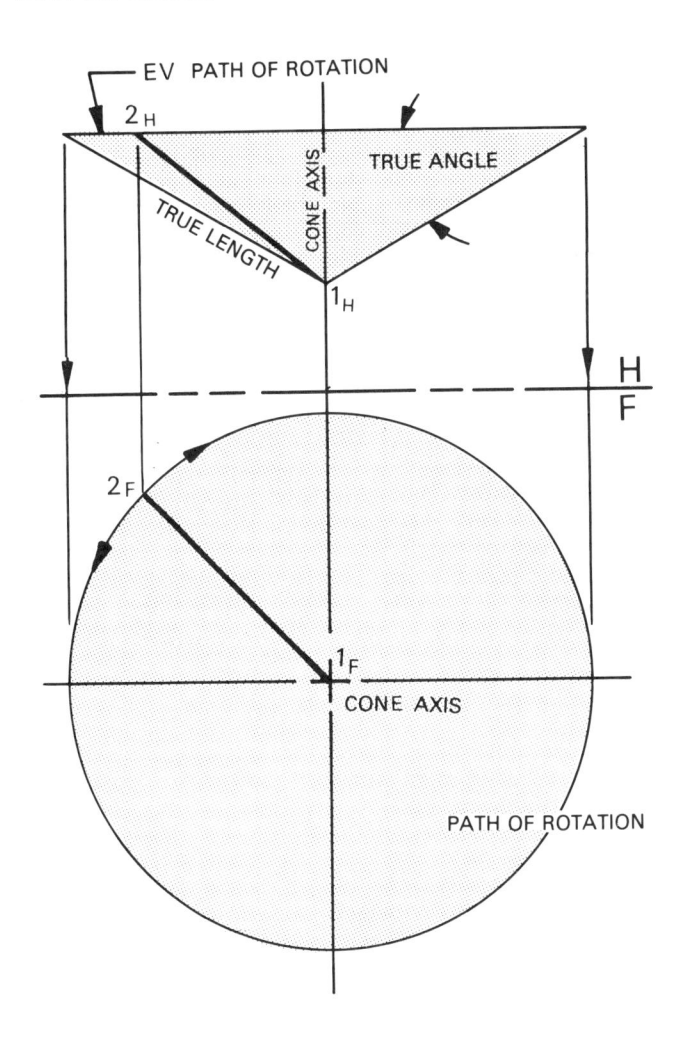

*The **locus** of a point or line is the set of all possible points (positions) formed by the movement of the line or point as determined by specified conditions.* A *circle* results if a point is revolved to all possible positions around a given line axis. If the axis is a point, the resulting movement of the point generates a sphere. When a line is revolved into all possible locations about a parallel axis line, the resulting generation produces a *cylinder* since all lines on the surface are the same distance from the center axis line. When an oblique line is revolved about an axis line which is not parallel to or intersecting the given line, a *hyperboloid of revolution* is formed.

The most commonly used locus of a line is generated by the revolving of a line about an axis that passes through (intersects) an end point of the given line and is called a *cone locus* of a line, Fig. 4-140. The cone represents all possible positions that the line could be in, given a specified angle and true length. The true angle that the line makes with the edge view of the path of rotation is called the *base angle* (*slope angle*). The angle formed by the axis line and the given line in a true length position is the *vertex angle*. The true length of the line is called the *slant height*.

In Fig. 4-140, line 1-2 is revolved about a horizontal axis that passes through point 1_F. The movement of point 2_F through a 360° path generates a *cone* which represents all possible positions of the line. In the frontal view the movement of 2_F is represented by a circle. In the adjacent horizontal view, the path of rotation appears as an edge and the line generates a cone (which appears as a triangle). The angle formed by the true length element and the edge view of the cones base is the slope or base angle (true angle).

131

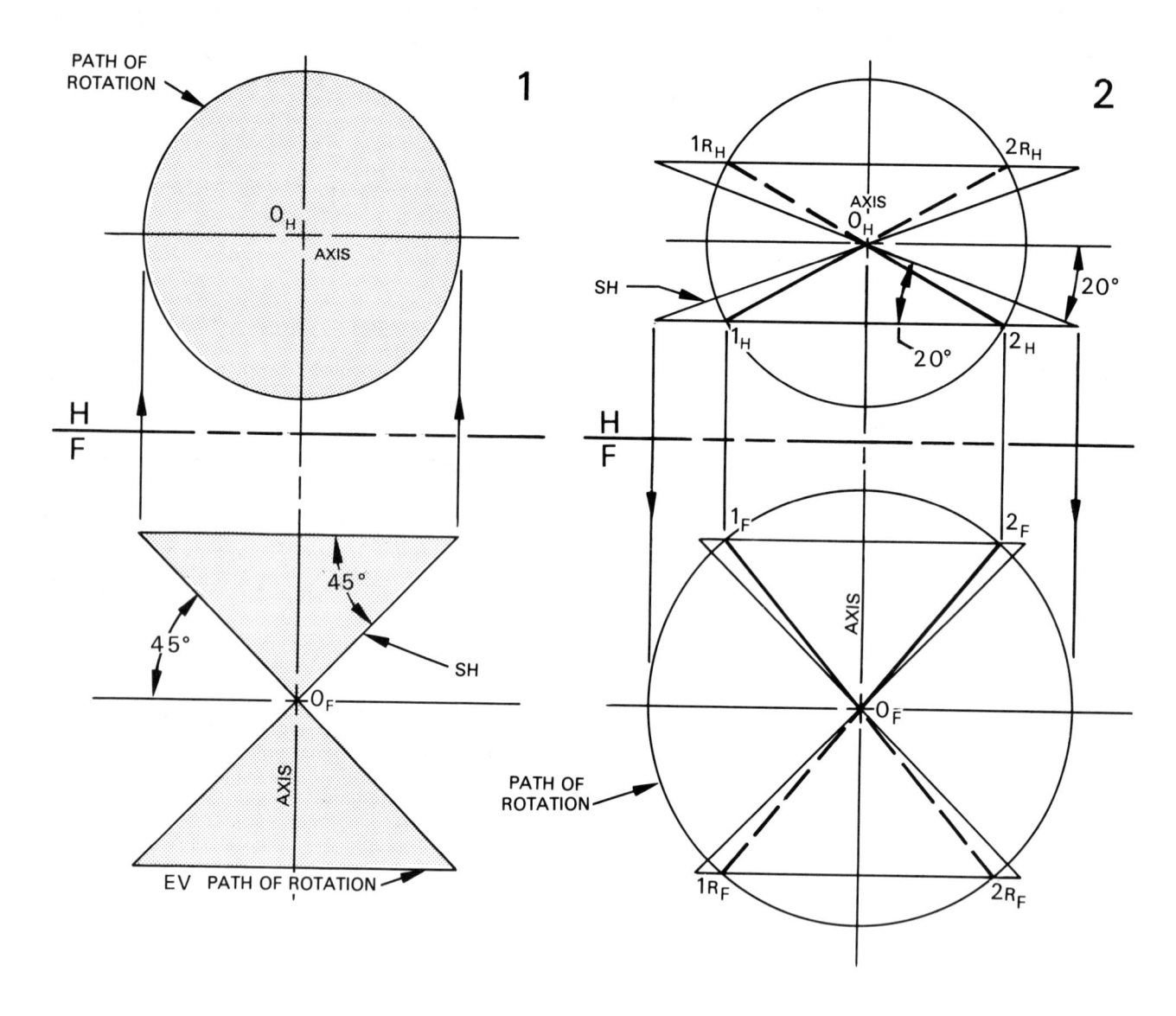

FIG. 4-142 Construction of a line at specified angles to two principal planes.

4.65 CONSTRUCTION OF A LINE AT SPECIFIED ANGLES TO TWO PRINCIPAL PLANES

To draw a line at a given angle to two principal planes, the line must be used to create a *cone of revolution* in both planes. The cones generated by the line have a common vertex and have axes lines that are perpendicular to one another with a slant height (SH) corresponding to the length of the given line. Each cone is generated by a given angle and line. Each cone has a specified angle (*base angle*) that it makes with the principal plane. If the sum of their angles total 90 degrees, the cones will be tangent. Note that the two cone angles cannot exceed 90 degrees in order to intersect or be tangent. There are eight possibilities for each set of angles.

In Fig. 4-142, a line is required that makes a 45° angle with the horizontal plane and a 20° angle with the frontal plane. The slant height (true length of line) is given.

1. Establish a convenient location for vertex point 0 in both views. Draw a right circular cone using the slant height and 45°. Note that two cones are drawn, since there are four possible positions in each view and the line could be slanting upward or downward to the right or left. The base is parallel to H/F. Complete the horizontal view by swinging the circle which results from the revolution of the line about the axis (1).

2. Draw a right circular cone using the given slant height and 20°. Again, two cones are drawn since the line could slant toward or away from the frontal plane. The base of the cone is parallel to the frontal plane and its base angle equals 20°. Complete the frontal view of the cone showing its path of rotation. The intersection of the two cones determines the position of the line at given angles to two principal planes. Eight possible positions are shown. Intersection line 0_F-1_F is one possible frontal position which could have a horizontal view of 0_H-1_H or 0_H-$1R_H$. Note that each line in each view can have two possible adjacent positions depending on the direction of the required line. Eight combinations are thus possible (2).

FIG. 4-143 Line at two given angles to two principal planes.

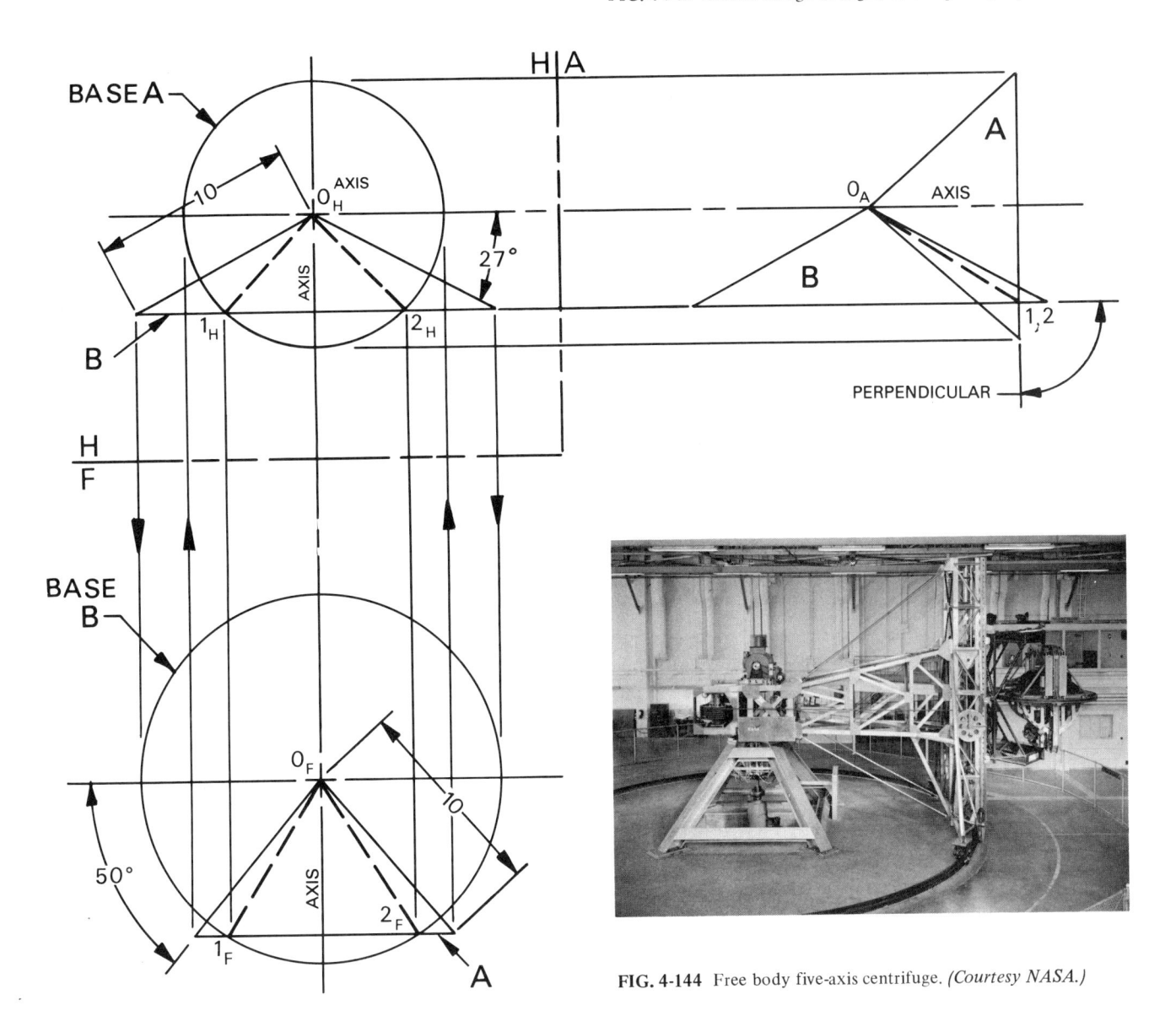

FIG. 4-144 Free body five-axis centrifuge. *(Courtesy NASA.)*

4.66 LOCUS OF A LINE AT GIVEN ANGLES WITH TWO PRINCIPAL PLANES

In Fig. 4-143, a line is required that makes a 50° angle with the horizontal plane and a 27° angle with the frontal plane. The line is 10 units long and slants down and forward. This is the same as section 4.65 except that only two positions are possible since the line direction was specified.

1. Locate vertex point 0 in the frontal and horizontal views.
2. Construct a cone in the frontal view with a slant height of 10 units and a base angle of 50°. The base edge is parallel to the horizontal plane. Since the line is to slant downward, only the lower cone need be drawn.
3. Complete the horizontal view by drawing a circle corresponding to the base of the cone, base A.
4. Construct a cone in the horizontal view using 10 units for the slant height and 27° for the base angle. Since the line is to slant forward, only the front cone need be drawn. The cone base is parallel to the frontal plane.
5. Complete the frontal view of base B.
6. Auxiliary view A is drawn to show the edge view of both cones in the same projection plane. Note that the bases are perpendicular and the intersection of the two bases fixes the position of the line. Line 0-1,2 represents the required line.
7. The intersection of the two cones in the frontal view establishes two possible locations for the line. 0_F-1_F slants downward, forward, and to the left. 0_F-2_F slants downward, forward, and to the right.
8. In the horizontal view the intersection of cones A and B fixes the two possible locations of the required line.

The revolution of the 5-axis free body centrifuge in Fig. 4-144 generates a cone with its angled brace. This testing devise actually generates a cone, two cylinders, and a sphere.

PROB. 4-43

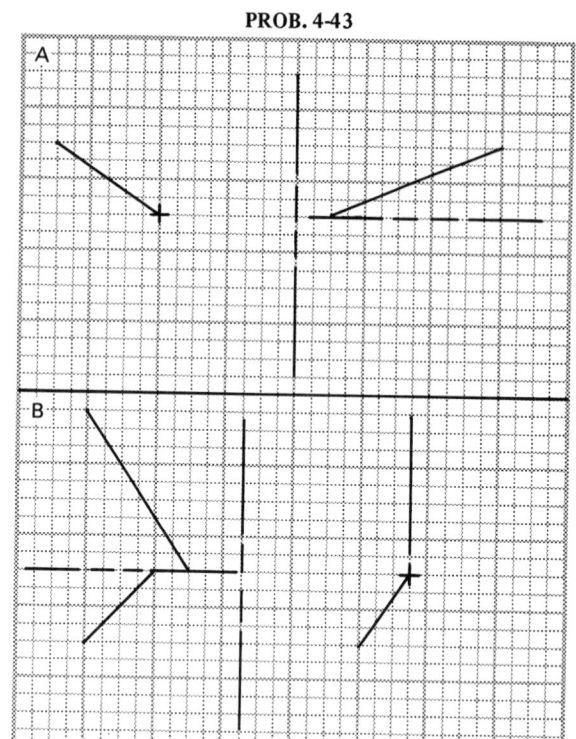

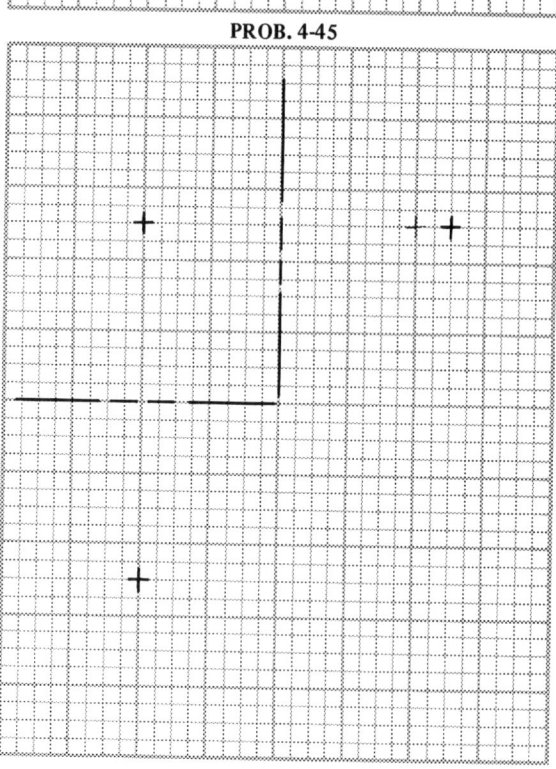

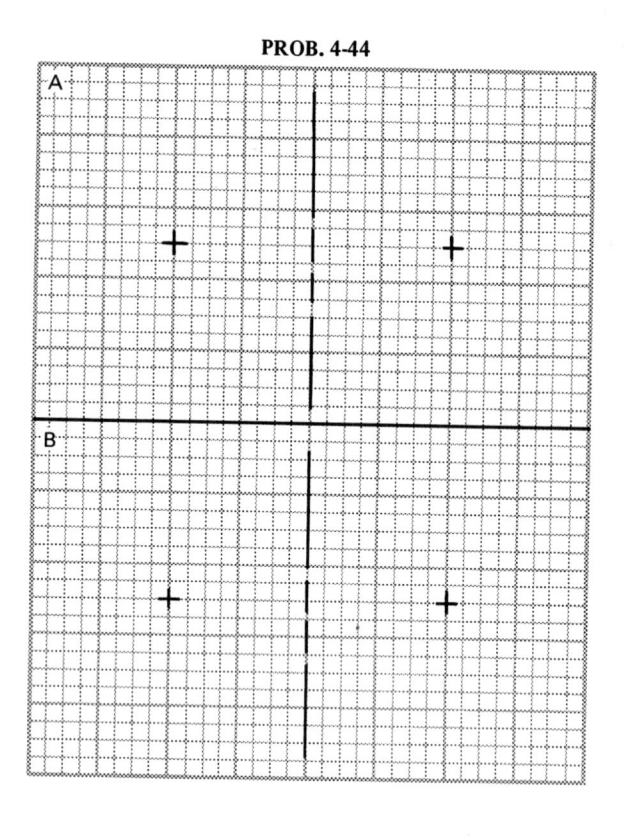

PROB. 4-44

PROB. 4-45

PROB. 4-43A. Show the complete cone locus of the given line and revolve the line 160° counterclockwise. Show its revolved position in all views.

PROB. 4-43B. Show the cone locus of both lines and revolve the short line 100° clockwise and the other line counterclockwise 200°. Show both in their revolved positions in each view.

PROB. 4-44A. Draw a line through the given point that makes a 45° angle with the profile plane and 55° with the horizontal plane. Show all possible solutions.

PROB. 4-44B. Construct a line through the given point which makes a 40° angle with the profile and 30° with the horizontal plane. Show all possible solutions.

PROB. 4-45. Draw a $1\frac{3}{4}''$ (44.4mm) line that makes an angle of 35° with the H plane and 50° with the F view. Show all three views.

PROB. 4-46

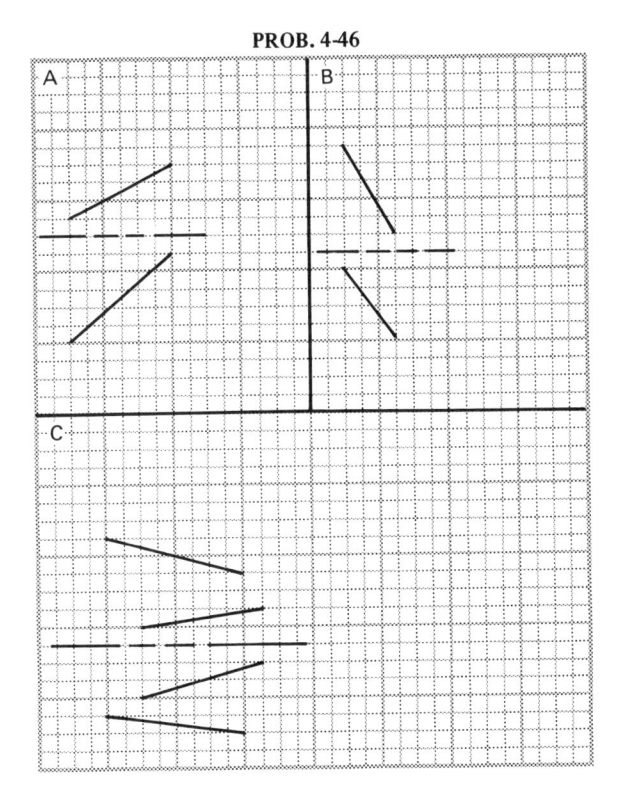

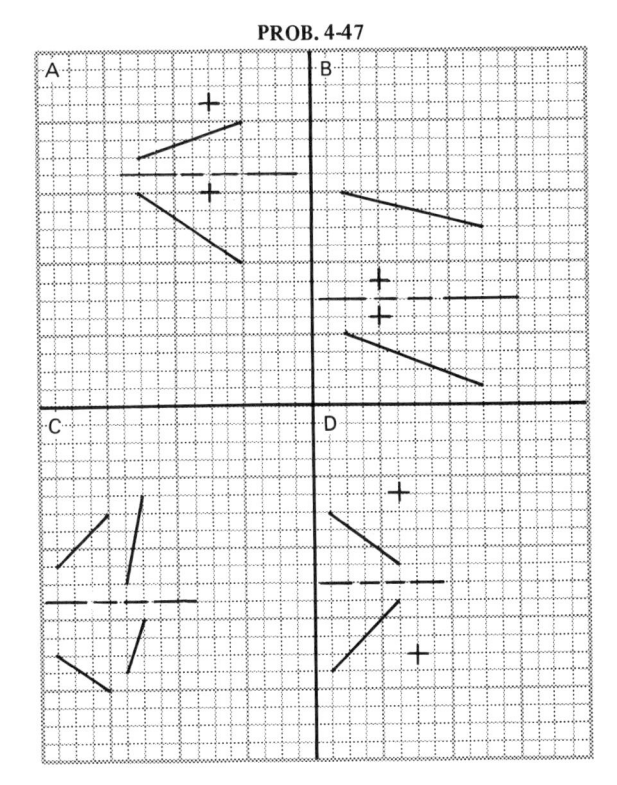

PROB. 4-47

TEST

PROB. 4-46A. Project an auxiliary view of the given line so that its projection is seen true length. What are the slope and bearing of the line? What is the angle that the line makes with the frontal plane?

PROB. 4-46B. Using rotation, what is the true length of the given line? Construct a true length diagram to check the answer.

PROB. 4-46C. Solve for the shortest perpendicular distance and the shortest horizontal connector between the skew lines. Show the two connectors in all views and calculate the bearing of each.

PROB. 4-47A. Project a 1" (25.4mm) long line that is perpendicular to the given line and passes through the point and is connected to the line.

PROB. 4-47B. A pipeline is to be laid from the given point and join another pipeline. It is to be of minimum length. Show in all views and note its bearing, length, and grade. Scale: 1" = 50 ft.

PROB. 4-47C. Project the shortest line between the skewed lines. Use the line method. What is the bearing of the line?

PROB. 4-47D. A line intersects the given line at an angle of 45°. The given point is one end of the line. What are the bearing and slope of the new line?

WORD PROBLEMS

PROB. 4-48. Draw line 1-2 that has a bearing of N 60°E and a slope angle of +38° and is 2.5" (63.5mm) in length. Point 1 is three units behind the frontal plane.

PROB. 4-49. Construct a line that has a grade of −25%, bearing of S 49°W, and is 3" (76.2mm) in length.

PROB. 4-50. Construct a line that forms a 23° angle with the frontal, and a 51° angle with the horizontal plane.

QUIZ

1. What is the difference between the bearing and the azimuth?
2. How many views are necessary to tell if a line is oblique?
3. What is a profile line?
4. What makes a line vertical?
5. In what views can the width dimension be measured?
6. What are skew lines?
7. What is the grade of a line?
8. How can you tell if two lines are perpendicular?
9. Parallel lines will appear _____ in _____ views.
10. If a line is vertical in the profile plane what type of line is it in the top and front view?
11. What is a secondary auxiliary view?
12. How many views are necessary to establish a line in space?
13. What is an inclined line?
14. How can you obtain a true length of an oblique line?
15. What is a point view of a line and how will it project in any subsequent view?

5

PLANES

FIG. 5-1 Infrared astronomical satellite with solar panels. *(Courtesy NASA.)*

5.1 PLANES

*A **plane** can be defined as a flat surface that is not curved or warped.* If any two points on it were connected to form a line, that line would be wholly in the surface of the plane. So far only points and lines have been discussed, but these elements have been drawn on *"projection planes"* representing the surfaces of the glass box as it is unfolded into the surface of the paper using orthographic projection. Horizontal, frontal, profile, and auxiliary projection planes have been used to complete the required views of points and lines. In this chapter, planes as objects in space are projected in the same manner onto the principal and auxiliary projection planes.

In general a plane will be considered a limited and defined shape as in Fig. 5-1, where the solar panels are planes, and Fig. 5-2, where much of the equipment is designed with plane surfaces. Though a plane may be limited, its borders can be extended indefinitely to solve for specific information concerning a problem. Just as a line can be extended to infinity, a plane is theoretically indefinite in size.

A ***plane*** can be fixed in space by locating any *three points* that lie in its surface and are not in a straight line. *A line and a point, two intersecting lines, or two parallel lines also define a plane.*

FIG. 5-2 Industrial model.
(Courtesy Engineering Model Associates, Inc.)

FIG. 5-3 Representing a plane with three points.

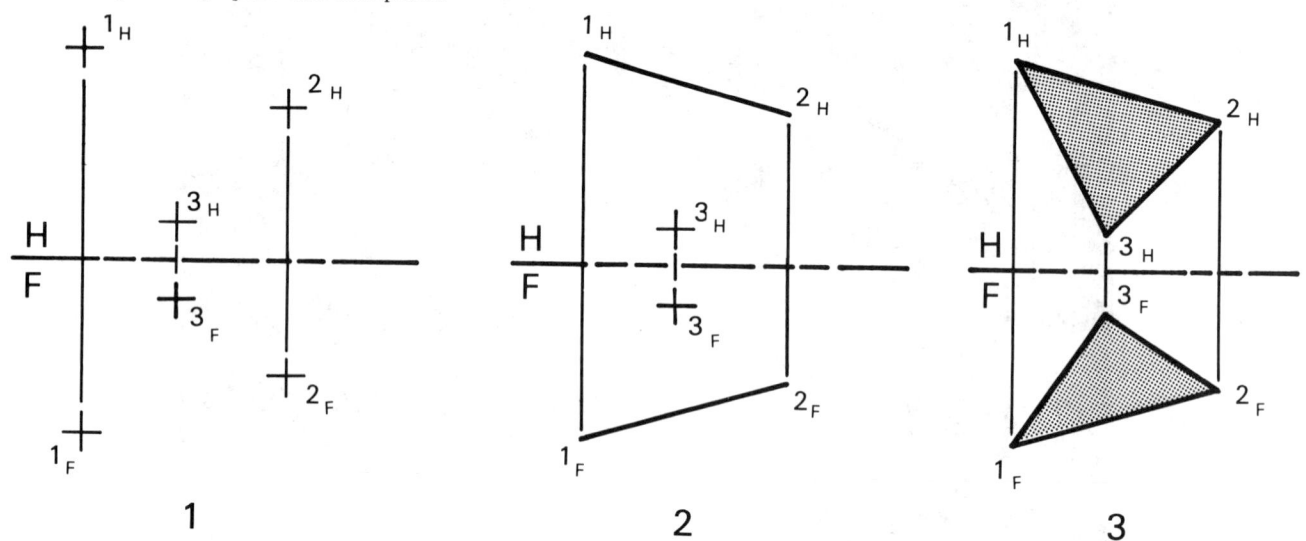

5.2 REPRESENTATION OF PLANES

A plane can be represented by four basic conditions:

1. *Three points not in a straight line,* Fig. 5-3(1).
2. *A point and a line,* Fig. 5-3(2).
3. *Two parallel lines,* Fig. 5-4(1).
4. *Two intersecting lines,* Fig. 5-4(2).

In Fig. 5-3, the same plane is defined by three points located in two views. The plane is identical in each case, only its method of representation is changed. The first method (1) is three individual unconnected points. In the second method (2), two of the points are connected, therefore the plane is defined by a point and a line. In method (3) the same three points are now connected to form plane 1-2-3. All three examples locate the end points of the same plane and define its surface when connected.

In Fig. 5-4, plane 1-2-3-4 is defined in three separate ways resulting in the same plane only different representations. In method (1) two parallel lines, line 1-2 and line 3-4, establish the plane. In the second method (2), two intersecting lines define the same plane, using line 1-4 and line 2-3. In example three (3) the two methods have been combined and enclosed to form plane 1-2-3-4. Lines 1-2 and 3-4 remain parallel and lines 1-4 and 2-3 remain intersecting. Note that not all planes form both parallel and intersecting lines with their end points. But *any two lines of a plane either intersect or are parallel if extended indefinitely.* Two intersecting lines have a common point (intersecting point) which is aligned in every adjacent view.

In general the example shown in Fig. 5-3(3) the enclosed triangular plane, and Fig. 5-4(3) the enclosed four-sided plane is the method used for representing a plane throughout the text.

FIG. 5-4 Representing a plane by parallel or intersecting line.

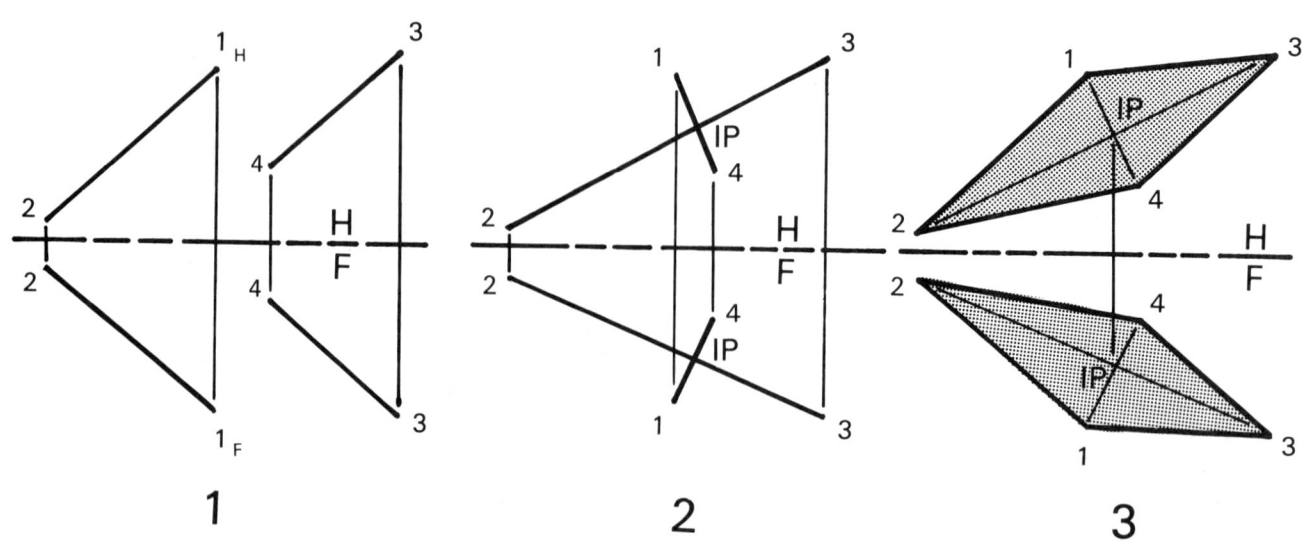

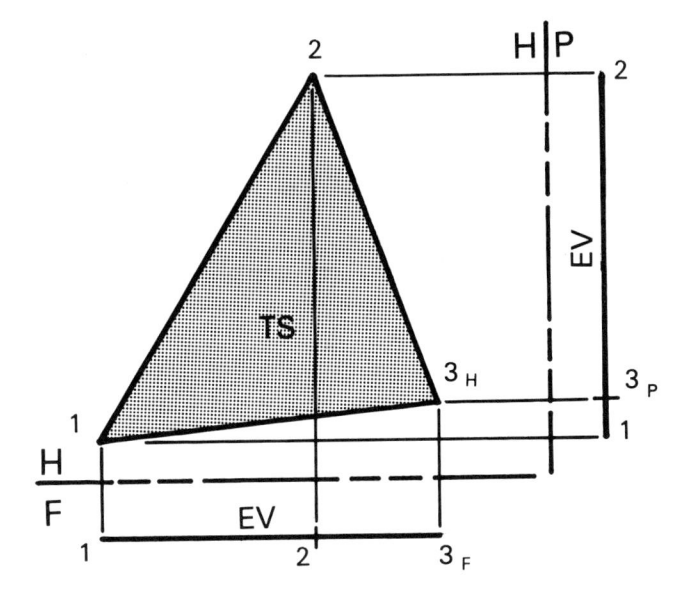

FIG. 5-5 Horizontal plane.

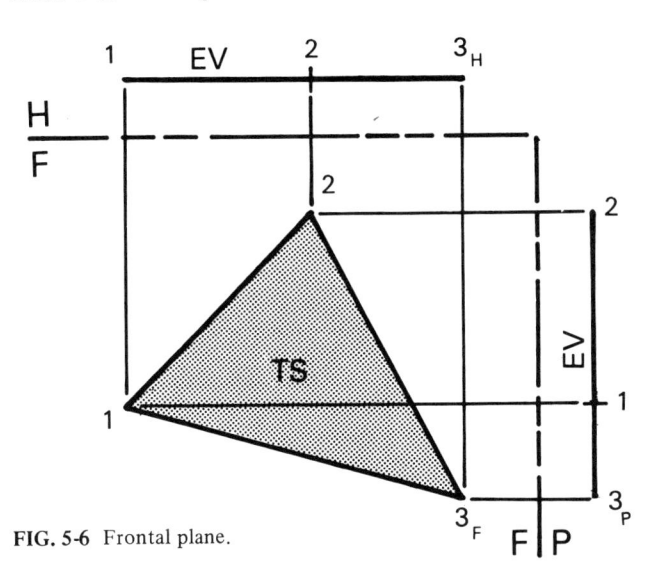

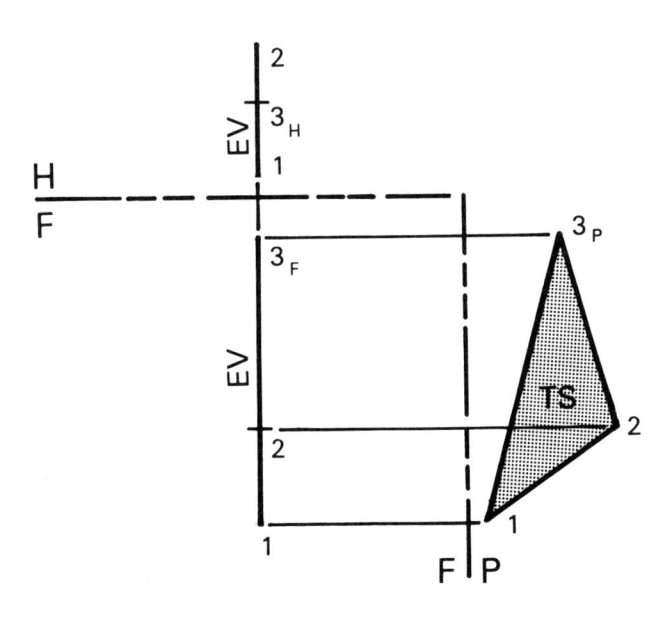

FIG. 5-6 Frontal plane.

FIG. 5-7 Profile plane.

5.3 PRINCIPAL PLANES

When a plane is parallel to a principal projection plane it is a **principal plane**. A principal plane can be a *horizontal plane,* a *frontal plane,* or a *profile plane* depending on its relationship to a *principal projection plane.* All lines in a horizontal plane, frontal plane, or profile plane are true length lines, therefore principal planes are made up of principal lines.

Remember, **principal projection planes** are imaginary sides of the unfolded glass box used to expedite the orthographic projection of an object (point, line, plane, solid) in descriptive geometry. On the other hand, *principal planes* are limited definite forms that happen to lie parallel to one principal projection plane.

To determine if a plane is a principal plane it is necessary to have at least two views, unless the given view shows the plane as parallel to a principal projection plane. In either case, two views are required to fix the position of any plane.

A *horizontal plane,* Fig. 5-5, is parallel to the horizontal projection plane. It is true size (true shape) in the horizontal view since all of its lines are principal lines, therefore they project true length. The frontal and profile view of a horizontal plane always shows the plane as an edge view. A horizontal plane is a level plane and shows as an edge in all elevation projections. Horizontal planes are perpendicular to the frontal and profile projection planes. In Fig. 5-5, the profile and frontal planes have been unfolded from the horizontal plane in order to show parallelism.

A *frontal plane,* Fig. 5-6, lies parallel to the frontal projection plane where it shows as true size. In the horizontal and profile views the plane appears as an edge view. All lines show true length in the frontal view, since they are principal lines (frontal lines). A frontal plane is perpendicular to the horizontal and profile projection planes. Frontal planes are vertical planes since they are always perpendicular to the horizontal projection plane.

A **profile plane**, Fig. 5-7, is true size in the profile view and appears as an edge in the frontal and horizontal views. Every line in the plane is true length in the profile view since they are profile lines. Profile planes are perpendicular to the frontal and horizontal projection planes. Profile planes are vertical planes since they are perpendicular to the horizontal projection plane.

FIG. 5-8 Vertical planes.

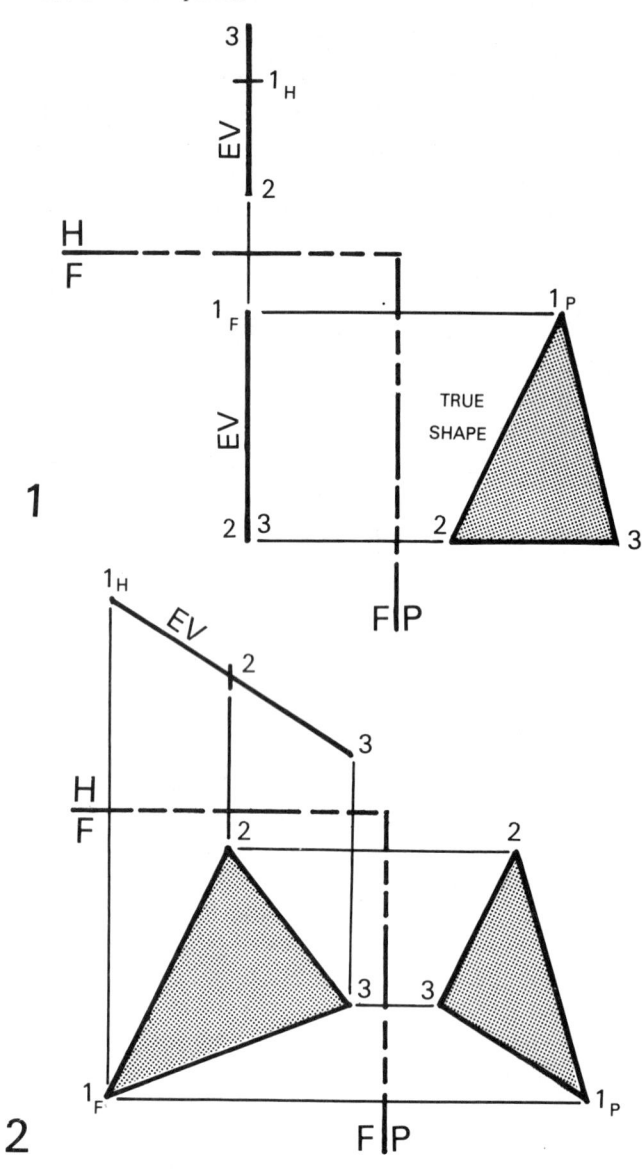

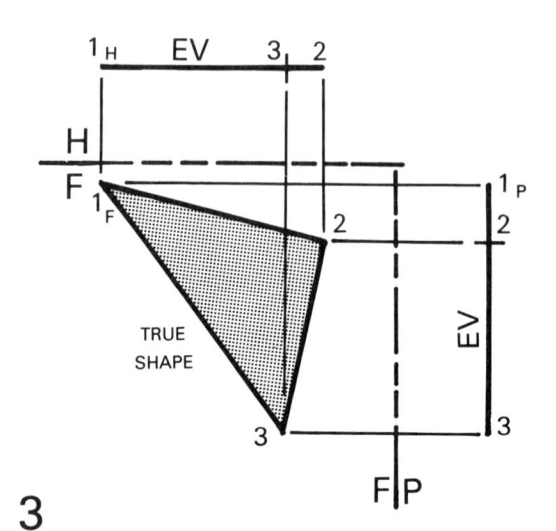

5.4 VERTICAL PLANES

Vertical planes are perpendicular to the horizontal projection plane. The horizontal view of all vertical planes shows the plane as an edge. There are three basic positions for a vertical plane, as shown in Fig. 5-8.

In (1) the vertical plane appears as an edge in the frontal and horizontal views. Plane 1-2-3 is perpendicular to the frontal and horizontal projection planes. This type of vertical plane is also a profile plane since it appears true shape in the profile view. The frontal and horizontal projections show the edge view of the plane parallel to the profile projection plane.

In (2) plane 1-2-3 is not parallel to a principal projection plane. Therefore it does not show as true size in any of the three principal views. The horizontal view of the plane establishes it as a vertical plane since it appears as an edge. The frontal and profile projections are foreshortened.

The third example of a vertical plane (3) is a frontal plane since it is true size in the frontal view. The horizontal and profile views show the plane as an edge and parallel to their adjacent projection planes.

The tail (vertical stabilizer) of the supersonic aircraft in Fig. 5-9 is an example of a vertical plane; the swept-back wings are horizontal planes.

FIG. 5-9 Supersonic aircraft. *(Courtesy NASA.)*

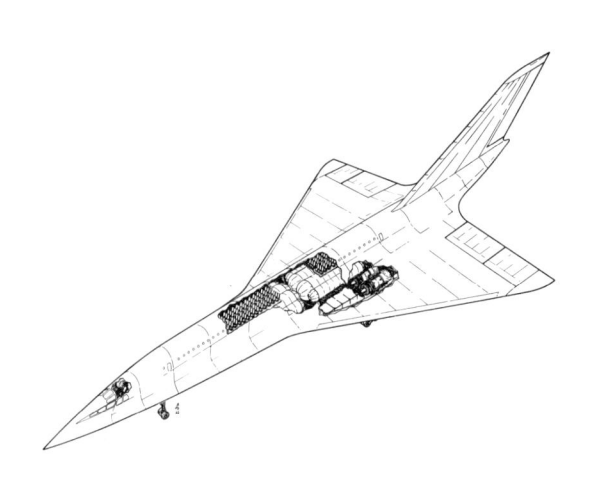

5.5 OBLIQUE AND INCLINED PLANES

The classification of planes is determined by their relationship to the three principal projection planes: frontal, horizontal, or profile. Principal planes (*normal planes*) appear as true size in one of the three principal projections and as edges in the other two. **Oblique planes** and **inclined planes** do not appear true size in any of the three principal views.

Oblique and inclined planes are not vertical or horizontal (level), and will not be parallel to a principal projection plane. In Fig. 5-10, examples of an oblique plane (1), and inclined planes (2 and 3) are given.

Oblique plane (1): An oblique plane is inclined to all three principal projection planes, which results in each view being foreshortened (distorted). The true size of an oblique plane cannot be seen in the three principal projections.

Inclined plane (2 and 3): An inclined plane does not appear true size in any of the principal projections. This form of plane is seen as an edge in the profile view (2) or in the frontal view (3) and foreshortened (distorted) in the other two views. An inclined plane will not appear as an edge in the horizontal view, though some texts refer to this situation as an inclined plane because it is "inclined" to the frontal and profile projection plane. A plane that is an edge in the horizontal view is a vertical plane. Note that an inclined plane is inclined (at an angle) to the horizontal projection plane; in other words, it is inclined in *elevation*, thus shows as an edge *only* in the frontal or profile views.

The solar paddles of the weather satellite in Fig. 5-11 are planes which rotate in order to orientate themselves to the suns rays based on the direction of orbit.

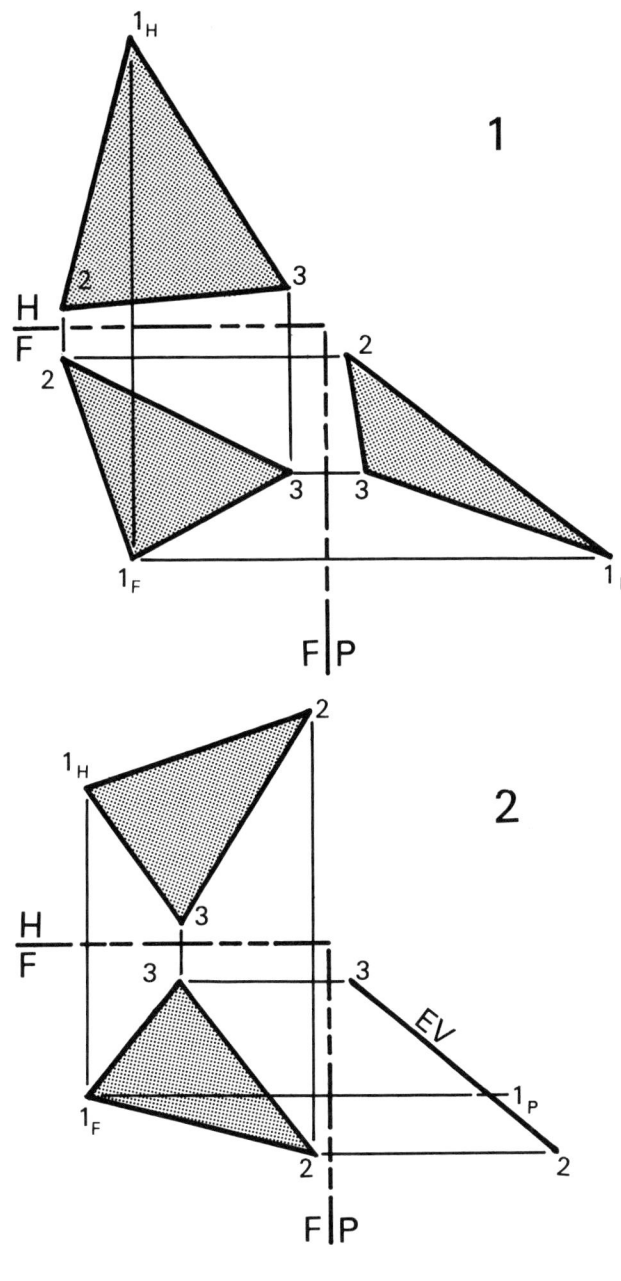

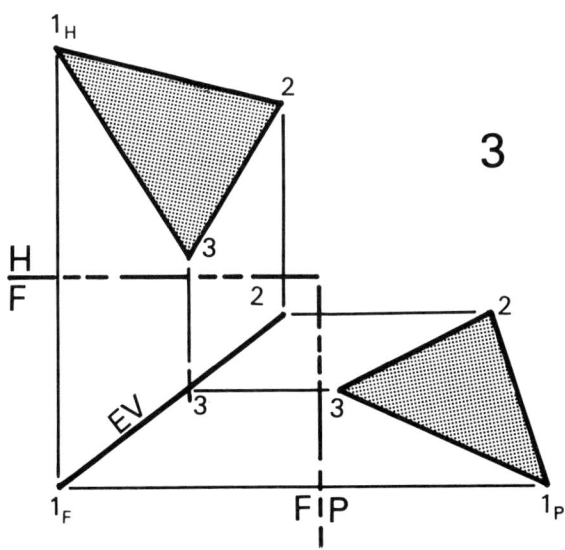

FIG. 5-10 Oblique (1) and inclined (2), (3) planes.

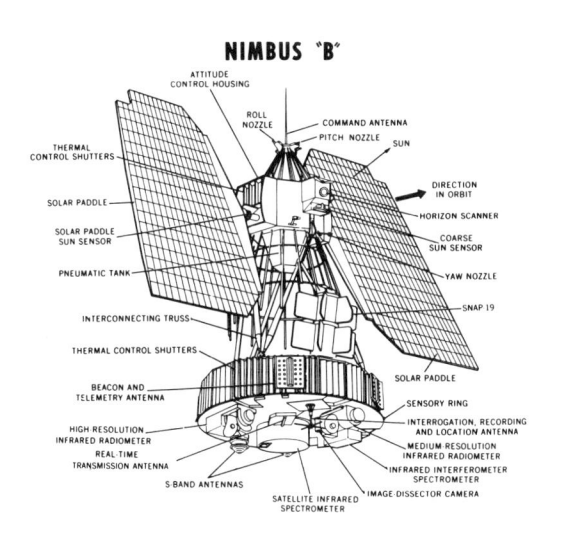

FIG. 5-11 Nimbus satellite. *(Courtesy NASA.)*

PROB. 5-1

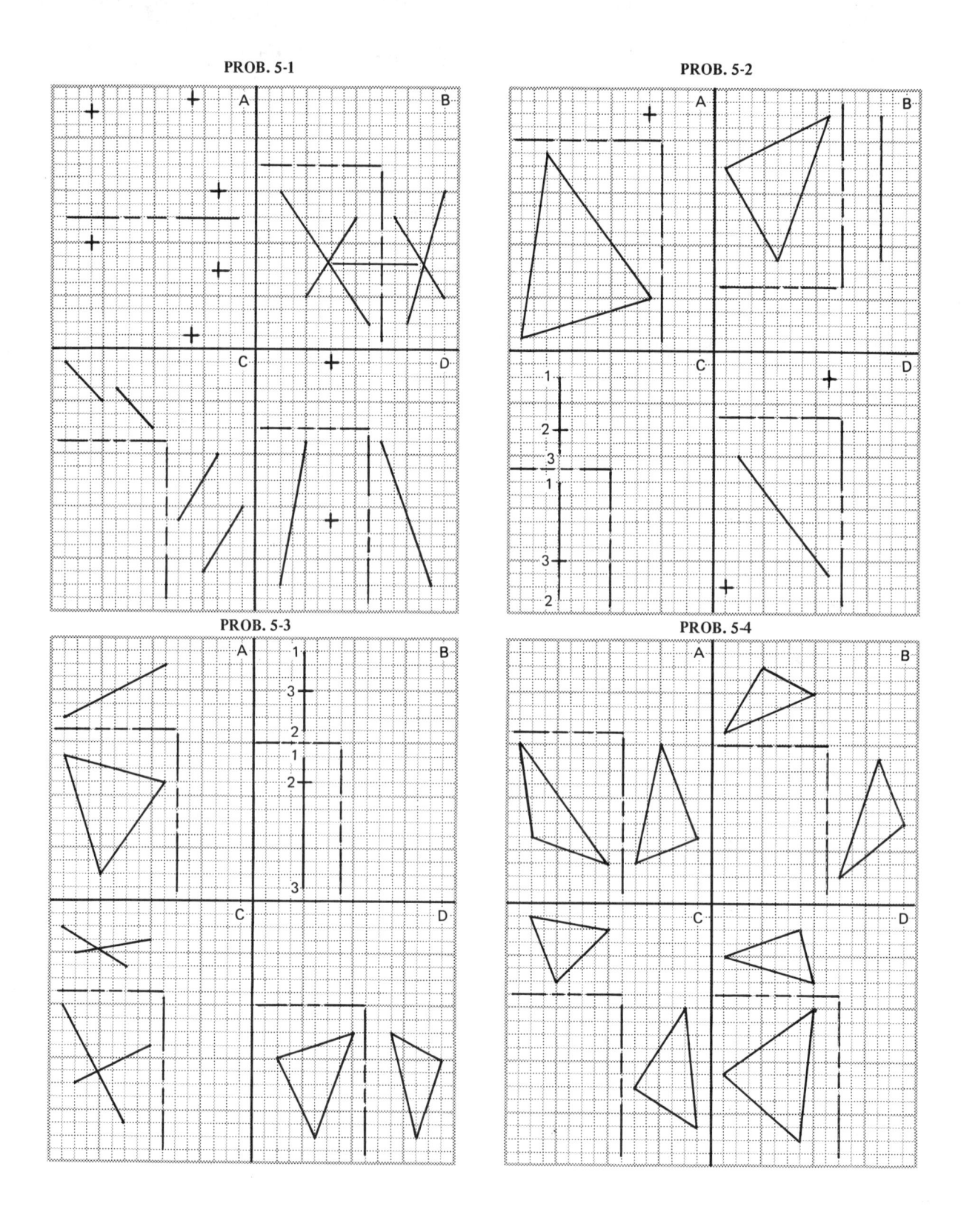

PROB. 5-2

PROB. 5-3

PROB. 5-4

PROBS. 5-1A, 5-1B, 5-1C, 5-1D. Complete the views of the planes for each problem. Connect the points and lines to form planes.

PROBS. 5-2A, 5-2B, 5-2C, 5-2D. Complete the views of the principal planes. Label for TS, TL, EV, and principal plane.

PROBS. 5-3A, 5-3B, 5-3C, 5-3D. Complete the views of the planes in each problem. Label for the type of plane, vertical, inclined, oblique, and principal plane.

PROBS. 5-4A, 5-4B, 5-4C, 5-4D. Same as 5-3.

The three principal views of oblique plane 1-2-3 in Fig. 5-12 are given and auxiliary views A and B are projected with a specific result in mind. Auxiliary view A is projected parallel to line 1_H-2_H by drawing fold line H/A parallel to line 1_H-2_H in the horizontal view. The resulting primary auxiliary view shows plane 1_A-2_A-3_A as oblique, but line 1_A-2_A is true length since it is parallel to the fold line in the adjacent view. Auxiliary A is constructed by drawing projection lines from each point on the plane in the horizontal view and transferring distances from the frontal view. Dimension D2 is transferred from the frontal view to locate point 3_A.

To locate a view where plane 1-2-3 would appear as an edge, auxiliary view B is projected perpendicular to frontal line 3_F-4_F (true length).

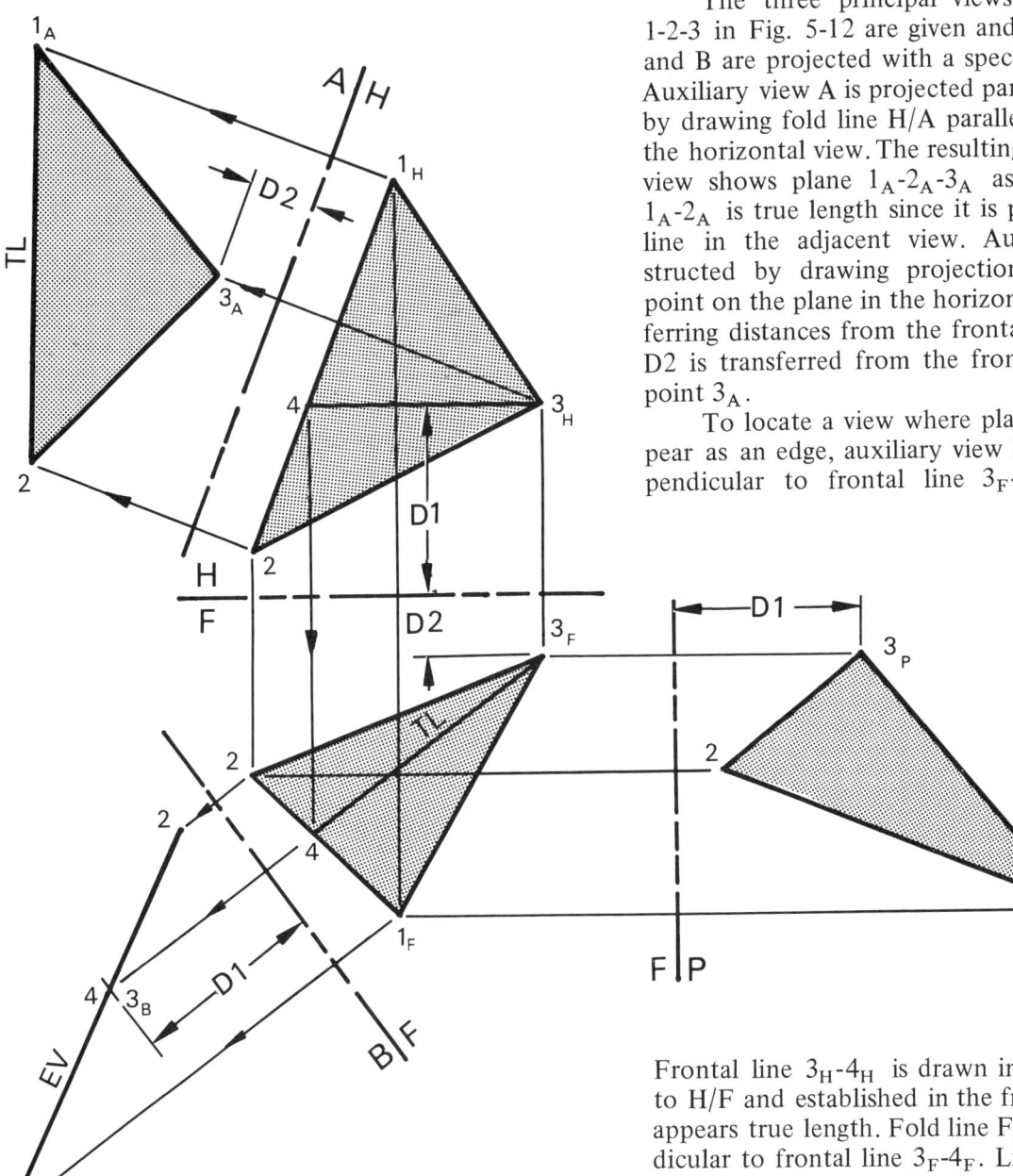

FIG. 5-12 Auxiliary projections of a plane.

Frontal line 3_H-4_H is drawn in the plane, parallel to H/F and established in the frontal view where it appears true length. Fold line F/B is drawn perpendicular to frontal line 3_F-4_F. Line 3_B-4_B is a point view in auxiliary B, therefore plane 1_B-2_B-3_B appears as an edge view projection. Dimension D1 is transferred from the previous view to locate the point view of line 3_B-4_B. D1 can be taken from the horizontal or profile views.

The preceding description is used to illustrate the fact that planes (and lines, solids) are located by projecting each point of the figure. It is extremely important to realize that all forms can be defined by their end points and therefore projected from view to view on adjacent projection planes using these points. Visualization of an object in an adjacent projection is not always possible, but projecting each individual point, connecting the points to form lines, and solving for proper visibility provides the solution without prior understanding of the end result. The final bounded form is as accurate as if it had been perfectly understood and visualized from the start.

5.6 AUXILIARY VIEWS OF A PLANE

Theoretically a plane is an unlimited flat surface. For practical purposes and to define the plane for graphical representations, planes are delineated by points and bounded by lines. In this text most planes are defined by three points and enclosed with lines. This is the easiest and most efficient method of representation. Planes can be projected from view to view by locating the end points of their lines and transferring each point separately.

FIG. 5-13 Points on planes.

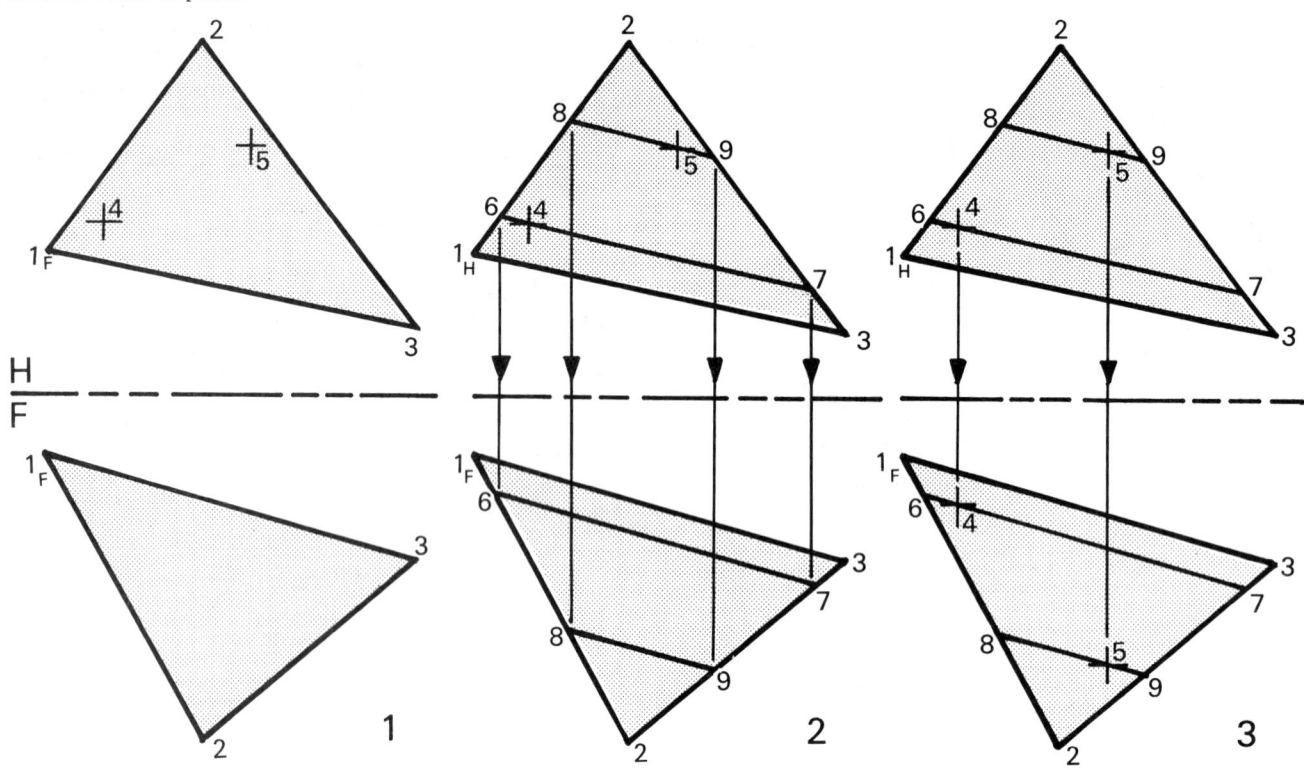

5.7 POINTS ON PLANES

To locate a point on a plane, the point need be given in only one view. A point on a plane lies on any line that passes through the point and lies in the plane. Where the point does not lie on a given line it is required to introduce a construction line that lies on the plane and through the point. There are an infinite number of lines that could be used, but the closer the line is to being parallel to the adjacent fold line the better. A line that is perpendicular or at too much of an angle to the fold line will not work or will project inaccurately. Note that subscripts are left off of many of the following examples.

In Fig. 5-13(1), plane 1-2-3 and points 4_H and 5_H are given. Both points lie on the plane. Find the position of the points in the frontal view.

1. Draw construction line 6_H-7_H through point 4_H and construction line 8_H-9_H through point 5_H. Using a line that is parallel to a given edge line expedites the solution.
2. Project the frontal view of line 6_F-7_F, and line 8_F-9_F.
3. Locate the frontal view of point 4_H and 5_H by drawing projection lines from the horizontal view until they intersect the construction lines in the frontal view.

In Fig. 5-14, point 4_H is located in the horizontal view on a line that is perpendicular to fold line H/F. In order to establish its frontal position, it is necessary to extend plane 1-2-3. Line 1_H-3_H is extended to point 5_H. Line 2_H-5_H is now at an angle to the fold line. Point 4_H is projected parallel to line 1_H-5_H until it intersects line 2_H-5_H at point 6_H. Point 6_F is located in the frontal view by projection, where it lies on 2_F-5_F. Point 6_F is projected parallel to Line 1_F-3_F to where it intersects line 2_F-3_F and establishes point 4_F. A number of lines could have been introduced on plane 1-2-5 and through point 4 to solve this problem.

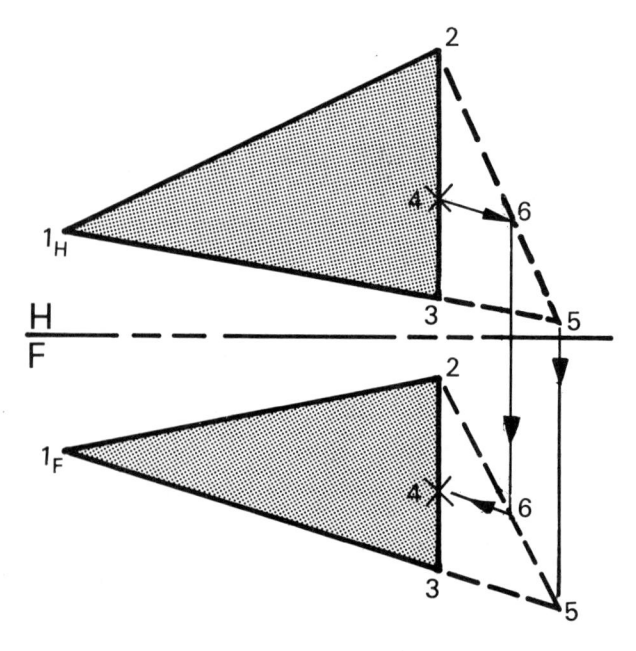

FIG. 5-14 Point on plane by plane extension.

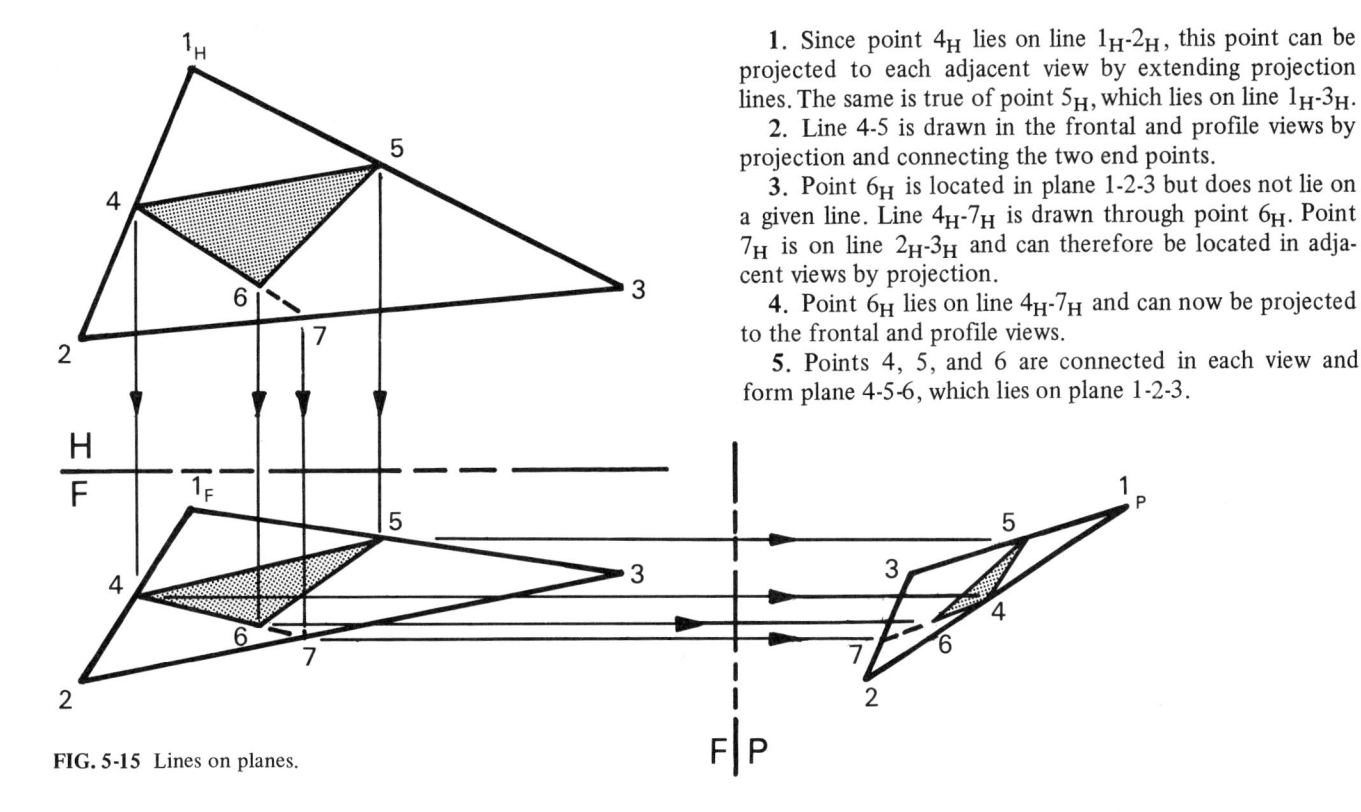

1. Since point 4_H lies on line 1_H-2_H, this point can be projected to each adjacent view by extending projection lines. The same is true of point 5_H, which lies on line 1_H-3_H.

2. Line 4-5 is drawn in the frontal and profile views by projection and connecting the two end points.

3. Point 6_H is located in plane 1-2-3 but does not lie on a given line. Line 4_H-7_H is drawn through point 6_H. Point 7_H is on line 2_H-3_H and can therefore be located in adjacent views by projection.

4. Point 6_H lies on line 4_H-7_H and can now be projected to the frontal and profile views.

5. Points 4, 5, and 6 are connected in each view and form plane 4-5-6, which lies on plane 1-2-3.

FIG. 5-15 Lines on planes.

5.8 LINES ON PLANES

A line which lies on a plane may be located in adjacent views by simple projection of its end points, or by projecting the point where it crosses a line on the given plane. In Fig. 5-15, three views of an oblique plane are given along with lines 4_H-5_H, 4_H-6_H, and 5_H-6_H in the horizontal view. In reality these lines form plane 4-5-6, which lies on plane 1-2-3. The frontal and profile projection of these three lines (plane 4-5-6) are required.

In Fig. 5-16, plane 1-2-3-4-5 is given in the horizontal view and partially complete in the frontal view (1). Using the above principles of lines and points on planes, the figure can be completed. Draw line 4_H-6_H from point 4_H through point 5_H until it intersects line 1_H-2_H at point 6_H. Locate point 6_F in the frontal view by projection where it lies on 1_F-2_F. Connect point 6_F to point 4_F. Point 5_F is located on line 4_F-6_F by projection. The figure can be completed by connecting the points.

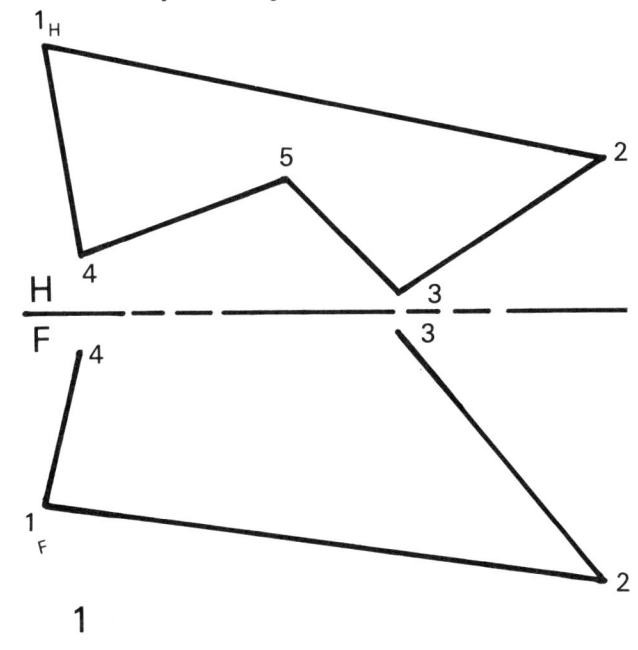

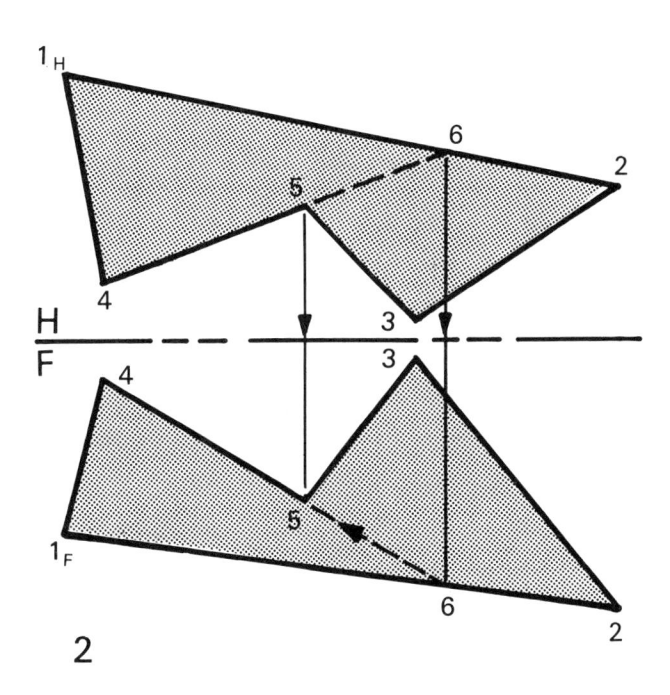

FIG. 5-16 Oblique lines on plane.

5.9 TRUE LENGTH LINES ON PLANES

Throughout the text it will be necessary to find a true length line which lies on an oblique plane in order to solve a particular problem, such as the edge view and true size of a plane, angle between two planes, or the strike and dip of a plane. A true length line can be found by drawing a line on the given plane parallel to the fold line. The adjacent projection shows the line as true length and in the given plane. In Fig. 5-17, lines have been located in each example so that they are parallel to the fold line in one view and project true length in the adjacent view.

Note that the examples are of oblique planes in the three principal projection planes: frontal, horizontal, and profile. Therefore these newly introduced lines are principal lines. A true length line can be found in any view using its adjacent projection to construct the line parallel to the fold line. When these views are not principal views, the lines will not be principal lines. The only requirement to finding a true length line in a plane is that the line be drawn parallel to the projection plane in one view and therefore project true length in the adjacent view.

In Fig. 5-17(1), line 3_H-4_H is drawn on the given oblique plane and parallel to H/F. The frontal projection of the line is on the plane and true length (a frontal line). In example (2), line 3_F-4_F is drawn parallel to H/F and on the given plane. The horizontal view shows the line as a horizontal line (true length) and on the plane. In example (3) line 2_F-4_F is drawn on the plane 1_F-2_F-3_F and parallel to F/P. Line 2_P-4_P appears true length in the profile view. It is a profile line.

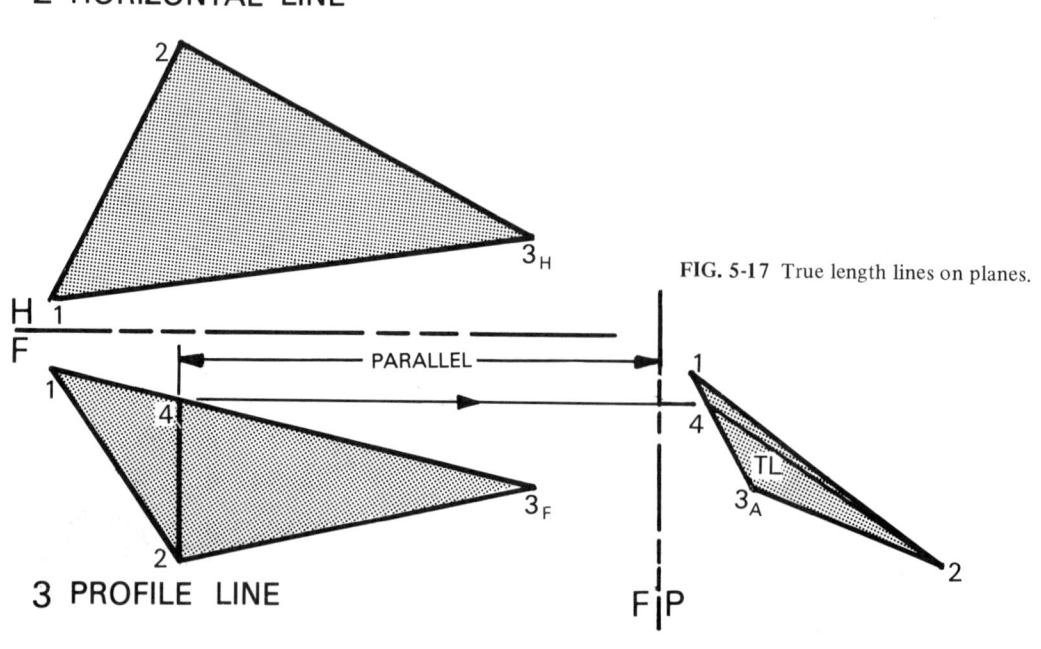

FIG. 5-17 True length lines on planes.

1 FRONTAL LINE

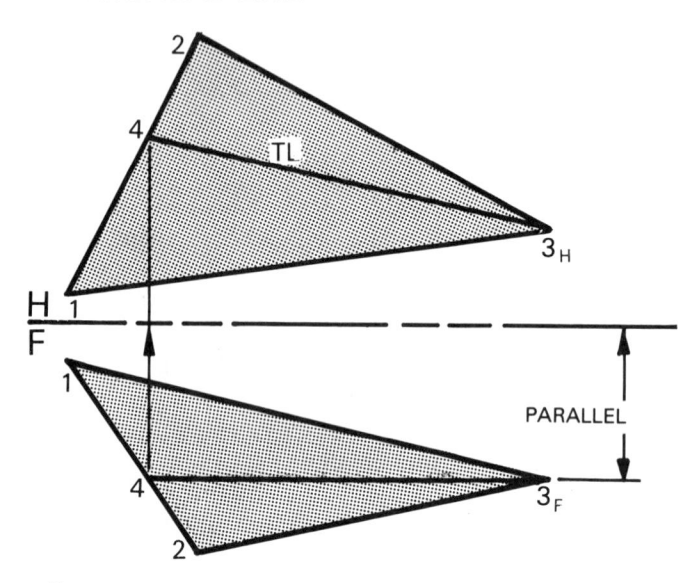

2 HORIZONTAL LINE

3 PROFILE LINE

FIG. 5-18 Line on a plane.

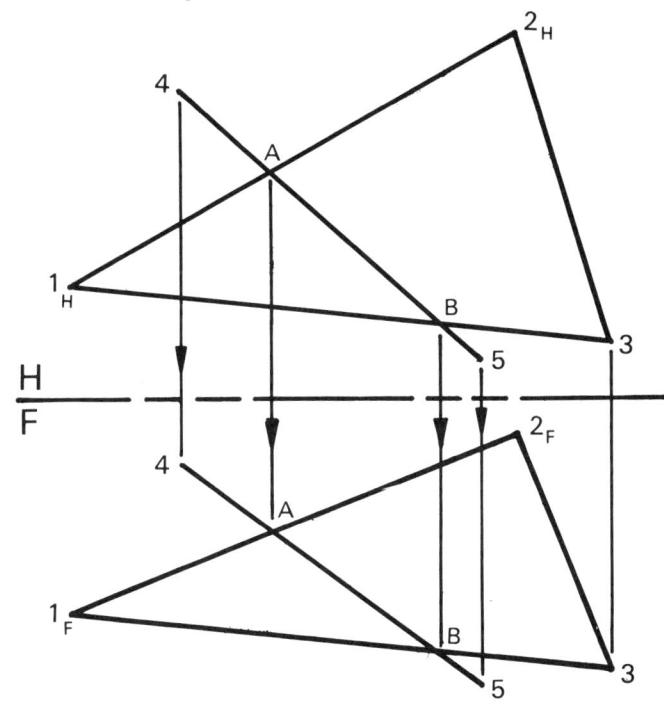

5.10 LOCATING PARALLEL AND NONPARALLEL LINES ON PLANES

Location of a line on a plane is a common requirement for solving a variety of descriptive geometry problems as they relate to actual industrial applications. In Fig. 5-18, plane 1-2-3 is given along with the horizontal view of line 4_H-5_H, which is said to lie on the plane. Note that the line could have been located in the frontal view and the horizontal view have been required. Line 4_H-5_H crosses line 1_H-2_H at point A_H and line 1_H-3_H at point B_H. Project points A_H and B_H to the frontal view where they still lie on their respective lines at points A_F and B_F. Draw a line through these points and extend it beyond the borders of the plane. Draw projection lines from points 4_H and 5_H in the horizontal view until they intersect extended line A_F-B_F. This fixes the position of points 4_F and 5_F and consequently line 4_F-5_F in the frontal view.

The wings and vertical stabilizer of the swing-wing supersonic aircraft are basically planes attached to the cylindrical fuselage, Fig. 5-19. Descriptive geometry concepts and procedures were used to design and model this experimental aircraft.

When a line is to lie on a given plane and parallel to one of its edges, the location process is simplified if one uses the theory of parallelism as applied to lines in space. *Parallel lines are parallel in all views.* Line 1 and line 2 are established in the horizontal view of Fig. 5-20. Plane 1-2-3 is given and the frontal projection of the lines is required. Line 1, line 2, and line 3-4 are parallel. Since parallel lines project parallel in all views, it is necessary to locate only one point where each line crosses an edge of the plane. Line 1 crosses line 1_H-4_H at point A_H, and line 2 crosses line 1_H-4_H at point C_H. Crossing points A and C are located along line 1_F-4_F in the frontal view by projection. Line 1 is drawn parallel to line 3_F-4_F and line 2 is drawn parallel to 3_F-4_F and line 1 in the frontal view. The end points of the lines are then determined by projection from the horizontal view.

FIG. 5-19 Model of swing wing aircraft. *(Courtesy NASA.)*

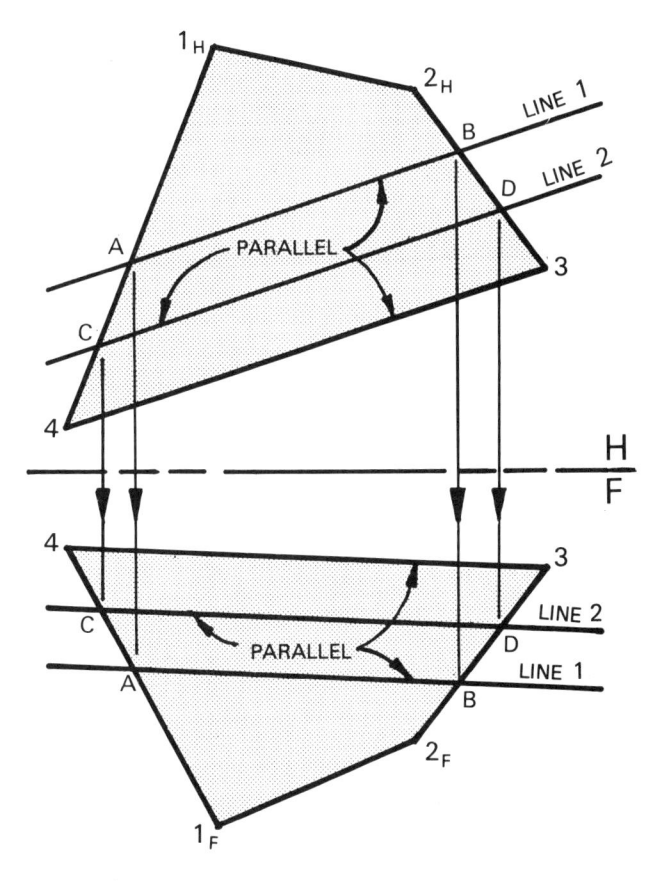

FIG. 5-20 Parallel lines on plane.

PROB. 5-5

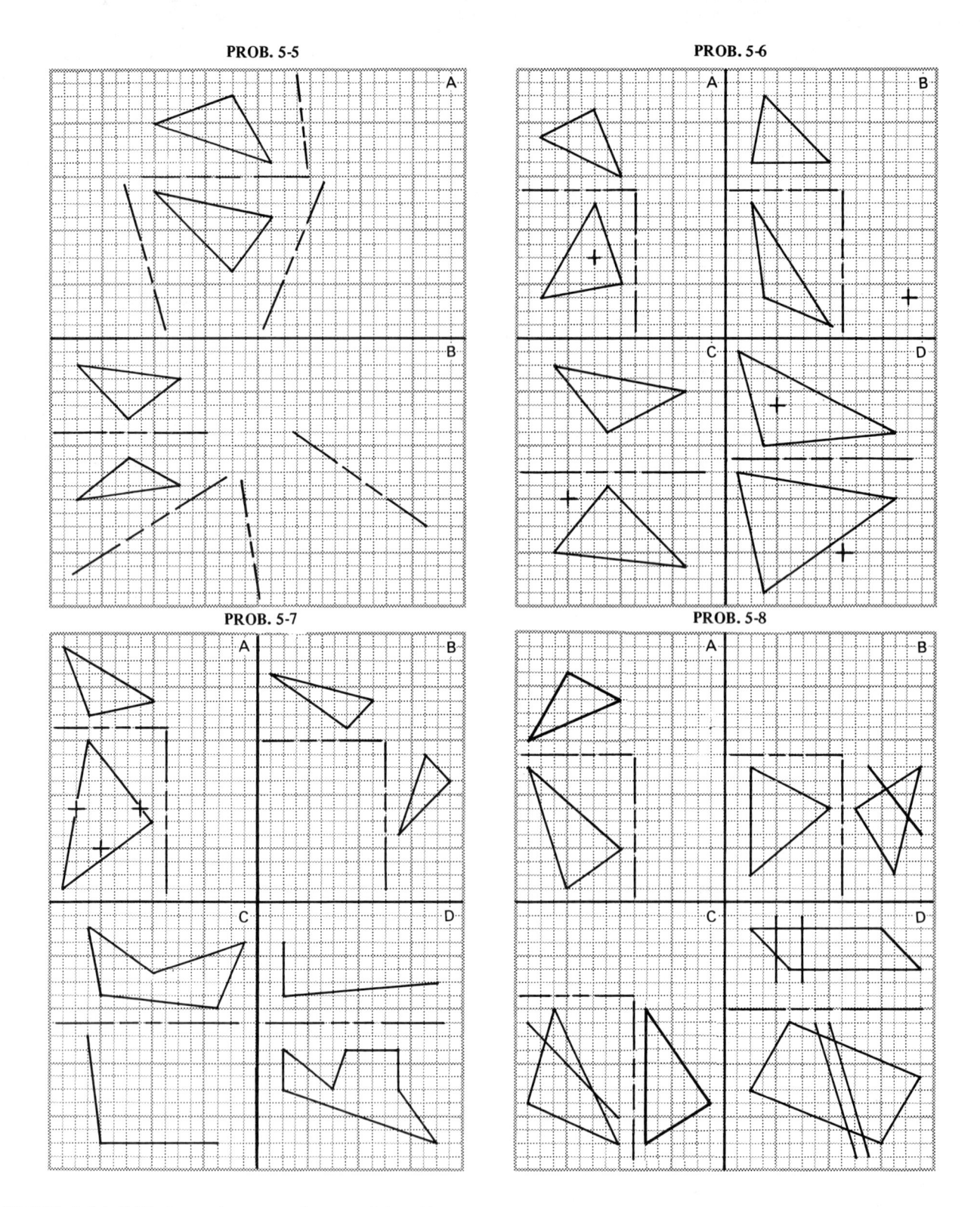

PROB. 5-6

PROB. 5-7

PROB. 5-8

PROBS. 5-5A, 5-5B. Draw the required views of the oblique planes.
PROB. 5-6A. The given point lies on the plane; complete all views.
PROB. 5-6B. Project the H and F views of the point so as to lie on the plane.
PROBS. 5-6C, 5-6D. Establish the point on the plane in each view using plane extension.
PROB. 5-7A. Show all three points on the plane and connect to form an inner plane.

PROB. 5-7B. Project the front view of the plane. Draw three evenly spaced frontal lines on the plane and show in all views.
PROBS. 5-7C, 5-7D. Complete the views of the planes.
PROB. 5-8A. Establish a profile, horizontal, and frontal line on the plane.
PROBS. 5-8B, 5-8C, 5-8D. Complete the views of the lines on the given planes.

5.11 EDGE VIEW OF A PLANE

The *edge view of a plane is seen in a view where the line of sight is parallel to the plane.* The line of sight is parallel to the plane when it is parallel to a true length line that lies on the plane. Since a projection plane is always perpendicular to the line of sight, it follows that a view drawn perpendicular to a plane (therefore perpendicular to a true length line that lies in the plane) shows the plane as an edge. This can be seen in a vertical plane which appears as an edge in the horizontal view, since it is perpendicular to the horizontal projection plane. A horizontal plane is perpendicular to the frontal and profile projection planes and thus appears as an edge in these two views.

When the given plane is oblique, an auxiliary projection is needed. In order to establish a line of sight parallel to the plane, a true length line needs to be drawn which lies on the plane. An auxiliary view where the line appears as a point view shows the plane as an edge. In Fig. 5-21, plane 1-2-3 is given and an edge view is required.

1. Draw line 1_H-4_H on plane 1_H-2_H-3_H, parallel to H/F, and complete the frontal view by projection. Line 1_F-4_F is true length.
2. Project auxiliary view A perpendicular to plane 1_F-2_F-3_F. The line of sight for this projection is parallel to the plane and parallel to true length line 1_F-4_F. Draw F/A perpendicular to 1_F-4_F and complete auxiliary view A by projection.
3. Auxiliary view A shows line 1_A-4_A as a point view and therefore plane 1_A-2_A-3_A appears as an edge view.

In Fig. 5-22, the edge view of plane 1-2-3 is shown in an auxiliary projection taken from the horizontal view. Normally this is the preferred practice since the slope (dip) angle is seen in a view where the plane is an edge and the horizontal projection plane appears as an edge. In this figure a horizontal line was drawn parallel to H/F and thus appears as true length in the horizontal view. H/A is drawn perpendicular to line 2_H-4_H. In auxiliary view A, line 2_A-4_A appears as a point view and plane 1_A-2_A-3_A as an edge view.

FIG. 5-21 Edge view of a plane.

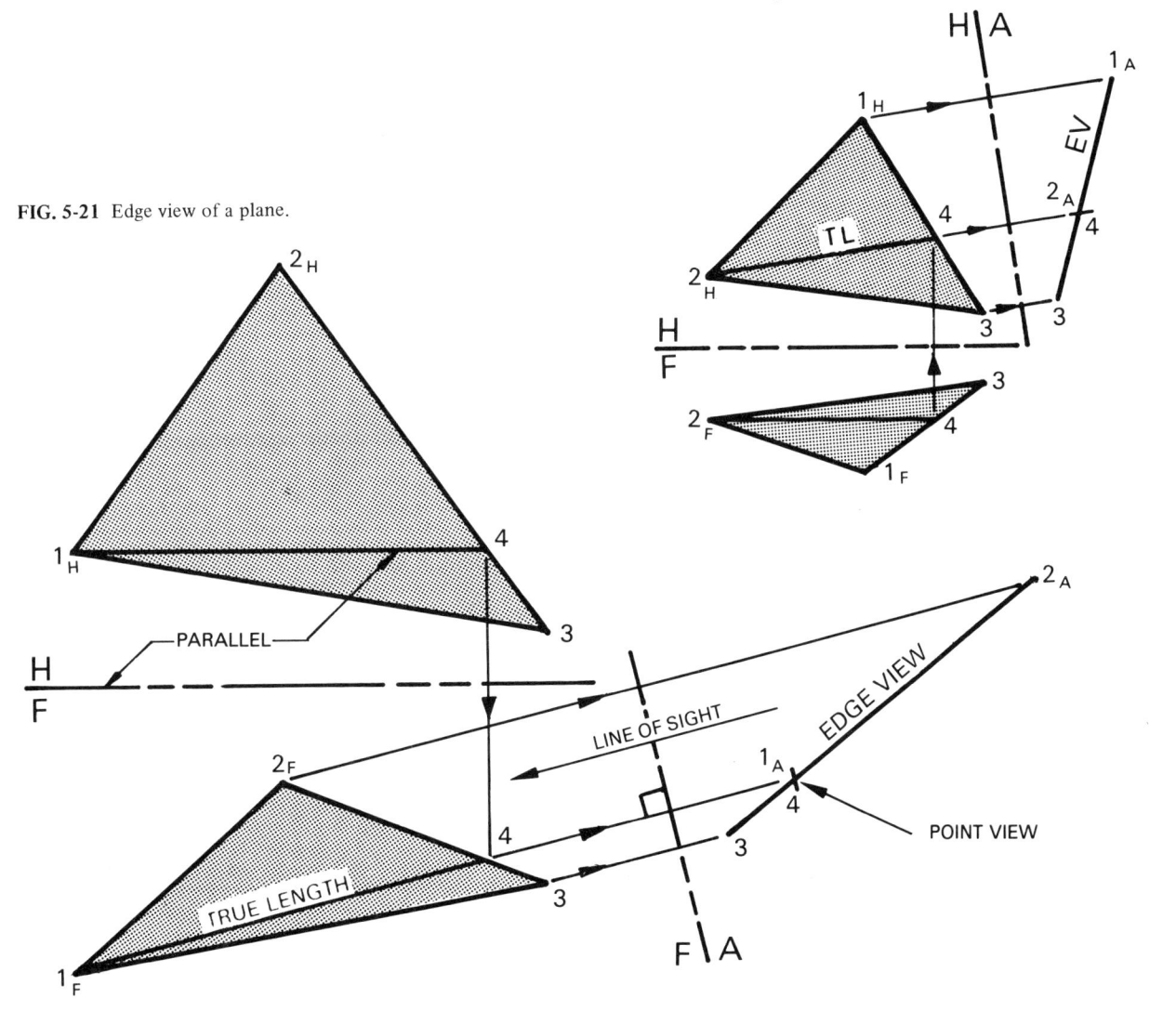

FIG. 5-22 Edge view.

FIG. 5-23 Edge view of a plane by primary auxiliary view.

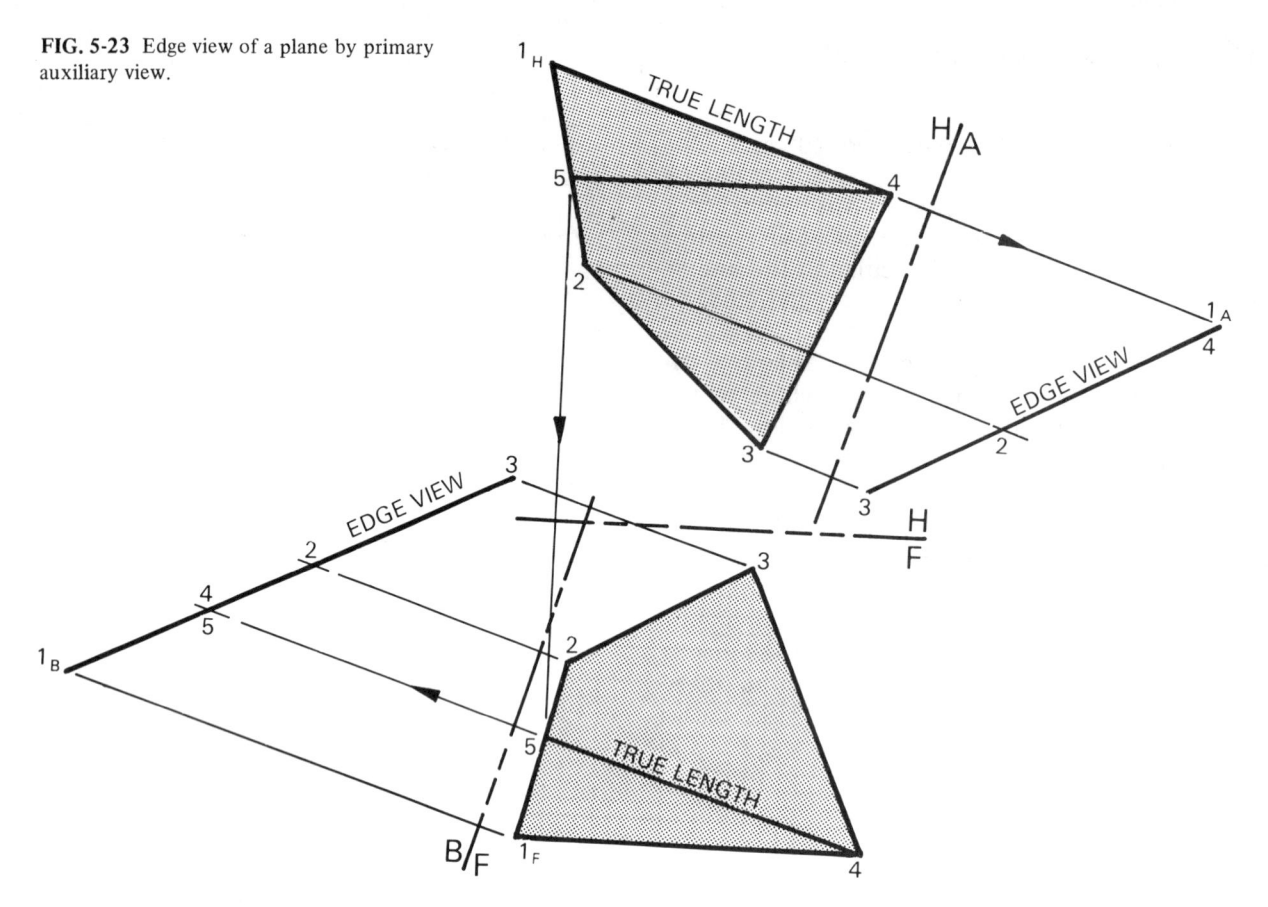

5.12 EDGE VIEW OF PLANE BY PRIMARY AUXILIARY VIEW

The edge view of a plane is seen in a view where one of its lines appears as a point. The point view of a line is obtained by projecting a view with a line of sight parallel to the true length line (the fold line is perpendicular to the line). In cases where one of the existing lines of a given plane is parallel to the fold line and therefore true length in the adjacent view, the construction of a true length line is unnecessary. In Fig. 5-23, both of these conditions are presented. Line 1_F-4_F is parallel to H/F and therefore is true length in the horizontal view. Fold line H/A is drawn perpendicular to line 1_H-4_H and auxiliary view A is projected with a line of sight parallel to the plane (and line). In auxiliary A, line 1_A-4_A appears as a point, thus plane 1_A-2_A-3_A is an edge.

In Fig. 5-23, an edge view auxiliary projection taken from the frontal view requires the construction of a true length line that lies on plane 1_F-2_F-3_F. Frontal line 4_H-5_H is drawn parallel to H/F and projected to the frontal view, where it appears true length. Auxiliary B is projected perpendicular to the plane. Its line of sight is parallel to the true length line and therefore is parallel to the plane. B/F is drawn perpendicular to line 4_F-5_F. Line 4_B-5_B is a point view in auxiliary B and plane 1_B-2_B-3_B appears as an edge.

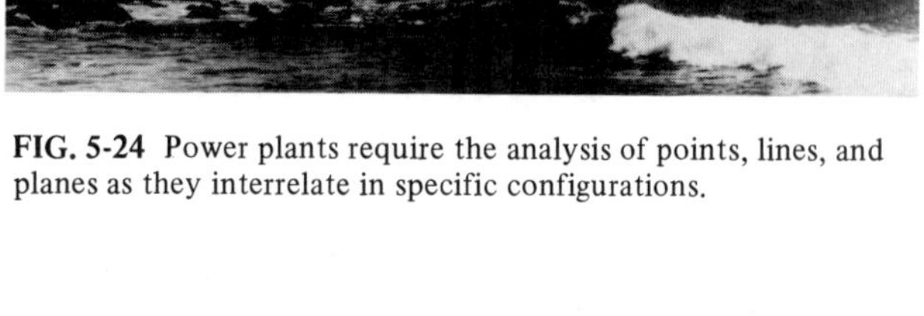

FIG. 5-24 Power plants require the analysis of points, lines, and planes as they interrelate in specific configurations.

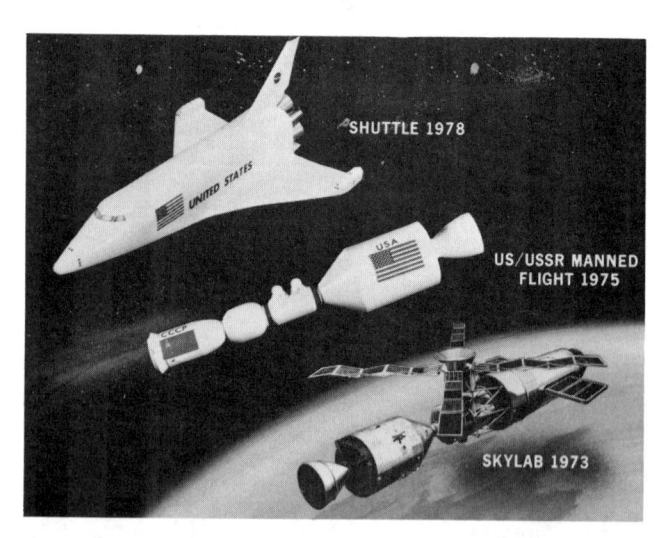

FIG. 5-25 The space program uses engineering data derived from mathematical and graphical spatial analysis. *(Courtesy NASA.)*

FIG. 5-26 True size of an oblique plane.

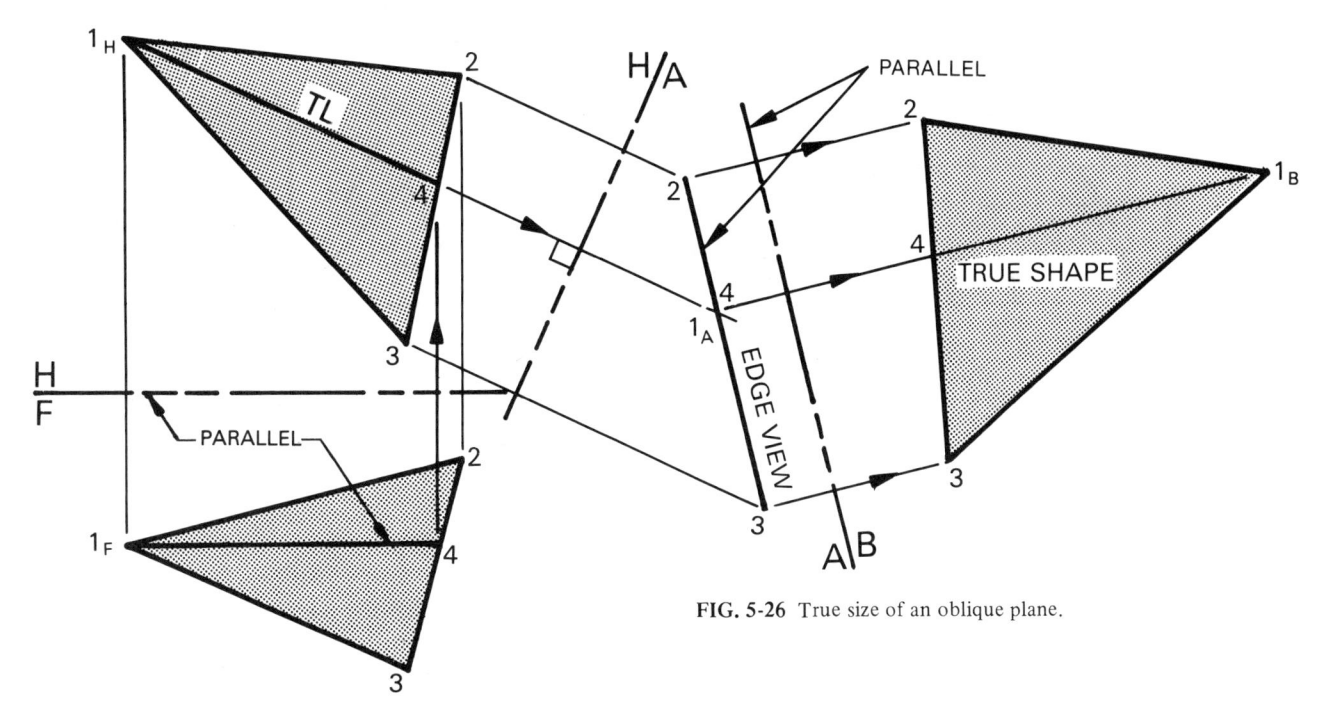

5.13 TRUE SIZE (SHAPE) OF AN OBLIQUE PLANE

*When the line of sight is perpendicular to the edge view of a plane it projects as **true size (shape)**.* The true size view is projected parallel to the edge view of the plane. Therefore the fold line between the views is drawn parallel to the edge view. An oblique plane does not appear as true size in any of the principal projection planes. Therefore a primary auxiliary and secondary auxiliary view is needed to solve for the true shape of an oblique plane.

In Fig. 5-26, oblique plane 1-2-3 is given and its true shape is required.

1. Draw horizontal line 1_F-4_F parallel to H/F and show it as true length in the horizontal view.

2. Draw H/A perpendicular to line 1_H-4_H and complete auxiliary view B. Line 1_A-4_A is a point view and plane 1_A-2_A-3_A an edge.

3. Project secondary auxiliary view B parallel to the edge view of plane 1_A-2_A-3_A. Draw A/B parallel to the edge view.

4. Complete auxiliary B; plane 1_B-2_B-3_B is true size (shape).

In Fig. 5-27, line 3_H-4_H of plane 1_H-2_H-3_H-4_H is parallel to H/F and therefore is true length (a frontal line) in the frontal view. To project an edge view of the plane, F/A is drawn perpendicular to 3_F-4_F. Auxiliary A shows line 3_A-4_A as a point and plane 1_A-2_A-3_A-4_A as an edge. A/B is drawn parallel to the edge view and auxiliary B completed. This secondary auxiliary view shows the plane as true shape. *Note that in a true shape/size view of a plane, all lines are true length and all angles are true angles.*

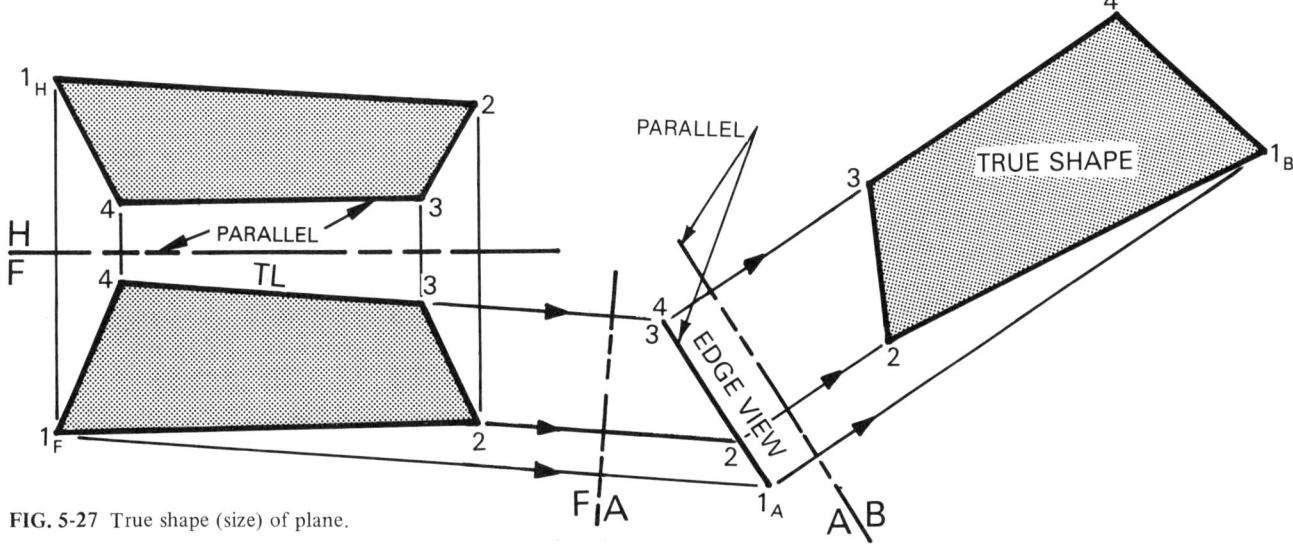

FIG. 5-27 True shape (size) of plane.

151

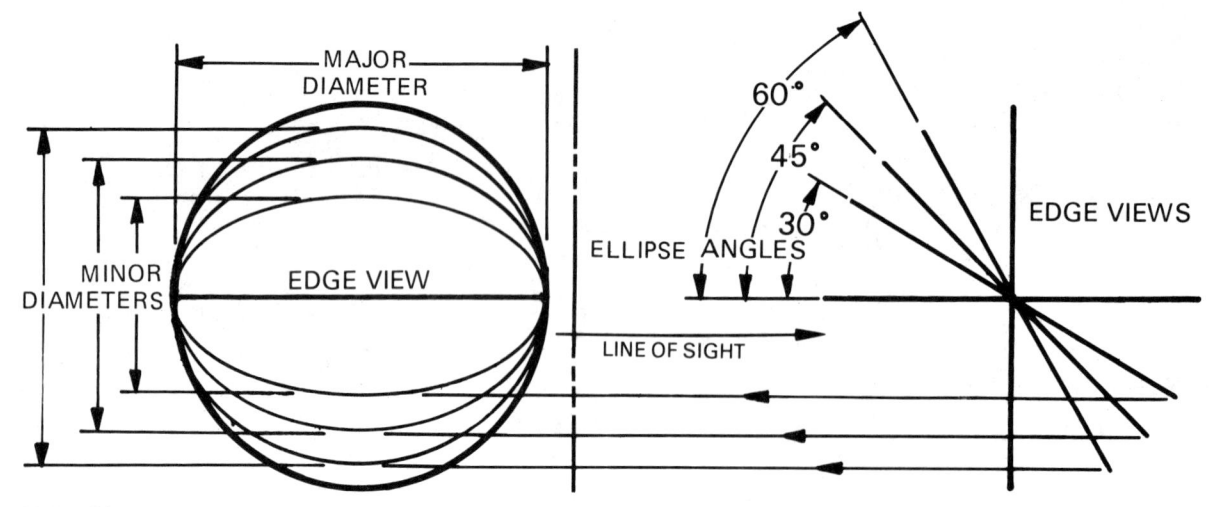

FIG. 5-28 Common ellipse angles.

5.14 VIEWS OF CIRCULAR PLANES

Circular planes are true size/shape in a view where the line of sight is perpendicular to the edge view of the plane. In the adjacent projection the plane appears as an edge and parallel to the fold line. The length of the edge view line is equal to the circle's diameter.

When a circular plane is oblique, it appears as an *ellipse*. In Fig. 5-28, the circular plane is shown at a variety of common angles to the line of sight. The *major diameter* remains the same as the circle is tilted, angled to the observer. The *minor diameter* is the smallest diameter in an elliptical view and changes as the ellipse angle changes. The minor diameter is always perpendicular to the major diameter. The angle formed by the edge of the plane and the line of sight is the *ellipse angle*. As the angle decreases, the minor diameter becomes smaller until both views are edge views. Whenever possible, one should use an ellipse template to draw the required elliptical form. Where the ellipse angle is an uncommon size or the circle diameter odd or oversize, one of the methods for ellipse construction described in Chapter 2 should be used.

An elliptical view of a circular plane along with each adjacent auxiliary view is plotted by locating a series of points along the outline of the circle in a true size view. These points are located in each adjacent view by projection and transferring distances to establish each individual point and connecting the series with a template or French curve. In Fig. 5-29, a normal view (true size) of the circular plane is

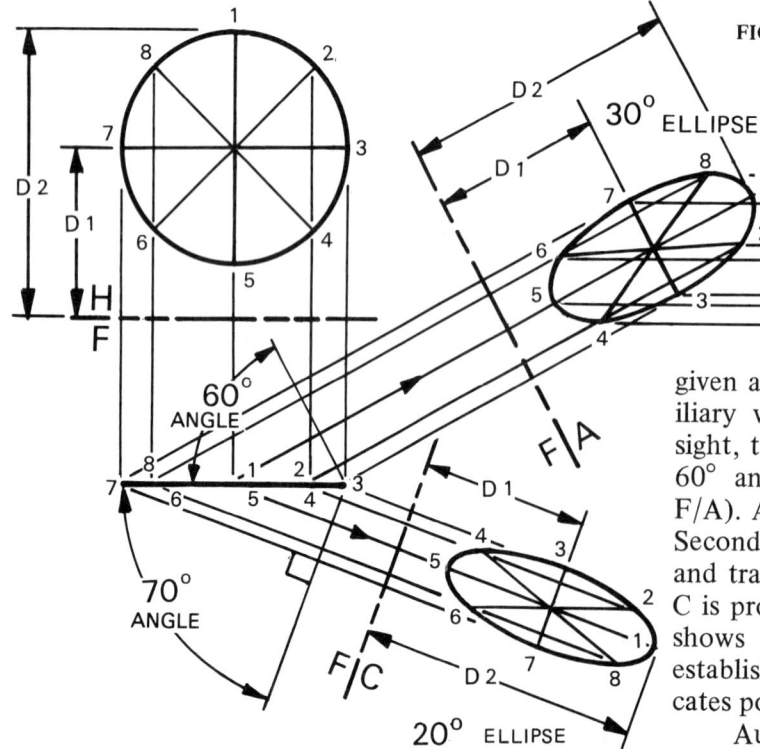

FIG. 5-29 Auxiliary views of a circle.

given along with its frontal edge view. Primary auxiliary view A forms a 30° angle with the line of sight, therefore the edge view of the plane forms a 60° angle with the adjacent view (and fold line F/A). Auxiliary A shows the plane as a 30° ellipse. Secondary auxiliary view B is drawn by projection and transferring distances for each point. Auxiliary C is projected at a 70° angle to the edge view and shows as a 20° ellipse. Note that dimensions D1 establishes points 3 and 7, and dimension D2 locates point 1, in auxiliary A and C.

Auxiliary B is a secondary auxiliary view.

FIG. 5-30 Plane figure on a given plane.

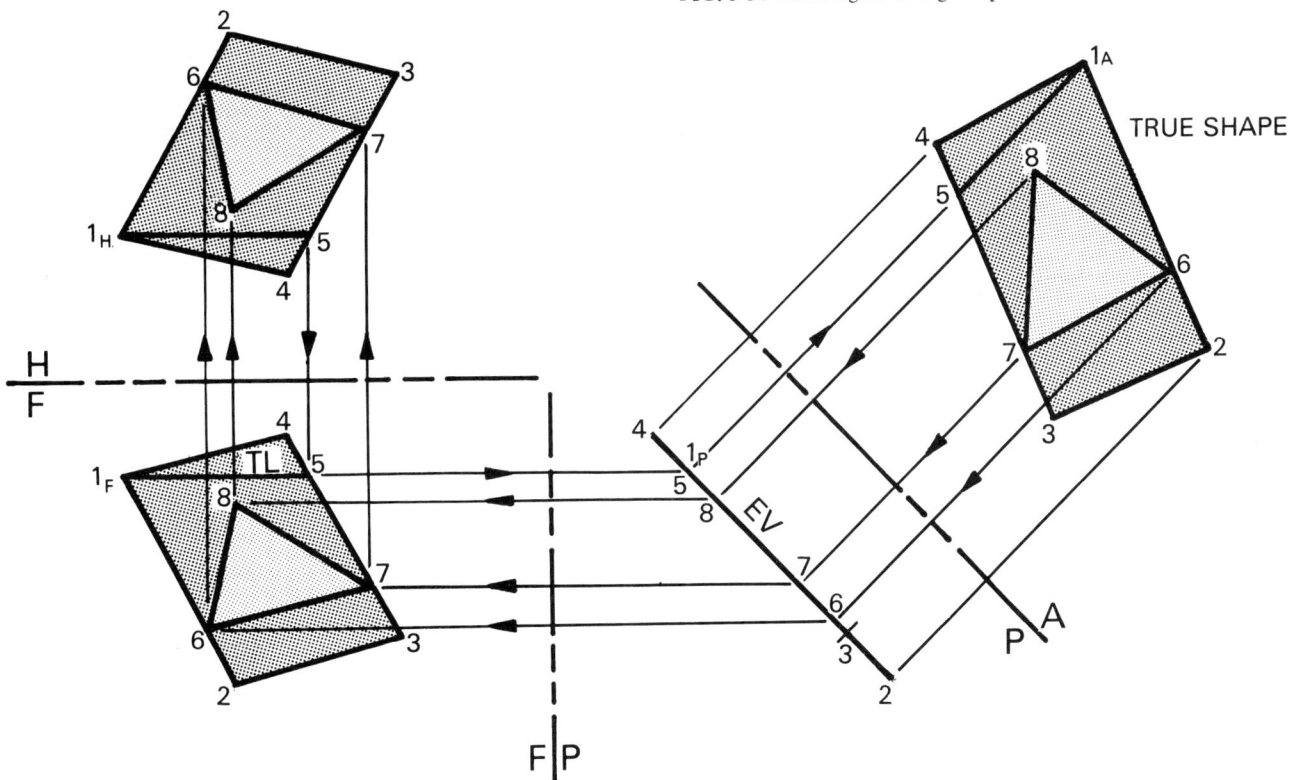

5.15 PLANE FIGURE ON A GIVEN PLANE

A given (plane) figure can be located on a plane in a view where the plane is shown as true size/shape. If the plane is not a principal plane, then a view must be projected where the plane is seen as true shape. The true size/shape of the plane is located in a projection plane that is parallel to an edge view of the plane. This normally involves the use of a primary and secondary auxiliary view. Note that specifications regarding the exact location, placement, and size of the figure must be given in order to establish its position on the plane.

In Fig. 5-30, the size of triangular plane 6-7-8 is given. This figure is to be centered on plane 1-2-3-4 with point 6 on edge line 1-2 and point 7 on edge 3-4. The frontal and horizontal views are given.

1. An edge view of plane 1-2-3-4 is needed before a true size view can be projected. Draw frontal line 1-5 parallel to H/F in the horizontal view and true length in the frontal view.
2. The primary auxiliary view is taken perpendicular to true length line 1_F-5_F. Note that plane 1-2-3-4 happens to be an inclined plane since this view is a profile projection, being perpendicular to the H/F fold line. Plane 1_P-2_P-3_P-4_P appears as an edge and line 1_P-5_P a point view.
3. The line of sight for a true size view is perpendicular to the edge view. P/A is drawn parallel to the edge view and auxiliary A completed showing plane 1_A-2_A-3_A-4_A true size/shape.
4. Figure 6-7-8 is positioned on the true size view of plane 1_A-2_A-3_A-4_A according to given specifications.
5. The profile, frontal, and horizontal views of the triangular plane are located by projection. Remember that plane 6-7-8 lies in plane 1-2-3-4 and is located by individually projecting each of its points back to all views where they fall on the plane.

Research and development of new ideas for technology, such as the design of the experimental spacecraft shown in Fig. 5-31, involves a combination of graphical and mathematical solutions to spatial relationships. Descriptive geometry as applied in orthographic projection is utilized to develop a variety of possible spatial alternatives for new products and engineering procedures.

FIG. 5-31 Original concept of a reusable space vehicle. *(Courtesy NASA.)*

FIG. 5-32 Circles on planes.

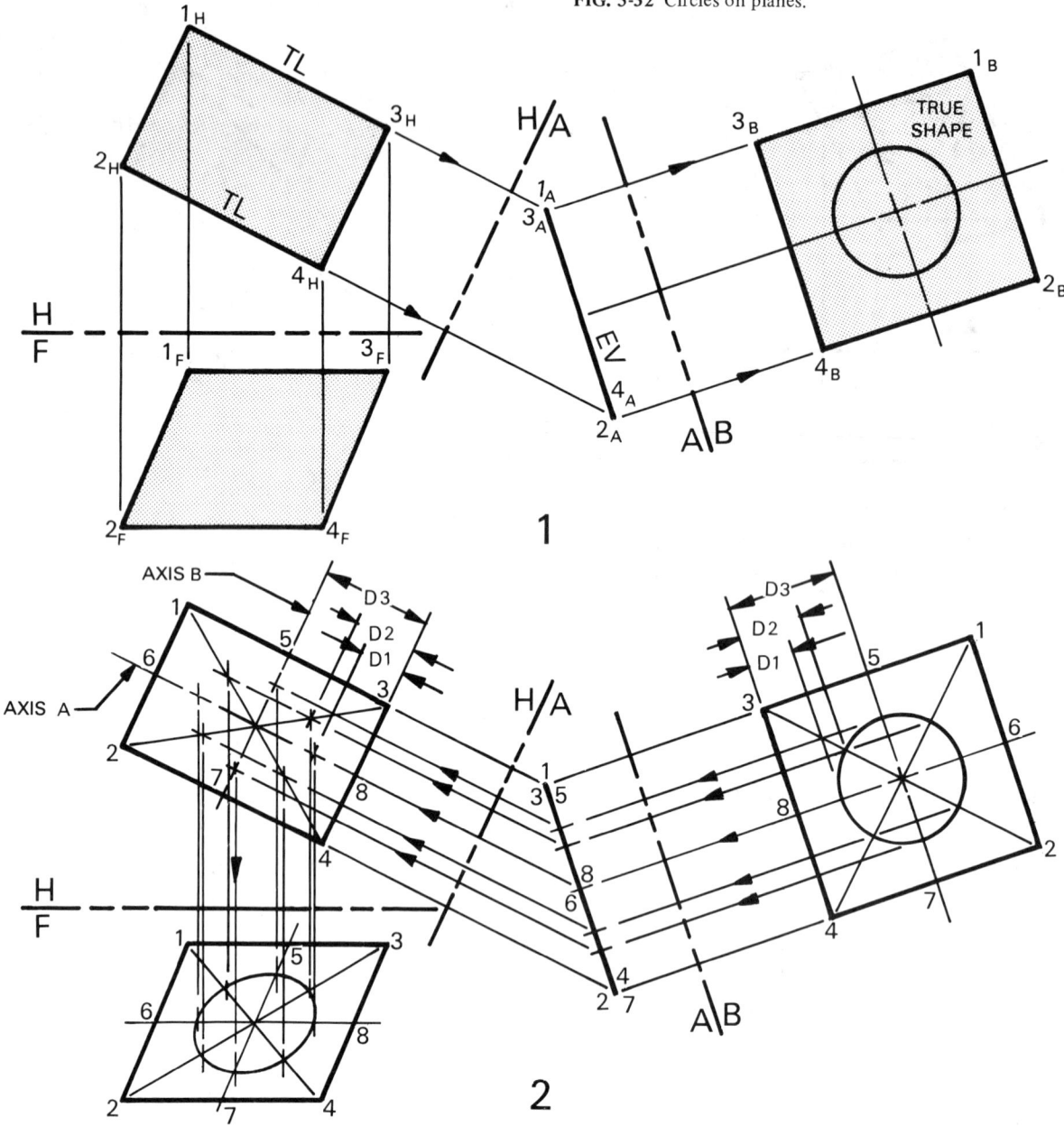

5.16 CIRCLES ON PLANES

To locate a given circle on a plane, a true shape view of the plane must be found. A typical problem found in industry is the location of a hole centered on a given surface. In Fig. 5-32, plane 1-2-3-4 is given and a hole/circle of a specific size is to drilled/drawn so that it is located in the exact center of the plane.

1. Line 1_H-3_H and line 2_H-4_H are horizontal lines (true length in the horizontal view). Therefore, a true length line need not be constructed to find the edge view. Draw H/A perpendicular to the horizontal lines and project auxiliary view A. Plane 1_A-2_A-3_A-4_A is an edge in this view (1).

2. Draw A/B parallel to the edge view of plane 1_A-2_A-3_A-4_A and project auxiliary B. This view shows the true size of the plane (1).

3. Locate the exact center of plane 1_R-2_R-3_R-4_R and draw the given circle (1).

4. To project the centered circle back to all previous views, a series of points needs to be located along its circumference. A convenient method to locate points on the circle is to divide the circle evenly by drawing lines from the corners of the plane (2).

5. Locate each point in auxiliary A by projection where they fall on the edge of plane 1_A-2_A-3_A-4_A. Dimensions D1, D2, and D3 are used to locate each point in the horizontal view by transferring them along their respective projection lines (2).

6. The frontal view of the circle is obtained by projection and transferring distances from auxiliary A (from H/A to each point on the edge view). Axis A (major diameter) and B (minor diameter) could also be used to locate and draw each view of the circle (2).

FIG. 5-33 Largest possible circle on a given plane figure.

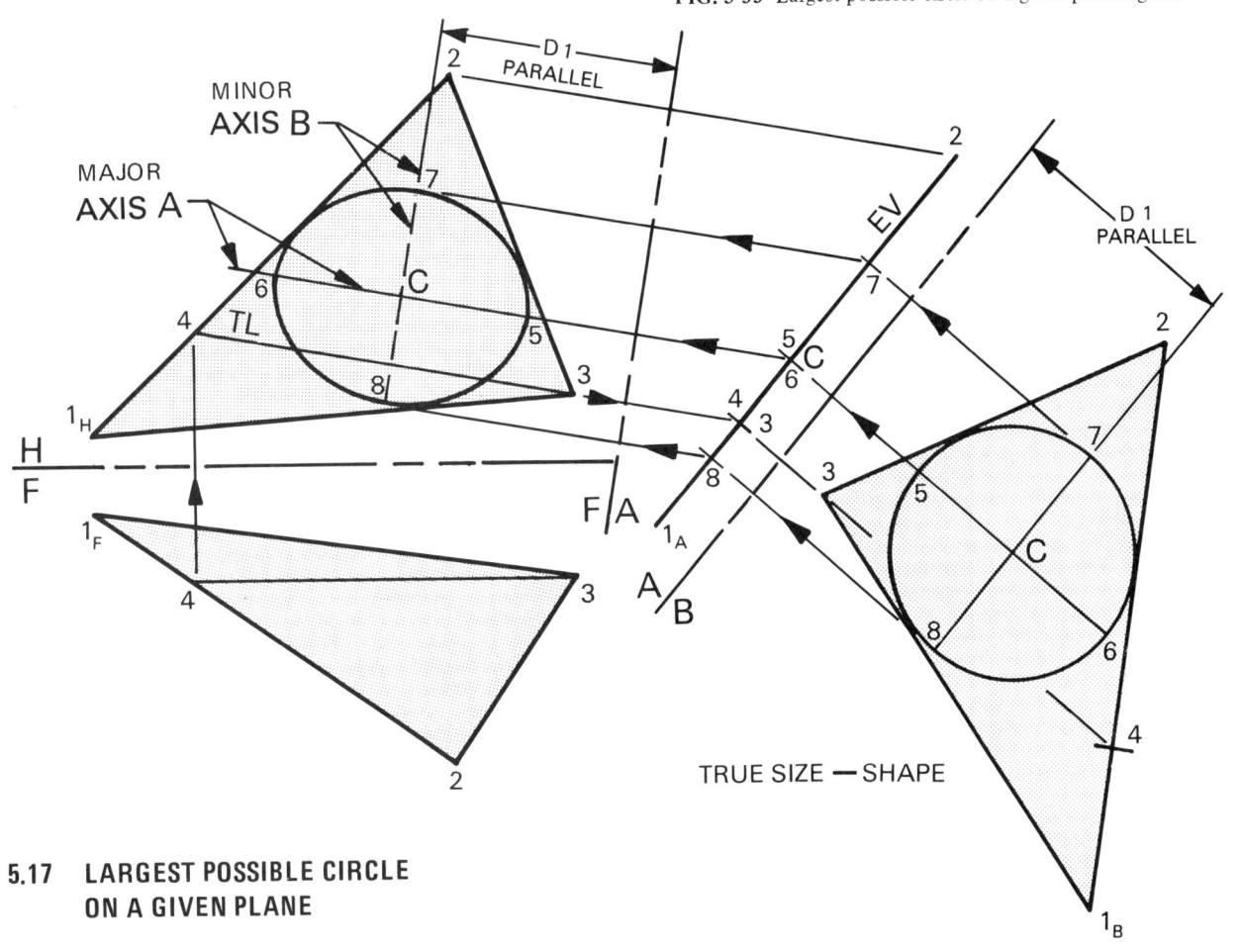

5.17 LARGEST POSSIBLE CIRCLE ON A GIVEN PLANE

The largest possible circle on a given oblique plane can be established in a view where the plane appears true size, Fig. 5-33. A circle can be drawn by using the method described in Chapter 2 for the construction of an inscribed circle of a triangle. This procedure establishes the location of the circle's center, C. The circumference of the circle is *tangent* to each leg of the triangle.

FIG. 5-34 Model builder detailing overhead crane. *(Engineering Model Associates, Inc.)*

After the true size view of the plane and circle is completed, the circle is projected back to the edge view. The major diameter, line 5_A-6_A, appears as a point view and the minor diameter is true length as measured from point 7_A to point 8_A along the edge view line. The horizontal view shows the major axis as a true length level line which is the diameter of the circle. Minor axis 7_H-8_H is perpendicular to major axis 5_H-6_H.

A circle projects as an ellipse in all views where it is not an edge view or true shape. If the major and minor axes are located in each oblique view, the ellipse can be constructed using the four-center, trammel, or any acceptable method as found in Chapter 2 or in a technical drawing text. This procedure eliminates the cumbersome practice of establishing a series of points along the circumference of the circle as outlined in the last section.

Note that the frontal view of the circle is not shown. This projection can be constructed by locating C (center) and both axes by projection, and transferring distances between F/A and the axes end points on the edge view to the frontal view.

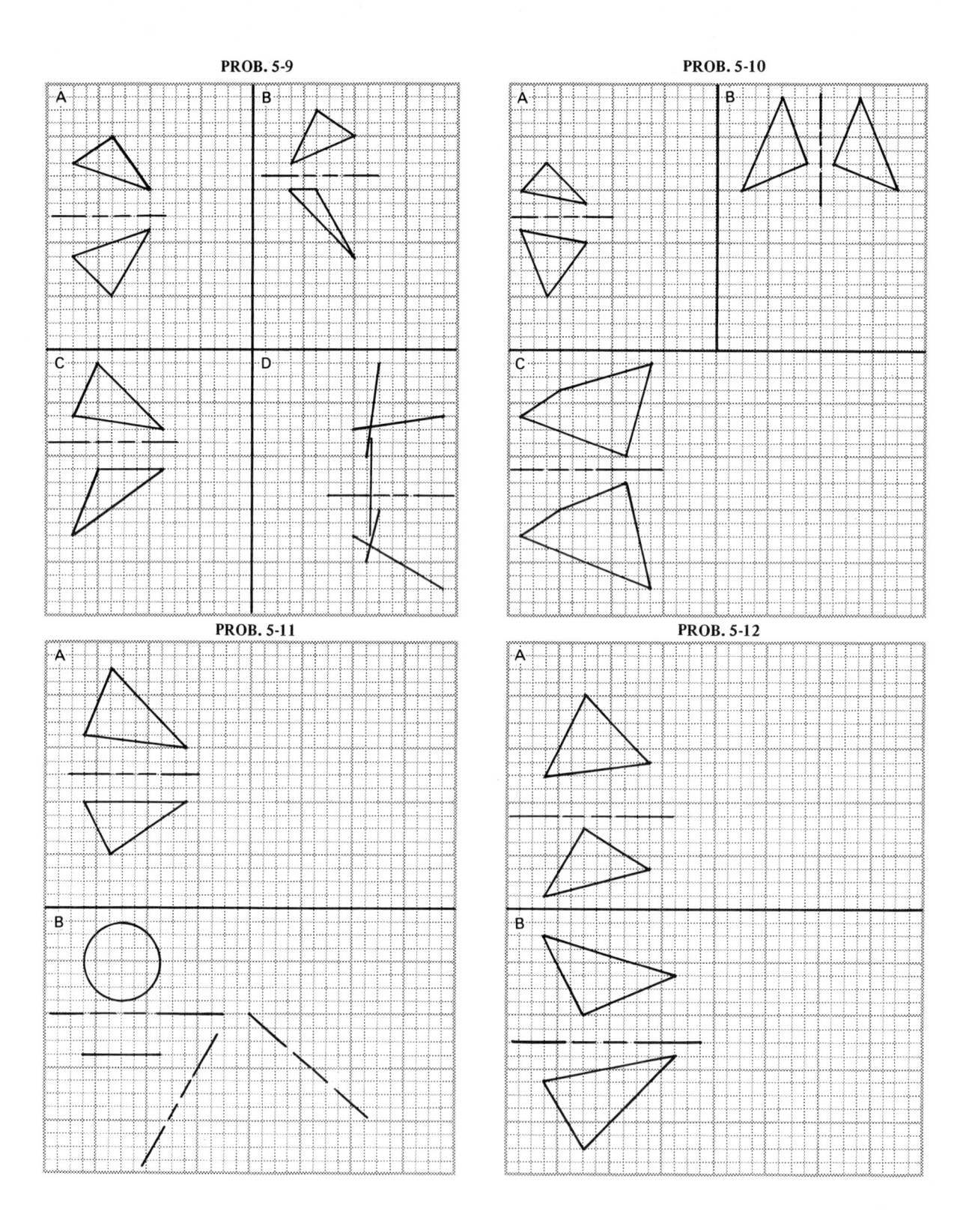

PROBS. 5-9A, 5-9B, 5-9C, 5-9D. Solve for the edge view of each plane.
PROBS. 5-10A, 5-10B, 5-10C. Project the edge view and true shape of each plane.
PROB. 5-11A. Solve for the edge view and true shape of the plane. Place the largest circle inside the plane and show in all views.

PROB. 5-11B. Complete the required views of the circular plane and draw the largest possible square inside the plane. Show the square in all views.
PROBS. 5-12A, 5-12B. Solve for the largest circle within the plane and show in all views.

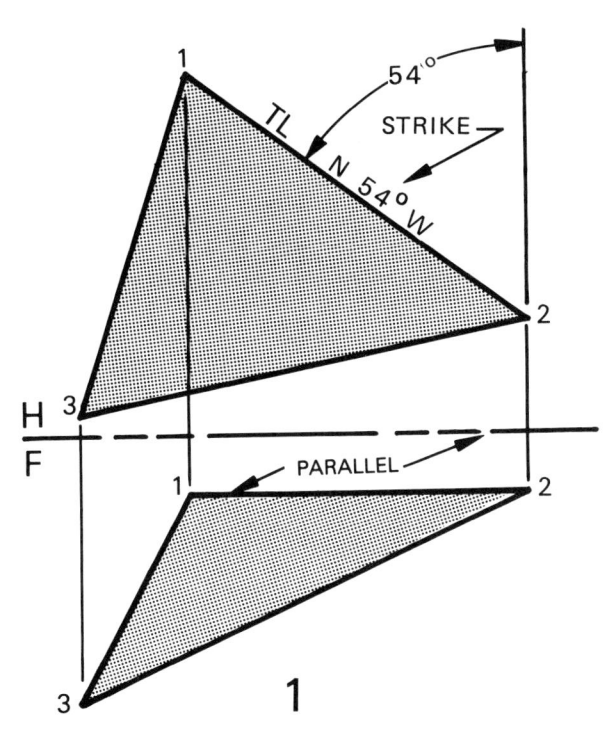

FIG. 5-35 Strike of a plane.

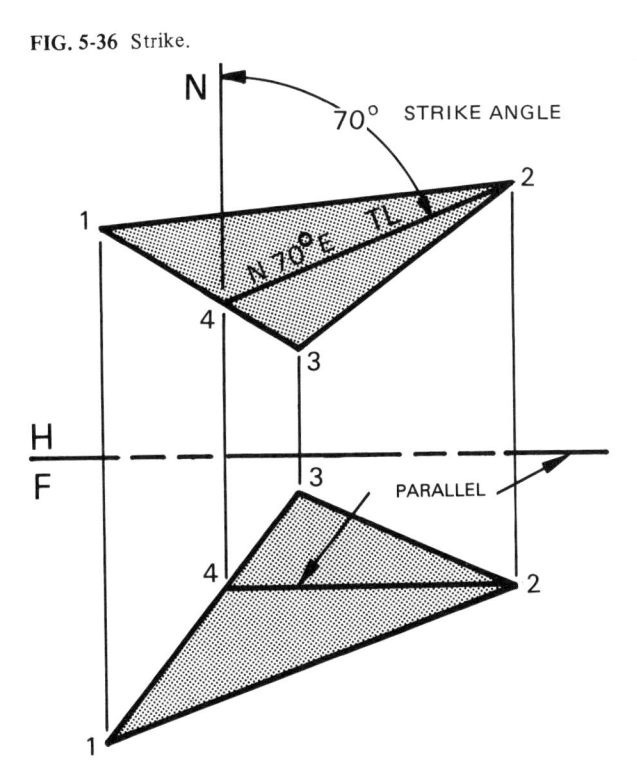

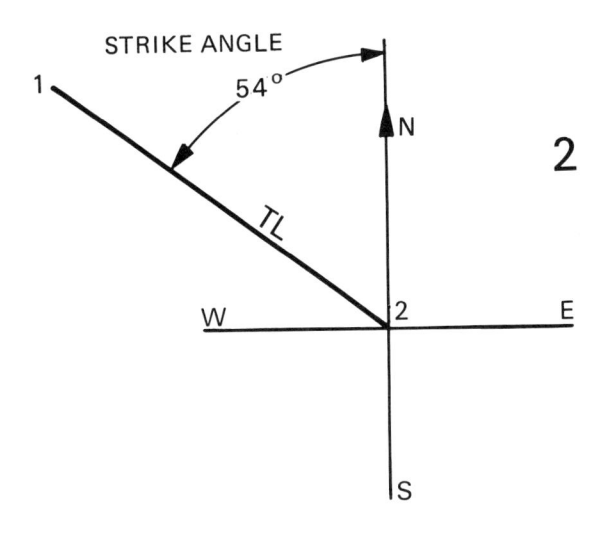

FIG. 5-36 Strike.

5.18 STRIKE OF A PLANE

*The **strike** of a plane is the bearing of a horizontal line which lies in the plane.* The strike is normally measured from the north as an acute angle. The strike line is a level line (horizontal line) and is therefore not slanted toward one of its end points. The strike is measured from a north-south meridian toward the east or west. Any true length line in the horizontal view can be used to measure the strike of the plane. The most common uses involving the strike of a plane involve mining and geological applications. This concept and its engineering applications are presented in detail in Chapter 8, Mining and Geology.

In Fig. 5-35(1), plane 1-2-3 is given and its strike is required. Line 1-2 is a horizontal line, being parallel to H/F in the frontal view and true length in the horizontal view. Note that line 1_F-2_F is "level" and therefore does not slope in the elevation view. The strike of the plane is equal to the bearing of horizontal line 1_H-2_H. The angle that line 1_H-2_H makes with the north-south meridian is measured as an acute angle from north (2). Line 1_H-2_H makes a 54° angle with the north-south reference and bears toward the west. The bearing of line 1_H-2_H is north 54 degrees west, N 54°W. An azimuth reading can also be used, N 306°.

In Fig. 5-36, plane 1-2-3 is oblique. In order to solve for the strike of the plane, a horizontal line must be constructed. Line 2-4 is drawn parallel to H/F. The horizontal projection of line 2_H-4_H shows the line as true length. The strike of plane 1-2-3 is measured from north as an acute angle. The strike equals the bearing of true length line 2_H-4_H. The bearing of horizontal line 2_H-4_H is north 70 degrees east, therefore the strike of plane 1-2-3 is written N 70°E (AZ 70°).

5.19 SLOPE OF A PLANE

*The angle that an edge view of a plane makes with the horizontal plane is the **slope angle**.* The slope angle can only be measured in an elevation view where the plane is an edge and the horizontal projection plane is an edge. The slope cannot be seen in a primary auxiliary taken from the frontal or profile views. The slope must be established in the frontal view or in a primary auxiliary view taken from the horizontal view (an elevation view). Both the plane and the horizontal projection plane must appear as edges in the same view in order to see the slope angle of the plane. In mining and geology the slope of a plane is referred to as the *dip*. See Chapter 8, Mining and Geology.

FIG. 5-37 Slope of a plane, 33° NE.

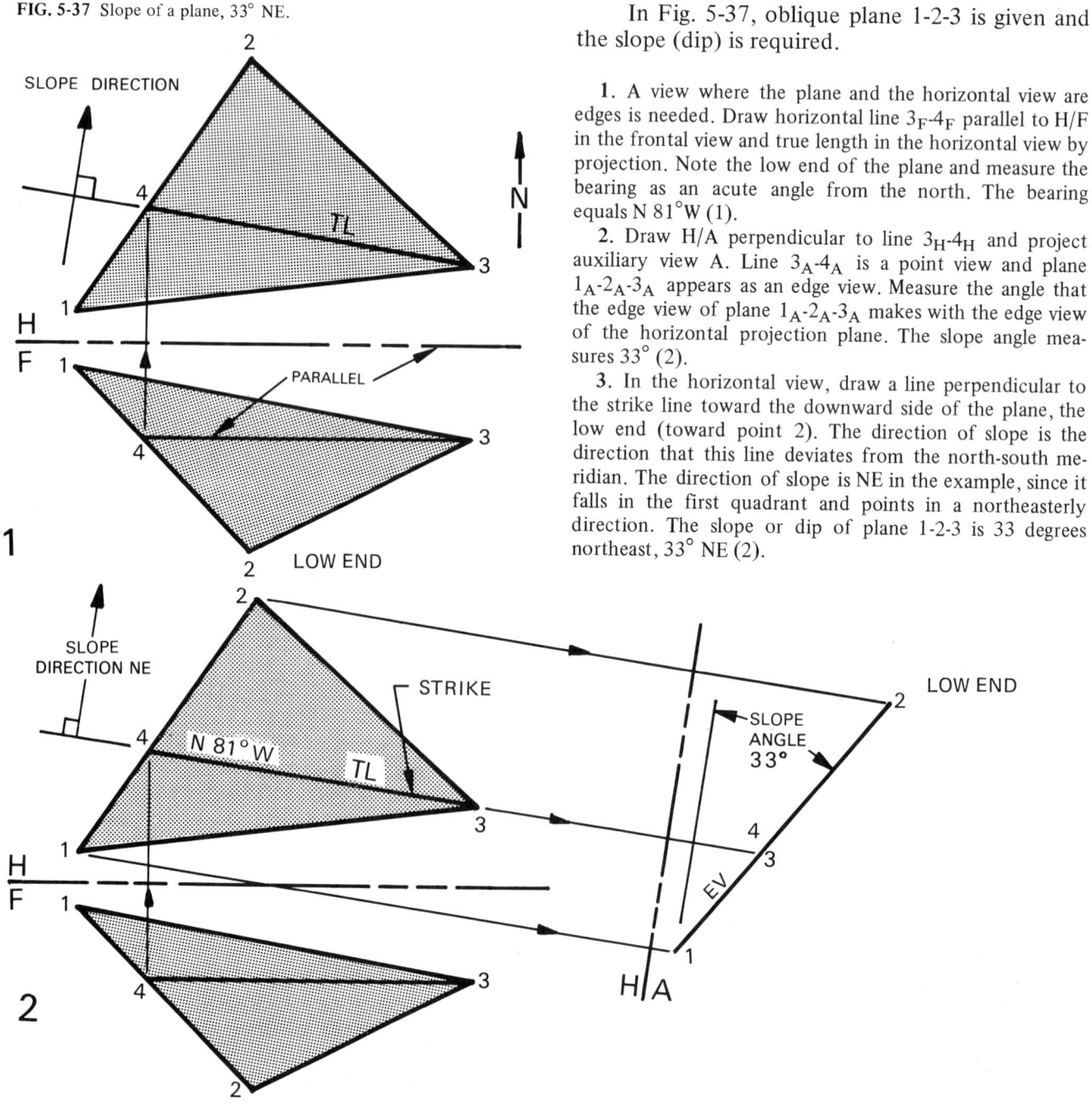

The **slope** or **dip** of a plane includes the *slope angle* (*dip angle*) and the general direction that the plane tips towards its downward end (low end). The direction of slope is established by drawing a line perpendicular to the strike line and towards the low end of the plane. The direction of the slope (dip) is read as the bearing of the line that is at a right angle to the strike line, giving only the compass directions not the degrees. Therefore the direction of slope is identified by NE, NW, SE, or SW, or by the four cardinal directions: N, S, E, or W.

When giving the slope or dip of a plane, the slope angle is stated first and the direction of slope (dip) second. An example of this is 45° NW: 45 degrees is the slope/dip angle and NW is the direction of slope/dip. This means that we have a plane which slopes 45° towards the northwest.

In Fig. 5-37, oblique plane 1-2-3 is given and the slope (dip) is required.

1. A view where the plane and the horizontal view are edges is needed. Draw horizontal line 3_F-4_F parallel to H/F in the frontal view and true length in the horizontal view by projection. Note the low end of the plane and measure the bearing as an acute angle from the north. The bearing equals N 81°W (1).

2. Draw H/A perpendicular to line 3_H-4_H and project auxiliary view A. Line 3_A-4_A is a point view and plane 1_A-2_A-3_A appears as an edge view. Measure the angle that the edge view of plane 1_A-2_A-3_A makes with the edge view of the horizontal projection plane. The slope angle measures 33° (2).

3. In the horizontal view, draw a line perpendicular to the strike line toward the downward side of the plane, the low end (toward point 2). The direction of slope is the direction that this line deviates from the north-south meridian. The direction of slope is NE in the example, since it falls in the first quadrant and points in a northeasterly direction. The slope or dip of plane 1-2-3 is 33 degrees northeast, 33° NE (2).

PROB. 5-13

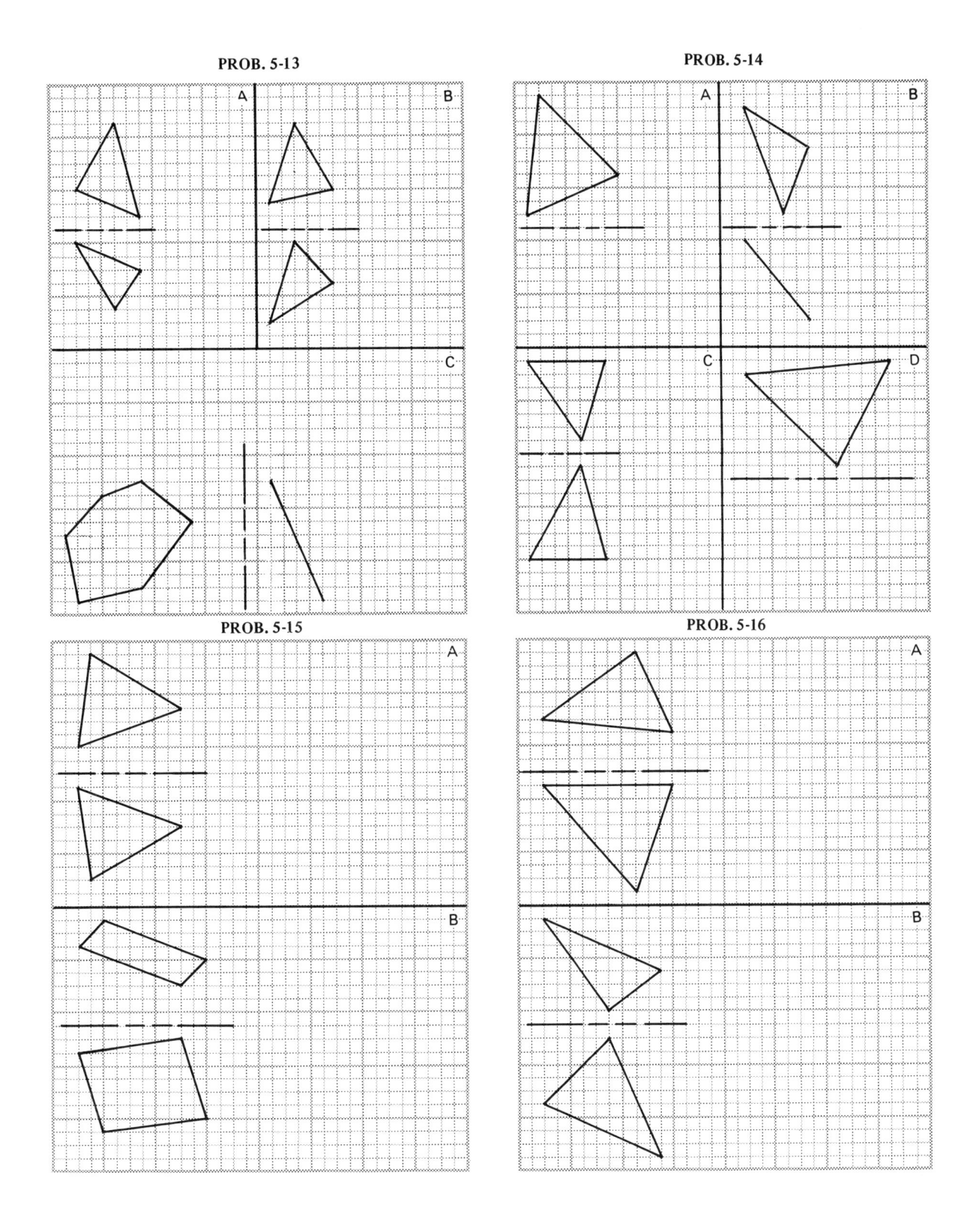

PROB. 5-14

PROB. 5-15

PROB. 5-16

PROBS. 5-13A, 5-13B. Determine the strike, direction of slope, and the slope angle.

PROB. 5-13C. Solve for the slope, strike, and true shape of the plane.

PROB. 5-14A. Determine the front view of the plane. It has a slope of 25° NE and a strike of N 85°W.

PROBS. 5-14B, 5-14C. What are the strike and slope of the plane?

PROB. 5-14D. Draw the missing view of the plane. The strike is due north, slope equals 30 degrees due east.

PROBS. 5-15A, 5-15B. Solve for the TS, EV, slope, and strike.

PROBS. 5-16A, 5-16B. Solve for the slope and strike of the plane. Construct an isosceles triangle with its apex at point 1 and base angles of 70° within the plane. Point 1 is the lowest point in (A) and the highest point in (B).

159

FIG. 5-38 Visibility of a line and a plane.

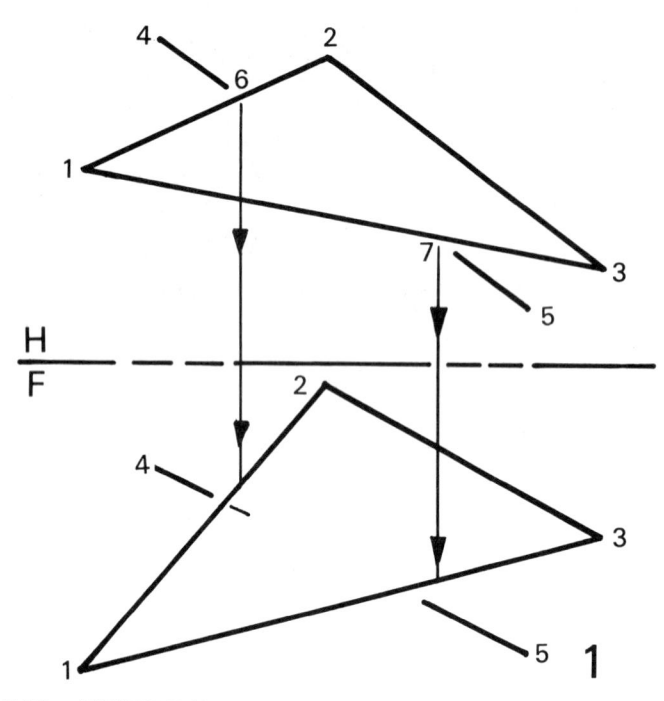

5.20 VISIBILITY

The visibility of a line and a plane can be determined by inspection of the relative location of each in relation to the projection planes. When this method is unsatisfactory, the visibility test should be given in each view to every separate place where the line and plane cross. This procedure was first outlined on page 83 in Chapter 4. It should first be determined if the two lines intersect—have a common point. When the crossing point is different in each adjacent view, then the lines do not intersect. To establish visibility, draw a sight line from the apparent point of intersection (crossing), perpendicular to the fold line between adjacent views. The first line that is met in the adjacent view will be visible in the view where the two lines crossed, where the visibility test originated.

In Fig. 5-38(1), plane 1-2-3 and line 4-5 are given; the proper visibility is required.

1. Line 4_H-5_H crosses line 1_H-2_H at point 6. Draw a sight line from point 6_H, perpendicular to the fold line, to the front view. The projector intersects line 1_F-2_F first. Therefore line 1_H-2_H is above line 4_H-5_H and visible in the horizontal view. Where line 4_H-5_H crosses line 1_H-3_H at point 7, draw a sight line to the frontal view. The projector intersects line 1_F-3_F first; therefore in the horizontal view, line 1_H-3_H is visible and line 4_H-5_H is hidden. Since the plane is above the line at both crossings it is solid (visible in the horizontal view) and the line is dashed (hidden).

2. The same procedure is used in the frontal view to solve for the visibility of the line and plane, only here the crossing points are projected to the horizontal view. Plane 1_H-2_H-3_H is encountered first by projectors extended from points 8_F and 9_F. Therefore, the plane is visible in the frontal view. Line 4_F-5_F appears hidden.

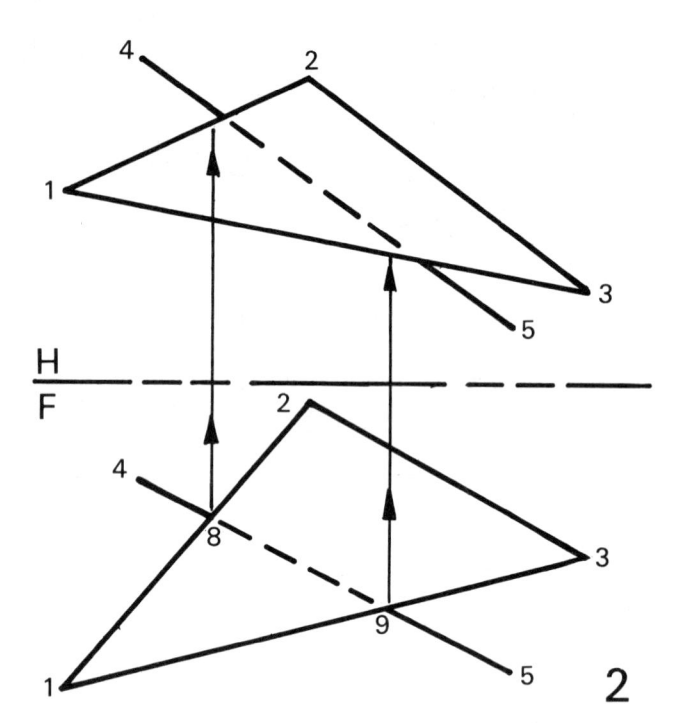

In Fig. 5-39, the visibility test establishes that the plane is above the line and thus is visible (solid) in the horizontal view. The line is in front of the plane so it appears solid (visible in the frontal view).

Note that *where a line is visible at one crossing and hidden at the other, the line and plane intersect.*

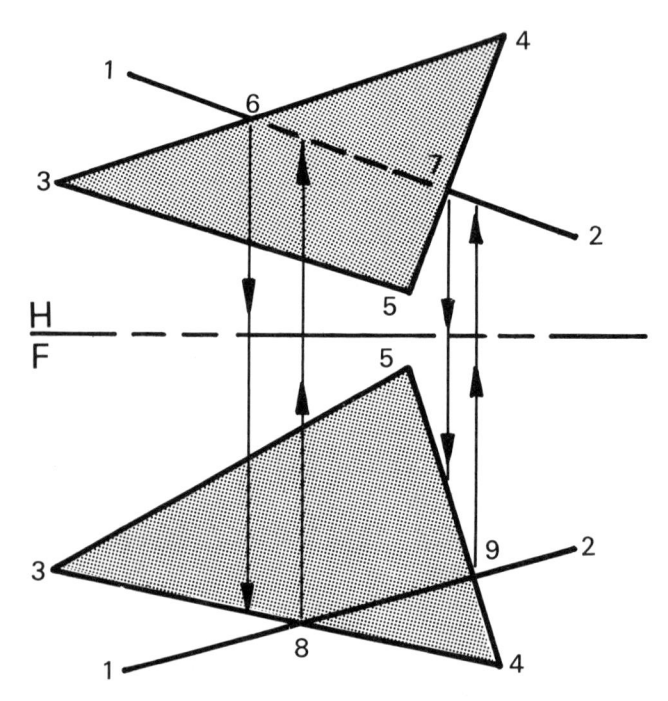

FIG. 5-39 Visibility.

FIG. 5-40 Construction of a plane parallel to a given line.

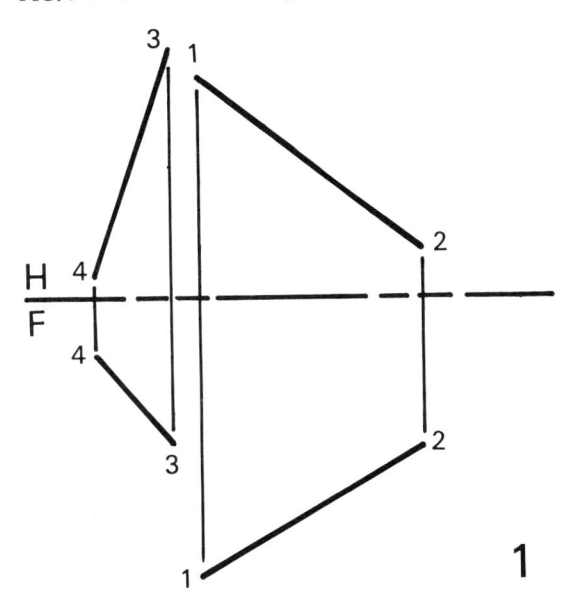

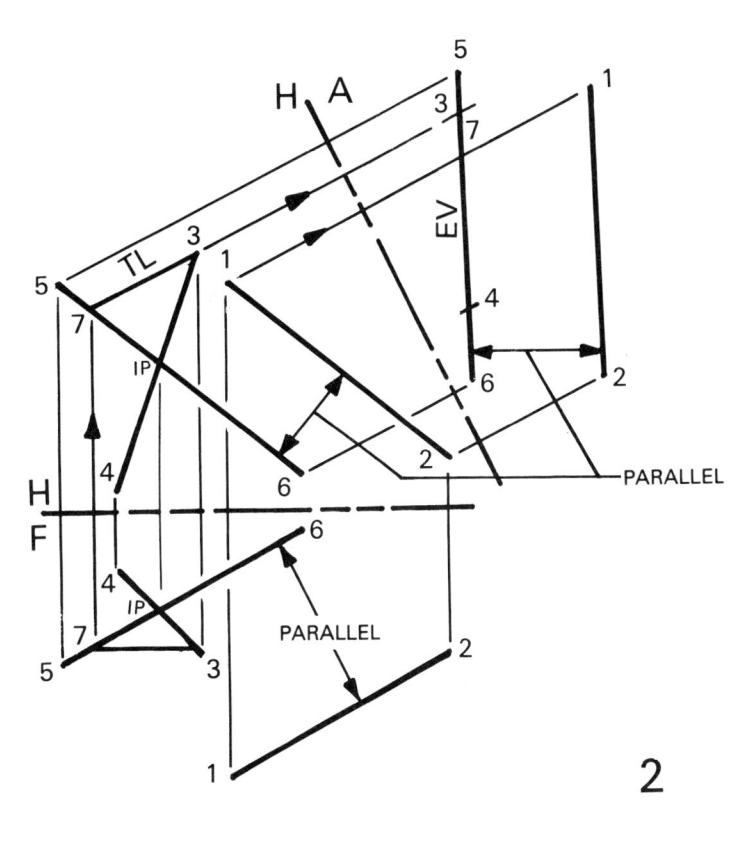

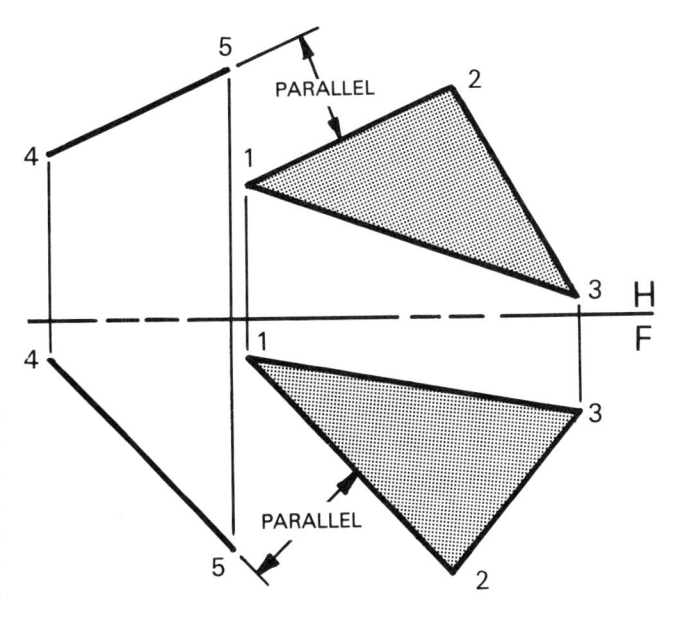

5.21 PARALLELISM OF LINES AND PLANES

A line and plane are parallel if the line is parallel to any line in the plane. To construct a plane using an existing line and parallel to a given line, it is necessary to draw a new line which intersects the existing line and is parallel to the given line. The newly formed plane is parallel to the given line, since the plane contains a line that is parallel to it. When forming a plane by drawing an intersecting line, the new line can be constructed anywhere along the existing line. The intersecting point (common point) of the intersecting lines is aligned in adjacent views. An auxiliary view showing the plane as an edge checks the solution, since the line appears parallel to the edge view of the plane in this projection.

A line is parallel to a plane if it is parallel to any line in the plane. This geometric theorem establishes the basis for the construction of a plane parallel to a given line. In Fig. 5-40(1), lines 1-2 and 3-4 are skew lines, nonintersecting and nonparallel. Construct a plane using line 3-4, which is parallel to line 1-2.

1. Draw line 5-6 a convenient length, through any point along line 3-4 and parallel to line 1-2. This procedure is used for both views. The end points of line 5-6 must line up in adjacent views. The common point of intersection (intersection point, IP) is also aligned in adjacent views (2).

2. To check the solution, project a view where the plane is an edge. Draw horizontal line 3_F-7_F parallel to H/F. Line 3-4 lies in plane 3-4-5-6 and appears true length in the horizontal view. Draw H/A perpendicular to horizontal line 3_H-7_H and project auxiliary A. Plane 3_A-4_A-5_A-6_A appears as an edge. Line 1_A-2_A projects as parallel to the plane (2).

To establish a line parallel to a given plane, draw the new line parallel to any line in the plane. In Fig. 5-41, plane 1-2-3 is given and a line parallel to it is required. Line 4-5 is drawn parallel to any line in the plane. Here, line 4-5 is drawn a convenient length and parallel to line 1-2 in both views.

FIG. 5-41 Construction of a line parallel to a given plane.

FIG. 5-42 Parallelism of planes.

In Fig. 5-42, planes 1-2-3 and 4-5-6 are parallel since intersecting lines in one plane are parallel to intersecting lines in the other. In this example, line 1-3 is parallel to line 4-5, and line 2-3 is parallel to line 5-6. Note that any line in one plane must be parallel to the other plane, since two planes in space must be parallel or intersect.

Normally, an auxiliary view is projected to prove parallelism of planes. Two planes are parallel when their edge views are parallel. An auxiliary view showing one of the planes as an edge shows the other as an edge and parallel, if they are in fact parallel planes. Note that two planes can be edges in a view and still not be parallel. The angle between the two planes can be measured in this type of view. The edge view of two parallel planes establishes the true distance (shortest perpendicular distance) between the two planes.

In Fig. 5-43, plane 1-2-3 and 4-5-6 are given. It is required to establish if the planes are parallel. If so, what is the true distance between them?

1. Draw horizontal line 6_F-7_F in plane 4-5-6, parallel to H/F and therefore true length in the horizontal view.
2. Project auxiliary A perpendicular to line 6_H-7_H by drawing H/A perpendicular to it.
3. In auxiliary A, both planes show as edges and also parallel to one another. The true distance between the planes is measured as the perpendicular distance between the two planes. The slope of the planes can also be measured in this view.

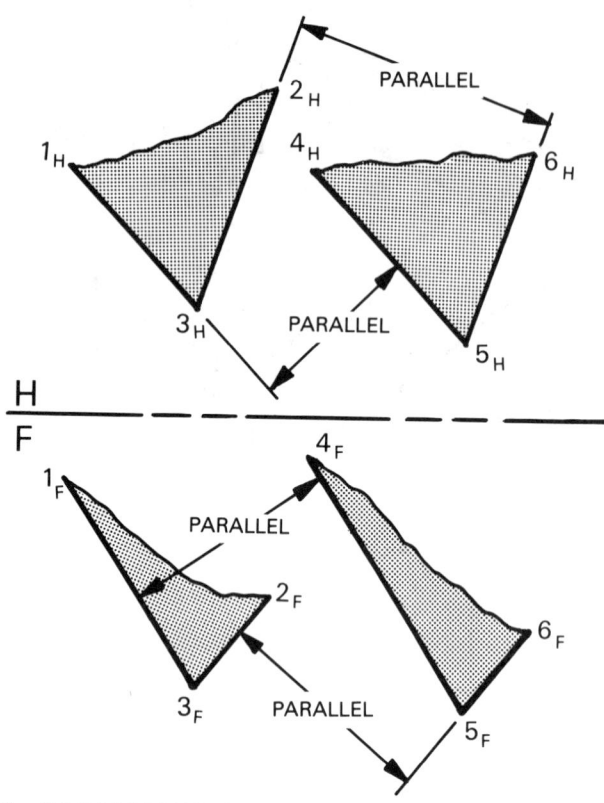

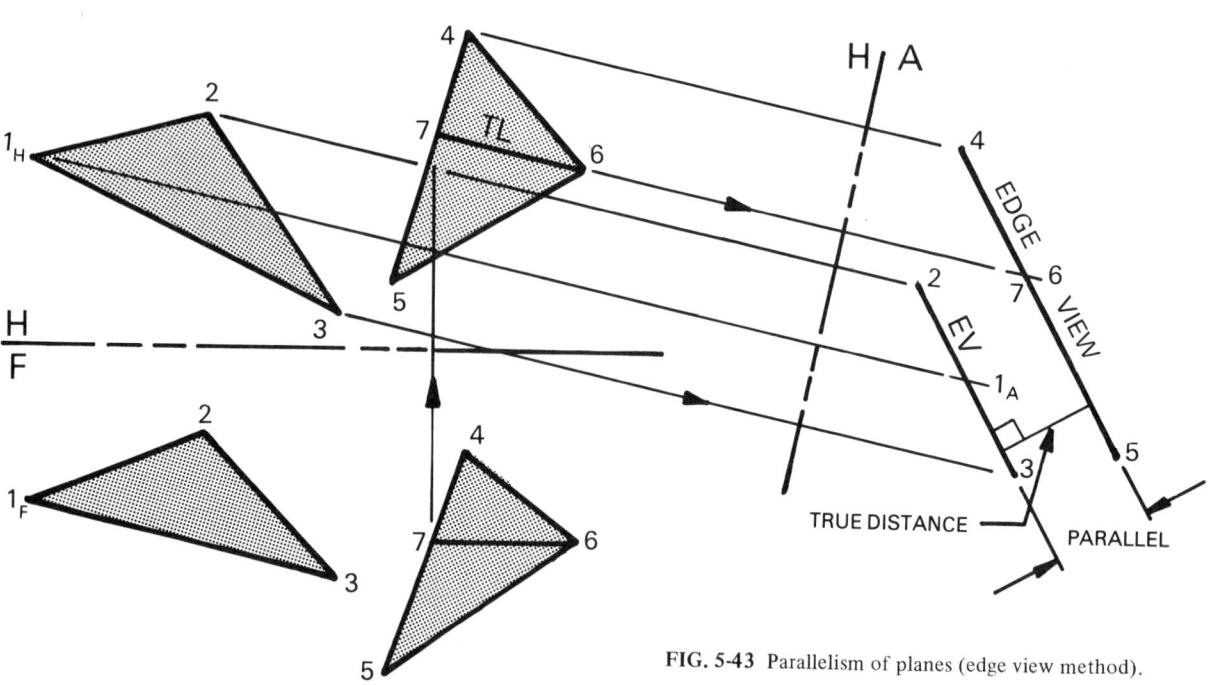

5.22 PARALLELISM OF PLANES

*Two planes are **parallel** if intersecting lines in one of the planes are parallel to intersecting lines in the other.* Parallelism is determined by attempting to draw a set of intersecting lines parallel to any two intersecting lines in the other plane. If the two sets of intersecting lines are parallel then the planes are parallel.

FIG. 5-43 Parallelism of planes (edge view method).

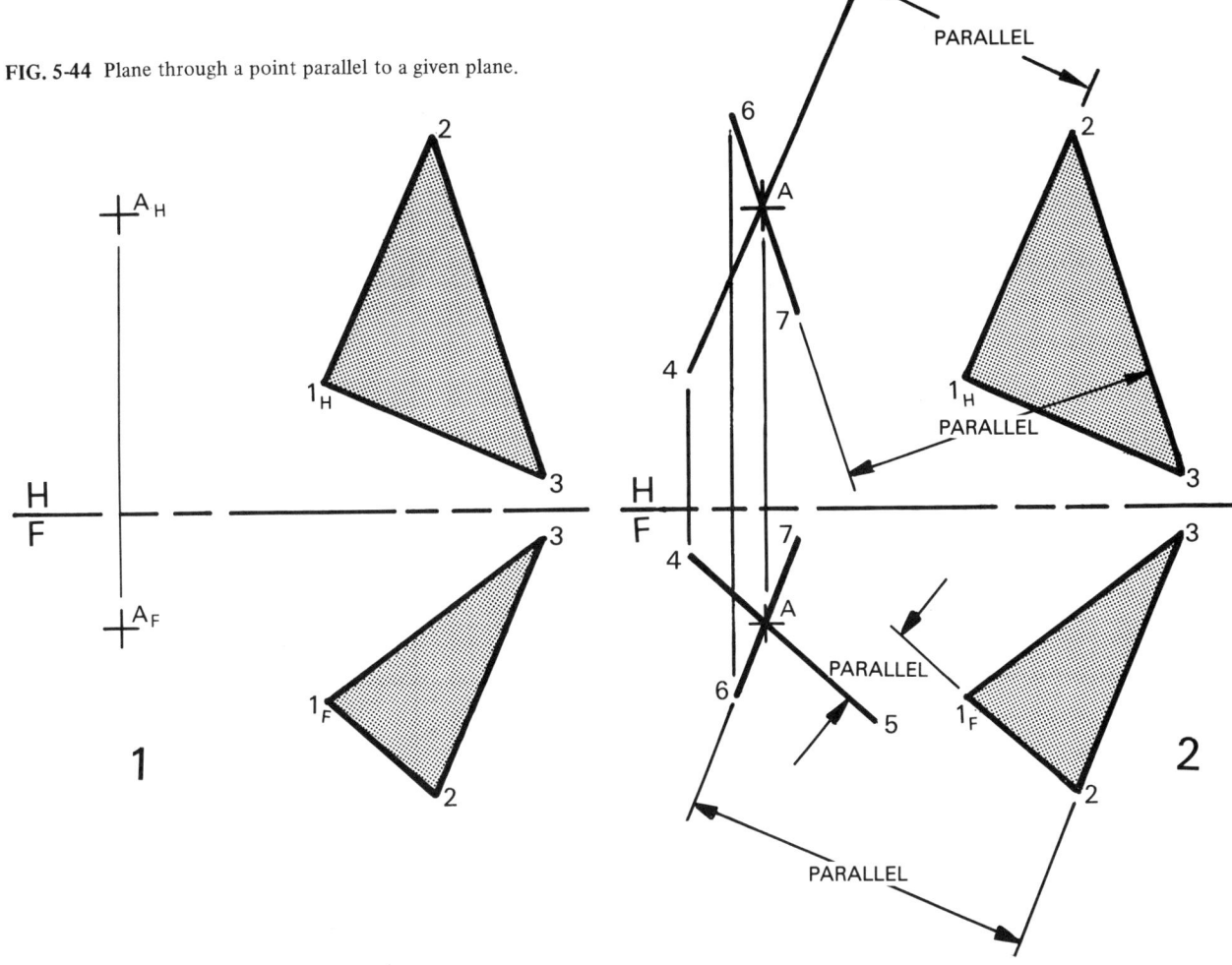

FIG. 5-44 Plane through a point parallel to a given plane.

5.23 PLANE THROUGH A POINT PARALLEL TO A GIVEN PLANE

A line is parallel to a plane if it is parallel to any line in that plane. Consequently, if two intersecting lines were drawn parallel to two existing lines of a given plane, the intersecting lines would form a plane parallel to the given plane. This statement can be used to construct a plane through an existing point parallel to a given plane. A view showing both planes as an edge will verify their parallelism.

In Fig. 5-44(1), point A and plane 1-2-3 are given. It is required to construct a plane through point A and parallel to plane 1-2-3.

1. Draw line 4-5 through point A and parallel to line 1-2 in both views. Make the new line a convenient length with its end points aligned in adjacent views (2).

2. Draw line 6-7 a convenient length, through point A and parallel to line 2-3 in both views (2).

3. Plane 4-5-6-7 and plane 1-2-3 are parallel. To check parallelism, a view showing the planes as edges could be projected. Both planes must be edges and parallel in the same view.

The concept of parallelism is used frequently in engineering work where a graphical solution is adequate. In Fig. 5-45, the Viking Lander is designed to land softly on Mars where it will perform a series of experiments, dispatching pictures and scientific information to earth. Graphical analysis using orthographic projection and descriptive geometry is an integral part of the design and production of space hardware.

FIG. 5-45 Model of a viking spacecraft. *(Courtesy NASA.)*

163

PROB. 5-17

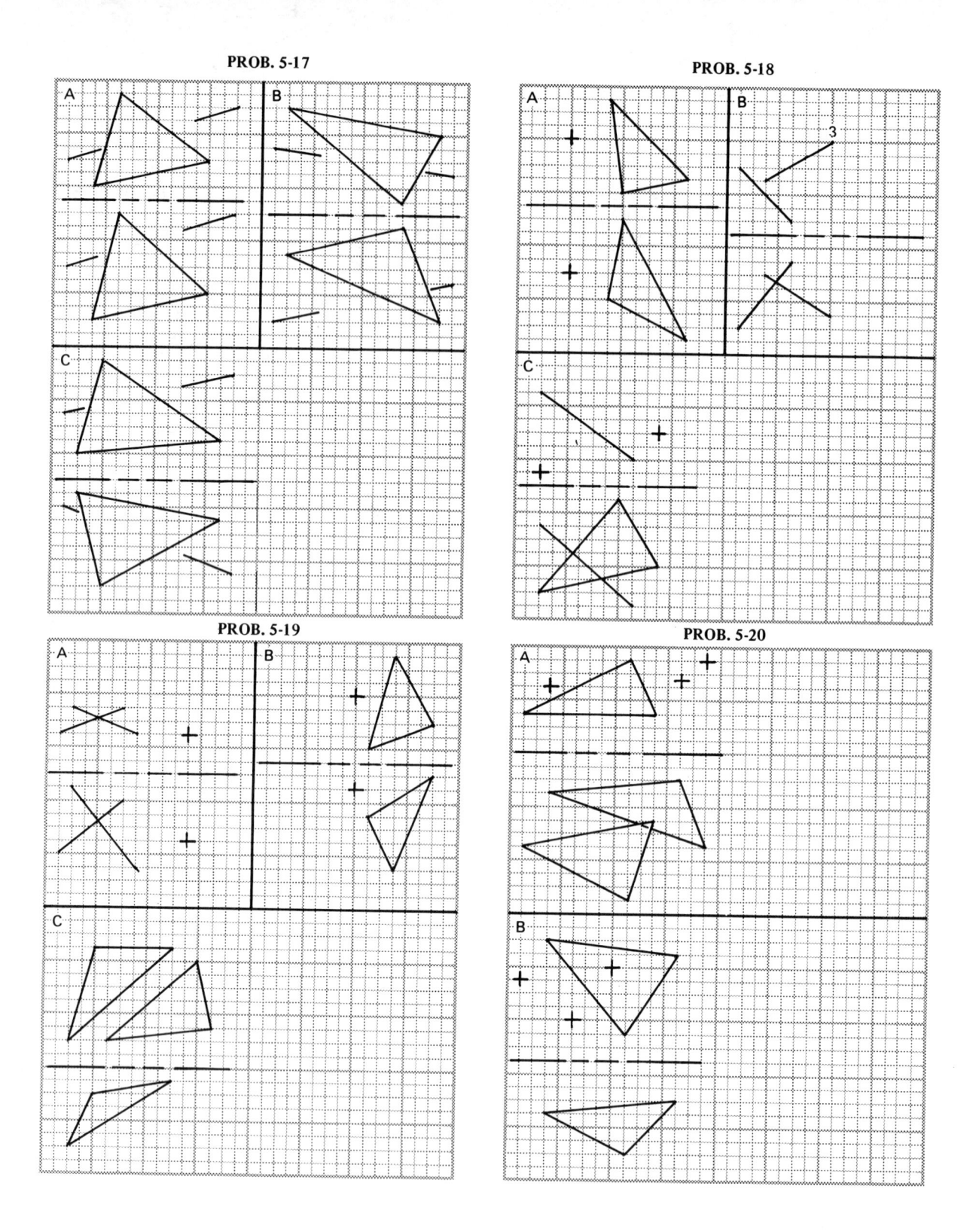

PROB. 5-18

PROB. 5-19

PROB. 5-20

PROBS. 5-17A, 5-17B, 5-17C. Complete the views and show correct visibility.

PROB. 5-18A. Draw a line through the given point and parallel to the plane.

PROB. 5-18B. Construct a plane with an intersecting line through line 3 and parallel to the other given line.

PROB. 5-18C. Complete the top view of the plane which is parallel to the line. Start in the F view and draw a line parallel to the given line and through one of the corners on the plane so that the line is on the plane.

PROBS. 5-19A, 5-19B. Construct a plane through the given point and parallel to the plane.

PROB. 5-19C. Finish the front view of the planes. The incomplete plane is parallel and somewhat below the other plane. They are .50″ (12.7mm) apart.

PROB. 5-20A. Are the two planes parallel? Solve for the edge view.

PROB. 5-20B. Construct a plane .625″ (15.8mm) above and parallel to the given plane and through the points. Solve for the TS and correct visibility. Use an auxiliary off of the frontal view. The planes are the same size.

FIG. 5-46 Perpendicularity of planes.

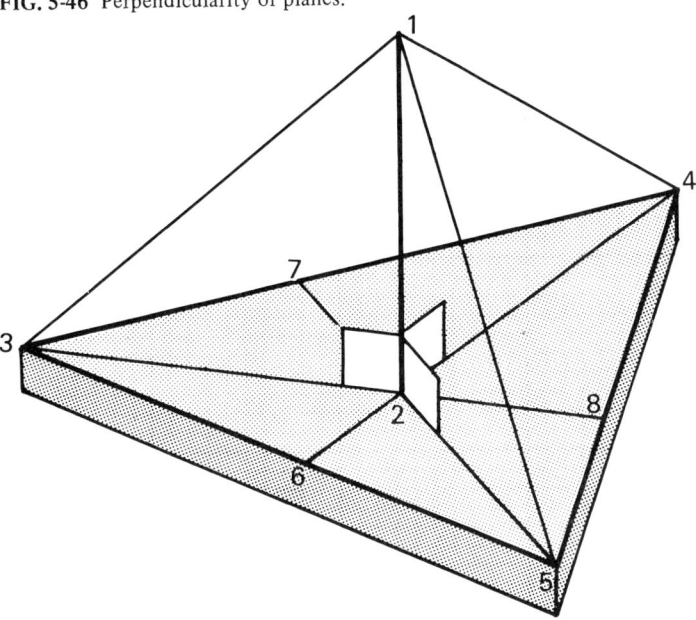

5.24 PERPENDICULARITY OF PLANES

*A line is **perpendicular** to a plane if it is perpendicular to two intersecting lines that lie on the plane.* If a line is perpendicular to a plane, it is perpendicular to all lines that fall on the plane and intersect the line, Fig. 5-46. To construct a line perpendicular to a given plane, a true length line must be established on the plane. A line can then be drawn perpendicular to the true length line. Each view of the perpendicular line can be drawn using the same method. In Fig. 5-47(1), line 5_F-6_F is perpendicular to frontal line (true length) 3_F-4_F, which lies in plane 1-2-3. Line 5_F-6_F is perpendicular to any true length line in the frontal view of the plane. Therefore line 5_F-6_F establishes the direction of all lines in the frontal view that are perpendicular to plane 1-2-3. Similarly, line 7_H-8_H is perpendicular to plane 1-2-3, since it is drawn perpendicular to horizontal line 2_H-4_H. Line 7_H-8_H establishes the direction of any line in the horizontal view that is perpendicular to the plane (2).

A plane is perpendicular to another plane if one line in it is perpendicular to the other plane. This can be seen in Fig. 5-46, where each vertical plane (1-2-3, 1-2-4, 1-2-5) is perpendicular to the horizontal plane since they all contain line 1-2, which is perpendicular to plane 3-4-5. Drawing a plane perpendicular to another plane is simply a matter of drawing one of its lines perpendicular to the given plane. This can easily be accomplished by establishing a true length line in the given plane as described above and drawing a new line perpendicular to it. The perpendicular line is part of the new plane and thus the planes are perpendicular.

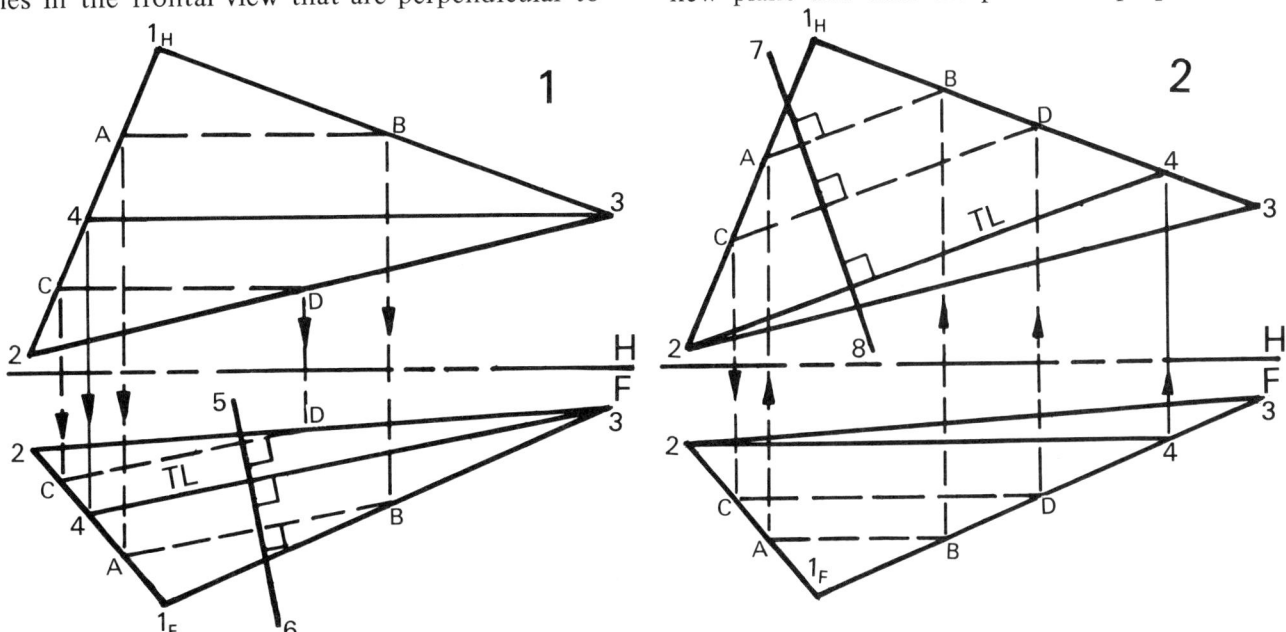

FIG. 5-47 Perpendicularity and true length lines.

FIG. 5-48 Line through a given point perpendicular to a plane.

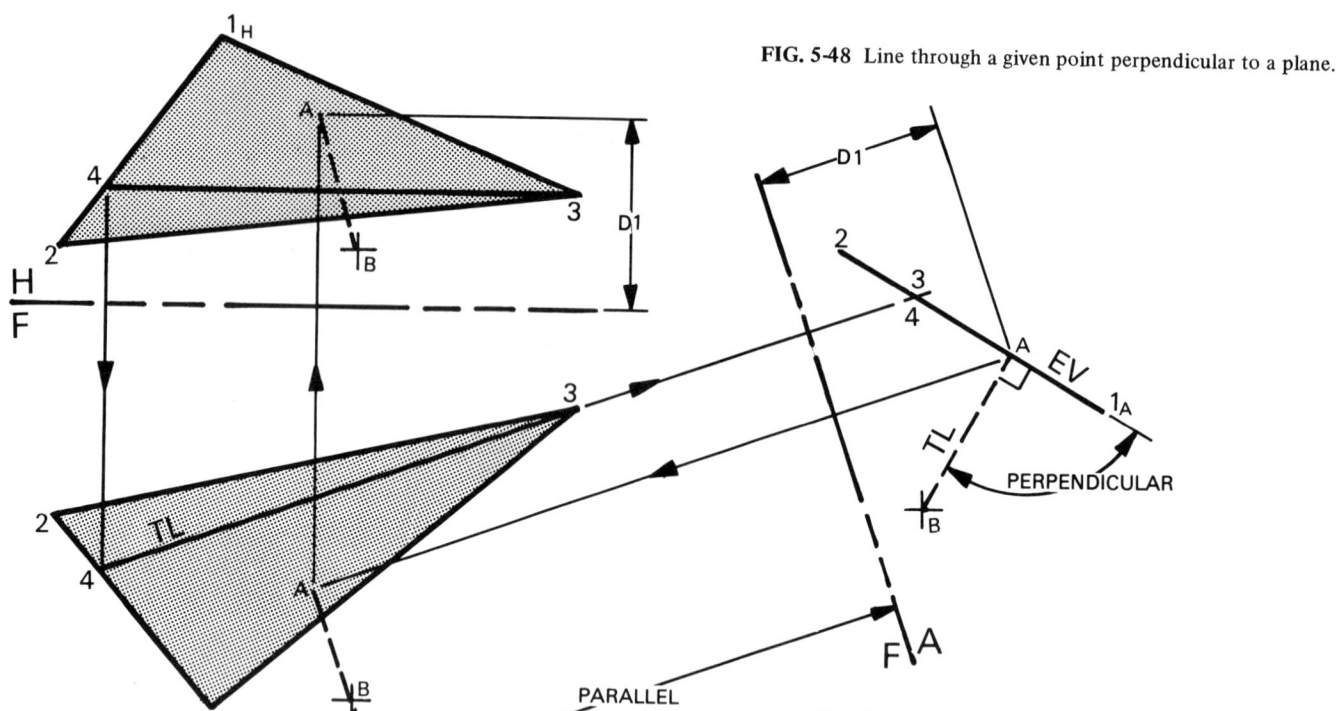

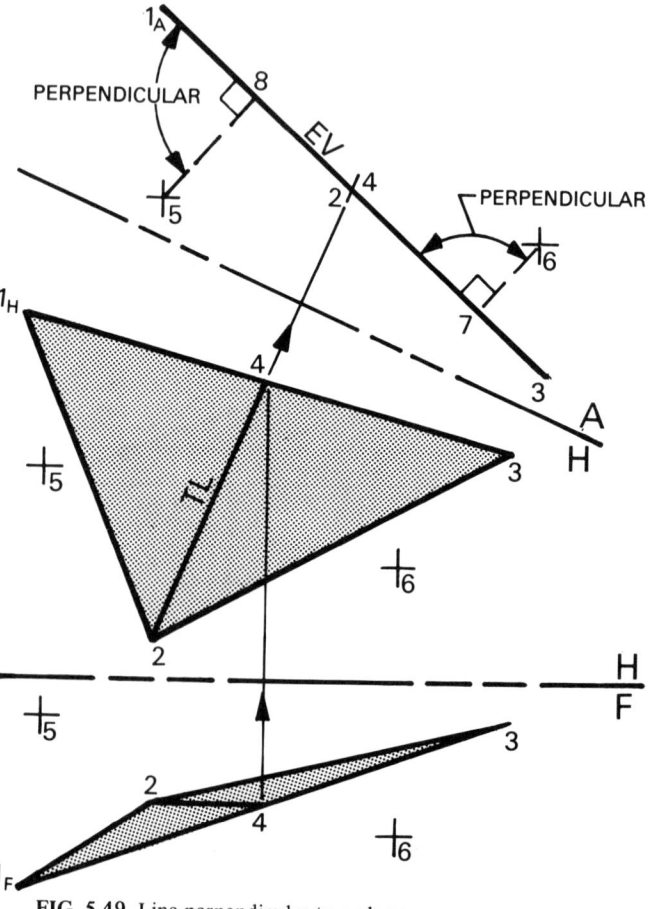

5.25 LINE PERPENDICULAR TO A PLANE (EDGE VIEW METHOD)

In a view where a plane appears as an edge, a line can be drawn through a given point and perpendicular to the plane. The point at which the line touches the plane is the *piercing* or *intersection point* of the line and the plane. This perpendicular line is a true length line in this view and projects parallel to the fold line in the previous adjacent view. The length of the perpendicular line is the shortest distance between the point and the plane. In Fig. 5-48, plane 1-2-3 and point B are given. A perpendicular line from the point to the plane is required.

1. Draw frontal line 3-4 in both views. Auxiliary A is projected perpendicular to line 3_H-4_H in order to show plane 1-2-3 as an edge. Complete auxiliary A.
2. Draw a line from point B_A perpendicular to plane 1_A-2_A-3_A. Line B_A-A_A is the shortest perpendicular distance from the point to the plane. Line B_A-A_A pierces plane 1_A-2_A-3_A at point A_A.
3. Project line B-A back to all previous views. Line B_F-A_F is parallel to F/A and perpendicular to true length line 3_H-4_H. Point A_F is fixed by projection from auxiliary A. The horizontal view of line B_F-A_F is located by the projection of point A_F and transferring dimension D1. Note that line B_H-A_H is perpendicular to any true length line which lies in plane 1_H-2_H-3_H.

In Fig. 5-49, a perpendicular line from each given external point to the plane is constructed by first finding the edge view of the plane. In auxiliary A, line 5_A-8_A is drawn perpendicular to plane 1_A-2_A-3_A, as is line 6_A-7_A. Points 7_A and 8_A represent the piercing points of their respective lines and the plane. Lines 5-8 and 6-7 are not projected back to the H and F views.

FIG. 5-49 Line perpendicular to a plane.

166

FIG. 5-50 Line perpendicular to a plane.

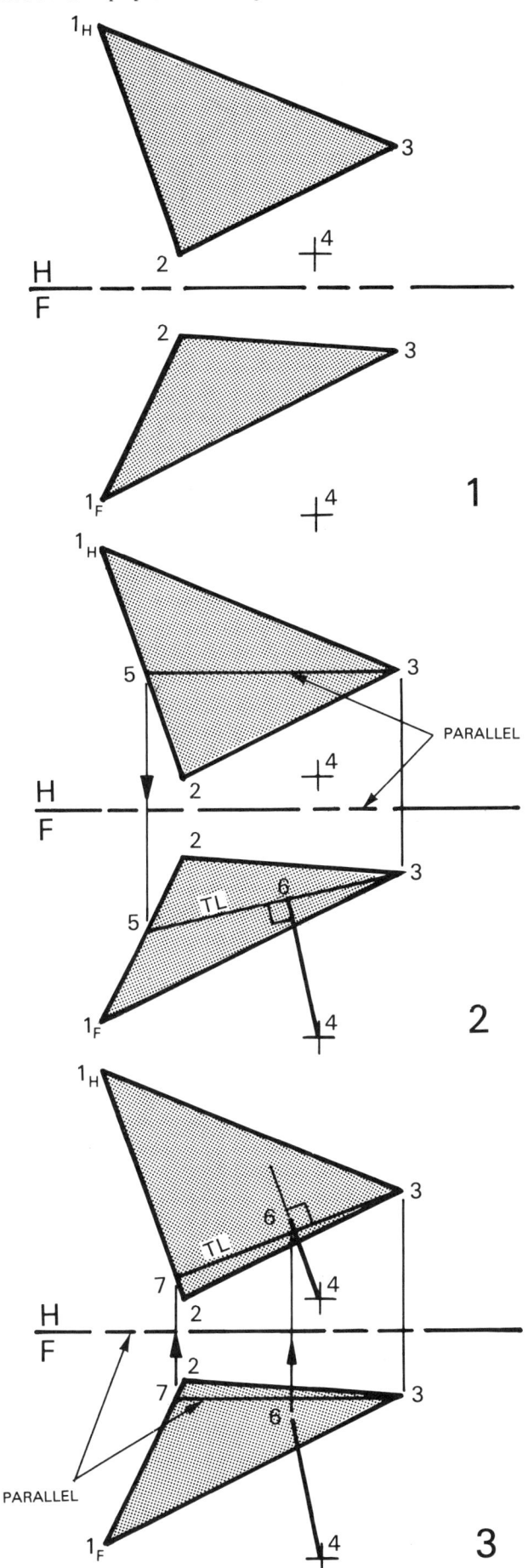

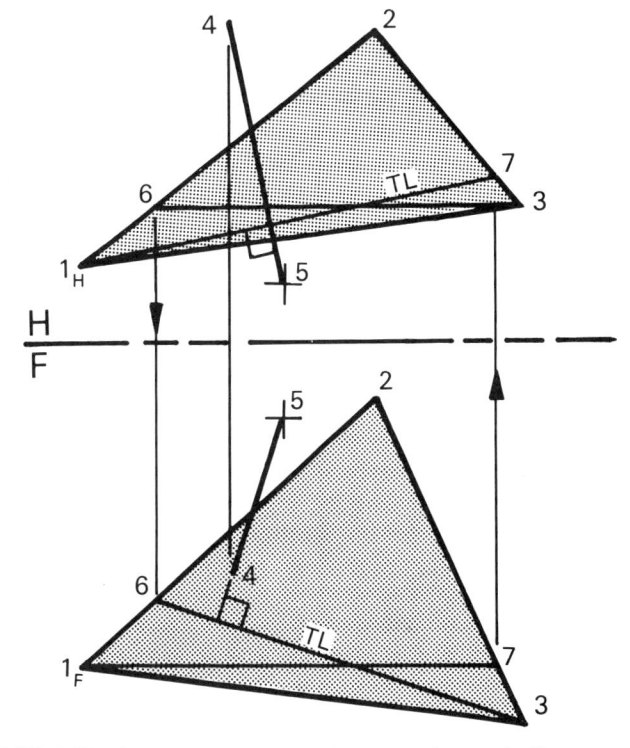

5.26 LINE PERPENDICULAR TO A PLANE (TWO VIEW METHOD)

When a line is perpendicular to a plane it is perpendicular to every line in the plane. Knowing this, a line can be drawn perpendicular to a plane by drawing it perpendicular to a true length line that lies in the plane. Note that in this section the piercing point of the line and the plane is not discussed and the shortest distance cannot be found without an edge view.

In Fig. 5-50, a line from point 4 perpendicular to plane 1-2-3 is required.

1. Plane 1-2-3 and point 4 are given.
2. Draw frontal line 3_H-5_H parallel to H/F, and true length in the frontal view. Draw a line from point 4_F perpendicular to 3_F-5_F. Label the end point 6_F. Note that line 4_F-6_F is any convenient length.
3. Draw horizontal line 3_F-7_F parallel to H/F and true length in the horizontal view. Draw a line from point 4_H perpendicular to line 3_H-7_H. Locate point 6_H by projection from the frontal view.

Line 4-6 is perpendicular to plane 1-2-3, since it is perpendicular to a true length line, in each view, which lies in the plane.

In Fig. 5-51, a line is drawn from a given point, perpendicular to the plane by establishing the direction of the line in each view. Line 4-5 is drawn perpendicular to frontal line 3_F-6_F in the frontal view and perpendicular to horizontal line 1_H-7_H in the horizontal view. The end points of the line are aligned in adjacent projections.

FIG. 5-51 Line perpendicular to a plane (two view method).

FIG. 5-52 Plane perpendicular to a line and through a given point.

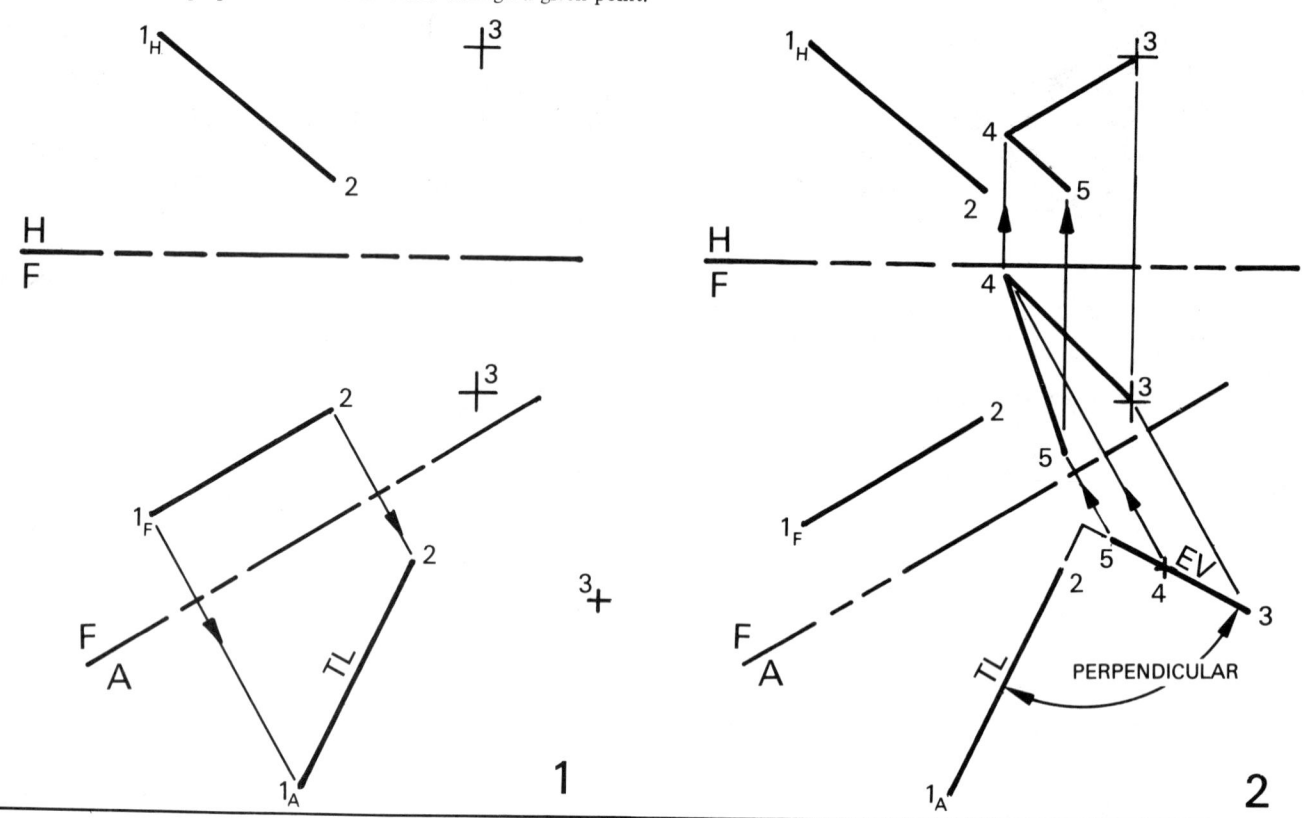

5.27 PLANE PERPENDICULAR TO A LINE AND THROUGH A GIVEN POINT

In Fig. 5-52, the edge view method of drawing a plane through a point and perpendicular to a line is illustrated.

1. Line 1-2 and point 3 are given. Draw F/A parallel to line 1_F-2_F. Line 1_A-2_A is true length in auxiliary A.
2. Construct an edge view of the required plane perpendicular to true length line 1_A-2_A. Points 4_A and 5_A are placed at convenient points along the edge view in order to define the plane. The frontal location of point 4 and 5 can be located anywhere along their projection lines when extended from auxiliary A. Points 4 and 5 are established in the horizontal view by projection and transferring distances from auxiliary A. Line 1-2 is true length and perpendicular to the edge view of plane 3-4-5. The line and the plane are therefore perpendicular.

In Fig. 5-53, plane 3-4-5 is drawn perpendicular to line 1-2 using the two-view method. Point 3 and line 1-2 are given. Frontal line 3_H-5_H is drawn a convenient length, parallel to H/F. Line 3_F-5_F is drawn true length and perpendicular to line 1_F-2_F, since a line is perpendicular to a plane if it is perpendicular to a true length line in that plane. Horizontal line 3_F-4_F is drawn parallel to H/F. Line 3_H-4_H is true length and perpendicular to line 1_H-2_H in the horizontal view. Intersecting lines 3-4 and 3-5 define a plane (3-4-5) perpendicular to line 1-2 and through point 3.

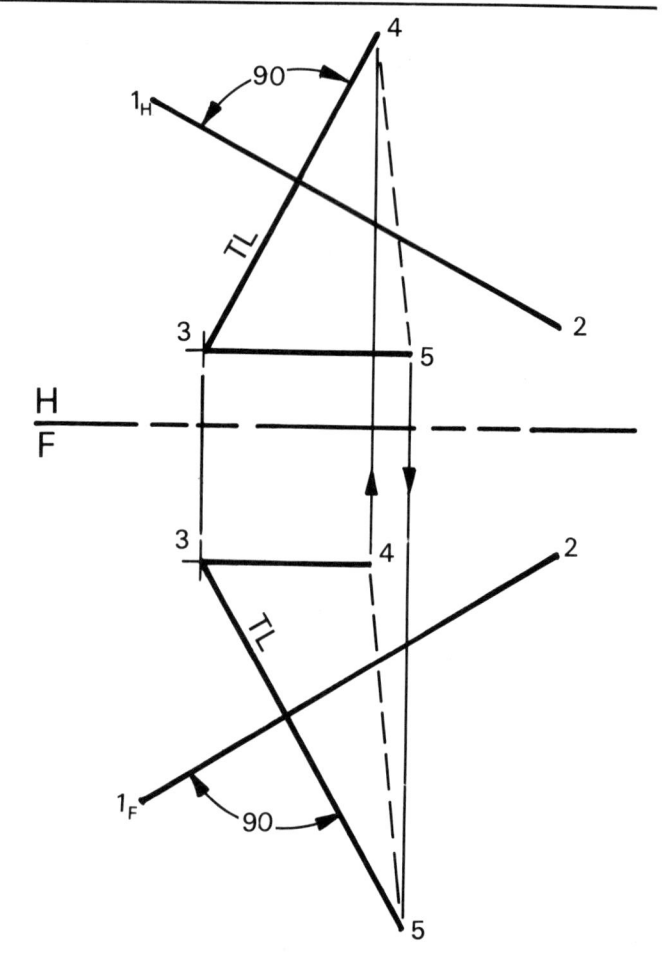

FIG. 5-53 Plane perpendicular to a line (two view method).

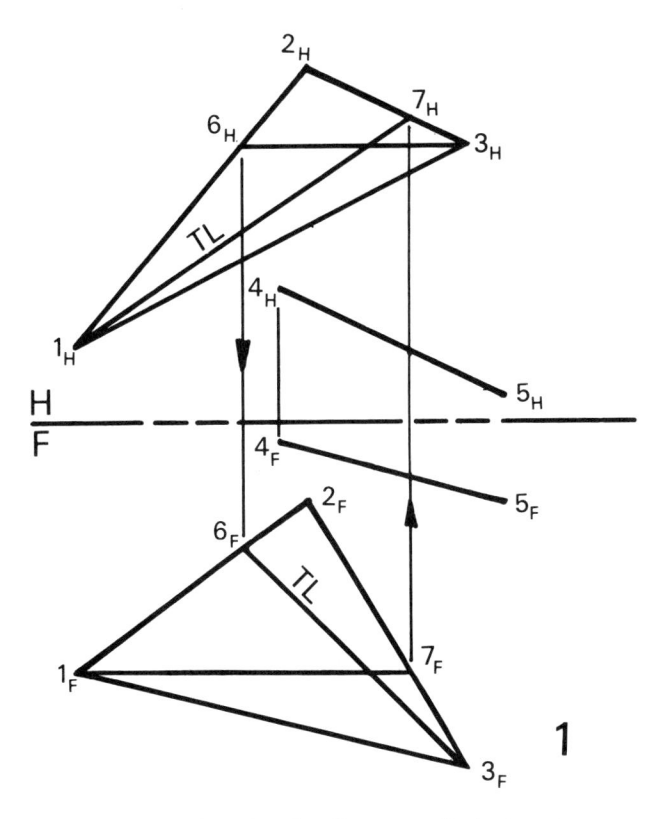

FIG. 5-54 Plane through a given line perpendicular to a plane.

5.28 PLANE THROUGH A GIVEN LINE PERPENDICULAR TO A PLANE

A line drawn perpendicular to a true length line that lies in a given plane, and intersecting an existing external line, forms a plane perpendicular to the given plane.

In Fig. 5-54, plane 1-2-3 and line 4-5 are given. A plane passing through the line and perpendicular to the plane is required. Note that the simplest way to construct a plane is by drawing two intersecting lines. Also, in order for a plane to be perpendicular to another plane it must contain a line which is perpendicular to the given plane. Since line 4-5 is given, a line intersecting it must be drawn perpendicular to a true length line that lies in the given plane.

1. Establish a true length line in plane 1-2-3 in each view. Draw frontal line 3_H-6_H parallel to H/F and project to the frontal view where it is true length. Draw horizontal line 1_F-7_F parallel to H/F and project to the horizontal view where it is true length.

2. In both views, from *any point* on line 4-5 draw an intersecting line perpendicular to the true length lines in plane 1-2-3. From point 5_F a line is drawn a convenient length and perpendicular to the frontal line which lies in plane 1_F-2_F-3_F. In the horizontal view draw a line from point 5_H perpendicular to the horizontal line which lies in plane 1_H-2_H-3_H. Locate end point 8 anywhere along this line and align in both views.

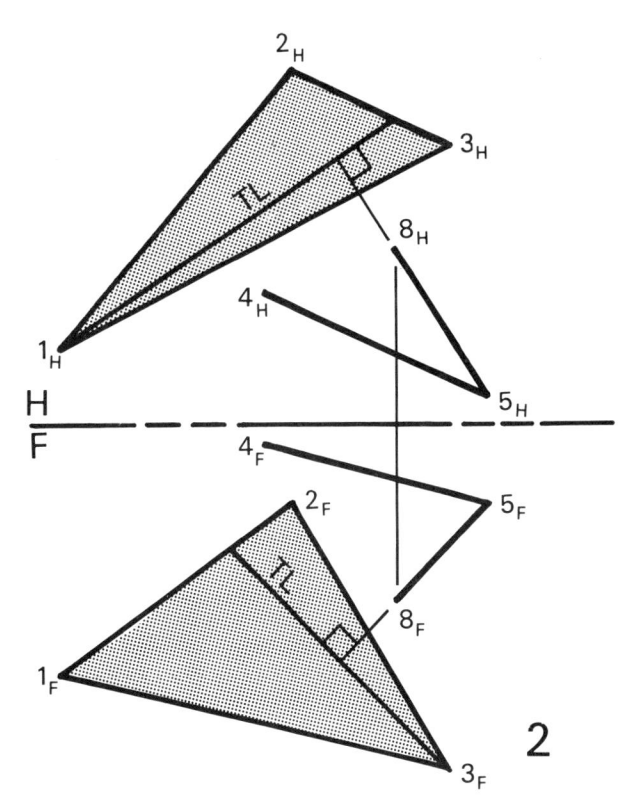

The theory of perpendicularity as used in descriptive geometry can be applied in the design of petrochemical facilities, Fig. 5-55. The design of pipe supports and structural bracing requires that the shortest perpendicular distance between a line and a surface (plane) be established.

FIG. 5-55 Piping model.

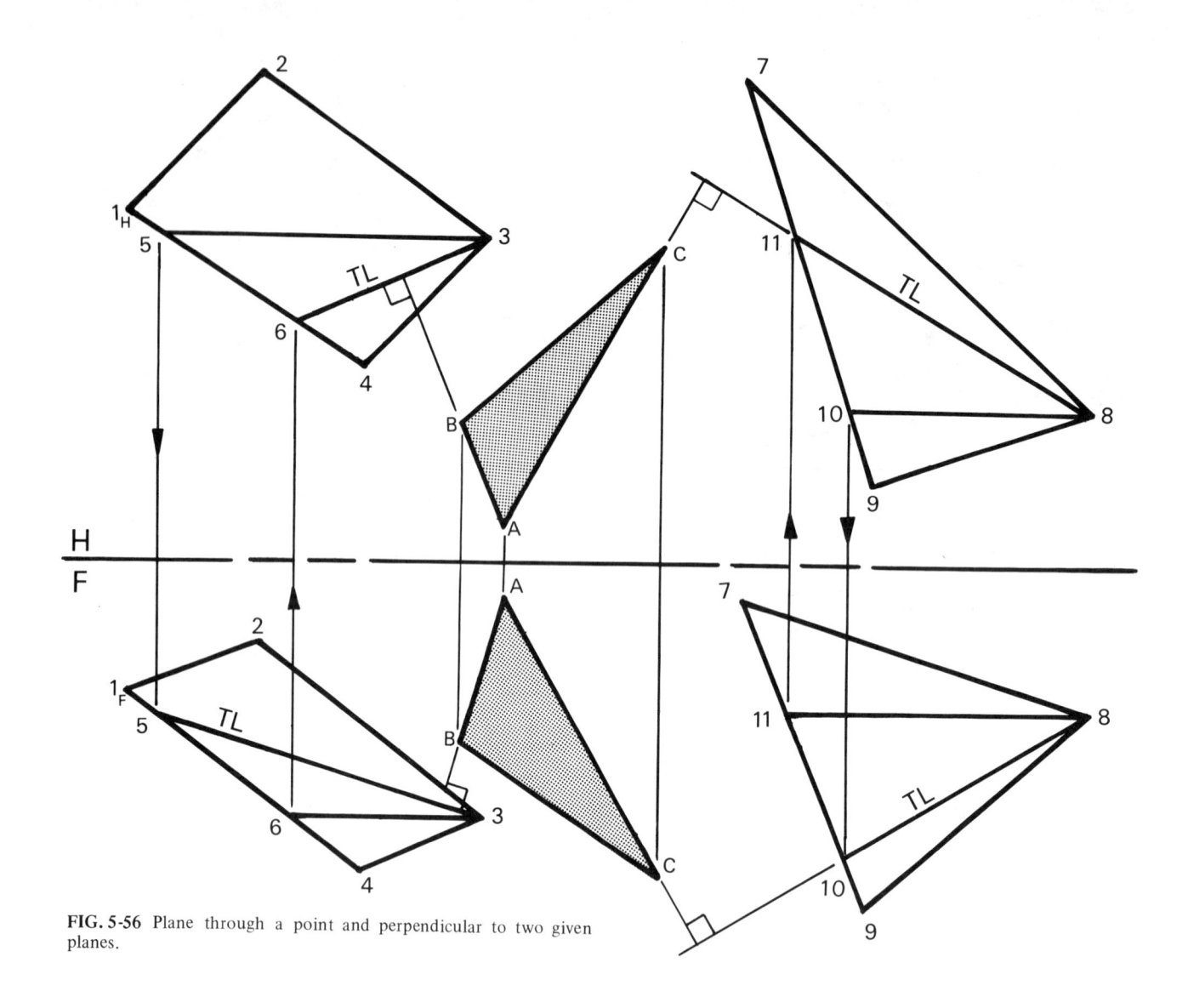

FIG. 5-56 Plane through a point and perpendicular to two given planes.

5.29 PLANE THROUGH A POINT AND PERPENDICULAR TO TWO GIVEN PLANES

A plane is perpendicular to another plane if one of its lines is perpendicular to the given plane. Two intersecting lines can be drawn through a point to create a plane. Therefore, a plane can be constructed through a given point and perpendicular to two given planes by drawing two intersecting lines from the point with each line perpendicular to one of the planes. The new plane formed from the two intersecting lines is perpendicular to both existing planes because each of its lines is perpendicular to a given plane.

In Fig. 5-56, plane 1-2-3-4 and plane 7-8-9 are given along with point A. A plane which passes through point A and is perpendicular to each of the two given planes is required. In order to draw a line perpendicular to a plane, it is necessary to establish a true length line which lies in the plane. Therefore a frontal and horizontal line must be located in both planes in each view.

1. Establish true length lines in both views of plane 1-2-3-4. Frontal line 3_H-5_H is drawn parallel to H/F, and true length in the frontal view. Horizontal line 3_F-6_F is drawn parallel to H/F, and true length in the horizontal view.

2. Establish true length lines in both views of plane 7-8-9. Draw frontal line 8_H-10_H parallel to H/F, and true length in the frontal view. Horizontal line 8_F-11_F is drawn parallel to H/F, and true length in the horizontal view.

3. From point A in both views draw a line perpendicular to plane 1-2-3-4. Point B can be located anywhere along this line. Line A_F-B_F is perpendicular to frontal line 3_F-5_F. Line A_H-B_H is perpendicular to horizontal line 3_H-6_H.

4. In both views, from point A, draw a line perpendicular to plane 7-8-9. Locate point C along this line. Line A_F-C_F is perpendicular to frontal line 8_F-10_F. Line A_H-C_H is perpendicular to horizontal line 8_H-11_H.

5. Enclose plane A-B-C by connecting the points.

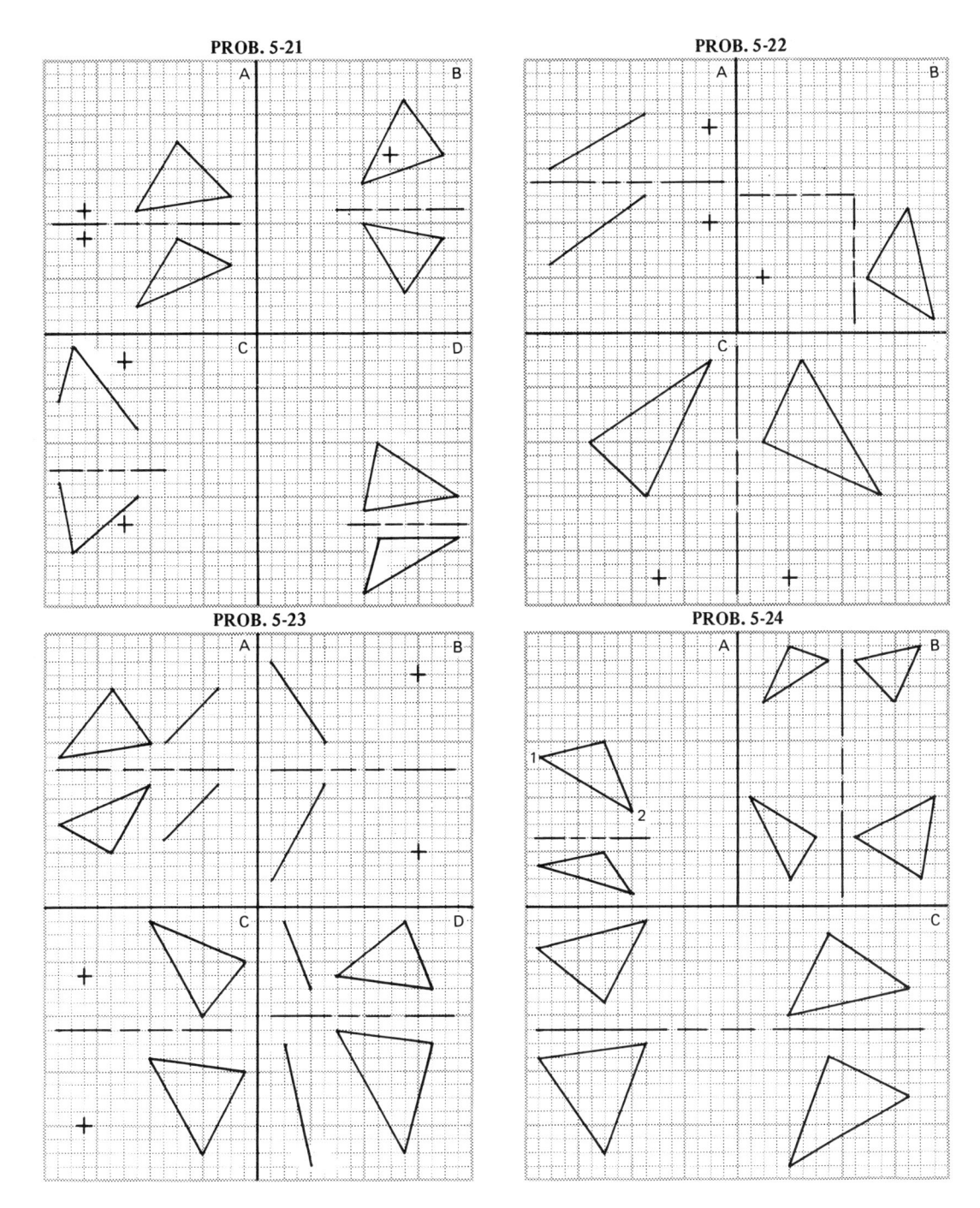

PROB. 5-21A. Construct a line that is through the given point and perpendicular to the plane.

PROB. 5-21B. Project a line that is 1" (25.4mm) long on the lower side of the plane and perpendicular to the plane. The given point is one end of the line and lies on the plane.

PROB. 5-21C. Draw a line .5" (12.7mm) long, perpendicular to the plane, and extending toward the plane from the given point.

PROB. 5-21D. Construct a line that is perpendicular to the plane. The line will extend up from the center of the plane.

PROB. 5-22A. Construct a plane that is perpendicular to the line. Represent the plane with two intersecting lines through the given point.

PROB. 5-22B. Show the profile plane in all views. Construct a 1" (25.4mm) frontal line that protrudes from the center of the plane toward the right.

PROB. 5-22C. From the point construct two lines, one perpendicular and one parallel to the plane.

PROB. 5-23A. Using the given line draw a plane perpendicular to the given plane.

PROB. 5-23B. Construct a plane through the point and perpendicular to the line.

PROB. 5-23C. Draw two planes through the point, one perpendicular and one parallel to the plane.

PROB. 5-23D. Construct a plane through the line and perpendicular to the plane.

PROB. 5-24A. Construct a plane that is equal in size and shape and perpendicular to the plane. The new plane lies below the given plane. Line 1-2 is the line of intersection between the planes.

PROBS. 5-24B, 5-24C. Construct a plane that is perpendicular to the two given planes.

171

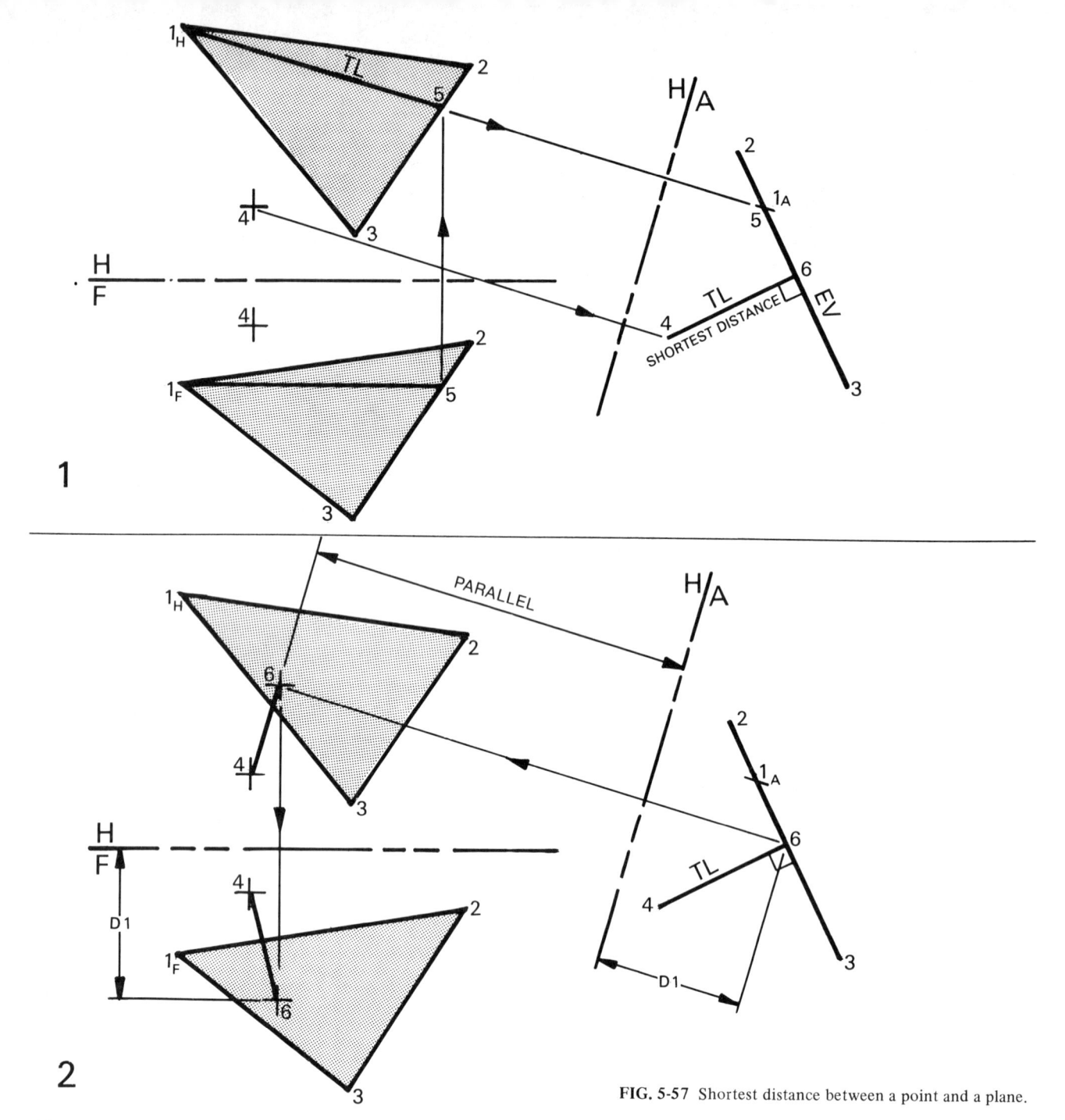

FIG. 5-57 Shortest distance between a point and a plane.

5.30 SHORTEST DISTANCE BETWEEN A POINT AND A PLANE

In Fig. 5-57(1), plane 1-2-3 and point 4 are given. The shortest connector is required.
The shortest distance between a point and a plane is a perpendicular line drawn between the point and the plane (page 166). A line drawn from a point to a plane is its shortest connector if drawn perpendicular to an edge view of the plane. In a view where the plane is an edge, the shortest distance is measured as the perpendicular distance between the point and the plane. If the edge view is established in an auxiliary projection from the horizontal view, it shows the slope (dip) of the plane and the slope of the line.

1. Draw horizontal line 1_F-5_F parallel to H/F, and true length in the horizontal view. H/A is drawn perpendicular to horizontal line 1_H-5_H. In auxiliary view A, plane 1_A-2_A-3_A appears as an edge. Draw a line from point 4_A perpendicular to the edge view of the plane. Point 6_A lies on the plane (at the point where the line pierces the plane). Line 4_A-6_A is the shortest distance between the point and plane. The acute angle formed by 4_A-6_A and the edge of the horizontal plane is the slope of the shortest connector.

2. Line 4_A-6_A is true length. Therefore it projects to the horizontal view as parallel to H/A. Point 6_H is fixed by projection from auxiliary A. Point 6_F is located by transferring dimension D1 from auxiliary A along its projection line in the frontal view.

172

FIG. 5-58 Shortest grade line between a point and a plane.

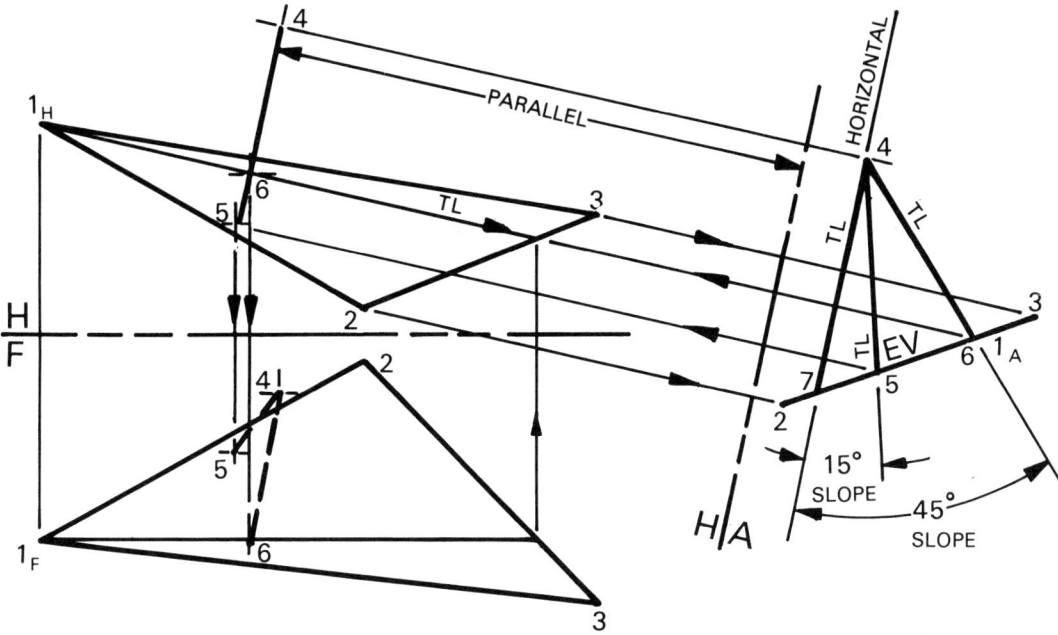

5.31 SHORTEST GRADE OR SLOPE LINE BETWEEN A POINT AND A PLANE

In many situations, the shortest connector between a point and plane will be too steep or form an unusable angle. In pipe support location, when the shortest connector interferes with an existing duct, cable tray, etc., it may be necessary to find the shortest connector at a given slope angle between the pipe and a structural surface. The shortest grade (slope) connector between two tunnels is also required in certain mining problems.

In Fig. 5-58, plane 1-2-3 and point 5 are given. Connectors with slope angles of 15° and 45° are required. Auxiliary A is projected perpendicular to a horizontal line. Plane 1_A-2_A-3_A appears as an edge and line 4_A-7_A is the shortest level distance. Line 4_A-5_A is drawn at 15° and line 4_A-6_A at 45° to the horizontal. All connectors are true length in auxiliary A and appear parallel to H/A in the horizontal view with their end points fixed by projection. The frontal view is established by projection and measurement.

In Fig. 5-59, auxiliary A shows plane 1_A-2_A-3_A as an edge. Line 4_A-5_A is the shortest perpendicular and line 4_A-6_A the shortest level (horizontal) line. Each is true length in auxiliary A and parallel to H/A in the horizontal view. Line 4_H-6_H is true length since its adjacent projections are parallel to their respective fold lines. The frontal view is established by projection and measurement. Note that horizontal connector 4_F-6_F is a level line and thus parallel to H/F.

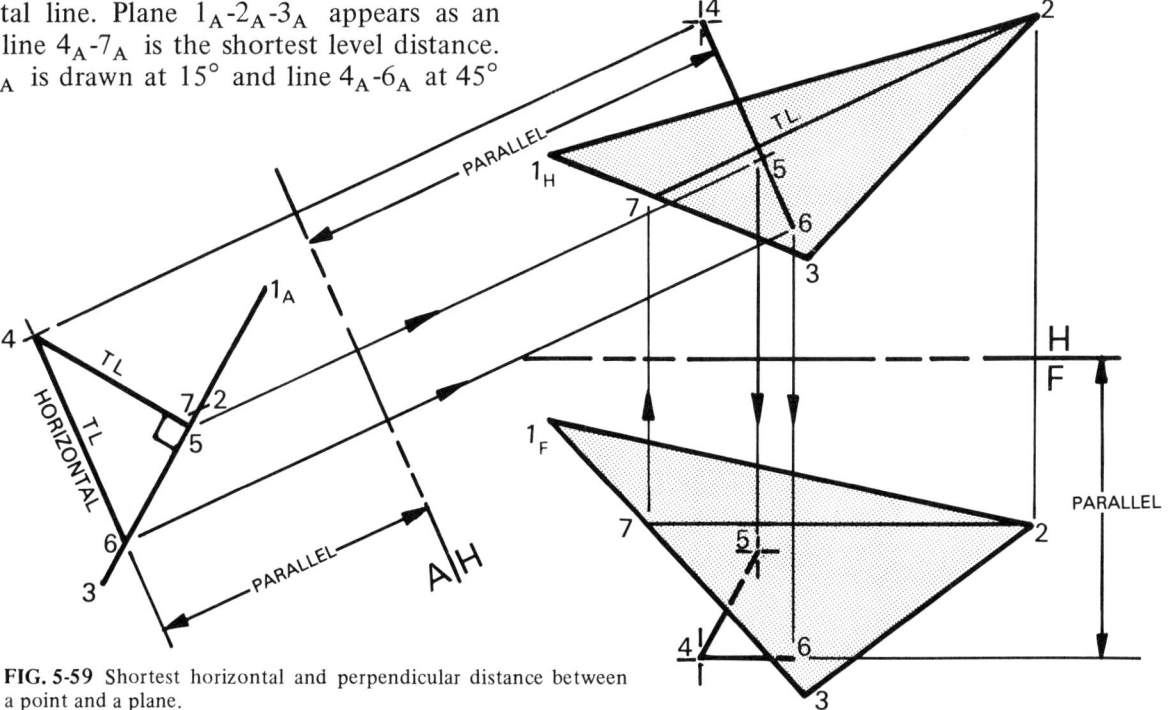

FIG. 5-59 Shortest horizontal and perpendicular distance between a point and a plane.

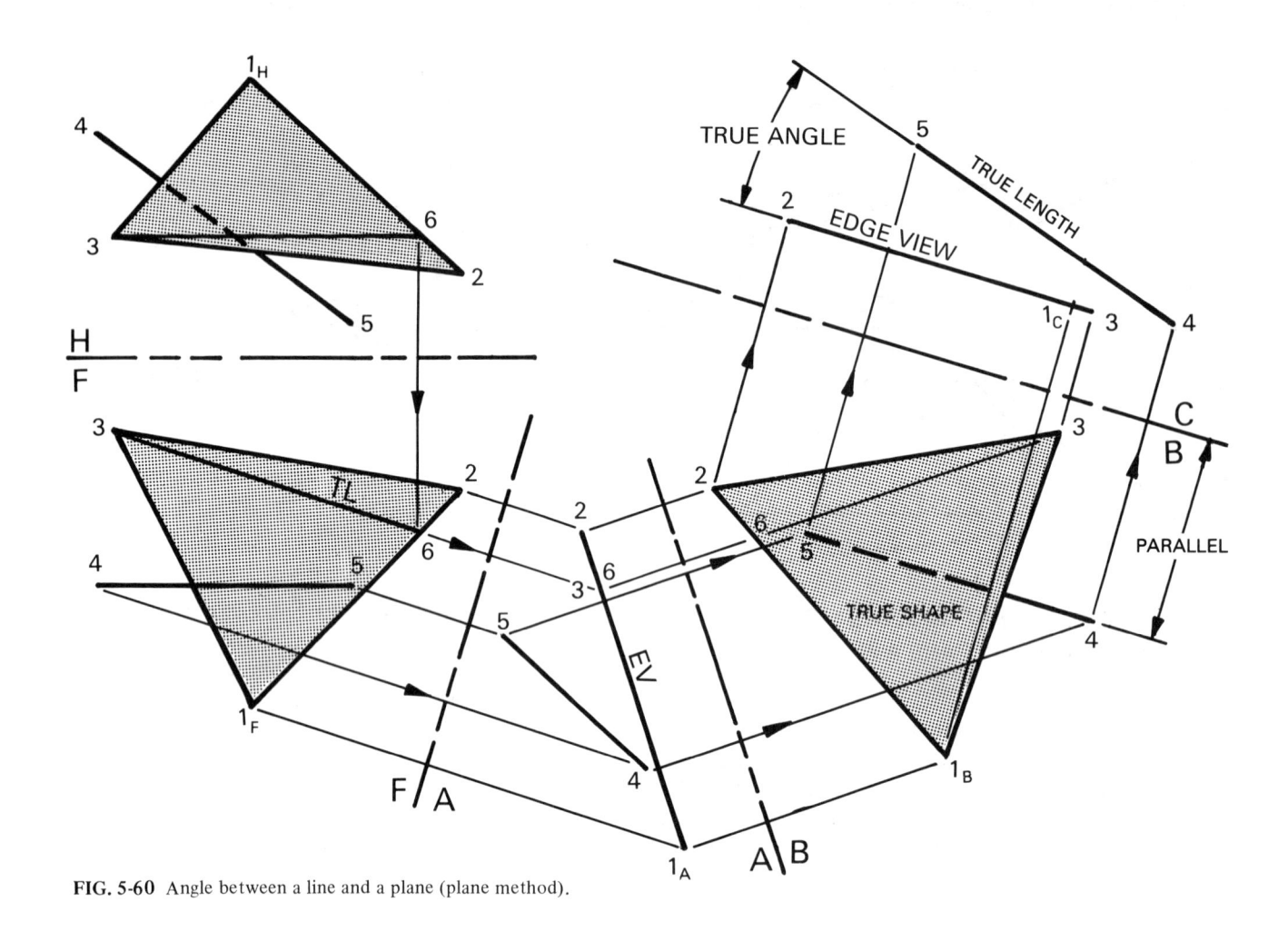

FIG. 5-60 Angle between a line and a plane (plane method).

5.32 ANGLE BETWEEN A LINE AND A PLANE (PLANE METHOD)

The angle between a line and a plane is measured in a view where the line is true length and the plane is an edge. Three successive projections are necessary to attain this situation, since the first auxiliary shows the planes as an edge with the line as oblique. When using the "*plane method*" the primary auxiliary, view A, is used to project an edge view of the plane. The secondary auxiliary, view B, is projected parallel to the edge view so that the plane appears as true size. A successive auxiliary, view C, is then taken parallel to the oblique line in auxiliary B. Auxiliary C shows the line as true length. The plane appears as an edge here since it was true size in the adjacent view. The true angle can be measured between the true length line and the edge view of the plane in this view.

In Fig. 5-60, plane 1-2-3 and line 4-5 are given. The true angle formed by the plane and the line are required.

1. Draw frontal line 3_H-6_H parallel to H/F, and project to the frontal view where it is true length.
2. Draw F/A perpendicular to 3_F-6_F and complete auxiliary A by projection and measurement. Plane 1_A-2_A-3_A projects as an edge view.
3. Draw A/B parallel to the edge view of the plane and complete auxiliary B by projection and transferring measurements from the frontal view. Plane 1_B-2_B-3_B is true size and line 4_B-5_B appears oblique.
4. Fold line B/C is drawn parallel to oblique line 4_B-5_B. Project auxiliary B with a line of sight perpendicular to line 4_B-5_B. Line 4_C-5_C is true length and plane 1_C-2_C-3_C shows as an edge view.
5. The true angle between the plane and line is measured as an acute angle formed by true length line 4_C-5_C and the edge view of plane 1_C-2_C-3_C.

174

5.33 ANGLE BETWEEN A LINE AND A PLANE (LINE METHOD)

In a view where a line is true length and a plane is an edge view, the true angle between them can be measured. The *"line method"* solves for the true length of the line in the primary auxiliary, view A. The secondary auxiliary, view B, establishes a view where the line is a point view. Any adjacent projection from this view shows the line in true length. Therefore, successive auxiliary, view C, is projected so that the plane is an edge view. The angle between the line and plane is measured as the acute angle formed by the true length line and the edge view of the plane in this view.

In Fig. 5-61, Plane 1-2-3-4 and line 5-6 are given. The true angle between them is required.

1. Draw F/A parallel to oblique line 5_F-6_F, and complete auxiliary A. Line 5_A-6_A appears true length and plane 1_A-2_A-3_A-4_A is oblique.
2. Establish A/B perpendicular to line 5_A-6_A, and complete auxiliary B. Line 5_B-6_B is a point view and plane 1_B-2_B-3_B-4_B is oblique.
3. An edge view of the plane is needed. Draw line 2_A-7_A on the plane and parallel to A/B in auxiliary A and project to auxiliary B where it shows true length.
4. Draw B/C perpendicular to true length line 2_B-7_B and complete auxiliary C. Line 5_C-6_C appears true length and plane 1_C-2_C-3_C-4_C shows as an edge view. The true angle is measured as shown.

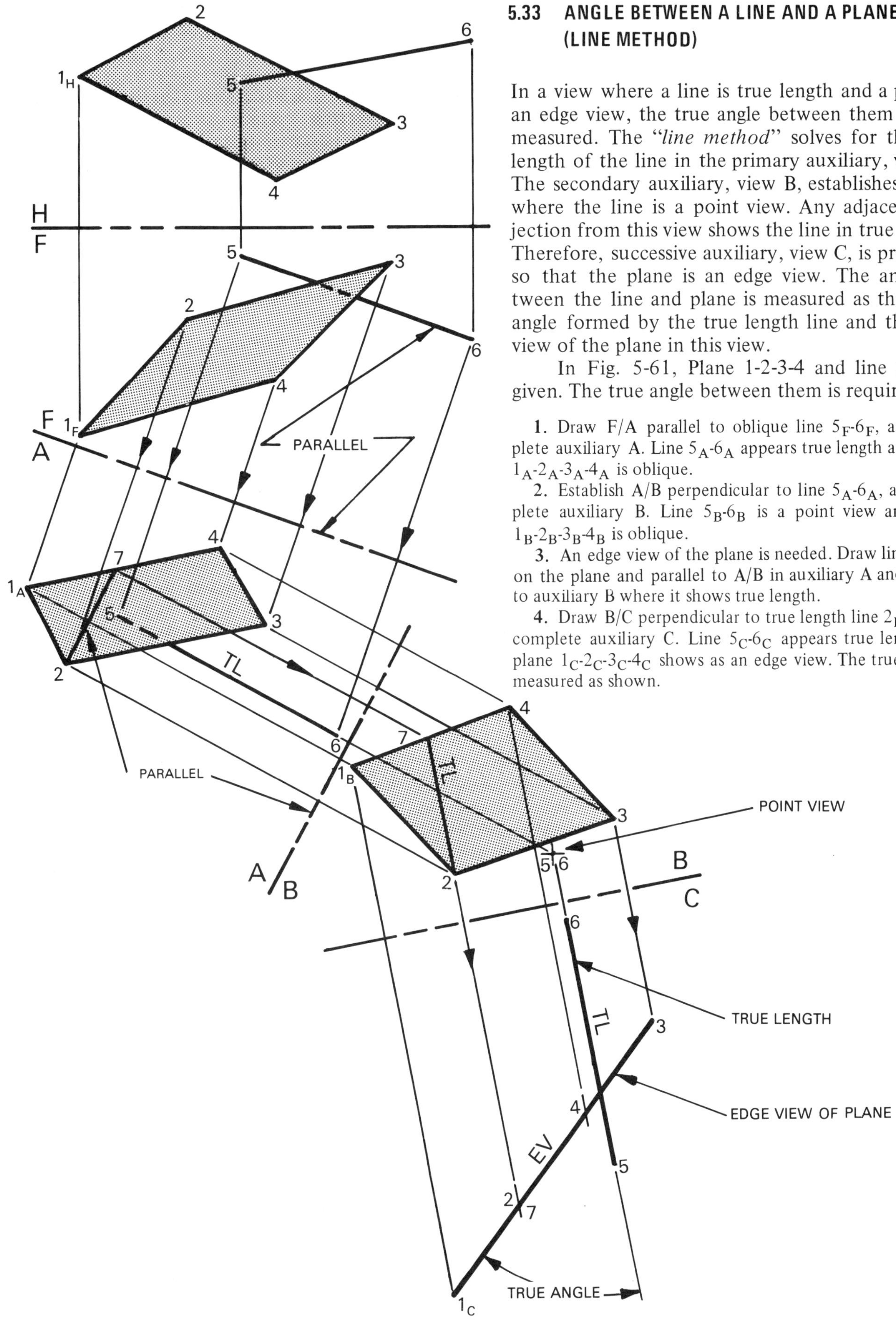

FIG. 5-61 True angle between a line and a plane (line method).

5.34 ANGLE BETWEEN A LINE AND A PLANE (COMPLEMENTARY ANGLE METHOD)

The complement of the angle between a line and plane equals 90° minus the true angle. The angle formed by a line drawn from a point on the line perpendicular to the plane is equal to the complement of the angle between the line and plane. A perpendicular connector is established by drawing a line from a point on the given line and perpendicular to a true length line that lies in the plane. In Fig. 5-62, plane 1-2-3 and line 4-5 are given; the angle between them is required.

1. Draw a line from any point on the given line and perpendicular to the plane in both views. Line 4_H-6_H is constructed perpendicular to horizontal line 3_H-8_H in the horizontal view and perpendicular to frontal line 2_F-7_F in the frontal view (1).

2. Form a plane by connecting points 4, 5, and 6. Draw horizontal line 4_F-9_F parallel to H/F and show as true length in the horizontal view (2).

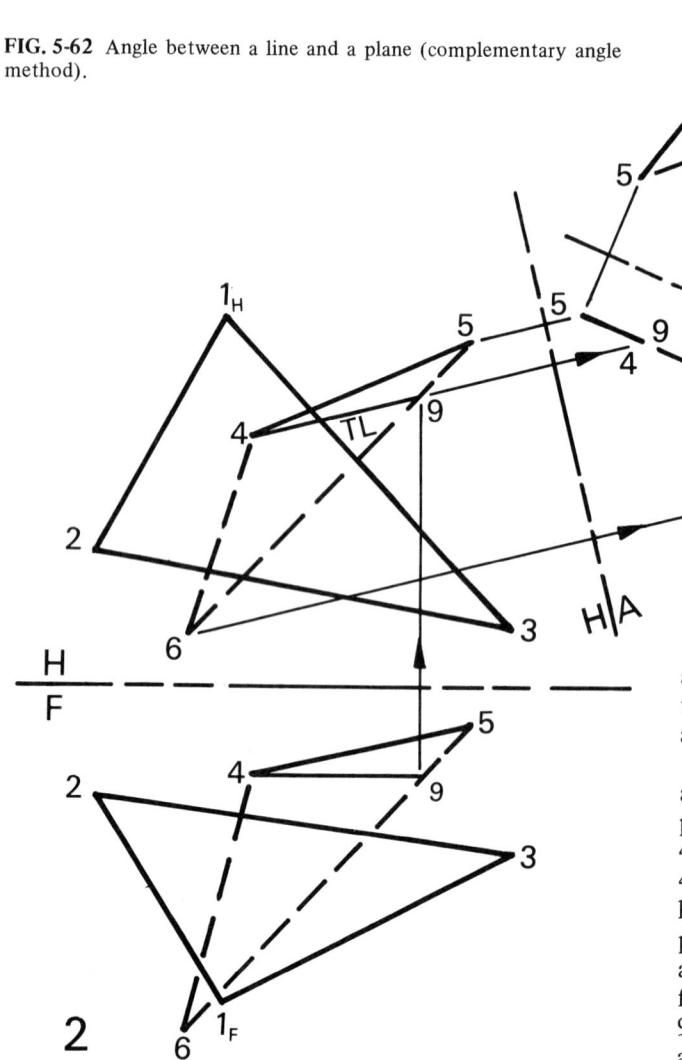

3. Draw H/A perpendicular to 4_H-9_H and complete auxiliary A. Plane 4_A-5_A-6_A is an edge view in this projection. Note that plane 1-2-3 need not be transferred to either auxiliary view (2).

4. Draw A/B parallel to the edge view of plane 4_A-5_A-6_A and project auxiliary B. Draw a line from point 4_B and perpendicular to line 4_B-6_B. Note that in this example line 4_B-6_B is extended in order to form an acute angle with line 4_B-5_B. This is not needed when the angle between them is less than 90 degrees, which would be the case if the original perpendicular lines were drawn from point 5 in the frontal and horizontal views. Angle A is the complementary angle formed by perpendicular line 4_B-6_B and line 4_B-5_B. The 90° angle minus the complementary angle equal the true angle between plane 1-2-3 and line 4-5 (2).

FIG. 5-62 Angle between a line and a plane (complementary angle method).

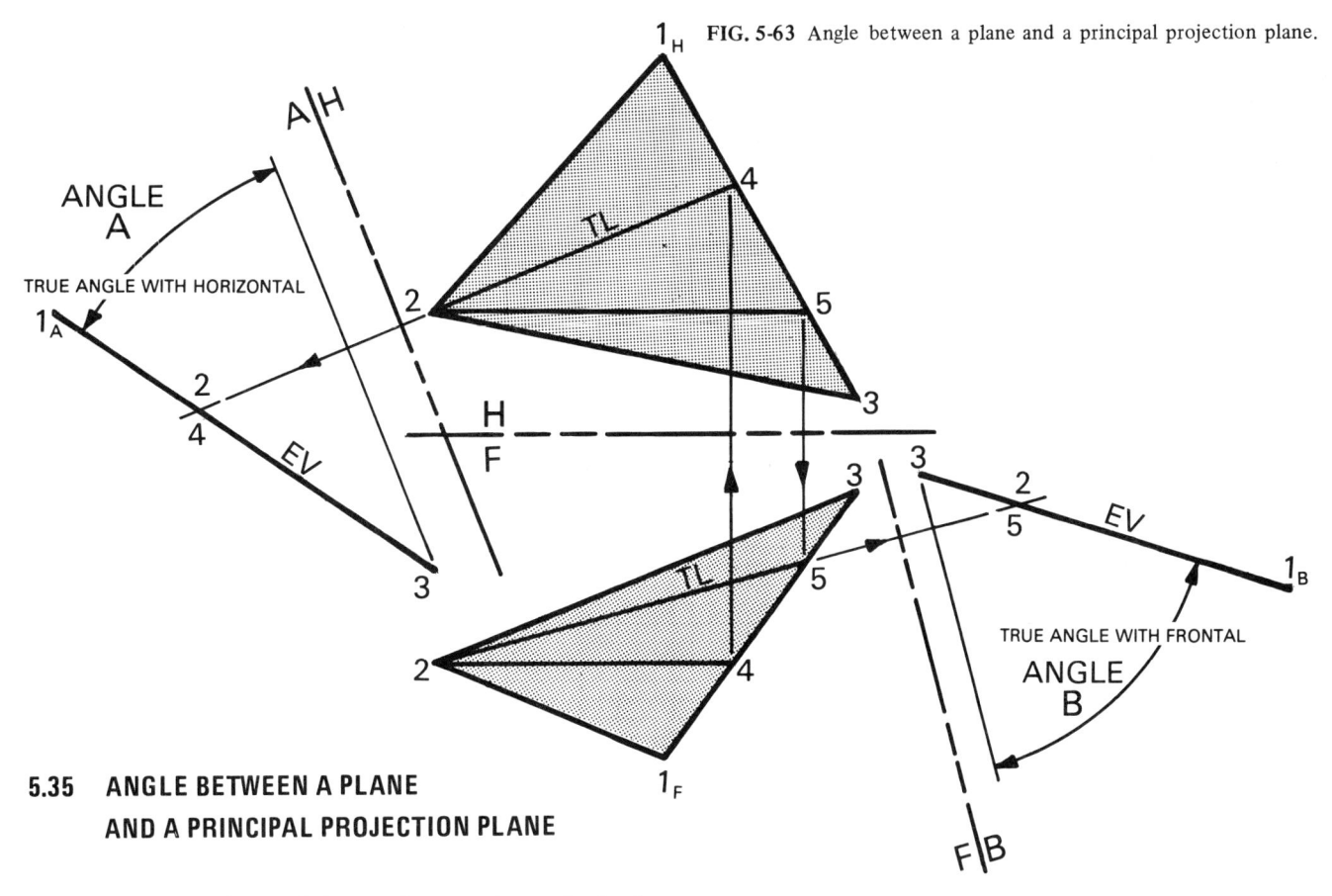

FIG. 5-63 Angle between a plane and a principal projection plane.

5.35 ANGLE BETWEEN A PLANE AND A PRINCIPAL PROJECTION PLANE

The true angle formed by a plane and a principal projection plane is measured in a view where both appear as edges. The angle between a plane and the frontal projection plane is seen in an auxiliary view taken from the frontal view where the plane projects as an edge view, auxiliary A in Fig. 5-63. The angle between a plane and the horizontal projection plane is seen in a view where the plane projects as an edge in an auxiliary view taken from the horizontal view, auxiliary B in Fig. 5-63.

A principal line is first established in each given view of the plane. Horizontal line 2_F-4_F is drawn parallel to H/F and true length in the horizontal view. Auxiliary A is projected perpendicular to line 2_F-4_F in order to establish an edge view of plane 1_A-2_A-3_A. The angle formed by plane 1_A-2_A-3_A and the edge of the horizontal projection plane (H/A) is the true angle between them (Angle A). Frontal line 2_H-5_H is drawn parallel to H/F, and true length in the frontal view. F/B is drawn perpendicular to frontal line 2_F-5_F. Auxiliary B shows plane 1_B-2_B-3_B as an edge. The true angle is measured as shown (Angle B).

In Fig. 5-64 the true angle between the plane and the profile projection plane is measured in auxiliary A. Profile line 3_F-4_F is drawn parallel to F/P and true length in the profile view. P/A is drawn perpendicular to 3_P-4_P. In auxiliary view A, plane 1_A-2_A-3_A appears as an edge view and the true angle between the plane and the profile projection plane is measured as shown (Angle C).

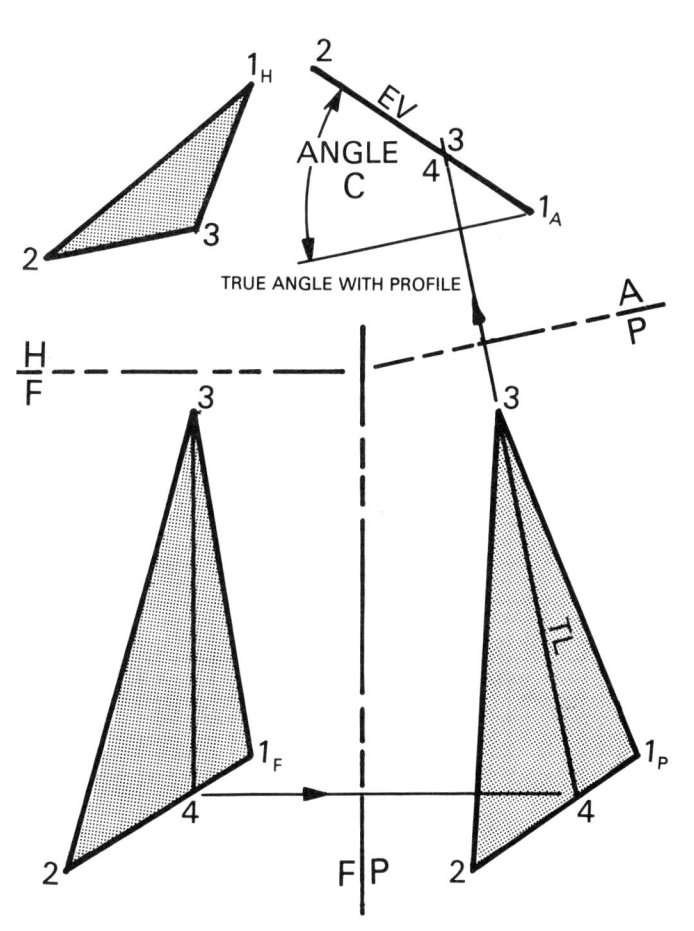

FIG. 5-64 Angle between a plane and the profile projection plane.

PROB. 5-25

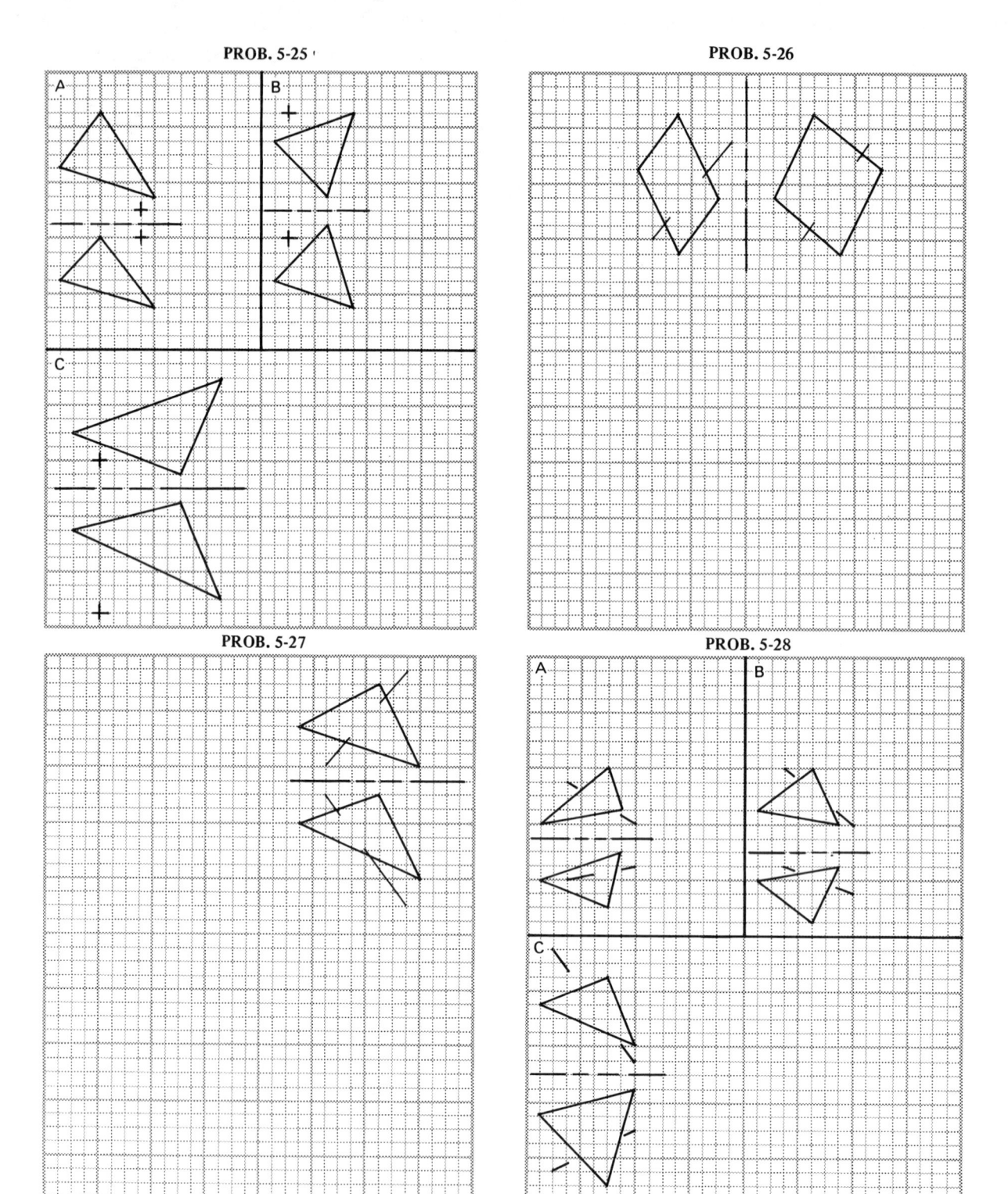

PROB. 5-26

PROB. 5-27

PROB. 5-28

PROB. 5-25A. Determine the shortest distance between the point and the plane. Show the connecting line in all views. What are the slope and bearing of the line?

PROB. 5-25B. Solve for the shortest distance, the shortest level distance, and the shortest 20% grade distance. Give the strike and slope of the plane.

PROB. 5-25C. Solve for positive grade distances of 15% and 25% between the point and the line. Show in all views. Also determine the shortest distance and shortest horizontal distance. Show only in the auxiliary view.

PROB. 5-26. Using the plane method solve for the angle between the line and the plane and determine proper visibility.

PROB. 5-27. Solve for the angle between the line and the plane using the line method and determine proper visibility.

PROB. 5-28A. Determine the angle between the line and plane utilizing the line method.

PROB. 5-28B. Solve for the angle between the line and the plane using the complementary angle method or the line method.

PROB. 5-28C. Using the complementary angle method determine the angle between the line and the plane.

FIG. 5-65 Angle between two planes.

5.36 ANGLE BETWEEN TWO PLANES

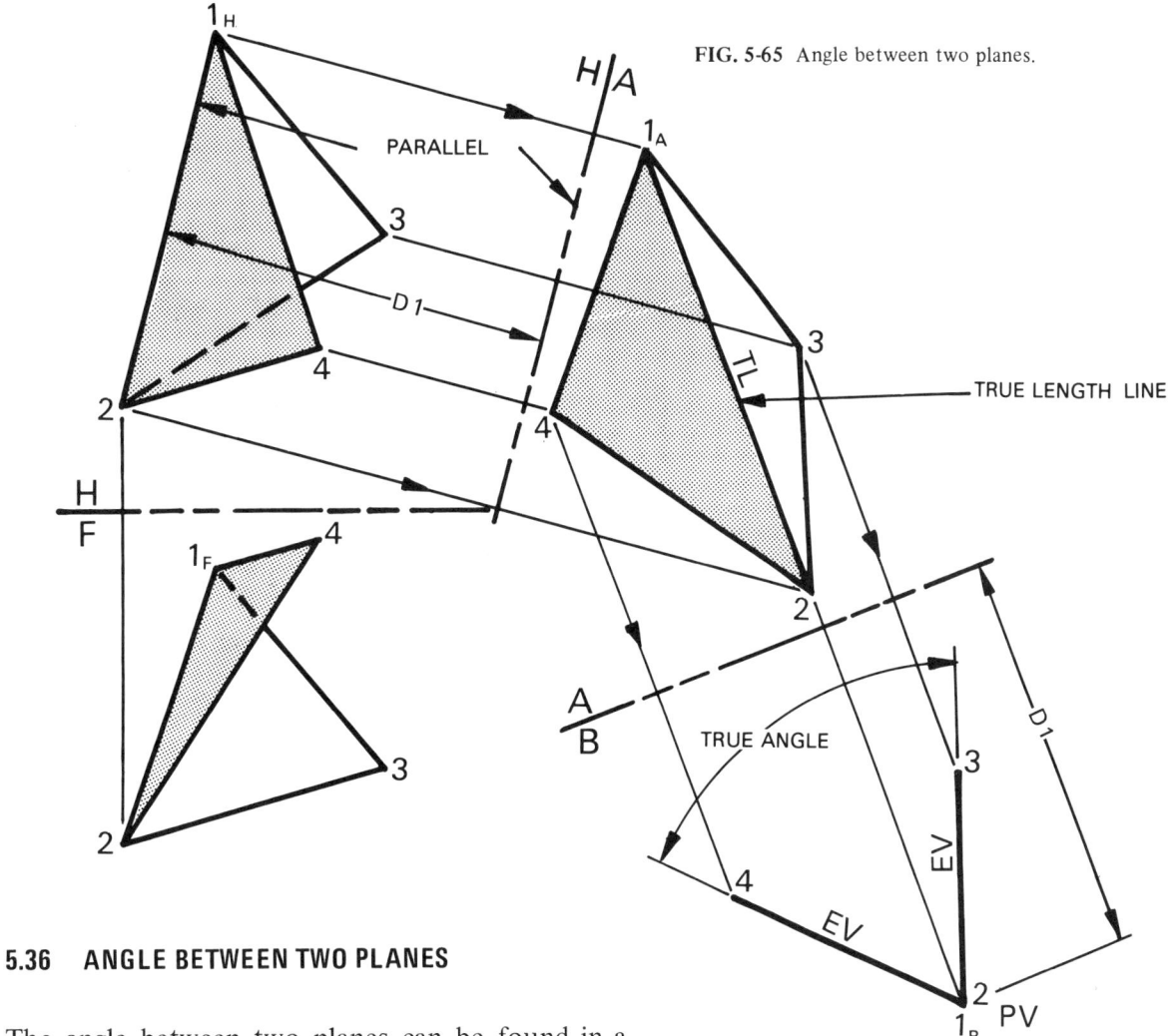

The angle between two planes can be found in a projection where both planes are seen as edge views. *The true angle between two intersecting planes is normally called a **dihedral angle**.* To solve for the angle between two intersecting planes, a view is necessary where the *common line* (intersection line) appears as a point view. In this view both planes show as edges and the angle between them can be measured. The first step in finding the angle between two planes involves projecting an auxiliary view where the common line is true length. An auxiliary view projected perpendicular to this true length intersection line shows the common line as a point and both planes as edges. The true angle between the planes is measured in this secondary auxiliary view.

In Fig. 5-65, two oblique planes with a common line are given. The dihedral angle formed by these two intersecting planes is required.

1. Draw H/A parallel to line 1_H-2_H. Line 1-2 is the common (intersection) line of the two oblique planes.
2. Complete auxiliary A. Line 1_A-2_A is true length in this projection.
3. Draw A/B perpendicular to true length line 1_A-2_A and complete auxiliary view B by projection and transferring dimensions from the horizontal view. Dimension D1 locates the point view of common line 1_B-2_B.
4. The true angle between the planes is measured in auxiliary B, since both intersecting planes appear as edges. This is the dihedral angle formed by the two planes.

In Fig. 5-66, the angle formed by welding two steel plates is an example of a dihedral angle between two intersecting planes. The supporting base for the airfoil machine has several intersecting steel plates which have been cut and welded at specific angles.

FIG. 5-66 Six-inch airfoil machine. *(Courtesy NASA.)*

FIG. 5-67 Angle between planes (dihedral angle).

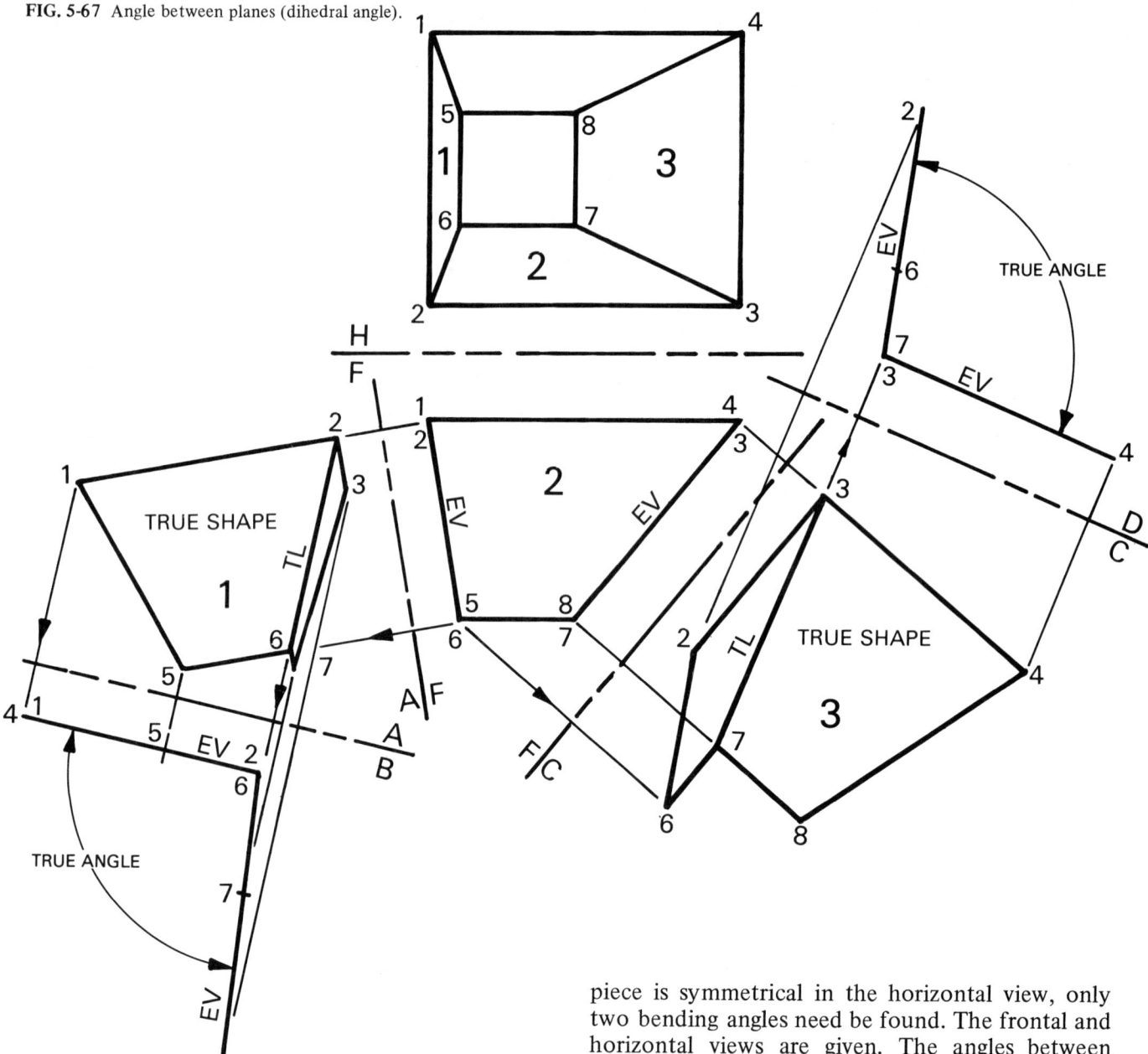

5.37 ANGLE BETWEEN TWO PLANES (DIHEDRAL ANGLE)

As stated in the previous section, *the angle formed by two planes is a **dihedral angle**.* To measure the true angle between two planes it is necessary to show their line of intersection as a point view. When a line on a plane appears as a point, the plane shows as an edge. If the common line between two planes is projected as a point, both planes appear as edges.

The sheet metal transition piece shown in Fig. 5-67 is formed by bending flat sheet stock at specific angles between each surface. The dihedral angle between each side must be known before the transition piece can be bent and welded. Since the piece is symmetrical in the horizontal view, only two bending angles need be found. The frontal and horizontal views are given. The angles between sides 1 and 2 and between sides 2 and 3 are necessary to manufacture the piece.

1. The frontal view of the transition piece shows side 1 and side 3 as edges. Draw F/A parallel to plane 1_F-2_F-5_F-6_F. In auxiliary A this plane is true shape and the line of intersection, 2_A-6_A, between plane 1 and 2 is true length.

2. Draw A/B perpendicular to line 2_A-6_A and complete auxiliary B. Line 2_B-6_B appears as a point view and both planes are edges. The true angle between plane 1 and plane 2 is measured as shown.

3. These same two steps are used to find the dihedral angle formed by plane 2 and 3. Auxiliary C is projected parallel to common line 3_F-7_F (and edge view 3_F-4_F-7_F-8_F). Line 3_C-7_C is a point view and plane 3_C-4_C-7_C-8_C shows as true shape.

4. D/C is drawn perpendicular to common line 3_C-7_C. In auxiliary D both planes appear as edges, since their line of intersection is a point view. The dihedral angle between plane 2 and plane 3 is measured as shown.

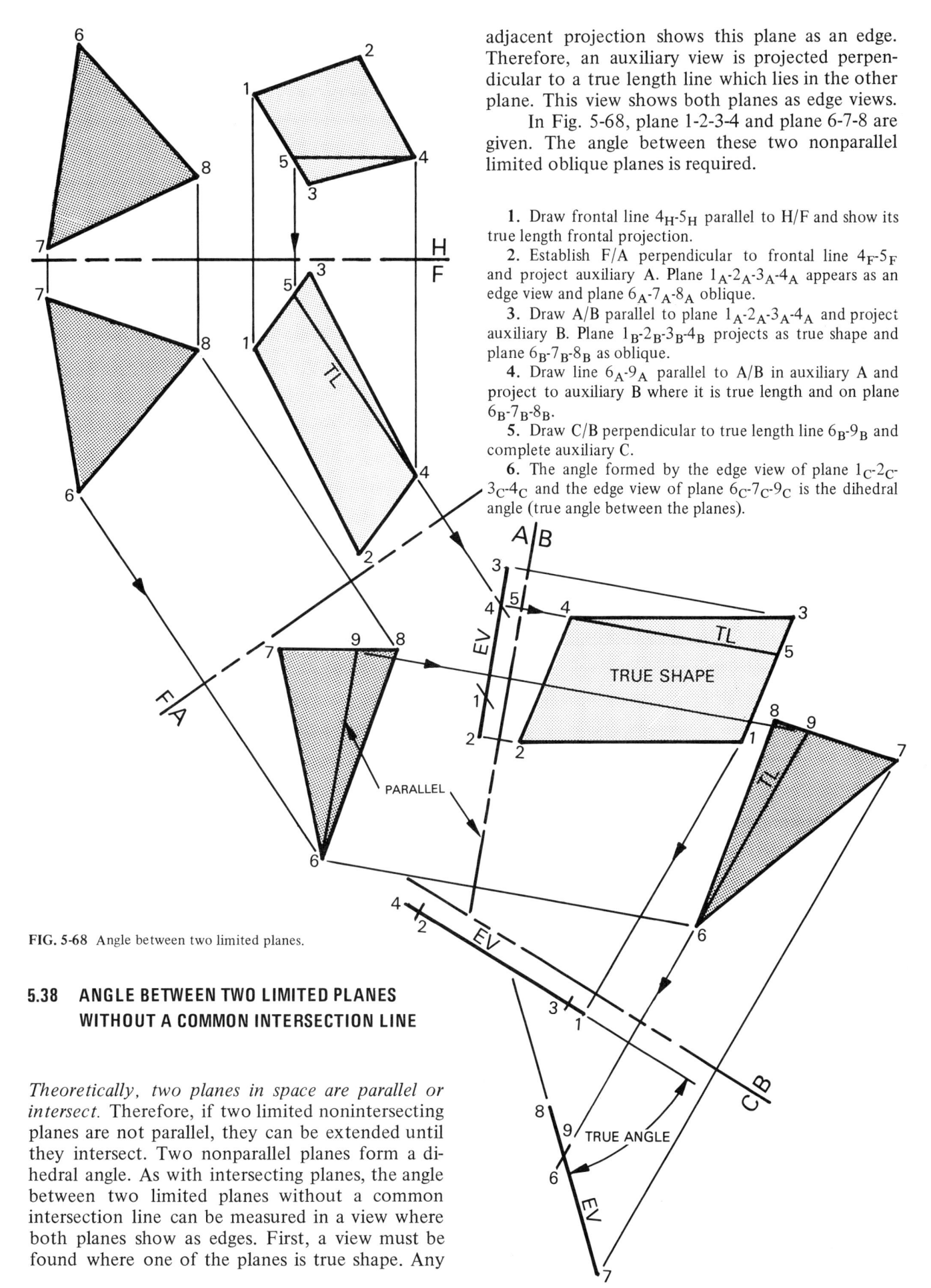

adjacent projection shows this plane as an edge. Therefore, an auxiliary view is projected perpendicular to a true length line which lies in the other plane. This view shows both planes as edge views.

In Fig. 5-68, plane 1-2-3-4 and plane 6-7-8 are given. The angle between these two nonparallel limited oblique planes is required.

1. Draw frontal line 4_H-5_H parallel to H/F and show its true length frontal projection.
2. Establish F/A perpendicular to frontal line 4_F-5_F and project auxiliary A. Plane 1_A-2_A-3_A-4_A appears as an edge view and plane 6_A-7_A-8_A oblique.
3. Draw A/B parallel to plane 1_A-2_A-3_A-4_A and project auxiliary B. Plane 1_B-2_B-3_B-4_B projects as true shape and plane 6_B-7_B-8_B as oblique.
4. Draw line 6_A-9_A parallel to A/B in auxiliary A and project to auxiliary B where it is true length and on plane 6_B-7_B-8_B.
5. Draw C/B perpendicular to true length line 6_B-9_B and complete auxiliary C.
6. The angle formed by the edge view of plane 1_C-2_C-3_C-4_C and the edge view of plane 6_C-7_C-9_C is the dihedral angle (true angle between the planes).

FIG. 5-68 Angle between two limited planes.

5.38 ANGLE BETWEEN TWO LIMITED PLANES WITHOUT A COMMON INTERSECTION LINE

Theoretically, two planes in space are parallel or intersect. Therefore, if two limited nonintersecting planes are not parallel, they can be extended until they intersect. Two nonparallel planes form a dihedral angle. As with intersecting planes, the angle between two limited planes without a common intersection line can be measured in a view where both planes show as edges. First, a view must be found where one of the planes is true shape. Any

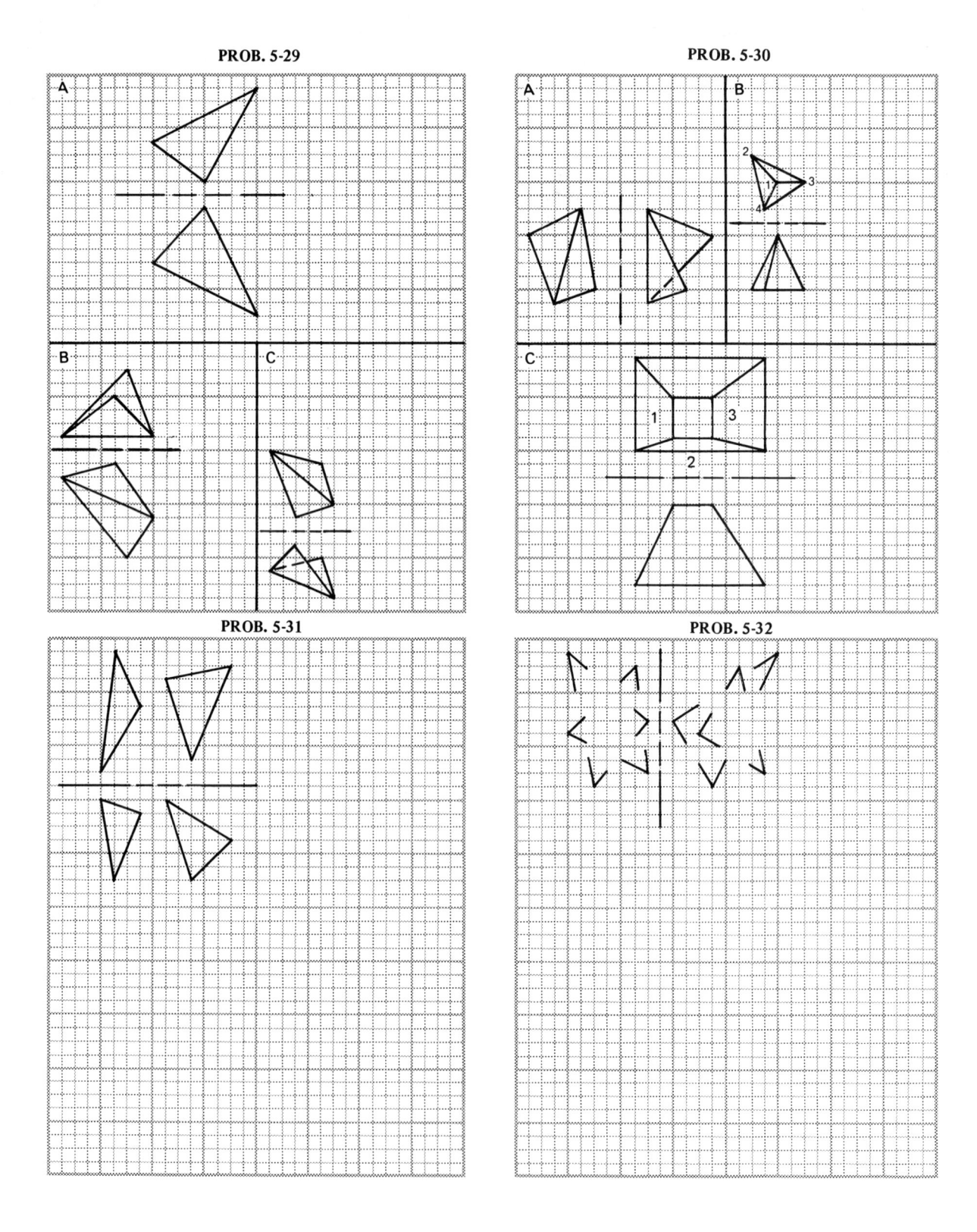

PROB. 5-29A. Determine the angle that the plane makes with the horizontal, frontal, and profile planes.

PROBS. 5-29B, 5-29C. Project and measure the true angle between the two connected planes.

PROB. 5-30A. Solve for the angle between the planes.

PROB. 5-30B. The solid form 1-2-3-4 is a tetrahedron. Determine the dihedral angle between surfaces 1-2-3 and 2-3-4 by projecting the auxiliary view that shows these two surfaces in edge view.

PROB. 5-30C. Solve for the angle between planes 1 and 2, and between the planes 2 and 3.

PROB. 5-31. Determine the angle between the two nonintersecting planes.

PROB. 5-32. Solve for the angle between the planes where the line of intersection is not given. Use the same procedure as when the planes are nonintersecting.

FIG. 5-69 Rotating cylinder wing flap system.

5.39 REVOLUTION OF PLANES

The *revolution of planes* involves the same basic principals as found in the revolution of points and lines; see Chapter 4, pages 121-33. Revolution is an excellent method for solving descriptive geometry problems in fewer views by eliminating time-consuming auxiliary projections. A variety of industrial applications involving revolving parts on machinery, aircraft, and for showing clearance between moving parts and mechanisms must be solved by graphical revolution. In Fig. 5-69, a rotating cylinder wing flap system is installed on a YOV-10A test bed aircraft. The cylinders are revolved in the direction of air flow. In Fig. 5-70 the rocket launcher is rotated about a horizontal axis line until the rocket is vertical.

Revolution of any object (point, line, plane, solid) requires that each individual point which makes up the form be revolved about an established axis line. Each point of a given form is revolved through the same number of specified degrees, in the same direction. The revolution of a point scribes a circular arc in the view where the axis line appears as a point. In the adjacent *perpendicular* projection, the axis line is true length and the circular path of rotation appears as an edge, always parallel to the related fold line.

FIG. 5-70 Portable rotating rocket launcher. *(Courtesy NASA.)*

FIG. 5-71 Edge view by revolution.

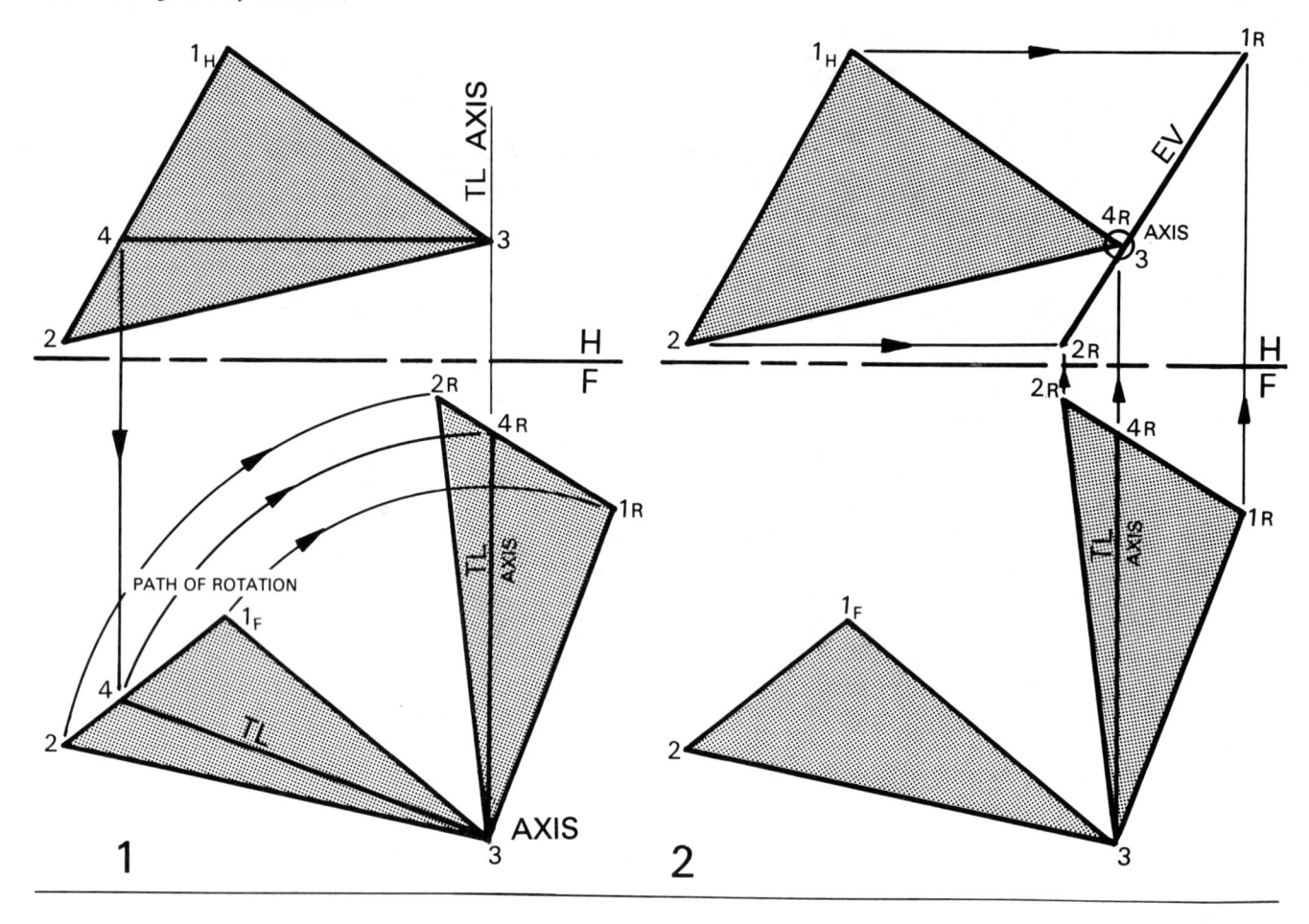

FIG. 5-72 Edge view of a plane by revolution.

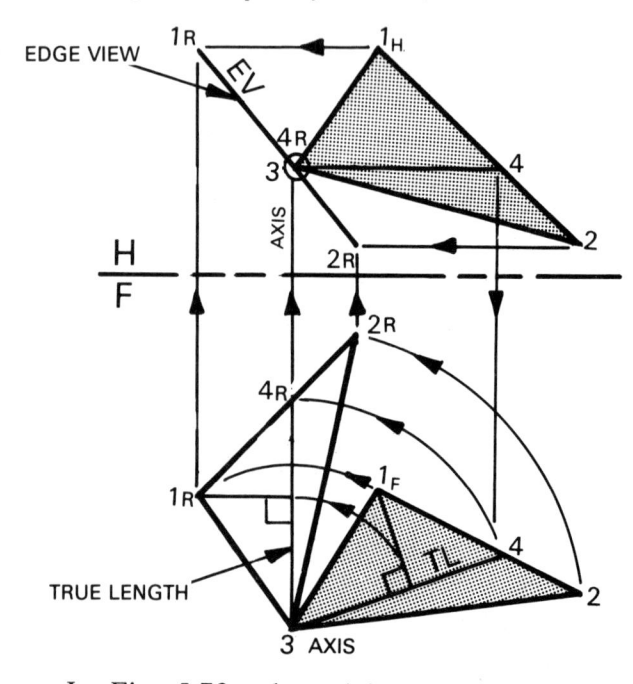

5.40 EDGE VIEW OF A PLANE USING REVOLUTION

The edge view of a plane is found in a view where a line in the plane appears as a point view. Normally an auxiliary projection is needed to solve for the edge view of a plane. Revolution can also be used to establish an edge view. By revolving a plane until it is perpendicular to a principal projection plane, its adjacent view will show the plane as an edge.

In Fig. 5-71, the edge view of oblique plane 1-2-3 is required.

1. Draw frontal line 3_H-4_H on the plane and parallel to H/F. Project as true length in the frontal view. Using point 3_F as an axis, revolve plane 1_F-2_F-3_F clockwise until frontal line 3_F-$4R_F$ is perpendicular to H/F. Plane $1R_F$-$2R_F$-3_F is now perpendicular to the horizontal projection plane. Line 3_F-$4R_F$ can be used as a TL axis line for revolving the plane parallel to the fold line if a true size view is required.

2. Project the revolved position of the plane to the horizontal view. Line 3-4R is a vertical line and appears as a point view. Point $1R_H$ and $2R_H$ are located by projection from the frontal view and by moving each point perpendicular to the axis (parallel to the adjacent fold line). Plane $1R_H$-$2R_H$-3_H is an edge view.

In Fig. 5-72, plane 1-2-3 is revolved until frontal line 3_F-$4R_F$ is perpendicular to H/F. The horizontal projection of plane $1R_H$-$2R_H$-3_H is an edge view.

FIG. 5-73 True size/shape of a plane by revolution.

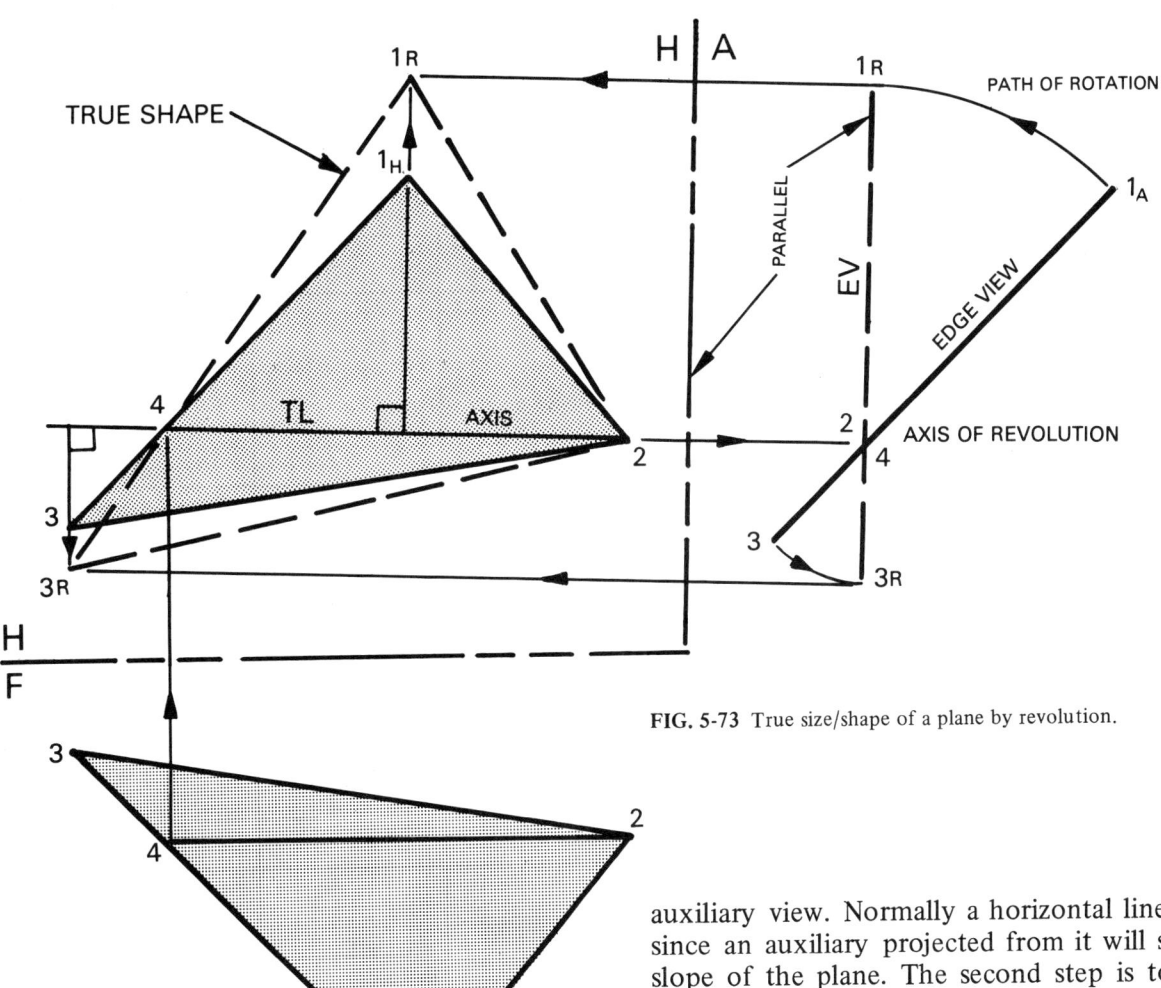

5.41 TRUE SIZE OF A PLANE BY REVOLUTION

The true size/shape of a plane can be determined by revolving the plane about a true length axis line which lies on the plane. The plane is revolved about the axis in a view where the plane appears as an edge and the axis line is a point view. The edge view of the plane is revolved until it is parallel to an adjacent projection plane. The revolved plane will then be parallel to the fold line and perpendicular to the line of sight for its adjacent projection.

Given two principal views of an oblique plane, the first step is to project the plane as an edge view. This can be done by establishing a principal (true length) line that lies on the plane in either view and projecting an auxiliary view perpendicular to it. The plane appears as an edge in this primary auxiliary view. Normally a horizontal line is used, since an auxiliary projected from it will show the slope of the plane. The second step is to revolve the edge view of the plane about the point view of the axis line (principal line) which lies on the plane, until the plane is parallel to the adjacent fold line. The revolved position of the plane is then projected back to the previous view, where it is true size in its revolved location.

In Fig. 5-73, the frontal and horizontal projections of oblique plane 1-2-3 are given. The true shape is required.

1. Draw horizontal line 2_F-4_F parallel to H/F and true length in the horizontal view. The strike of plane 1-2-3 is the bearing of horizontal line 2_H-4_H.

2. Fold line H/A is drawn perpendicular to horizontal line 2_H-4_H and auxiliary A is completed. Plane 1_A-2_A-3_A is an edge view in auxiliary A. The slope of the plane can be measured between the plane and H/A.

3. Using the point view of horizontal line 2_A-4_A as the axis of revolution, rotate the plane until it is parallel to H/A. Point 1_A revolves to position $1R_A$ and point 3_A revolves to position $3R_A$. Both points move counterclockwise through the same angular displacement.

4. Project the revolved position of plane $1R_A$-2_A-$3R_A$ back to the horizontal view. Locate point $1R_H$ and $3R_H$ by moving point 1_H and 3_H perpendicular to the true length axis line 2_H-4_H, until they intersect projection lines extended from the revolved points in auxiliary A.

5. Connect points $1R_H$, $3R_H$, and 2_H. Plane $1R_H$-2_H-$3R_H$ is the true shape of the plane.

5.42 TRUE SHAPE OF A PLANE USING REVOLUTION

In Fig. 5-74, the edge view of plane 1-2-3-4 is revolved so that its resulting true size projection will not be on top of the given view of the plane. This procedure allows the revolved true shape plane to be shown without the confusion caused by the overlapping of projections. In Fig. 5-74, plane 1-2-3-4 is given and the true shape of the plane is required. Plane 5-6-7 lies on the plane.

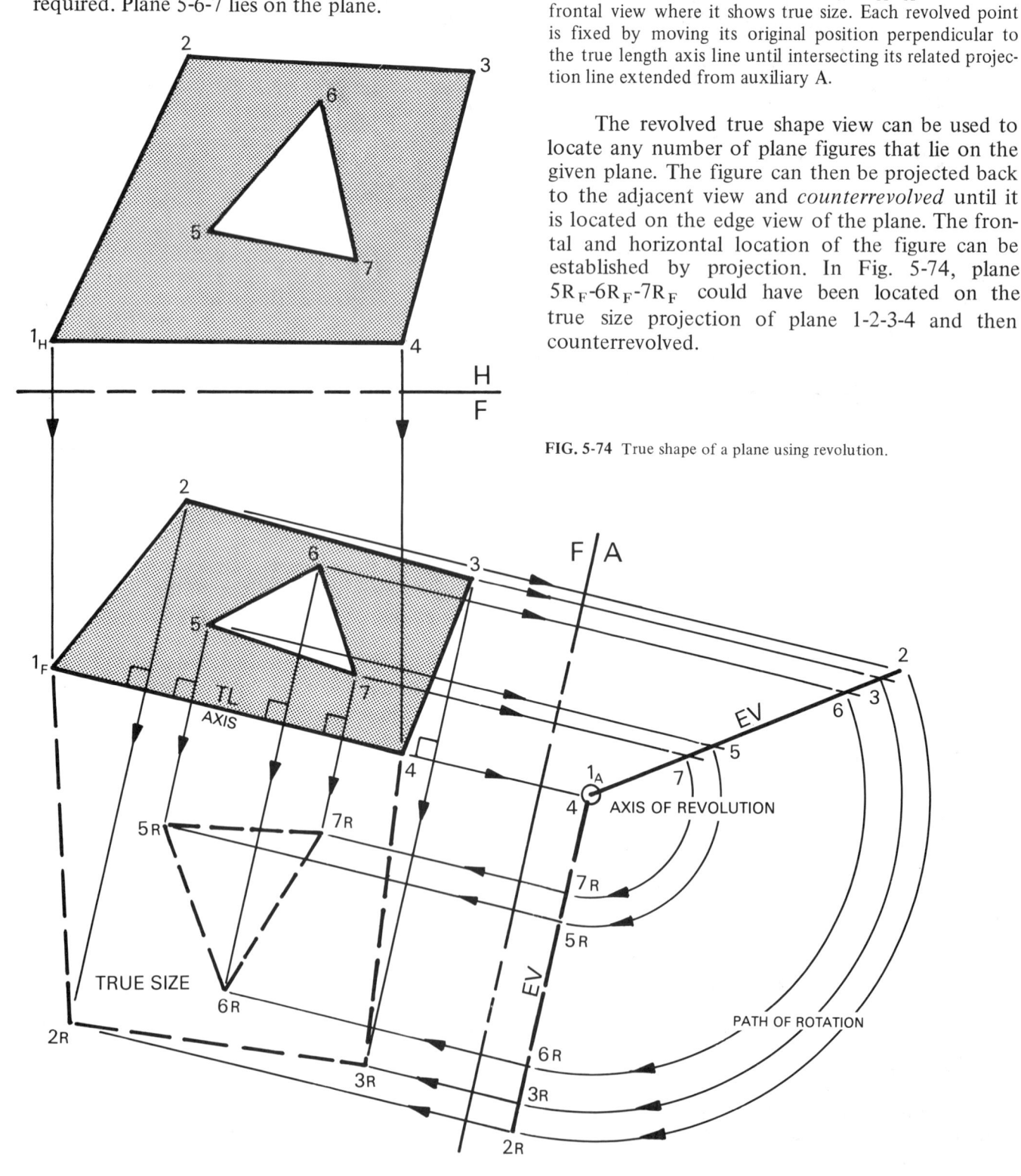

1. Line 1_H-4_H is parallel to H/F, therefore it is a true length line in the frontal view. Draw F/A perpendicular to 1_F-4_F and project auxiliary A. Line 1_F-4_F is the true length axis line.
2. Plane 1_A-2_A-3_A-4_A is an edge view in auxiliary A. Using the point view of line 1_A-4_A as the axis of revolution, rotate the edge view of the plane until it is parallel to F/A. Plane 1_A-2_A-3_A-4_A is revolved clockwise to position 1_A-$2R_A$-$3R_A$-4_A.
3. Project revolved plane 1_A-$2R_A$-$3R_A$-4_A back to the frontal view where it shows true size. Each revolved point is fixed by moving its original position perpendicular to the true length axis line until intersecting its related projection line extended from auxiliary A.

The revolved true shape view can be used to locate any number of plane figures that lie on the given plane. The figure can then be projected back to the adjacent view and *counterrevolved* until it is located on the edge view of the plane. The frontal and horizontal location of the figure can be established by projection. In Fig. 5-74, plane $5R_F$-$6R_F$-$7R_F$ could have been located on the true size projection of plane 1-2-3-4 and then counterrevolved.

FIG. 5-74 True shape of a plane using revolution.

186

FIG. 5-75 Edge view and true size of a plane by double revolution.

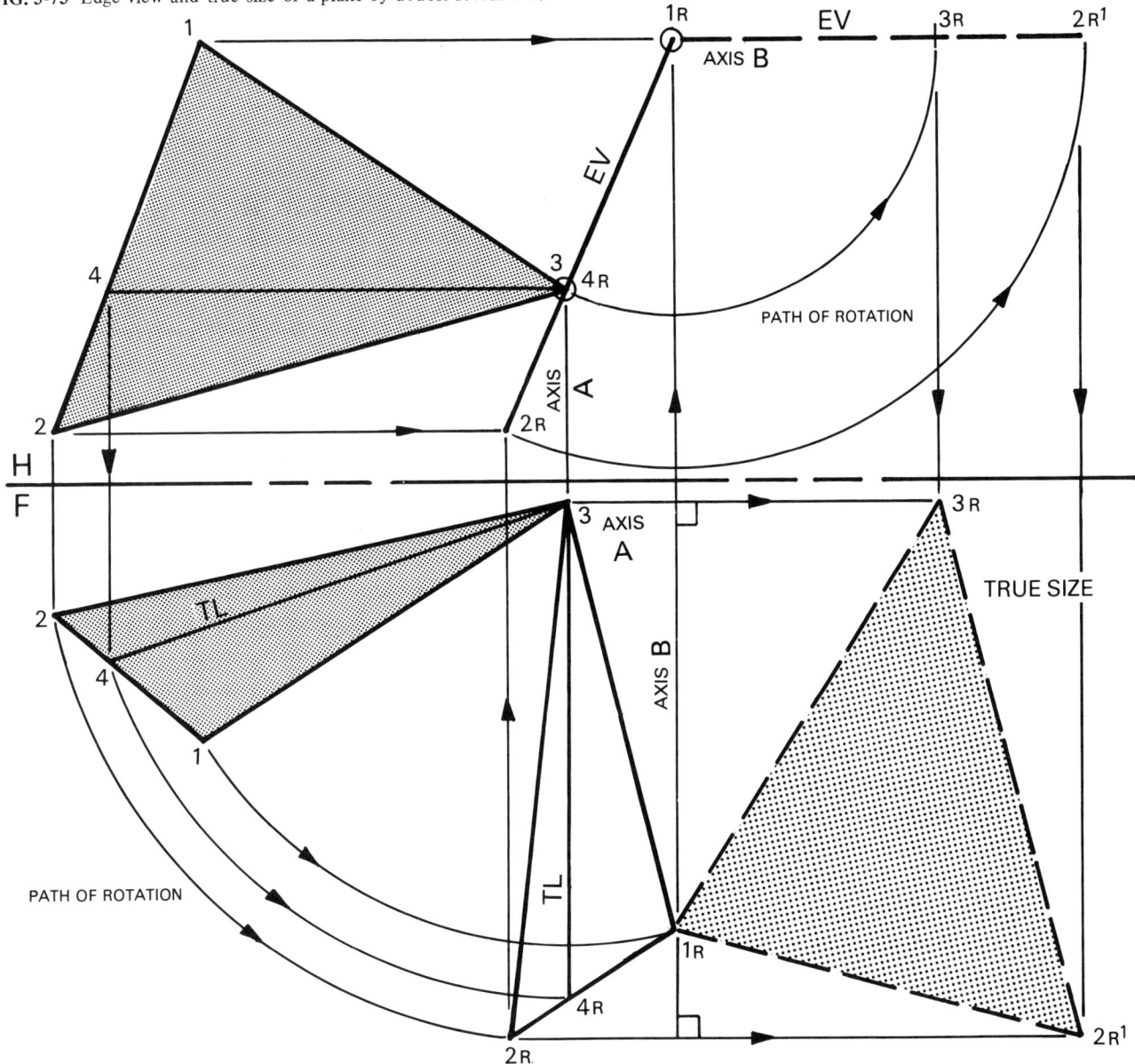

5.43 DOUBLE REVOLUTION OF A PLANE

The edge view and true shape of an oblique plane can be solved for in the two given views by double revolution. An auxiliary projection is unnecessary when using this method. *Double revolution* can cause confusion and clutter the drawing if the plane is revolved so as to overlap the existing views. This confusion can be eliminated by revolving the plane away from the given projections, as in Fig. 5-75.

The first step is to establish a true length line on the plane. Second, the plane is revolved until the true length line is perpendicular to the fold line and projected to the adjacent view where it appears as an edge. Third, the edge view is revolved until parallel to the fold line and then projected to the adjacent view where it shows true size/shape.

In Fig. 5-75, oblique plane 1-2-3 is given. Using only the existing views, find the true size.

1. Draw frontal line 3_H-4_H parallel to H/F and project as true length in the frontal view.

2. Revolve 1_F-2_F-3_F about axis A, which passes through point 3_F, until true length line 3_F-$4R_F$ is vertical (perpendicular to H/F).

3. Project the revolved position of the plane to the horizontal view. Plane $1R_H$-$2R_H$-3_H appears as an edge view.

4. Revolve the edge view of the plane about axis B, which passes through point $1R_H$, until the plane is parallel with H/F. Plane $1R_H$-$2R_H^1$-$3R_H$ is parallel to the frontal projection plane and remains as an edge in the horizontal view.

5. Project the revolved position to the frontal view where plane $1R_F$-$2R_F^1$-$3R_F$ is true size/shape.

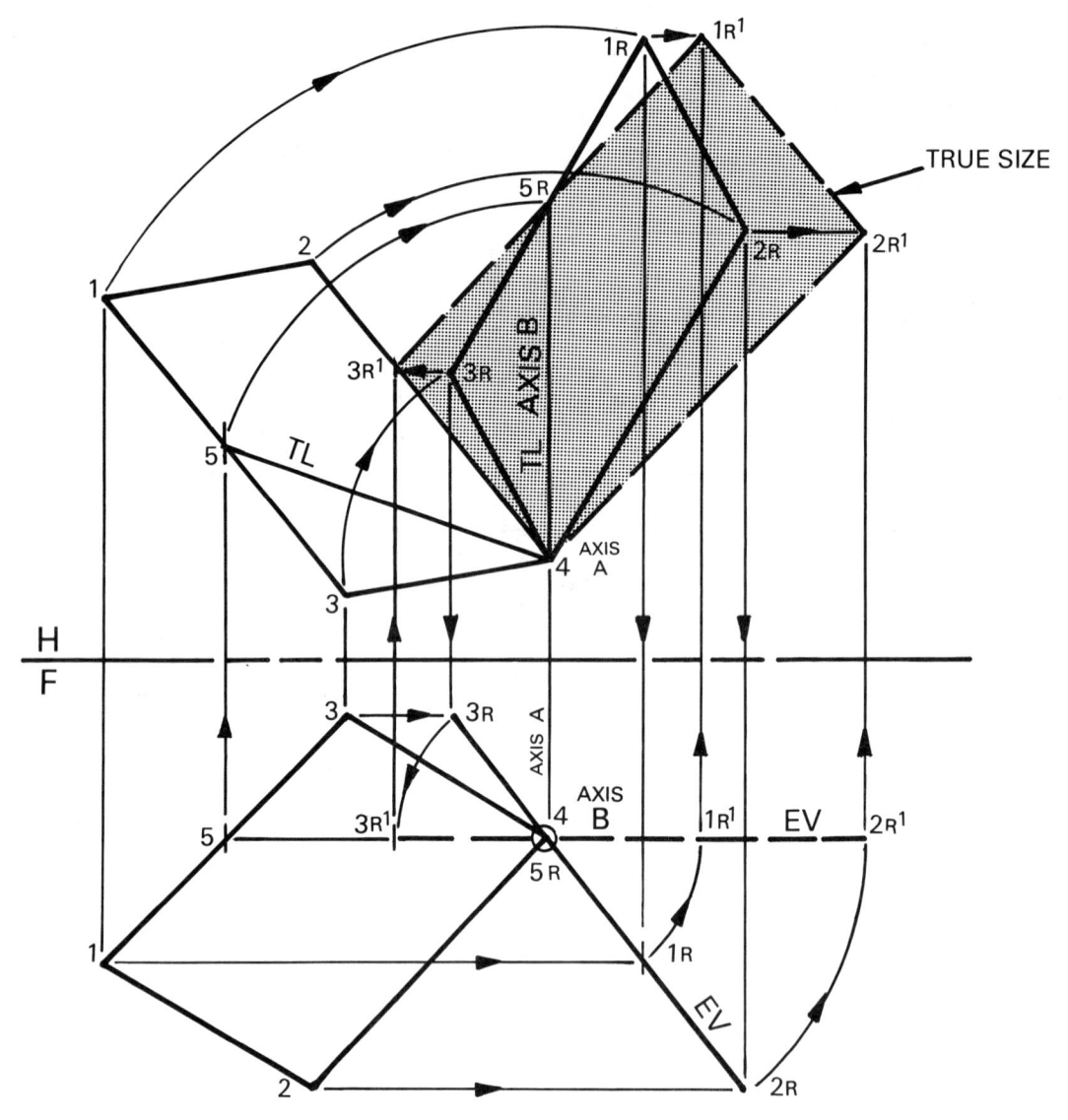

FIG. 5-76 True size of a plane by double revolution.

5.44 TRUE SIZE OF A PLANE BY DOUBLE REVOLUTION

Double revolution can save time and space by eliminating one or two auxiliary views normally associated with the projection of an oblique plane as true size. Whether or not the final graphical solution is adequate depends on the choice of axes, drafting quality, and accuracy of projection. In Fig. 5-76, plane 1-2-3 is given and a view showing it as true size is required. Note how the choice of axis has determined the final placement of the true size view. Without the shading, this illustration might prove somewhat more cluttered and confusing. An axis which revolves the plane away from the given projection is preferred, as in Fig. 5-75. However, space restrictions may require that the final solution overlap the given and first revolution views as in Fig. 5-76.

1. Draw horizontal line 4_F-5_F parallel to H/F and true length in the horizontal view.
2. Rotate plane 1-2-3-4 about axis A (point 4_H) until line 4_H-5_H is perpendicular to H/F.
3. Project plane $1R_H$-$2R_H$-$3R_H$-4_H to the frontal view. Plane $1R_F$-$2R_F$-$3R_F$-4_F appears as an edge view.
4. Using the point view of line 4_F-$5R_F$ as axis B, revolve the edge view of plane $1R_F$-$2R_F$-$3R_F$-4_F until it is parallel with the horizontal plane (H/F).
5. Project plane $1R_F^1$-$2R_F^1$-$3R_F^1$-4_F to the horizontal view. Plane $1R_H^1$-$2R_H^1$-$3R_H^1$-4_H is a true size projection.

PROB. 5-33

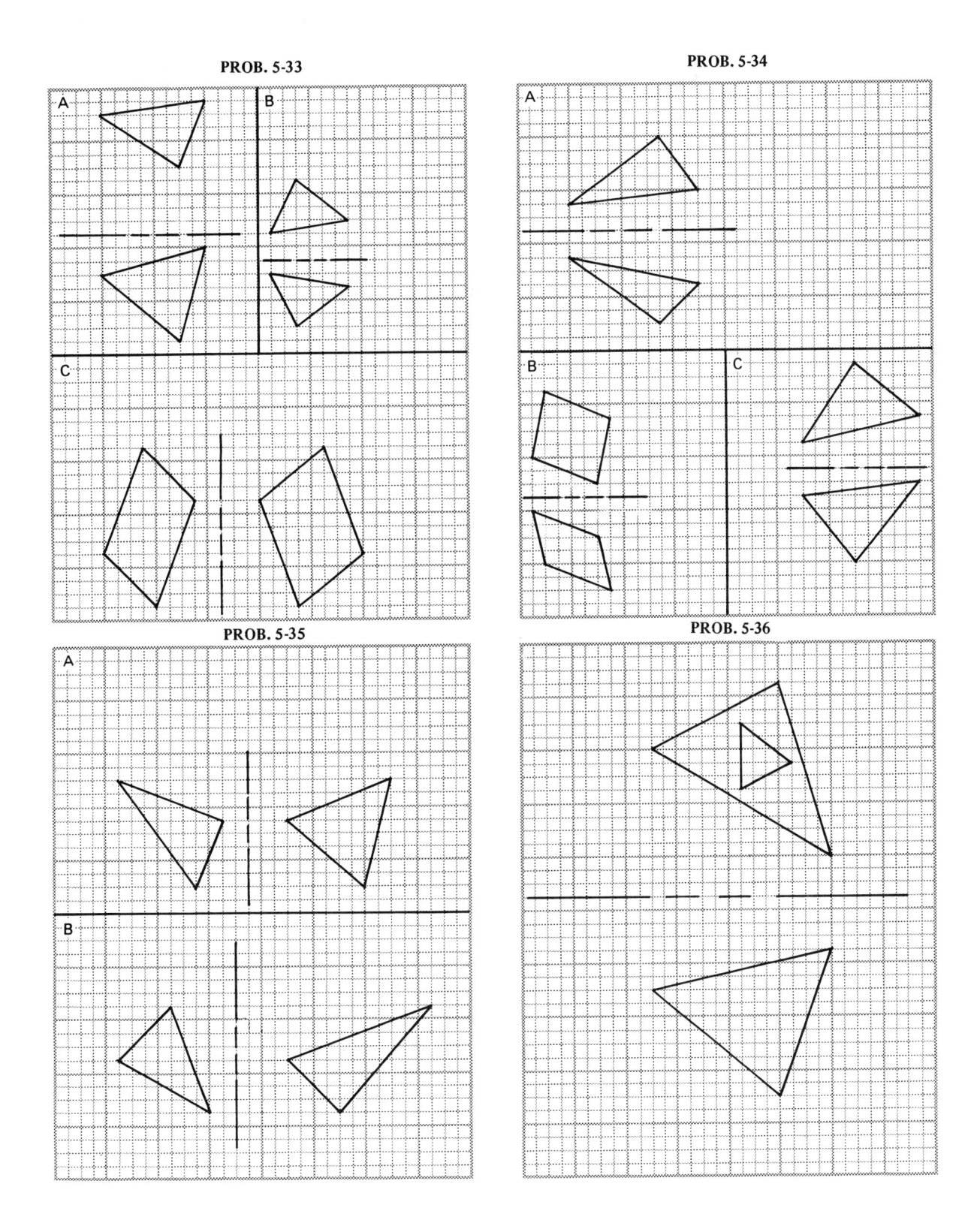

PROB. 5-34

PROB. 5-35

PROB. 5-36

PROB. 5-33A. Solve for the edge view of the plane using revolution.

PROB. 5-33B. Using rotation, project a true size view of the plane. Verify by projecting an auxiliary view.

PROB. 5-33C. Solve for the edge view by revolving the plane in the profile view.

PROB. 5-34A. Project an edge view off of the H view, then using revolution solve for the true shape.

PROBS. 5-34B, 5-34C. Solve for the true shape of the given plane by projecting an edge view and then using revolution.

PROB. 5-35A. Solve for the true shape of the plane by double rotation. The edge view will show in the P view.

PROB. 5-35B. Using double revolution, solve for the true shape of the plane. The edge view appears in the F view. Construct the largest circle that can be inscribed in the triangle using geometric construction.

PROB. 5-36. Using double revolution, solve for the true shape of the plane. Project the figure that lies on the plane into all views and show its true shape.

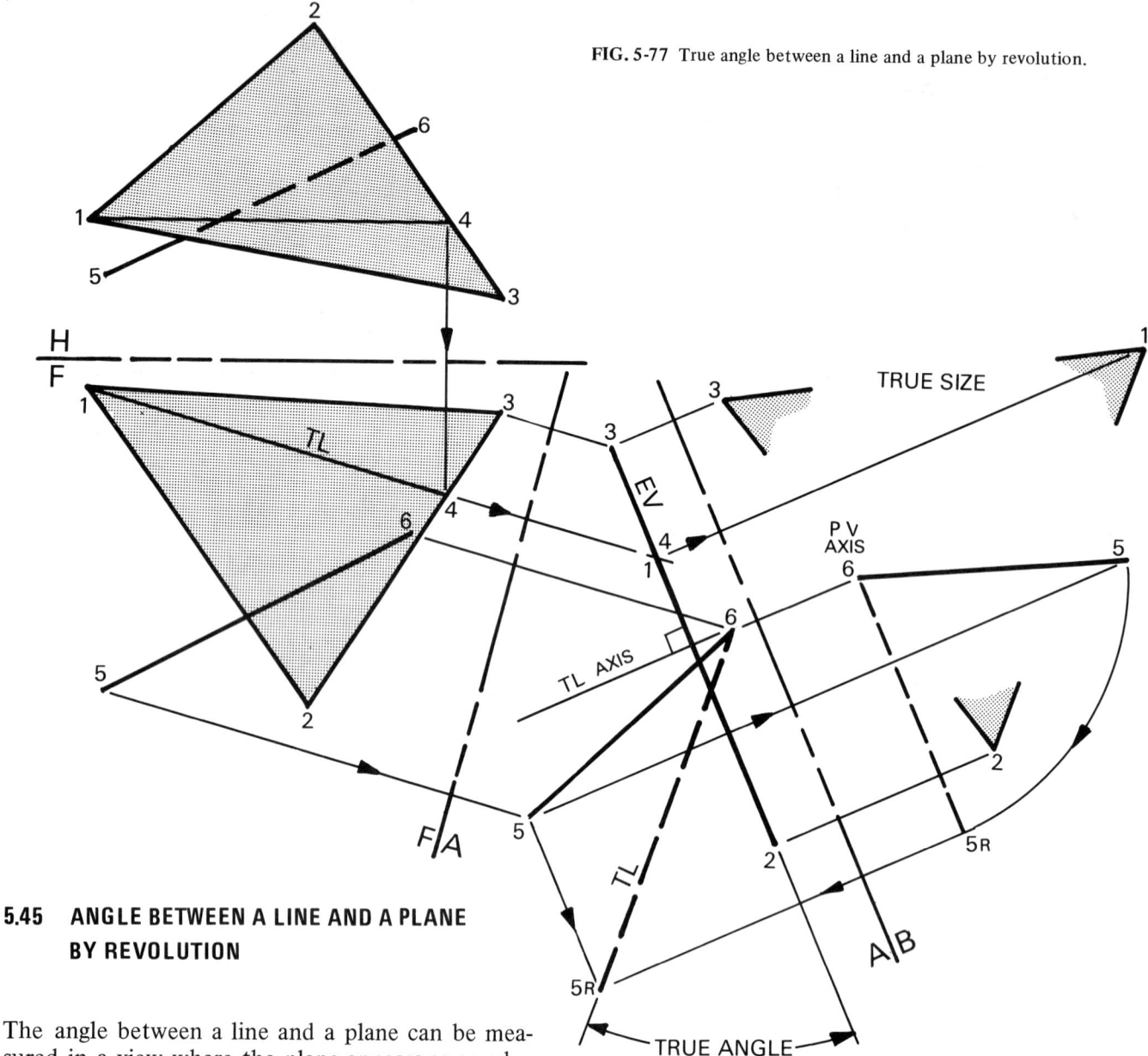

FIG. 5-77 True angle between a line and a plane by revolution.

5.45 ANGLE BETWEEN A LINE AND A PLANE BY REVOLUTION

The angle between a line and a plane can be measured in a view where the plane appears as an edge and the line is true length. Revolution can be used to solve for the true angle. Revolution eliminates one auxiliary projection. The "plane method" is used as shown on page 174 except that the third auxiliary view is not required. The given plane is projected as an edge. A second auxiliary is then taken showing the plane as true shape. In this view the line is revolved parallel to the adjacent fold line and then is projected back to the previous view where it will be true length. Since this view shows the plane as an edge and the line as true length in its revolved position the angle between them can be measured. Note that the axis appears as a point in a view where the plane is true size. The true length view of the axis line will be perpendicular to the edge view of the plane. In other words, the line is revolved about an axis line which is perpendicular to the edge view of the plane.

In Fig. 5-77, plane 1-2-3 and line 5-6 are given. The angle between the line and plane is required.

1. Draw frontal line 1_H-4_H parallel to H/F and project as true length in the frontal view.

2. Draw F/A perpendicular to 1_F-4_F and project auxiliary A. Line 1_A-4_A is a point view and plane 1_A-2_A-3_A is an edge view. Line 5_A-6_A is oblique.

3. Fold line A/B is drawn parallel to the edge view of plane 1_A-2_A-3_A, and auxiliary B completed. Plane 1_B-2_B-3_B is true size and line 5_B-6_B appears oblique. The true shape view of the plane need not be shown since this view is only used to revolve the line.

4. Axis 6_B is a point view of the axis of revolution in auxiliary view B. The true length of the axis line will be perpendicular to the edge view of the plane in auxiliary A. Revolve line 6_B-5_B about point 6_B until it is parallel to A/B.

5. Line 6_B-$5R_B$ is parallel to the adjacent projection plane and therefore projects to auxiliary A as true length. Point 5_A moves perpendicular to the axis line to position $5R_A$.

6. Line 6_A-$5R_A$ is true length. Plane 1_A-2_A-3_A is an edge view. The true angle between the line and plane is measured as shown.

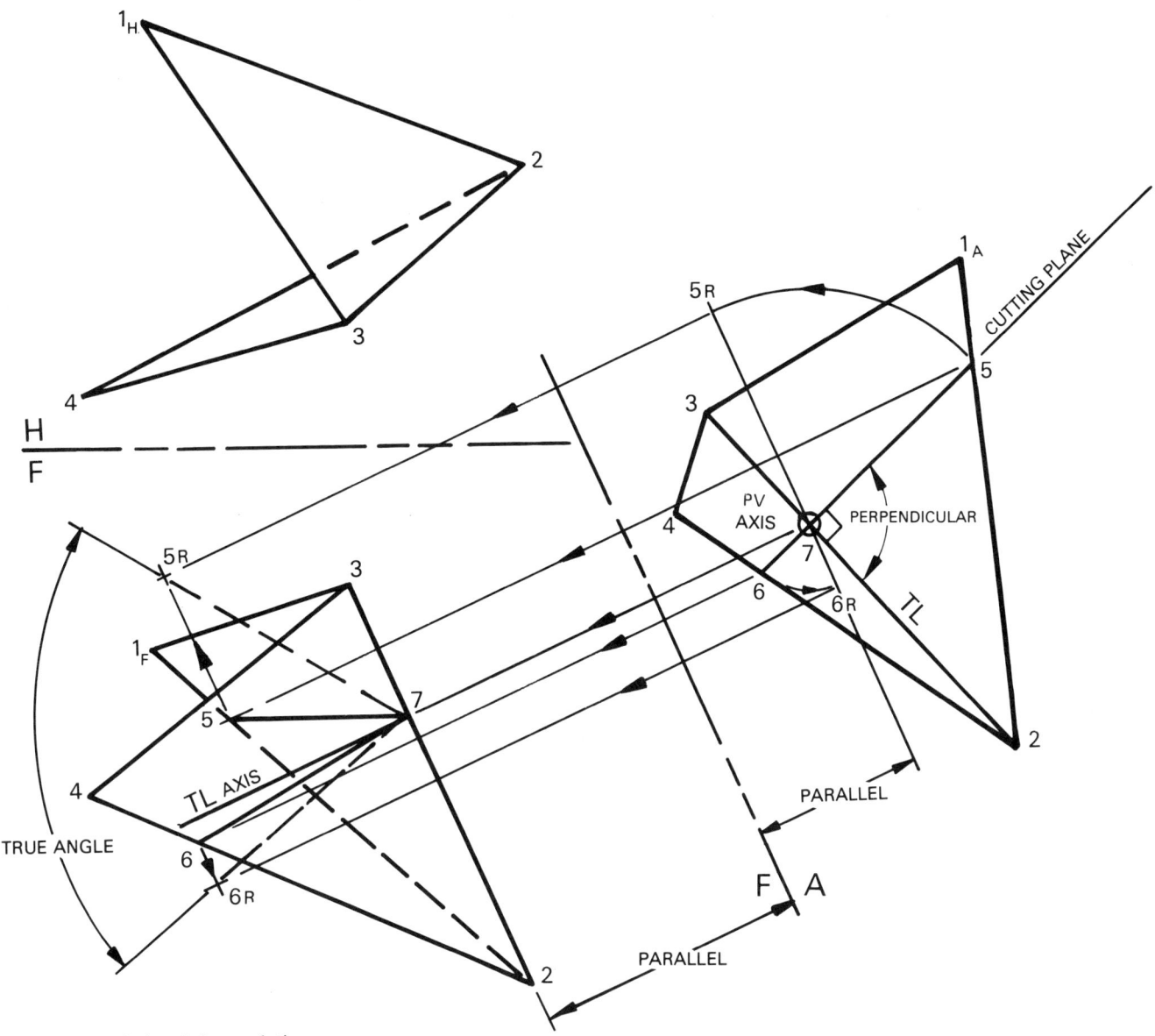

FIG. 5-78 Dihedral angle by revolution.

5.46 ANGLE BETWEEN TWO PLANES BY REVOLUTION

Revolution can be used to find the dihedral angle of two planes. A cutting plane is passed perpendicular to the intersection (common) line between the two planes in a view where the line appears as true length. The plane formed by the cutting plane as it cuts the two intersecting planes is revolved until parallel to the adjacent fold line. This new plane is projected back to the previous view, where it appears true size. The true angle can be measured in this view as the angle formed by the cutting plane intersecting the given planes.

In Fig. 5-78, two intersecting planes are given and their dihedral angle is required.

1. Draw F/A parallel to intersection line 2_F-3_F and project auxiliary view A. Intersection line 2_A-3_A is true length in auxiliary A. Both planes are oblique.

2. Pass a cutting plane perpendicular to true length intersection line 2_A-3_A. The cutting plane can be established anywhere along the true length intersection line. For greater accuracy, locate the cutting plane where the given planes are shown clearly, preferably through a wide portion of each plane. This is the edge view of the cutting plane.

3. Where the cutting plane intersects (cuts) the two given planes, label these points 5_A, 6_A, and 7_A. The cutting plane is now defined by plane 5_A-7_A-6_A.

4. The new plane formed by the cutting plane and the given intersecting planes is revolved about one of its points until parallel to the fold line. Here, plane 5_A-7_A-6_A is revolved about point 7_A until parallel to F/A, $5R_A$-7_A-$6R_A$. Point 7_A is the point view of the axis line.

5. The axis line is drawn true length and perpendicular to line 2_F-3_F and F/A in the frontal view. Points 5_F and 6_F move perpendicular to the axis line (parallel to F/A) to position $5R_F$ and $6R_F$ respectively. Cutting plane $5R_F$-7_F-$6R_F$ is true size in this view.

6. The dihedral angle $5R_F$-7_F-$6R_F$ is measured as shown.

5.47 DIHEDRAL ANGLE BY REVOLUTION

As in the previous section the angle between planes, or dihedral angle, can be found by rotating a cutting plane passed perpendicular to the line of intersection (common line) between the two planes. In Fig. 5-79, two intersecting oblique planes are given and their dihedral angle is required.

1. A true length view of intersection line 1-3 must be found first. Draw A/F parallel to 1_F-3_F and project auxiliary A. Common line 1_A-3_A is true length and both planes are oblique.
2. Pass a cutting plane perpendicular to line 1_A-3_A. The cutting plane "cuts" line 1_A-2_A at point 5_A, line 3_A-4_A at point 6_A, and line 1_A-3_A at point 7_A.

FIG. 5-79 Dihedral angle.

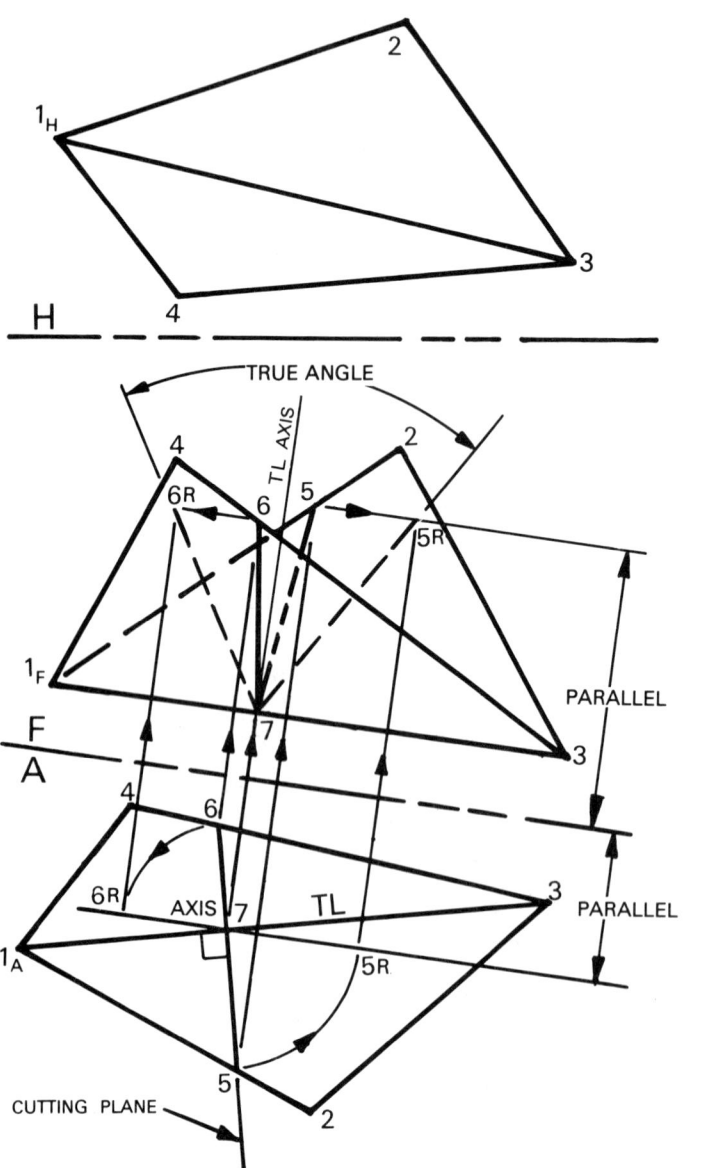

3. Plane 5_A-7_A-6_A represents the edge view of the cutting plane and when revolved equals the dihedral angle. Project plane 5_A-7_A-6_A back to the frontal view.
4. Using point 7_A as the point view of the axis, revolve plane 5_A-7_A-6_A until parallel to F/A and project to the frontal view. Points 5_F and 6_F move perpendicular to the axis line (parallel to F/A) to positions $5R_F$ and $6R_F$.
5. The dihedral angle is measured as angle $5R_F$-7_F-$6R_F$ in the frontal view.

In Fig. 5-80, only two views are necessary to solve for the dihedral angle. Common line 1-2 is a horizontal line (as are lines 3-4 and 5-6) and thus true length in the horizontal view. A cutting plane is passed perpendicular to line 1_H-2_H. Where the cutting plane cuts plane 1-2-3-4 it forms line 7-8, and where it cuts plane 1-2-5-6 it forms line 7-9. Intersecting lines 7-8 and 7-9 form a plane whose angle equals the dihedral angle between the two given planes, angle 8-7-9. Plane 8_H-7_H-9_H is revolved parallel to H/F and projected to the frontal view. The true angle between the planes is measured as shown.

FIG. 5-80 Dihedral angle by revolution.

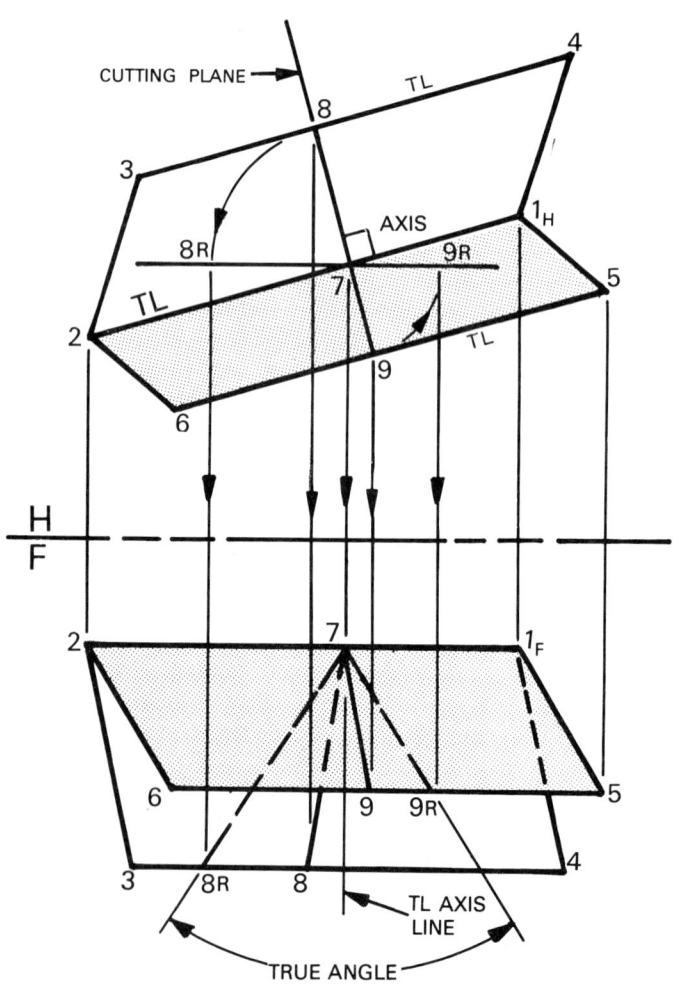

PROB. 5-37

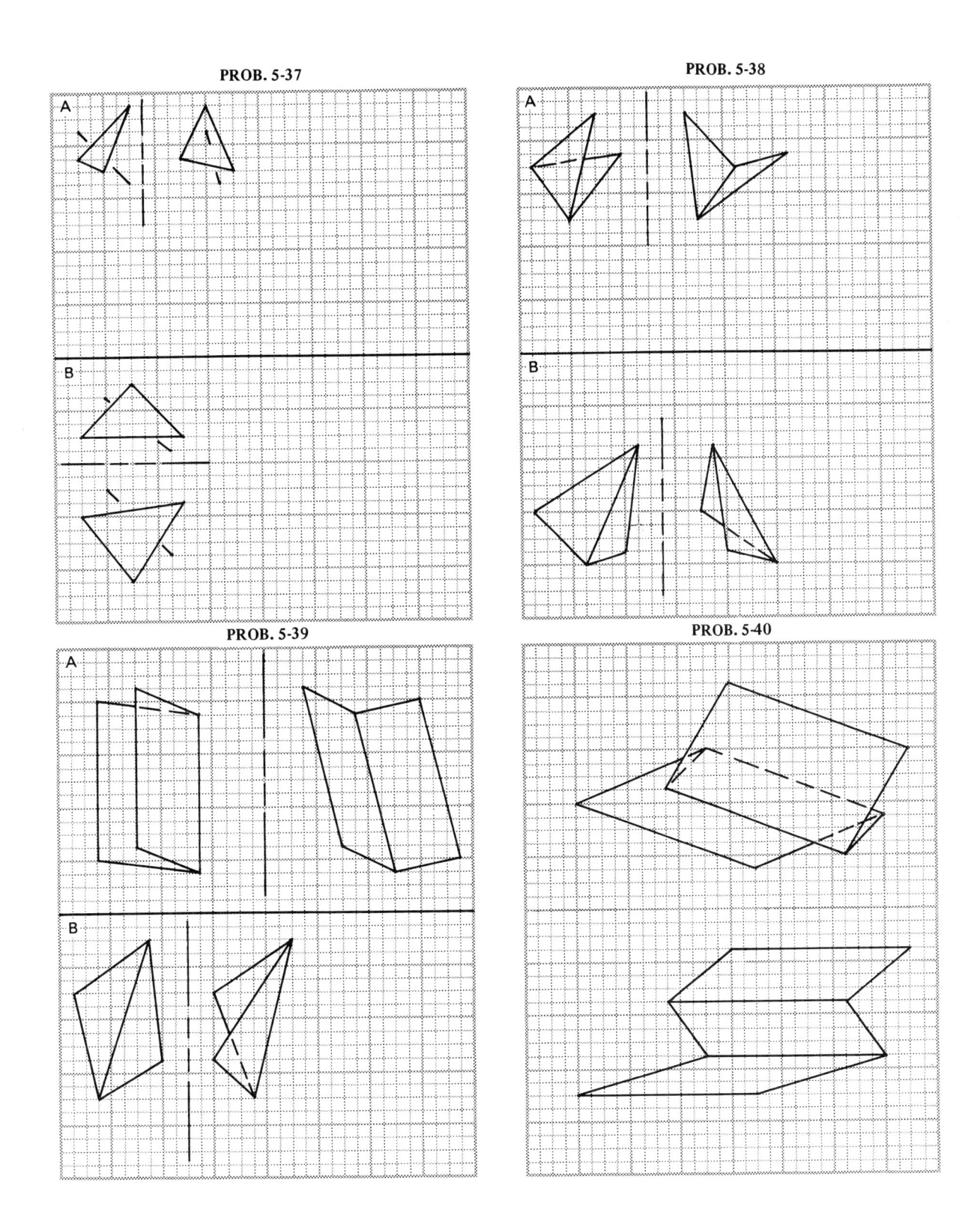

PROB. 5-38

PROB. 5-39

PROB. 5-40

PROBS. 5-37A, 5-37B. Solve for the angle between the line and the plane using revolution.

PROBS. 5-38A, 5-38B. Using rotation, solve for the angle between the connected planes.

PROBS. 5-39A, 5-39B. Determine the dihedral angle between the planes using revolution.

PROB. 5-40. Using rotation, solve for the angle between the planes. There will be two angles to solve for in this problem.

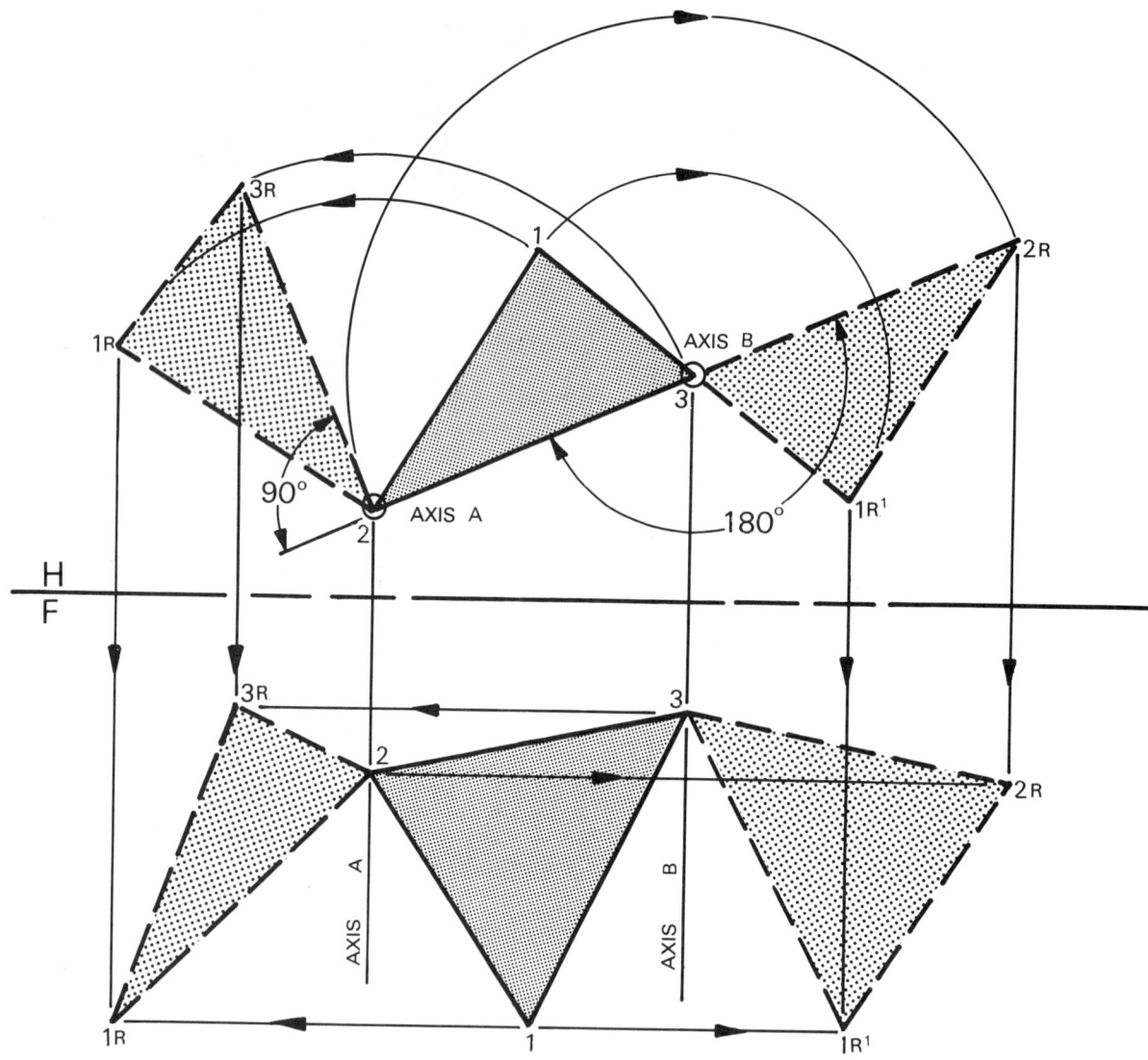

FIG. 5-81 Revolution of a plane about a vertical axis at a specified angle.

5.48 REVOLUTION OF PLANES AT SPECIFIED ANGLES

Engineering problems may require the revolution of an object or plane at a specified angle. A view showing the revolution of a mechanism may be necessary to establish sufficient space for the object to swing through its angular displacement. The clearance between the revolving form and adjacent machinery, etc., may also be required.

In Fig. 5-81, plane 1-2-3 is given. Revolve the plane 90° counterclockwise about a vertical axis which passes through point 2_H. Through point 3_H, rotate the plane 180° clockwise about a vertical axis. Show the frontal view of both revolved positions.

1. Revolve plane 1_H-2_H-3_H 90° counterclockwise about axis A, which passes through point 2_H. Points 1 and 3 are both revolved 90°. Plane $1R_H$-2_H-$3R_H$ represents the revolved position of the plane in the horizontal view. Note that revolution never changes the shape of the form being revolved in the view where the axis appears as a point.

2. Project the revolved position of the plane to the frontal view. Point 3_F moves to position $3R_F$ and point 1_F moves to position $1R_F$, perpendicular to the axis line (parallel to H/F).

3. Revolve plane 1_H-2_H-3_H 180° clockwise about a vertical axis which passes through point 3_H, axis B. Plane $1R_H^1$-$2R_H$-3_H is the revolved position.

4. Project the revolved position of the plane to the frontal view. Point 1_F and point 2_F are moved perpendicular to the axis line until they intersect the projection lines extended from the horizontal view. Plane $1R_F^1$-$2R_F$-3_F is the frontal view of the revolved plane.

FIG. 5-82 Revolution of a plane about a given line.

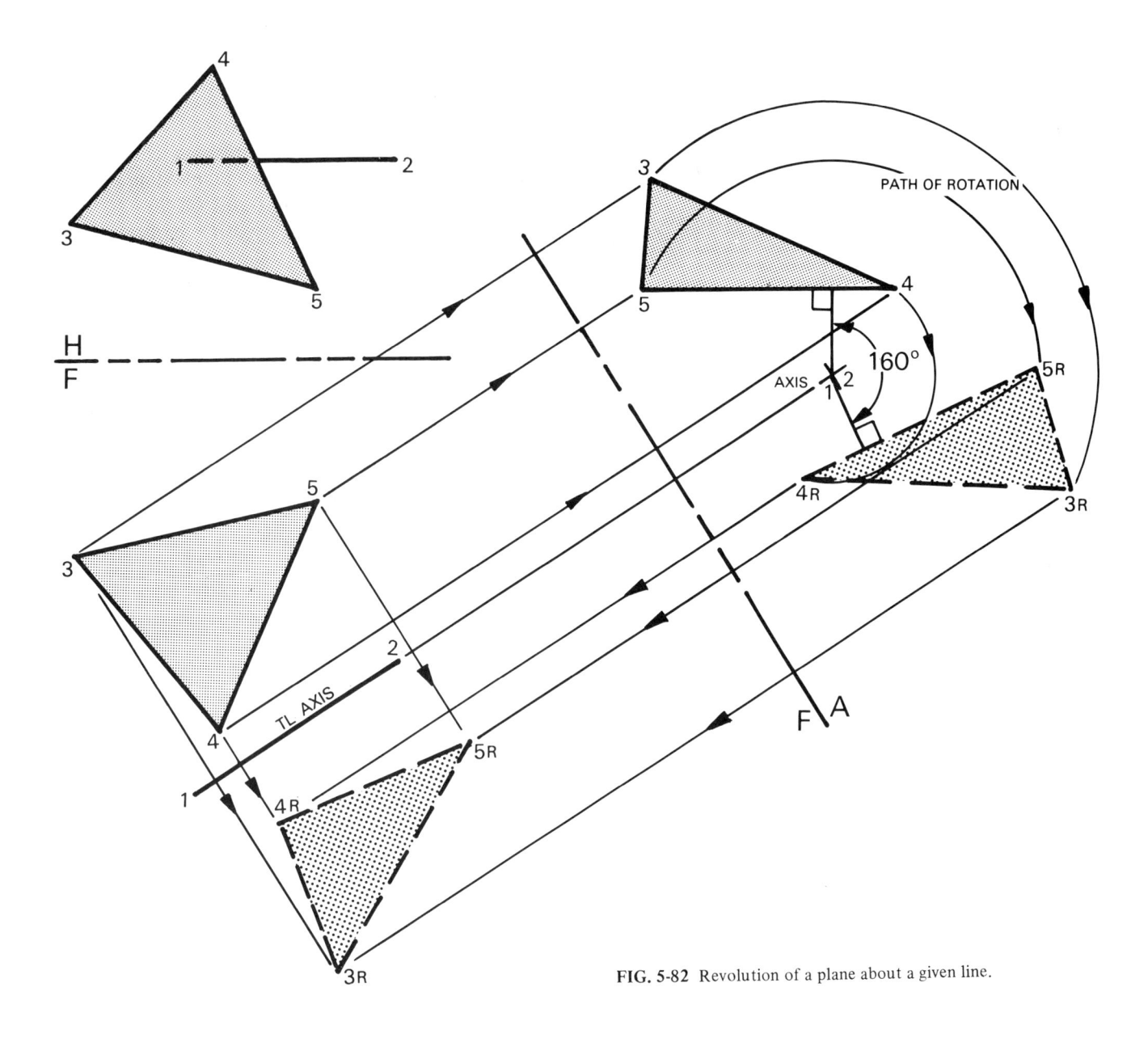

5.49 REVOLUTION OF A PLANE ABOUT A GIVEN LINE

A plane is revolved about a given external line in a view where the line appears as a point view. The true length of the line must be found before the point view can be established. The point view of the line is projected in a view taken perpendicular to the true length line.

In Fig. 5-82, line 1-2 and plane 3-4-5 are given. The plane is to be revolved 160° clockwise about the line. Line 1-2 is a frontal line, therefore true length in the frontal view. Plane 3-4-5 is oblique.

1. Draw fold line F/A perpendicular to frontal line 1_F-2_F. The line of sight for auxiliary view A is parallel to frontal line 1_F-2_F.

2. Project auxiliary view A. Line 1_A-2_A is a point view. Plane 3_A-4_A-5_A appears oblique.

3. Using line 1_A-2_A as the axis of revolution, revolve every point in the plane clockwise 160°. Since all points of the plane are external to the axis line, they move through the same angular displacement. Plane $3R_A$-$4R_A$-$5R_A$ is the revolved position of the given plane.

4. The revolved position of the plane is projected back to the frontal view where each of its points moves perpendicular to the axis line 1_F-2_F, parallel to F/A. Plane $3R_F$-$4R_F$-$5R_F$ is the frontal position of the revolved plane. Note that the horizontal position of the plane is not shown, but could be located by projection and measurement.

FIG. 5-83 Restricted revolution and clearance.

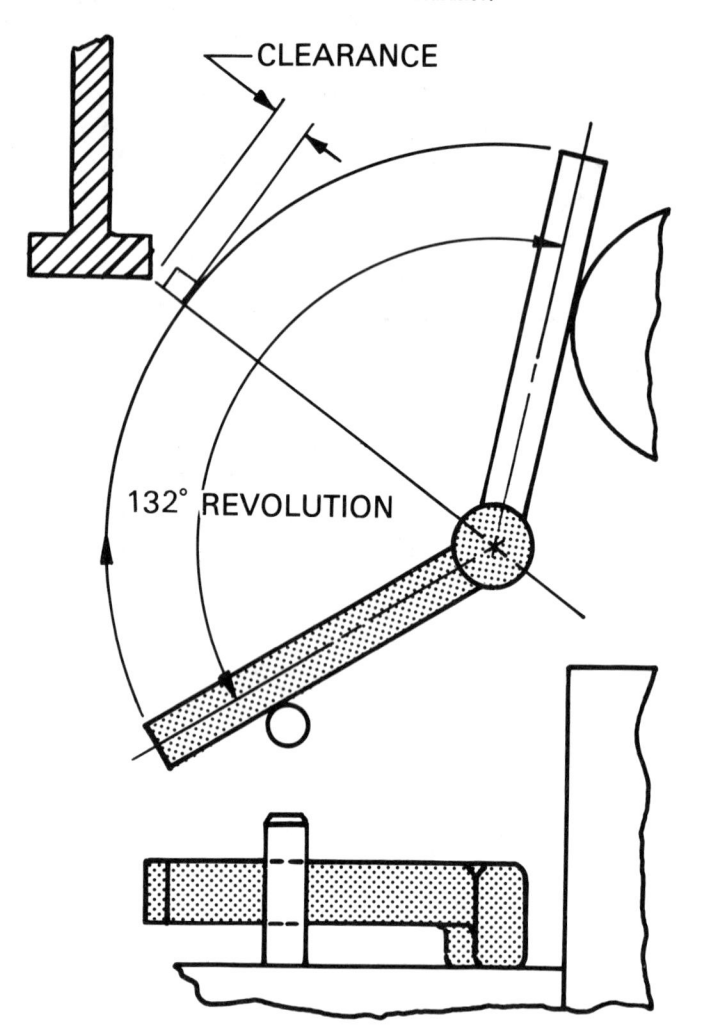

5.50 RESTRICTED REVOLUTION AND CLEARANCE

When a machine part or any type of mechanical device must be free to rotate about an axis through a circle or circular arc, revolution is used to find the extent of the piece's movement. The circular arc created by the extreme point of the revolving part determines its clearance with surrounding surfaces. The circular plane created by the rotating part scribes an arc which equals the extreme line of intersection between the part and an external obstruction. In order for the object to move through its prescribed angular displacement, all surrounding surfaces must lie outside the circular plane of intersection.

The lever in Fig. 5-83 can move in a circular arc of 132° before it hits an obstruction. Its forward position is fixed by a dowel pin. The clearance between the moving lever and the steel beam is measured as the perpendicular distance between the beam and the circular line of intersection, along a line passing through the axis.

In Fig. 5-84, each of the cages on the continuous-motion animal centrifuge rotates about a horizontal axis. The centrifuge itself revolves using an electric motor and a belt drive.

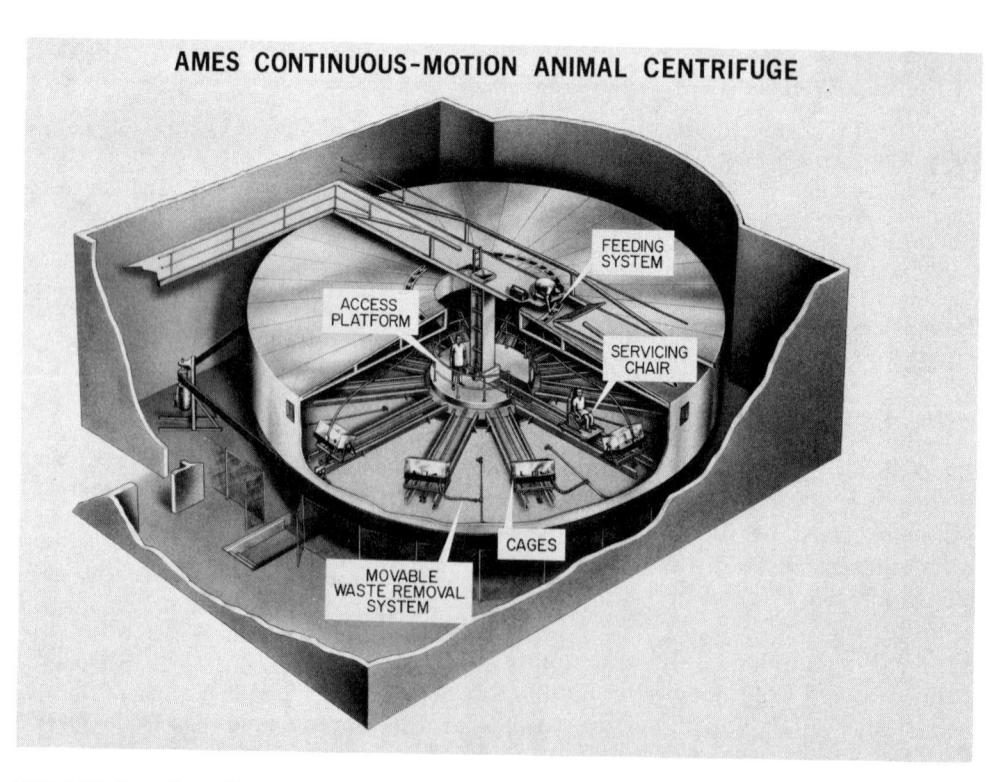

FIG. 5-84 Centrifuge. *(Courtesy NASA.)*

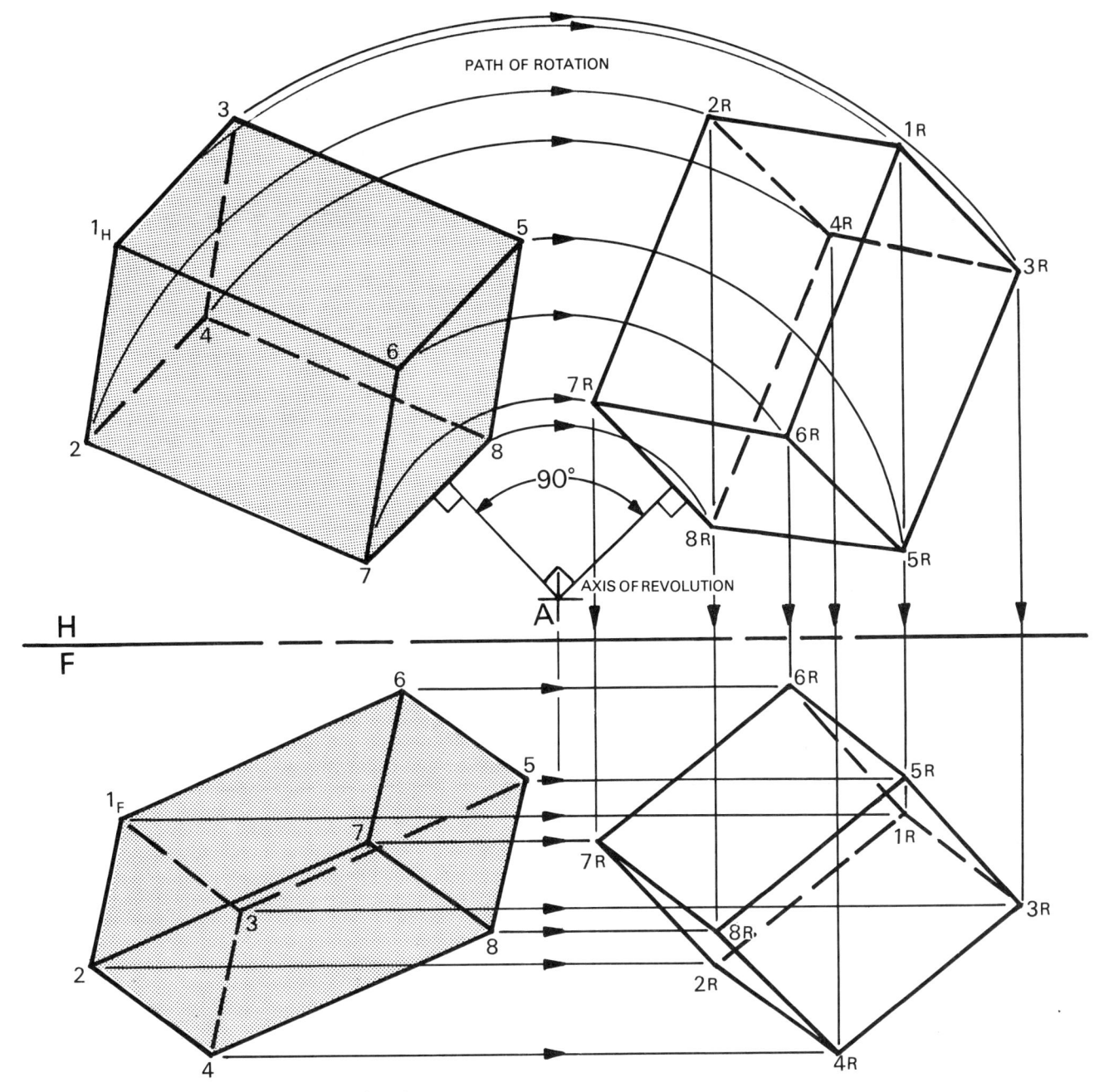

FIG. 5-85 Revolution of a solid at a specified angle.

5.51 REVOLUTION OF A SOLID

A designer may find it necessary to revolve an object/solid about a given axis line. Most often the object will be revolved to a new position where some operation can be performed on it or with it. Usually, the revolved position shows the object in a normal (true shape) or edge view perspective. A more complicated situation arises when the given object is oblique and its rotated position is also oblique.

In Fig. 5-85, a rectangular prism is revolved 90° about a given external axis. Note that the revolution of a given object requires that each point on the form revolve through the same angle. In the example, all points in the horizontal view revolve clockwise through 90°, since axis A is given as external to the prism. In the view where the axis is a point, the prism retains the exact shape it had before it was revolved, only its *position* has changed.

The revolved frontal view is located by drawing projection lines from each point on the rotated prism in the horizontal view. Every point of the given projection in the frontal view moves perpendicular to the axis line (parallel to fold line H/F), until intersecting the projection lines extended from the revolved prism in the horizontal view. Note that the frontal position of the revolved prism changes *shape* as well as position.

FIG. 5-86 Double revolution of a solid.

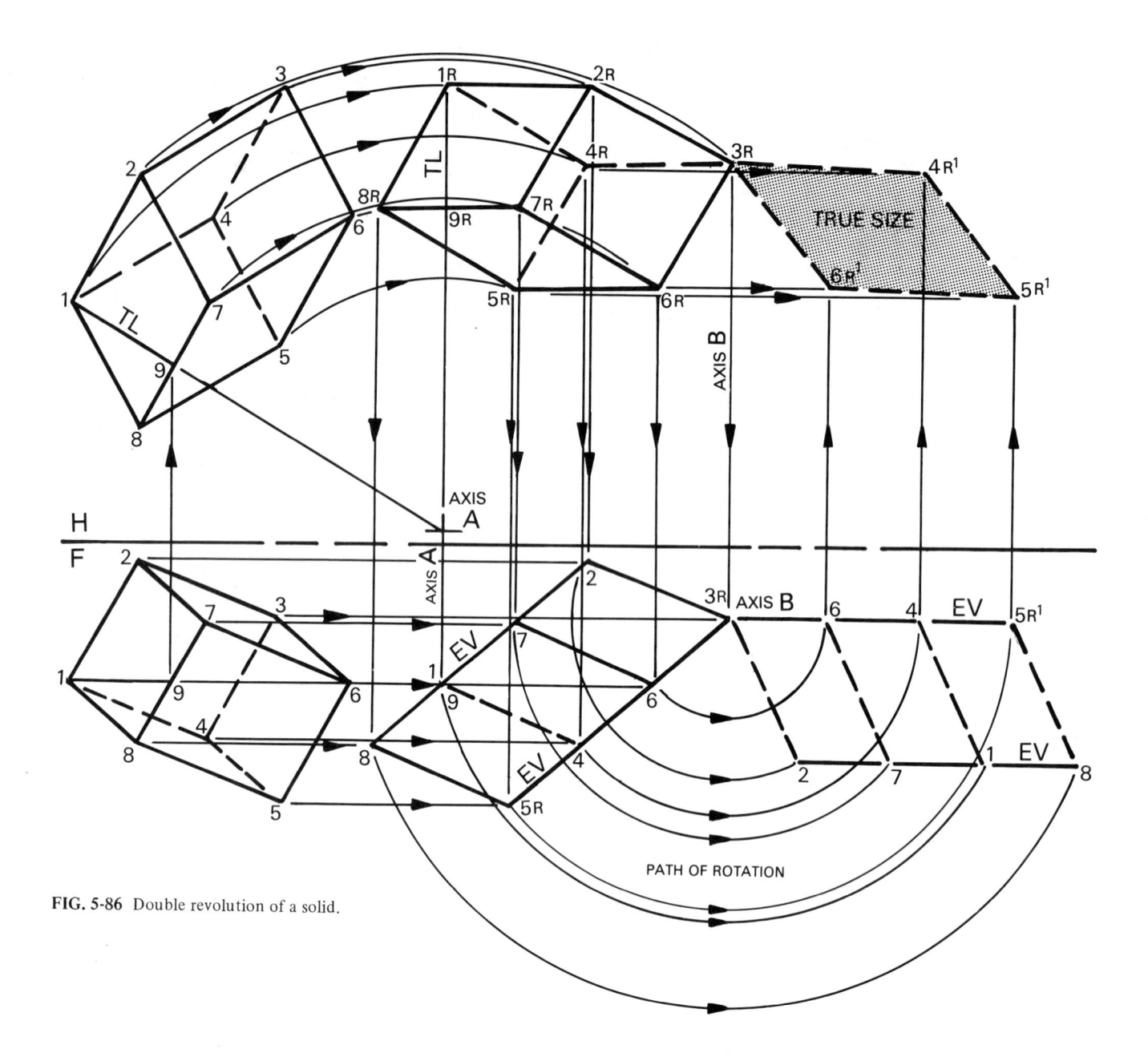

5.52 DOUBLE REVOLUTION OF A SOLID

The prism in Fig. 5-86 is revolved twice in order to show the true shape of one of its surfaces. The first revolution establishes the edge view of the surface, the second shows the plane as true shape. Each new projection is revolved so as not to overlap the preceding view.

1. Horizontal line 1_F-9_F is drawn parallel to H/F and projected to the horizontal view where it appears true length.

2. Axis A is established by extending line 1_H-9_H. The axis is located so that the revolved position of the prism does not overlap the existing view. This may take a couple of trial placements of the axis and revolutions of point 8_H, since the point closest to the existing view is needed.

3. The prism is revolved about axis A until horizontal line 1_H-9_H is perpendicular to the frontal plane. Line $1R_H$-$9R_H$ is perpendicular to H/F.

4. The frontal view of the prism in its first revolved position is drawn by moving all points perpendicular to the axis line and drawing projection lines from the horizontal view. The prism has two of its plane surfaces shown as edges in this position.

5. A horizontal axis is passed through point $3R_F$, axis B. All points of the prism are revolved about axis B until the edge views of the two surfaces are parallel to the fold line. Only the top surface need be projected to the horizontal view.

6. Plane $3R_H$-$4R_H^1$-$5R_H^1$-$6R_H^1$ is true size.

198

PROB. 5-41

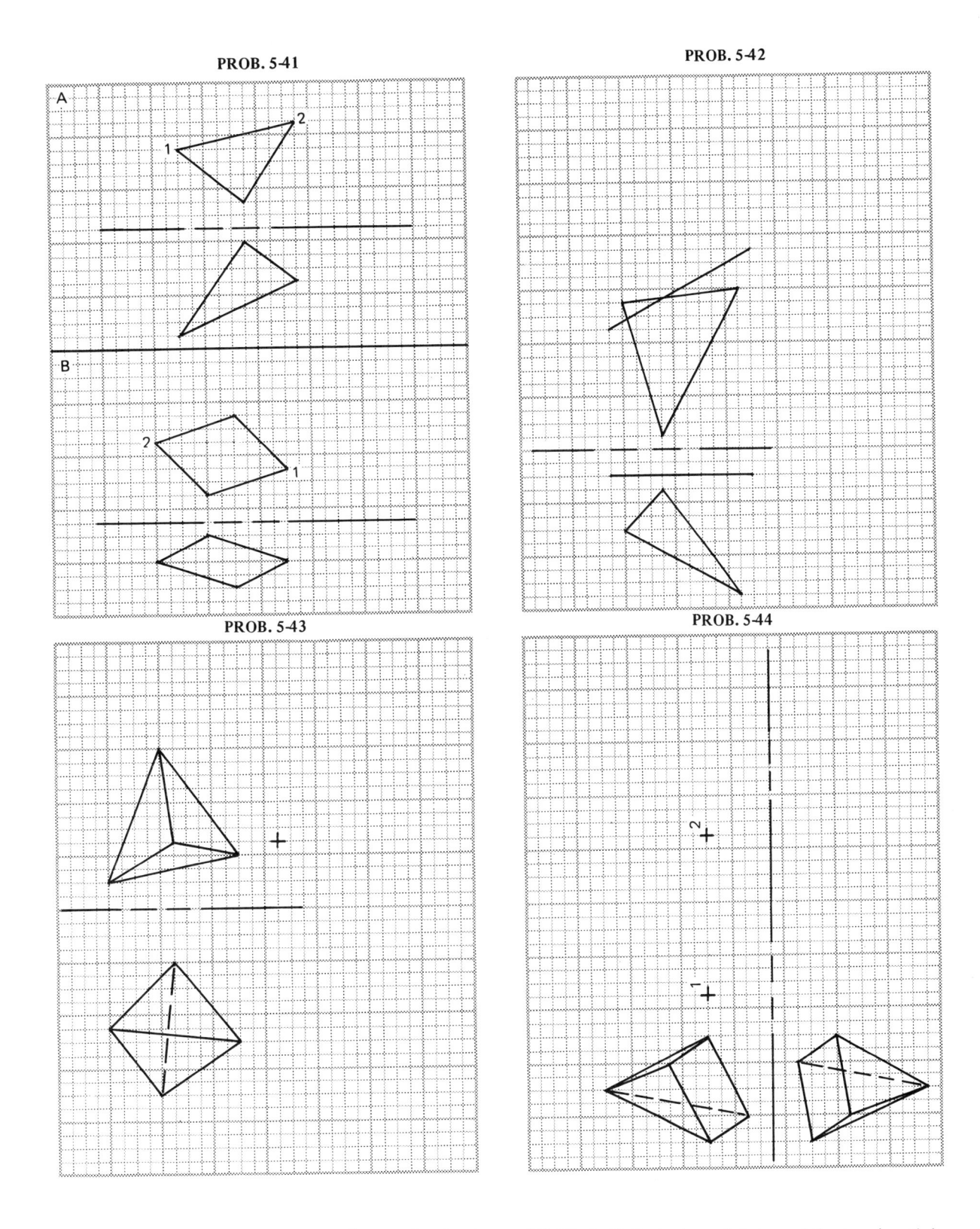

PROB. 5-42

PROB. 5-43

PROB. 5-44

PROB. 5-41A. Revolve the given plane 180° around point 1 and 130° counterclockwise around point 2. Both are to be revolved in the top view and projected to the front view.

PROB. 5-41B. Rotate the plane 151° counterclockwise around point 2 and 151° clockwise around point 1. Rotate in the H view and show in the F view.

PROB. 5-42. Project a point view of the true length line and revolve the plane 90° downward around the line. Show the revolved position in all views.

PROB. 5-43. Rotate the figure 140° around the given point and show in the front view.

PROB. 5-44. Revolve the pyramid 120° clockwise around point 1 and then 110° clockwise around point 2. Show all projections in both views.

FIG. 5-87 Piercing point of a line and a plane.

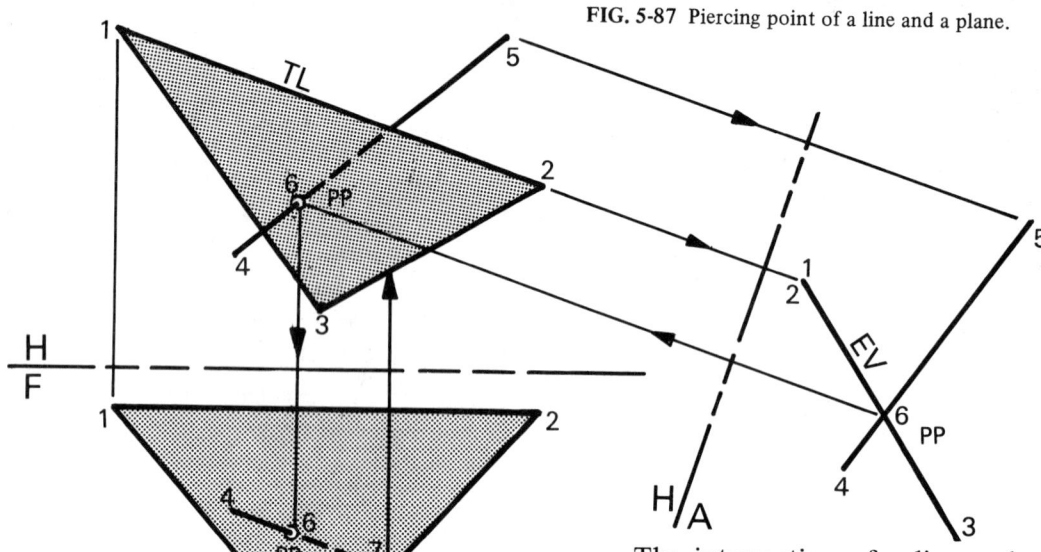

5.53 PIERCING POINTS (EDGE VIEW METHOD)

A line and a plane have three possible relationships.

1. *A line can lie on a plane.*
2. *A line can be parallel to a plane.*
3. *A line can intersect (pierce) a plane.*

The intersection of a line and a plane is possibly the most important and frequently required solution in descriptive geometry. The procedure for finding the intersection of a line and a plane can be applied to intersections in all categories: line and a solid, two planes, a plane and a solid, two solids. If broken down into specific types of intersections: line and a sphere, plane and cone, cylinder and pyramid; the list is endless. In this chapter only the intersection of a line and a plane is presented. All other types are covered in Chapter 6, Intersections.

The intersection of a line and a plane forms the basis of intersections of all forms, since objects are composed of lines and planes. Both the line and the plane can be extended to solve for theoretical intersections—those which lie outside the given bounded plane or beyond the given length of the line.

The point at which a line intersects (pierces) a plane is its "piercing point." This piercing point can be obtained by the *"edge view (auxiliary view) method"* or the *"cutting plane method."* In Fig. 5-87, plane 1-2-3 and line 4-5 are given. Their piercing point and proper visibility are required.

1. Draw H/A perpendicular to horizontal line 1_H-2_H and project auxiliary A. Plane 1_A-2_A-3_A appears as an edge view and line 4_A-5_A as oblique.
2. The piercing point (point 6_A) is where the line crosses the edge line of the plane, point 6_A.
3. Project point 6_A to the horizontal view and frontal views as shown.
4. Proper visibility is determined by inspection of auxiliary A for the horizontal view and by the visibility test for the frontal view.

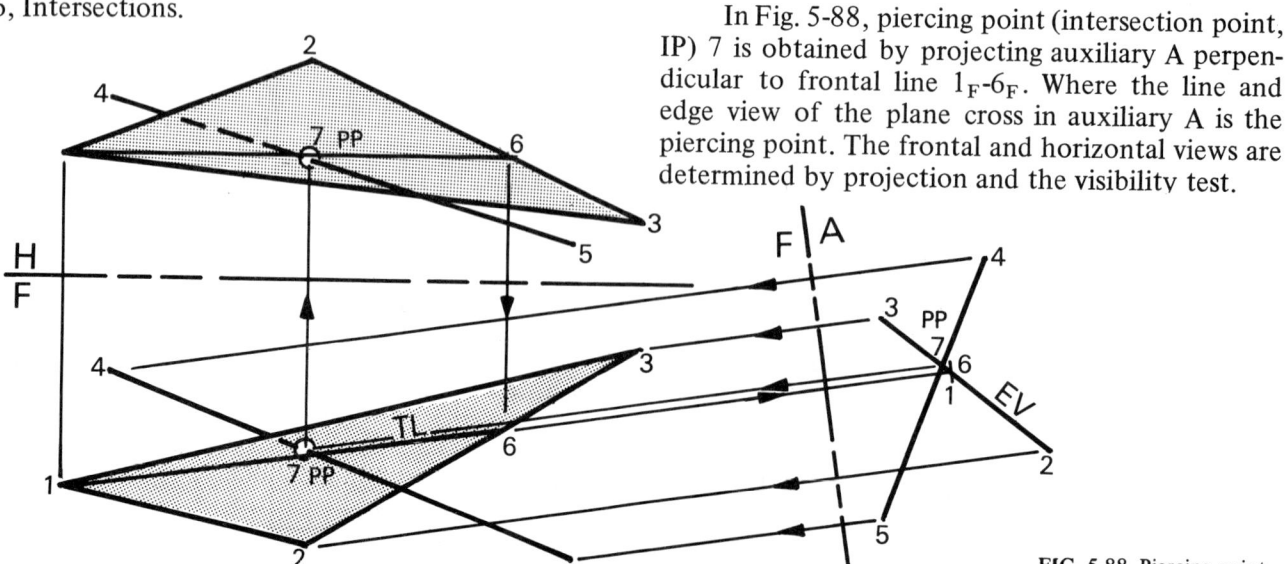

In Fig. 5-88, piercing point (intersection point, IP) 7 is obtained by projecting auxiliary A perpendicular to frontal line 1_F-6_F. Where the line and edge view of the plane cross in auxiliary A is the piercing point. The frontal and horizontal views are determined by projection and the visibility test.

FIG. 5-88 Piercing point.

FIG. 5-89 Piercing point (edge view method).

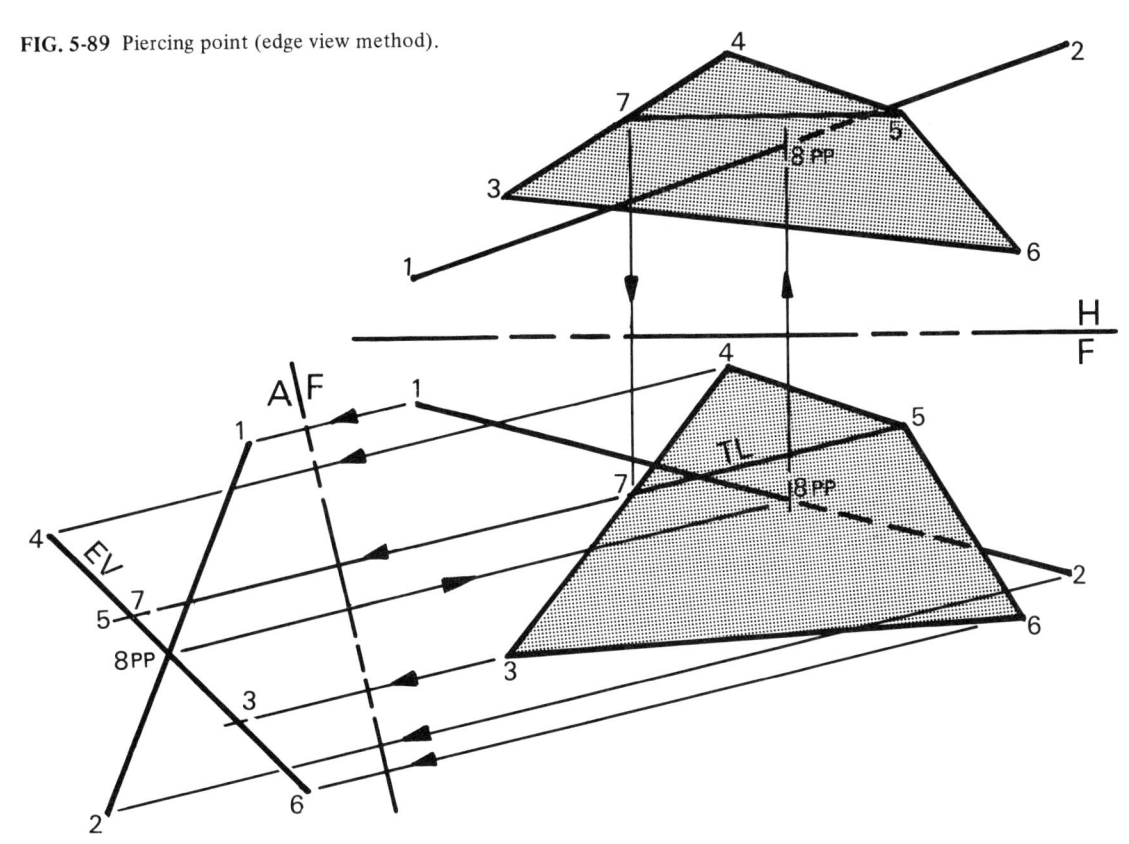

5.54 PIERCING POINT OF A LINE AND A PLANE (EDGE VIEW METHOD)

As in the previous section, the edge view method is the most accurate method for finding the intersection of a line and a plane. Piercing points (PP) can also be established by projecting a view where the line is a point view. The "cutting plane method" is useful when time and space are considerations, since it requires only two views. The "edge view method" can be used as check on all other methods since it shows such a clear and concise answer.

FIG. 5-90 Piercing point of a line and a vertical plane.

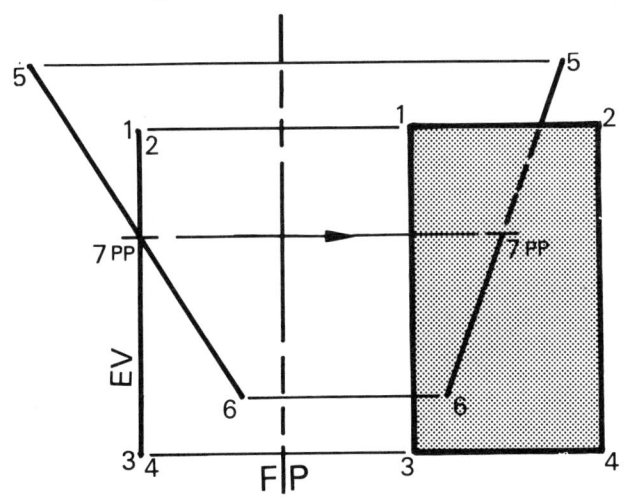

In Fig. 5-89, line 1-2 and plane 3-4-5-6 are given. The piercing point and visibility of the line and plane are required.

1. Project an edge view of the plane. Draw frontal line 5_H-7_H parallel to H/F and true length in the frontal view.

2. Draw F/A perpendicular to frontal line 5_F-7_F and project auxiliary A. Plane 3_A-4_A-5_A-6_A appears as an edge view and line 1_A-2_A as oblique.

3. Where line 1_A-2_A and the edge line of the plane cross is the piercing point, point 8_A. Project point 8 to the frontal view where it lies on the line and plane at position 8_F. Project point 8_F to the horizontal view where it lies on the line and plane at position 8_H.

4. Solve for the proper visibility by inspection and using the visibility test.

When a plane is a principal plane it appears as an edge in one of the three principal projection planes. Therefore, an auxiliary projection need not be taken. In Fig. 5-90, profile plane 1-2-3-4 and line 5-6 are given. Since plane 1-2-3-4 is a profile plane it appears true size in the profile view and as an edge in the frontal view (and horizontal view). The intersection of the line and the plane is apparent in the frontal view. Piercing point 7_F is projected to the profile view where it lies on the line. *Note that all piercing/intersecting points are common points of the plane and the line.* Visibility is determined by inspection of the frontal view.

201

FIG. 5-91 Piercing points (cutting plane method).

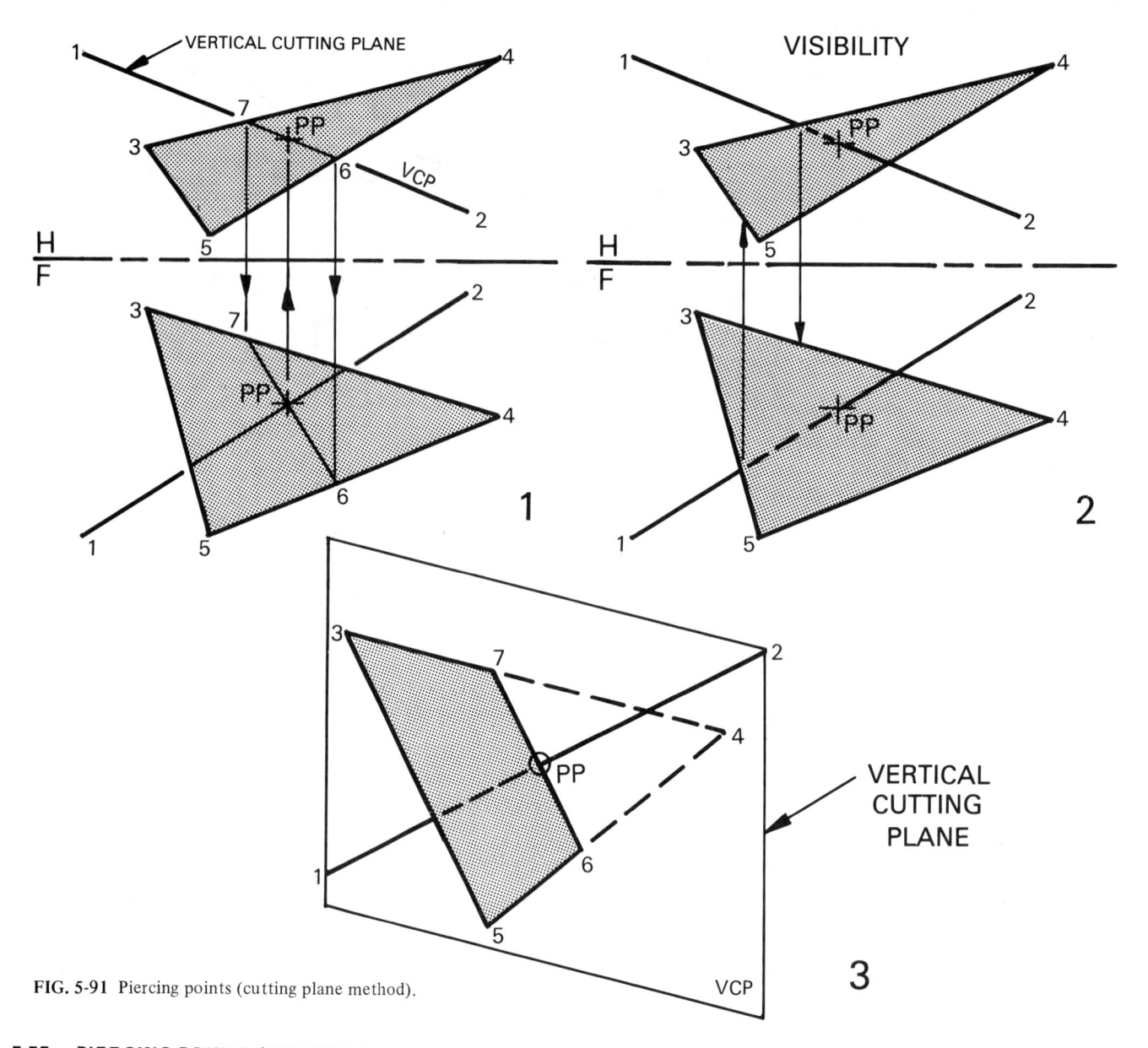

5.55 PIERCING POINTS (CUTTING PLANE METHOD)

If two planes intersect, their line of intersection contains all lines which lie on one plane and pierce the other. The *"cutting plane method"* involves the forming of a new plane which contains the given line. A cutting plane is used which shows as an edge view in one of the principal projection planes. In Fig. 5-91, a vertical cutting plane (VCP) was formed by passing a plane through line 1_H-2_H. Line 1_H-2_H represents the edge view of the VCP. Where this VCP "cuts" plane 3_H-4_H-5_H it forms a line of intersection which is common to both planes, line 6_H-7_H. This line of intersection is projected to the adjacent view, where it lies on both planes, line 6_F-7_F (1). The line of intersection between the two planes must be parallel or intersect the given line. If the line of intersection intersects the given line in the adjacent projection, then it will establish the piercing point of the line and the plane, PP. When using a vertical cutting plane the piercing point is established by projecting the line of intersection from the horizontal view. Point 6_H and point 7_H are projected to the frontal view where they form line 6_F-7_F. Line 6_F-7_F crosses line 1_F-2_F at PP. The horizontal view of PP can be located by projection. Visibility is determined using the visibility test, since inspection of an edge view is not possible (2).

If the line of intersection does not cross the given line, the line and plane do not intersect. In this case the given line lies in front of or behind and above or below the plane.

In Fig. 5-91(3), a pictorial view of this problem is provided. The vertical cutting plane contains line 1-2. VCP cuts plane 3-4-5 along a line of intersection, line 6-7. Line 6-7 lies on both planes, is a common line. Where line 6-7 crosses line 1-2 they will intersect at PP. PP is the piercing point of line 1-2 and plane 4-5-6.

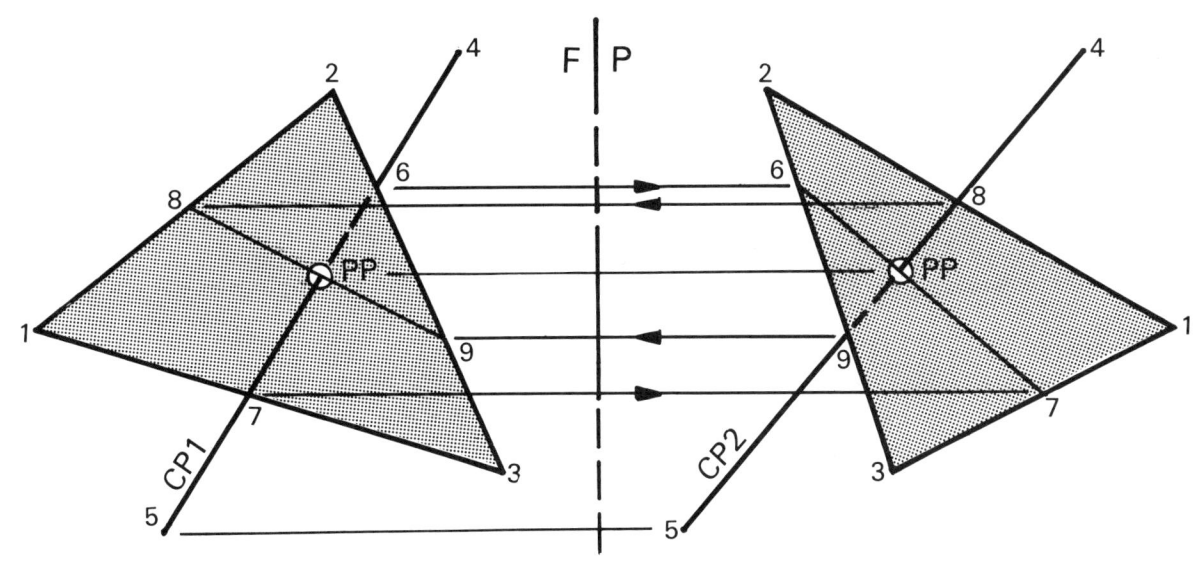

FIG. 5-92 Piercing point of a line and a plane (individual line method).

5.56 PIERCING POINT OF A LINE AND A PLANE (INDIVIDUAL LINE METHOD)

When using the cutting plane method it must be noted that a cutting plane can be introduced in both views (*individual line method*). This procedure provides a check since the piercing point must be aligned in all adjacent projections. In Fig. 5-92, plane 1-2-3 and line 4-5 are given. The line is not parallel to the plane, therefore they intersect. Using the cutting plane method, solve for the piercing point of the line and the plane.

1. A cutting plane is drawn containing line 4_F-5_F, which appears as an edge in the frontal view, CP 1.
2. Where CP 1 crosses line 1_F-3_F and line 2_F-3_F, it forms line 6_F-7_F. Line 6-7 is the line of intersection.
3. Project point 6_F and 7_F to the profile view. Line 6_P-7_P is the line of intersection (common line) between plane 1_P-2_P-3_P and CP 1. The piercing point is located where line 4_P-5_P and line 8_P-9_P cross, PP.
4. The frontal location of PP can be established by projection, but in this example another method is used. The "*individual line method*" solves for the intersection of a line and plane separately in each view by forming a new cutting plane with the line in each projection. CP 2 is passed through line 4_P-5_P in the profile view. CP 2 is an edge view and coincident with line 4_P-5_P.
5. Project the crossing points, 8_P and 9_P, of CP 2 and the plane to the frontal view.
6. Line 8_F-9_F is the line of intersection between plane 1-2-3 and CP 2. Line 8_F-9_F crosses line 4_F-5_F at PP.

In many cases the line and plane will not intersect, given the limited length of the line or size of the plane. The theoretical or *extended point of intersection* (piercing point) can be found by extending the line or plane. In Fig. 5-93, plane 1-2-3-4 and line 5-6 are not parallel. Since the given views do not show the line and plane crossing, it is necessary to extend the line. Line 5_F-6_F is extended to point 8_F. CP 1 is introduced as a plane containing line 5_F-8_F, which appears as an edge in the frontal view. CP 1 intersects plane 1-2-3-4 at points 7_F and 8_F. Line 7_F-8_F is projected to the horizontal view where it represents the line of intersection between the two planes. Line 5_H-6_H is extended until it crosses line 7_H-8_H and locates the piercing point of the line and the plane. PP is located in the frontal view by projection.

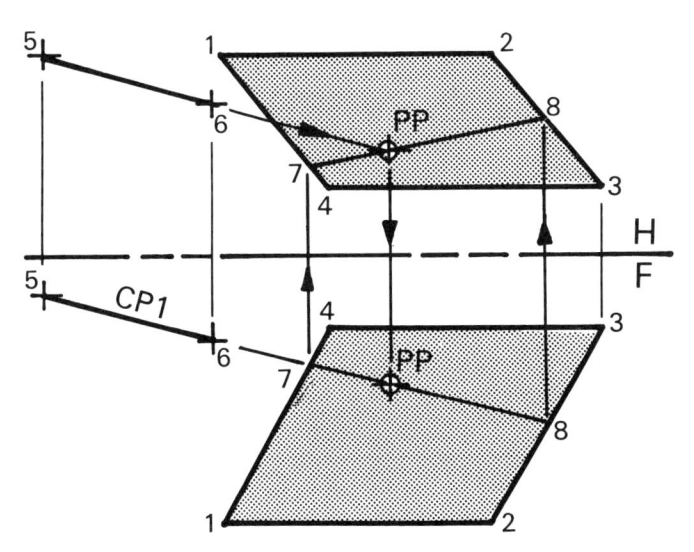

FIG. 5-93 Piercing point by line extension.

In Fig. 5-95, plane 3-4-5-6 and vertical line 1-2 are given. The piercing point and proper visibility are required. Line 7_H-8_H is drawn through the point view of line 1_H-2_H. Line 7_H-8_H is projected to the frontal view. This line is on the plane and contains a point on line 1-2. Line 7_F-8_F crosses line 1_F-2_F at PP. PP is a common point to line 1-2, line 7-8, and plane 3-4-5-6.

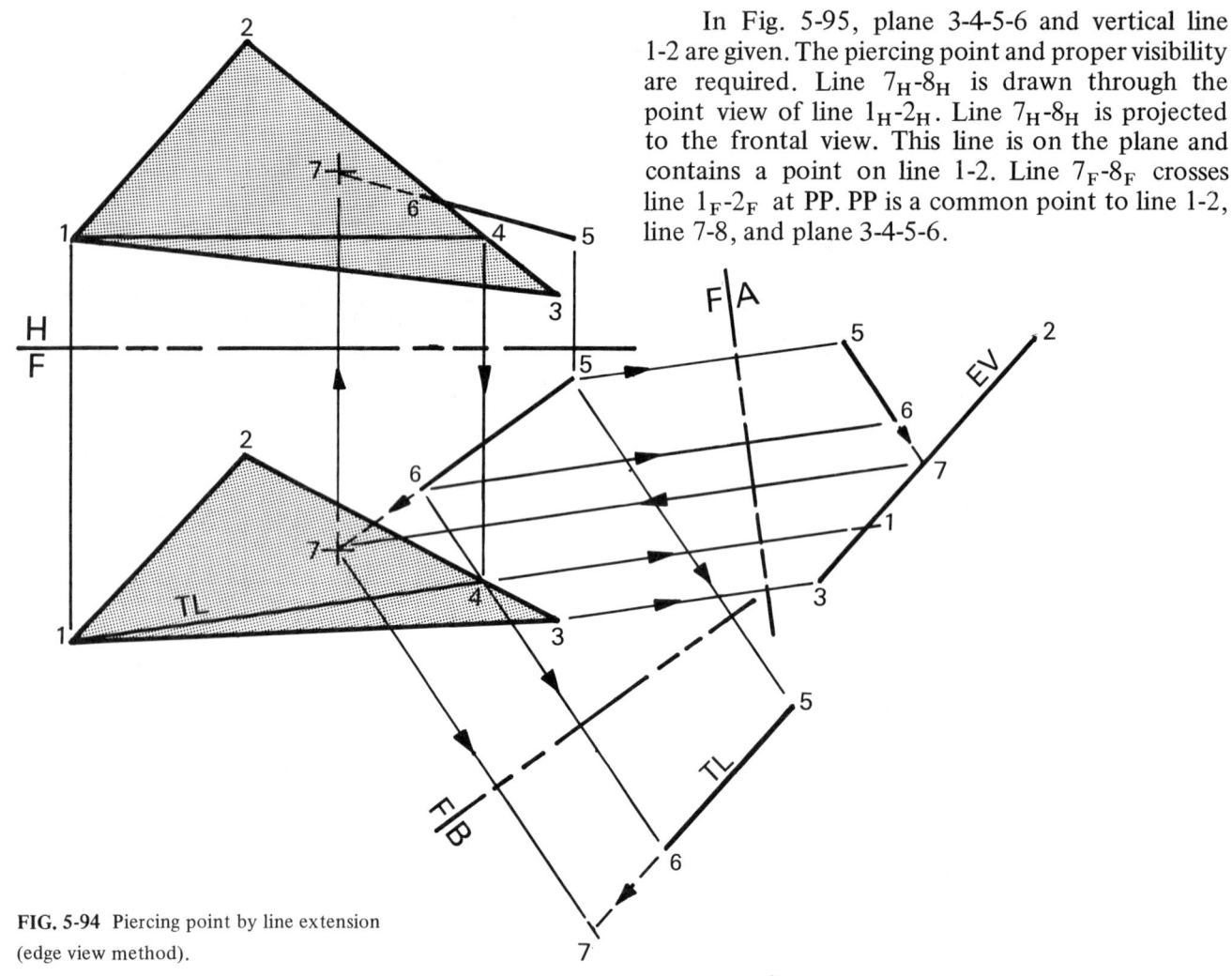

FIG. 5-94 Piercing point by line extension (edge view method).

5.57 PIERCING POINT BY LINE EXTENSION (EDGE VIEW METHOD)

The piercing point of a line and a plane can be found by line extension using the edge view method. In Fig. 5-94, plane 1-2-3 and line 5-6 are given. The plane could represent an ore vein and the line a tunnel. The point at which the tunnel first hits the ore vein is required, along with the length of the tunnel.

1. Draw frontal line 1_H-4_H parallel to H/F and true length in the frontal view.
2. Establish A/B perpendicular to frontal line 1_F-4_F and complete auxiliary A. Plane 1_A-2_A-3_A is an edge view and line 5_A-6_A is oblique.
3. Extend line 5_A-6_A until it intersects the edge view of plane 1_A-2_A-3_A. Point 7_A is the piercing point.
4. Locate point 7 in the frontal and horizontal views by projection and line extension as shown.
5. Draw F/B parallel to 5_F-7_F and project auxiliary B. The true length of the tunnel is shown as line 5_B-7_B.

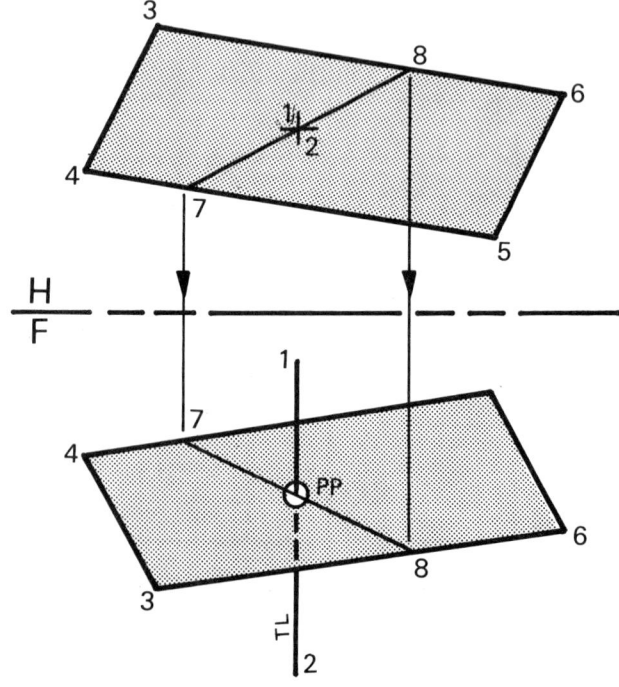

FIG. 5-95 Piercing point of a vertical line and oblique plane.

204

PROB. 5-45

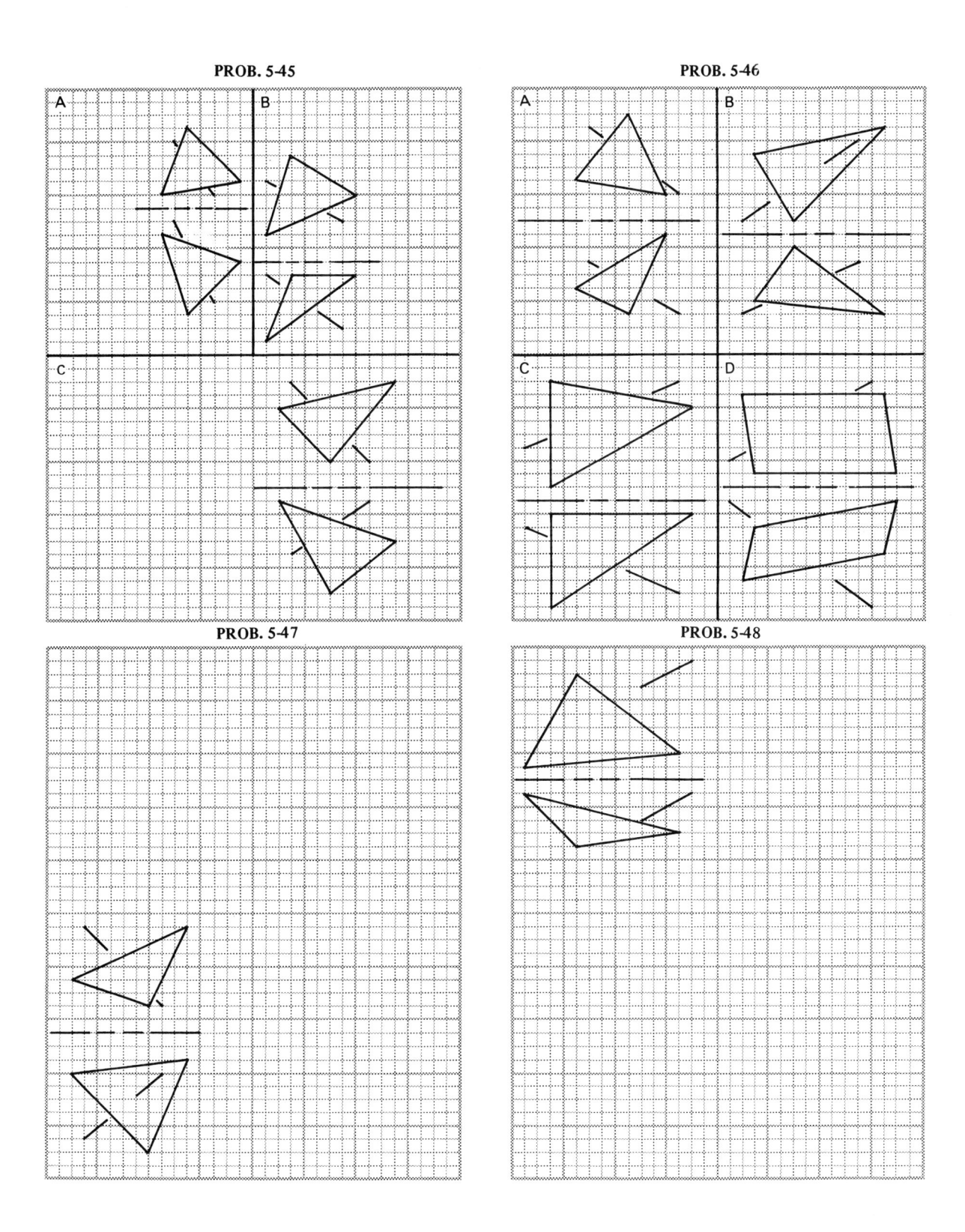

PROB. 5-46

PROB. 5-47

PROB. 5-48

PROBS. 5-45A, 5-45B, 5-45C. Using the edge view method, determine the piercing point and show in all views along with the proper visibility.

PROBS. 5-46A, 5-46B, 5-46C, 5-46D. Using the cutting plane method, solve for the piercing point of the line and the plane. Complete all views and show the proper visibility.

PROB. 5-47. Determine the piercing point and the angle between the line and the plane. Note the angle and complete all views.

PROB. 5-48. Find the true length of the line and the angle that the line makes with the plane. Solve for the piercing point by the edge view method and line extension.

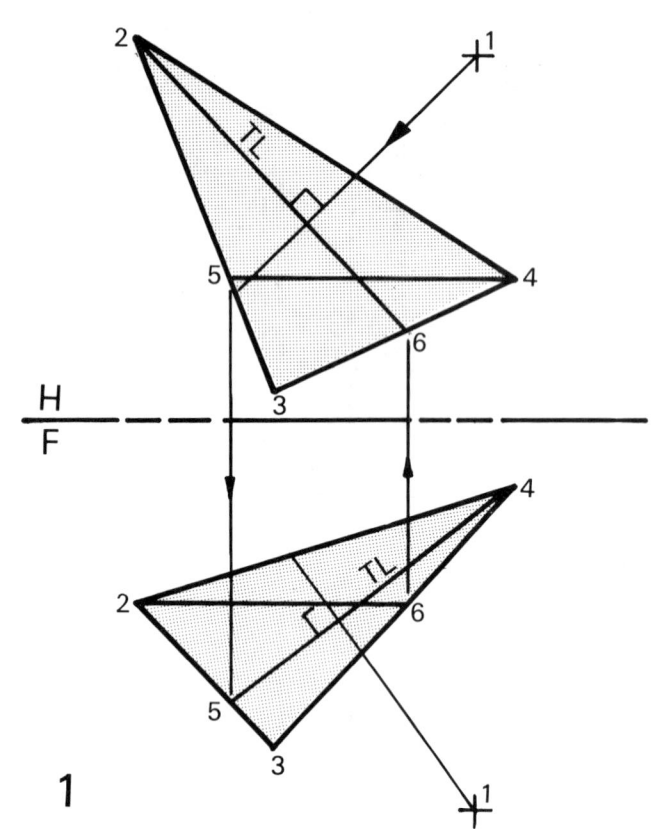

FIG. 5-96 Projection of a point on a plane.

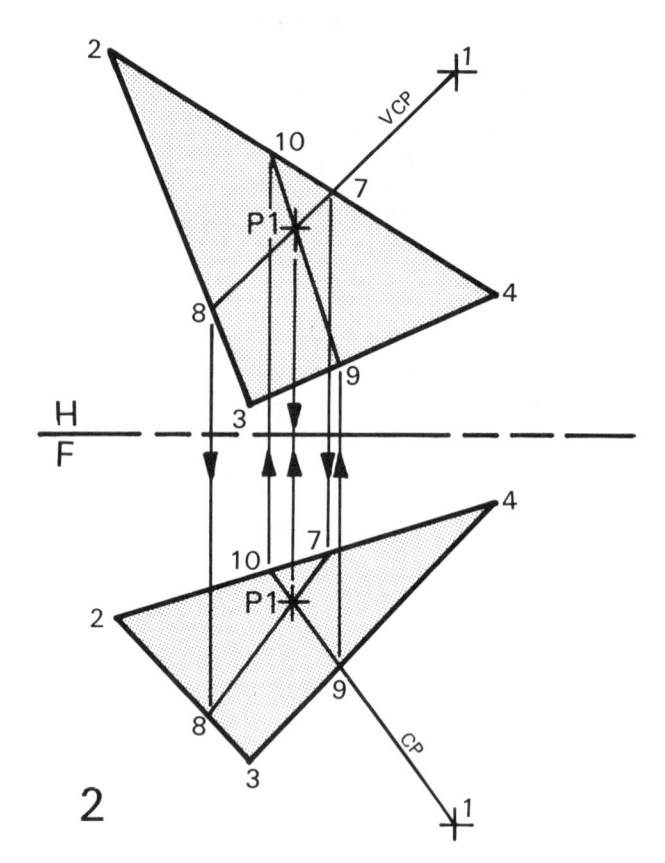

5.58 PROJECTION OF A POINT ON A PLANE

The projection of a point on a plane can be found by using the cutting plane method for finding an intersection point and the theory of perpendicularity. A point is projected onto a plane where a perpendicular line extended from the point intersects the plane. The point of intersection or piercing point is the points projection on the plane. This procedure is applied in Chapter 8, Shades and Shadows. Each point (corner) of an object is projected onto the horizontal surface on which it sets. When connected, these projected points represent the shadow area of the object. The shadow that the aircraft makes with the floor can be seen in Fig. 5-97.

In Fig. 5-96, the plane and the point are given. The projection of the point on the plane is required.

1. Frontal line 4-5 and horizontal line 2-6 are drawn in both views. Draw a line from point 1_H perpendicular to horizontal line 2_H-6_H. Draw a line from point 1_F perpendicular to frontal line 4_F-5_F.
2. Using the cutting plane method solve for the piercing point, P1. Line 1_H-8_H represents the edge view of a vertical cutting plane; where it cuts plane 2_H-3_H-4_H it forms a line on the plane, line 7_H-8_H. Project line 7-8 to the frontal view. Where line 7_F-8_F crosses the perpendicular line extended from point 1_F, it locates point P1. P1 can be projected to the horizontal view or found by the cutting plane method using line 9-10.

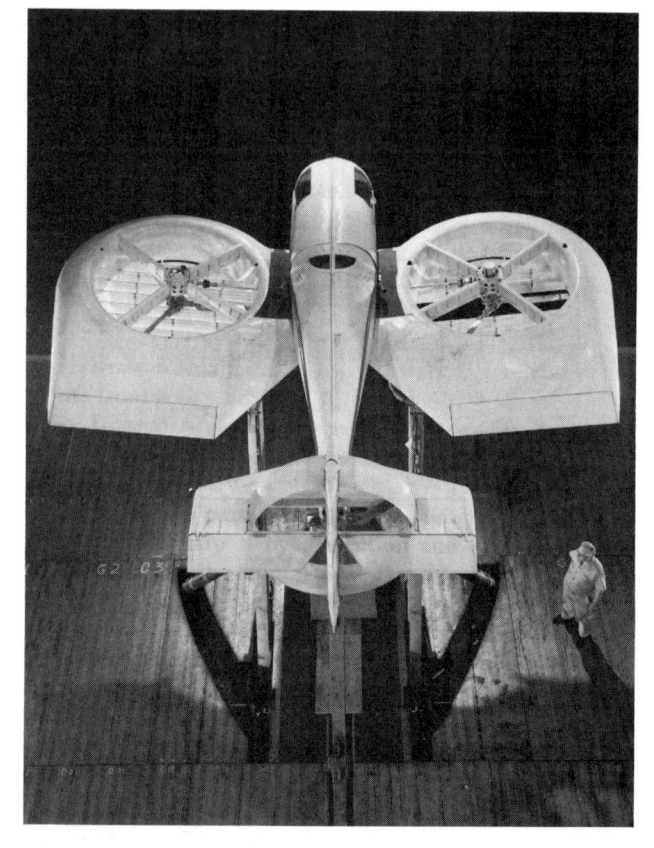

FIG. 5-97 Vanguard VTOL. *(Courtesy NASA.)*

206

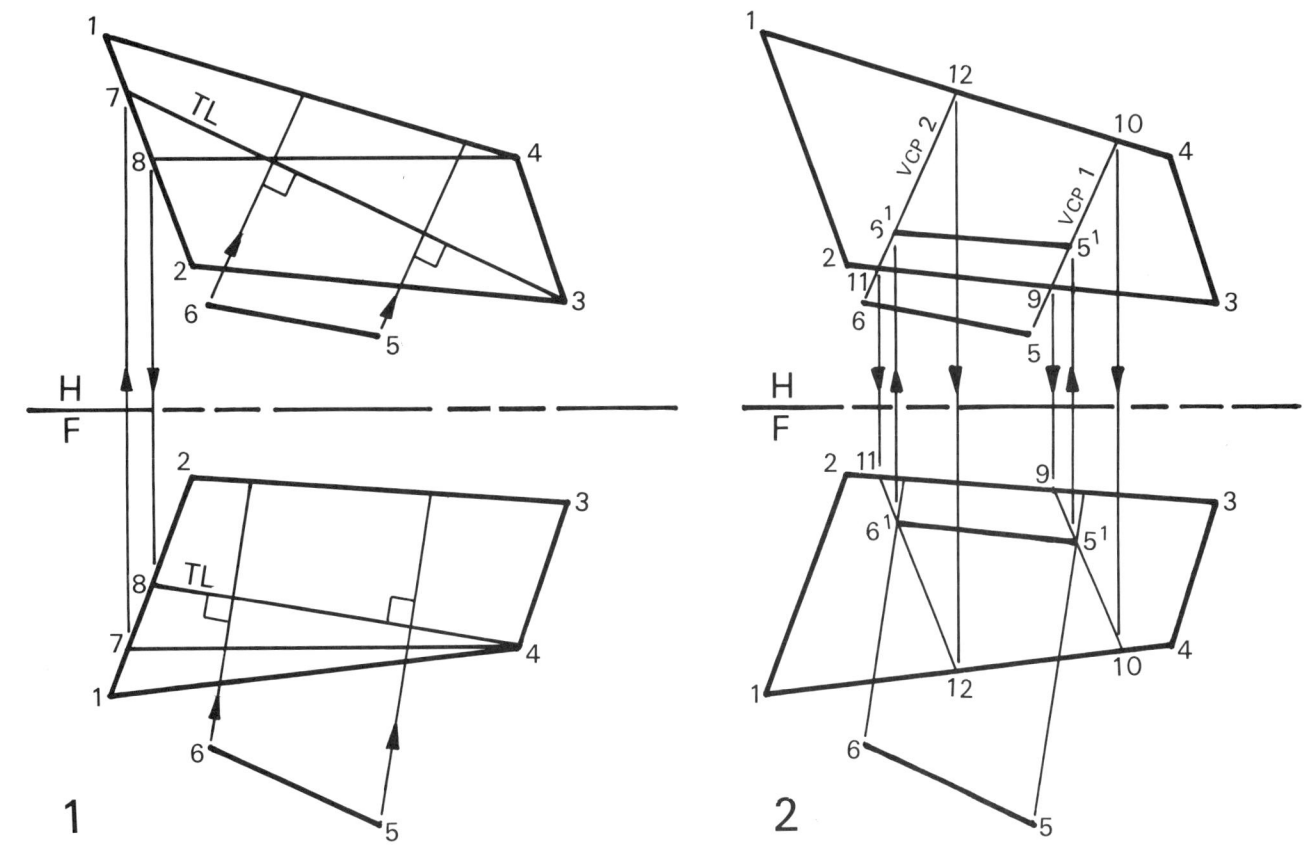

FIG. 5-98 Projection of a line on a plane (cutting plane method).

5.59 PROJECTION OF A LINE ON A PLANE (CUTTING PLANE METHOD)

A line can be projected on a plane by projecting each of its end points onto the plane. This procedure is described in the previous section. In Fig. 5-98, plane 1-2-3-4 and line 5-6 are given. The projection of the line on the plane is required. Since a line is defined by its end points, it is only necessary to project these two points onto the plane.

1. Draw a frontal line and a horizontal line which lie on the plane, and show their projection in both views. Extend a line from point 5_H and a line from point 6_H, perpendicular to horizontal line 3_H-7_H. Extend a line from point 5_F and a line from point 6_F, perpendicular to frontal line 4_F-8_F.

2. Two vertical cutting planes are established in the horizontal view using the two perpendicular lines drawn from point 5_H and point 6_H. VCP1 cuts plane 1_H-2_H-3_H-4_H at point 9_H and point 10_H. Line 9_H-10_H is projected to the frontal view. Line 9_F-10_F crosses the perpendicular line extended from point 5_F at piercing point 5^1_F. This point represents the projection of point 5 onto the plane in the frontal view. The horizontal view of point 5^1 is located by projection as shown. This same procedure is repeated using VCP2. Line 5^1-6^1 is the projection of the line 5-6 onto plane 1-2-3-4.

The structural configuration of the tracking antenna in Fig. 5-99 illustrates many of the spatial relationships of points, lines, and planes in space. The angle between lines, intersecting lines and planes, true length, and true size relationships must be determined in the design stage using the theory of descriptive geometry.

FIG. 5-99 Eighty-five-foot tracking antenna. *(Courtesy NASA.)*

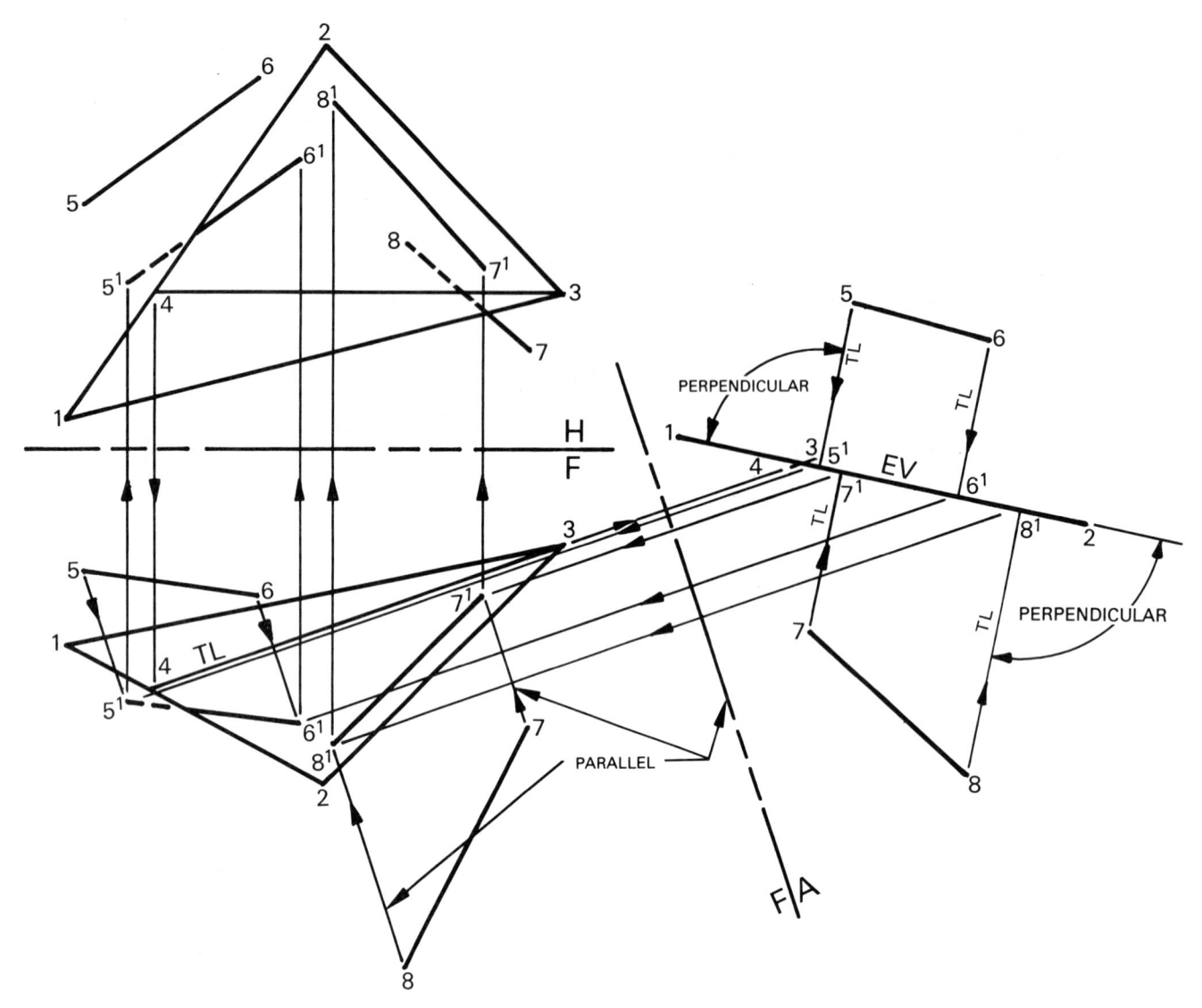

FIG. 5-100 Projection of a line on a plane (edge view method).

5.60 PROJECTION OF A LINE ON A PLANE (EDGE VIEW METHOD)

When more than one line is to be projected onto a given plane the edge view method is useful. This method is also better suited for locating a line on an infinite plane or where the line will not project onto the given plane as defined by its limited boundaries.

In Fig. 5-100, the projection of lines 5-6 and 7-8 on plane 1-2-3 is established by the edge view method. Note that the projection of line 5-6 will not lie totally within the given bounded plane.

1. Draw frontal line 3_H-4_H parallel to H/F and true length in the frontal view.
2. Draw F/A perpendicular to frontal line 3_F-4_F and project auxiliary A. Plane 1_A-2_A-3_A is an edge view and lines 4_A-5_A and 7_A-8_A are both oblique.
3. Draw projectors from point 5_A, 6_A, 7_A, and 8_A perpendicular to the edge view of plane 1_A-2_A-3_A. Where these perpendicular projectors intersect the edge view of the plane at piercing points 5_A^1, 6_A^1, 7_A^1, and 8_A^1 they determine the projection of the given lines onto the plane.
4. Projector lines 5_A-5_A^1, 6_A-6_A^1, 7_A-7_A^1, and 8_A-8_A^1 are true length in auxiliary A. Therefore, they will be parallel to F/A in the frontal view. Each of the four piercing points can be projected to the frontal view as shown. The point at which each crosses its corresponding projection line will fix the frontal location of its respective piercing point.

As an example, point 7_A^1 is projected to the frontal view. A projection line is drawn from point 7_A^1 to the frontal view. Where it crosses the projector extended from point 7, parallel to F/A, it locates point 7_F^1. Point 7_F^1 is the frontal location of the piercing point. Therefore, one of the end points of the projection of line 7-8 onto plane 1-2-3.

5. All four piercing points are projected to the horizontal view. Line 5^1-6^1 and line 7^1-8^1 are located in the horizontal view by transferring distances and projection. Note that point 5^1 does not lie on the given plane as defined by points 1-2-3, but lies on an extended (infinite) plane of which plane 1-2-3 is only a part.

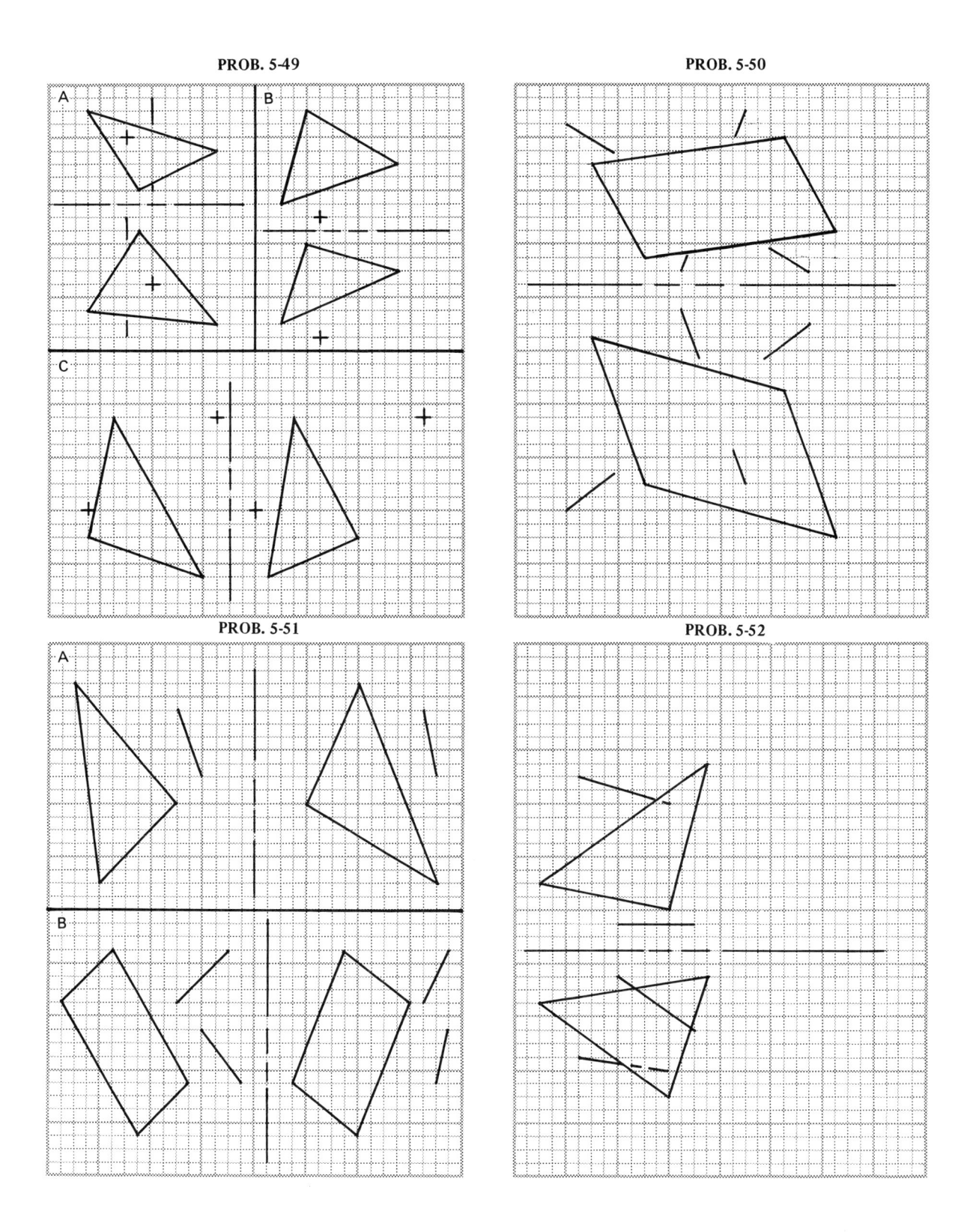

PROB. 5-49A. Solve for the piercing points of the two lines and plane.
PROB. 5-49B. Project the point onto the plane.
PROB. 5-49C. Project the two points onto the plane.
PROB. 5-50. Solve for the piercing points of the two lines and the plane. Use the cutting plane method.
PROB. 5-51A. Project the line onto the plane. Use the cutting plane method.
PROB. 5-51B. Using the cutting plane method, project the two lines onto the plane.
PROB. 5-52. Using the edge view method, project the lines onto the plane.

PROB. 5-53

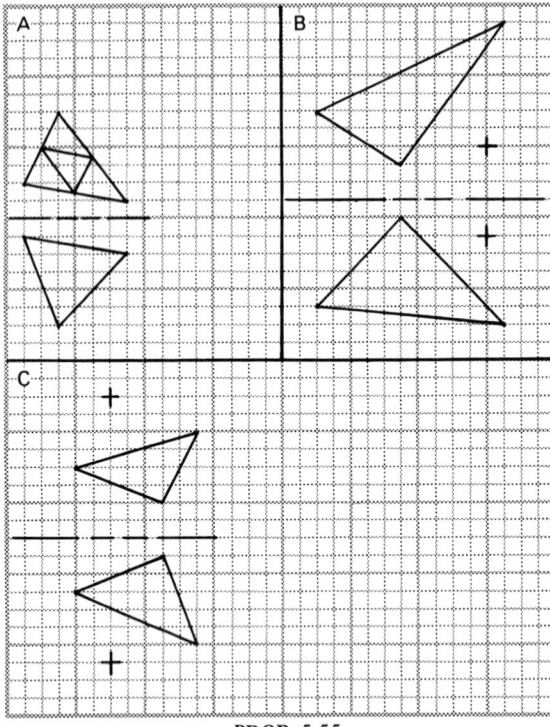

PROB. 5-54

PROB. 5-55

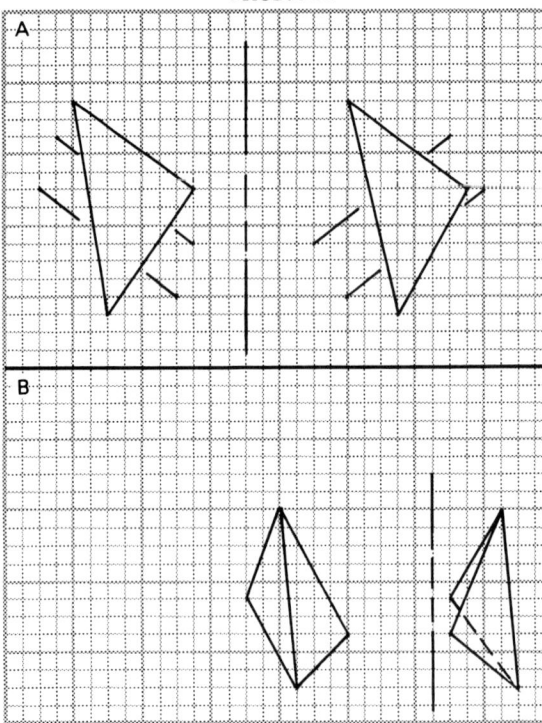

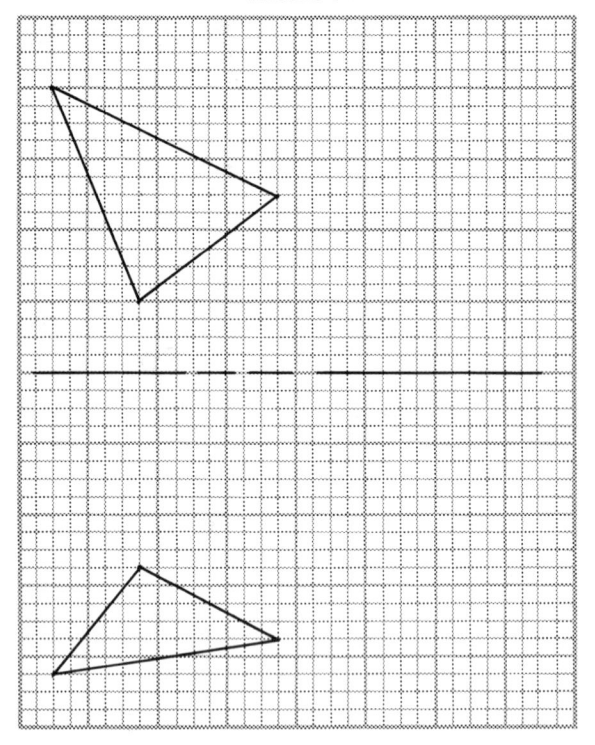

TEST

PROB. 5-53A. Complete the views of the plane and the figure that lies on the plane. Solve for the EV, TS, strike, and slope.

PROB. 5-53B. Establish two lines from the given point. One line will be parallel and the other perpendicular to the plane. What is the relationship of the two lines and the given plane?

PROB. 5-53C. Draw a line from the given point and perpendicular to the plane. Solve for the TS, slope, and strike of the plane. What is the diameter of a circle that circumscribes the plane?

PROB. 5-54. Using double rotation, solve for the TS. Use geometric construction to draw the largest inscribed circle.

PROB. 5-55A. Find the piercing points of the two lines and the unlimited plane using the cutting plane method.

PROB. 5-55B. Using revolution, solve for the TS of each plane and the angle between the planes.

QUIZ

1. Define frontal, horizontal, and profile plane.
2. What is the difference between an inclined and an oblique plane?
3. What type of projection precedes a perpendicular view of the TS of a plane?
4. What is the slope of a plane?
5. How does bearing differ from strike?
6. What is a dihedral angle?
7. Name four ways to show a plane.
8. Describe how to solve for the edge view of an oblique plane.
9. What view shows the slope of a plane?
10. Define the piercing point of a line and a plane.

6

INTERSECTIONS

FIG. 6-1 Power plant model, large diameter scrubber ducting has many intersections of different shapes.

6.1 INTERSECTIONS

The design of practically all products will involve lines, planes, and solids that intersect. Cubes, prisms, pyramids, cylinders, cones, spheres, etc., and their intersecting variations are just a few of the many forms used in engineering design work. Part of the responsibility of a designer is to establish these forms in such a manner that the result is a functional, produceable object. A necessary step in this process is the determination of the lines of intersection of the various shapes so that they can be graphically described and economically manufactured. A further step in the manufacturing process—development of patterns—is discussed in Chapter 7.

*The intersection of two shapes forms a **line of intersection**, or **common line**.* A basic step in finding the line of intersection between two geometric shapes is the determination of the intersection of a line and a plane, the piercing point, which was presented in the last chapter. The points of intersection are located by the projection of an *edge view* of the plane and/or the introduction of *cutting planes* of known orientation. These two methods may be utilized separately or together, depending on the requirements of the problem.

FIG. 6-2 X-15 shows configuration featuring short, square-tip swept-back wings and wedge-shaped upper vertical. *(Courtesy NASA.)*

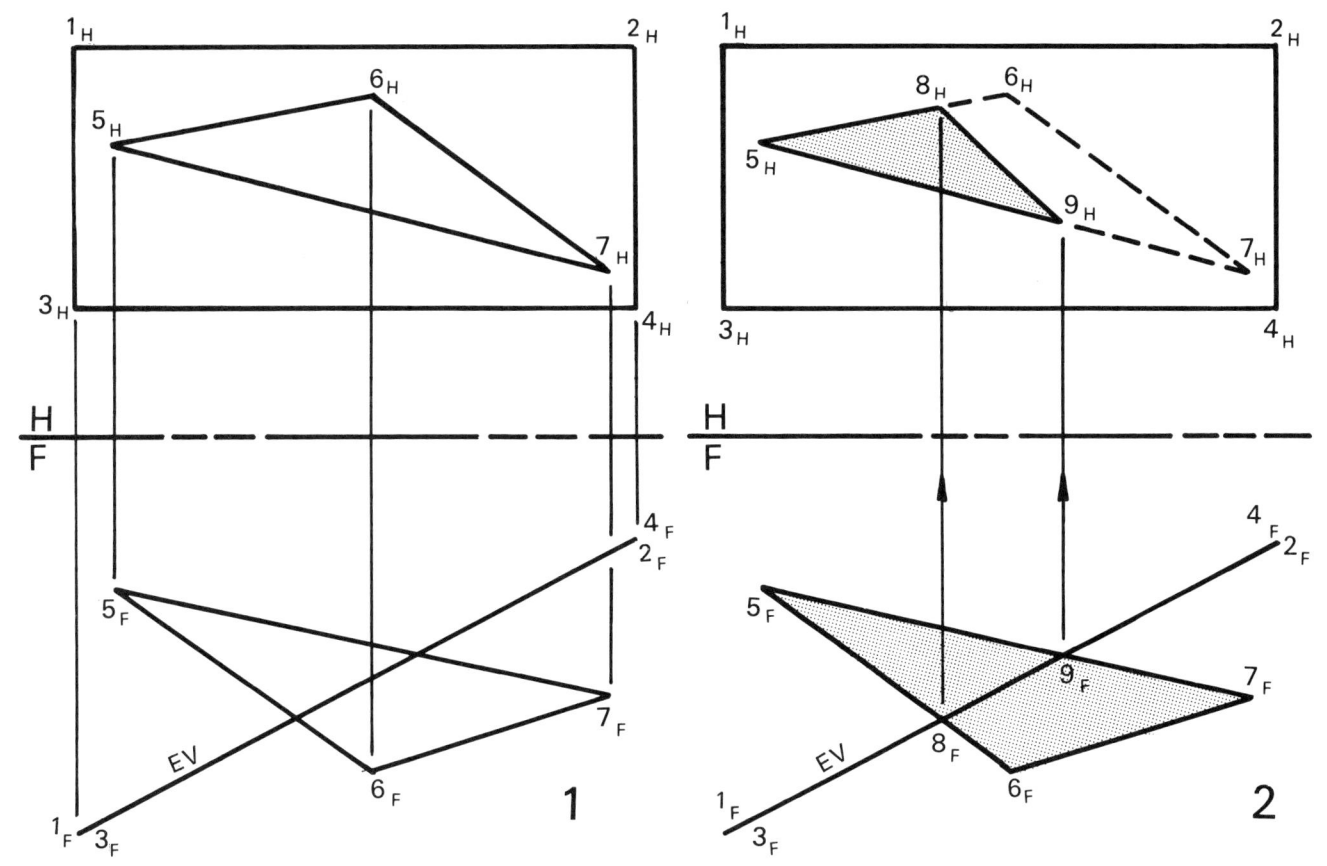

FIG. 6-3 Intersection of two planes (edge view method).

6.2 INTERSECTION OF TWO PLANES (EDGE VIEW METHOD)

The intersection of two planes can be determined by finding the edge view of one of the planes. Where any two lines on one plane pierce the edge view of the other plane, it determines their line of intersection or common line.

Both lines and planes are considered unlimited for construction purposes. Therefore both given planes and their line of intersection can be extended as required. The actual intersection of two defined planes will of course have a limited line of intersection which must be common to both planes.

To establish the intersection of two planes, it is necessary to find two points common to both planes. These points of intersection form a straight line. In Fig. 6-3, the line of intersection and proper visibility are required.

1. Inclined plane 1-2-3-4 and oblique plane 4-5-6 are given. Plane 1-2-3-4 appears as an edge in the frontal view.
2. Line 5_F-6_F and line 5_F-7_F pierce the edge view of plane 1_F-2_F-3_F-4_F at points 8_F and 9_F, respectively. Project these two piercing points to the horizontal view where they form line 8_H-9_H, which is the line of intersection. Visibility is determined by inspection: The portion of the plane formed by points 5-8-9 is above plane 1-2-3-4 in the frontal view, therefore it is visible in the horizontal view.

In Fig. 6-4, line 1_F-4_F pierces the edge view of plane 5_F-6_F-7_F at point 8_F. Line 2_F-3_F pierces plane 5_F-6_F-7_F at point 9_F. These two points are projected to the profile view where they establish the *extended line of intersection*. Where line 8_P-9_P crosses plane 5_P-6_P-7_P at point A and B determines the true line of intersection between the two limited planes.

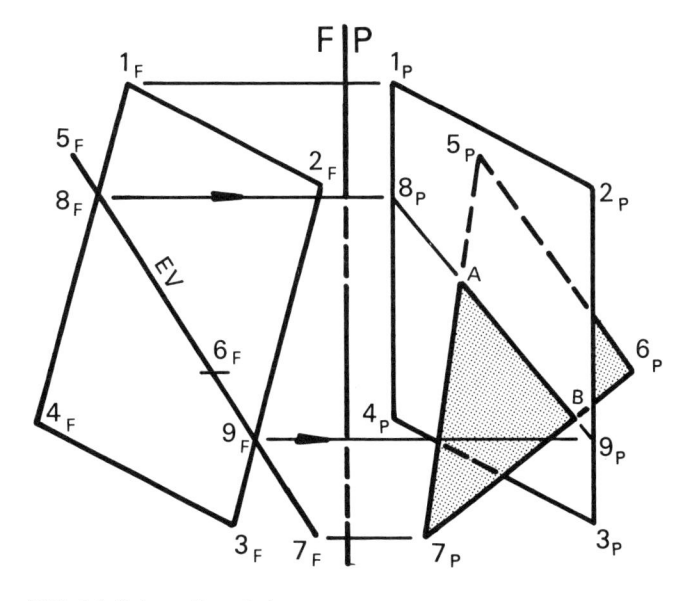

FIG. 6-4 Intersection of planes.

FIG. 6-5 Intersection of two oblique planes (edge view method).

6.3 INTERSECTION OF TWO OBLIQUE PLANES (EDGE VIEW METHOD)

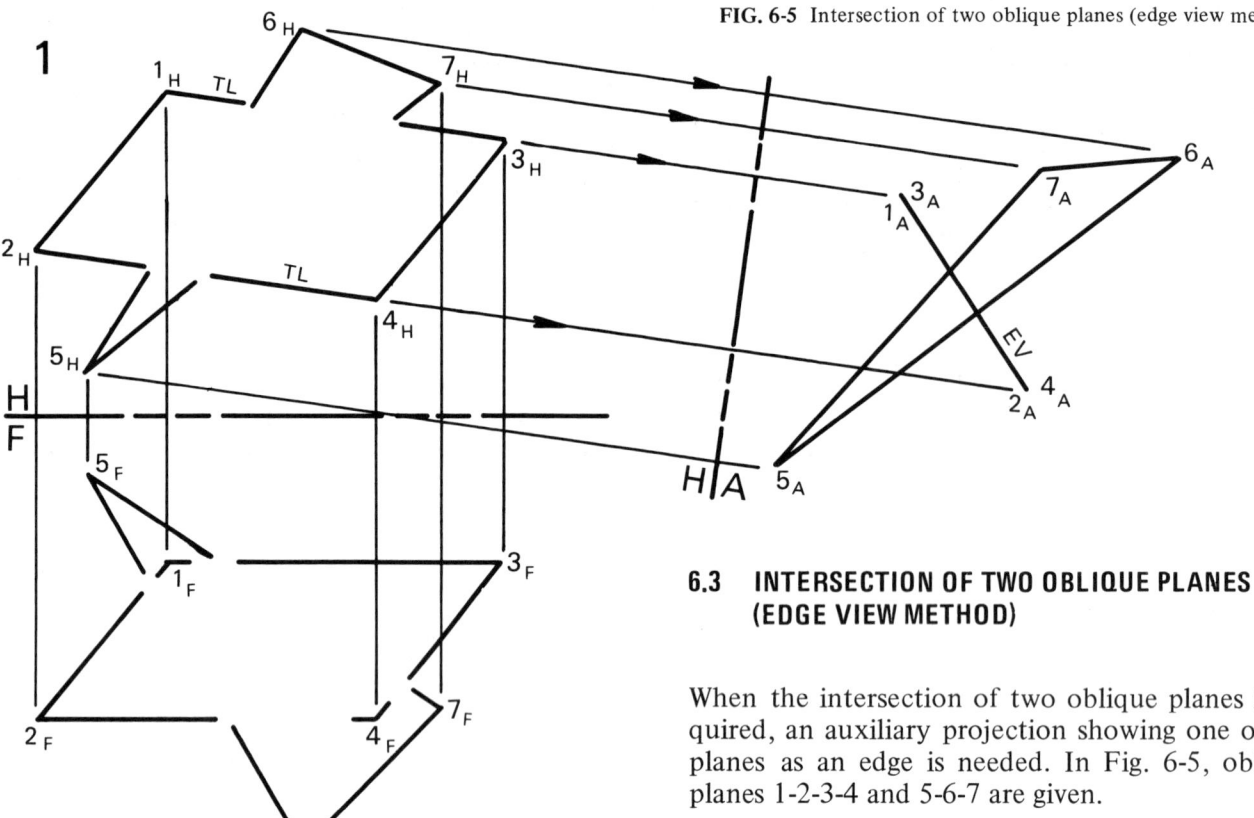

When the intersection of two oblique planes is required, an auxiliary projection showing one of the planes as an edge is needed. In Fig. 6-5, oblique planes 1-2-3-4 and 5-6-7 are given.

1. Line 1-3 and line 2-4 are horizontal lines—true length in the horizontal view. Draw H/A perpendicular to line 2_H-4_H and project auxiliary A. Plane 1_A-2_A-3_A-4_A appears as an edge view and plane 5_A-6_A-7_A is oblique.

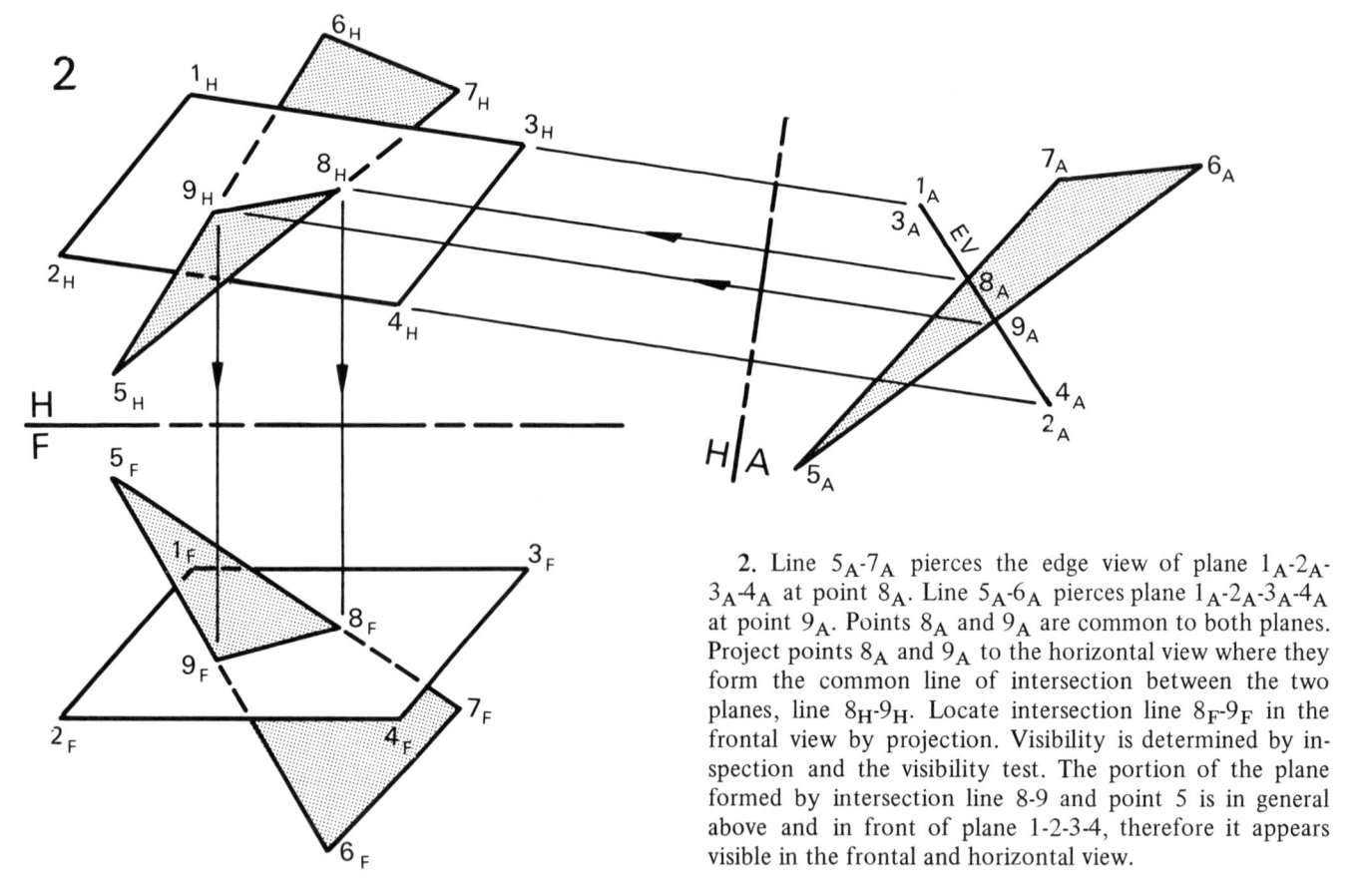

2. Line 5_A-7_A pierces the edge view of plane 1_A-2_A-3_A-4_A at point 8_A. Line 5_A-6_A pierces plane 1_A-2_A-3_A-4_A at point 9_A. Points 8_A and 9_A are common to both planes. Project points 8_A and 9_A to the horizontal view where they form the common line of intersection between the two planes, line 8_H-9_H. Locate intersection line 8_F-9_F in the frontal view by projection. Visibility is determined by inspection and the visibility test. The portion of the plane formed by intersection line 8-9 and point 5 is in general above and in front of plane 1-2-3-4, therefore it appears visible in the frontal and horizontal view.

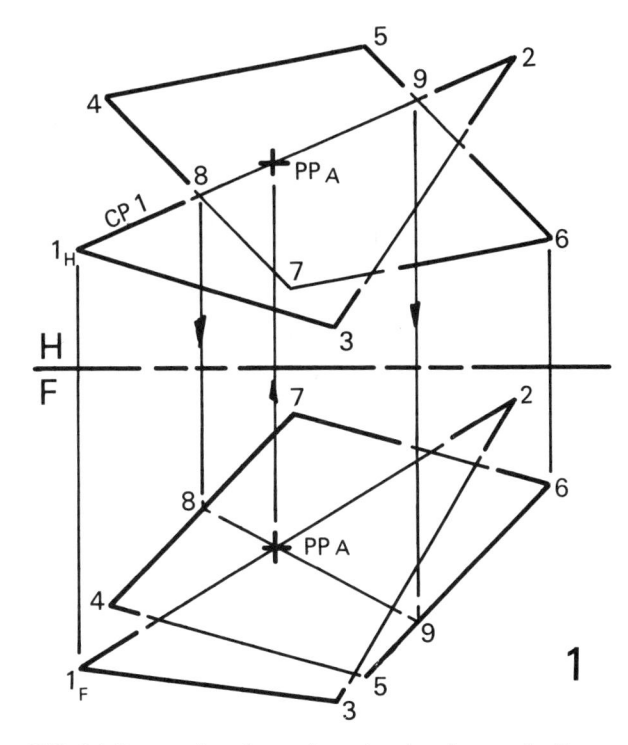

FIG. 6-6 Intersection of two planes (cutting plane method).

6.4 INTERSECTION OF TWO PLANES (CUTTING PLANE METHOD)

The line of intersection between two planes is found by locating two points common to both planes. The *"cutting plane method"* for finding the intersection of a line and a plane can be used to establish these two common piercing points, page 202. Each piercing point must be found individually and then projected to the adjacent view. Using cutting planes to solve for the piercing point of a line and a plane in each view, instead of projecting located points from view to view, is called the *"individual line method."*

In Fig. 6-6, oblique planes 1-2-3 and 4-5-6-7 are given. Their line of intersection is to be determined using the cutting plane method. Note that some lines make better cutting planes than others. Suitability is determined by trial and error. Some lines will obviously not cross the other plane and therefore can not be used (unless extended). Others cross only very small parts of the other plane and likewise may not be adequate. It may be necessary to extend a line in some cases. Also, cutting planes can be established using lines of different planes in the same view. In Fig. 6-6, cutting planes are passed through different lines on the same plane.

1. Pass a vertical cutting plane through line 1_H-2_H in the horizontal view. Line 1_H-2_H represents the edge view of CP 1. CP 1 cuts line 4_H-7_H at point 8_H and line 5_H-6_H at point 9_H. Project points 8 and 9 to the frontal view, where they form line 8_F-9_F. Line 8_F-9_F crosses (intersects) line 1_F-2_F at PP A. PP A is the piercing point of line 1-2 and plane 4-5-6-7. Project piercing point PP A to the horizontal view. PP A is a common point shared by both planes.

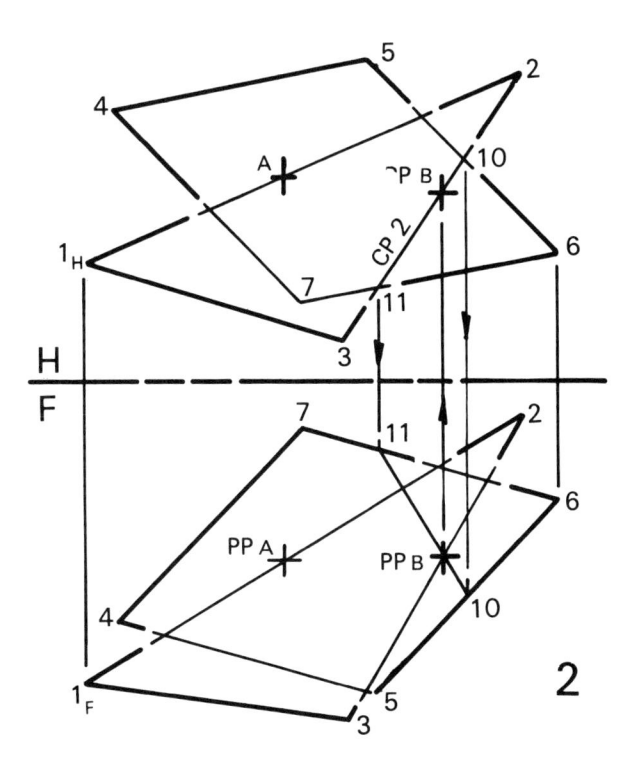

2. Pass vertical cutting plane CP 2 through line 2_H-3_H. Line 2_H-3_H represents the edge view of CP 2. CP 2 crosses line 5_H-6_H and line 6_H-7_H at point 10_H and point 11_H respectively. Project points 10 and 11 to the frontal view, where they form line 10_F-11_F. Line 10_F-11_F crosses (intersects) line 2_F-3_F at piercing point PP B. Project PP B to the horizontal view.

3. Connect point PP A and point PP B to form intersection line A-B. Visibility is determined by applying the visibility test as shown.

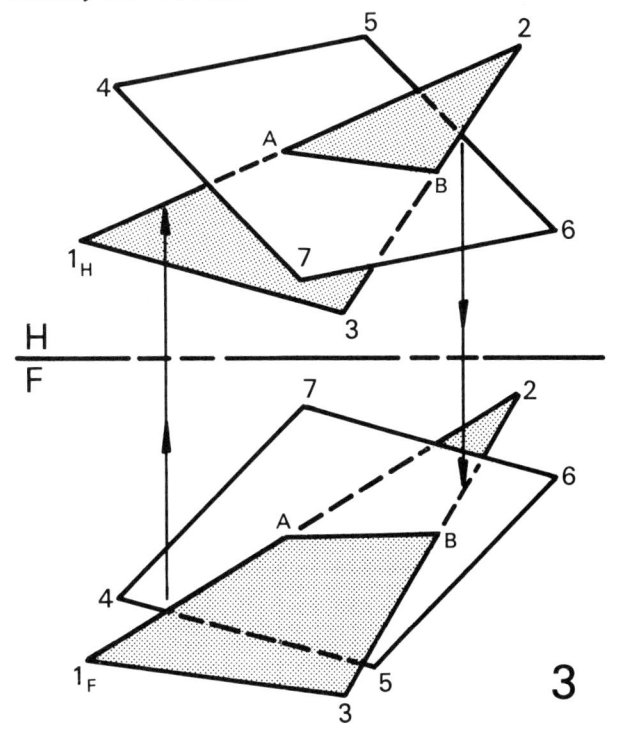

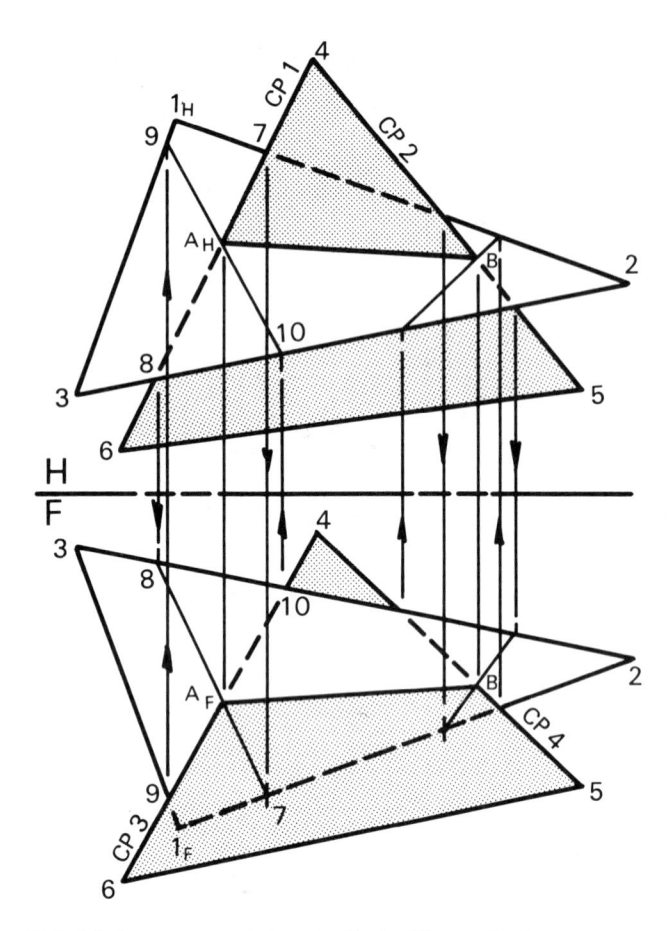

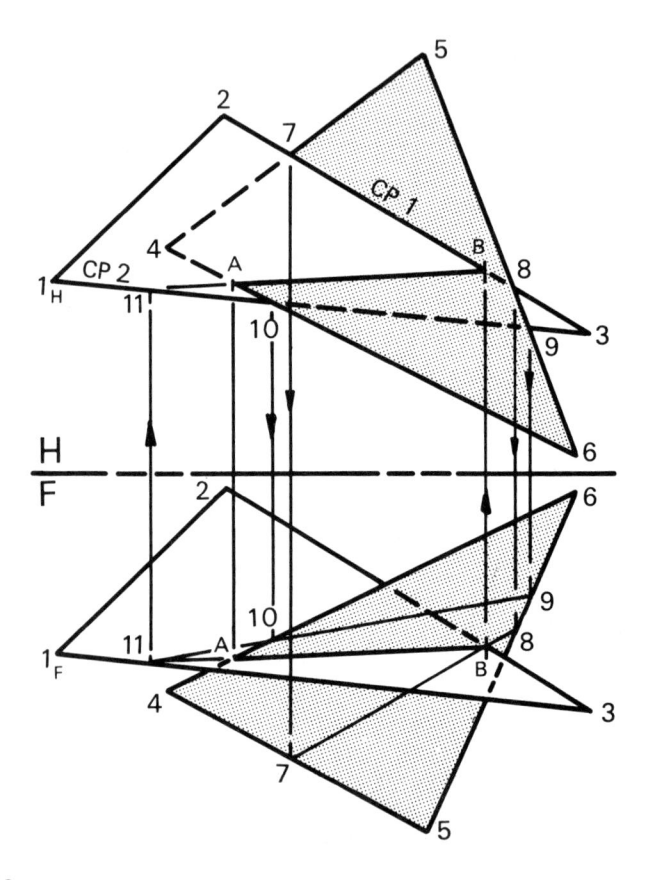

FIG. 6-7 Intersection of planes (individual line method).

2. Pass another vertical cutting plane (CP 2) through line 4_H-5_H and repeat the process described above to locate point B_F.
3. Pass cutting plane CP 3 through line 4_F-6_F. The edge view of CP 3 is represented as line 4_F-6_F. CP 3 intersects plane 4_F-5_F-6_F at points 9_F and 10_F which are then projected to the horizontal view. Line 9_H-10_H lies on plane 1_H-2_H-3_H and intersects line 4_H-6_H at point A_H. Therefore line 4_H-6_H pierces plane 1_H-2_H-3_H at point A_H. Piercing points A_H and A_F must be inline between views.
4. Repeat step 3 using line 4_F-5_F as the edge view of CP 4. Piercing point B must be aligned in adjacent views.
5. Connect piercing points A and B. Line A-B is the line of intersection between the two planes. Visibility is established by the visibility test.

In Fig. 6-8, CP 1 is passed through line 2_H-3_H and cuts plane 4_H-5_H-6_H along line 7_H-8_H. Line 7_F-8_F intersects line 2_F-3_F at piercing point B_F, which is then projected to the horizontal view. CP 2 is passed through line 1_H-3_H and cuts plane 4_H-5_H-6_H along line 9_H-10_H. The frontal projection of line 9_F-10_F does not cross line 1_F-3_F and therefore must be extended to point 11_F on line 1_F-3_F. Connecting point 11_F and point B_F forms the *extended line of intersection* between the two planes. Line 11_F-B_F crosses line 4_F-6_F at piercing point A_F. Piercing points A and B form the line of intersection between the two planes and can be projected to the horizontal view as shown.

FIG. 6-8 Intersection of planes (using cutting planes).

6.5 INTERSECTION OF PLANES (CUTTING PLANE METHOD)

The intersection of two planes can be determined by finding two lines in one plane which pierce the other plane. Two (or more) piercing points will establish a straight line that is common to both planes.

The *"cutting plane method"* is used to find the piercing point of a line on one plane that intersects the other plane. A piercing point located in one view may be projected to the adjacent view. The *"individual line method"* locates two piercing points of a line and a plane in each view instead of projecting established points between views. This method is time consuming but provides its own check, since piercing points must be aligned between adjacent views.

The individual line method is illustrated in Fig. 6-7. Planes 1-2-3 and 4-5-6 are given and their line of intersection is required. Note that the choice of cutting planes is arbitrary.

1. Pass a vertical cutting plane through line 4_H-6_H. Line 4_H-6_H is the edge view of CP 1. CP 1 intersects plane 1_H-2_H-3_H at point 7_H and point 8_H. Project line 7-8 to the frontal view. Line 7_F-8_F intersects line 4_F-6_F at point A_F. Piercing point A_F is one point on the line of intersection.

FIG. 6-9 Intersection of two infinite planes.

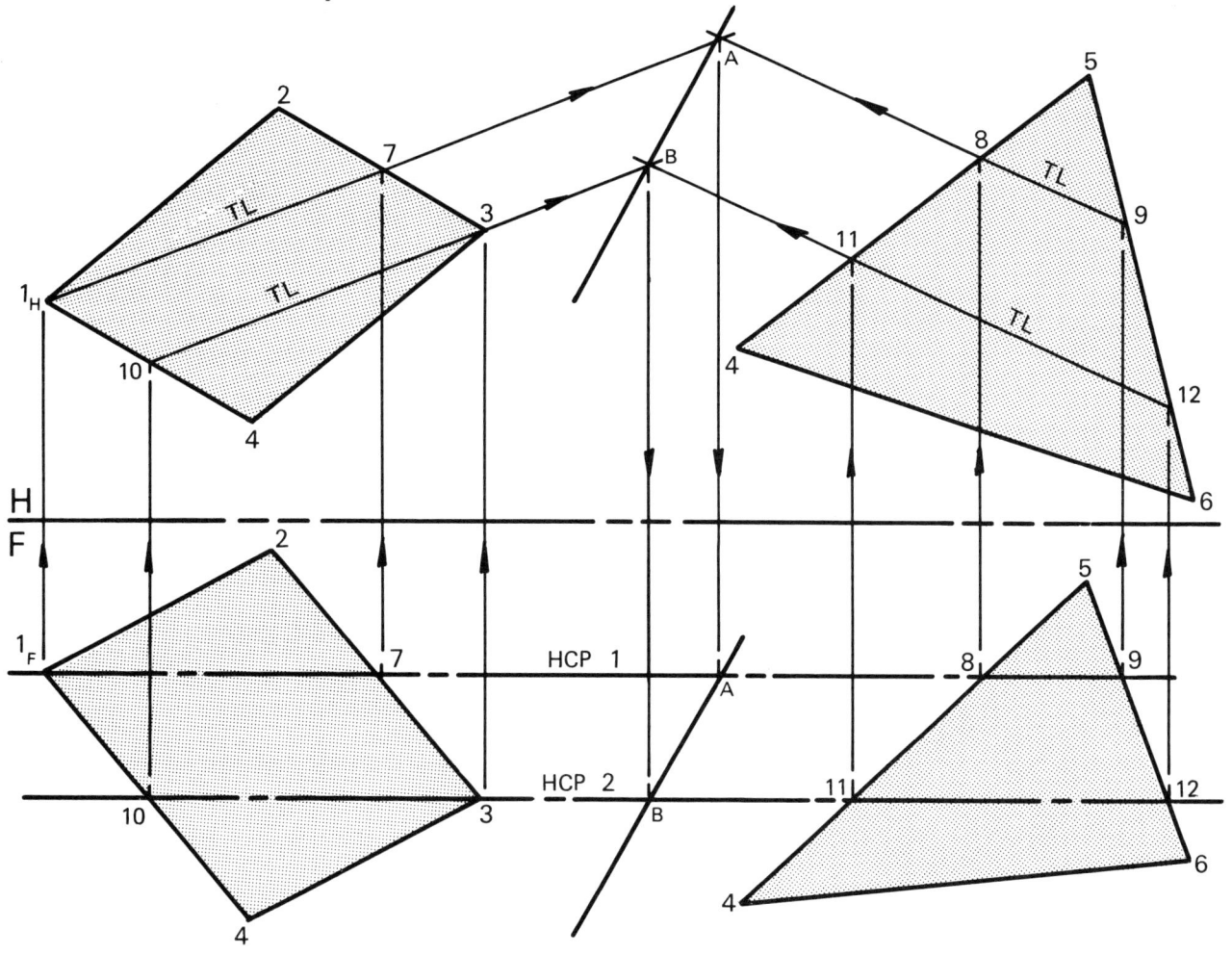

6.6 INTERSECTION OF TWO INFINITE PLANES (CUTTING PLANE METHOD)

The intersection line between two infinite planes can be found by passing a third plane through the given two planes to locate one piercing point. A minimum of two cutting planes are required, since each will locate only one piercing point. This method does not work if the line of intersection happens to be parallel to the newly introduced cutting plane, which is rarely the case.

A more accurate solution can be found if the two cutting planes are introduced through the widest part of the planes and a sufficient distance apart. The line of intersection established in one view is projected to the adjacent view, or two more cutting planes are passed through the planes in the adjacent view and the process repeated.

Each cutting plane is established as an edge in one view; normally two parallel vertical or two parallel horizontal cutting planes are used. Note that a minimum of two cutting planes are required; each locates one piercing point shared by all three planes. Two such piercing points are needed to establish a line of intersection.

In Fig. 6-9, plane 1-2-3-4 and plane 4-5-6 are given. The line of intersection is required.

1. Pass a horizontal cutting plane through point 1_F. HCP 1 is parallel to H/F and appears as an edge in the frontal view. HCP 1 intersects plane 1_F-2_F-3_F-4_F at points 1_F and 7_F, and plane 4_F-5_F-6_F at points 8_F and 9_F.
2. Project points 1 and 7, and points 8 and 9 to the horizontal view where they form lines 1_H-7_H and 8_H-9_H. Lines 1_H-7_H and 8_H-9_H are extended in the horizontal view until they intersect at piercing point A_H. Point A is projected to the frontal view where it lies on the edge view of HCP 1. Point A is a point common to all three planes. Note that line 1-7 and line 8-9 are horizontal lines, true length in the horizontal view.
3. Steps 1 and 2 are repeated using HCP 2. Note that lines 3-10 and 11-12 are horizontal lines and are parallel to lines 1-7 and 8-9 respectively.
4. Piercing points A and B are connected to form the line of intersection between the two extended planes.

Mining and geology applications of locating the intersection of two nonparallel infinite planes are presented in Chapter 8.

PROB. 6-1

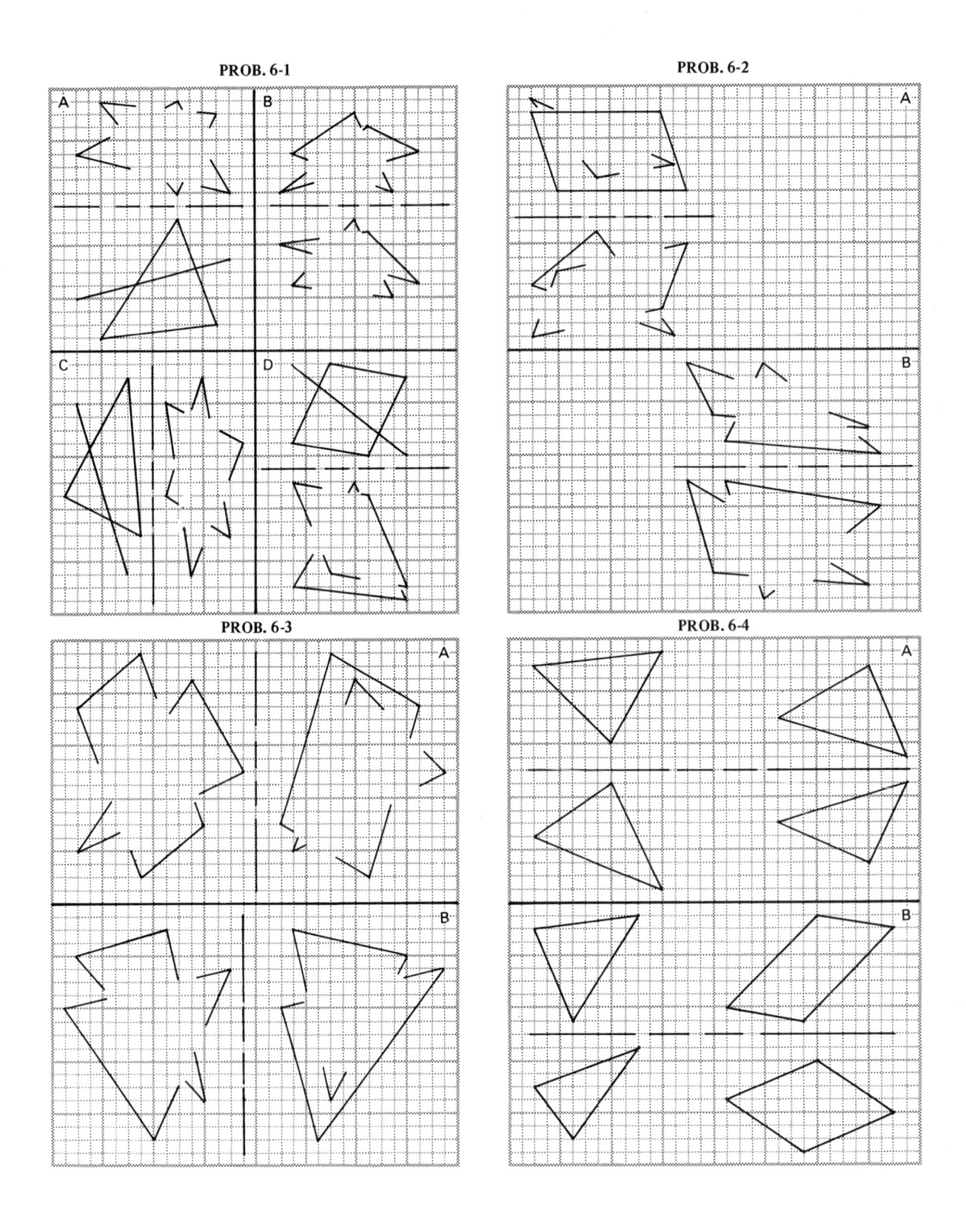

PROB. 6-2

PROB. 6-3

PROB. 6-4

PROBS. 6-1A, 6-1B, 6-1C, 6-1D. Solve for the intersection of the two planes. In A, C, and D one of the planes shows as an edge view.

PROBS. 6-2A, 6-2B. Using the edge view method, solve for the intersection of the two planes.

PROBS. 6-3A, 6-3B. Complete the two views of each problem and solve for the intersection using the cutting plane method.

PROBS. 6-4A, 6-4B. Solve for the intersection of the unlimited planes. Note the bearing of the intersection line.

218

FIG. 6-10 Intersection of a line and a prism.

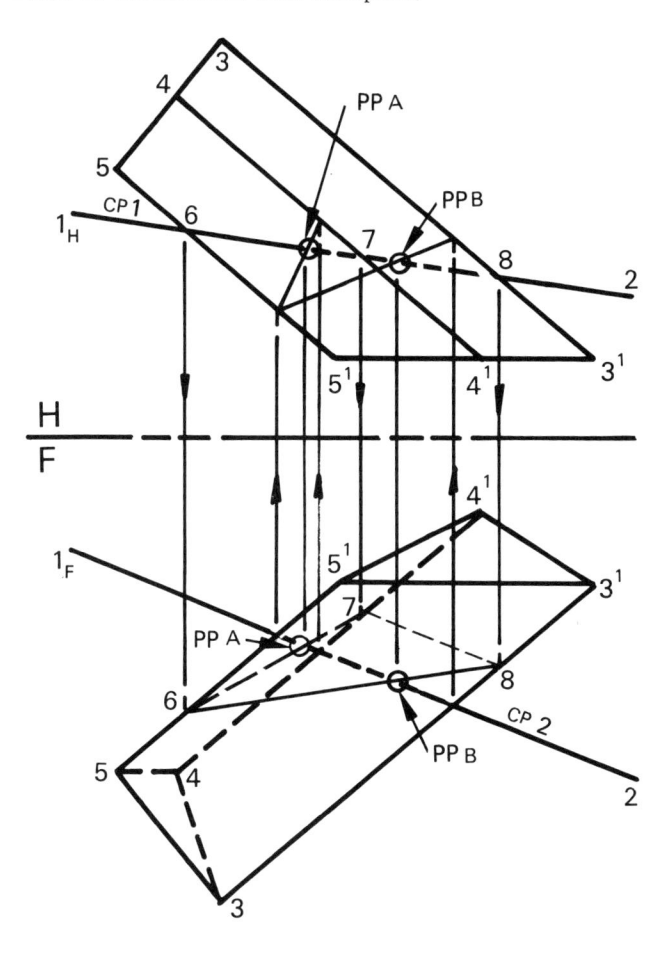

1. A vertical cutting plane is passed through line 1_H-2_H, CP 1. CP 1 intersects plane 4_H-4_H^1-5_H-5_H^1 along line 6_H-7_H. Line 6-7 is projected to the frontal view. Line 6_F-7_F intersects line 1_F-2_F at piercing point PP A. The horizontal position of PP A is established by projection.

2. CP 1 intersects plane 3_H-3_H^1-5_H-5_H^1 along line 6_H-8_H, which is then projected to the frontal view. Line 6_F-8_F intersects line 1_F-2_F at piercing point PP B. PP B is located in the horizontal view by projection.

Note that CP 1 cuts a plane section from the prism. Plane section 6_F-7_F-8_F intersects line 1_F-2_F at PP A and PP B.

3. CP 2 is passed through line 1_F-2_F and the first two steps repeated if the individual line method is employed.

4. The visibility test is used to establish proper visibility. Note that where a line lies "within" the prism (between points A and B in this case) it will always be *hidden* (dashed).

In Fig. 6-11, a vertical cutting plane is passed through line 1_H-2_H in the horizontal view. VCP intersects the prisms edges at points 7_H, 8_H, 9_H, and 10_H, which are then projected to the frontal view and establish a plane section. Plane section 7_F-8_F-9_F-10_F intersects line 1_F-2_F at PP A and PP B. PP A and PP B are located in the horizontal view by projection and visibility is determined by the visibility test.

6.7 INTERSECTION OF A LINE AND A PRISM (CUTTING PLANE METHOD)

The intersection of a line and a prism can be found by locating the piercing point of the line and each plane with which it will intersect. A prism is a *polyhedron* whose sides are composed of parallelograms. The bases of a prism may be parallel or oblique. When the prism has two parallel bases at right angles to its sides, it is called a *right prism*. All faces (sides and bases) are composed of plane surfaces. Therefore a line intersects the prism by piercing a plane face. The piercing point of the line and plane face can be determined by the edge view method or the cutting plane method. An edge view of the prism will easily show the location of the intersection of the line and the prism. Note that the line must pierce the prism through two of its plane faces. Therefore two piercing points must be established. The line can be thought of as an arrow which is shot into the prism. The line/arrow enters one of the plane faces and exits through another plane face.

In Fig. 6-10, the prism and the line are given. The intersection and proper visibility are required.

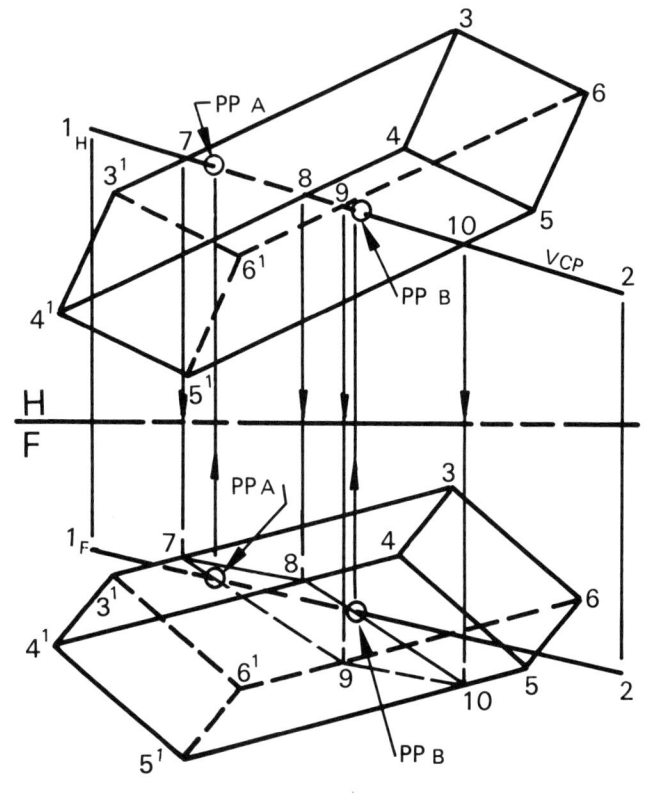

FIG. 6-11 Intersection of a line and an oblique prism.

219

FIG. 6-12 Intersection of a plane and a right prism (edge view method).

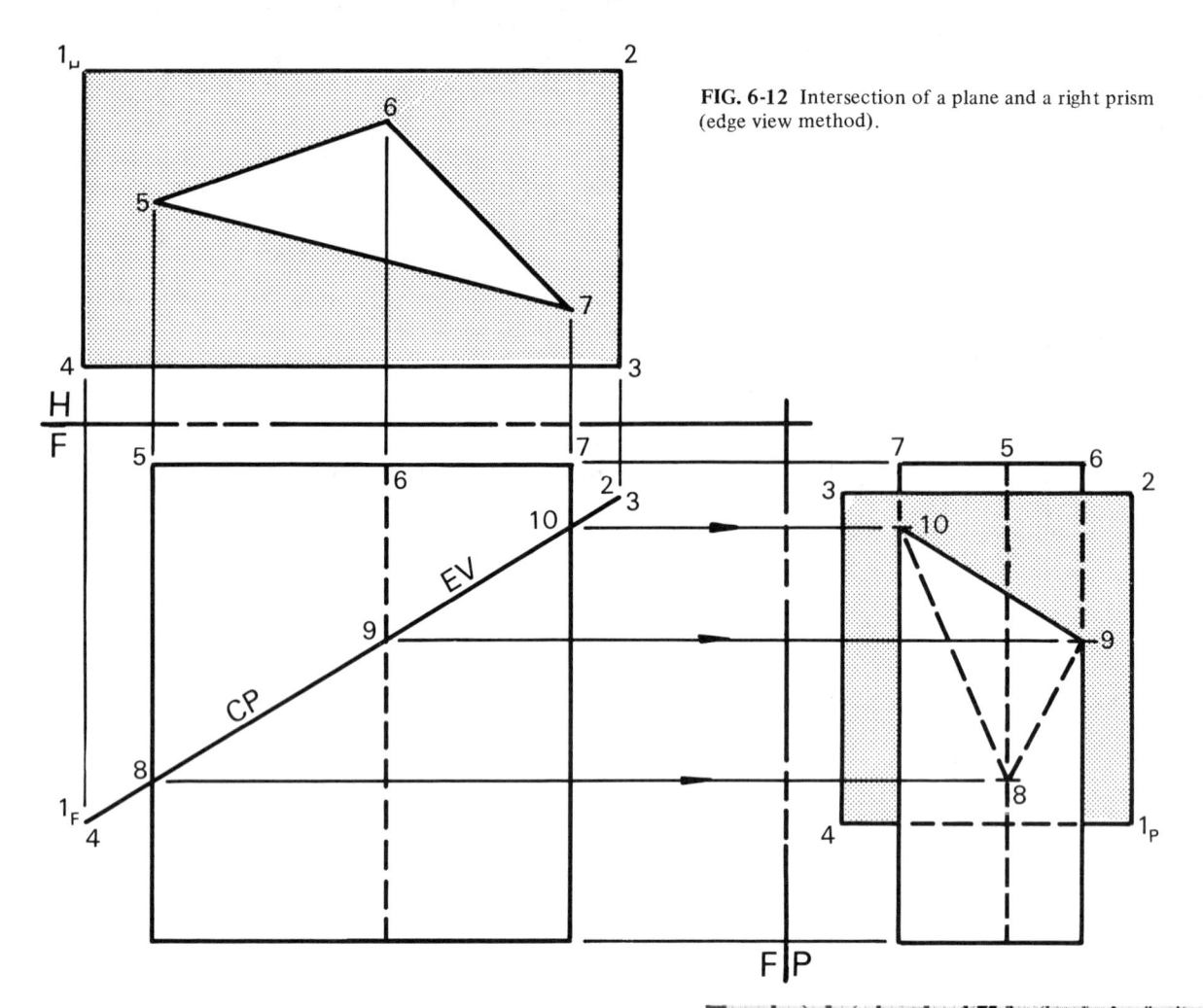

6.8 INTERSECTION OF A PLANE AND A RIGHT PRISM (EDGE VIEW METHOD)

The intersection of a plane and any prism is easily established in a view where the plane appears as an edge, Fig. 6-12. The edge view of the plane can be used as the edge view of a cutting plane which cuts a plane section from the prism in the adjacent projection. This cutting plane cuts a straight line of intersection on each plane face of the prism with which it intersects. An edge view of the prism could also be used to form cutting planes and solve for the intersection of each plane face of the prism with the given plane figure.

In Fig. 6-12, the right prism and plane are given. Their intersection lines and proper visibility are required.

1. The edge view of plane 1_F-2_F-3_F-4_F intersects the edges of the prism at points 8_F, 9_F, 10_F in the frontal view. Project intersection points 8, 9, and 10 to the profile view. Prism 5-6-7 is a right prism. All surfaces of the prism appear as edges in the horizontal view.
2. Points 8_P, 9_P, and 10_P form plane section 8_P-9_P-10_P in the profile view. This plane section determines the intersection lines of the plane and prism.
3. Visibility is determined by inspection of the relationship of the plane and the prism in the frontal and horizontal views. Line 2_F-3_F is the closest to F/P and therefore visible in the profile view. Line 1_P-4_P is hidden in the profile view. Edge line 5 in the profile view is hidden. All lines which are drawn to this edge line are hidden. Therefore, intersection lines 8_P-9_P and 8_P-10_P are hidden.

FIG. 6-13 Hydroplane. *(Courtesy NASA.)*

The intersection of surfaces is an integral part of engineering design work. The hydroplane shown in Fig. 6-13 is designed using a variety of shapes and surfaces that intersect to form the desired structural form of the vessel.

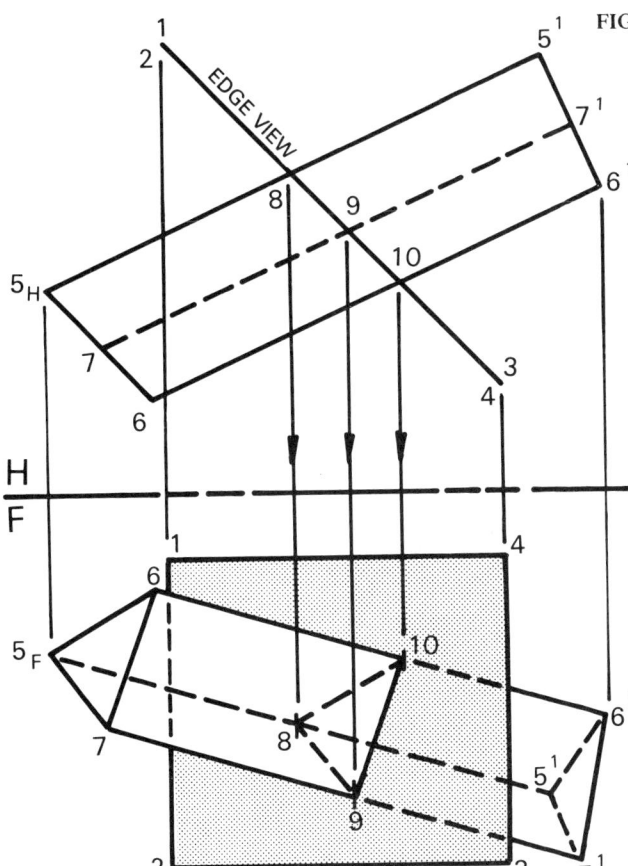

FIG. 6-14 Intersection of a vertical plane and an oblique prism.

FIG. 6-15 The shape of the supersonic transport is composed of intersecting plane, curved, and warped surfaces. *(Courtesy NASA.)*

FIG. 6-16 Intersection of an inclined plane and an oblique prism.

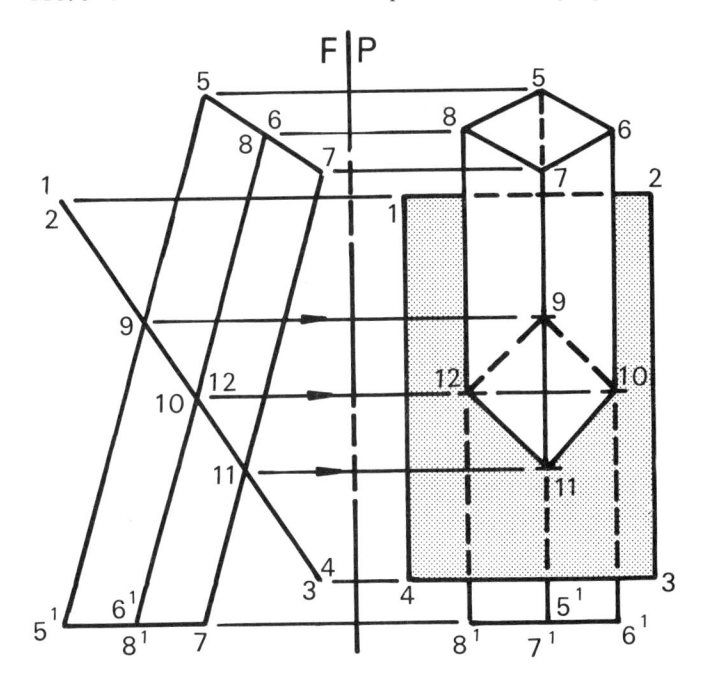

6.9 INTERSECTION OF A PLANE AND AN OBLIQUE PRISM (EDGE VIEW METHOD)

The intersection of an oblique prism and a plane can be found in any view showing the plane as an edge. The edge view of the plane will *"cut"* a plane section of the prism. The lines of intersection can be established by projecting the points at which the plane intersects each edge line of the prism to the adjacent view. These lines of intersection determine the plane section. The plane section establishes the lines of intersection of the plane and the prism.

In Fig. 6-14, the intersection of vertical plane 1-2-3-4 and the prism is required.

1. Use the edge view of plane 1_H-2_H-3_H-4_H as the edge of a cutting plane which intersects the prism at points 8_H, 9_H, and 10_H. Points 8, 9, and 10 are the points at which the lateral edges of the prism pierce the plane.

2. Project points 8, 9, and 10 to the frontal view where they establish triangular plane section 8_F-9_F-10_F. Lines 8_F-9_F, 8_F-10_F, and 9_F-10_F are the lines of intersection between the plane and the prism.

3. Visibility is established by inspection of the plane and prism in the horizontal view. Line 1_H-2_H is behind the prism and is therefore hidden in the frontal view. Point 8_F lies on edge line 5_F-5_F^1 which is hidden in the frontal view. All lines which intersect at point 8_F are hidden.

In Fig. 6-16, the inclined plane and the prism are given. The edge view of inclined plane 1_F-2_F-3_F-4_F is used to cut a plane section from the prism. When projected to the profile view points 9_P, 10_P, 11_P, and 12_P determine the plane section and therefore the lines of intersection. Visibility is established by inspection of the frontal view.

The supersonic transport shown in Fig. 6-15 is an example of intersection of surfaces. The triangular wing surfaces intersect the cylindrical fuselage and the prism-shaped air intake duct on the underside of the plane.

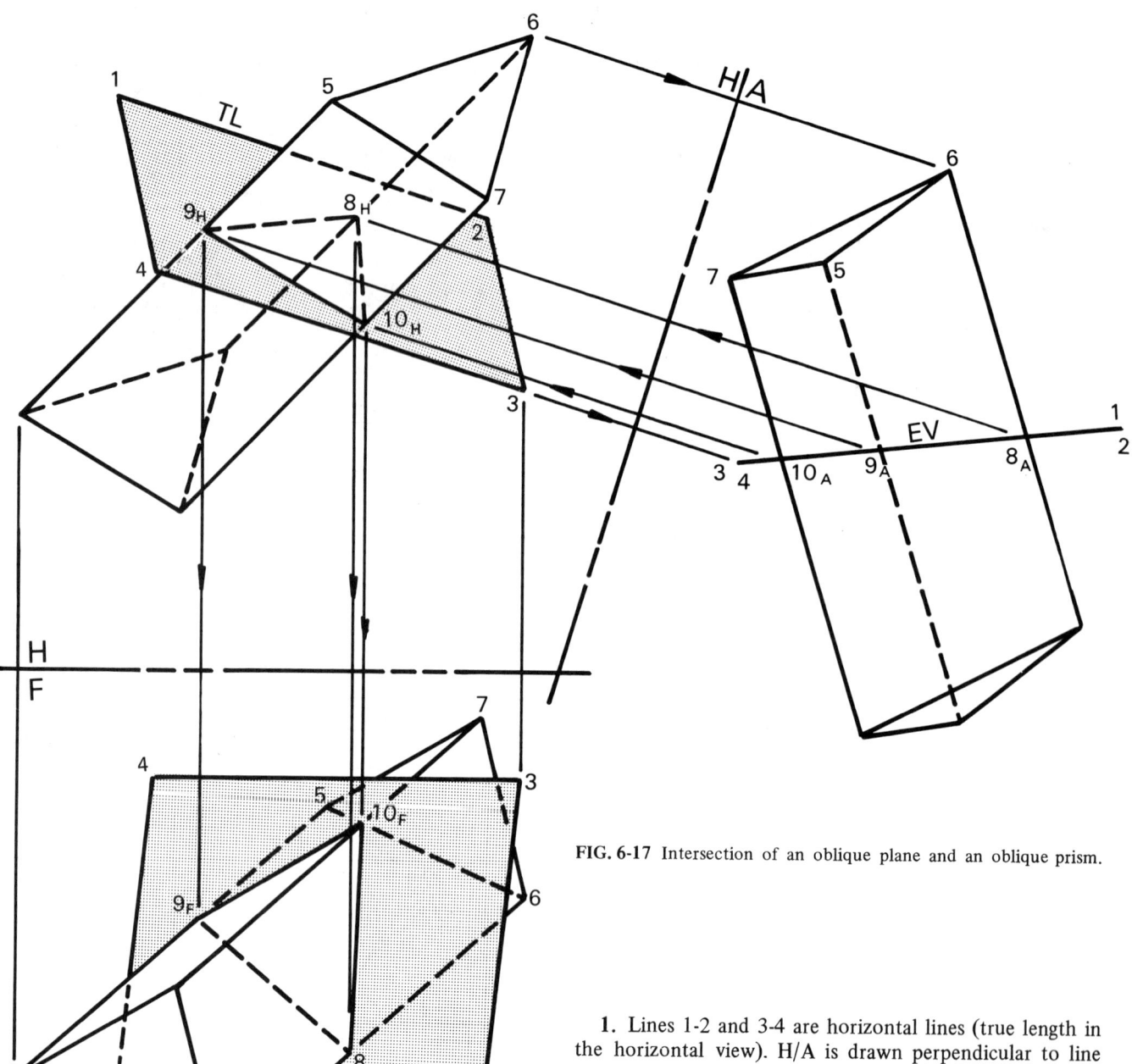

FIG. 6-17 Intersection of an oblique plane and an oblique prism.

6.10 INTERSECTION OF AN OBLIQUE PLANE AND AN OBLIQUE PRISM (EDGE VIEW METHOD)

The intersection of a plane and a prism can be established in a view where the plane appears as an edge. An auxiliary view showing the plane as an edge is required if the plane is oblique in the given views.

In Fig. 6-17, plane 1-2-3-4 and the prism are both oblique in the given frontal and horizontal views. The intersection and correct visibility are required.

1. Lines 1-2 and 3-4 are horizontal lines (true length in the horizontal view). H/A is drawn perpendicular to line 1_H-2_H. Complete auxiliary A. Plane 1_A-2_A-3_A-4_A appears as an edge view and the prism as oblique.

2. Plane 1_A-2_A-3_A-4_A represents the edge view of a cutting plane. Plane 1_A-2_A-3_A-4_A intersects the prism at points 8_A, 9_A, and 10_A. In other words, the edge lines of the prism pierce the plane at points 8_A, 9_A, and 10_A. Project all three piercing points to the horizontal view. The horizontal view of the piercing points determines the plane section cut from the prism, which in turn corresponds to the intersection of the plane and prism.

3. Project points 8, 9, and 10 to the frontal view. Points 8_F, 9_F, and 10_F determine the frontal view of the line of intersection (triangular plane section). The frontal location of each piercing point can also be fixed by transferring distances from auxiliary A along projection lines drawn from each point in the horizontal view. Note that this last method will insure the accurate location of the intersection points. This method should always be used to check the placement of the piercing points in an oblique view.

4. Visibility is determined by inspection of auxiliary A and or the visibility test.

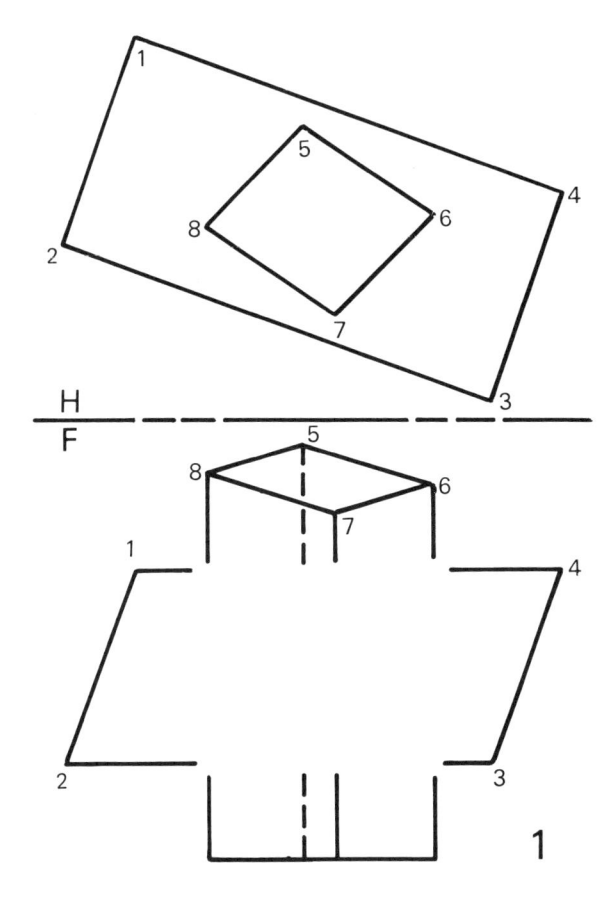

FIG. 6-18 Intersection of a plane and a right prism (cutting plane method).

6.11 INTERSECTION OF A PLANE AND A RIGHT PRISM (CUTTING PLANE METHOD)

The line of intersection between two surfaces is a common line defined by connected piercing points. A line on one surface pierces the other surface along the line of intersection. Two or more piercing points define this line of intersection. In Fig. 6-18, the intersection of plane 1-2-3-4 and the right prism is required.

1. Plane 1-2-3-4 and a prism defined by edge lines 5, 6, 7, and 8 are given. Their line of intersection is to be determined by the cutting plane method.

2. Pass a vertical cutting plane (CP 1) through the vertical plane represented by edge lines 5 and 8, and CP 2 through edge lines 6 and 7. CP 1 intersects line 1_H-4_H and line 2_H-3_H at point 9_H and point 10_H respectively. CP 2 intersects line 1_H-4_H at point 11_H and line 2_H-3_H at point 12_H. Project all four points to the frontal view where they form line 9_F-10_F and line 11_F-12_F. Line 9_F-10_F intersects edge lines 5 and 8 at points A and B, and line 11_F-12_F intersects edge lines 6 and 7 at points C and D. A, B, C, and D are the piercing points of the edge lines of the prism and the plane. Piercing points A, B, C, and D are connected to establish the lines of intersection between the plane and prism.

3. Vertical cutting planes CP 3 and CP 4 could be used instead of CP 1 and CP 2 or as a check.

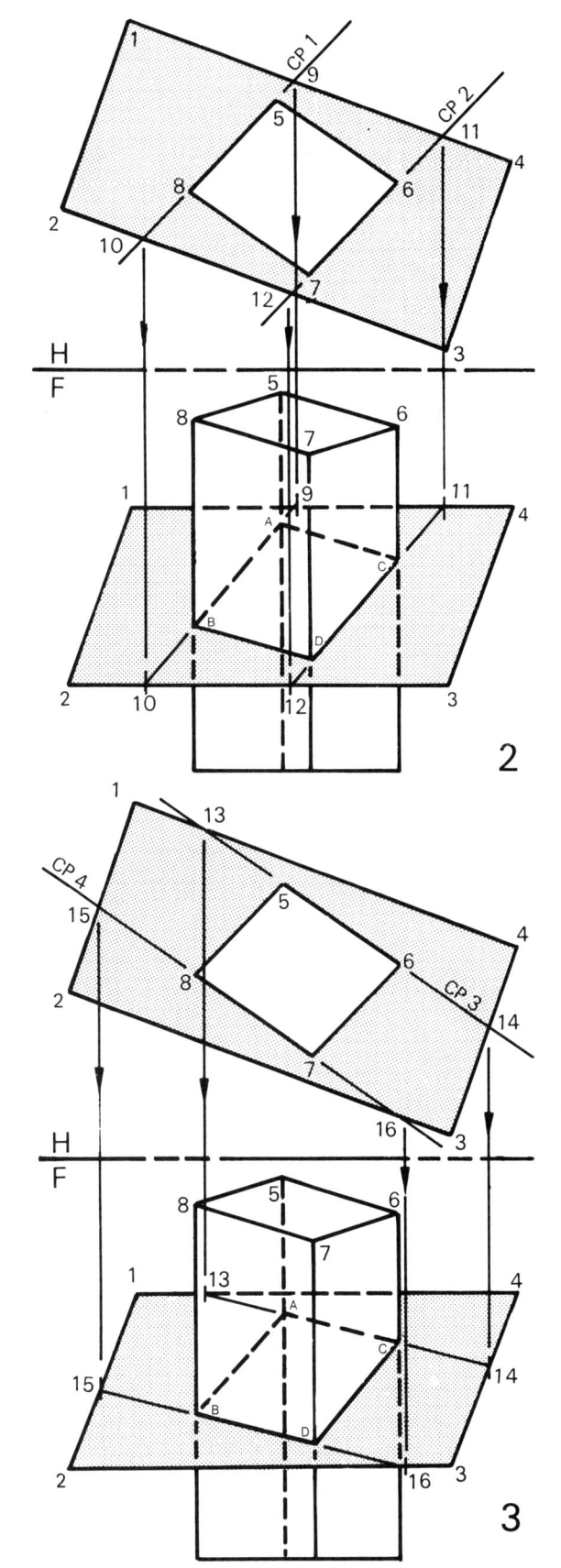

FIG. 6-19 Intersection of a plane and a prism.

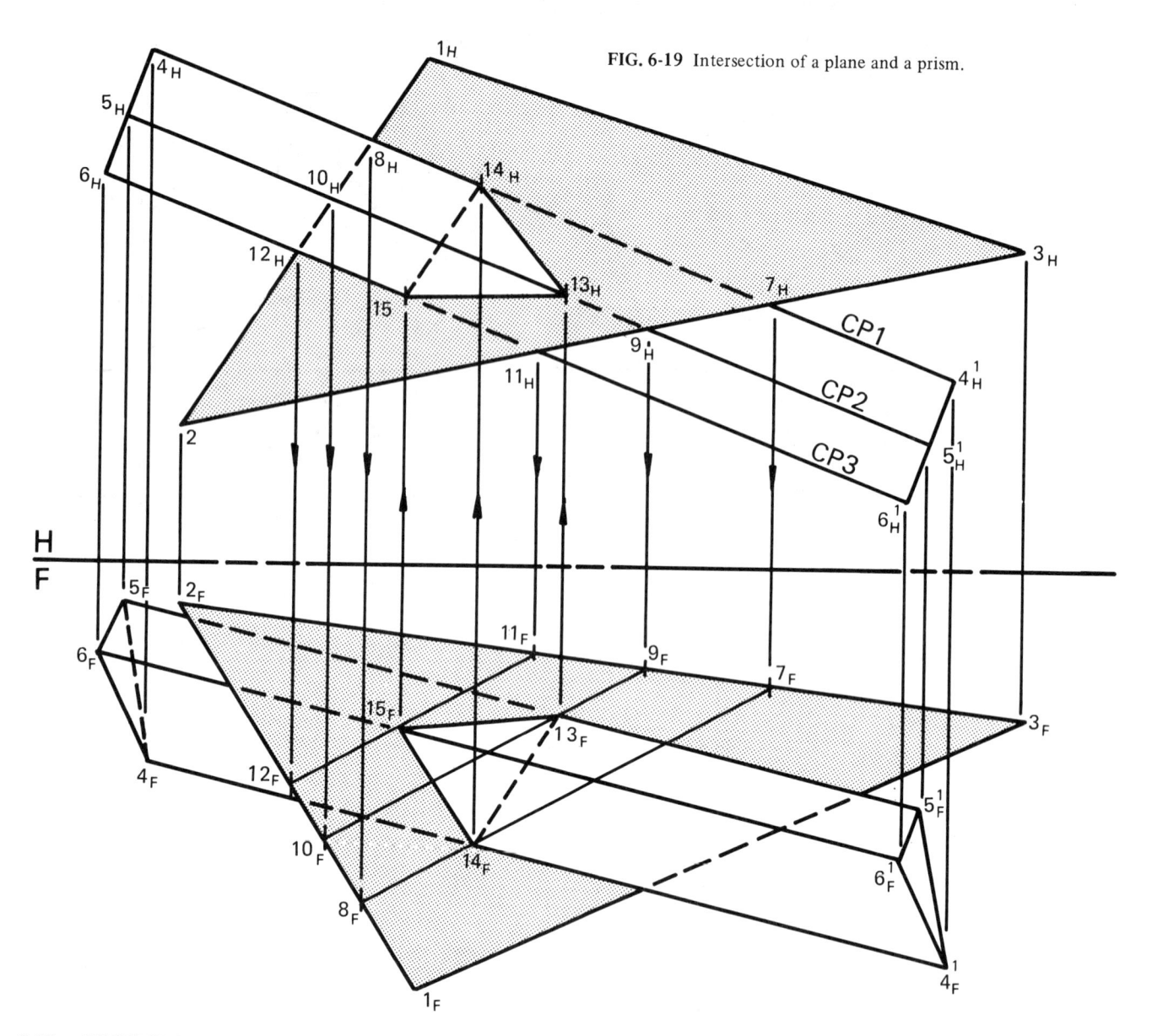

6.12 INTERSECTION OF AN OBLIQUE PLANE AND AN OBLIQUE PRISM (CUTTING PLANE METHOD)

The line of intersection of an oblique prism and oblique plane can be readily located by projecting an auxiliary view where the plane is an edge or the prism is a right section edge view. Note that a right section of a prism is a view projected perpendicular to its edge lines where they appear as true length. The cutting plane method can also be used.

Cutting planes can be introduced at any angle, in any view. Vertical, horizontal and front edge view cutting planes passed through existing lines are the most convenient. In Fig. 6-19 the plane and the prism are given. The line of intersection is required.

1. Pass vertical cutting plane CP 1 through line 4_H-4_H^1 in the horizontal view.
2. CP 1 intersects line 1_H-2_H at point 8_H, and line 2_H-3_H at point 7_H. Project points 7 and 8 to the frontal view where they form line 7_F-8_F.
3. Line 7_F-8_F intersects line 4_F-4_F^1 at point 14_F. Point 14_F is the piercing point of line 4_F-4_F^1 and plane 1_F-2_F-3_F.
4. Project piercing point 14 to the horizontal view.
5. Pass vertical cutting planes CP 2 and CP 3 through lines 5_H-5_H^1 and 6_H-6_H^1 respectively.
6. CP 2 intersects line 2_H-3_H at point 9_H and line 1_H-2_H at point 10_H. CP 3 intersects line 2_H-3_H and line 1_H-2_H at points 11_H and 12_H respectively. Project points 9 and 10, and points 11 and 12 to the frontal view where they form lines 9_F-10_F and 11_F-12_F.
7. Line 9_F-10_F intersects line 5_F-5_F^1 at piercing point 13_F. Line 11_F-12_F intersects line 6_F-6_F^1 at piercing point 15_F. Project piercing points 13 and 15 to the horizontal view.
8. Connect all three piercing points in both views to establish the line of intersection between the plane and the prism.

Note that all three vertical cutting planes are parallel in the horizontal view and therefore they cut parallel lines on the plane in the frontal view.

224

PROB. 6-5

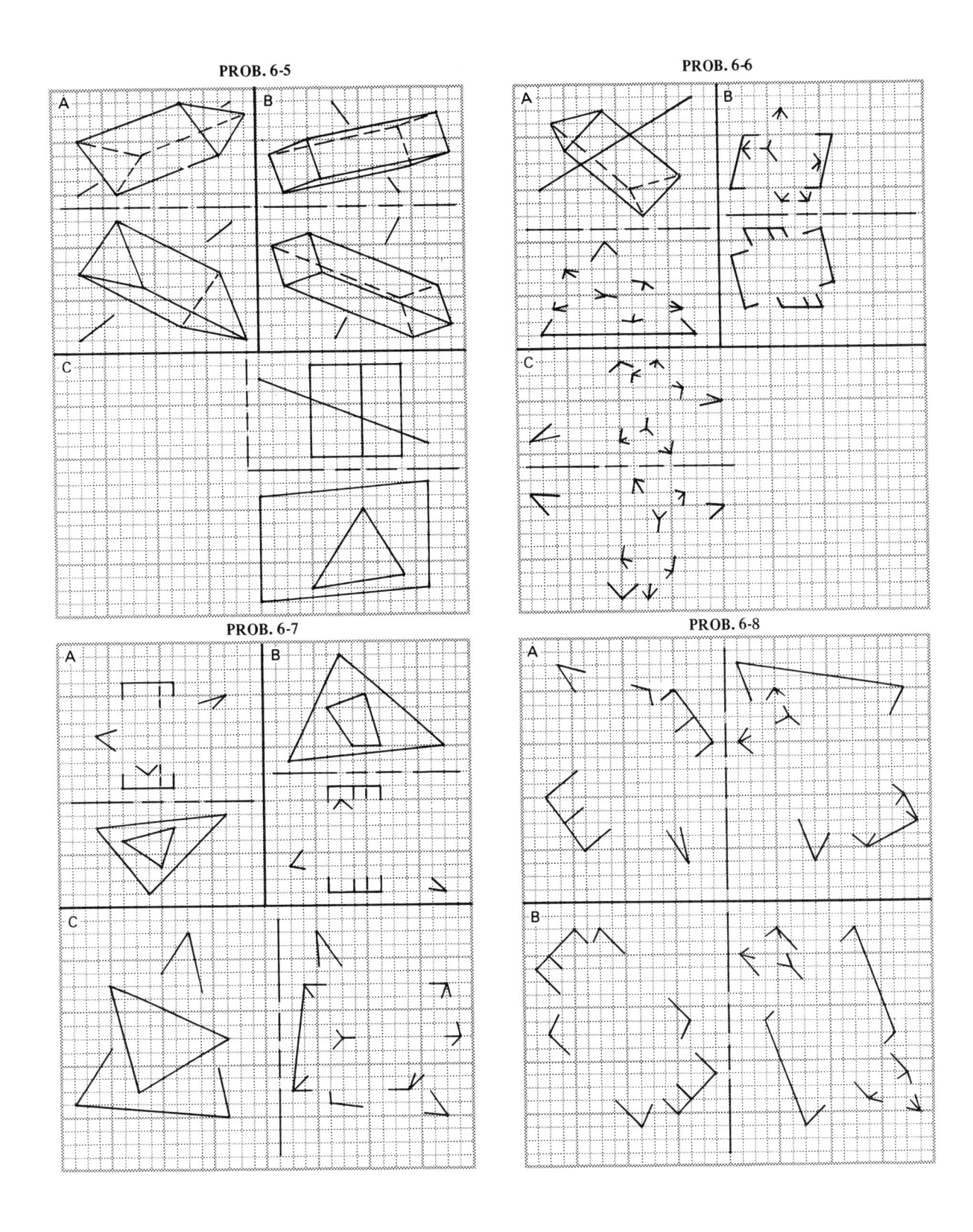

PROB. 6-6

PROB. 6-7

PROB. 6-8

PROBS. 6-5A, 6-5B. Find the piercing points of the line and the prism and show the proper visibility.
PROB. 6-5C. Project an auxiliary view from the horizontal view. Solve for the intersection of the plane and the prism.
PROBS. 6-6A, 6-6B, 6-6C. Use the edge view method to solve for the intersection of the plane and prism.

PROBS. 6-7A, 6-7B, 6-7C. Using the cutting plane method, determine the intersection between the oblique plane and right prism.
PROBS. 6-8A, 6-8B. Determine the intersection between the oblique plane and oblique prism.

225

6.13 INTERSECTION OF A PLANE AND A PYRAMID (EDGE VIEW METHOD)

The intersection of a plane and a pyramid can be established in a view where the plane appears as an edge. Where the edge view of the plane intersects the pyramid determines the points of intersection (piercing points) of each edge line of the pyramid and the plane. The piercing points thus established can be projected back to the adjacent views. If the given plane is oblique, an auxiliary view where the plane projects as an edge is necessary.

In Fig. 6-20, oblique plane 1-2-3-4 and right pyramid 5-6-7-8 are given. The intersection line and proper visibility are required.

1. Line 1-2 and 3-4 are frontal lines (true length in the frontal view). Draw F/A perpendicular to line 1_F-2_F and project auxiliary view A. Plane 1_A-2_A-3_A-4_A appears as an edge and the pyramid as oblique.

FIG. 6-20 Intersection of a plane and a pyramid.

2. Plane 1_A-2_A-3_A-4_A intersects edge lines 5_A-8_A, 6_A-8_A, and 7_A-8_A at points 9, 10, and 11 respectively. Project the intersection/piercing points to the frontal and horizontal views.

3. Connect piercing points 9, 10, and 11 to form the line of intersection.

4. Visibility is determined by inspection of auxiliary A. The portion of the pyramid represented by points 8-9-10-11 is above and somewhat in front of the plane. Therefore this portion is visible in the frontal and horizontal views.

The Viking Lander in Fig. 6-21 is composed of a variety of intersecting surfaces and lines, including cylinders, prisms, spheres, cones, and plane surfaces.

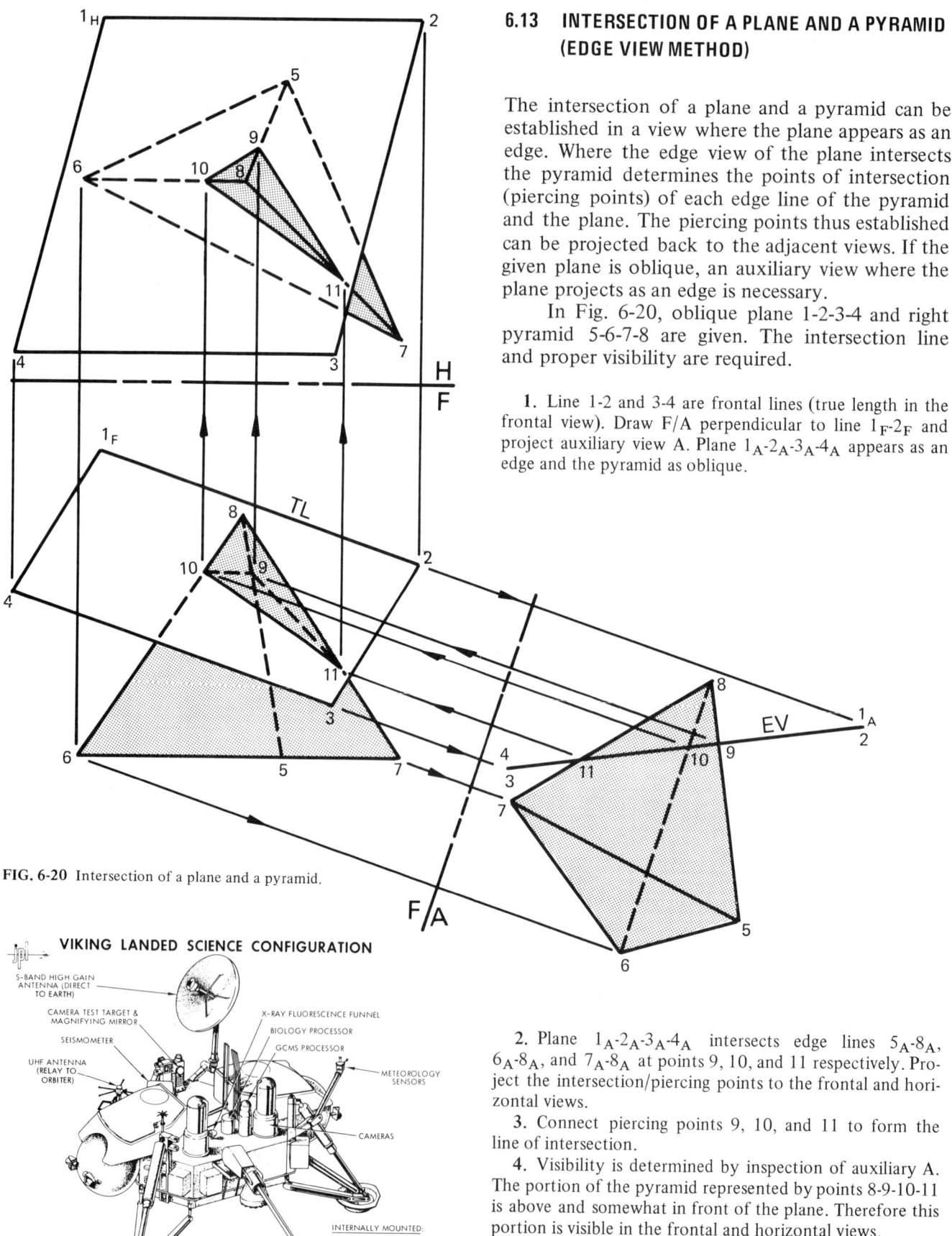

FIG. 6-21 Viking. *(Courtesy NASA.)*

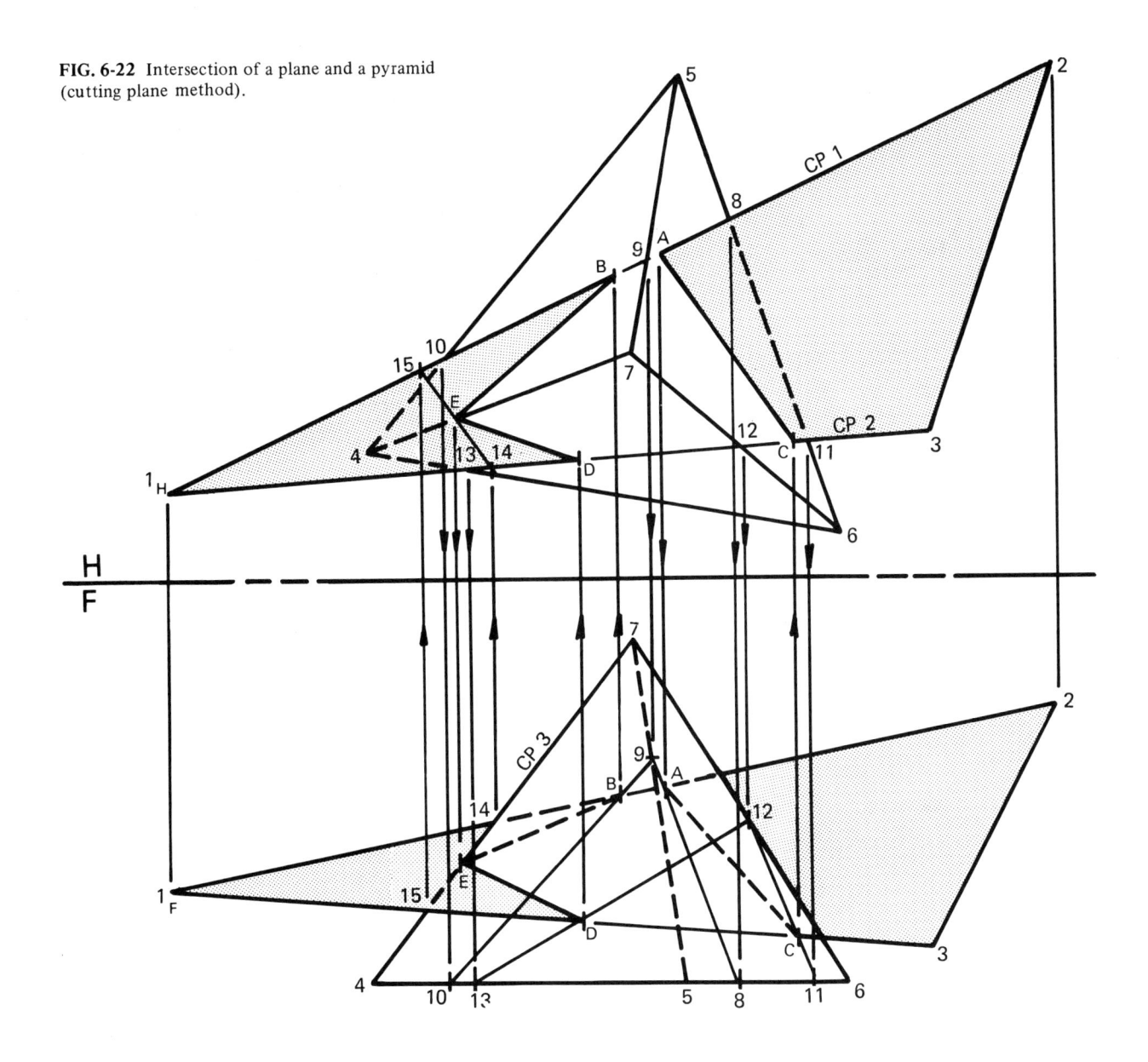

FIG. 6-22 Intersection of a plane and a pyramid (cutting plane method).

6.14 INTERSECTION OF A PLANE AND A PYRAMID (CUTTING PLANE METHOD)

The intersection of a plane and a pyramid can be determined by passing a series of cutting planes through the edge lines of the pyramid, the lines that form the plane or a combination of the two.

In Fig. 6-22, oblique plane 1-2-3 and pyramid 4-5-6-7 are given.

1. Pass vertical cutting plane CP 1 through line 1_H-2_H. CP 1 intersects edge lines 5_H-6_H at point 8_H, 5_H-7_H at point 9_H and 4_H-5_H at point 10_H.
2. Project points 8, 9, and 10 to the frontal view where they form lines 8_F-9_F and 9_F-10_F. Line 8_F-9_F intersects line 1_F-2_F at point A. Line 9_F-10_F intersects line 1_F-2_F at point B. Project points A and B to the horizontal view. Line 1-2 pierces surface 5-6-7 at point A and surface 4-5-7 at point B.

3. Pass CP 2 through line 2_H-3_H. CP 2 intersects lines 5_H-6_H, 6_H-7_H, and 4_H-6_H at points 11_H, 12_H, and 13_H respectively.
4. Project points 11, 12, and 13 to the frontal view where they form lines 11_F-12_F and 12_F-13_F. Line 11_F-12_F intersects line 1_F-3_F at point C. Line 12_F-13_F intersects line 1_F-3_F at point D. Piercing points C and D are projected to the horizontal view. Line 1-3 pierces surface 5-6-7 at C and surface 4-6-7 at point D.
5. Points A and C form intersection line A-C. Plane 1-2-3 intersects pyramid 4-5-6-7 on surface 5-6-7 at intersection line A-C. Line 1-2 intersects the pyramid at point B and line 1-3 at point D. Line B-D lies inside the pyramid and therefore does not represent the intersection of the plane and pyramid.
6. Where edge line 4-7 pierces plane 1-2-3 needs to be determined. Pass CP 3 through edge line 4_F-7_F. CP 3 intersects line 1_F-2_F at point 14_F and line 1_F-3_F at point 15_F. Project points 14 and 15 to the horizontal view where they form line 14_H-15_H, which intersects edge line 4_H-7_H at point E. Project point E to the frontal view.
7. Line B-E and line D-E complete the intersection.

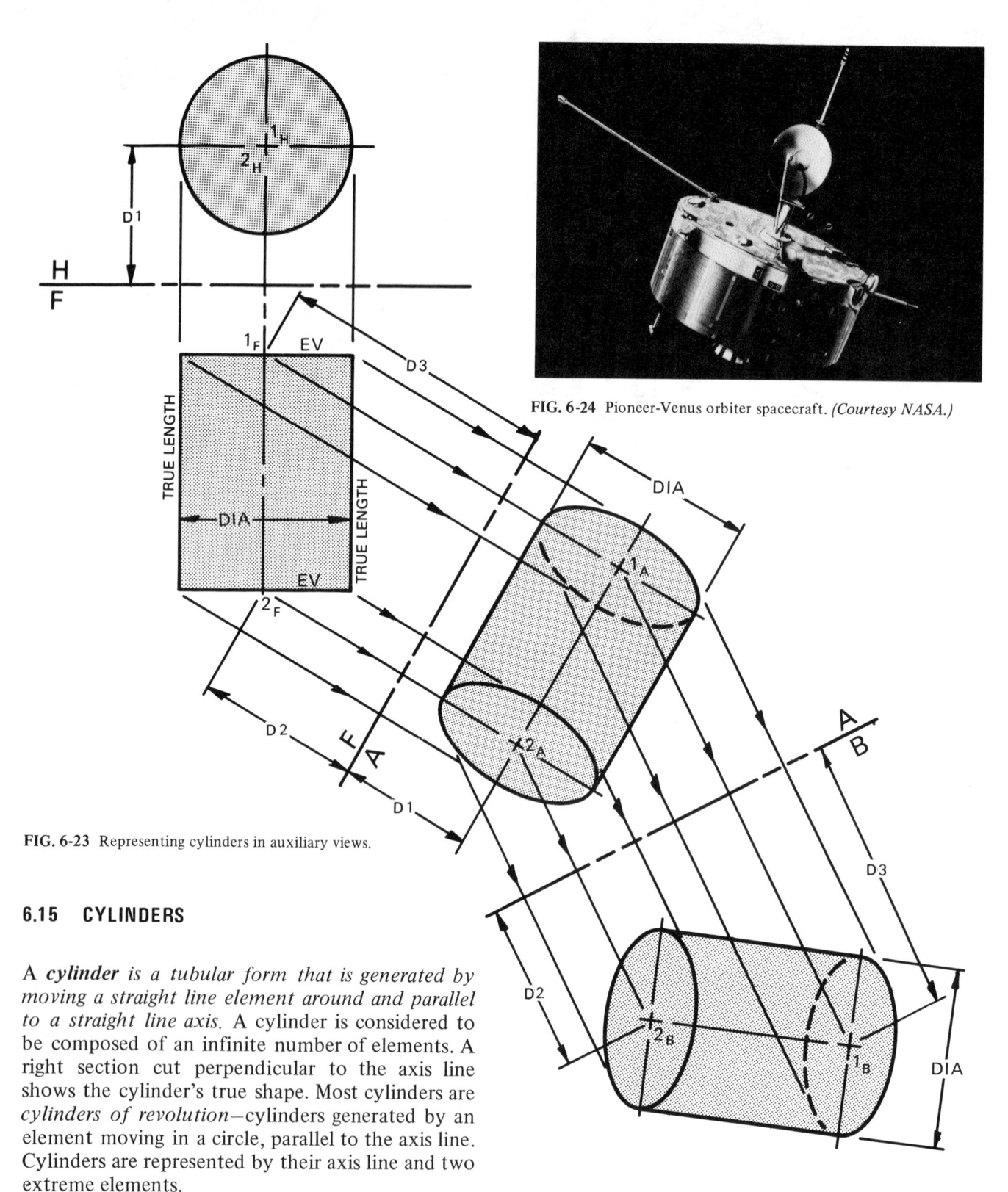

FIG. 6-23 Representing cylinders in auxiliary views.

FIG. 6-24 Pioneer-Venus orbiter spacecraft. *(Courtesy NASA.)*

6.15 CYLINDERS

A *cylinder is a tubular form that is generated by moving a straight line element around and parallel to a straight line axis.* A cylinder is considered to be composed of an infinite number of elements. A right section cut perpendicular to the axis line shows the cylinder's true shape. Most cylinders are *cylinders of revolution*—cylinders generated by an element moving in a circle, parallel to the axis line. Cylinders are represented by their axis line and two extreme elements.

In Fig. 6-23, cylinder 1-2 appears as a right section in the horizontal view where its cylindrical surface is represented as a circle with its axis as a point view. The frontal view shows the two extreme elements as true length and the bases as edges. Auxiliaries A and B are projected so that the cylinder's bases appear as ellipses. Note that the distance between extreme elements remains the same in all projections, since this is a cylinder of revolution.

The major axis of an ellipse will always be perpendicular to the axis line and equal to the diameter, if the cylinder is a cylinder of revolution.

Most cylindrical forms are circular in shape. The Pioneer spacecraft shown in Fig. 6-24 is an example of a cylinder of revolution. Pipes and most vessels are right circular cylinders.

FIG. 6-25 Oblique cylinder.

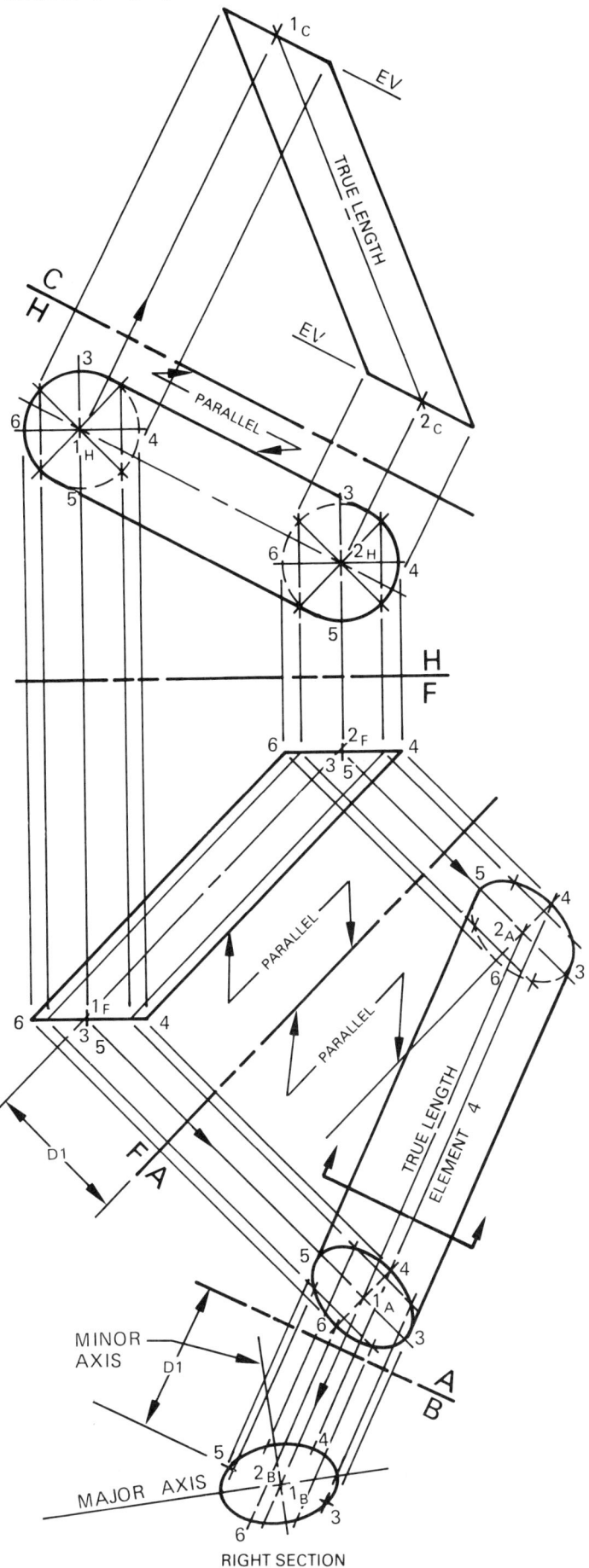

FIG. 6-26 All pipes are cylinders of revolution.

6.16 OBLIQUE CYLINDERS

Where a cylinder appears as *oblique*, every element on its surface and the axis line shows as foreshortened. The true shape of an oblique cylinder must be projected from a view where the axis is true length. A view where the axis appears as a point view and the cylinder's surface is an edge shows the true shape of the cylinder. The true shape can be a circle, as when the cylinder is an oblique cylinder of revolution or an elliptical form. A view projected perpendicular to the axis line (end view or right section), where it appears as true length, shows the true shape of the cylinder. The true length of the cylinder equals the true length of the axis.

In Fig. 6-25, oblique cylinder 1-2 is given in the frontal and horizontal views. The true shape of the base plane can be seen in the horizontal view since the base appears as an edge, and parallel to the fold line in the frontal view. An auxiliary projection taken parallel to the axis line in any view will show its true length. In the example, F/A is drawn parallel to 1_F-2_F. 1_A-2_A is true length. The elliptical view of the base is drawn by establishing a series of points along the perimeter of its base and transferring them from view to view. The major and minor axis lines could also be located and geometric construction applied. Fold line A/B is drawn perpendicular to 1_A-2_A. In auxiliary B the cylinder appears true shape, since this is an end view or "right section."

Auxiliary C is projected parallel to 1_H-2_H. Since the base is true shape in the horizontal view, it projects as an edge. The true length is measured along axis line 1_A-2_A. Space permitting, a right section could also be projected from this view.

229

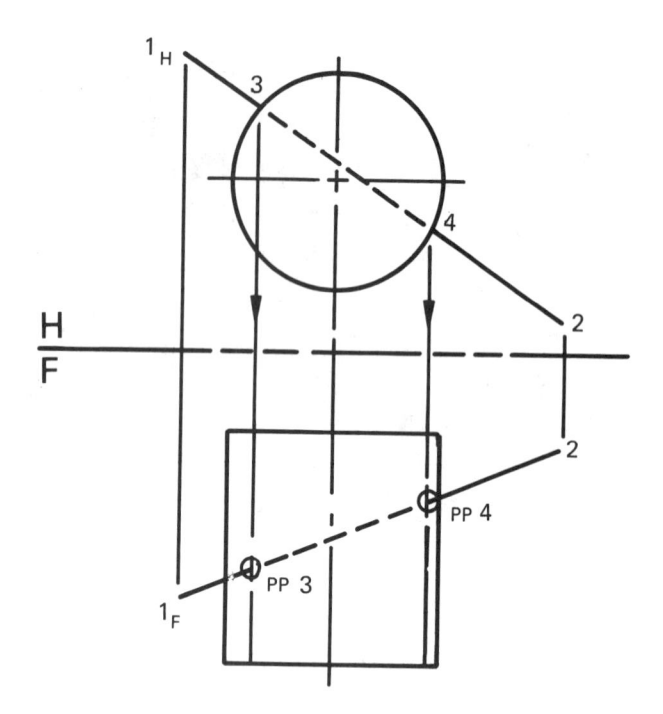

FIG. 6-27 Intersection of a line and a cylinder.

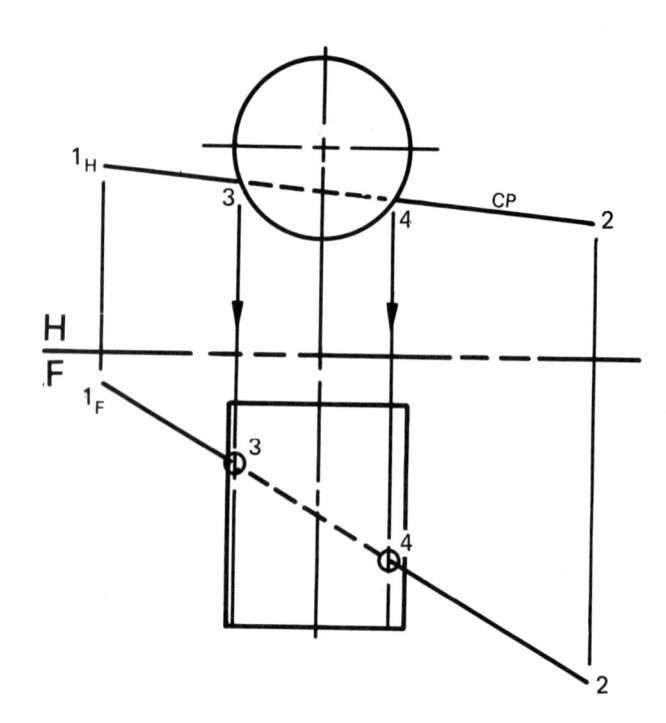

FIG. 6-28 Piercing points of a line and a cylinder.

6.17 INTERSECTION OF A LINE AND A CYLINDER

The intersection points of a line and a cylinder are apparent in a view where the cylinders axis appears as a point. In Fig. 6-27 line 1_H-2_H pierces the cylinder at points 3_H and 4_H. These piercing points are projected to the frontal view where they lie on the line and establish PP 3 and PP 4.

In Fig. 6-28, a vertical cutting plane is passed through line 1_A-2_A and parallel to the axis. The intersection of the cutting plane and the cylinder cuts two straight-line elements on the cylinder in the adjacent view. The intersection of the two elements and line 1_F-2_F locate points 3 and 4.

Visibility is established by inspection of the horizontal view, for both examples.

In Fig. 6-29, the intersection of the line and the oblique cylinder is required. A cutting plane passed through the line and parallel to the axis intersects the cylinder as two parallel elements. The intersection of the elements and the line establishes the two piercing points.

1. Construct a cutting plane containing line 1-2 and parallel to the axis line (cylinder's wall). Draw line 2_F-4_F parallel to the cylinder's axis.
2. Draw line 3_F-4_F parallel to H/F. For convenience it is passed through the edge view of the cylinder upper base.
3. Draw a line from point 2_H parallel to the axis. Point 4_H is located at the intersection of this line and a projection line drawn from point 4_F in the frontal view.

4. Connect point 3_H and point 4_H. Line 3_H-4_H intersects the base of the cylinder at points 5_H and 6_H. Points 5_H and 6_H are the uppermost points of two elements that lie on the plane and cut elements on the surface of the cylinder. The intersection of the elements and line 1-2 establish piercing points 7 and 8.

FIG. 6-29 Intersection of a line and oblique cylinder.

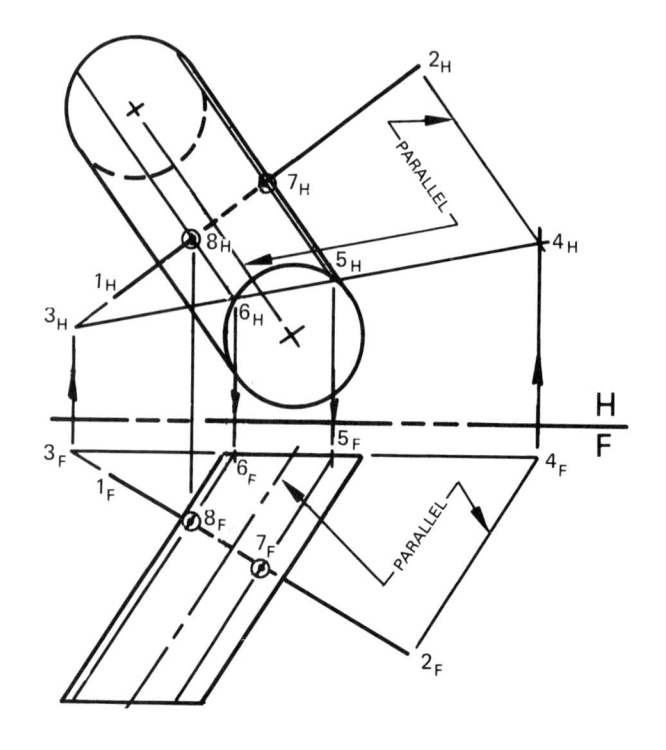

230

6.18 INTERSECTION OF A PLANE AND A CYLINDER

The line of intersection of a plane and a cylinder can be determined by passing a series of cutting planes parallel to the cylinder's axis. Each CP cuts elements on the cylinder which pierce the plane to form an elliptical line of intersection. Accuracy increases with the number of cutting planes used.

In Fig. 6-30, the inclined plane and the right cylinder are given.

1. Pass three vertical cutting planes parallel to the axis and through the cylinder. CP 1 cuts two elements on the cylinder which intersect the edge view of the plane at piercing points 2_P and 8_P. Project point 2 and 8 to the frontal view.
2. Repeat this process using CP 2 and CP 3.
3. Locate piercing points 1 and 5 by projection from the profile view.
4. Connect the piercing points showing proper visibility.

In Fig. 6-31, a series of vertical cutting planes are passed parallel to the axis and through the cylinder. Each cutting plane establishes an element on the cylinder and a line on the plane. Where these related lines and elements intersect, they establish the required piercing points.

1. Draw CP 1 and CP 2 parallel to the axis line. CP 1 intersects line 1_H-3_H at point 4_H and line 2_H-3_H at point 5_H. CP 2 intersects line 1_H-2_H at point 6_H and line 2_H-3_H at point 7_H. Both CPs establish an element on the cylinder. Project the elements to the frontal view along with lines 4-5 and 6-7. Line 4_F-5_F intersects its element at piercing point A and line 6_F-7_F intersects its corresponding element at point B.
2. Repeat step 1 using CP 3, CP 4, and CP 5. Note that each of these cutting planes cuts two elements on the cylinder. Therefore, each locates two piercing points. Since point 2_H is front of the cylinder, lines 1-2 and 2-3 are visible, as is point B.

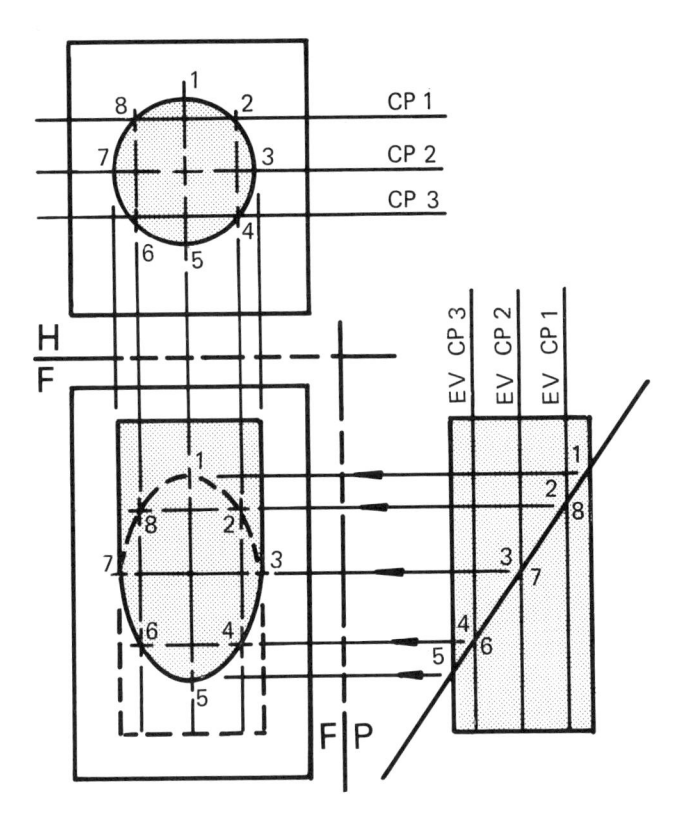

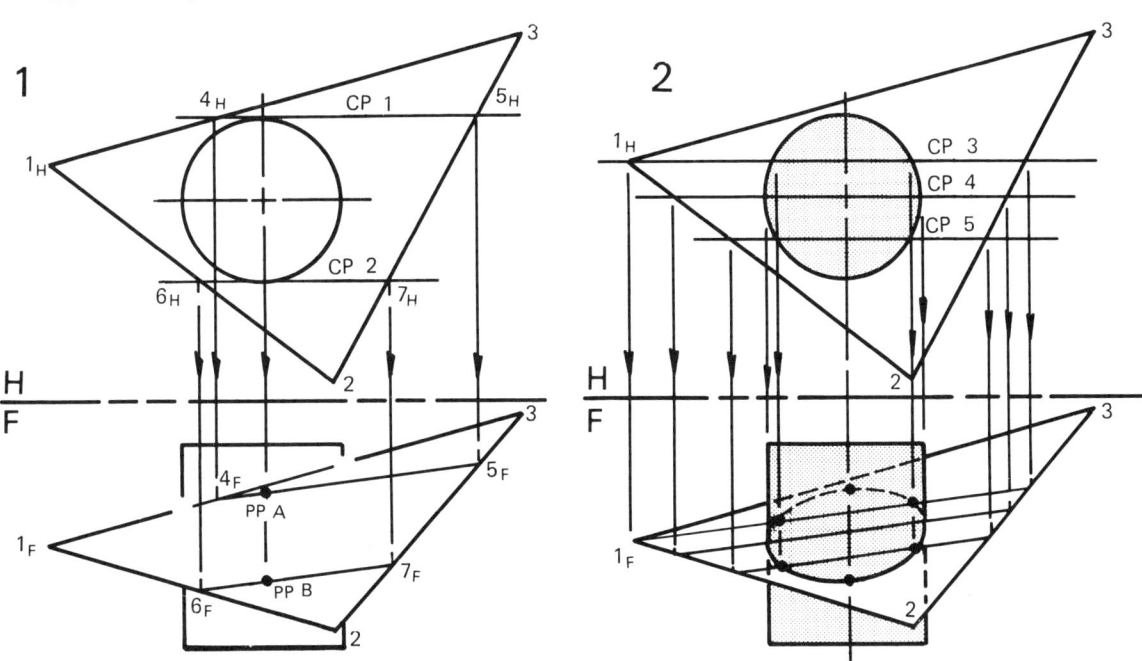

FIG. 6-30 Intersection of a plane and a cylinder.

FIG. 6-31 Intersection of a plane and a cylinder (cutting plane method).

6.19 INTERSECTION OF AN OBLIQUE PLANE AND AN OBLIQUE CYLINDER (EDGE VIEW METHOD)

The intersection of a plane and a cylinder, both of which are oblique in their given views, can be solved by projecting an auxiliary view where the plane appears as an edge. This auxiliary view can be projected from either given view, but a view where the cylinder bases show as edges is the most convenient. A series of cutting planes are introduced parallel to the axis of the cylinder in a view where the plane is an edge view. The cutting planes scribe elements along the surface of the cylinder. The intersection of the elements and the plane establish a series of corresponding piercing points, which are then located in all previous views by projection and transferring distances. The piercing points are connected in sequence to form the line of intersection.

In Fig. 6-32, the horizontal and frontal views of the oblique plane 1-2-3 and the oblique cylinder are given. Their intersection line and proper visibility are required.

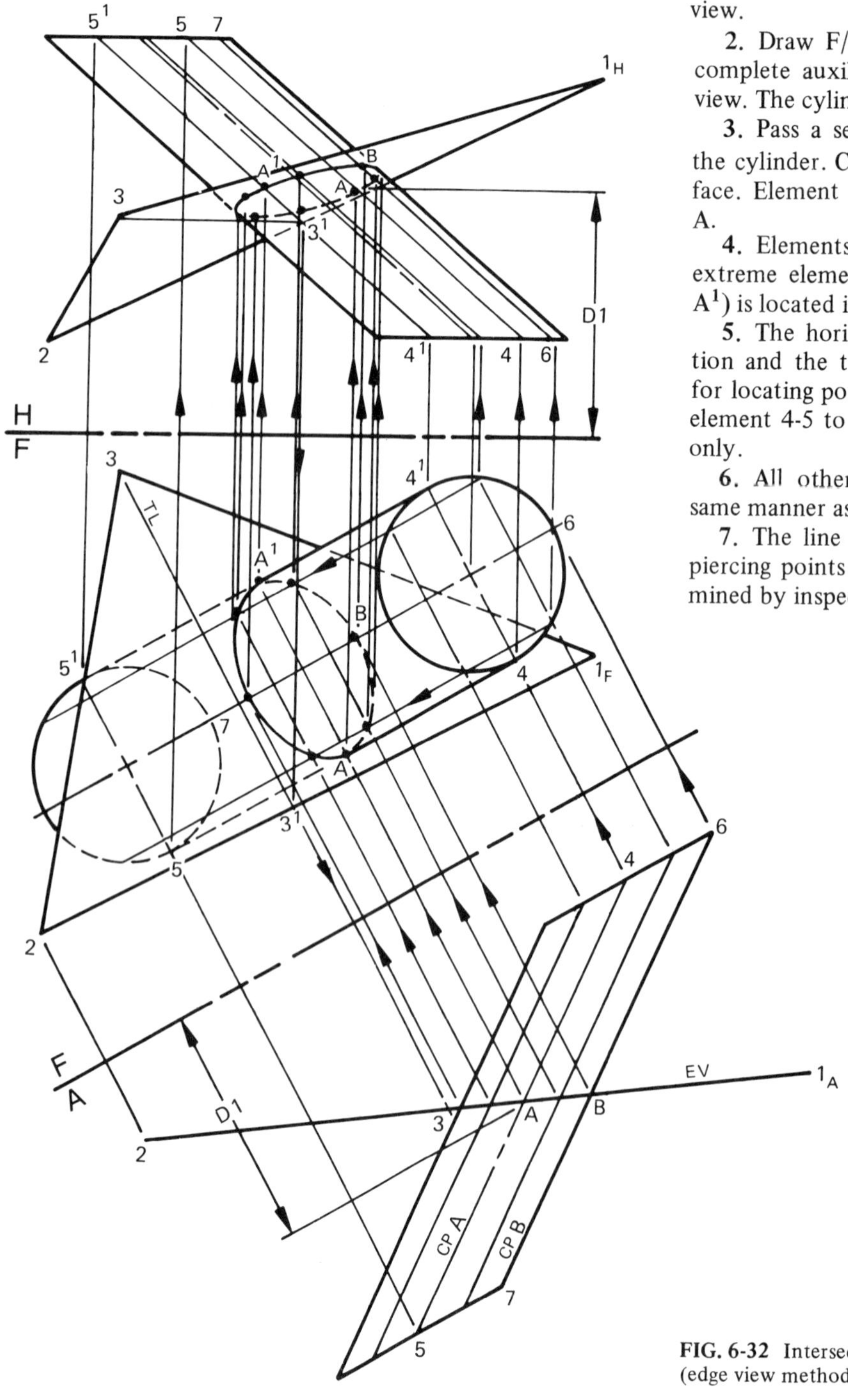

1. An auxiliary view projected from the frontal view shows the cylinder bases as edges since they are true shape in the frontal view. Frontal line $3\text{-}3^1$ is drawn parallel to H/F in the horizontal view and true length in the frontal view.

2. Draw F/A perpendicular to frontal line $3_F\text{-}3_F^1$, and complete auxiliary A. Plane $1_A\text{-}2_A\text{-}3_A$ appears as an edge view. The cylinder is oblique, with its bases as edges.

3. Pass a series of cutting planes parallel to the axis of the cylinder. CP A cuts two elements on the cylinder's surface. Element $4_A\text{-}5_A$ intersects the plane at piercing point A.

4. Elements 4 and 5 (and $4^1\text{-}5^1$) correspond to the two extreme elements in the frontal view. Point A (and point A^1) is located in the frontal view by projection.

5. The horizontal view of point A is located by projection and the transferring of distance D1. Another method for locating point A_H would be to project its corresponding element 4-5 to the horizontal view and locate by projection only.

6. All other piercing points can be established in the same manner as described in the above steps.

7. The line of intersection is drawn by connecting the piercing points in their proper sequence. Visibility is determined by inspection of auxiliary A.

FIG. 6-32 Intersection of an oblique plane and an oblique cylinder (edge view method).

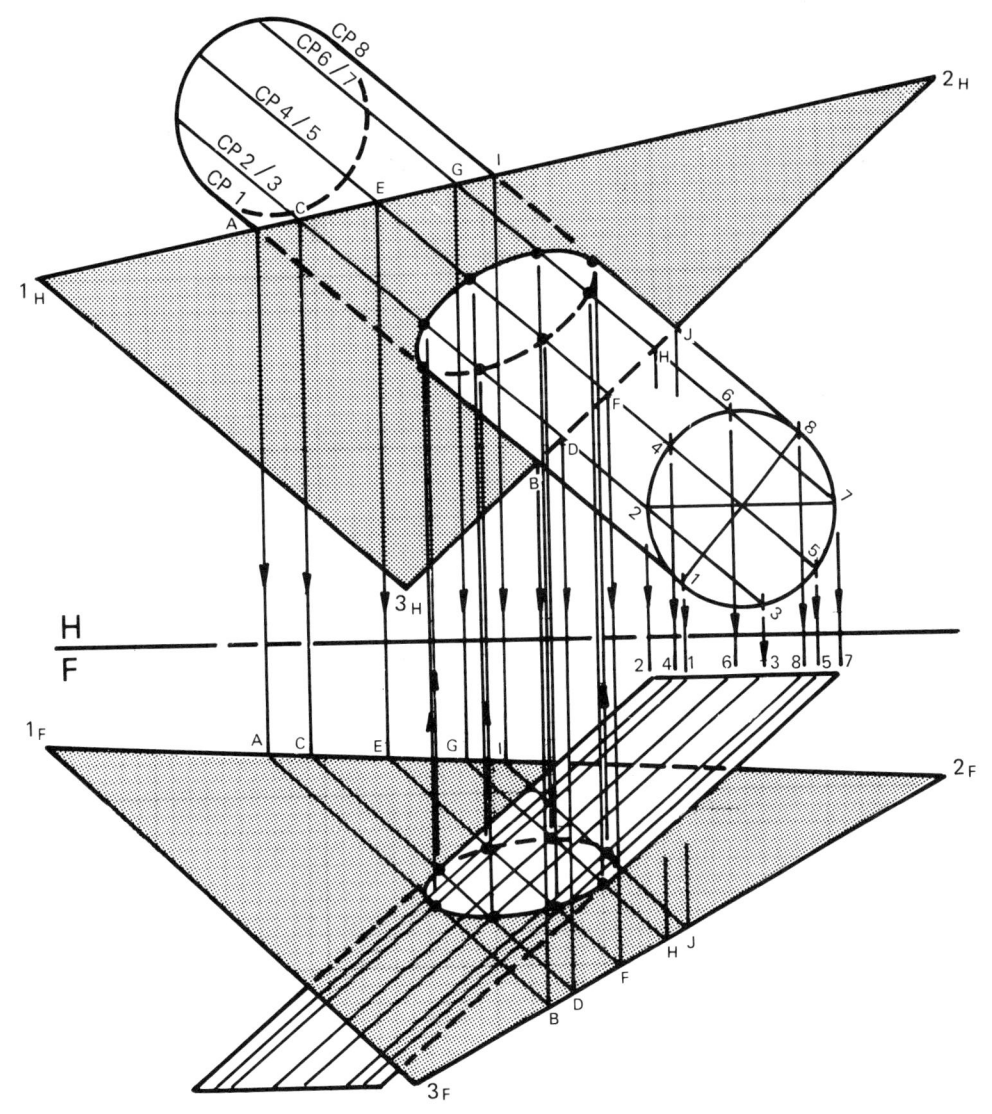

FIG. 6-33 Intersection of an oblique plane and an oblique cylinder (cutting plane method).

6.20 INTERSECTION OF AN OBLIQUE PLANE AND AN OBLIQUE CYLINDER (CUTTING PLANE METHOD)

By passing a series of cutting planes parallel to the cylinder's axis, the corresponding elements on the cylinder's surface pierce the given plane and establish a line of intersection.

In Fig. 6-33, plane 1-2-3 and the cylinder are given. The line of intersection and proper visibility are required.

1. Establish a series of cutting planes through the cylinder and parallel to its axis line. Project their corresponding elements to the frontal view and label as shown.

2. CP 1 intersects line 1_H-2_H of the plane at point A and line 2_H-3_H at point B. Project points A and B to the frontal view. Points A and B form line A_F-B_F.

3. Line A_F-B_F intersects element 1 and locates the piercing point of the element and the plane.

4. CP 2/3 intersects plane 1_H-2_H-3_H at points C and D, which are then projected to the frontal view and form line C_F-D_F.

5. Line C_F-D_F intersects element 2 and element 3 thus establishing the piercing points of elements 2 and 3 and the plane.

6. This process is repeated for each cutting plane.

7. Connect the piercing points and determine visibility. Where element 1 crosses line 1_H-2_H in the horizontal view at point A, draw a sight line to the frontal view. Since the projector from point A intersects line 1_F-2_F first, line 1_H-2_H is solid in the horizontal view. Repeat this procedure at each crossing of the plane and the extreme elements of the cylinder in both views.

233

PROB. 6-9

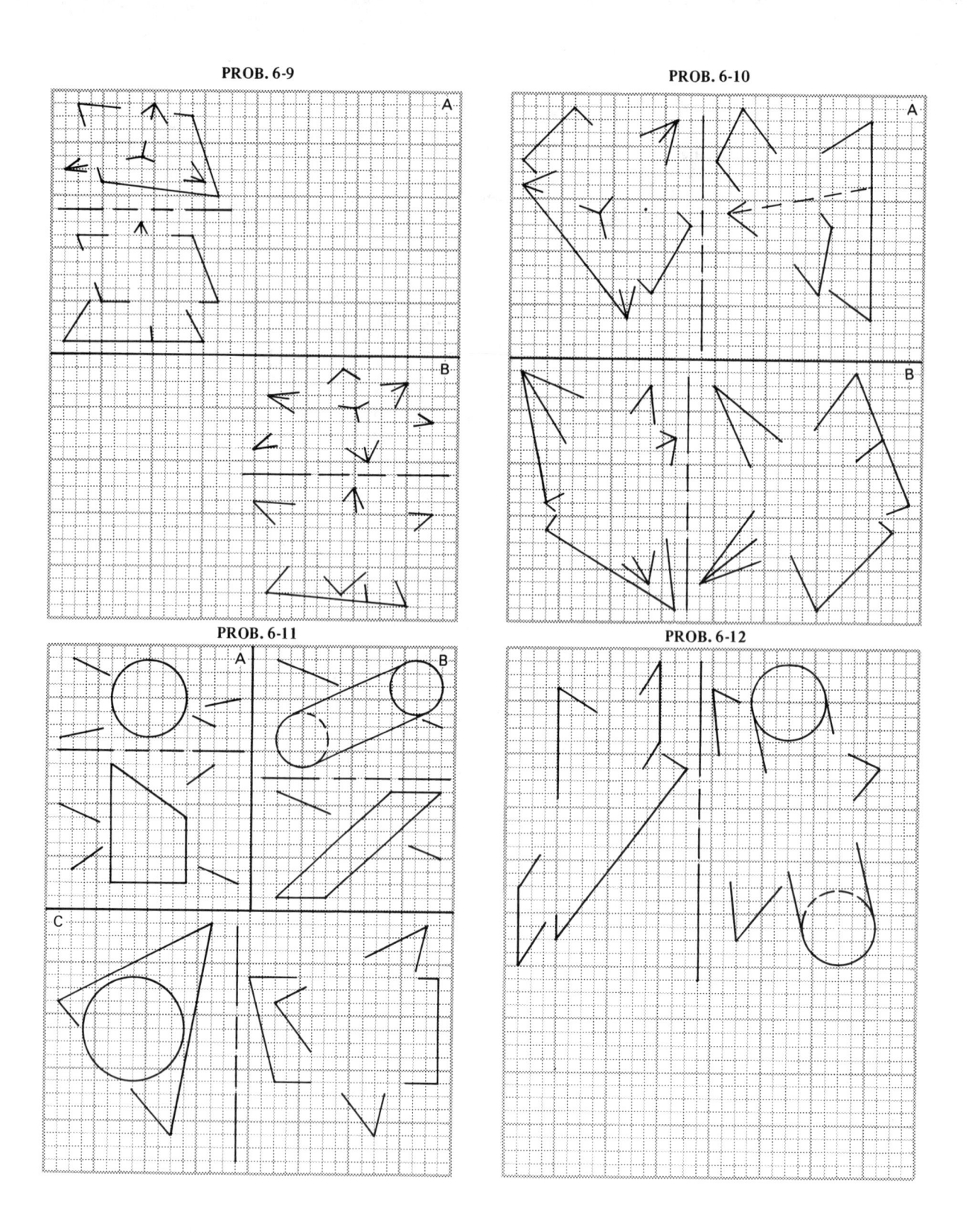

PROB. 6-10

PROB. 6-11

PROB. 6-12

PROBS. 6-9A, 6-9B. Solve for the intersection of the pyramid and the plane by use of the edge view method.

PROBS. 6-10A, 6-10B. Determine the intersection of the oblique plane and pyramid using the cutting plane method.

PROBS. 6-11A, 6-11B. Complete the two views of the line and cylinder and determine the piercing points and visibility.

PROB. 6-11C. Complete the intersection of the plane and the cylinder using the cutting plane method.

PROB. 6-12. Determine the intersection of the plane and the oblique cylinder. Use the cutting plane method and then check the answer by projecting an auxiliary edge view.

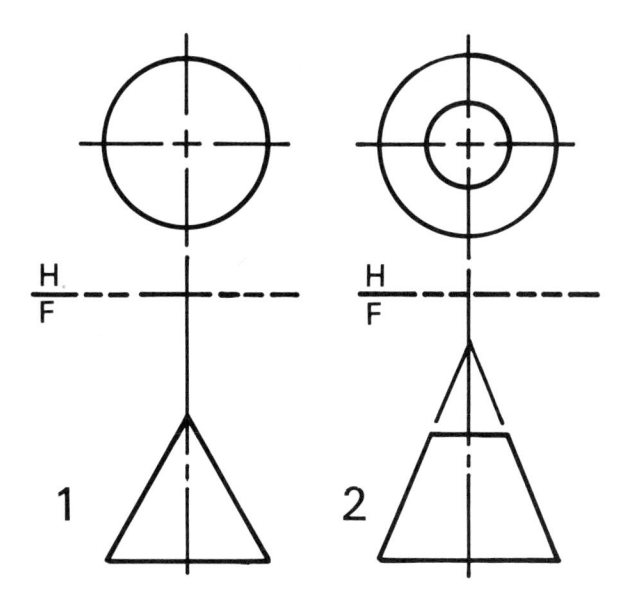

FIG. 6-34 Right circular cones (cones of revolution).

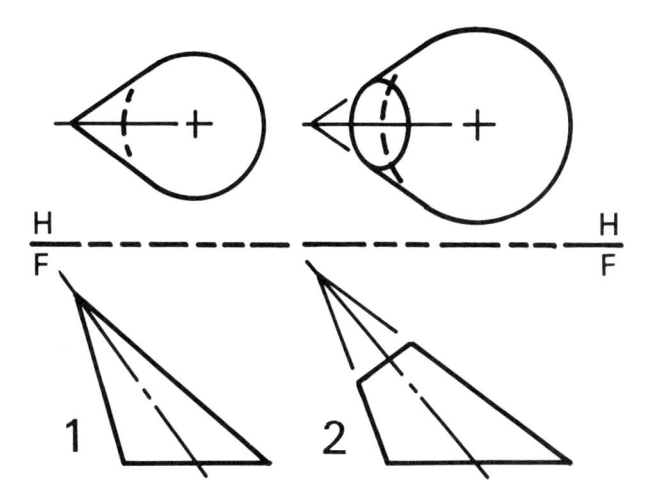

FIG. 6-35 Elliptical cone.

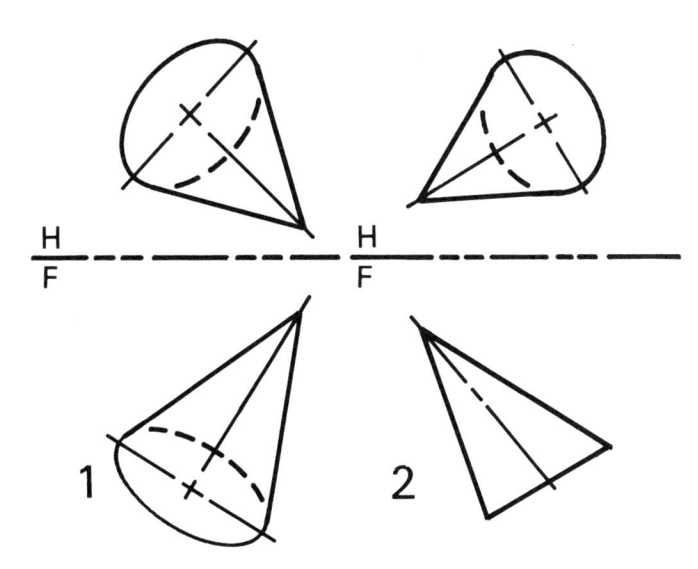

FIG. 6-36 Oblique views of cones.

6.21 CONES

*A **cone** is a single-curved surface* formed by line segments/elements connecting the vertex with all points on the perimeter of the base. *A **cone** is generated by the movement of a straight line element passed through the vertex and moving around the boundary of the base.* A cone generated by a right triangle rotating about one of its legs is a *right cone* or *cone of revolution,* Fig. 6-34 and Fig. 6-37. If a right section cut from the cone is an ellipse, the cone is an *elliptical cone.* A cone with a circular base whose right section is an ellipse is sometimes referred to as an *oblique circular cone,* Fig. 6-35. If a cone is cut below its vertex, it is termed a *truncated cone,* Fig. 6-34(2) and Fig. 6-35(2).

Cones are represented by two extreme elements, an axis line and a base of a defined shape. Where the cone is viewed perpendicular to its axis, Fig. 6-34, the axis line appears as a point and the base as a circle or ellipse. Normally, elements are not shown when the projection is perpendicular to the axis. Only the two extreme elements are needed in any given view which is not perpendicular to the axis line. A cutting plane passed parallel to the axis of a cone cuts two elements on the cones surface. Any number of useful elements may be added to the cone in order to find a solution to a problem. An oblique view of a cone results in the foreshortening of any element not parallel to the projection plane, Fig. 6-36.

FIG. 6-37 This delta rocket is composed of cylindrical and conical shapes. *(Courtesy NASA.)*

FIG. 6-38 Point on a cone.

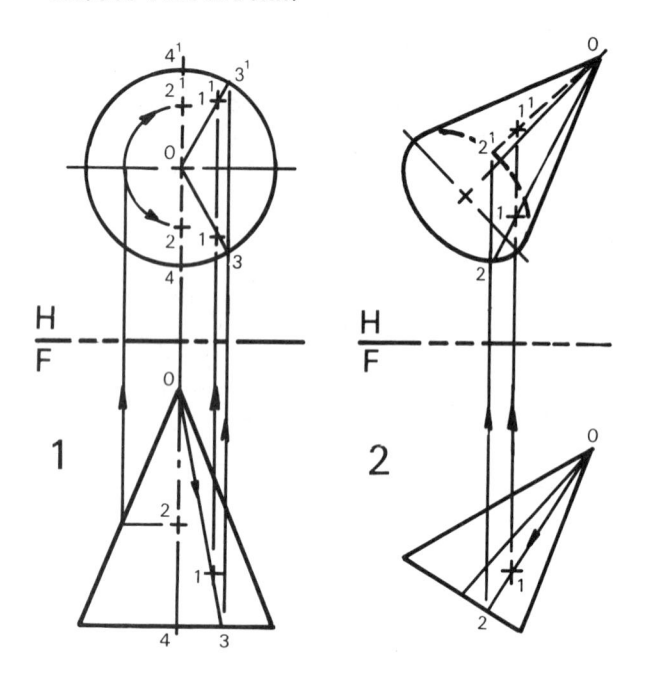

6.22 INTERSECTION OF A LINE AND A CONE

In order to understand how to locate the piercing point of a line and a cone, it is necessary to be able to locate a point on a cone. In Fig. 6-38(1), points 1 and 2 are given in the frontal view. A point which lies on a cone also lies on a line/element of the cone. Elements 0_F-3_F and 0_F-4_F are drawn through point 1_F and point 2_F respectively and projected to the horizontal view. Points 1 and 2 are projected to the horizontal view and located at the intersection of their projection lines and their corresponding element. Point 2_F is projected to the extreme element and then to the horizontal view since it lies inline with the vertical axis line. Two possible positions for each point exist, depending on whether it is hidden or visible in the frontal view. In Fig. 6-38(2), element 0_F-2_F is passed through point 1_F and projected to the horizontal view. Point 1 is located by projection.

The intersection of a line and a cone can be located by passing a cutting plane containing the line, through the cone's vertex and its base. This cutting plane cuts two straight line elements on the cone, which intersect the line where the line pierces the cone. In Fig. 6-39, line 1-2 and the right cone are given.

1. Construct a convenient-sized cutting plane containing line 1_F-2_F and the vertex 0_F. Extend the plane until it intersects the plane of the cone's base. Cutting plane 0_F-3_F-4_F intersects line 1_F-2_F at points 5_F and 6_F.

2. Project points 5 and 6 to the horizontal view. Draw a line from point 0_F through point 5_H until it intersects the projection line extended from point 3_F in the frontal view. Repeat this process to complete cutting plane 0_H-3_H-4_H.

3. Line 3_H-4_H lies in the plane of the cone's base and intersects the curve of the base at points 7_H and 8_H. Project points 7 and 8 to the frontal view. Elements 0-7 and 0-8 intersect line 1-2 at PP1 and PP2 respectively. Project PP 1 and PP 2 to the H view.

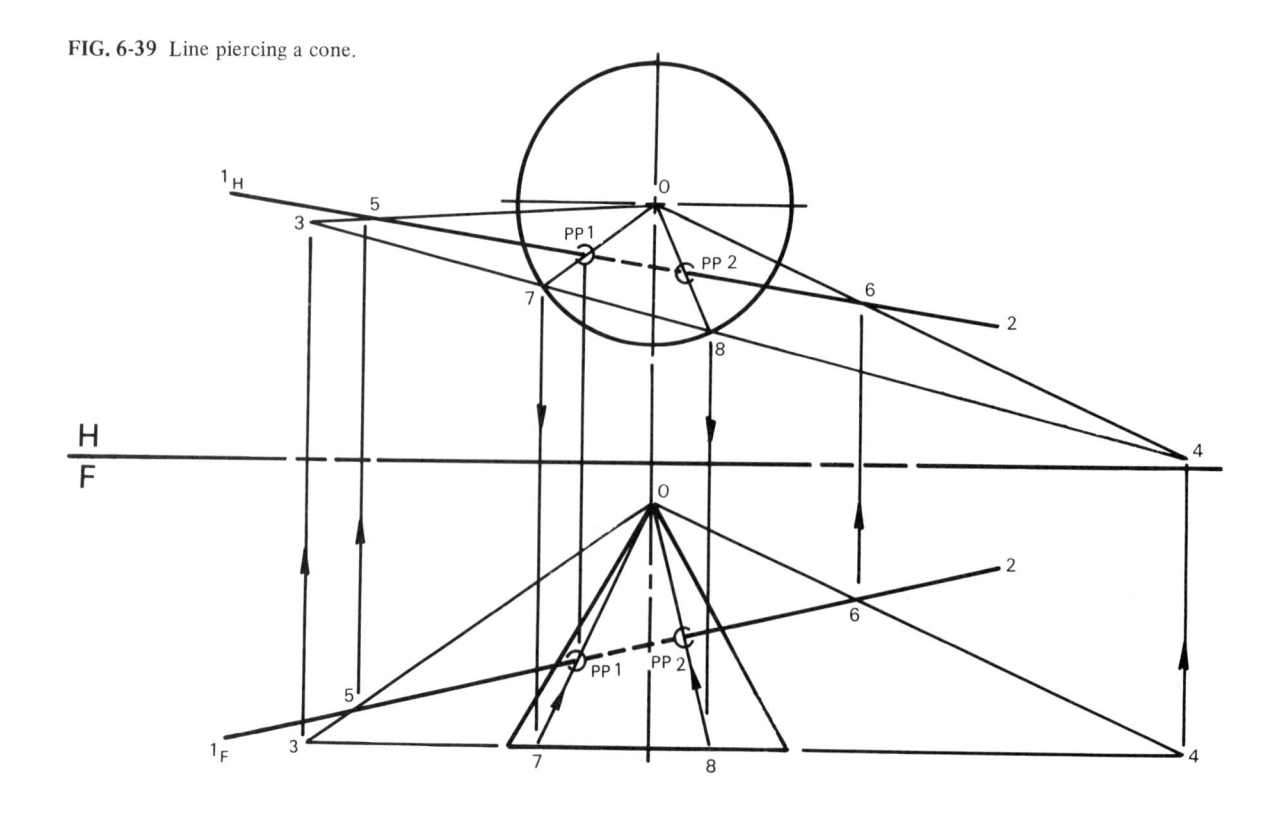

FIG. 6-39 Line piercing a cone.

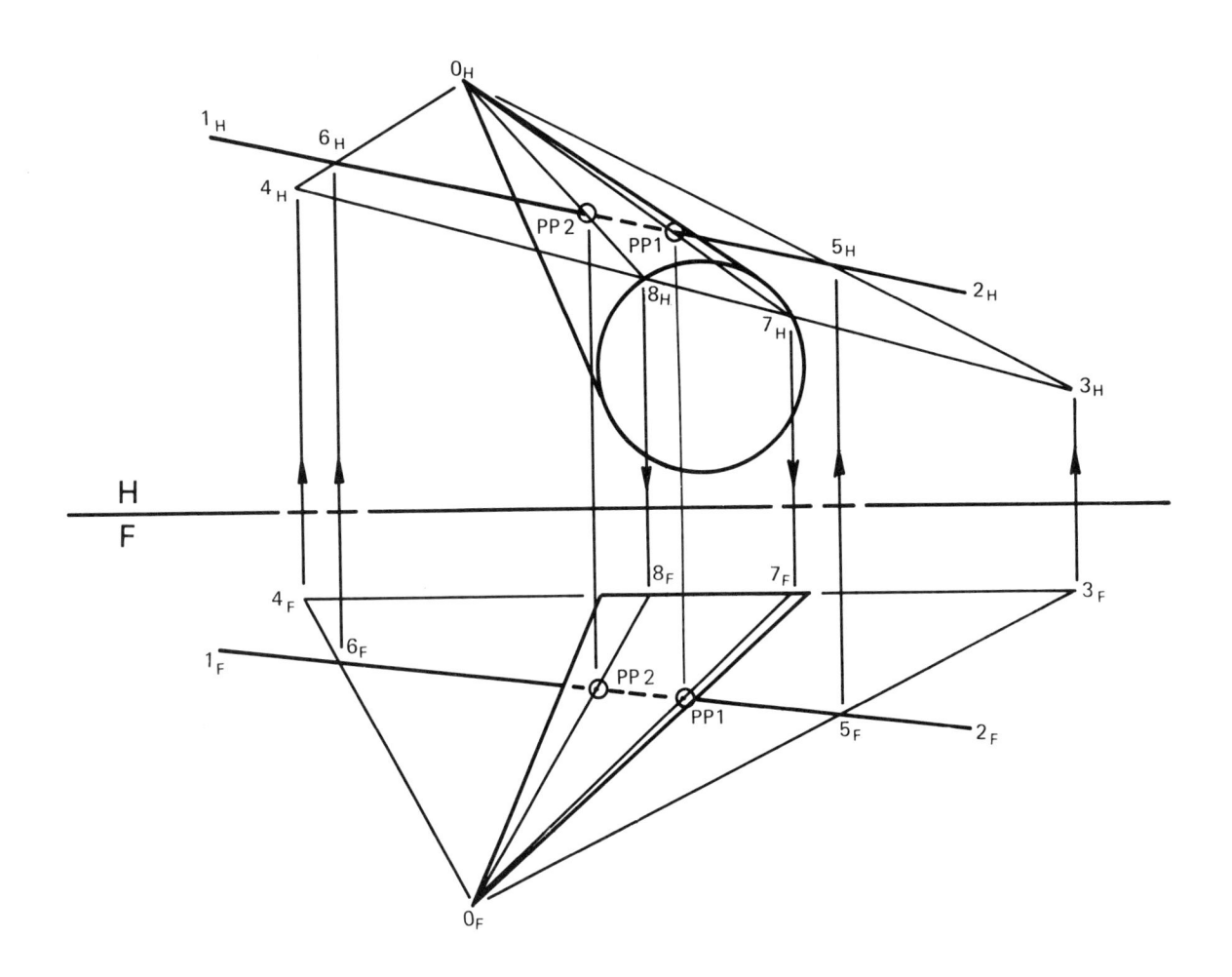

FIG. 6-40 Intersection of a line and an oblique cone.

6.23 INTERSECTION OF A LINE AND AN OBLIQUE CONE

The intersection of a line and a cone can be determined by passing a cutting plane containing the line and through the vertex and base of the cone. This cutting plane cuts two straight line elements on the cone. The piercing points of the line and the cone are located at the intersection of each element and the line. If a cutting plane containing the line and through the vertex does not pass through (intersect) the base, then the line does not intersect the cone.

In Fig. 6-40, line 1-2 and the oblique cone are given. The points at which the line pierces the cone are required.

1. Construct a convenient-size cutting plane containing line 1_F-2_F and through vertex 0_F. Extend the plane until it intersects the plane of the cone's base in the frontal view. Note that the cutting plane can be any size as long as it contains the line. If space permits, the plane can be constructed through the end points of the line. Normally the plane will be drawn as in the example, through two convenient positions on the line so that its base will lie on the base of the cone.

2. Plane 0_F-3_F-4_F contains line 1_F-2_F and intersects it at points 5_F and 6_F. Project points 5 and 6 to the horizontal view. Form the cutting plane by drawing lines from the vertex through points 5 and 6 until they intersect projection lines extended from points 3_F and 4_F in the frontal view.

3. Line 3_H-4_H lies on the plane of the cone's base and intersects the curve of the base at points 7_H and 8_H. Project points 7 and 8 to the frontal view.

4. Connect the vertex 0 with points 7 and 8 in both views. Lines 0-7 and 0-8 are elements of the cone. The intersection of each element and line 1-2 determines the piercing points of the line and the cone.

5. Element 0-7 intersects line 1-2 at piercing point PP 1 and element 0-8 intersects line 1-2 at PP 2. Visibility is determined by the visibility test.

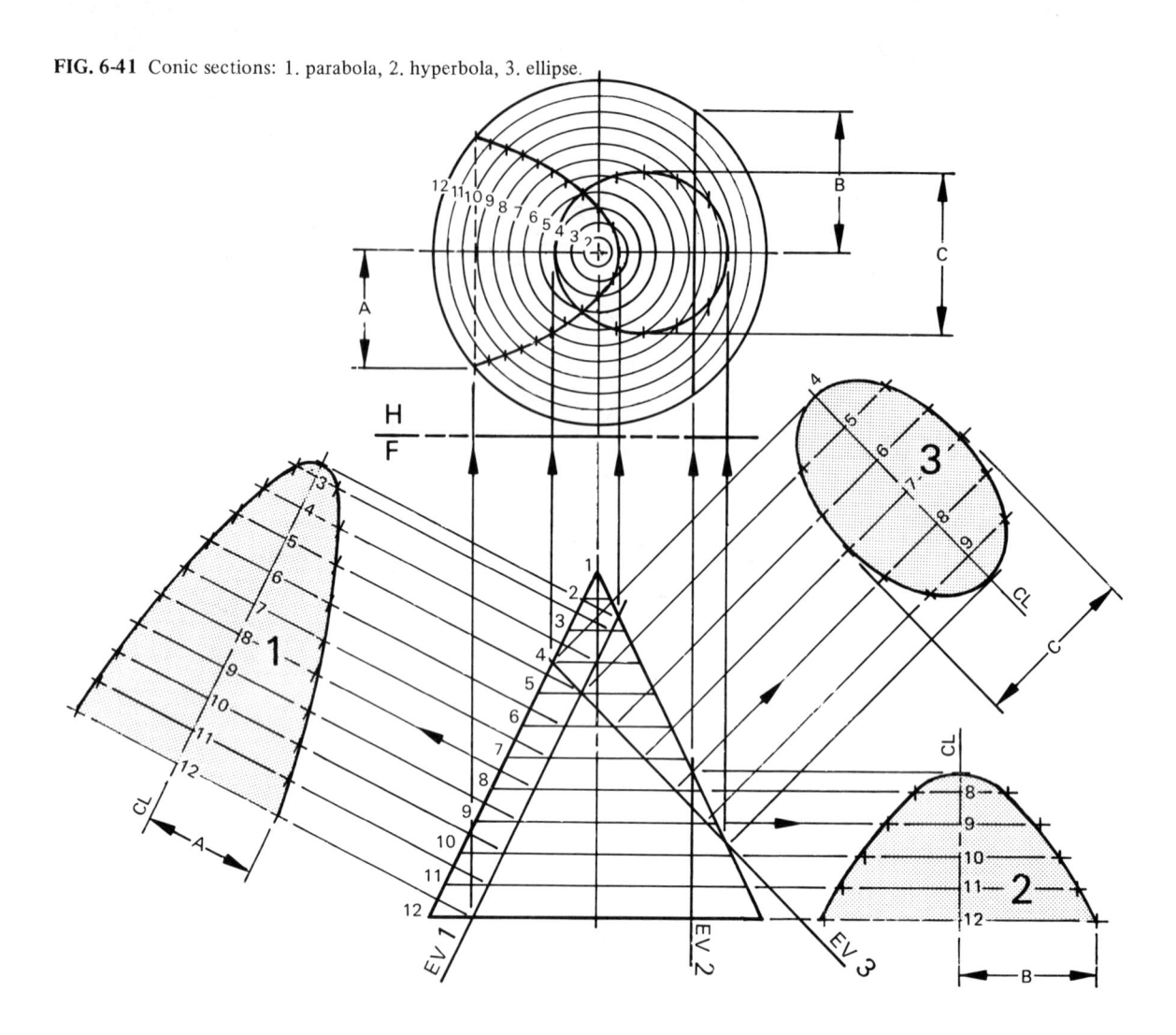

FIG. 6-41 Conic sections: 1. parabola, 2. hyperbola, 3. ellipse.

6.24 CONIC SECTIONS
(INTERSECTION OF A PLANE AND A CONE)

*The intersection of a plane and a right cone is called a **conic section**.* Five types of shapes can result from this intersection, Fig. 6-41.

1. **Parabola**: A plane passed parallel to an extreme element of the cone, therefore forming the same base angle, cuts a parabola (1).
2. **Hyperbola**: A plane passed through the cone, at a greater angle than the base angle results in a hyperbola (2).
3. **Ellipse**: A plane which cuts all the elements of the cone, but is not perpendicular to the axis, forms a true ellipse (3).
4. **Isosceles triangle**: A plane passed through the vertex cuts an isosceles triangle (the frontal view).
5. **Circle**: A plane passed perpendicular to the axis forms a circular intersection. A series of horizontal cutting planes have been introduced in the frontal view, which project as circles in the horizontal view.

The intersection of a cone and a plane is established by passing a series of horizontal cutting planes (perpendicular to the axis of the cone). In Fig. 6-41, the frontal and horizontal views of the cone are given along with the edge view of three planes which intersect the cone. The horizontal view and the true shape of each intersection are required.

1. Pass a series of evenly spaced horizontal cutting planes through the cone, CP 1 through CP 12.
2. Each cutting plane projects as a circle in the horizontal view.
3. EV 1 intersects cutting planes 3 through 12 in the frontal view. Project each intersection point to the horizontal view. The intersection of EV 1 and the cone forms a parabola.
4. The true shape of the parabola is seen in a view projected parallel to EV 1. The centerline of the parabola is drawn parallel to EV 1 and the intersection points of the plane (EV 1) and each cutting plane are projected from the frontal view. Distances are transferred from the horizontal view, as is dimension A.
5. Repeat steps 3 and 4 to establish the intersection of EV 2 and EV 3 with the cone. EV 2 projects as a line in the horizontal view and as a hyperbola in a true shape view (2). EV 3 forms an ellipse in the horizontal view and projects as a true ellipse in a true shape view (3).

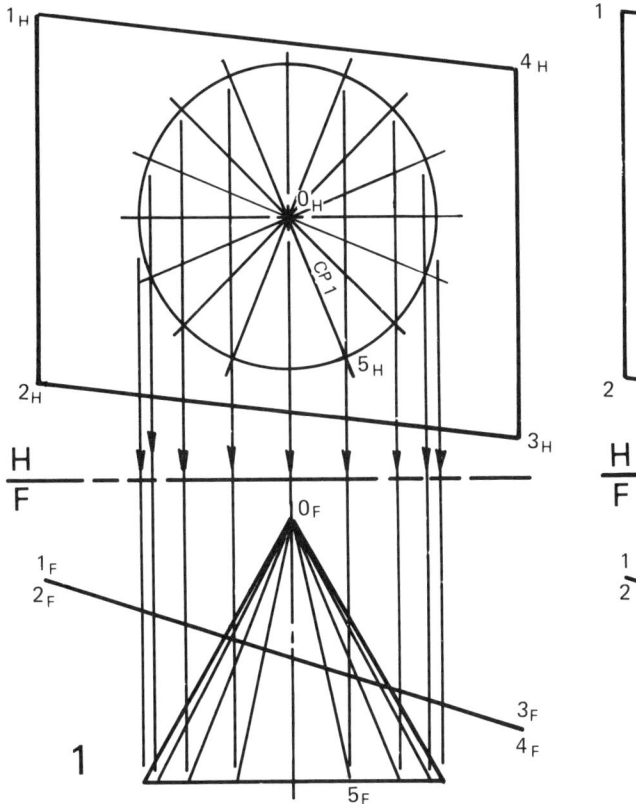

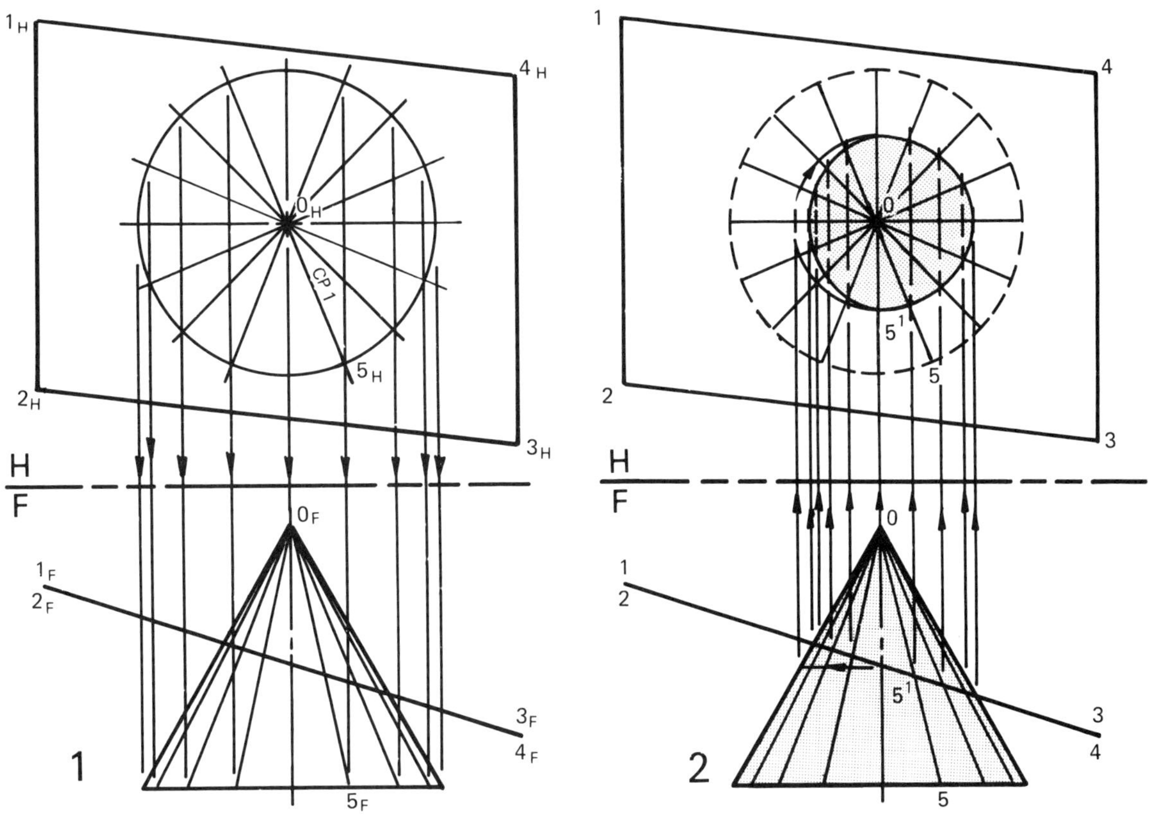

FIG. 6-42 Intersection of a plane and a cone.

6.25 INTERSECTION OF A PLANE AND A CONE

The intersection of a plane and a cone can be determined by passing a series of vertical cutting planes through the vertex of the cone. Each cutting plane intersects the cone as an element on its surface. The intersection of each element and the edge view of the given plane establish a series of piercing points which when connected form the line of intersection.

In Fig. 6-42, inclined plane 1-2-3-4 and the cone are given. The line of intersection is required.

1. Pass a series of vertical cutting planes through the vertex of the cone. Each cutting plane intersects the surface of the cone as a straight line element. Project each element to the frontal view where it forms a line on the cone, O_F-5_F.
2. The intersection of an element and the edge view of plane 1-2-3-4 establishes a piercing point, point 5_F^1. Project each piercing point to the horizontal view. Connect the piercing points of the elements and the plane to form the line of intersection. The portion of the cone above the plane in the frontal view is visible in the horizontal view.

The resulting horizontal view of the plane and the cone forms an ellipse. The true shape of this elliptical conic section can be seen in a view projected parallel to the edge view of the plane where it shows true shape.

The parabolic petal-type solar collector in Fig. 6-43 is an example of an applied conical design.

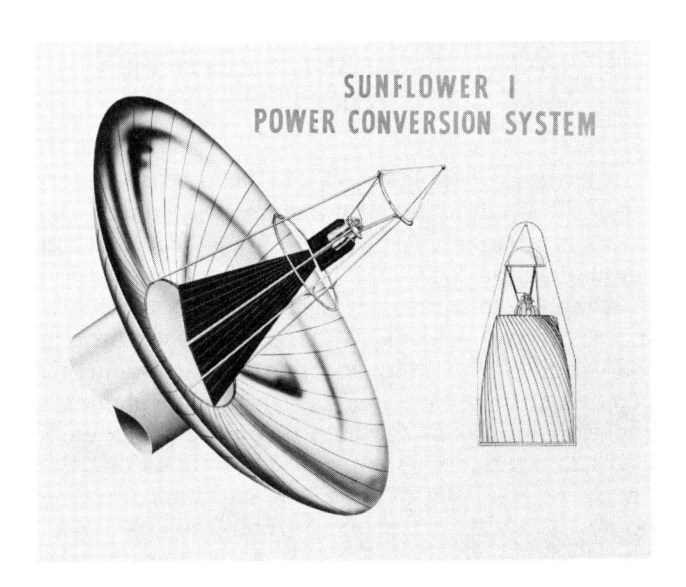

FIG. 6-43 Parabolic petal-type solar collector.

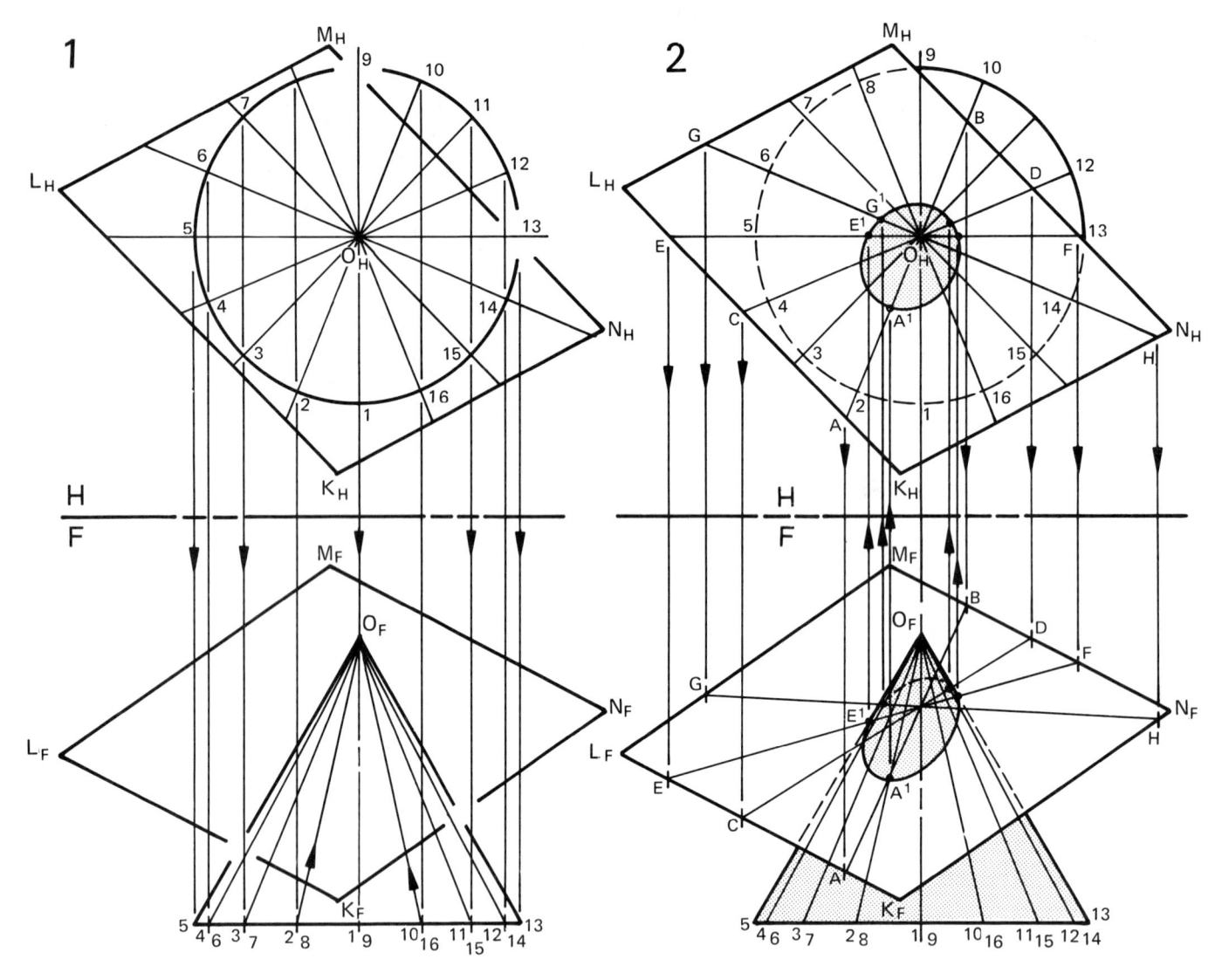

FIG. 6-44 Intersection of an oblique plane and a cone (cutting plane method).

6.26 INTERSECTION OF AN OBLIQUE PLANE AND A CONE (CUTTING PLANE METHOD)

The cutting plane method can be used to determine the intersection of a plane and a cone if the plane is oblique in its given views. This method eliminates the need for an auxiliary projection showing the plane as an edge. A series of evenly spaced vertical cutting planes are passed through the vertex of the cone, and the plane. The cutting planes intersect the cone as straight line elements and the plane as lines which lie on the plane. Each element intersects its corresponding line along the line of intersection of the plane and the cone. The point at which an element intersects its corresponding line on the plane locates a point on the line of intersection. This intersection point lies on the plane and on the cone's surface.

In Fig. 6-44, oblique plane K-L-M-N and the right cone are given and their intersection line is required.

1. Pass a series of evenly spaced vertical cutting planes through the cone's vertex. Project the elements to the frontal view and label as shown.

2. Each element corresponds to a cutting plane which intersects the cone and the plane. As an example, a cutting plane passed through the cone and intersecting the plane forms elements 0_H-2_H and 0_H-10_H, and also intersects the plane at points A_H and B_H. Points A_H and B_H form line A_H-B_H, which lies on the plane and represents the intersection of the cutting plane and the given plane. Line A-B is projected to the frontal view where it intersects element 0_F-2_F at piercing point A_F^1. Each line formed by the intersection of the cutting plane and the plane intersects two corresponding elements on the cone. The line of intersection is determined by connecting the piercing points. The piercing points are projected to the horizontal view to locate the line of intersection in that view. Visibility is then determined for both views.

PROB. 6-13

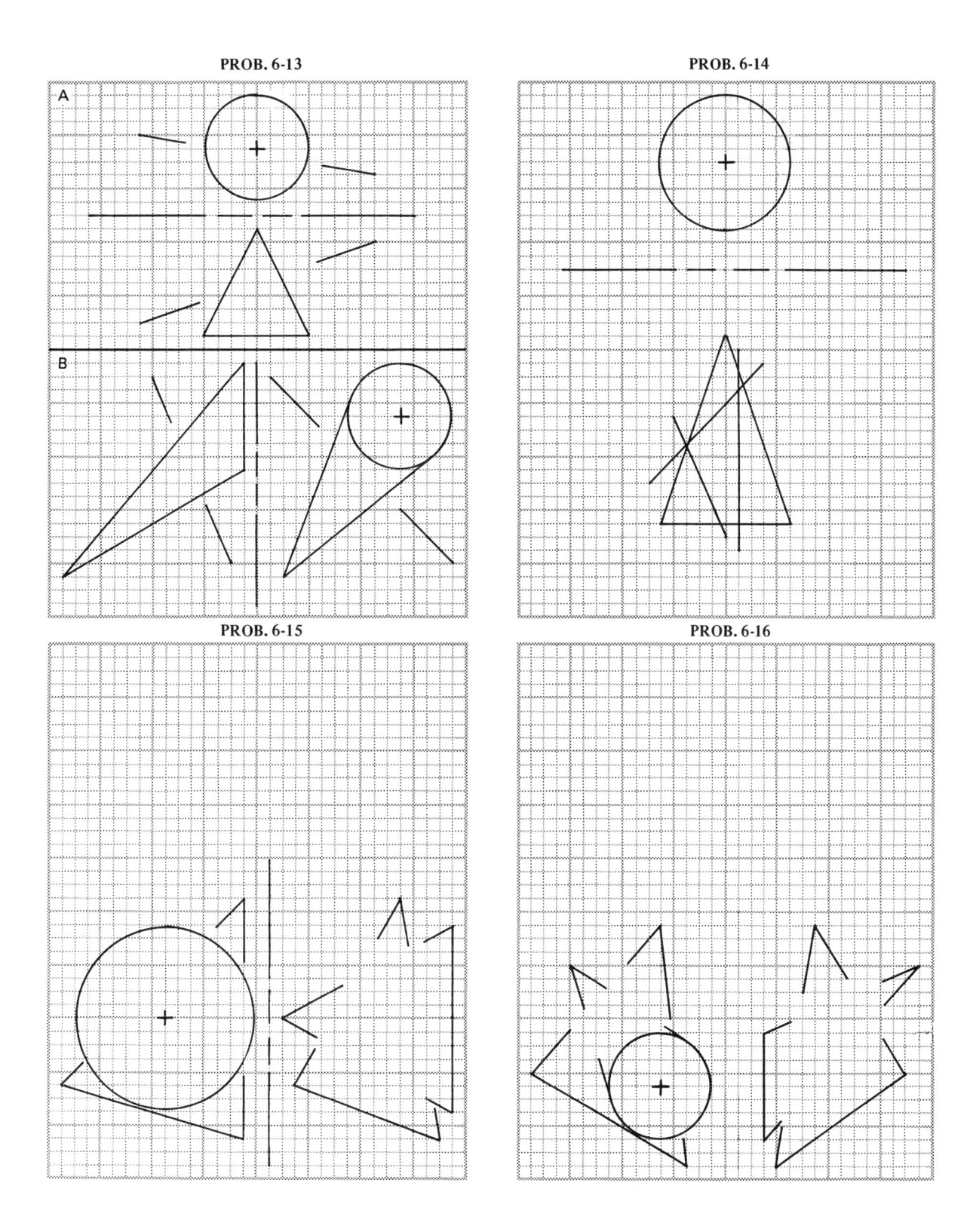

PROB. 6-14

PROB. 6-15

PROB. 6-16

PROBS. 6-13A, 6-13B. Solve for the piercing points and visibility of the line and cone.

PROB. 6-14. The front view of the right cone is cut by three separate planes. Show the resulting true shape conic sections and the intersection lines in the H view.

PROB. 6-15. By means of an auxiliary edge view solve for the intersection of the plane and the cone. Check the answer by using the cutting plane method.

PROB. 6-16. Using the edge view method determine the intersection between the plane and the cone. Project the auxiliary edge view so that the cone's base appears as an edge.

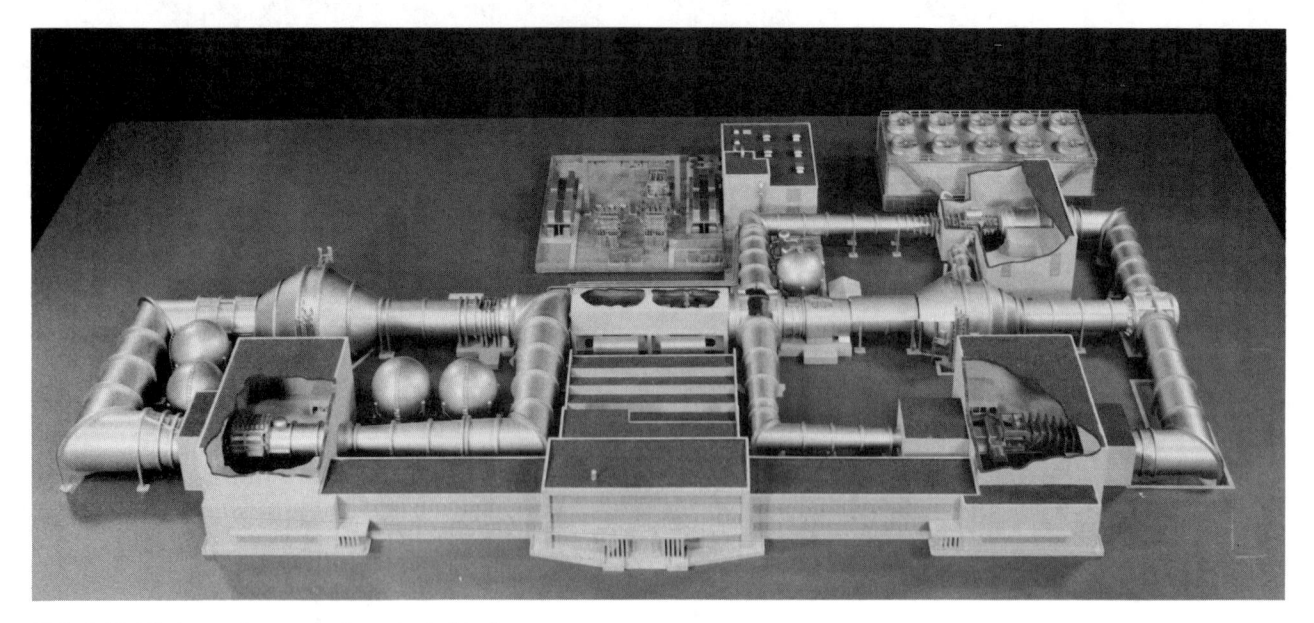

FIG. 6-45 Wind tunnel, note the five spherical tanks. *(Courtesy NASA.)*

6.27 SPHERES

A *sphere* can be defined as geometric form bounded by a surface containing all possible points at a given distance from its center. *A sphere is generated by moving a circle around an axis line which passes through its center.* Spheres are *double-curved surfaces* and contain no straight lines. Spheres are represented as circles equal to their diameter in all projections. The centerlines of the circles are normally shown in all views.

Spheres or portions of spheres are found in the design of a variety of industrial products, buildings, and vessels. In Fig. 6-45, the model of the wind tunnel shows five spherical fuel tanks. The dome of the nuclear containment building shown in Fig. 6-46 is a half-sphere (hemisphere).

A plane passed through the center of a sphere and at an angle to the adjacent projection plane creates an elliptical line of intersection, A and B in Fig. 6-47(1). This type of intersection is known as a *"great circle"* of a sphere.

A plane passed parallel to the adjacent projection plane and not through its center cuts a *"small circle,"* C and D in Fig. 6-47(2).

FIG. 6-46 Containment building with a hemisphere dome shape.

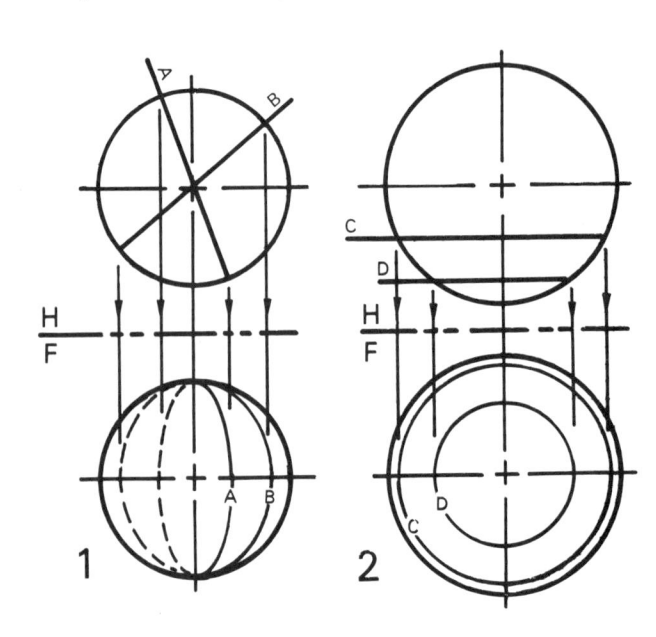

FIG. 6-47 Great and small circles of a sphere.

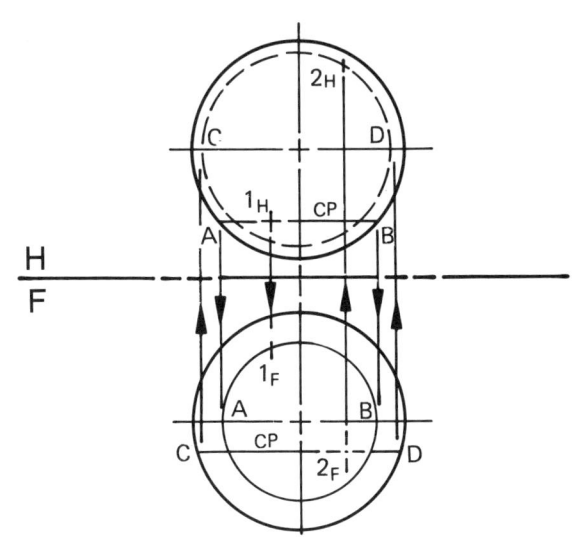

FIG. 6-48 Point on a sphere.

6.28 INTERSECTION OF A LINE AND A SPHERE

Before discussing the intersection of a line and a sphere it is important to know how to locate a point which lies on a sphere. The location of the point must be given in one view. A point lies on the sphere and on a line that lies on the sphere. Therefore a line which lies on the sphere is drawn containing the point. This line is a circle if a cutting plane containing the point is passed parallel to the adjacent fold line. In Fig. 6-48, point 1_H is given as visible in the horizontal view. Point 2_F is given as hidden in the frontal view.

1. Draw A-B parallel to H/F and through point 1_H. Project A-B to the frontal view where it appears as a "small circle". Project point 1 to the frontal view. Point 1_F is at the first intersection of the projection line and the small circle since it is visible in the horizontal view.
2. Draw cutting plane C-D through point 2_F and parallel to H/F, and project to the horizontal view. Point 2_H is located at the second intersection of small circle C-D and the projection line.

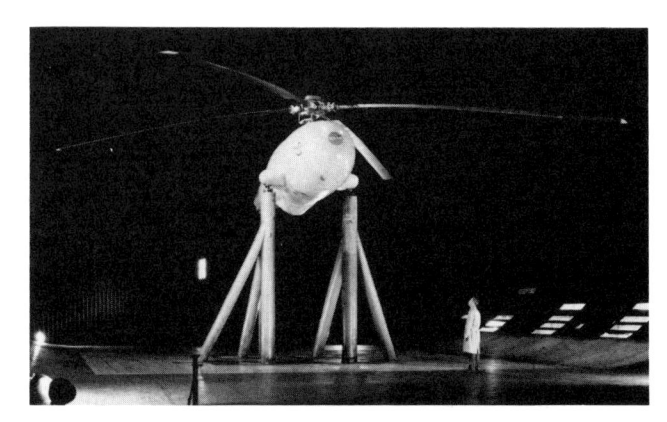

FIG. 6-49 The nose of this aircraft has a spherical shape. *(Courtesy NASA.)*

The intersection of a line and a sphere can be located by passing a cutting plane through the given line and projecting a view where the line is true length and the cutting plane appears as a small circle (true size). The required piercing points are located at the intersection of the small circle and the true length line.

In Fig. 6-50, line 1-2 and the sphere are given. The piercing points of the line and sphere, and proper visibility are required.

1. Pass a CP through line 1_H-2_H. The CP intersects the sphere at points A and B.
2. Draw H/A parallel to line 1_H-2_H and project auxiliary A. Line 1_A-2_A is true length and the cutting plane appears as small circle A-B.
3. Line 1_A-2_A intersects small circle A-B at piercing points 3_A and 4_A.
4. Project piercing points 3 and 4 to the horizontal and frontal views and show proper visibility.

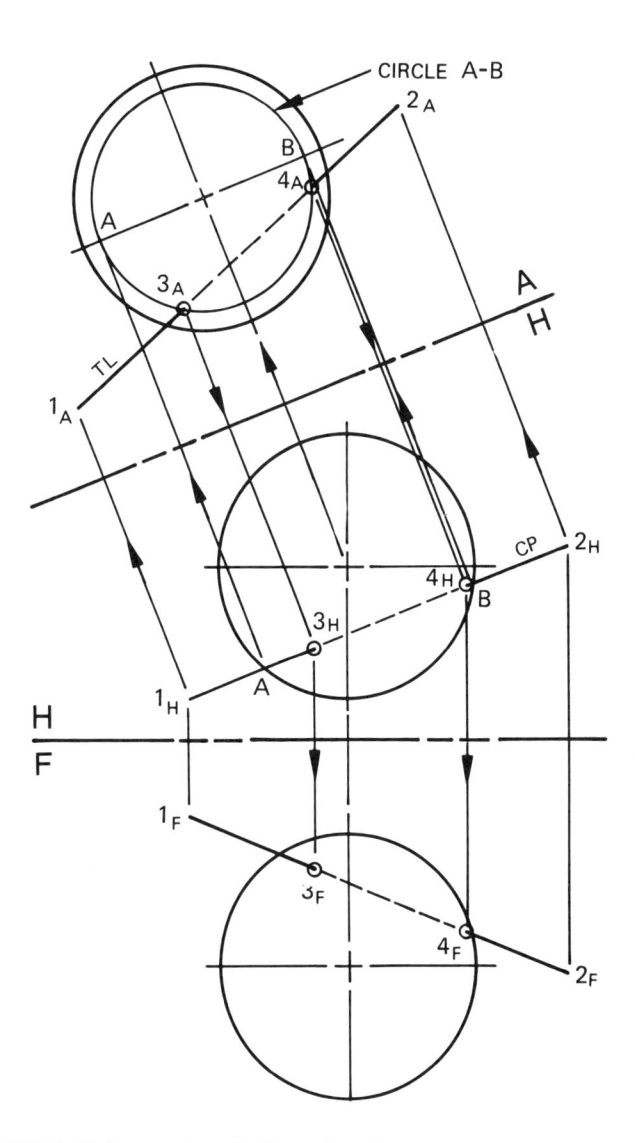

FIG. 6-50 Intersection of a line and a sphere.

243

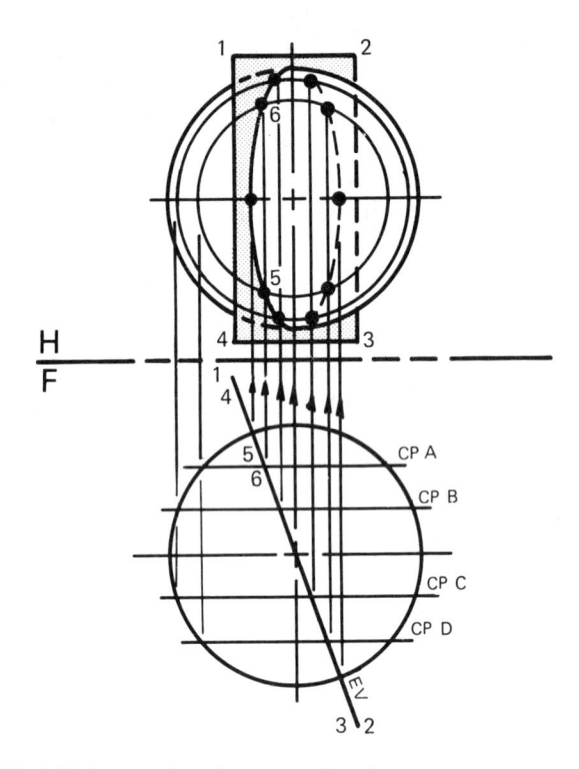

FIG. 6-51 Intersection of a plane and a sphere.

6.29 INTERSECTION OF A PLANE AND A SPHERE

The intersection of a plane and a sphere results in a *circular line of intersection*. If the plane is oblique, the line of intersection appears as an ellipse. The extreme piercing points and therefore the major and minor axes must first be found using the edge view or cutting plane method. The actual ellipse can be constructed by means of an ellipse template using the major and minor axes (1), plotting a series of piercing points established by cutting planes in a view showing the plane as an edge (2), or passing a series of cutting planes through the sphere and the plane where the plane is oblique (3).

In Fig. 6-51, the intersection of the sphere and plane is required.

1. Pass a series of evenly spaced horizontal cutting planes through the sphere and project to the horizontal view. Each CP cuts a small circle section.
2. Each CP intersects the edge view of the plane and locates two piercing points which are projected to the horizontal view to establish the line of intersection.

In Fig. 6-52, the cutting plane and edge view methods are shown. CPA through CPD are introduced in the frontal view, or in auxiliary A if the edge view method is used. Each cutting plane cuts small circle sections on the sphere in the horizontal view. Points A and F establish the minor axis of the ellipse in the horizontal view. The piercing points are projected from the intersection of the CPs and the edge view to the horizontal view as shown. The line of intersection in the frontal view is established by transferring distances and projection.

The cutting plane method establishes lines in the plane at the intersection of the CP and the oblique plane in the frontal view. Piercing points are determined in the horizontal view by the intersection of each small circle section and its corresponding line on the plane.

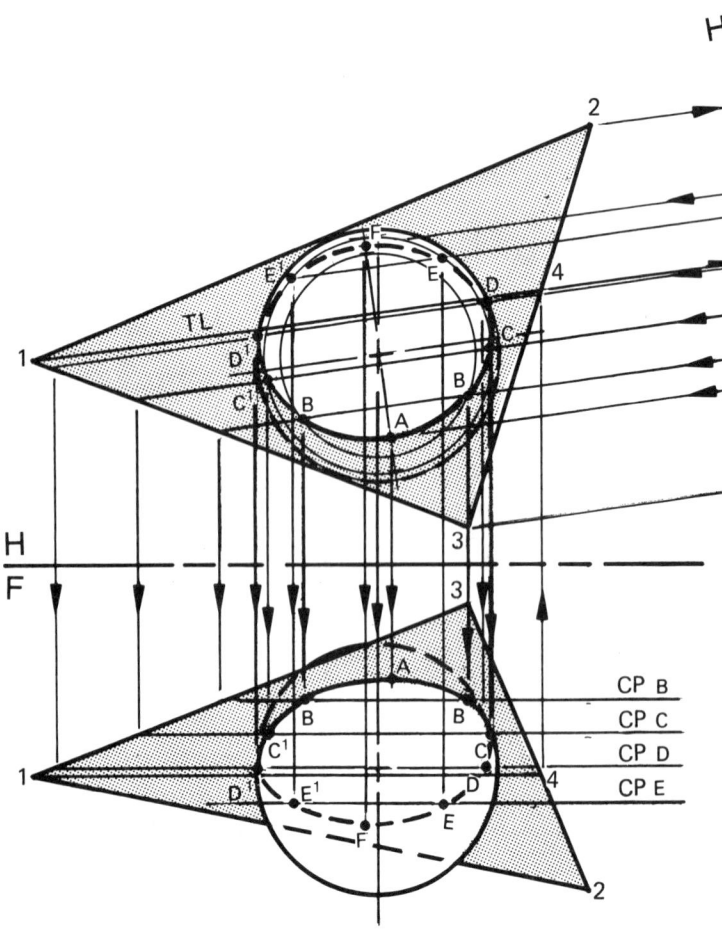

FIG. 6-52 Intersection of an oblique plane and a sphere.

FIG. 6-53 Intersection of a plane and a torus.

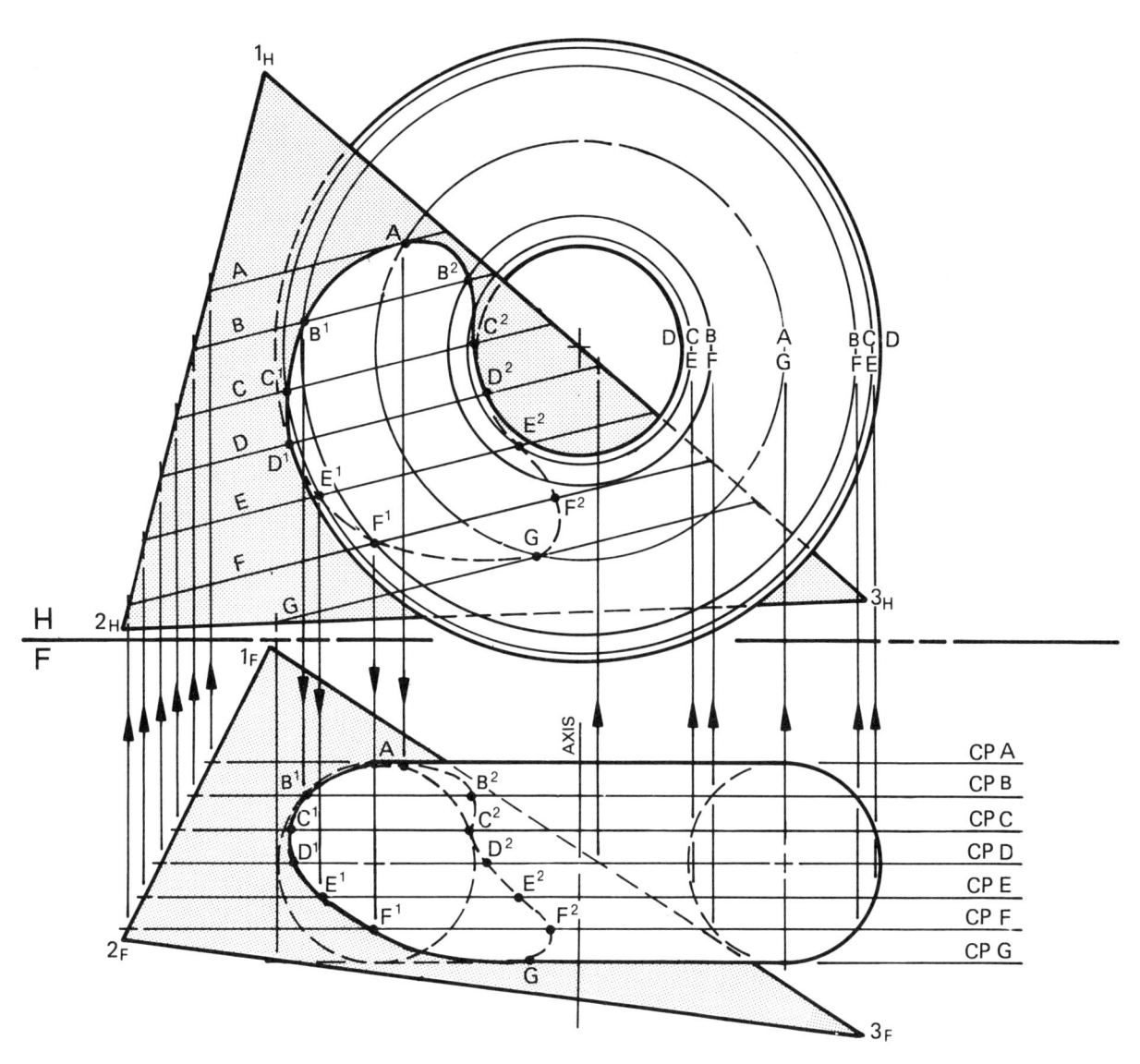

6.30 INTERSECTION OF A PLANE AND A TORUS

A torus is a double-curved surface of revolution generated by revolving a curve about an external axis. It commonly appears as a doughnut-shaped surface generated by a circle revolved about an axis in its plane which does not intersect the circle. An *annular torus* as shown in Fig. 6-53 has a closed circular generating line and a circle cross-sectional shape.

The intersection of a plane and a torus is established by passing a series of cutting planes perpendicular to its axis line. Cutting planes passed perpendicular to the axis line intersect the torus as circles and the plane as straight lines. The intersection of the circle elements and its corresponding straight line elements cut from the plane determines two piercing points along the line of intersection. This intersection represents a point common to both the plane and the torus, thus it lies on the line of intersection. Connecting the piercing points and determining proper visibility completes the problem.

In Fig. 6-53, plane 1-2-3 and the torus are given. The line of intersection is determined by passing a series of evenly spaced cutting planes perpendicular to the axis line in the frontal view (CP A through CP G).

Each intersection of a cutting plane and the torus is projected to the horizontal view, where it appears as a circle element. The intersection of each cutting plane and the given plane is also projected to the horizontal view where it appears as straight line element on the plane.

The intersection of a straight line element on the plane with its corresponding circle element on the torus establishes two piercing points along the line of intersection. CP A and CP G establish only one piercing point apiece since they lie along the upper and lower surface of the torus. Connect the piercing points as shown. Visibility is determined by the visibility test.

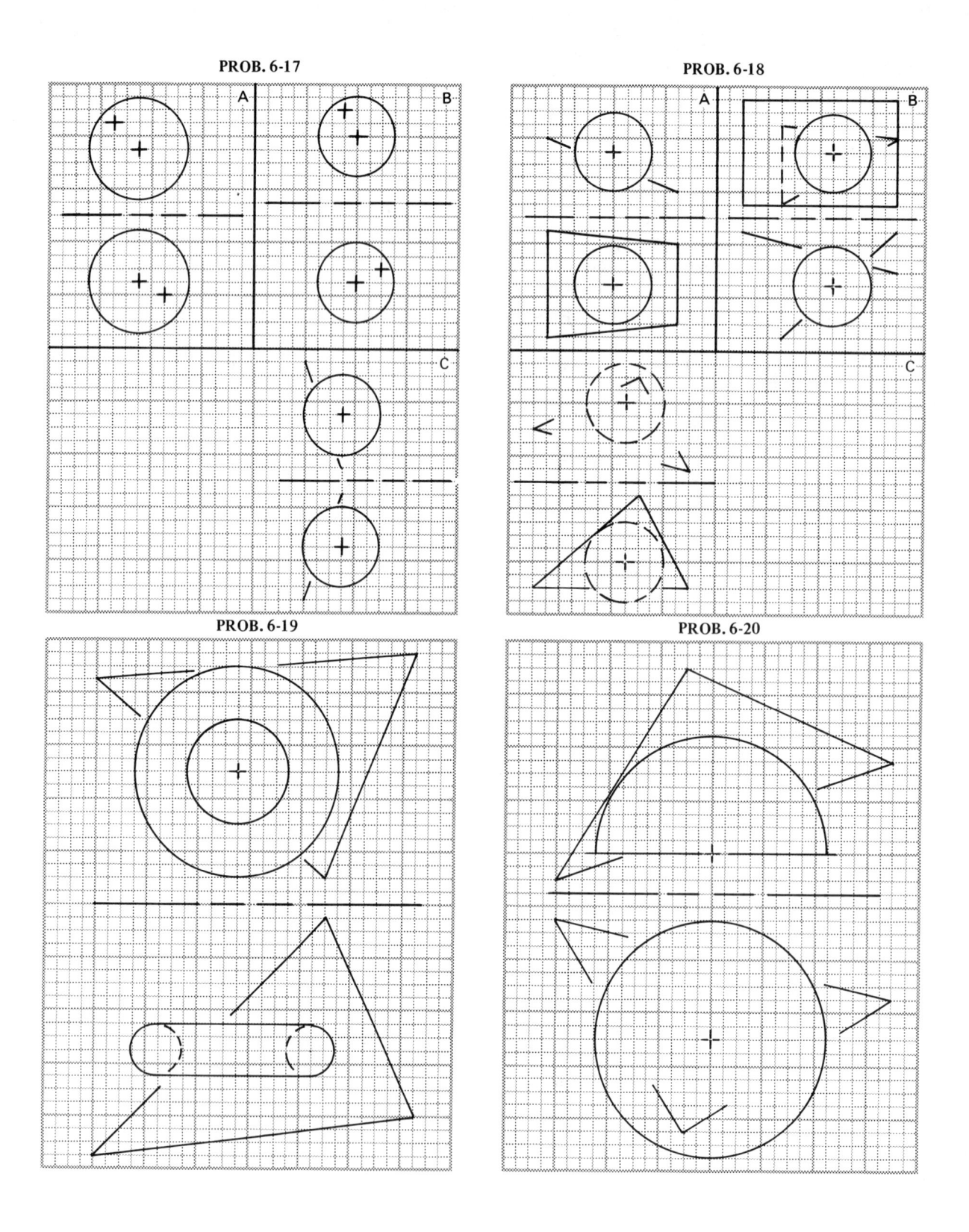

PROBS. 6-17A, 6-17B. Determine the position of the given points on the sphere. In problem A the given points are visible. In problem B the points shown are both hidden.
PROB. 6-17C. Using the edge view method, solve for the intersection of the line and the sphere.
PROB. 6-18A. Solve for the intersection of the plane and sphere.
PROB. 6-18B. Determine the visibility and intersection of each plane and the sphere.

PROB. 6-18C. Complete the views and solve for the intersection of the plane and sphere by means of an edge view projected from the horizontal view.
PROB. 6-19. Using multiple cutting planes determine the intersection between the oblique plane and torus and establish the correct visibility.
PROB. 6-20. Complete the two views and solve for the intersection between the plane and half sphere and establish the correct visibility.

FIG. 6-54 Intersection of two prisms (edge view method).

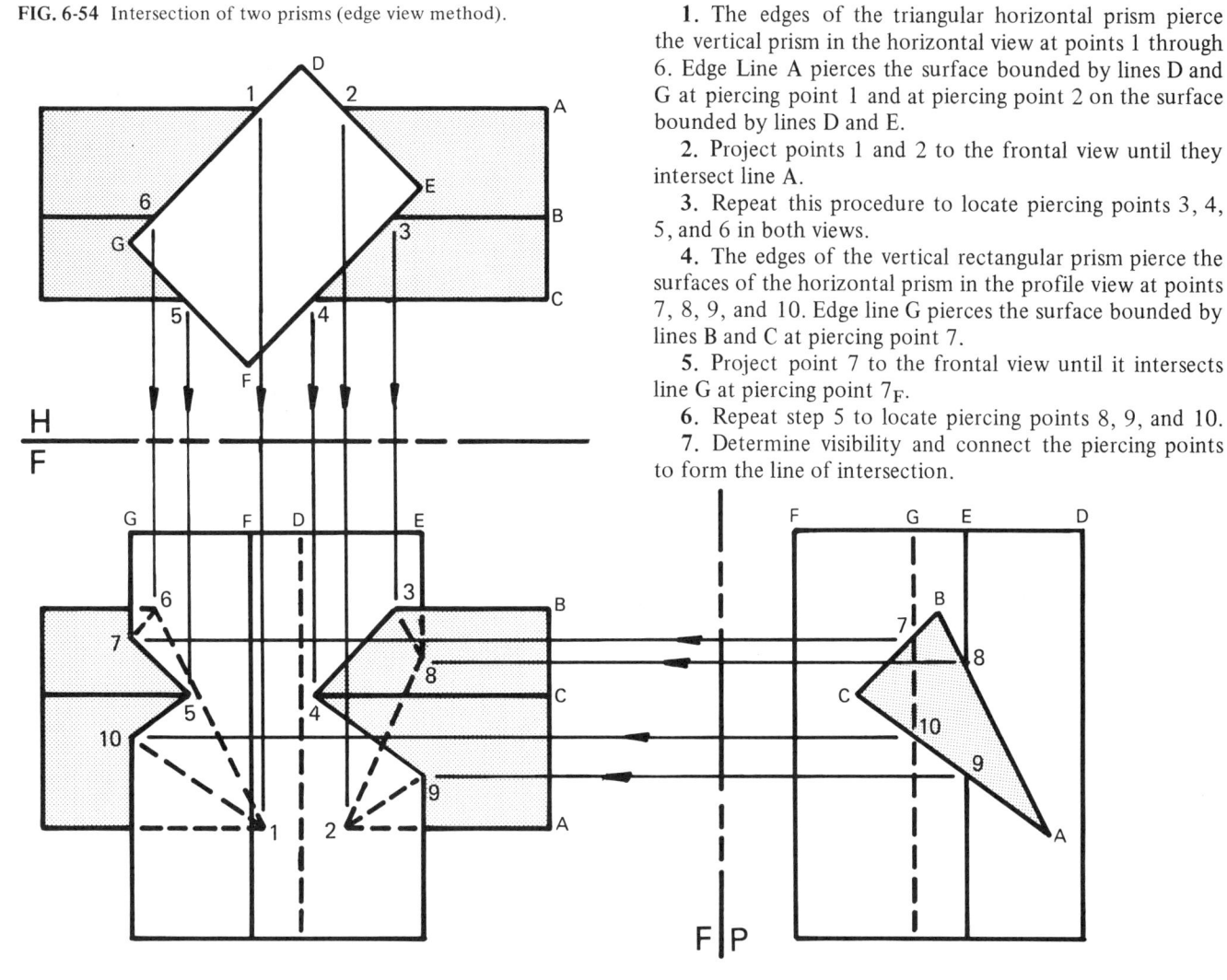

1. The edges of the triangular horizontal prism pierce the vertical prism in the horizontal view at points 1 through 6. Edge Line A pierces the surface bounded by lines D and G at piercing point 1 and at piercing point 2 on the surface bounded by lines D and E.

2. Project points 1 and 2 to the frontal view until they intersect line A.

3. Repeat this procedure to locate piercing points 3, 4, 5, and 6 in both views.

4. The edges of the vertical rectangular prism pierce the surfaces of the horizontal prism in the profile view at points 7, 8, 9, and 10. Edge line G pierces the surface bounded by lines B and C at piercing point 7.

5. Project point 7 to the frontal view until it intersects line G at piercing point 7_F.

6. Repeat step 5 to locate piercing points 8, 9, and 10.

7. Determine visibility and connect the piercing points to form the line of intersection.

6.31 INTERSECTION OF PRISMS

The intersection of two prisms can be determined by the edge view method. Each prism must appear as an edge in a given or auxiliary view. The piercing point of an edge line of one prism and a surface of the other prism can be established in a view where one of the prisms is an edge view. Basically, this type of problem can be reduced to finding the piercing point of a line and a plane. Where each edge line of a prism pierces a surface (plane) of the other prism, a point (piercing point) on the line of intersection is established. The line of intersection includes only surface lines of intersection, not those which will be "inside" the prisms.

In Fig. 6-54, two right prisms intersect at right angles. The horizontal view shows the edge view of rectangular prism and the profile view shows the triangular prism as an edge view.

The model of the telescope in Fig. 6-55 was designed and constructed using a variety of intersecting shapes.

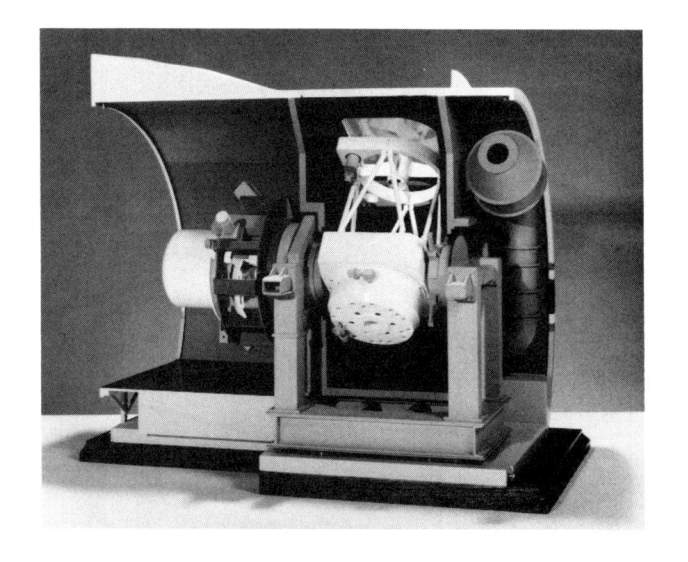

FIG. 6-55 Model of aircraft telescope. *(Courtesy NASA.)*

6.32 INTERSECTION OF TWO PRISMS (EDGE VIEW METHOD)

The line of intersection of two prisms is established by finding the piercing points of the edge lines of one prism with each surface of the other prism. This process is repeated using the lines of the second prism and the surfaces of the first. The intersection of prisms is theoretically the intersection of individual lines and planes or the intersection of two planes. Each prism must be shown as an edge view. If only one prism is given in edge view, an auxiliary view must be projected showing the other prism as an edge view. The edge view of a prism shows all of its surfaces as edges. The intersection of a line of one prism and the edge view of a surface of the other prism locates a piercing point. This piercing point is a common point for both prisms. Two properly connected piercing points form a line of intersection (common line). By properly connecting the piercing points of each line and surface, the line of intersection of the two prisms is established.

In Fig. 6-56, the horizontal and frontal views of the two prisms are given. The line of intersection is required.

1. Draw an auxiliary view showing the triangular prism as an edge view. Each of the edge lines of the triangular prism are frontal lines (true length in the frontal view). Therefore, draw F/A perpendicular to line 1 and project auxiliary A.

2. In the horizontal view, edge line 1 pierces the surface bounded by lines A and D at point 1_H^1. The surface bounded by lines A and B is pierced by line 2 at point 2_H^1 and line 3 at point 3_H^1.

3. Project points 1^1, 2^1, 3^1 to the frontal view to establish the end points of line 1, line 2 and line 3.

4. In auxiliary A, corner line A-A^1 intersects two of the surfaces of the triangular prism at point 4 and point 5.

5. Project piercing points 4 and 5 to the frontal view until they intersect corner line A_F-A_F^1.

6. Visibility is determined by inspection of the profile and horizontal view. Connect the piercing points in the proper sequence. In the frontal view intersection lines 3^1-2^1 and 2^1-4 are visible, all others are hidden.

FIG. 6-56 Intersection of two prisms.

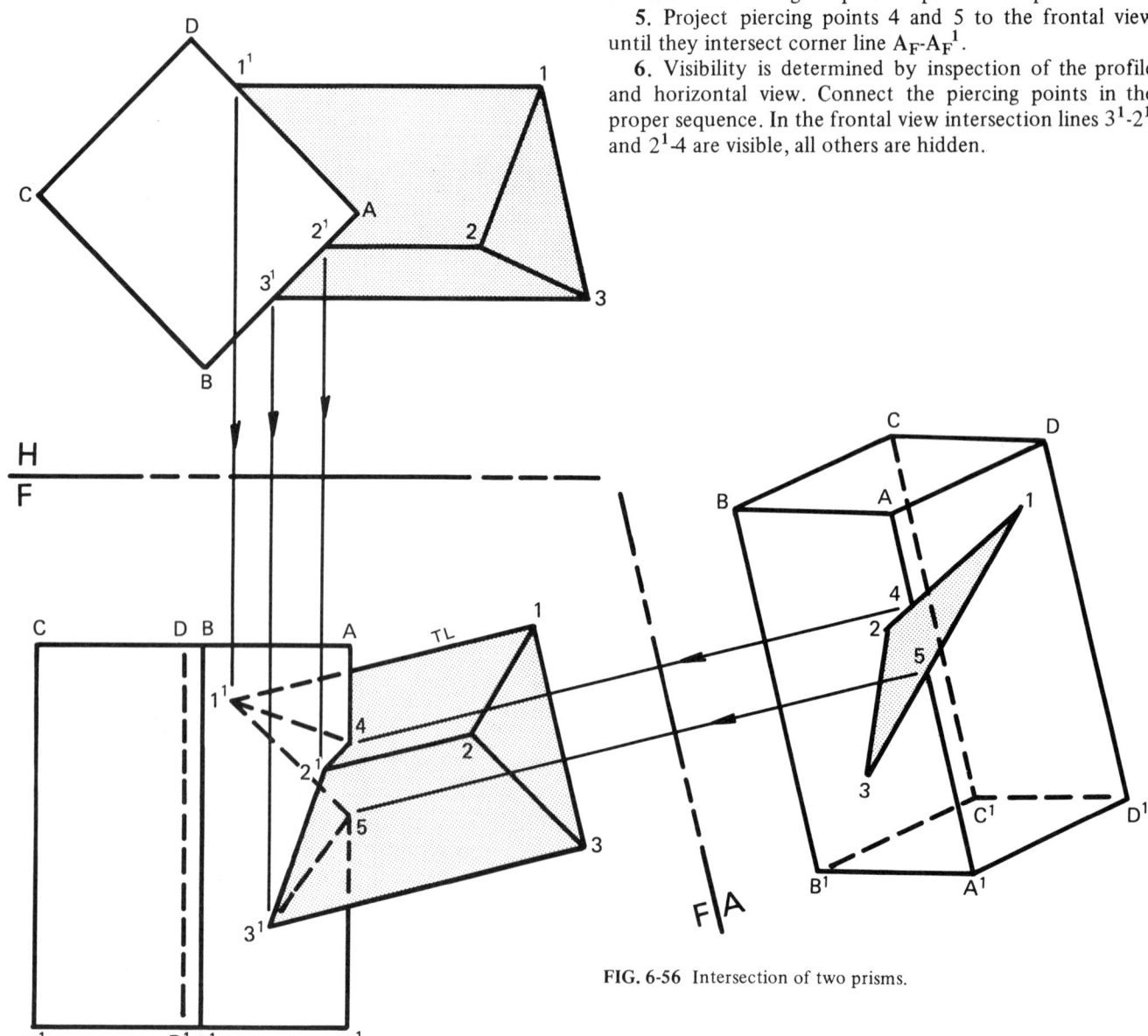

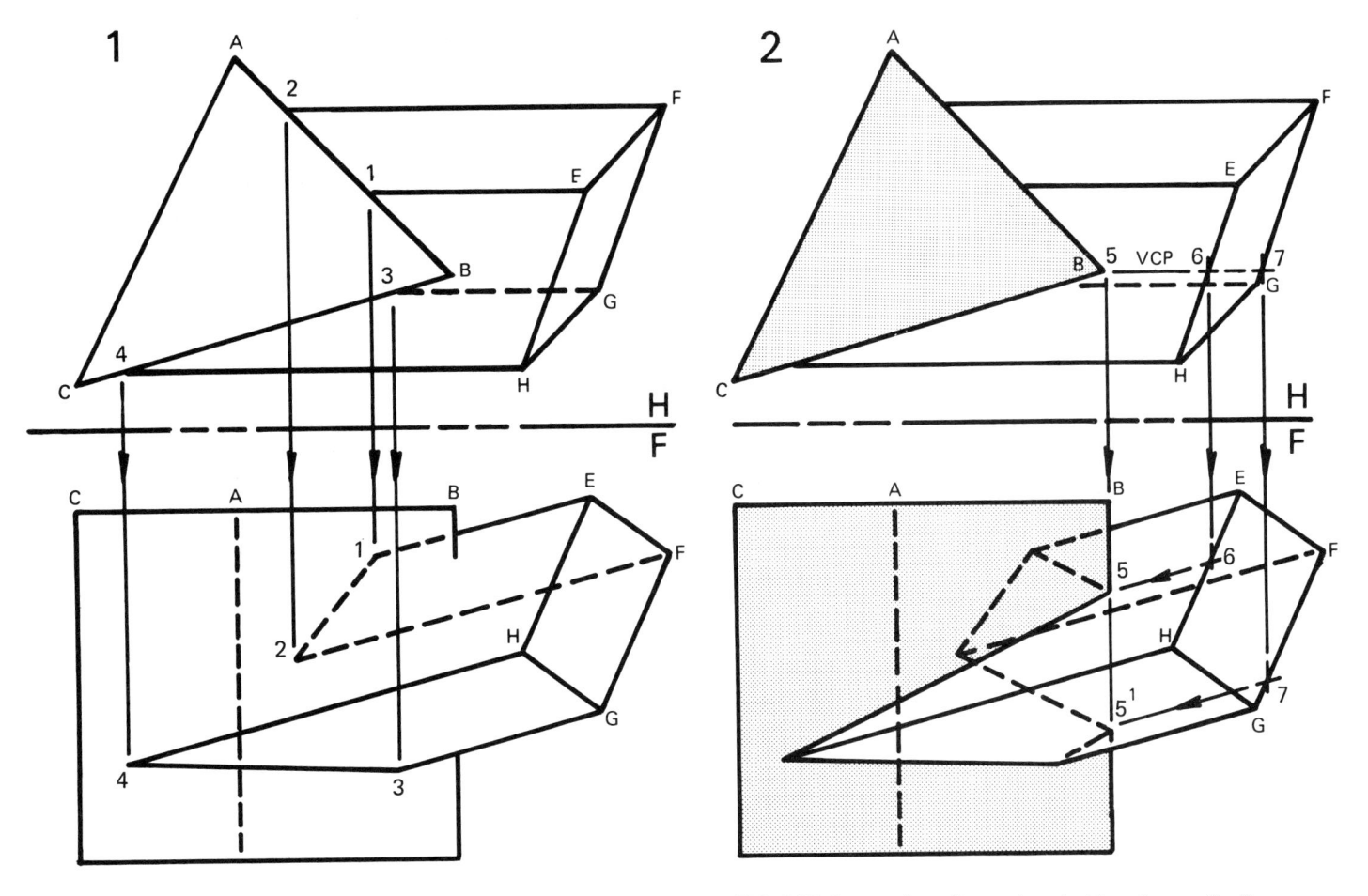

FIG. 6-57 Intersection of two prisms (cutting plane method).

6.33 INTERSECTION OF TWO PRISMS (CUTTING PLANE METHOD)

In Fig. 6-57, the frontal and horizontal view of the two prisms is given. The edge view of the vertical triangular prism appears in the horizontal view. Using the cutting plane method determine the line of intersection and proper visibility.

1. In the horizontal view locate the intersection of lines E, F, G, H, and the surfaces of the vertical prism. Line E intersects the vertical prism at piercing point 1, line F at point 2, line G at point 3, and line H at point 4. Project piercing points 1, 2, 3, and 4 to the frontal view. Lines 1-2 and 3-4 establish two of the intersection lines.

2. In the horizontal view pass a vertical cutting plane through corner line B and parallel to the edge lines of the rectangular prism. VCP intersects corner line B at point 5 and line E-H at point 6_H and line F-G at point 7_H. Project points 6 and 7 to the frontal view. Line 5-6 and line 5^1-7 are drawn parallel to the rectangular prism lines and intersect corner line B at points 5 and 5^1 respectively. Visibility is determined and the frontal view completed as shown.

In Fig. 6-58, a sheet metal model of the intersection of a rectangular prism and an octagonal prism is shown.

FIG. 6-58 Intersection of an octagonal prism and a rectangular prism.

PROB. 6-21

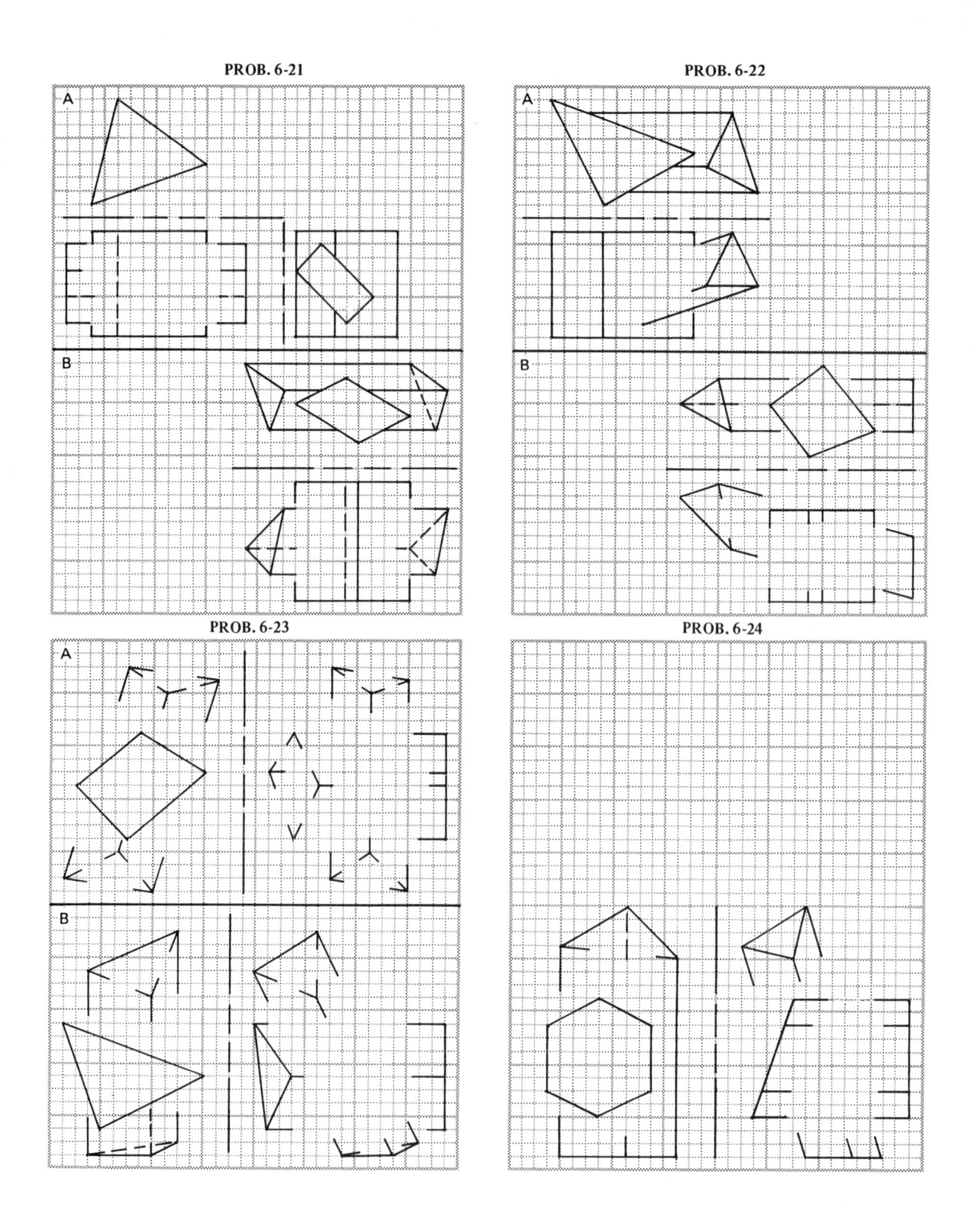

PROB. 6-22

PROB. 6-23

PROB. 6-24

PROB. 6-21A. Complete the views and determine the intersection of the two prisms.
PROB. 6-21B. Draw a left profile view and complete the intersection and visibility of the two prisms.

PROBS. 6-22A, 6-22B. Using the edge view method, complete the views and solve for the intersection.
PROBS. 6-23A, 6-23B. Determine the intersection between the prisms using the cutting plane method.
PROB. 6-24. Solve for the intersection between the prisms using either method.

6.34 INTERSECTION OF A PRISM AND A PYRAMID

The intersection of a prism and a pyramid can be determined using the same methods as for the intersection of two prisms. An edge view of the prism provides a view where the lines of the pyramid intersect surfaces of the prism. Similarly, a view where a portion of the pyramid appears as an edge determines a piercing point of a line of the prism and a surface of the pyramid.

In Fig. 6-59, the intersection of the vertical prism and the right pyramid is required.

1. In the horizontal view, the lines forming the inclined surfaces of the pyramid intersect the edge view surfaces of the prism. Label these piercing points 1, 2, 3, and 5, and project to the frontal view. Line 1_F-2_F is parallel to the base of the pyramid. Points 3_F and 5_F determine points of intersection along the edges of the pyramid.

2. Point 4 can be found by passing a vertical cutting plane through the edge view of the prism as shown. The cutting plane intersects the front face of the pyramid along line 3-6. Project point 6 to the frontal view and draw line 3_F-6_F. Line 3_F-6_F intersects the front vertical edge of the prism and establishes piercing point 4_F. Note that point 7 could have been used to establish piercing point 4.

3. Connect the piercing points and determine visibility as shown.

The intersection of a horizontal prism and a right pyramid is required in Fig. 6-60.

1. Pass horizontal cutting plane CP A through the top edge of the prism in the frontal view. CP A intersects the edges of the pyramid at points 1 and 2.

2. Project points 1 and 2 to the horizontal view. CP A intersects the pyramid as a triangular section whose lines are parallel to the base lines of the pyramid. Points 1_H and 2_H determine the location of this triangular section.

3. Pass horizontal cutting plane CP B through the lower edge of the prism in the frontal view. CP B intersects the edges of the pyramid at points 3_F and 4_F.

4. Project points 3 and 4 to the horizontal view. CP B intersects the pyramid as a triangular section whose lines are the parallel to the base lines of the pyramid.

5. Where CP A and CP B intersect *both* the prism and the pyramid the line of intersection is established.

FIG. 6-59 Intersection of a vertical prism and a pyramid.

FIG. 6-60 Intersection of a pyramid and a horizontal prism.

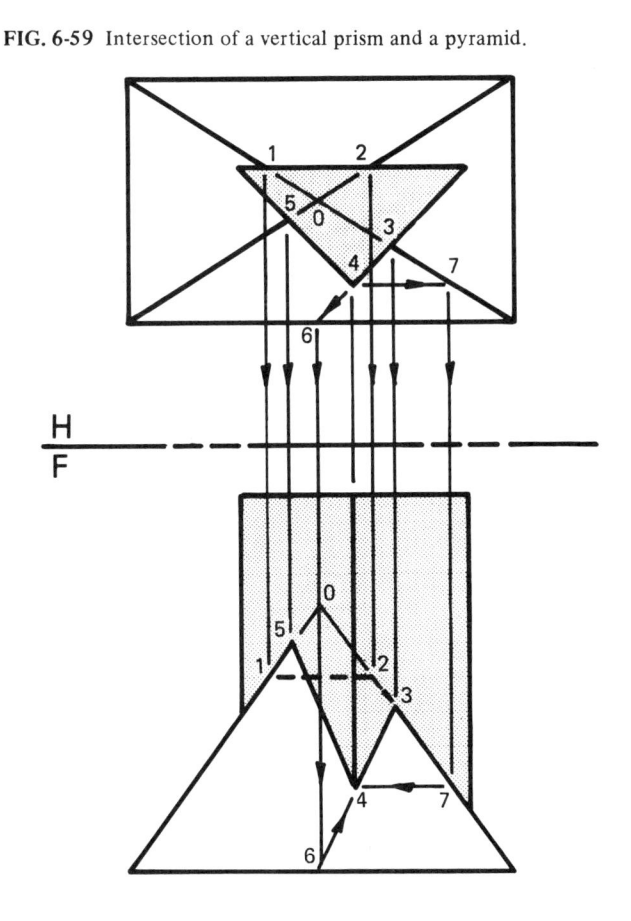

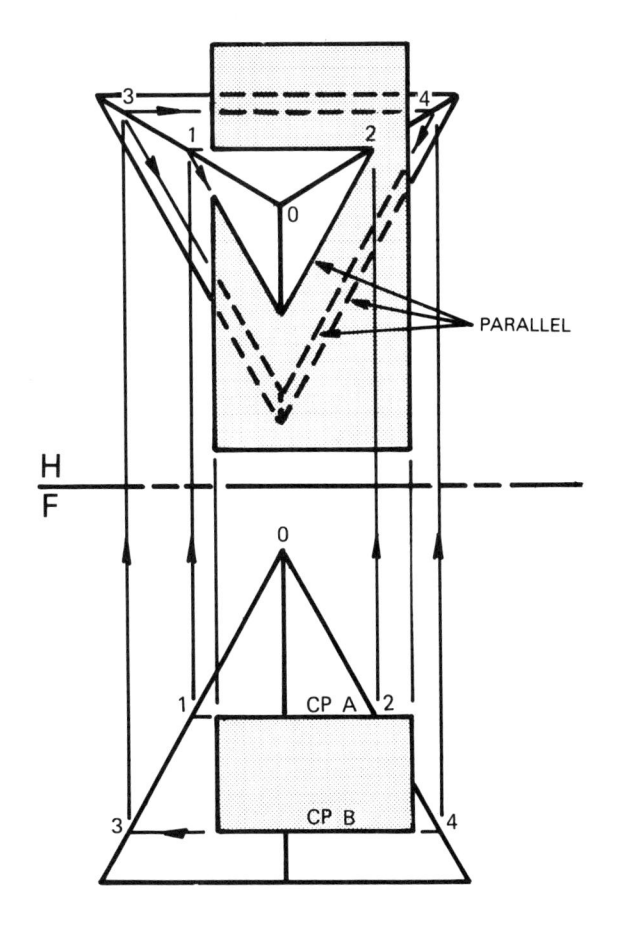

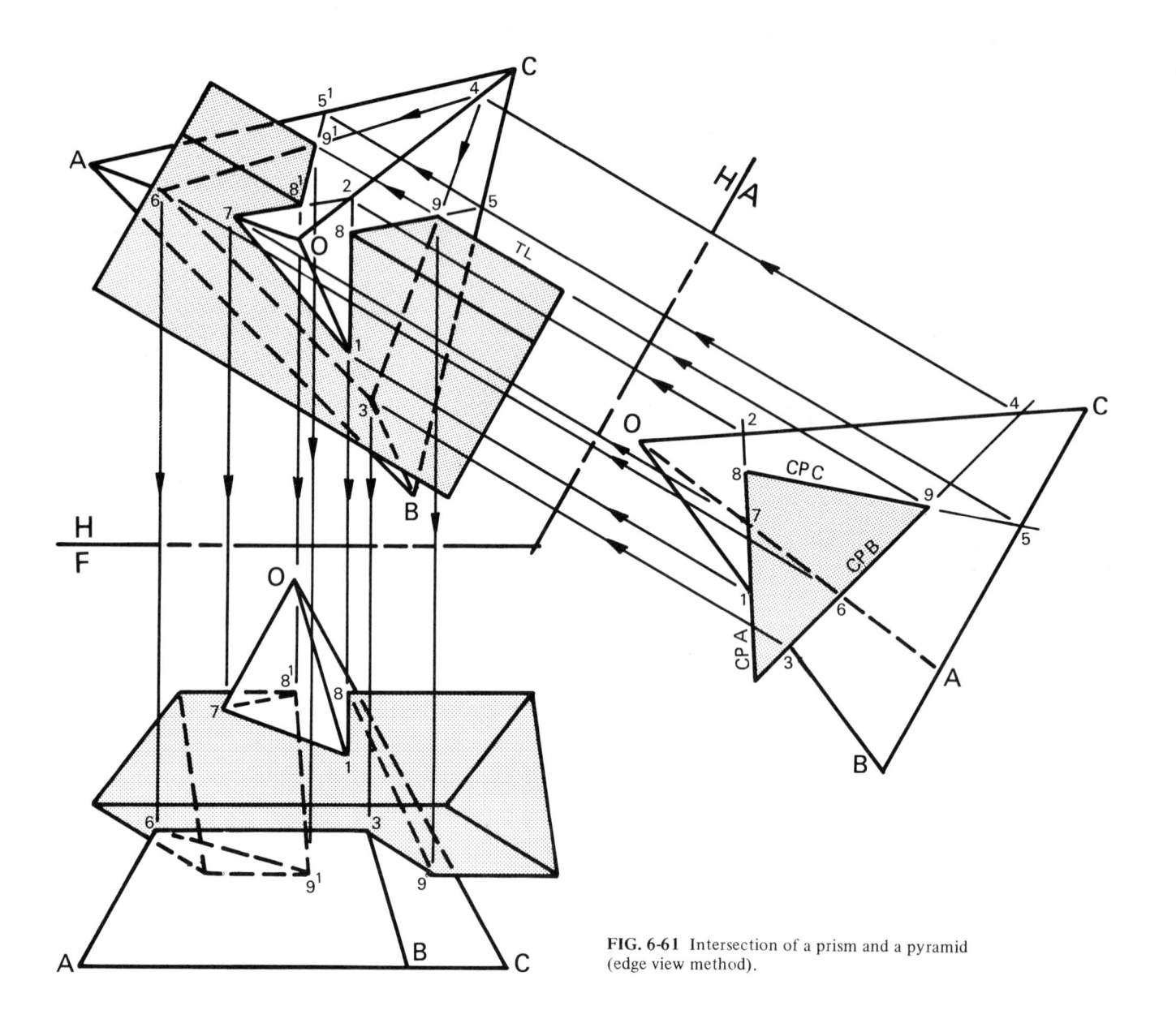

FIG. 6-61 Intersection of a prism and a pyramid (edge view method).

6.35 INTERSECTION OF A PRISM AND A PYRAMID (EDGE VIEW METHOD)

In Fig. 6-61, the given frontal and horizontal projections of the intersecting prism and pyramid are not shown in edge view. An auxiliary view must be projected where the prism appears as an edge.

1. The given prism is a horizontal right prism. Each of its edge lines are true length in the horizontal view. Draw H/A perpendicular to the lateral edge lines of the prism in the horizontal view and project auxiliary A. The prism appears as an edge.

2. Pass cutting planes CP A, CP B, and CP C through the edge view surfaces of the prism in auxiliary A.

3. CP A intersects line 0-B at point 1, line 0-C at point 2 and line 0-A at point 7. Project points 1, 2, and 7 to the horizontal view where they establish a triangular section cut from the pyramid. Plane section 1-2-7 locates piercing points 1_H, 7_H, 8_H, and 8_H^1 in the horizontal view. The frontal view of these piercing points is established by projection.

4. CP B intersects line 0-B at point 3, line 0-C at point 4, and line 0-A at point 6. CP B cuts a plane section from the pyramid and establishes piercing points 3_H, 4_H, 9_H, and 9_H^1 in the horizontal view. Project each piercing point to the frontal view.

5. CP C is not necessary for the completion of the problem but provides a reliable check for points 8, 8^1, 9, and 9^1.

6. Connect the piercing points in proper sequence and establish visibility as shown.

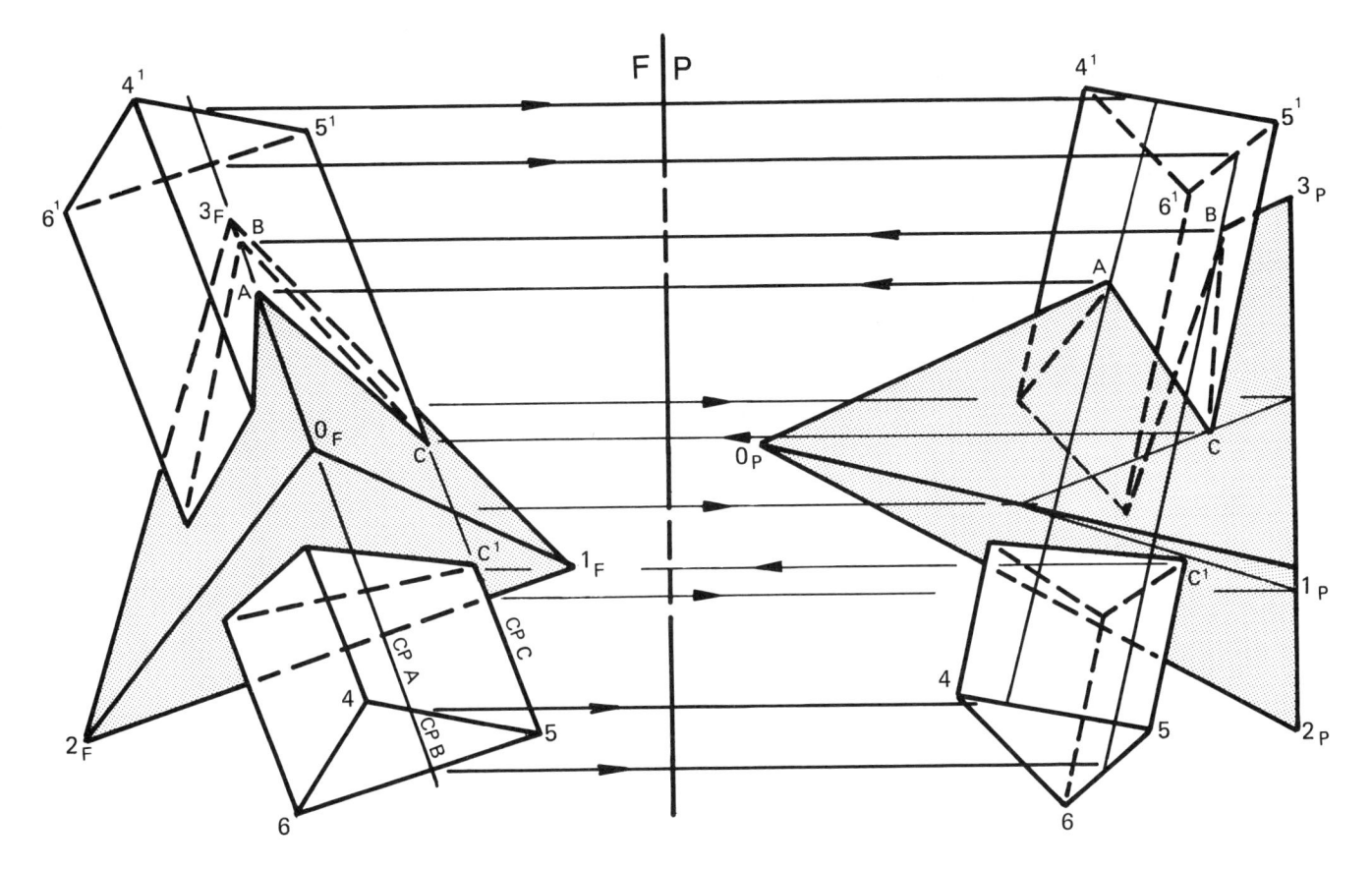

FIG. 6-62 Intersection of a prism and a pyramid (cutting plane method).

6.36 INTERSECTION OF A PRISM AND A PYRAMID (CUTTING PLANE METHOD)

The cutting plane method can be applied to the intersection of a prism and a pyramid when only two principal views are used. In Fig. 6-62, prism 4-5-6 and pyramid 0-1-2-3 are given. Neither form appears as an edge in the given frontal and profile views. The line of intersection is required using only the given projections.

1. Pass a series of cutting planes through convenient lines of the prism or pyramid in either view.
2. CP A and CP B are actually the same cutting plane but have been defined separately so as to illustrate locating points A and B. CP A and CP B are passed through edge line 0_F-3_F of the pyramid.
3. CP A intersects the prism as a line on its surface. The intersection of this cutting plane line and line 0_P-3_P determine piercing point A. Project point A to the frontal view. CP B intersects line 0-3 at piercing point B.
4. CP C is passed through line 5 of the prism in the frontal view. Note that CP C is **not** parallel to the other two cutting planes. CP C intersects two surfaces of the pyramid and locates points C and C^1.
5. Cutting planes can be passed through lines 4 and 6 of the prism in the frontal view to locate all other piercing points required to determine the line of intersection.

Fig. 6-63 is a sheet metal development and welded intersection of a prism and a pyramid.

FIG. 6-63 Intersection of a prism and a pyramid.

253

PROB. 6-25

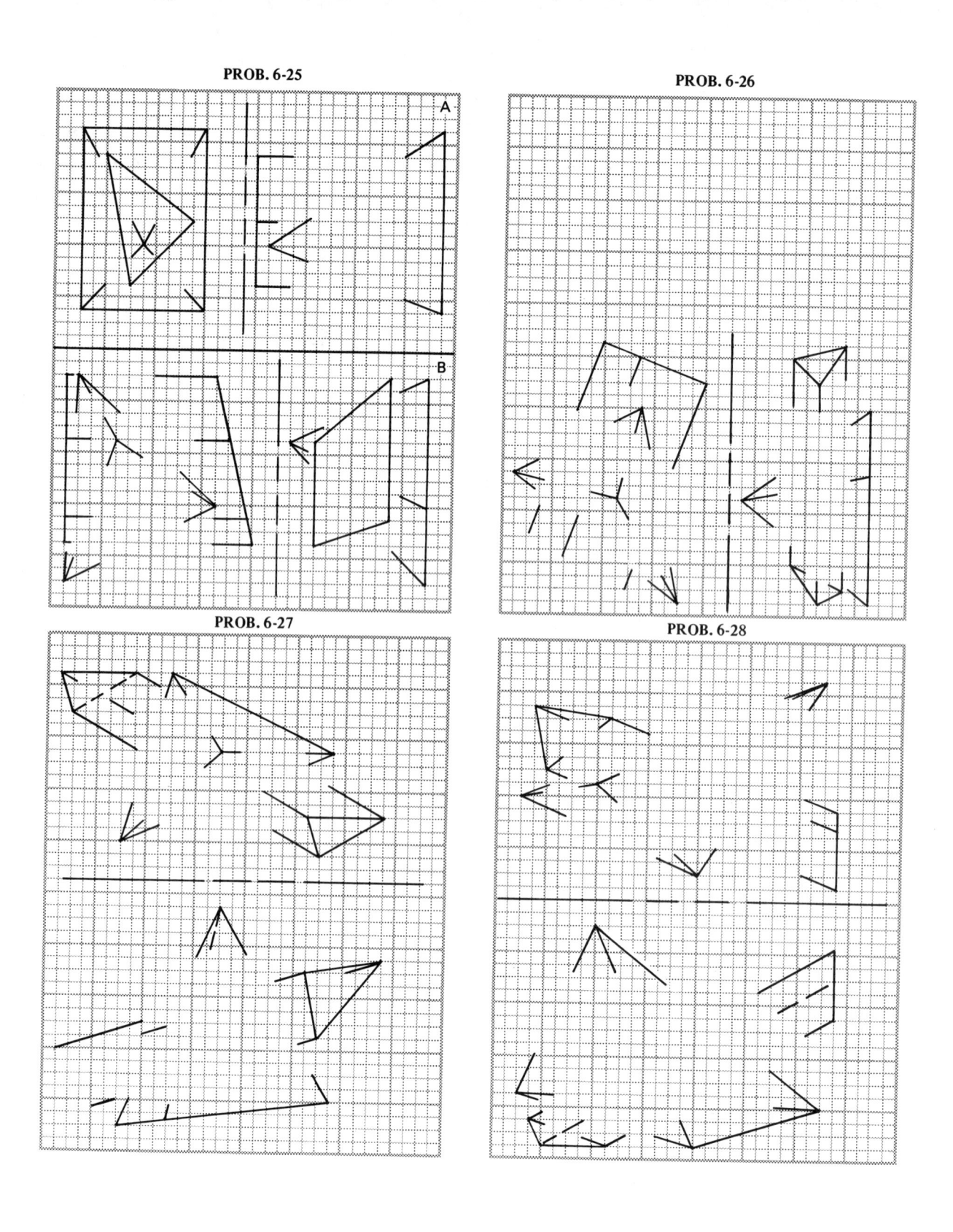

PROB. 6-26

PROB. 6-27

PROB. 6-28

PROBS. 6-25A, 6-25B. Solve for the intersection of the pyramid and prism. Use the cutting plane method.

PROB. 6-26. Complete the given views and solve for the intersection by means of a complete edge view.

PROB. 6-27. Use the cutting plane method to determine the intersection of the given figures.

PROB. 6-28. Solve for the intersection between the oblique prism and the oblique pyramid. Use the cutting plane method.

254

FIG. 6-64 Heat exchangers and a manifold. *(Courtesy NASA.)*

6.37 INTERSECTION OF CYLINDERS

The intersection of two cylinders is a common industrial problem. In Fig. 6-64 the piping manifold and the heat exchangers for the wind tunnel turbines are composed of a number of intersecting shapes of various sizes.

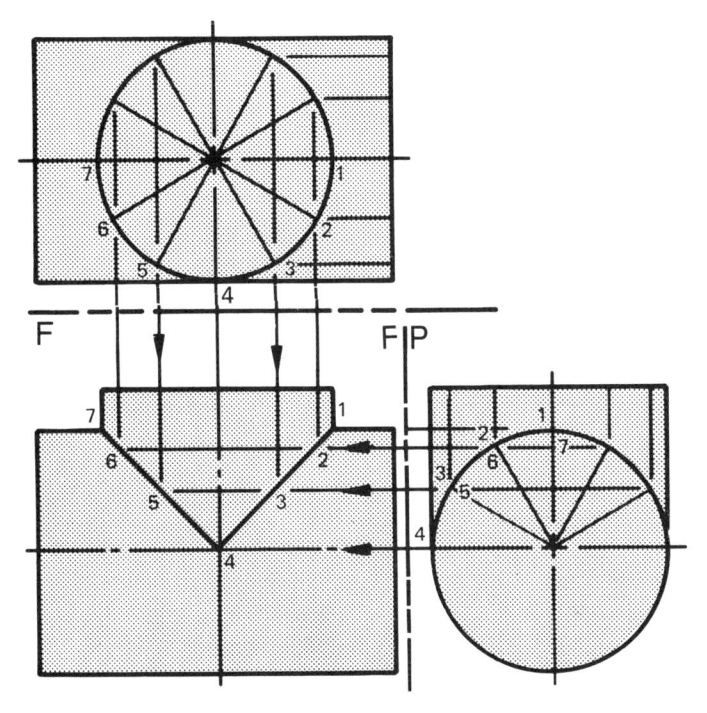

FIG. 6-65 Intersection of two cylinders.

Two intersecting perpendicular right cylinders of the same diameter intersect as in Fig. 6-65. The line of intersection can be determined by showing each cylinder as an edge view and passing a series of equally spaced cutting planes through both cylinders. Each cutting plane intersects a cylinder as an element on its surface. The intersection of related elements determines the line of intersection. Each intersection point is actually the piercing point of an element of one cylinder and the surface of the other cylinder.

In Fig. 6-65 both cylinders are the same diameter and intersect one another at right angles. The resulting curved line of intersection appears as straight lines in the frontal view. Therefore, in this case the line of intersection could have been determined by simply drawing the straight lines from point 1 to point 4 to point 7.

To illustrate the edge view method as applied to all perpendicular intersections of two cylinders, regardless of their diameters, the following description is provided: A series of elements are drawn on the surface of one cylinder by equally dividing the edge view of the vertical cylinder as shown. Each vertical cutting plane passes parallel to the cylinder's axis and cuts a straight-line element on both surfaces. Points 1 through 7 represent the intersection of related elements established by the intersection of a cutting plane and each cylinder. The profile view can also be used to equally divide the horizontal cylinder and establish vertical cutting planes as shown.

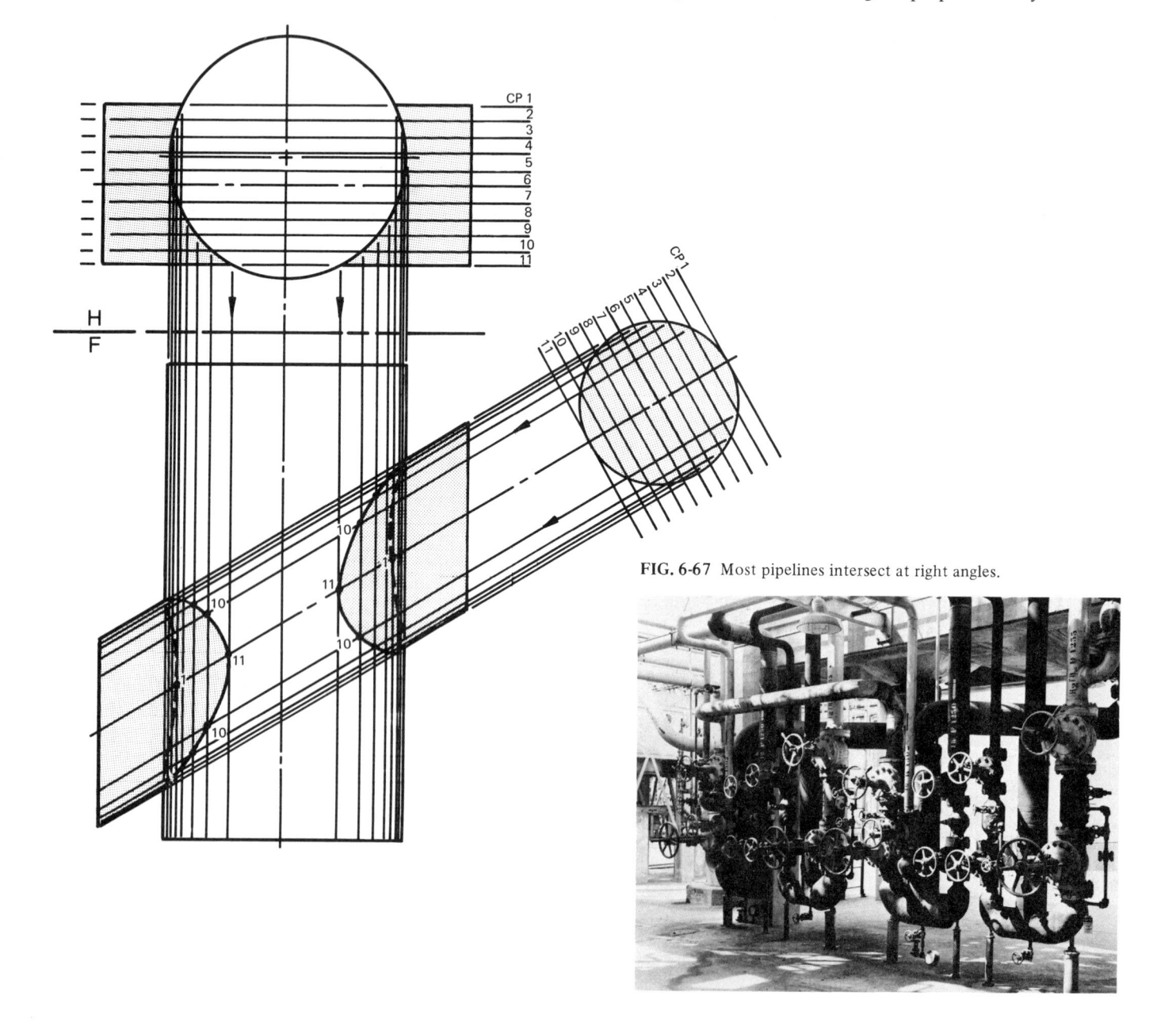

6.38 INTERSECTION OF TWO CYLINDERS (NOT AT RIGHT ANGLES)

An edge view of both cylinders is necessary. Project an auxiliary view of the cylinder which does not appear as an edge in a given view. Pass a series of equally spaced cutting planes parallel to the axis line of both cylinders. Cutting planes parallel to both axes appear as edges in views where the cylinders show as edge views, as in Fig. 6-66 where CP 1 through CP 11 appear as edges in horizontal and auxiliary views. Each cutting plane intersects both cylinders as elements on their surface. Related elements intersect along the line of intersection of the two cylinders. Accuracy increases proportionally to the number of cutting planes and therefore piercing points. Piercing points are connected by means of a french curve.

In Fig. 6-66 the intersection of two pipes, not at right angles is required.

1. Project an edge view of the small-diameter pipe. It is unnecessary to establish a fold line, since only the pipe's diameter is shown.
2. Pass a series of cutting planes through the small pipe, parallel to the axes of **both** cylinders and show in the edge views.
3. Each CP cuts two elements from each cylinder except CP 1 and CP 11, which establish only one element each on the small pipe. The frontal view shows the intersection of related elements. Elements cut by CP 10 intersect at point 10. Since there are two elements cut from each pipe, four points of intersection result.
4. Establish the intersection of all elements and connect the points after determining the proper visibility.

FIG. 6-66 Intersection of two cylinders (not at right angles).

FIG. 6-67 Most pipelines intersect at right angles.

FIG. 6-68 Intersection of two oblique cylinders.

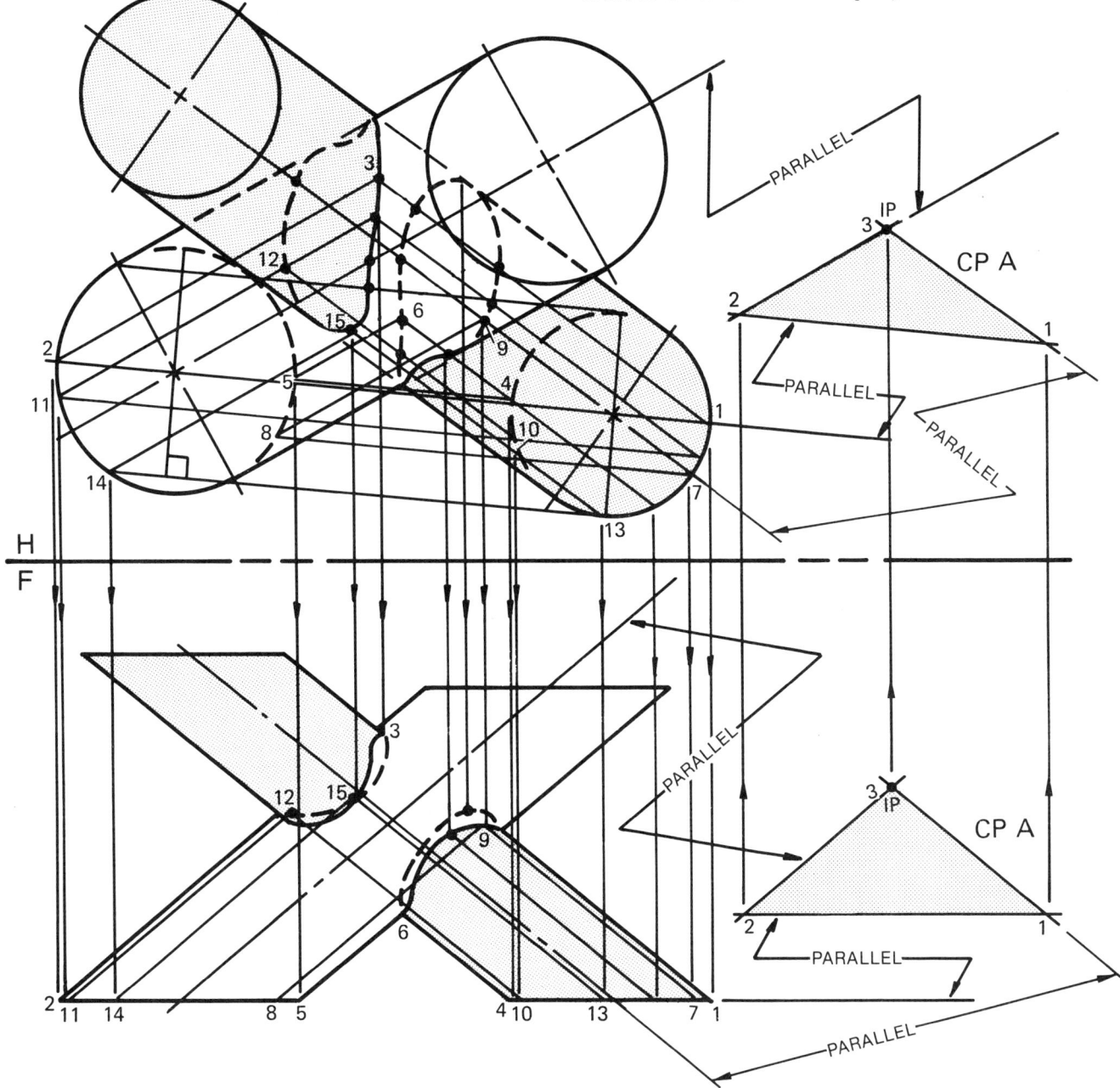

6.39 INTERSECTION OF TWO OBLIQUE CYLINDERS

The intersection of two oblique cylinders can be determined by passing a series of cutting planes parallel to the axes of both cylinders. *The cutting planes are oblique and parallel to one another.* A cutting plane cuts an element on each cylinder's surface. The intersection of two related elements establish one intersection point along the line of intersection.

In Fig. 6-68, the intersection of the two oblique cylinders is required.

1. A triangular *setup* cutting plane (CP A) is constructed next to the cylinders. In the frontal view, draw horizontal line 1-2 parallel to and lying on the base plane of the cylinders where they show as an edge. Draw line 1-3 parallel to one cylinder axis and line 2-3 parallel to the other.

2. Construct CP A in the horizontal view. Project points 1 and 2 to the horizontal view. Draw line 1-2 parallel to a line drawn through the center points of each cylinder's circular base plane. Lines 1-3 and 2-3 are drawn parallel to their respective axis line as shown.

3. Pass a series of oblique cutting planes parallel to CP A and through the two cylinders as shown. Two lines of intersection are required. Therefore use both sides of each cylinder's base circle to establish elements for the cutting planes. As an example, CP 1-2-3 locates point 3 along the upper line of intersection. CP 4-5-6 locates point 6 along the lower line of intersection.

4. Project the cutting planes' elements to the frontal view and locate intersection points by projection.

5. Establish visibility and connect the points.

257

6.40 INTERSECTION OF A PRISM AND A CYLINDER

The line of intersection between a prism and a cylinder can be determined by passing a series of cutting planes parallel to the axis of the cylinder and through the prism. Each cutting plane intersects the prism and the cylinder as elements on their surfaces. A cutting plane cuts related elements on the surface of the cylinder and on the surface of the prism. The intersection of related elements establish intersection points along the line of intersection. An adequate number of cutting planes should be used to locate and draw a smooth line of intersection. Normally, vertical or horizontal cutting planes are drawn parallel to the axis line in a view where the cylinder appears as an edge (right section). For convenience, evenly space the cutting planes about the point view of the axis line.

In Fig. 6-69, the intersection of the given horizontal cylinder and the vertical prism is required.

1. Project a profile view showing only the right section of the cylinder.
2. Pass a series of evenly spaced vertical cutting planes through the edge view of the cylinder in the profile view. CP 1 through CP 11 are drawn in the profile view and the horizontal view.
3. With the exception of CP 1 and CP 11, each cutting plane cuts two elements along the surface of the cylinder and one on the surface of the prism, thus establishing two intersection points per cutting plane. Project the elements of the cylinder established in the profile view to the frontal view. Project the elements of the prism established in the horizontal view to the frontal view.
4. The intersection of related elements in the frontal view determines the line of intersection. CP 1 and CP 11 cut one element each along the extreme front and back of the cylinder in the profile view and along the prism in the horizontal view. The intersection of related elements locates point 1_F and 11_F.
5. Examination of the horizontal view establishes the proper visibility.

In Fig. 6-70, the petrochemical engineering design model exhibits a variety of geometric shapes. The intersection of shapes and development of patterns for vessels, ducting, piping, etc., is an integral part of design work in the industrial modeling and piping design field.

FIG. 6-69 Intersection of a prism and a cylinder.

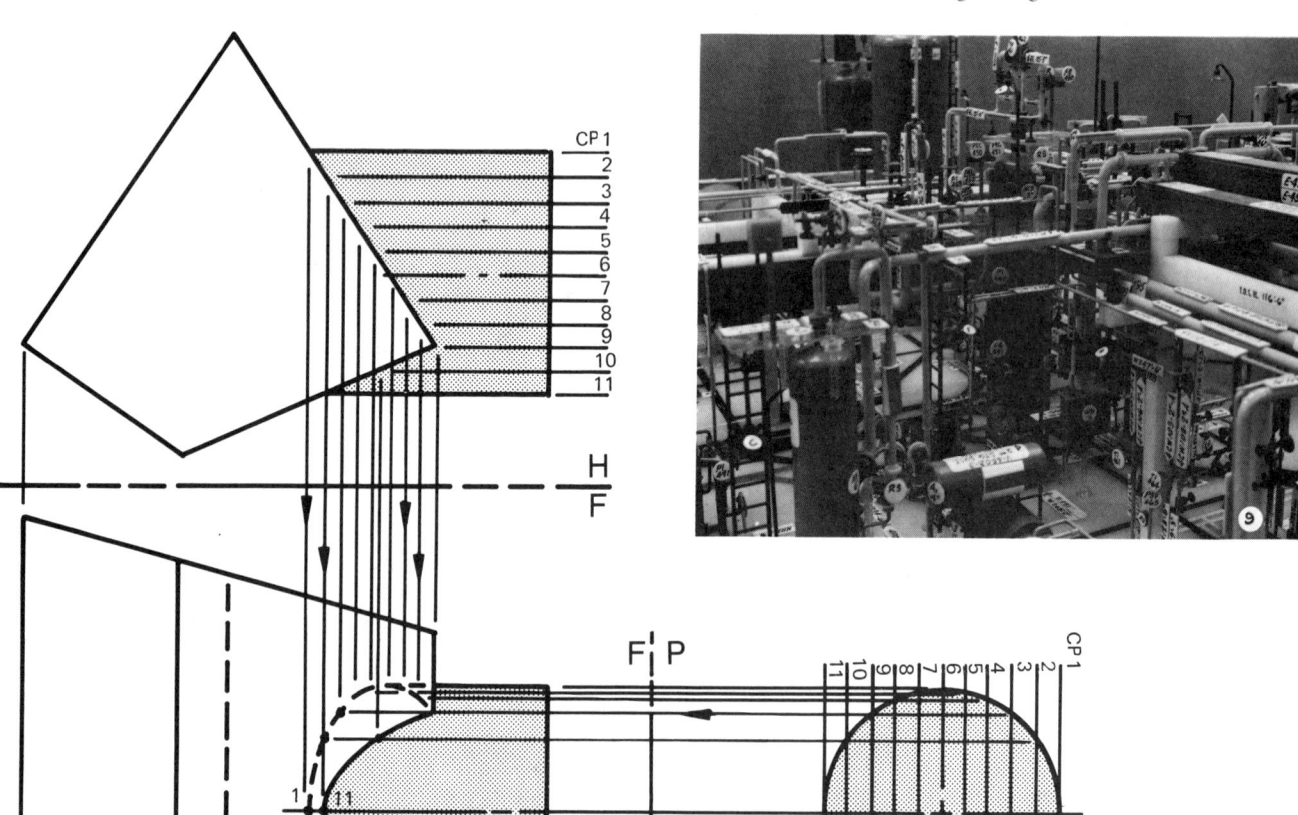

FIG. 6-70 Petrochemical engineering model.

FIG. 6-71 Intersection of a cylinder and a prism (edge view method).

6.41 INTERSECTION OF A CYLINDER AND A PRISM (EDGE VIEW METHOD)

In Fig. 6-71, the vertical right cylinder and the inclined prism are given in the frontal and horizontal views. A series of cutting planes are drawn through an end view of the prism (right section) and parallel to the axis of the cylinder. Each cutting plane cuts an element on the surface of the cylinder and prism. Intersecting related elements determine the line of intersection.

1. Draw F/A perpendicular to the true length lines of the prism in the frontal view. Project auxiliary A. The cylinder need not be shown.
2. Pass a series of evenly spaced vertical cutting planes parallel to the cylinder's axis and through the right section of the prism in auxiliary A. Show the cutting planes in the horizontal view.
3. The edge lines of the prism intersect the cylinder in the horizontal view. Project piercing points A^1, B^1, C^1, and D^1 to the frontal view.
4. Project elements established on the prism in auxiliary A and elements established on the cylinder in the horizontal view to the frontal view. The intersection of related elements determines points on the line of intersection. Note that each cutting plane cuts two elements on the prism and one on the cylinder. Therefore each cutting plane locates two points on the line of intersection.
5. Connect the intersection points in proper sequence after determining visibility.

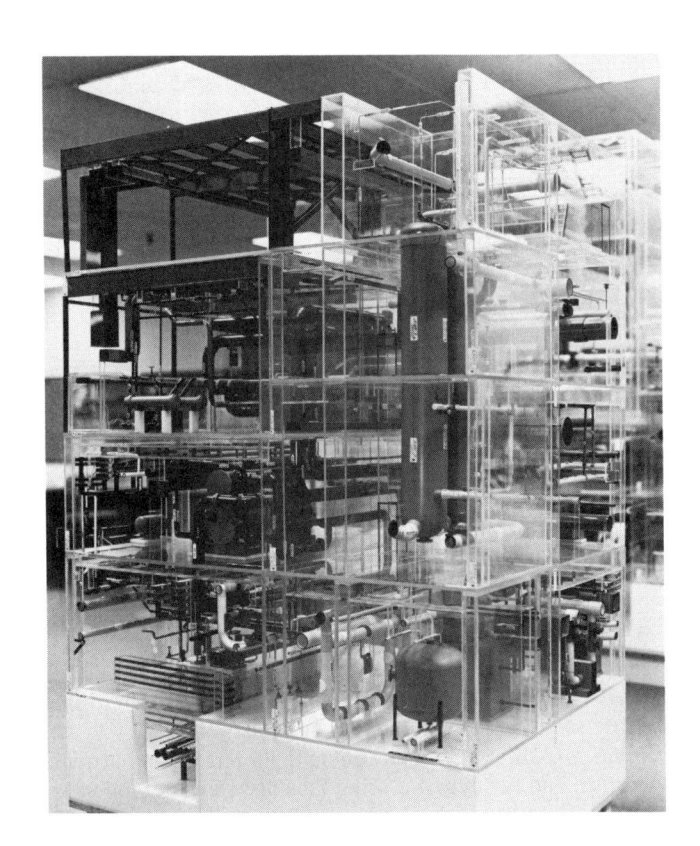

FIG. 6-72 A variety of intersecting geometric forms can be seen on this power plant model. *(Courtesy EMA.)*

PROB. 6-29

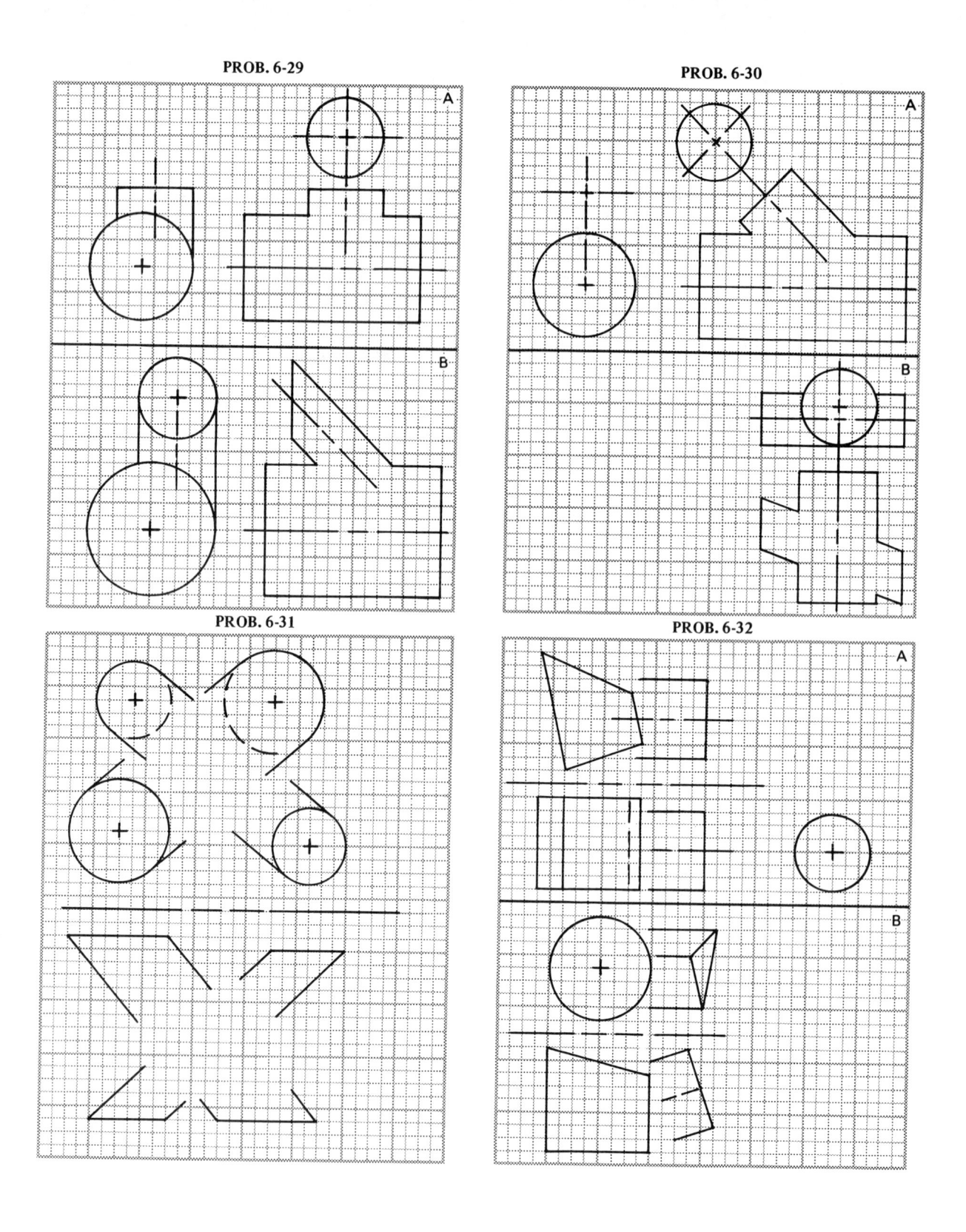

PROB. 6-30

PROB. 6-31

PROB. 6-32

PROBS. 6-29A, 6-29B. Solve for the intersection between the given cylinders.

PROBS. 6-30A, 6-30B. Complete the views and solve for the intersection between the cylinders.

PROB. 6-31. Use the cutting plane method to determine the intersection between the two oblique cylinders.

PROBS. 6-32A, 6-32B. Determine the intersection between the prism and the cylinder using the edge view method.

FIG. 6-73 Cone shaped Apollo spacecraft. *(Courtesy NASA.)*

FIG. 6-74 Intersection of a cone and a horizontal prism.

6.42 INTERSECTION OF CONES

Conical shapes are used in the design of a wide variety of industrial products, structures and commercial applications. Cones used as they are or in combination with other geometric shapes can be found throughout all engineering fields. In general, the right circular cone and frustrum of a right circular cone are the most common. Oblique cones with circular bases are sometimes used as transition pieces and in ducting HVAC designs. The Apollo 8 spacecraft shown in Fig. 6-73 is a conical shape which merges into an elliptical dish bottom to form the heat shield. In Fig. 6-74, the sheet metal model of the intersection of a right cone and a horizontal prism was drawn and developed using descriptive geometry. Figure 6-75, from left to right, shows model of an intersection of a right vertical circular cylinder with a right circular cone, a horizontal cylinder with a right circular cone, and a rectangular prism with a right circular cone.

The following pages cover the intersection of cones with a variety of geometric forms.

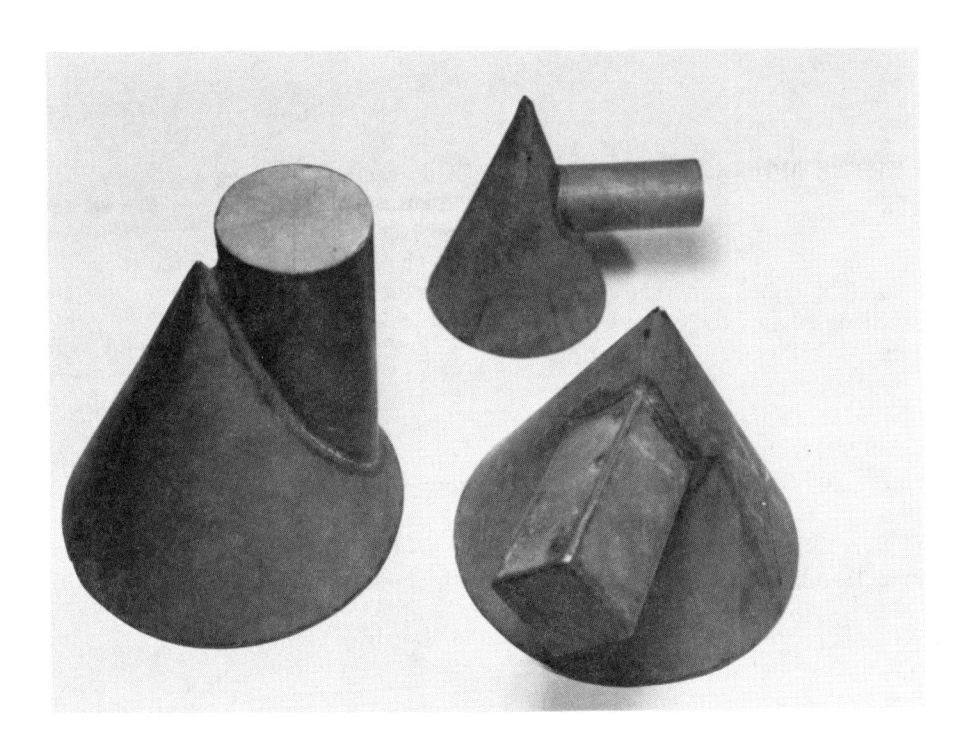

FIG. 6-75 Sheet metal models of conical intersections.

FIG. 6-76 The launch vehicle is composed of intersecting cones and cylinders. *(Courtesy NASA.)*

FIG. 6-77 Intersection of two right circular cones.

6.43 INTERSECTION OF TWO RIGHT CIRCULAR CONES

The intersection of two right cones can readily be determined by using cutting planes to locate common points along the line of intersection. A cutting plane passed parallel to the bases of two intersecting right cones cuts circular elements on the surface of each cone. The intersection points of related circle elements determine points along the line of intersection of the cones.

In Fig. 6-77, the two right circular cones are given. The line of intersection is required.

1. Pass a series of evenly spaced horizontal cutting planes parallel to the base planes of the two cones.

2. Project the cutting planes to the horizontal view. Each cutting plane appears as a true shape circle element on the surface of both cones.

3. The intersection of related circle elements determines two common points along the line of intersection between the cones.

4. Project each intersection point to the frontal view as shown.

5. The highest elevation point of the intersection line is located by passing a vertical cutting plane through the vertices of both cones in the horizontal view. Cutting plane A-B cuts straight-line elements along the surface of each cone from the vertex to base circle. Line VA-A intersects line VB-B at point 1_F in the frontal view. Project point 1 to the horizontal view. Point 1 is the highest point along the line of intersection.

6. Determine visibility and connect the points in proper sequence in both views.

FIG. 6-78 Intersection of two oblique cones.

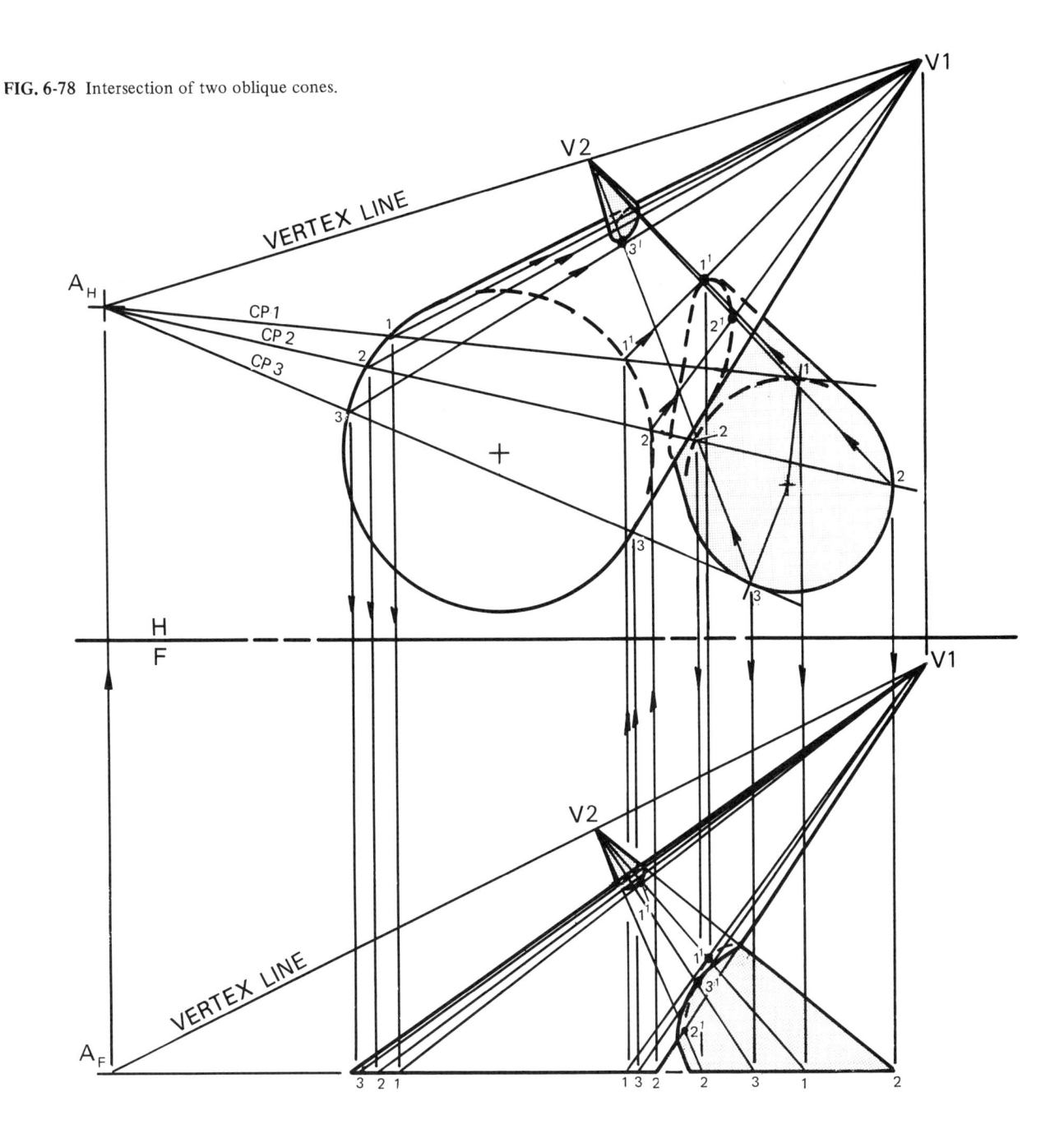

6.44 INTERSECTION OF TWO OBLIQUE CONES

The intersection of two oblique cones can be determined by passing cutting planes through the vertices of both cones. A line formed by connecting the vertices is the **vertex line,** which when extended intersects the base plane at the *"point of convergence."* Each cutting plane contains the vertex line. The intersection of a cutting plane and the cones will cut elements along the surface of each cone. The intersection of related elements determines a point on the line of intersection. In Fig. 6-78, the intersection of the two cones is required.

1. In the frontal view draw a line through the vertices of both cones. Extend the vertex line until it intersects the base plane at point A_F (point of convergence). Project point A to the horizontal view and draw the vertex line.
2. Draw CP 1 containing the vertex line and tangent to the base circle of cone 2. CP 3 is passed through the vertex line and tangent to the other side of the base circle of cone 2. Any number of CPs can be passed between the two limiting cutting planes, 1 and 2.
3. Each CP cuts elements along the surface of the cones. CP 2 cuts each cone twice, thus locating two intersection points along each line of intersection.
4. Project all elements to the frontal view and locate their corresponding intersection points.

6.45 INTERSECTION OF A CONE AND A HORIZONTAL CYLINDER

The intersection of a cone and a cylinder can be determined by passing a series of horizontal CPs parallel to the cylinder's axis in a view where the cylinder appears as a right section. Each CP cuts elements along the surface of both forms. The intersection of related elements locates a point on the line of intersection. In Fig. 6-79, the intersection of the cone and cylinder is required.

1. Project the profile view to show the right section (edge view) of the cylinder.
2. Evenly divide the cylinder as shown. Each division corresponds to *a horizontal cutting plane,* CP 1 through CP 7. Extend the cutting planes to the frontal view.
3. The highest and lowest points of the intersection are established in the frontal view where CP 1 intersects the cone at point 1 and CP 7 at point 7. Project points 1 and 7 to the horizontal view as shown.
4. Project the cutting planes to the horizontal view. Each CP appears as a circle element on the cone and a straight line element on the surface of the cylinder.
5. The intersection of related elements determine a point on the line of intersection. Locate the points in both views.
6. With the exception of points 1 and 7, each common point is used to plot a line of intersection which is symmetrical to the axis of the cylinder in the horizontal view. Determine visibility and connect the points as a smooth curve representing the line of intersection.

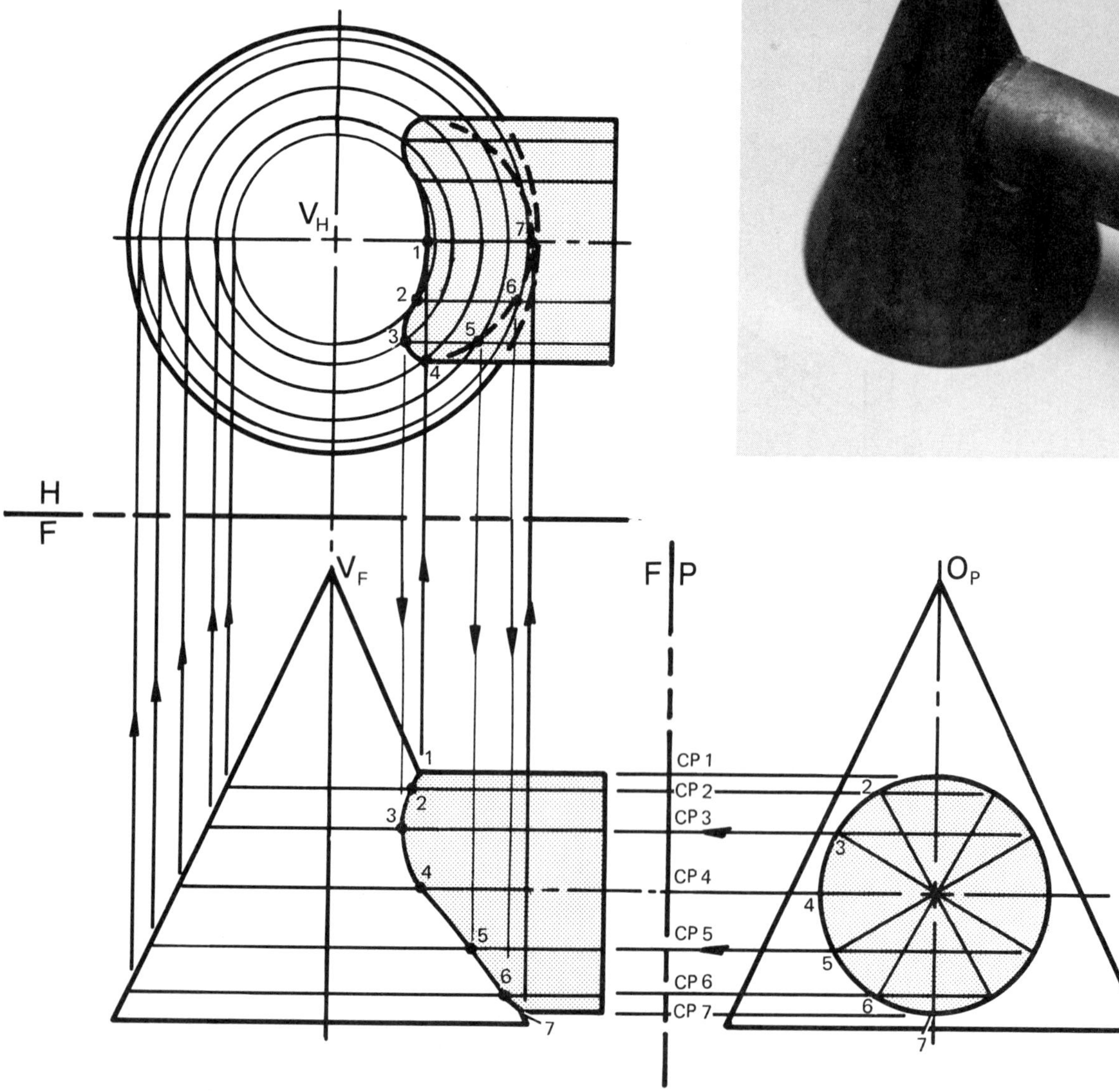

FIG. 6-80 Intersection of a cone and a cylinder.

FIG. 6-79 Intersection of a right cone and a horizontal cylinder.

FIG. 6-81 Intersection of a cone and a cylinder.

6.46 INTERSECTION OF A CONE AND A CYLINDER

A vertical cutting plane passed through the vertex of the cone and parallel to its axis intersects both the cone and cylinder as straight-line elements on their surfaces. A right section view of the cylinder is required to fix the position of the elements along its surface. The intersection of an element on the cone with a related element on the cylinder establishes a point on the line of intersection.

In Fig. 6-81, the intersection of the cone and cylinder is required.

1. Project auxiliary A perpendicular to the cylinder. The cylinder appears as right section.
2. Evenly divide one half of the circumference of the cone's base in the horizontal view. Since the intersection is symmetrical about the cylinder's axis, only the front divisions need be used as cutting planes. Each division corresponds to a vertical CP passed through the vertex of the cone and parallel to its axis.
3. Each CP cuts a straight line element along the surface of the cone. Locate the CPs in each view by projection of the cone's elements.
4. The intersection of the CPs and the cylinder in auxiliary A establish related elements along the surface of the cylinder. Project the cylinder's elements to the frontal view.
5. The intersection of related elements in the frontal view determines points along the line of intersection. CP 1 locates points 1 and 6 at the extremes of the intersection line. CP 2 locates points 2 and 5 as shown.

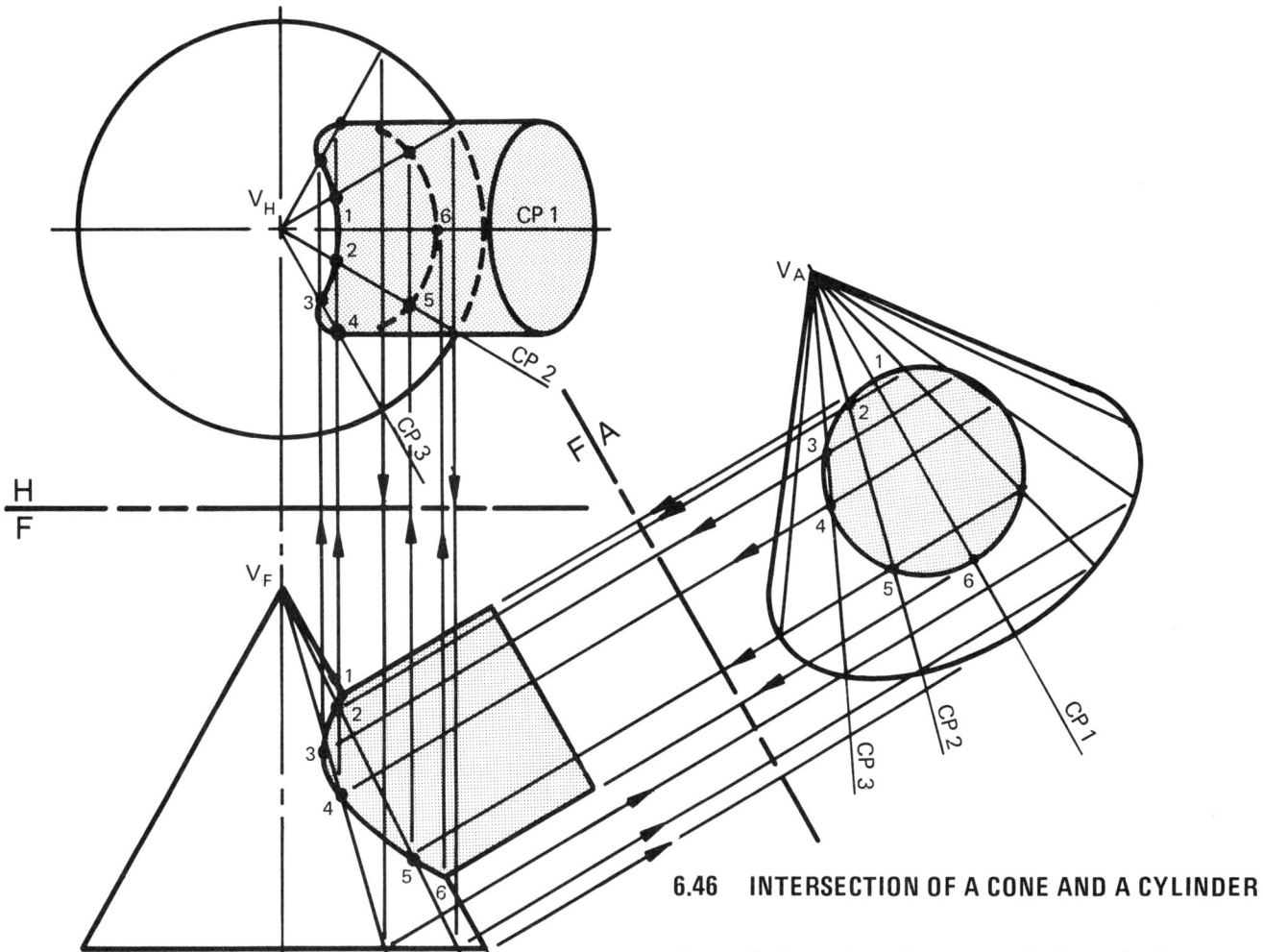

FIG. 6-82 Three cylinders, two frustrums of a cone and a half sphere, were used in the design of the satellite. *(Courtesy NASA.)*

PROB. 6-33

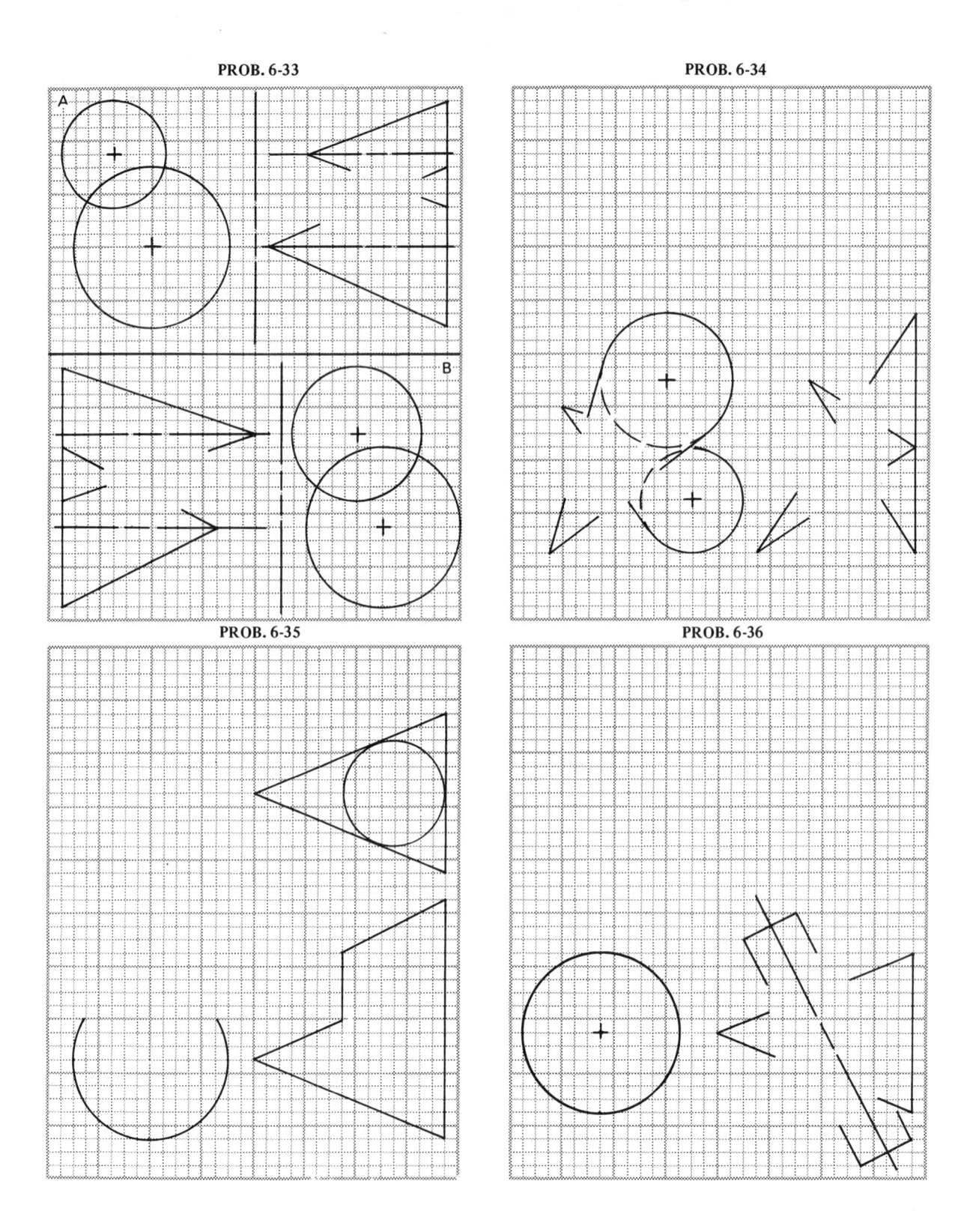

PROB. 6-34

PROB. 6-35

PROB. 6-36

PROBS. 6-33A, 6-33B. Complete the views of the intersecting cones.

PROB. 6-34. Determine the intersection between the oblique cones.

PROB. 6-35. Complete the three views of the cone intersected by the cylinder.

PROB. 6-36. Use an auxiliary edge view to determine the intersection between the cone and cylinder. The cylinder is true length in the frontal view.

FIG. 6-83 Intersection of a right circular cylinder and a right vertical cone.

6.47 INTERSECTION OF A CONE AND A VERTICAL CYLINDER

By passing a series of radial cutting planes parallel to the axes of an intersecting cone and vertical cylinder, the lines of intersection may be determined. Each cutting plane cuts straight-line elements along the surface of the cone and the cylinder. The intersection of related elements establishes points along the line of intersection. In Fig. 6-83, the intersection of the cone and vertical cylinder is required.

1. Evenly divide the circumference of the cone's base, 1 through 12. Each division corresponds to a vertical cutting plane passed parallel to the axis of the cone and cylinder. The CPs intersect both the cone and cylinder as straight-line elements along their surfaces.

2. Project all elements to the frontal view as shown. The intersection of related elements locates a point along the line of intersection. As an example, a cutting plane passed through V_H and point 6 cuts an element on both geometric forms. The intersection of these two related elements locates point 6^1 on the line of intersection in the frontal view.

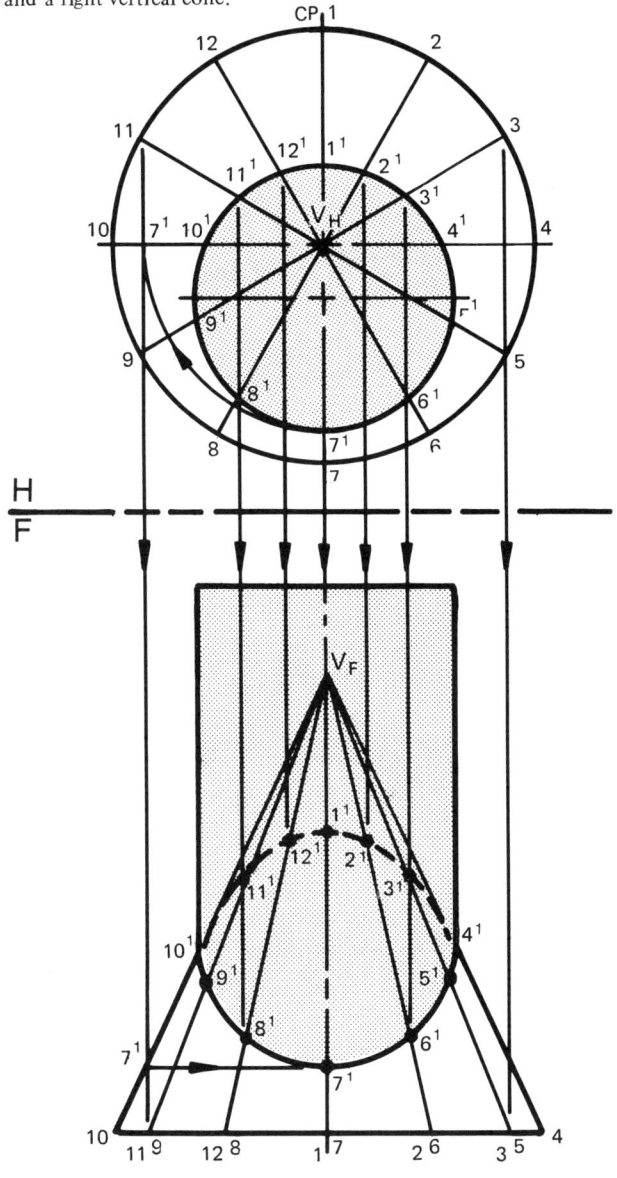

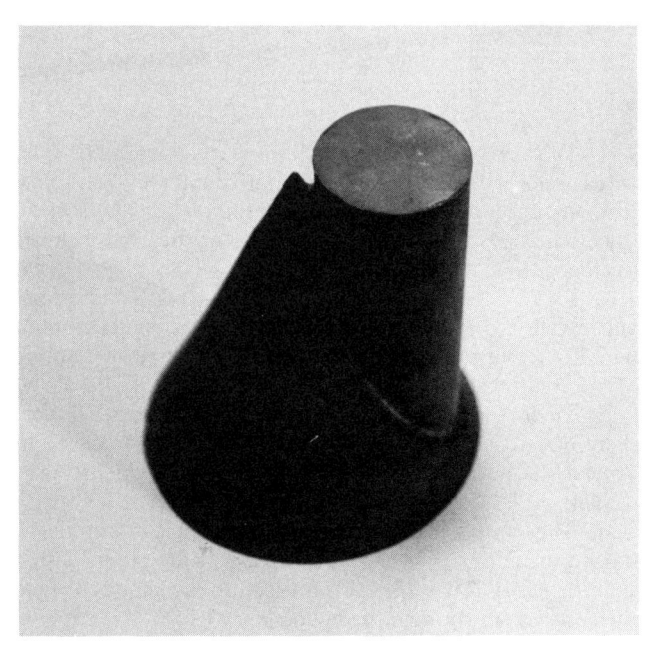

FIG. 6-84 Intersection of a cone and a cylinder with parallel axes.

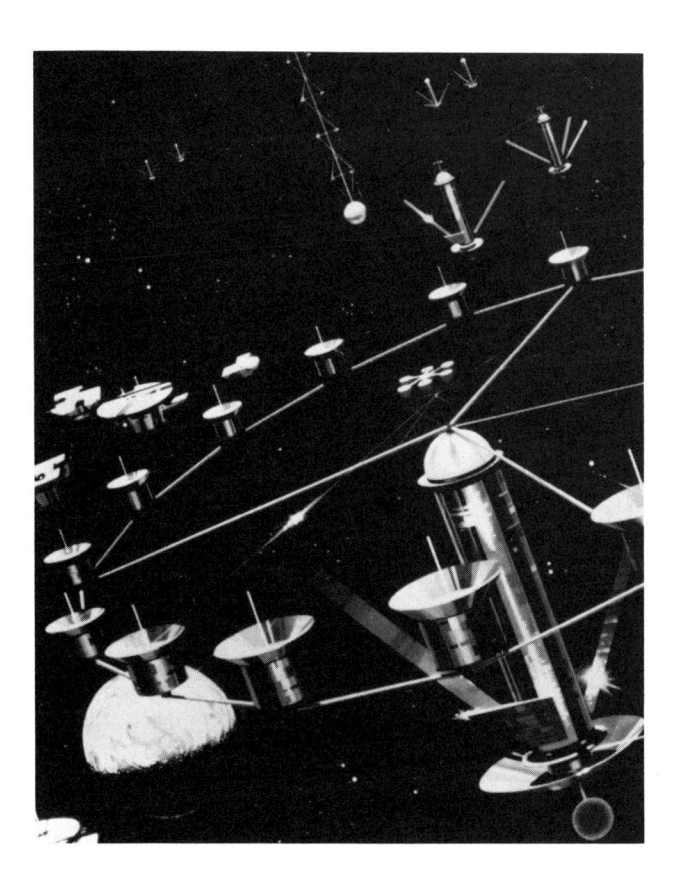

FIG. 6-85 Space colony designed using cylinders and conical shapes. *(Courtesy NASA.)*

FIG. 6-86 Intersection of an oblique cone and a horizontal circular cylinder.

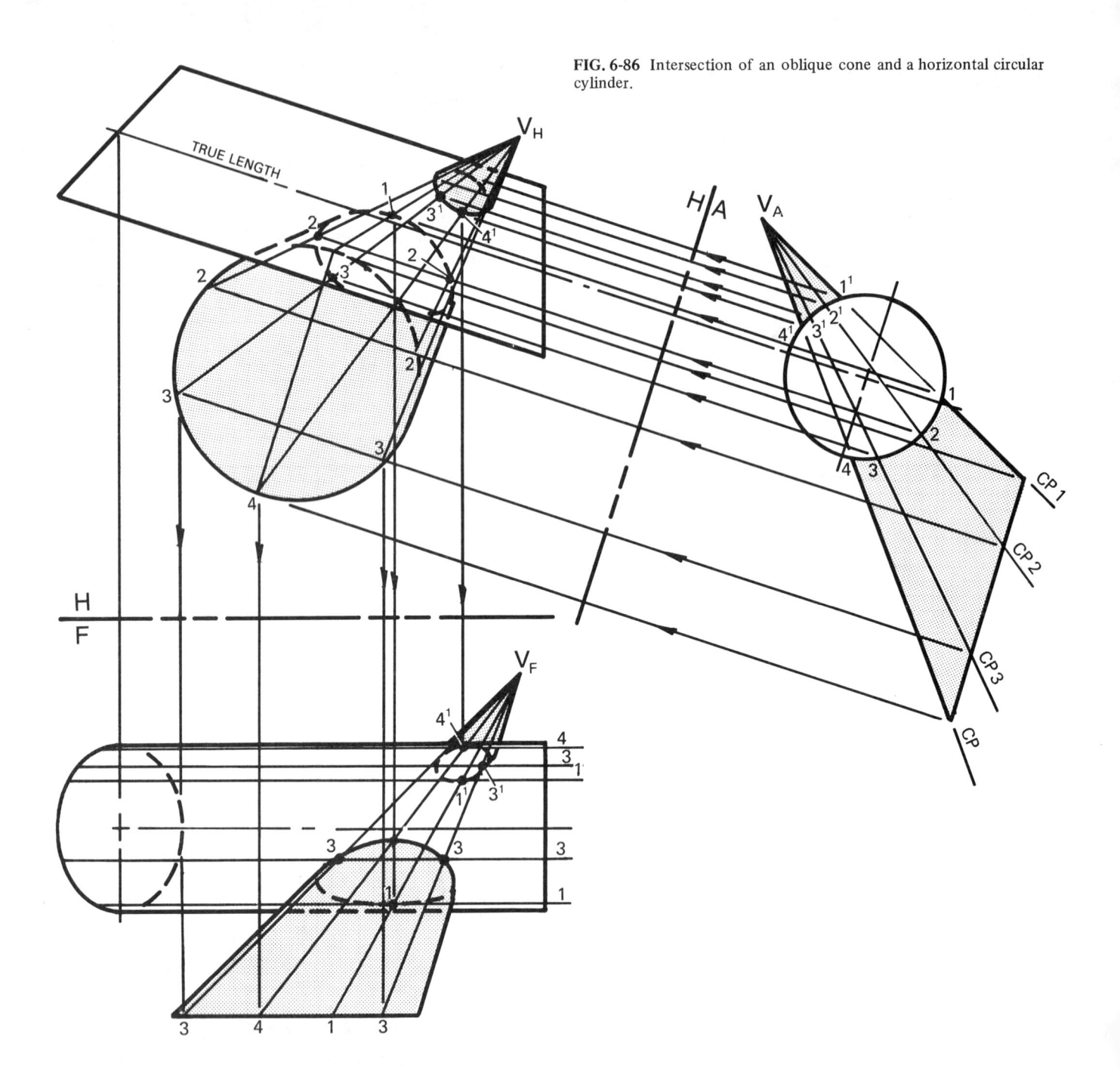

6.48 INTERSECTION OF AN OBLIQUE CONE AND A CYLINDER

Cutting planes which cut related elements from an oblique cone and a cylinder can be used to determine their line of intersection. Cutting planes are passed through the cone and cylinder in a view where the cylinder's axis appears as a point (right section view). Each cutting plane cuts two elements along the surface of the cone and cylinder. Related elements intersect at a point common to both forms thus locating the line of intersection.

In Fig. 6-86, the frontal and horizontal view of the oblique cone and the horizontal cylinder is given. The line of intersection is required.

1. Draw H/A perpendicular to the cylinder's axis line, which appears true length in the horizontal view, and project auxiliary A. The cylinder appears as a right section since its axis is a point view. The base of the cone projects as an edge.
2. Pass CP 1 through 4 through the vertex of the cone and parallel to the axis of the cylinder. Increasing the number of cutting planes will increase the accuracy of the line of intersection.
3. Each CP intersects the cone and cylinder twice, thereby cutting two elements along the surface of each. Project the horizontal and frontal views of the elements as shown.
4. The intersection of related elements determines a point on the line of intersection. Connect the points with a smooth curve. Two lines of intersection are required since the cone passes through the cylinder.

FIG. 6-87 Intersection of an oblique cone and cylinder.

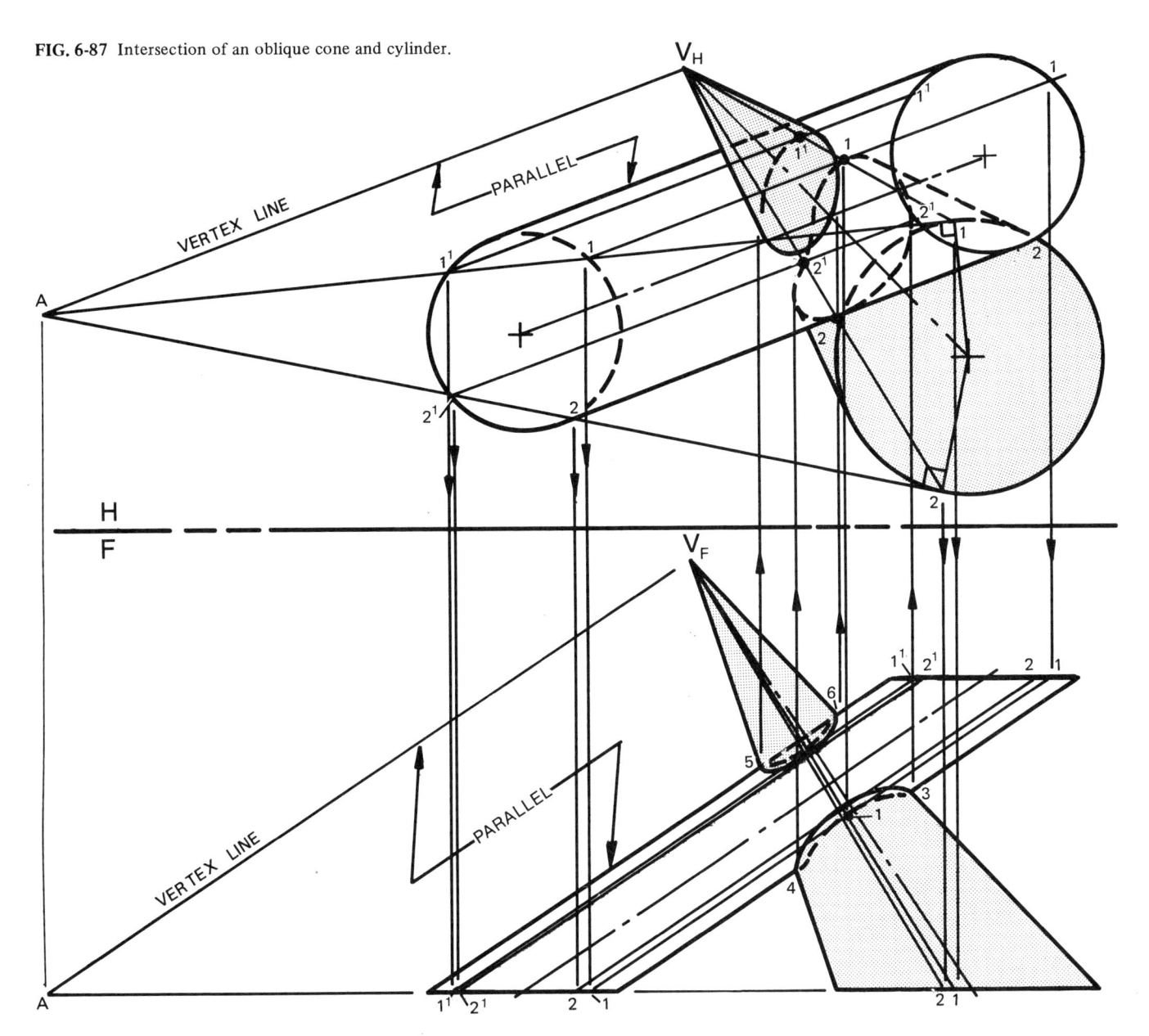

6.49 INTERSECTION OF AN OBLIQUE CONE AND AN OBLIQUE CYLINDER

As in the case of two intersecting oblique cones, the intersection of an oblique cone and an oblique cylinder requires the use of *oblique cutting planes*. In Fig. 6-87 the cylinder and cone have parallel bases. Oblique cutting planes are passed through the vertex of the cone and parallel to the axis of the cylinder. This is accomplished by drawing a *vertex line* through the vertex of the cone and parallel to the axis of the cylinder. This vertex line is extended until it intersects the base plane of the two figures. All cutting planes are drawn using the vertex line and a horizontal line lying in the plane of the base and intersecting both base curves. Each CP cuts an element on the surface of cone and cylinder. Related elements intersect along the line of intersection. Use as many CPs as necessary to locate enough points to form a smooth curve along the lines of intersection.

1. In the frontal view draw the vertex line through the vertex of the cone and parallel to the axis of the cylinder until it pierces the base plane at point A. Point A is the *point of convergence*. Project the vertex line to the horizontal view.

2. Draw cutting planes V-A-1 and V-A-2 tangent to the base circle of the cone. These two cutting planes are the limiting planes since they determine the extent of the intersection line. Use as many CPs as necessary.

3. Each shown CP cuts one element along the cone's surface and two along the surface of the cylinder. Intersecting elements determine the line of intersection. V-A-1 cuts elements from the cone and cylinder as shown. Points 1 and 1¹ are located at the intersection of the related elements and establish one point along each line of intersection.

4. Connect the points after determining visibility. A visible point has two visible elements.

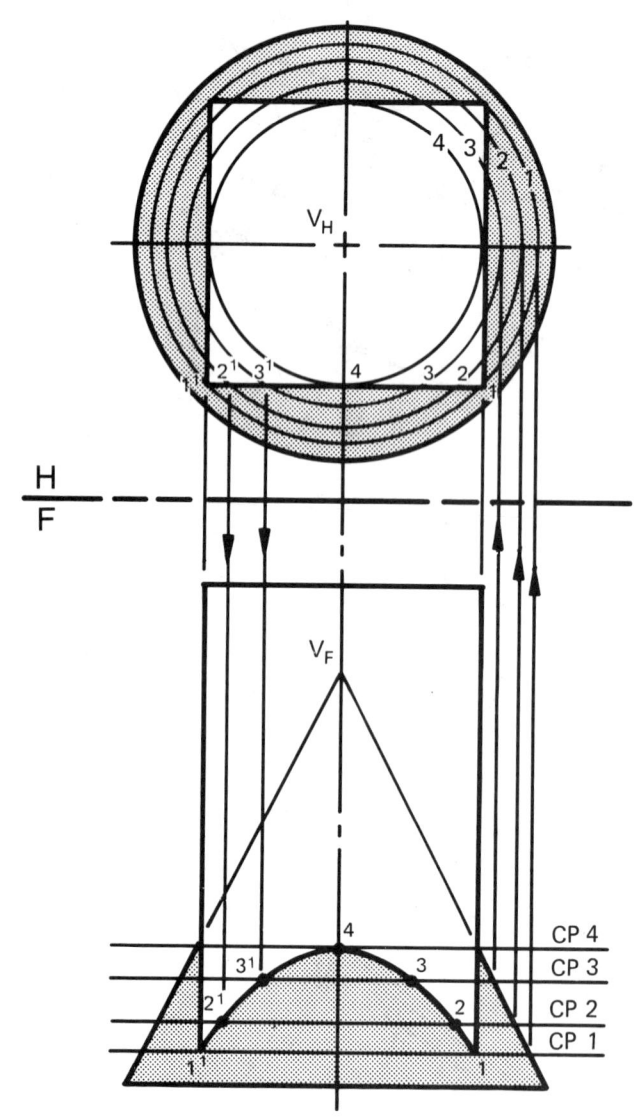

FIG. 6-88 Intersection of a cone and a vertical prism.

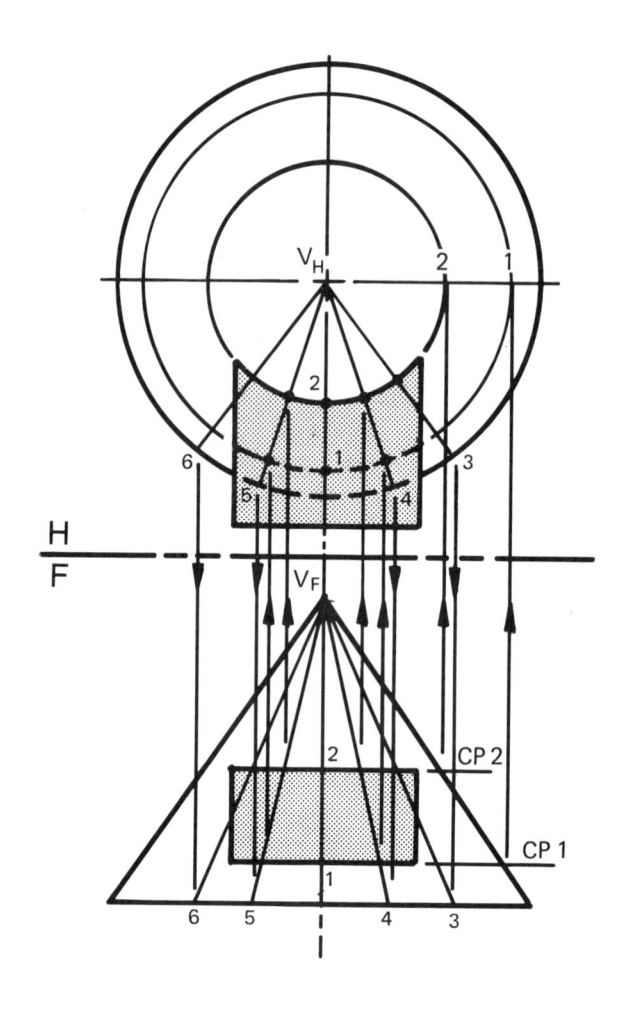

FIG. 6-89 Intersection of a cone and a horizontal prism.

6.50 INTERSECTION OF A CONE AND A PRISM

The intersection of a cone and a prism can be established by passing a series of cutting planes through the figures. When the prism is a vertical prism, horizontal cutting planes perpendicular to the cone's axis should be used, Fig. 6-88. When the prism is horizontal, vertical CPs passed through the axis of the cone or horizontal CPs perpendicular to the axis of the cone can be used, Fig. 6-89.

In Fig. 6-88, the intersection of a vertical prism and a right circular cone is required.

1. Draw horizontal CP 1 through CP 4 perpendicular to the axis of the cone. Project the CPs to the horizontal view where they appear as circle elements on the surface of the cone. CP 1 is drawn first so that it intersects the corners of the prism in the horizontal view, therefore locating the extreme intersecting point of the edge lines and the cone.

2. The intersection of the circle elements and the prism's edges in the horizontal view locate piercing points along the line of intersection. Project the points to the frontal view.

3. Determine visibility and connect the points as shown to form the line of intersection.

In Fig. 6-89, the intersection of the horizontal prism and the right circular cone is required.

1. Pass horizontal CP 1 and CP 2 through the upper and lower horizontal planes of the prism. Project the CPs to the horizontal view where they appear as circle elements.

2. Since the upper and lower surfaces of the prism are horizontal planes, the line of intersection coincides with the circle elements cut by the CPs.

3. When the prism's surfaces are not horizontal planes, vertical CPs passed through the axis of the cone in the horizontal view are used. CPs 3, 4, 5, and 6 cut elements along the cone's surface. Their intersection with the prism in the frontal view cuts two elements each on the prism. Intersecting elements determine the line of intersection in the horizontal view.

270

PROB. 6-37

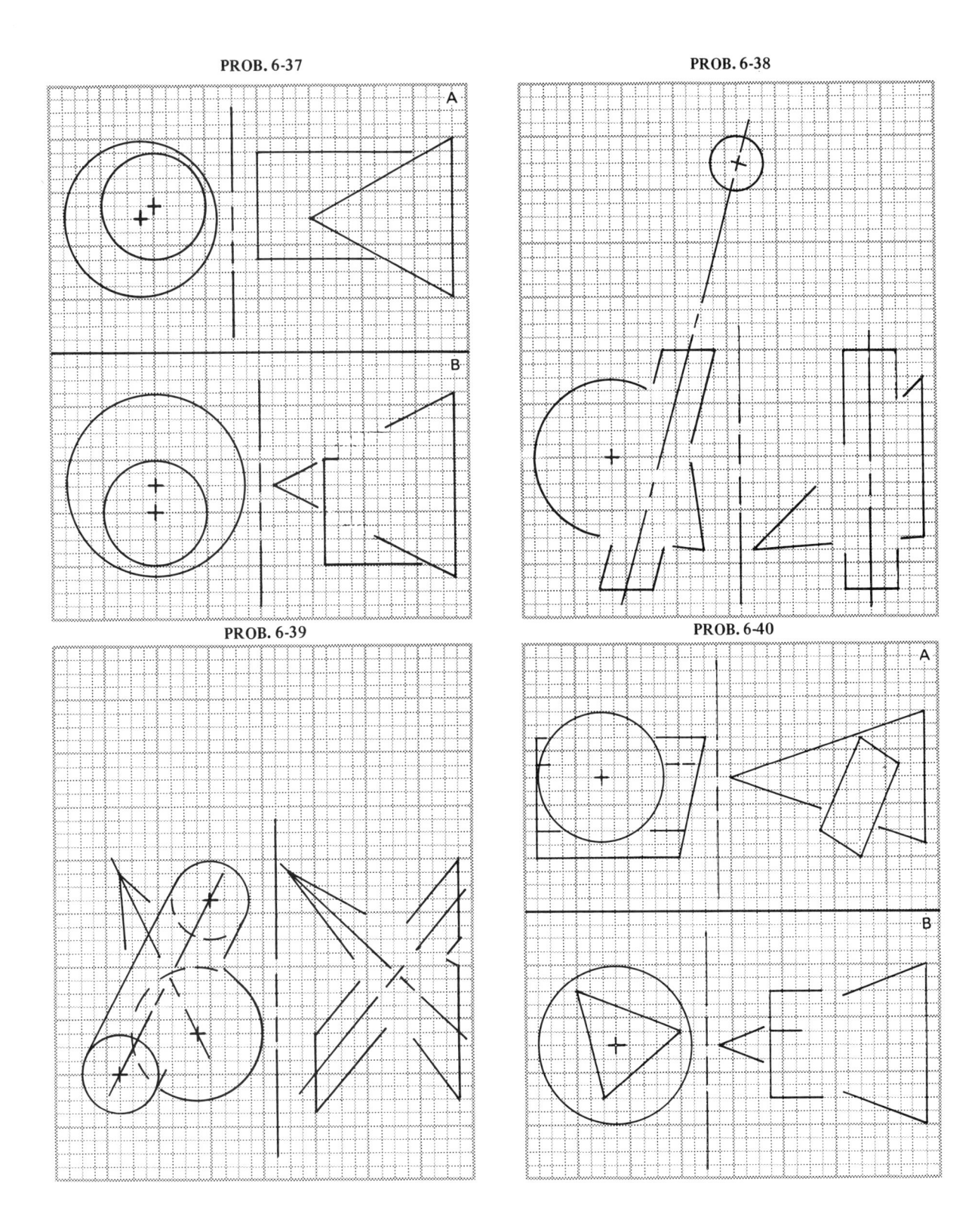

PROB. 6-38

PROB. 6-39

PROB. 6-40

PROBS. 6-37A, 6-37B. Complete the views of the intersecting shapes.

PROB. 6-38. Using the edge view method, determine the intersection between the figures.

PROB. 6-39. Solve for the intersection of the oblique cone and oblique cylinder.

PROBS. 6-40A, 6-40B. Complete the views of the intersecting shapes.

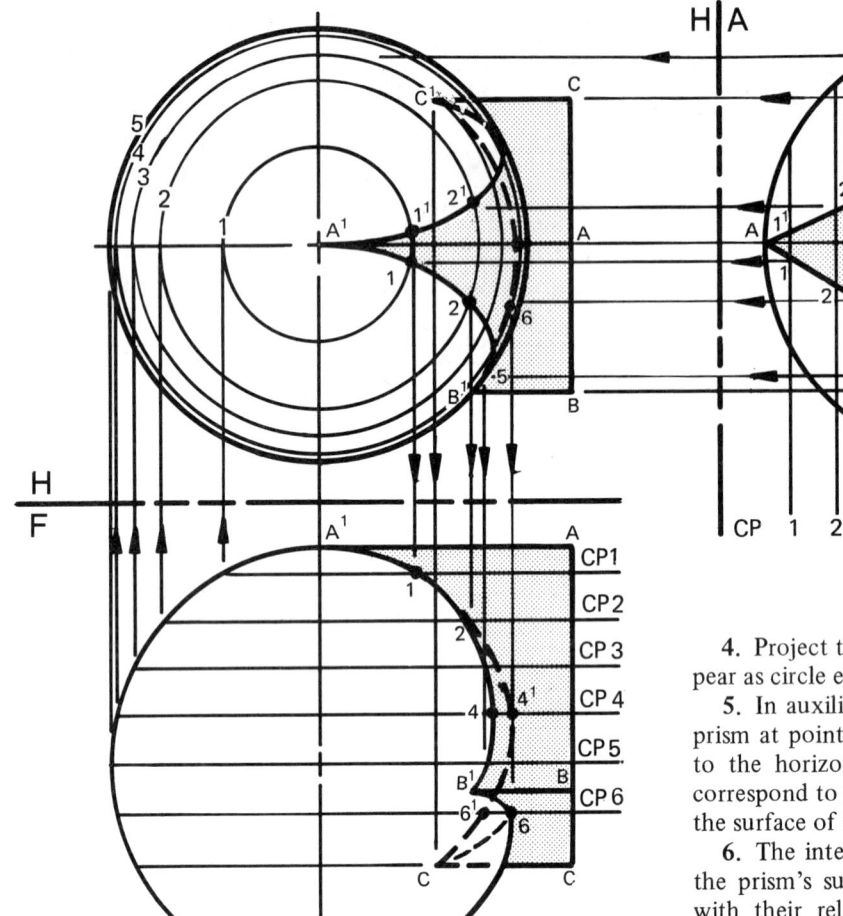

FIG. 6-90 Intersection of a sphere and a prism.

6.51 INTERSECTION OF A SPHERE AND A PRISM

A cutting plane passed through a sphere cuts a circle element on its surface. To find the intersection of a sphere and a prism, pass a series of evenly spaced cutting planes through an edge view of the prism. Each cutting plane cuts a circle element on the sphere and intersects the prism as a straight line element. Related circle elements and straight-line elements intersect at a point on the line of intersection.

In Fig. 6-90, the intersection of the sphere and the triangular prism is required. The frontal and horizontal views are given.

1. Draw auxiliary A perpendicular to the prism. The prism appears as a right section (edge view).
2. Edge line A of the prism is tangent to the sphere. Draw line A tangent to the sphere in the frontal view and project to the horizontal view as shown.
3. Pass a series of evenly spaced horizontal cutting planes through the prism and sphere in auxiliary A. Show the cutting planes in the frontal view.

4. Project the CPs to the horizontal view, where they appear as circle elements on the surface of the sphere.
5. In auxiliary A the CPs intersect the edge view of the prism at points 1 through 6. Project each intersection point to the horizontal view. Note that the intersection points correspond to piercing points of elements cut by the CPs on the surface of the prism with the surface of the sphere.
6. The intersection of the projection lines (elements on the prism's surface) drawn from the points in auxiliary A with their related circle elements establish points on the line of intersection. Note that the prism is composed of three plane surfaces, therefore there are three curved lines of intersection.
7. Project the intersection points to the frontal view. Determine visibility and connect the points in proper sequence to form the line of intersection.

In Fig. 6-91, the space probe was designed using interrelated and intersecting spheres, cones, cylinders, and prisms.

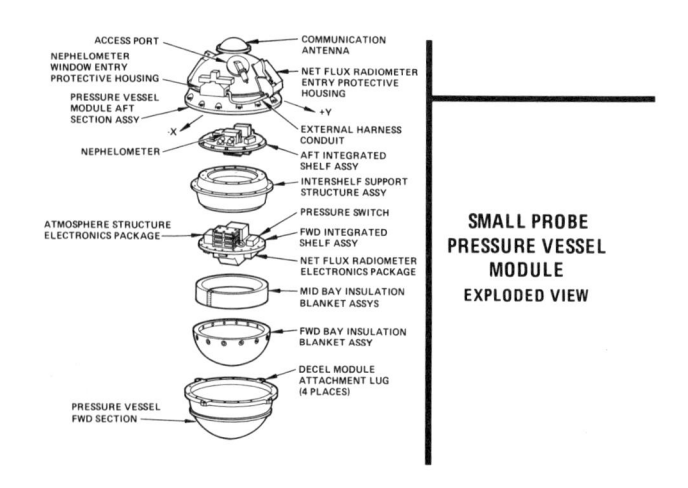

FIG. 6-91 Spherical space probe. *(Courtesy NASA.)*

272

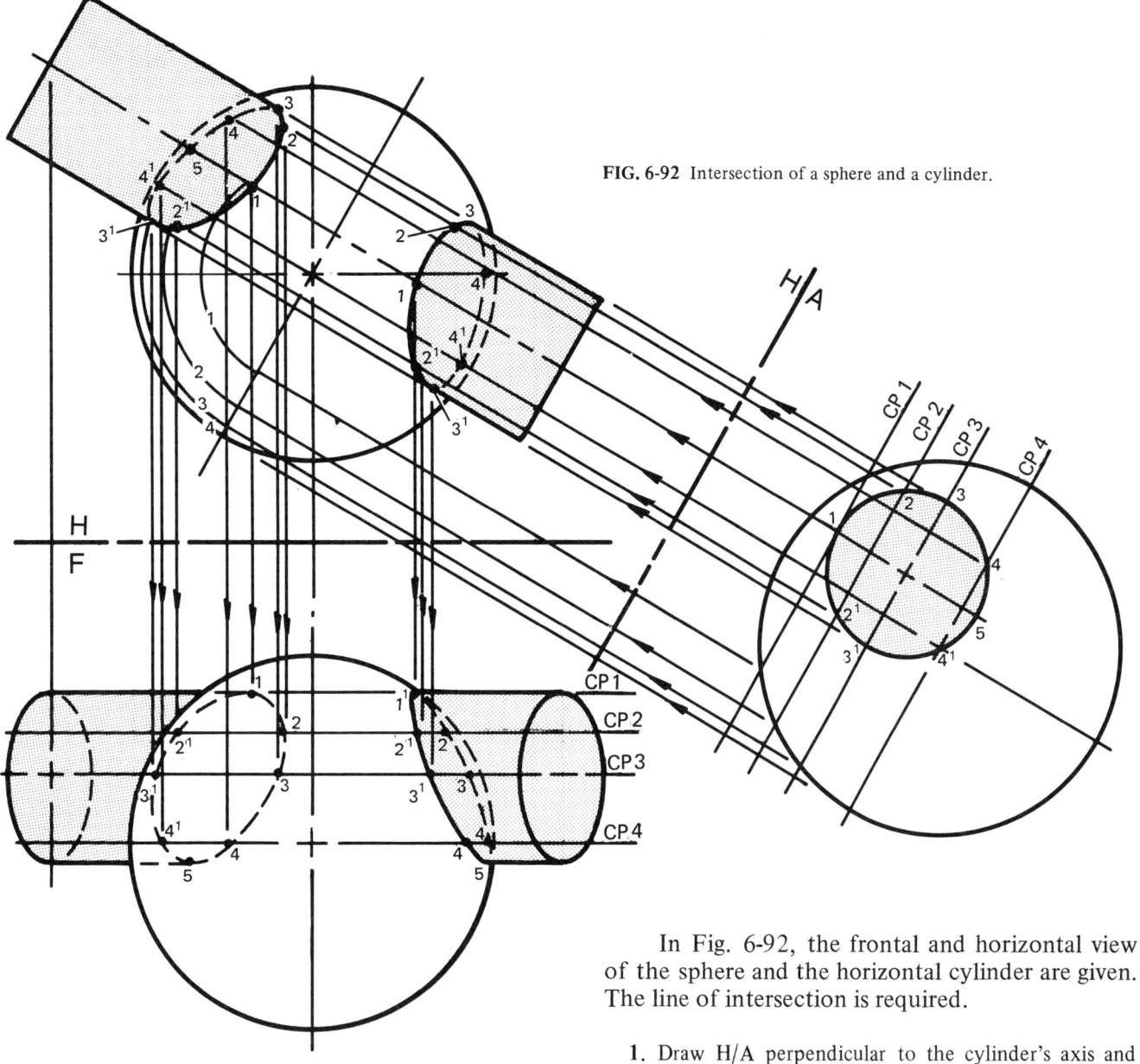

FIG. 6-92 Intersection of a sphere and a cylinder.

In Fig. 6-92, the frontal and horizontal view of the sphere and the horizontal cylinder are given. The line of intersection is required.

1. Draw H/A perpendicular to the cylinder's axis and project auxiliary A. The cylinder appears as a right section with its axis line as a point view.
2. Pass a series of conveniently spaced horizontal cutting planes through the edge view of the cylinder and the sphere in auxiliary A. Show the cutting planes in the frontal view. Project the cutting planes to the horizontal view, where they appear as circles on the sphere.
3. CP 1 cuts a straight-line element along the upper surface of the cylinder and a circular element on the surface of the sphere. The intersection of the sphere's circle element and the cylinder's straight-line element in the horizontal view locates a point on the line of intersection, point 1. Point 1 is the uppermost point on the line of intersection.
4. CP 2 through CP 3 intersect the sphere as circle elements and cut two straight-line elements each on the surface of the cylinder. The intersection of related elements in the horizontal view locates points 2, 2^1, 3, and 3^1.
5. Project all points to the frontal view to their corresponding cutting planes.
6. Determine visibility and connect the points in proper sequence to establish the line of intersection. Since the cylinder goes through the sphere (pierces it), two curved lines of intersection result.

6.52 INTERSECTION OF A SPHERE AND A CYLINDER

By passing a series of cutting planes parallel to the axis of a cylinder and through a sphere, points along the line of intersection can be determined. A cutting plane drawn parallel to the axis of a cylinder will cut straight-line elements along its surface. Any cutting plane passed through a sphere cuts a circle element on its surface. Therefore, the intersection of related circle and straight-line elements establish points on the line of intersection. Each point represents a point common to both the sphere and the cylinder. Cutting planes are conveniently passed parallel to the axis of a cylinder where the cylinder appears as an edge (right section). A right section shows the cylinder's axis line as a point view.

PROB. 6-41

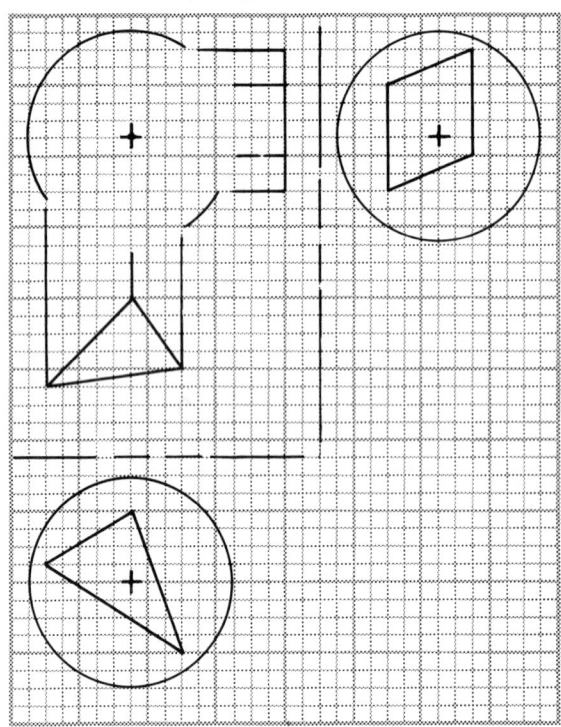

PROB. 6-42

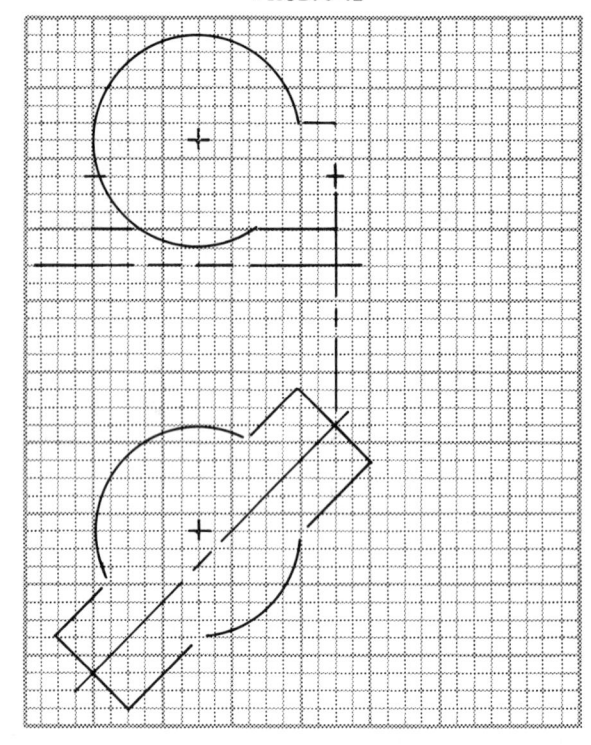

PROB. 6-43

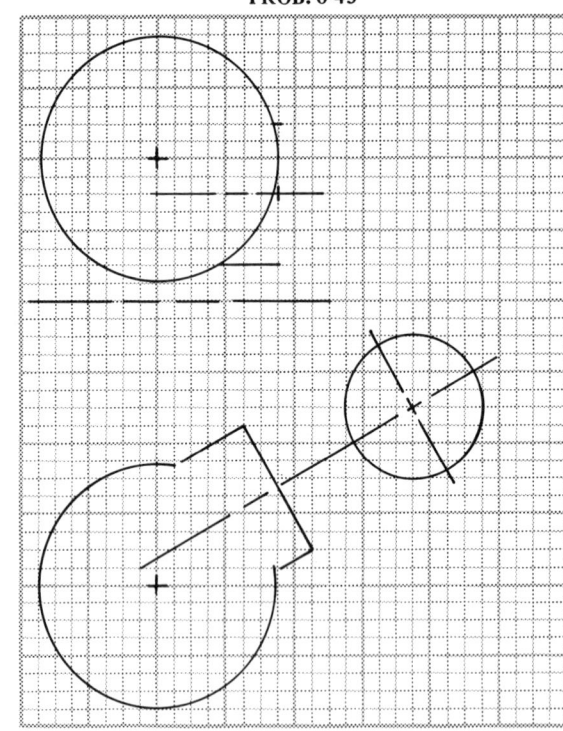

PROB. 6-41. Complete all three views of the intersecting shapes.

PROB. 6-42. Use the edge view method to complete the intersection of the sphere and cylinder.

PROB. 6-43. Solve for the intersection between the sphere and cylinder.

274

6.53 PICTORIAL INTERSECTIONS

The use of *pictorial intersections* will aid in understanding descriptive geometry problems traditionally completed as orthographic projections. The following sections should be studied in conjunction with corresponding concepts presented in the first part of the chapter. Each pictorial example is drawn as an isometric projection of a typical intersection: line piercing plane, intersection of two planes, etc. The examples are partially enclosed in an isometric box. Concepts presented throughout the text can be applied to the pictorial intersections with the understanding that they must be transferred to isometric projections. Therefore, the relationship of lines and planes, parallelism, and cutting planes are an essential part of any solution to a pictorial problem.

In Fig. 6-93, the intersection of the line and plane is required.

1. Line 1-2 and plane 3-4-5 are given. Line 4-5 may be extended to locate where the plane intersects the isometric box.

2. A vertical cutting plane is drawn containing line 1-2.

3. Vertical CP A-B-C-D intersects plane 3-4-5 at points 6 and 7. Line B-C lies on the same vertical isometric box plane as line 3-4, they intersect at point 6. Line C-D lies on the same lower horizontal isometric box plane as line 4-5 and they intersect at point 7. Line 6-7 lies on plane 3-4-5 and intersects line 1-2 at the piercing point of line 1-2 and plane 3-4-5.

4. An alternative method passes an inclined cutting plane through line 1-2. CP E-F-G-H intersects plane 3-4-5 at points 3 and 8. Line 3-8 intersects line 1-2 at the piercing point of line 1-2 and plane 3-4-5.

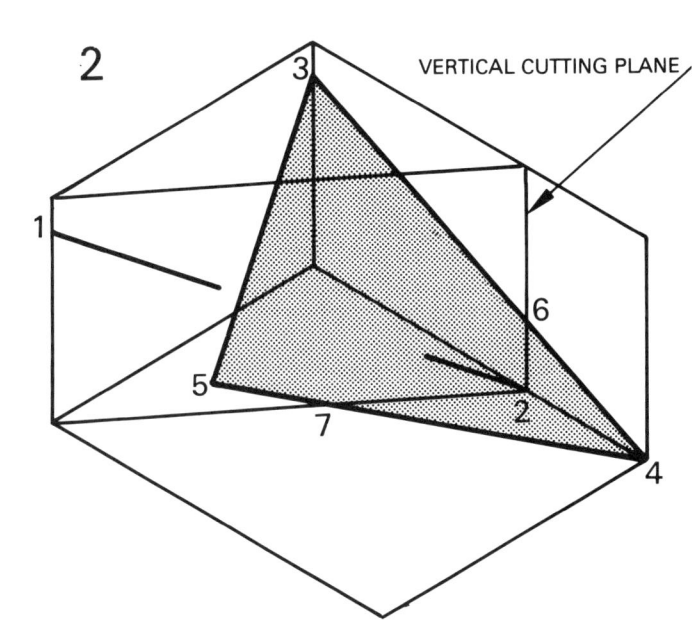

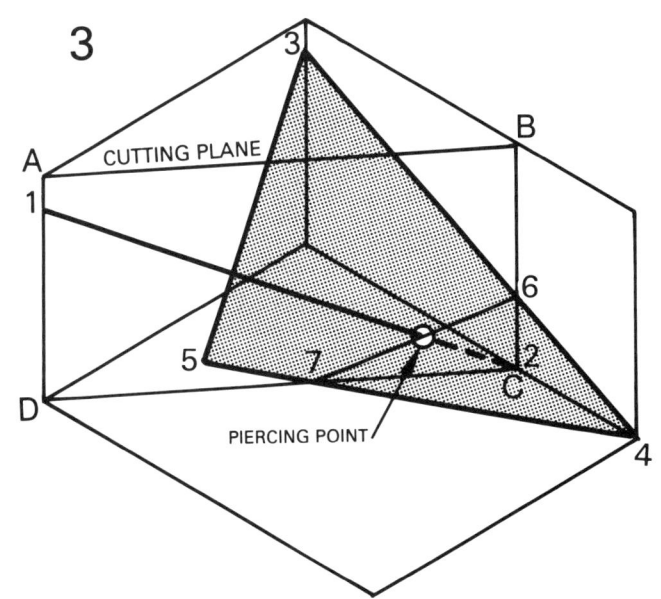

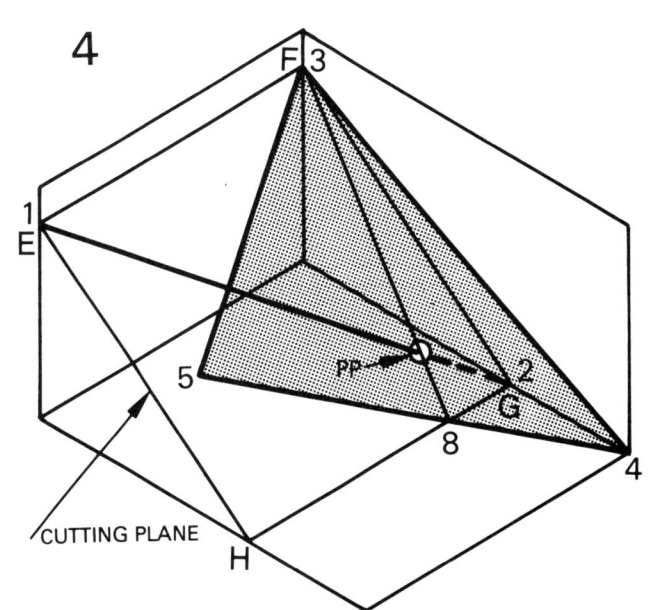

FIG. 6-93 Pictorial intersection of a line and a plane.

FIG. 6-94 Pictorial intersection.

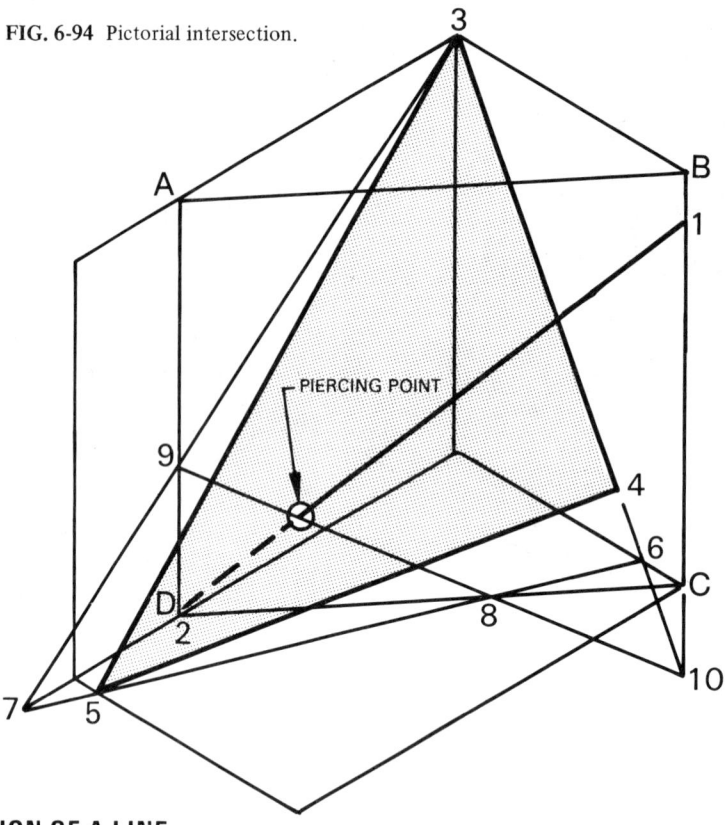

6.54 PICTORIAL INTERSECTION OF A LINE AND A PLANE

Planes and lines can be extended. Lines lying in the same plane must be parallel or intersect. Cutting planes cut straight-line elements on plane surfaces where they intersect. These and other concepts may be applied to pictorial intersection problems.

The intersection of a plane and a line can be determined by passing a cutting plane containing the line through the plane. The intersection of the element cut by the CP on the plane and the given line locates the piercing point of the line and the plane.

In Fig. 6-94, the intersection of line 1-2 and plane 3-4-5 is required. Note line 3-4 lies on the vertical isometric box plane.

1. Pass vertical cutting plane A-B-C-D containing line 1-2 through the plane.
2. Extend line 3-4 to point 6 on the lower horizontal plane. Extend line 5-6 to point 7 on the vertical isometric plane. Plane 3-6-7 now intersects the isometric box along two of its vertical and one of its horizontal planes.
3. CP A-B-C-D intersects plane 3-6-7 at point 8 and 9. Line A-D intersects line 3-7 at point 9. Line C-D intersects line 6-7 at point 8.
4. Connect points 8-9. Line 8-9 lies on plane 3-4-5 and intersects line 1-2 at the piercing point of the given line and plane.
5. To check the answer, extend line 3-4 until it intersects the vertical isometric plane at point 10. Connect points 9-10. Line 9-10 must pass through the piercing point as shown.

In Fig. 6-95, vertical cutting plane A-B-C-D containing given line 1-2 is passed through plane 3-4-5 as shown. Line 4-5 is extended to point 6. Point 6 is at the intersection of the vertical and horizontal isometric box planes. CP A-B-C-D intersects plane 3-5-6 at points 7 and 8. Line 7-8 intersects line 1-2 at the piercing point of the given line and plane.

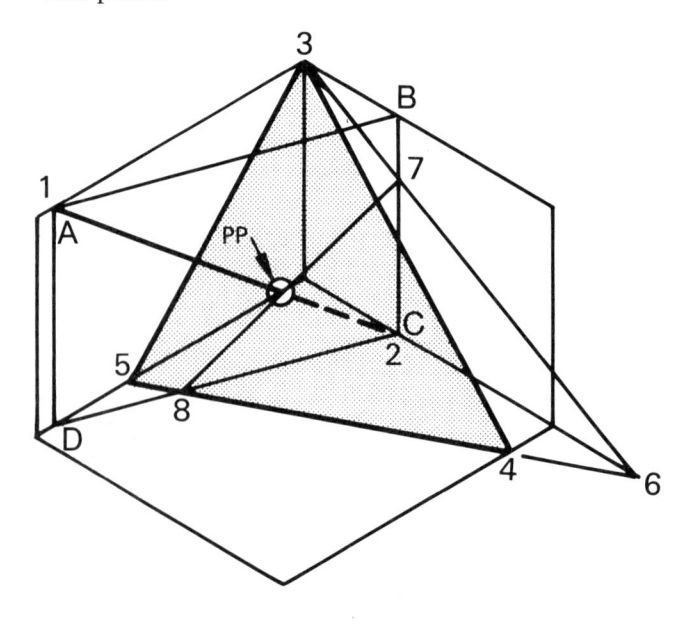

FIG. 6-95 Intersection of a line and a plane.

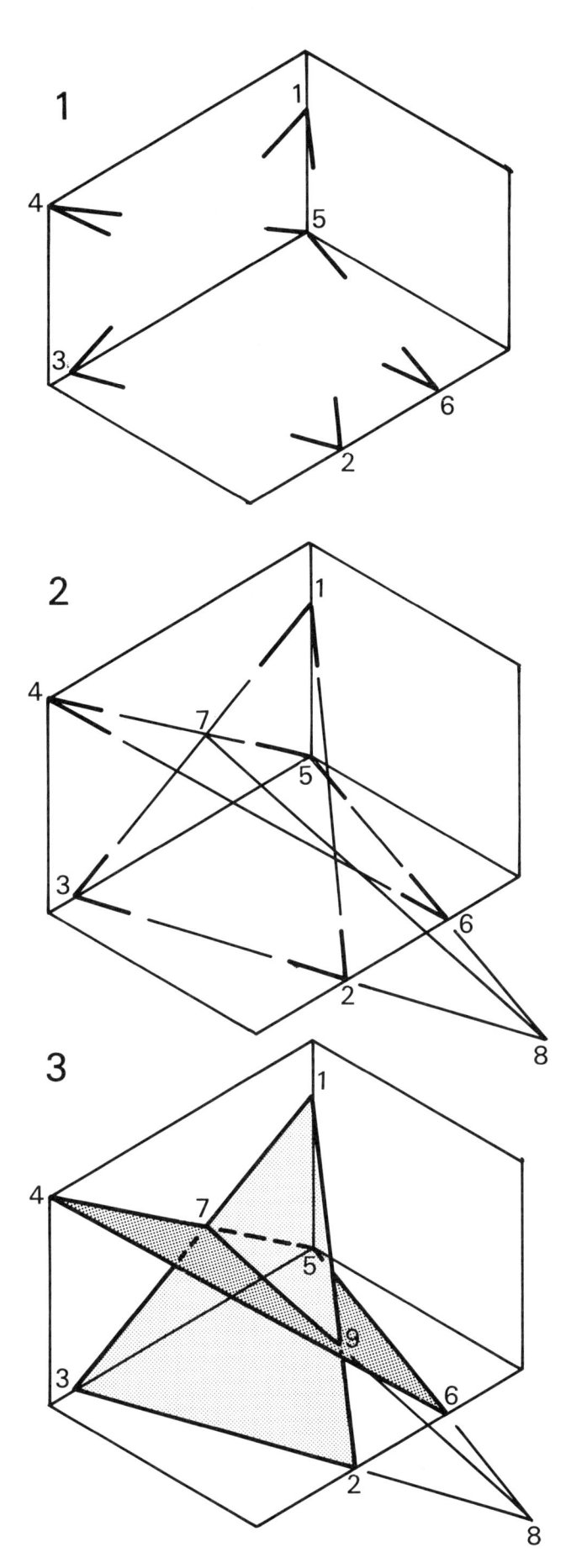

6.55 PICTORIAL INTERSECTION OF TWO PLANES

Two planes in space either intersect or are parallel. The intersection of two planes can be determined by the cutting plane method or by locating the intersection of a line on one plane with a surface of the other plane. Two such intersection points establish two points along the line of intersection between the two planes. By connecting the points (and extending the line if necessary) the line of intersection may be located. Visibility is established by observing the relationship of the planes with the enclosing isometric box.

In Fig.6-96 the intersection of plane 1-2-3 and 4-5-6 is required.

1. Plane 1-2-3 and 4-5-6 are given. Note that line 2-3 and 5-6 lie in the same lower horizontal plane of the isometric box. Line 4-5 and 1-3 lie in the same vertical isometric box plane (1).

2. The cutting plane method is not required, since two points along the line of intersection can be located with the given information. Line 1-3 intersects line 4-5 (and therefore plane 4-5-6) at point 7 (2).

3. Line 2-3 and 5-6 lie in the same plane and are obviously not parallel. Extend line 2-3 and 5-6 until they intersect at point 8. Point 8 is a common point to both planes. Therefore point 8 is on the line of intersection of the two planes (2).

4. Connect points 7 and 8. Line 7-8 is the extended line of intersection between the two limited planes. Line 7-8 intersects line 1-2 at point 9. Point 9 locates the piercing point of line 1-2 and plane 4-5-6. Line 7-9 is the line of intersection between the two given planes (3).

5. Visibility is established by noting that line 5-6 is below and behind the line of intersection. The line of intersection is visible, as is the upper part of plane 1-2-3 as represented by plane 1-7-9. Point 4 is above and in front of plane 1-2-3. Therefore, from the line of intersection to point 4 is visible.

FIG. 6-96 Pictorial intersection of two planes.

PROB. 6-44A, B, C

A

B

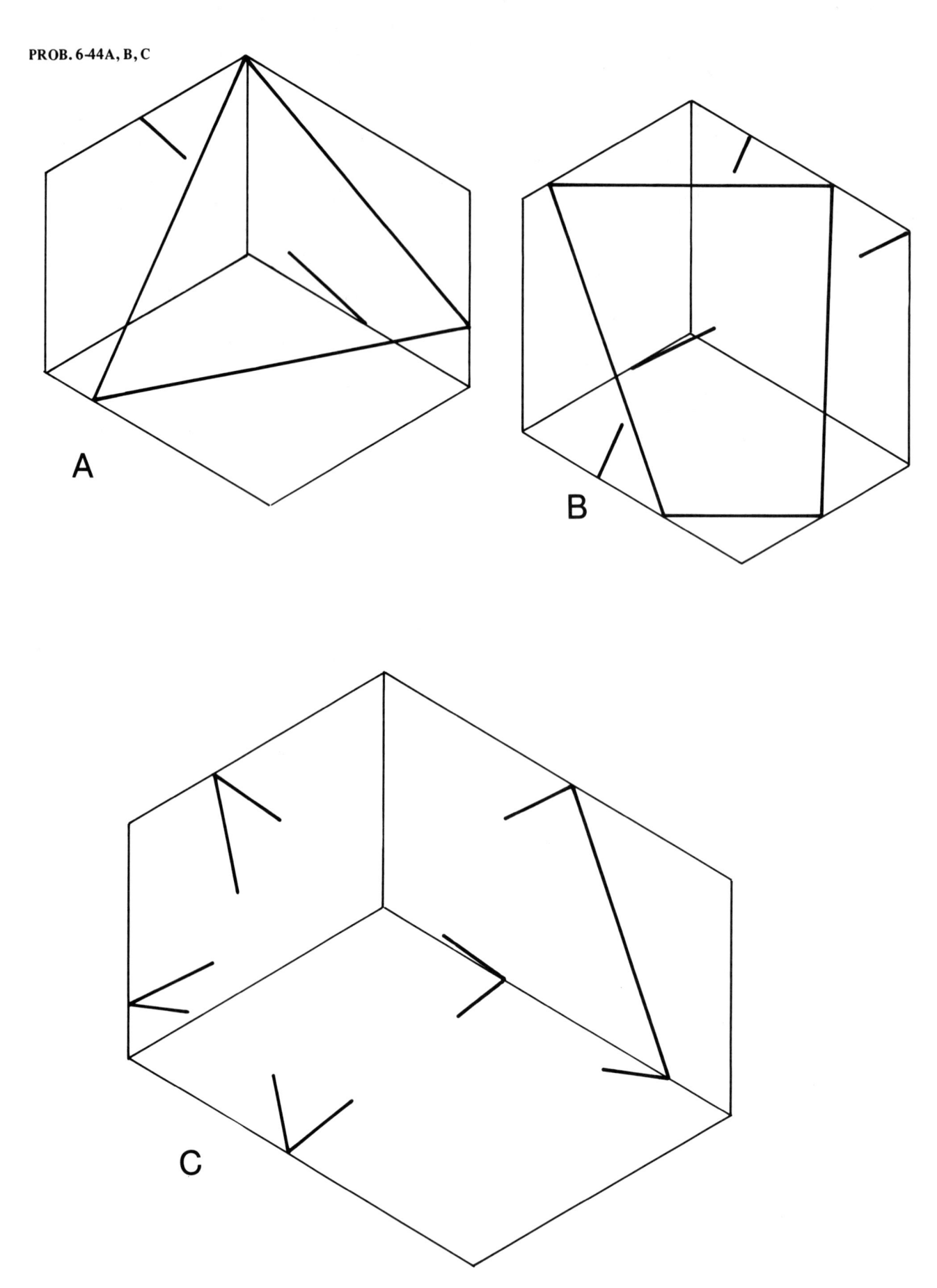

C

PROBS. 6-44 A, B, C. Complete the pictorial intersection problems. These problems can be completed in text if desired.

278

6.56 PICTORIAL INTERSECTION OF PLANES

The intersection of planes can be determined by applying the concept of parallelism to the problem. In Fig. 6-97, planes 1-2-3 and 4-5-6 are given. Their line of intersection is required.

1. Lines 2-3 and 4-5 lie in the same lower horizontal plane. Lines 2-3 and 4-5 intersect at point 7. Point 7 is one point along the line of intersection.
2. Draw a line from point 6 parallel to line 4-5. This line lies in the upper horizontal isometric box plane since it was drawn parallel to line 4-5.
3. Draw a line from point 1, parallel to line 2-3, until it intersects the line drawn from point 6 at point 8. Line 1-8 is in the upper horizontal plane, since it is parallel to line 2-3.
4. Point 8 is a common point for both given planes. Point 8 lies on the extended line of intersection. Connect point 7 and 8. Line 7-8 intersects line 4-6 at point 9. Line 7-9 is the line of intersection.

In Fig. 6-98, the intersection of planes 4-5-6 and 7-8-9 with plane 1-2-3 is required.

1. Plane 1-2-3 intersects plane 4-5-6 at points A and B. Line A-B is the line of intersection.
2. Extend line 7-8 to point 10. Line 9-10 lies on the lower horizontal plane. The intersection of line 9-10 and line 2-3 locates point 13. Point 13 is one point on the extended line of intersection.
3. Extend line 1-3 to point 11. Line 2-11 intersects line 7-8 on the vertical plane at point 12.
4. Connect points 12 and 13 to form the extended line of intersection. Line 12-13 intersects line 1-2 and line 8-9 at points C and D respectively. Line C-D is the line of intersection.

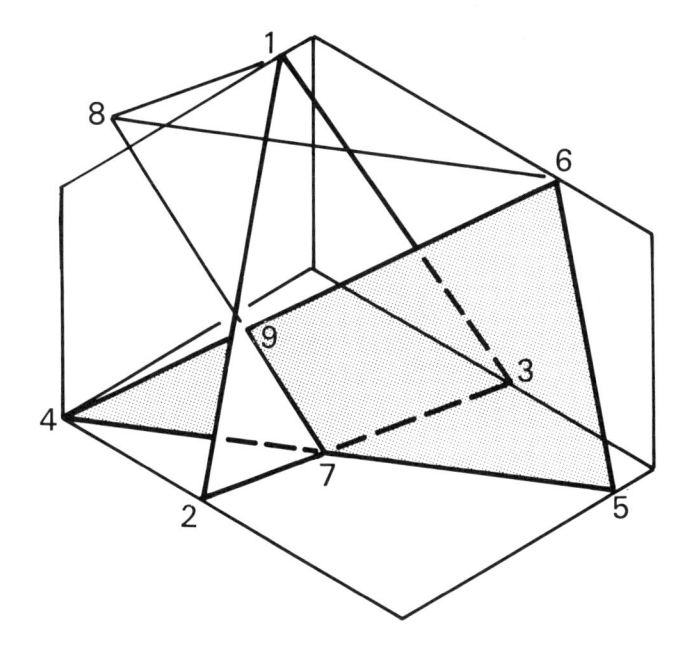

FIG. 6-97 Intersection of two planes.

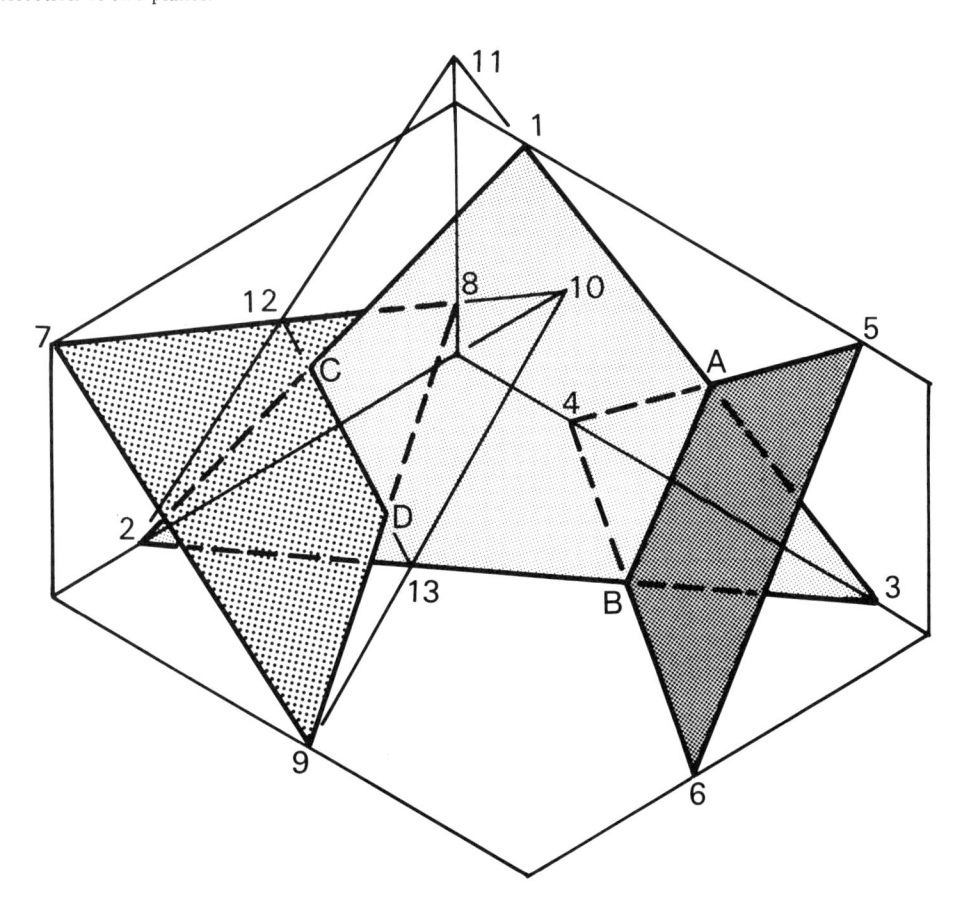

FIG. 6-98 Pictorial intersection of three planes.

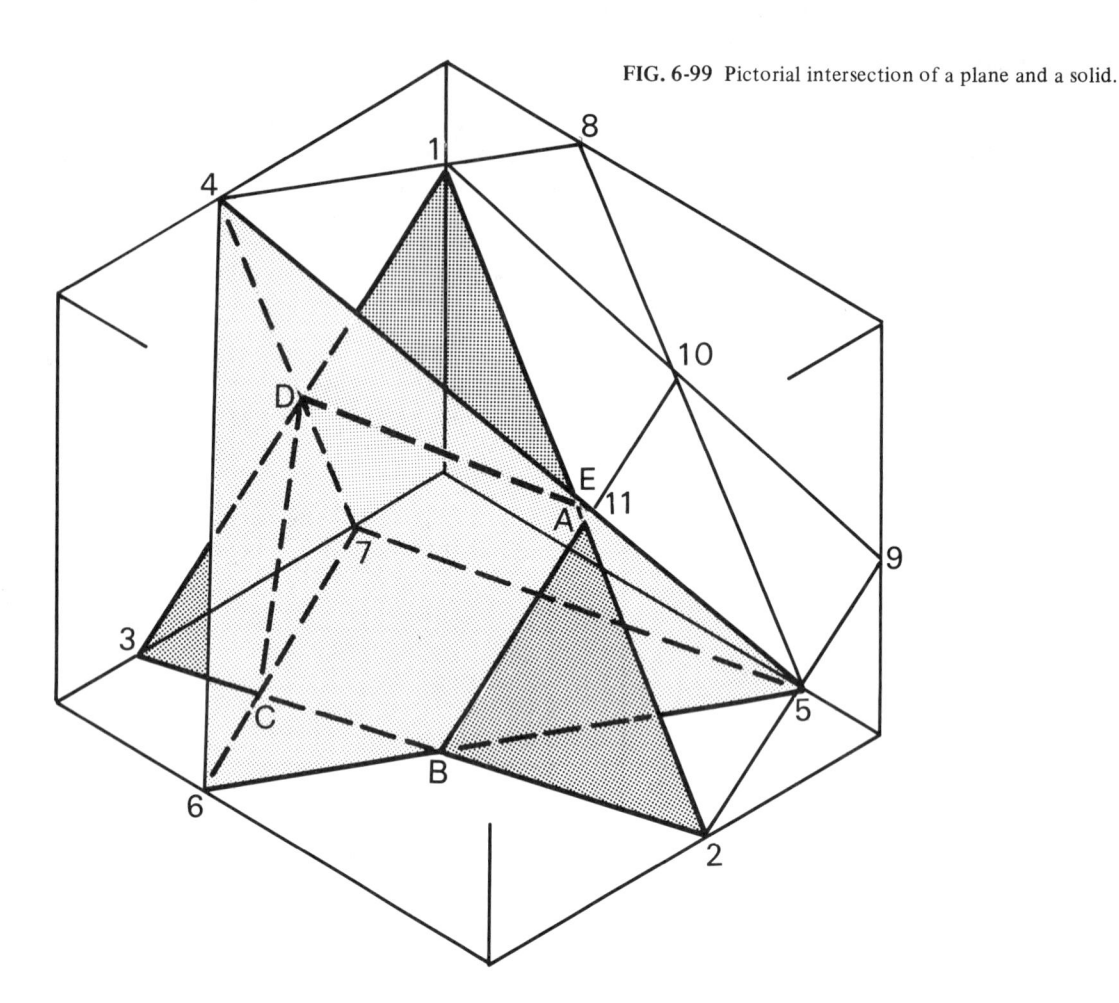

FIG. 6-99 Pictorial intersection of a plane and a solid.

6.57 PICTORIAL INTERSECTION OF A PLANE AND A SOLID

The pictorial intersection of a plane and a solid can be established by determining the intersection of the plane and each surface of the solid. In Fig. 6-99, the line of intersection between plane 1-2-3 and figure 4-5-6-7 is required.

1. Plane 5-6-7 and line 2-3 lie in the lower horizontal plane. Line 2-3 intersects line 5-6 at point B and line 6-7 at point C. Connect points B and C. Line B-C is a portion of the total line of intersection.
2. Line 1-3 and line 4-7 lie in the same vertical plane. Line 1-3 intersects line 4-7 at point D. Connect point C and D. Line C-D is another portion of the line of intersection.
3. Draw line 4-8 parallel to line 5-6. Connect point 8 and point 5 to form plane 4-8-5-6. Line 4-8 is in the upper horizontal plane. Note that line 4-8 is above and in front of point 1.
4. Draw line 2-9 parallel to line 1-3. Line 2-9 is in a vertical plane. Note that line 2-9 is in front of and above point 5. Connect point 9 and point 1 to form plane 1-9-2-3.
5. Lines 1-9 and 5-8 lie on the same plane and intersect at point 10. Point 10 is one point on the extended line of intersection between plane 1-2-3 and plane 4-5-6.
6. Connect point 10 and point B. Line B-10 pierces plane 4-5-6 at point A. Line A-B is a portion of the line of intersection.
7. Point E is the piercing point of line 1-2 and plane 4-5-7. The construction required to locate this point is not shown. Note that the line drawn between points A and E lies within the solid. Line D-E is the last part of the line of intersection.

The petrochemical project modeled in Fig. 6-100 exhibits a variety of intersecting geometric forms.

FIG. 6-100 Petrochemical model.

280

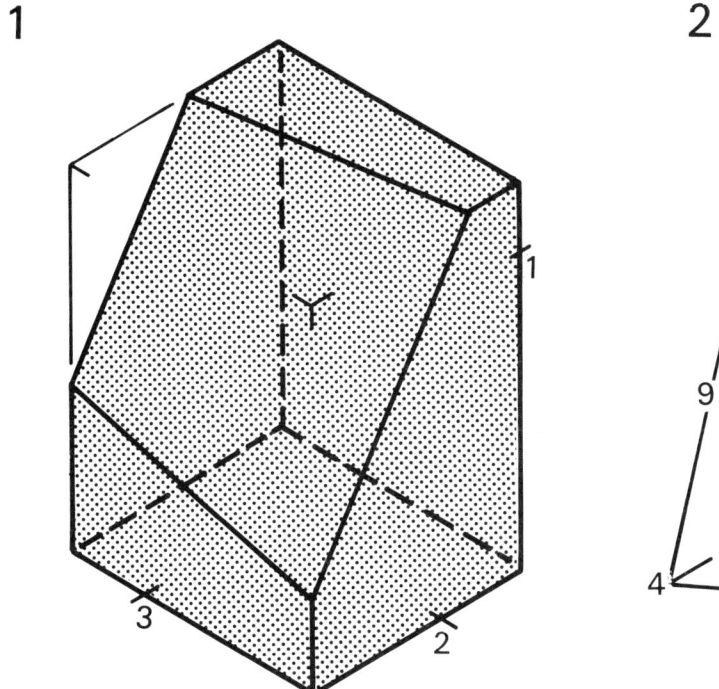

FIG. 6-101 Pictorial intersection of a plane and a solid object.

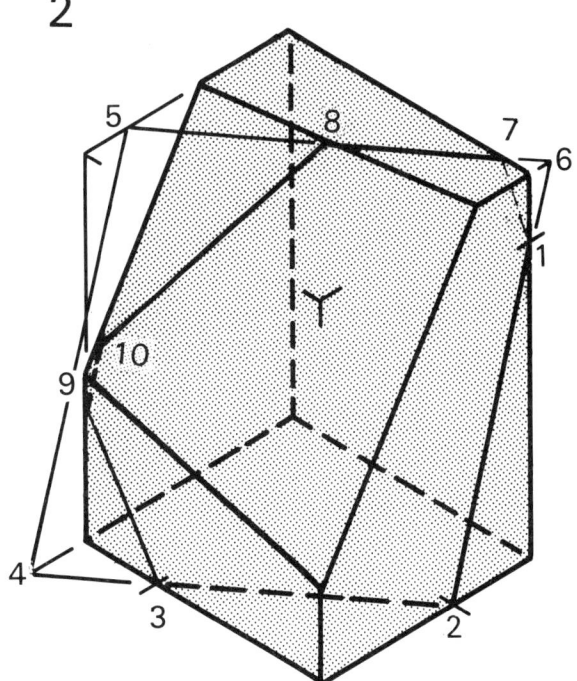

6.58 PICTORIAL INTERSECTION OF A PLANE AND A SOLID OBJECT

The intersection of a plane and a given solid object can be established by applying previous concepts of parallelism, lines on planes, and intersection of lines and planes. In Fig. 6-101, the intersection of the solid block and the plane is required.

1. Plane 1-2-3 and the object are given in isometric projection (1).
2. Point 1 and point 2 lie on the same vertical plane. Line 1-2 is the line of intersection of the plane and the solid on the right vertical surface. Points 2 and 3 lie on the base plane of the object and are connected to form line 2-3. Line 2-3 is a portion of the required line of intersection (2).
3. Extend line 2-3 until it intersects the left vertical surface of the solid at point 4. Extend line 1-2 to point 6 where it intersects the upper horizontal face of the object (2).
4. Draw a line from point 4 parallel to line 2-6 until it intersects the upper horizontal surface at point 5. Connect point 5 and point 6. Line 2-6 is parallel to line 4-5. Line 5-6 is parallel to line 2-4. Line 5-6 intersects two lines on the upper horizontal surface at points 7 and 8. Line 7-8 is a portion of the line of intersection, as is line 1-7 (2).
5. Draw line 3-9 parallel to line 1-7 and line 9-10 parallel to line 1-2. Line 8-10 completes the line of intersection.

In Fig. 6-102, plane 1-2-3 and the object are given. Extend line 1-2 to point 4. Draw line 5-6 parallel to line 1-4 and through point 3. Line 1-6 and line 4-5 are parallel. Plane 1-4-5-6 intersects the object as shown and establishes the line of intersection, 1-10, 10-9, 9-8, 8-5, 5-7, 7-12, and 12-1.

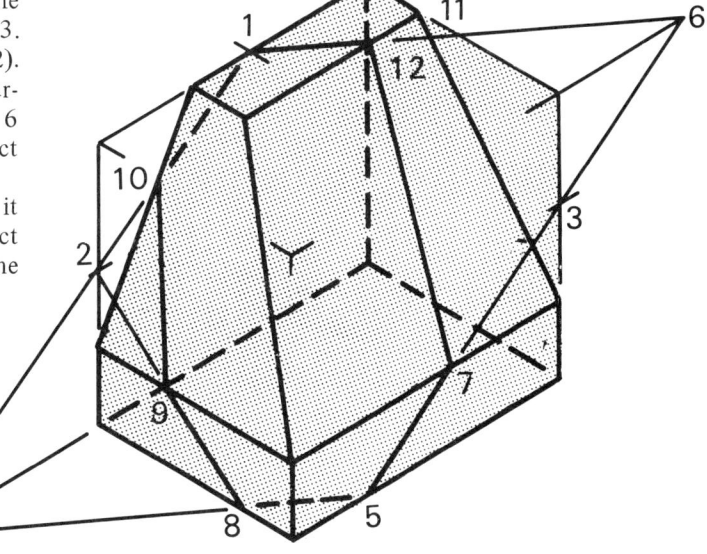

FIG. 6-102 Pictorial intersection.

PROB. 6-45A, B, C

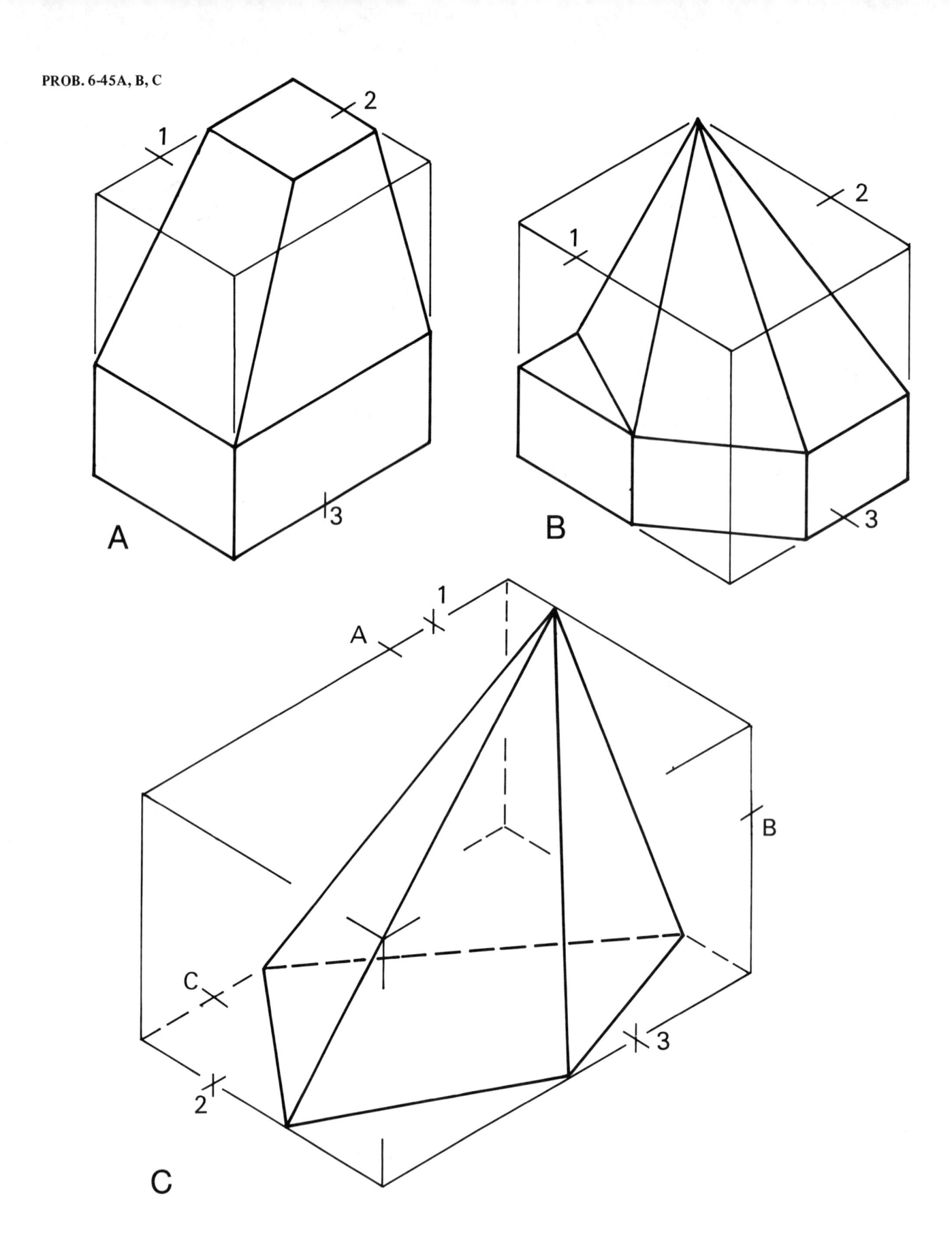

PROBS. 6-45 A, B, C. For each pictorial problem, solve for the intersection of the plane and the solid. A choice of planes is given in problem C.

282

PROB. 6-46A through F

A

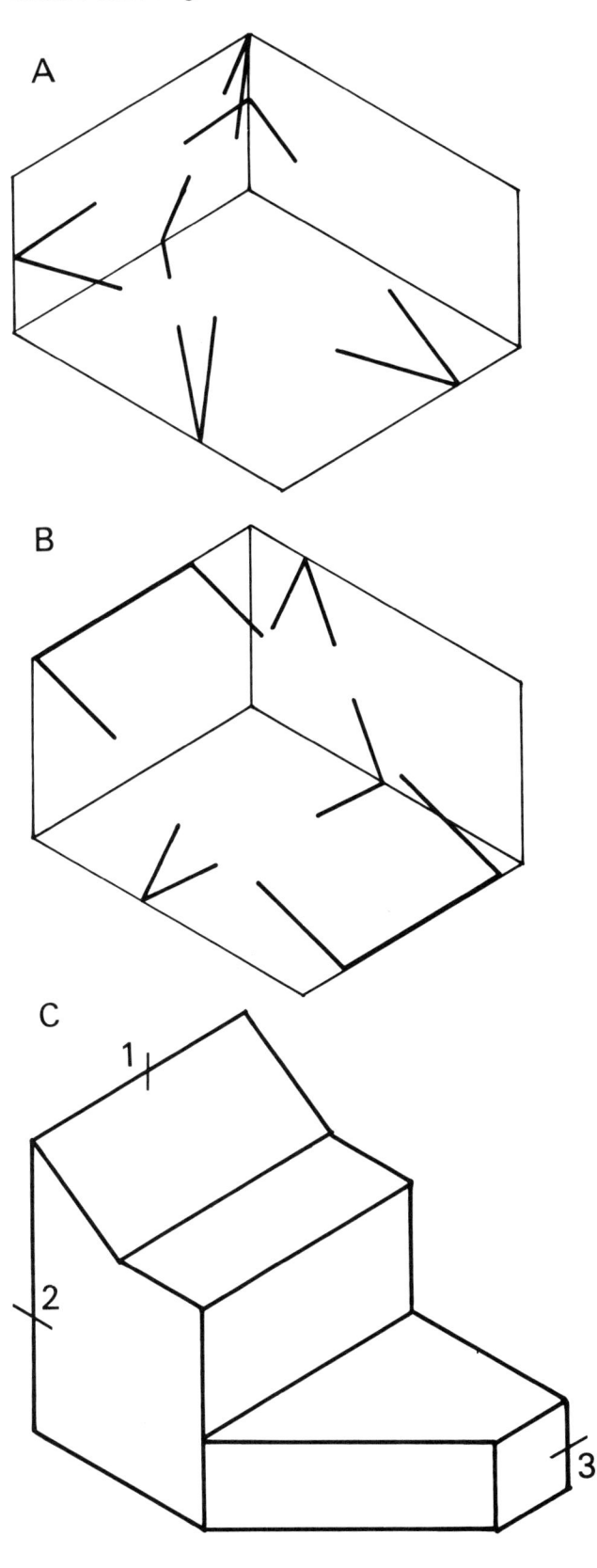

B

C

D

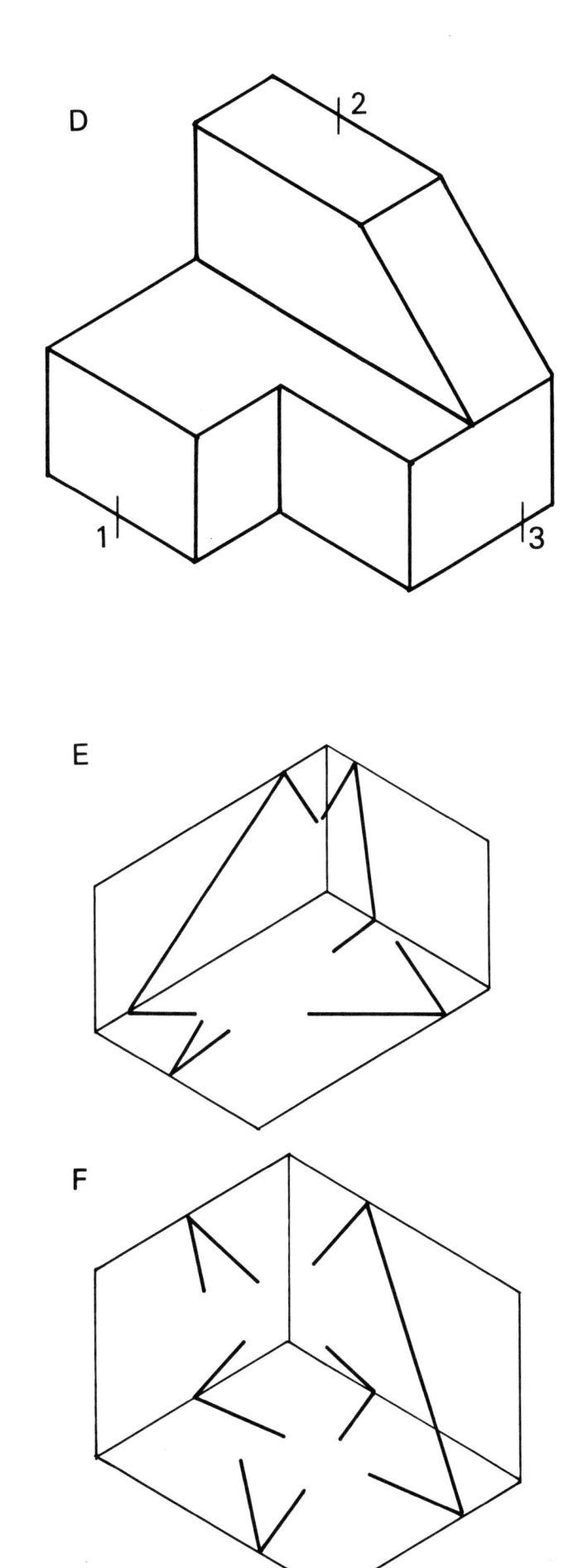

E

F

PROBS. 6-46 A through F. Complete the pictorial intersection problems as assigned. Note that these problems as well as problems 6-45 A through C can be drawn in the book.

283

6.59 ONE-POINT PERSPECTIVE (PICTORIAL INTERSECTION)

One-point perspective intersection problems can be solved by applying the concept of parallelism to lines lying in the front and rear surfaces of the object. Extending lines of the plane to intersect the

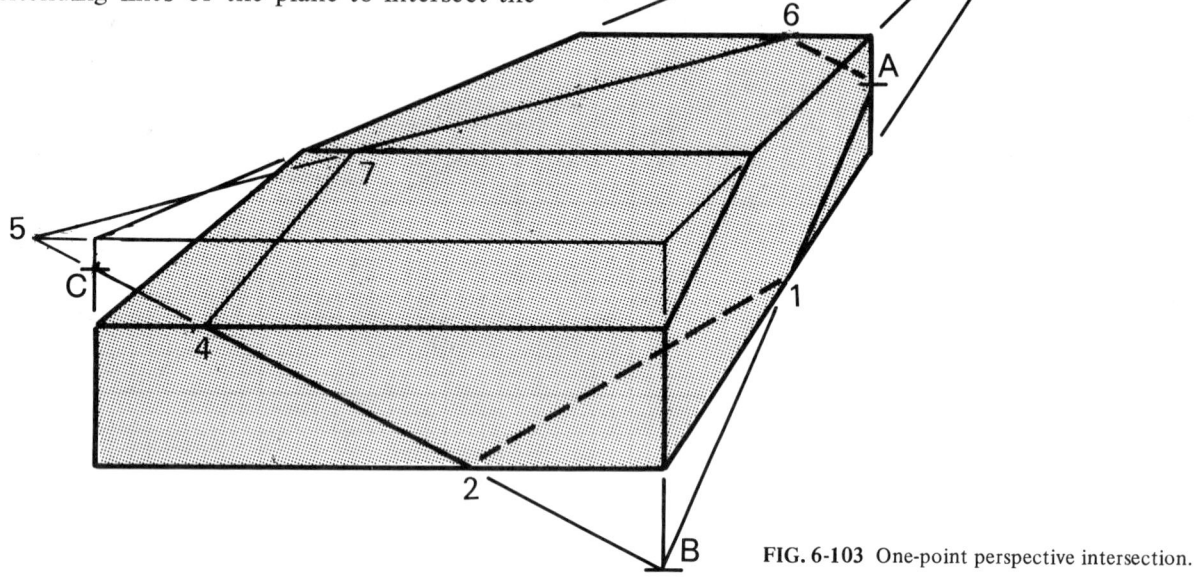

FIG. 6-103 One-point perspective intersection.

object's edge lines helps establish the line of intersection on receding surfaces. In Fig. 6-103 the intersection of plane A-B-C and the object shown in perspective projection is required.

1. Line A-B lies on the right vertical face of the object. Line A-B intersects the base line of the object at point 1. Line A-1 is a visible portion of the total line of intersection.
2. Line B-C lies in the front vertical face of the object. The intersection of line B-C with lines of the object at points 2 and 4 locates the line of intersection on the front face.
3. Connect points 1 and 2. Line 1-2 determines the line of intersection on the lower base plane.
4. The front and rear vertical faces of an object drawn as a one-point perspective are parallel to the projection plane. Extend line B-C to point 5. Draw line A-6 parallel to line B-5. The intersection of line 5-6 and the object establishes point 7. Line 6-7 and line 7-4 complete the line of intersection.

In Fig. 6-104, the intersection of plane A-B-C and the object is required.

1. Line A-B intersects the front face along points 1 and 4 and establishes part of the line of intersection.
2. Line B-C intersects a line in the left vertical face at point 2. Line C-2 is a portion of the line of intersection.
3. Draw line C-3 and line 2-5 parallel to line A-B. Line 2-5 and C-3 establish two more parts of the line of intersection.
4. Connect points 3 and 4 to form intersection line 3-4. Connect points 1 and 5 to form the last part of the line of intersection.

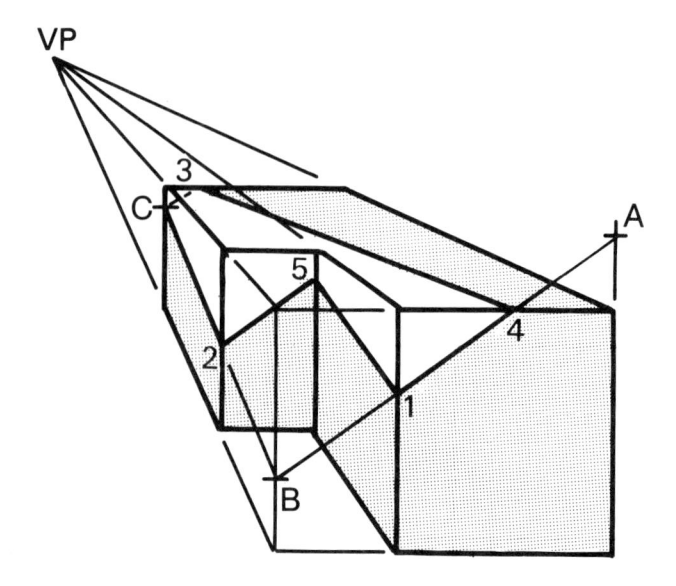

FIG. 6-104 Perspective intersection.

PROB. 6-47

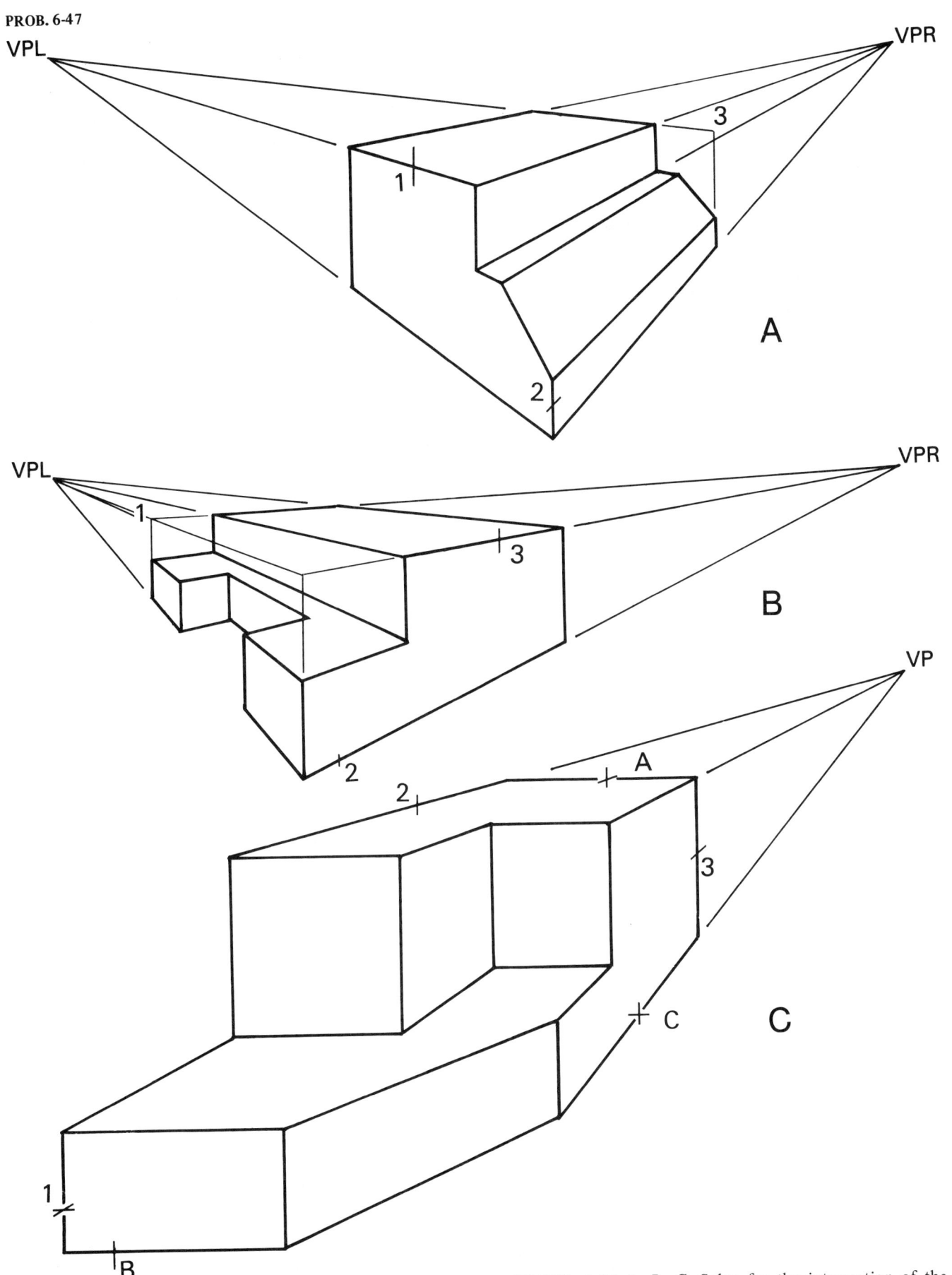

PROBS. 6-47 A, B, C. Solve for the intersection of the plane and the perspective figure. Problem C has two planes to choose from. These projects can be completed in the text if desired.

285

FIG. 6-105 Pipelaying ship model. *(Courtesy Magee-Bralla, Inc.)*

FIG. 6-106 F8 aircraft. *(Courtesy Magee-Bralla, Inc.)*

6.60 FAIR SURFACES

Fairing** is shaping a member or structure for the primary function of establishing a smooth, continuous outline.* Fairing a surface of a boat's hull reduces drag caused by contact with the water, Figs. 6-105 and 6-107. Fairing a car's body surface reduces air resistance and increases gasoline mileage. The contour surfaces of an airplane require fairing to eliminate drag and air resistance, Fig. 6-106. Each of the above-mentioned surfaces, car, ship, and airplane, is made up of ***double-curved surfaces. *A double-curved surface is generated by the movement of a curved line.* Ship hulls, automobile bodies, airplane fuselages are double-curved surfaces designed by drawing curved contour lines representing lines of the surface. Of course, graphical design is not used to originate streamlined shapes or advanced aerodynamic aircraft, efficient swift ships, etc. Computer calculations, computer modeling, and scale modeling all have their places in the total design effort. When a large double-curved surface such as a ship is being constructed, lines and contours are laid out full scale. This process is called *lofting*.

The actual process of fairing a surface is done by passing a series of vertical and horizontal sections (cutting planes) through the given structure to be contoured. Each section position is called a *station*. **Horizontal sections** are called *water lines* (WL) when the surface is a ship's hull as in Fig. 6-108. ***Transverse*** (perpendicular to the hull) ***vertical sections*** are called *frame lines* (FL). ***Vertical sections*** passed through the length of the hull cut *buttock lines* (BL). BL, FL, and WL stations are numbered or lettered and generally are evenly spaced, as shown in Fig. 6-108.

Frame lines appear as straight lines in the breadth plane (top view) and the body plan, and as contour lines in the sheer plan (side view). Note that the choice of principal projection plane positions is different when drawing a ship's hull and fairing its surface. Buttock lines are straight lines in the breadth plan and body plans, and contours in the sheer plan. Water lines are straight lines in the body and sheer plans and contours in the breadth plan.

Fairing is accomplished by the drawing of water lines and buttock lines in accordance with the given frame lines. The resulting surface must be a smooth and continuous one. To check the fairness of a surface, the intersection of an assumed frame line and a series of plane sections representing cutting planes passed through a water line or a buttock line must result in a smooth curved line. A series of points located by the intersection of frame lines and station sections can be plotted along the surface of the given ship's hull. These points must agree with one another between the two *adjacent* projections.

FIG. 6-107 Oceangoing barge. *(Courtesy NASA.)*

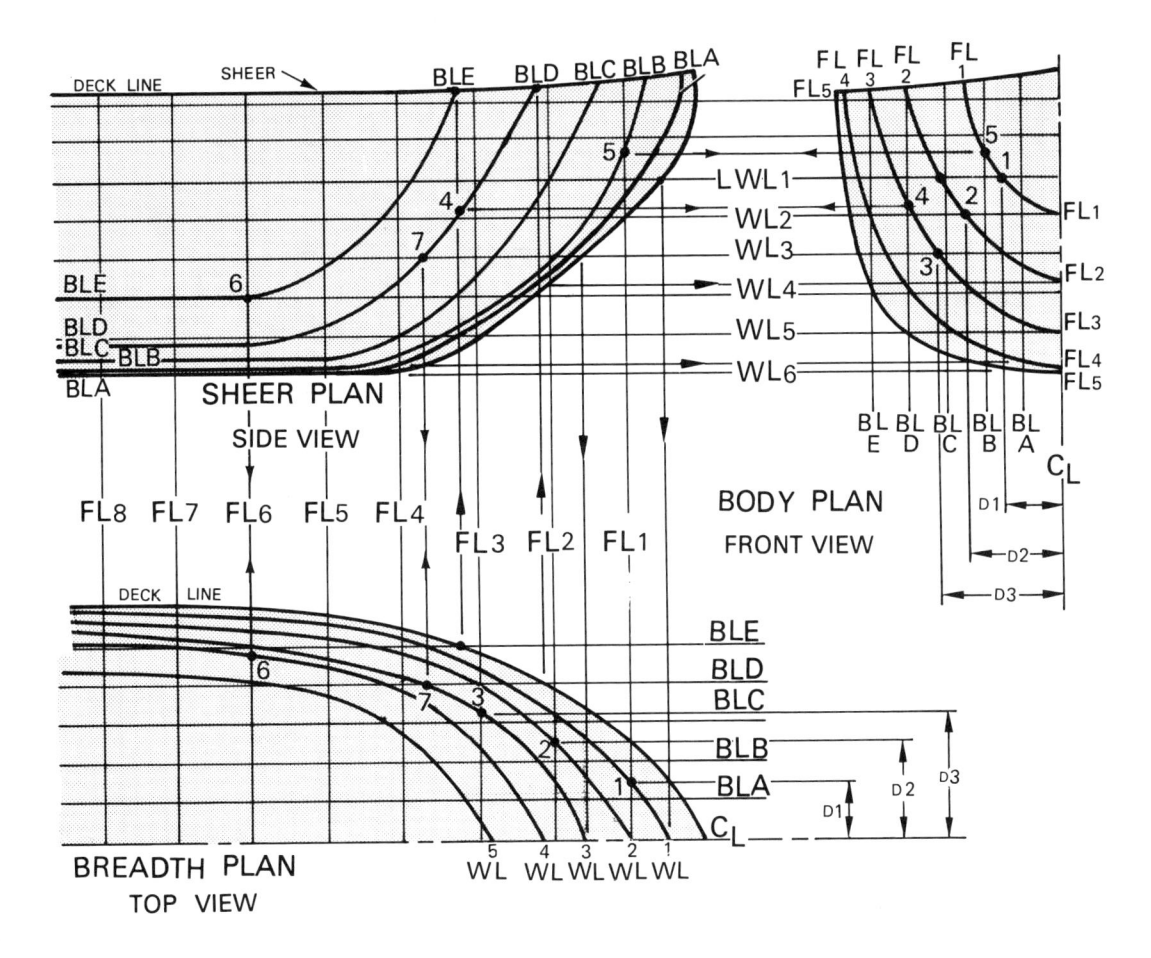

FIG. 6-108 Fairing of a ship's hull.

6.61 FAIRING

The bow (forward portion) of a ship's hull is shown in Fig. 6-108 in breadth, sheer, and body plan views. The frame lines are given and the fairness of the contour must be determined. Note that when a particular set of given frame lines do not pass a fairing test they must be *repositioned*. The process of adjustment or repositioning of the contour of the surface is called *fairing*. In the given example, frame lines 1 through 5 are seen. Each frame line represents a section cut by a transversal cutting plane passed through the stations in the sheer and breadth plans. The water lines are evenly spaced horizontal cutting planes. LWL is the *load water line*.

The intersection of a water station section and a frame line results in a point on the surface of the hull. This point must be aligned between the two views as is point 5. A series of such points must result in a smooth continuous curved line. The intersection of a buttock station section with a frame line results in a point on the surface of the hull.

The intersection of LWL and FL1 results in point 1 as seen in the body plan. Points 2 and 3 result from the intersection of WL2 and WL3 with FL2 and FL3 respectively. The breadth plan location of point 1 must be located on FL1 and dimension D1 from the centerline of the hull, likewise for points 2 and 3. As can be seen, all three points fulfill this requirement.

The intersection of the buttock station section BLB and frame line FL results in point 5. Point 5 must lie on BLB at FL1 in the sheer plan. The intersection of station BLB and FL3 results in point 4. Point 4 should lie on FL3 and BLD in the sheer plan. As can be seen, it does not. In this case the buttock line needs to be adjusted.

The intersection of a buttock station section and a water line in the breadth plan must result in a common point. BLD intersects WL3 at point 7. Point 7 must lie on WL3 and BLD in the sheer plan, as it does. A final test is made where WL4 intersects BLE in the sheer plan. Point 6 must be on WL4 and BLE in the breadth plan. Again it can be seen that the contour line needs to be slightly adjusted to fulfill this requirement.

PROB. 6-48

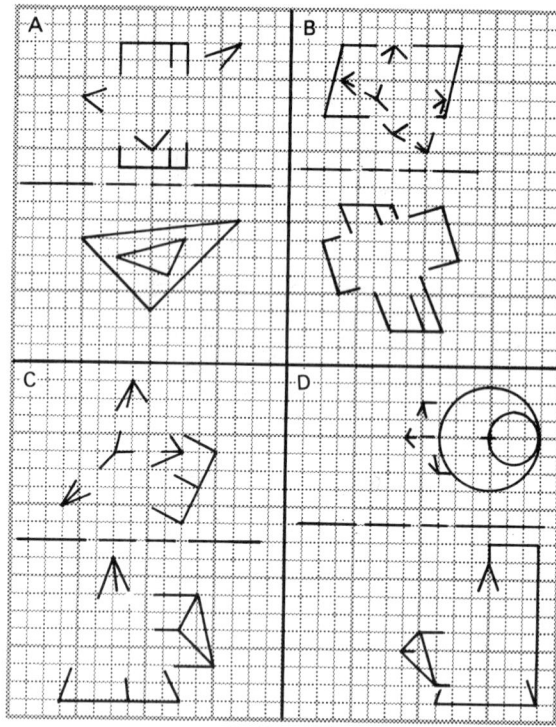

PROB. 6-49

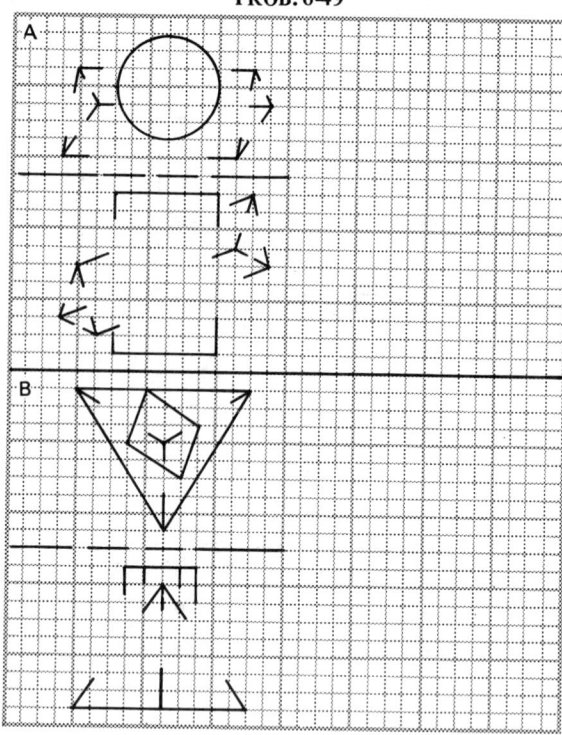

PROB. 6-50

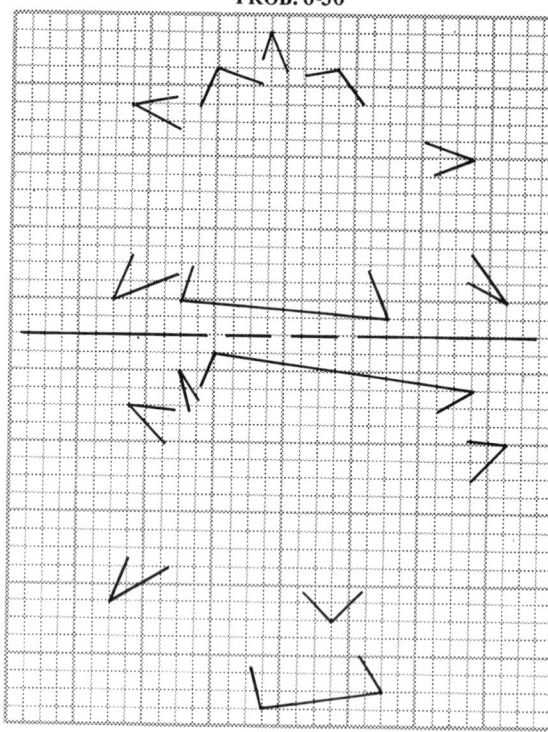

PROB. 6-48D. Project a left profile view and solve for the intersection of the cone, cylinder, and prism.
PROB. 6-49A. Use the edge view method to determine the intersection.
PROB. 6-49B. Draw a side view and show the intersection of the two shapes.
PROB. 6-50. Determine the intersection between planes 1-2-3, A-B-C, and L-M-N.

QUIZ

1. Define *prism*.
2. Briefly describe the difference between the two methods of solving for intersections.
3. When a plane intersects a cylinder at an angle to its axis, the resulting intersection will be what type of shape?
4. What is a torus?
5. Define *conic section*.
6. What is the individual line method of intersection?
7. Intersecting cylindrical shapes are common in what engineering area?
8. Define *fairing*.
9. What is lofting and where is it used?
10. Define *point of convergence*.

TEST

PROB. 6-48A. Determine the intersection between the plane and prism.
PROB. 6-48B. By auxiliary view, determine the intersection of the plane and prism.
PROB. 6-48C. Solve for the intersection between the two shapes.

PROB. 6-51A through H

EXTRA ASSIGNMENTS

PROBS. 6-51 A through H. Instructor can use these problems as alternative assignments. They can be scaled directly from the text and drawn in a specified scale using SI or English units.

PROB. 6-52A through E

A

B

C

D

E

EXTRA ASSIGNMENTS

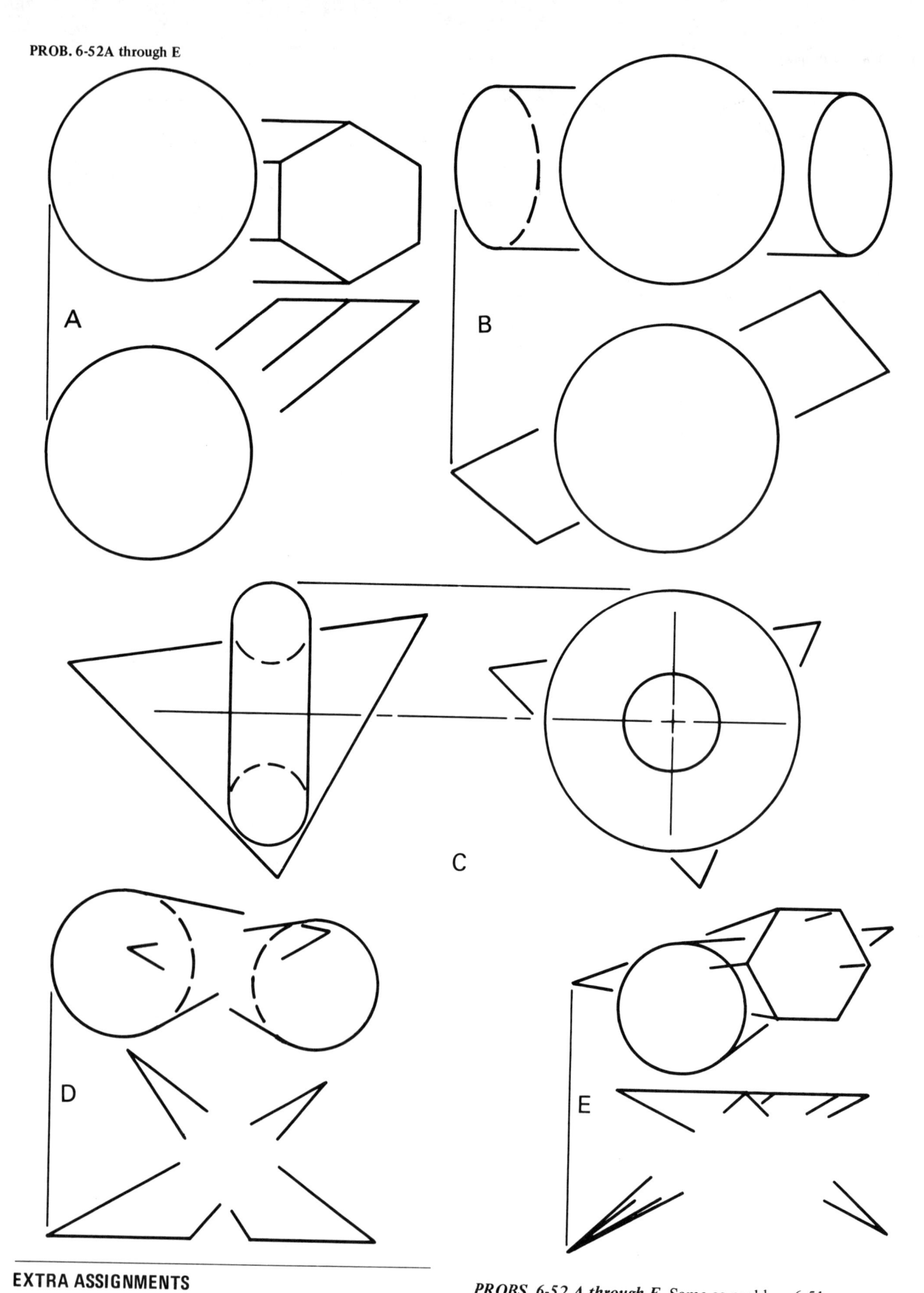

PROBS. 6-52 A through E. Same as problem 6-51.

290

7

DEVELOPMENTS

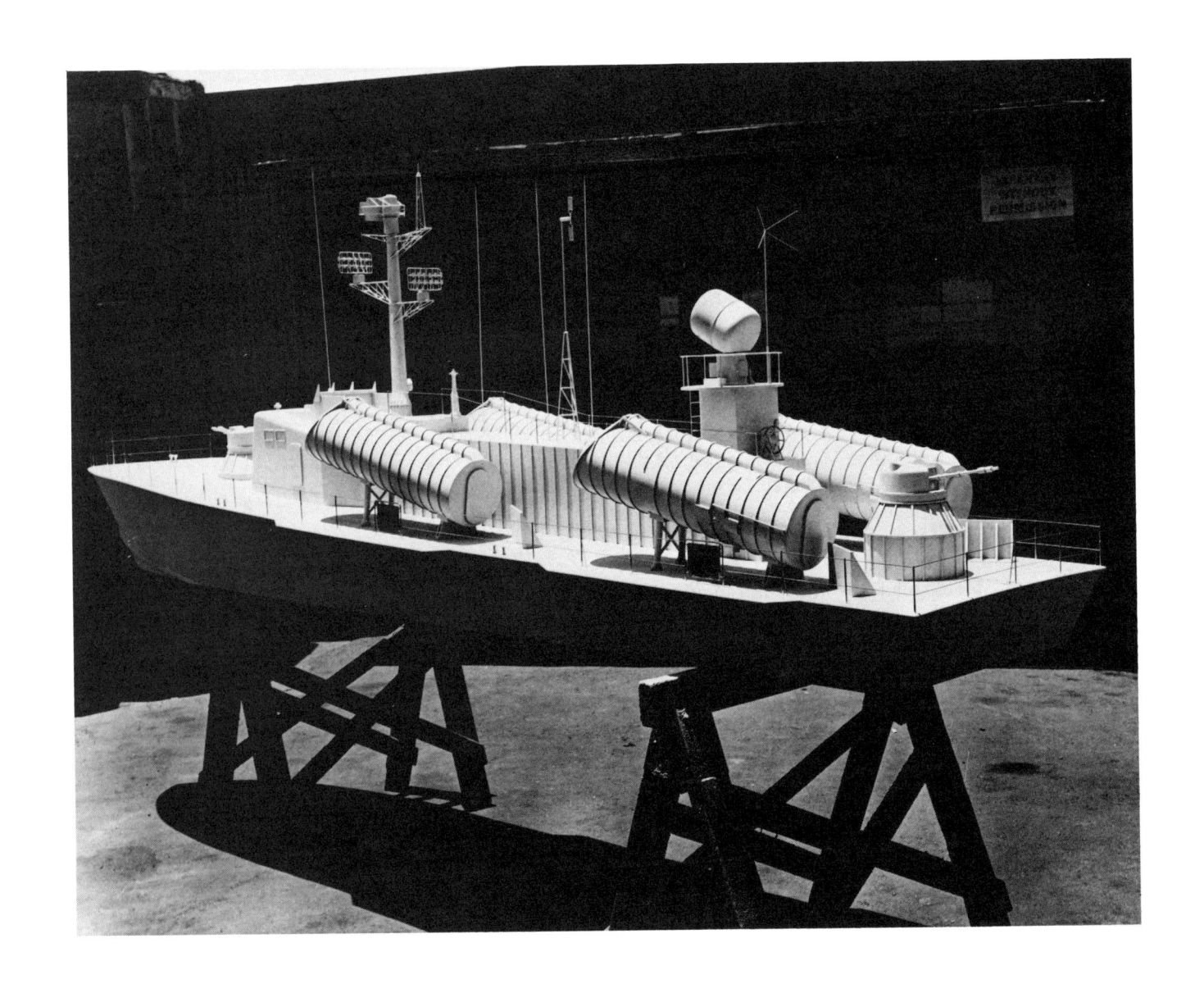

FIG. 7-1 Wind tunnel. *(Courtesy NASA.)*

7.1 DEVELOPMENTS

A variety of industrial structures, products, and manufactured parts are made from flat sheet stock material. Parts designed to be produced of flat materials are cut from a pattern that is drawn as a development from the orthographic projection. The complete layout drawing of a part showing its total surface area in one view is constructed using true length dimensions. This flat plane drawing shows each surface of the part as true shape, therefore all lines in a development are true length. All surfaces of the object are connected along their adjacent bend lines. Sheet metal objects, cardboard packaging, large diameter cylindrical vessels/piping, funnels, cans, and ducting are just a few of the many types of objects made from developments. In Fig. 7-1, many of the interconnecting cylindrical and conical shapes used on the wind tunnel were designed and produced using a development. Each section of the wind tunnel was separately laid out and then rolled to produce the required shape and diameter. In Fig. 7-2, the vessel bridge of the destroyer model is made of flat sheet brass plate bent to the required configuration.

A *pattern* is made from the original development drawing and used in the shop to scribe or set up the true shape configuration of a part to be produced. The actual developed flat sheet configuration is then cut according to its pattern. The final operations include bending, folding or rolling, and stretching the part to its required design. Welding, glueing, soldering, bolting, seaming, or riveting can be used to join the piece's seam edge.

FIG. 7-2 Brass model of destroyer used for reflectricity studies. *(Courtesy Magee-Bralla, Inc.)*

7.2 BASIC DEVELOPMENTS

*The four most common shapes that are developable are the **prism**, **pyramid**, **cylinder**, and **cone**,* in all of their variations. The development of an object is normally accomplished by unfolding or unrolling its surfaces onto the plane of the paper. The actual drawing of the object consists of showing each successive surface as true shape and connected along common edges. One edge line serves as a ***seam*** for a shape composed of plane surfaces. The seam or break line for a curved shape will be along a line/*element* on its surface.

Throughout this chapter each of the objects is developed as an ***inside-up*** pattern drawing. That is to say, the object is unfolded/unrolled so that the inside surface is face up. In some cases a pattern may be required to show as an outside-up development. The difference in drawing this variation will not cause any problems to one who has mastered the traditional inside-up method.

In Fig. 7-3(1) and (2), the prism and pyramid have been unfolded inside-up so that each surface is laid flat and connected along common edges. The first and last line of any development is the same line, because they are joined together along the seam. A right prism unfolds as a rectangle. The length of the rectangle is equal to the perimeter of the base and its width is equal to its altitude.

In Fig. 7-3(3) and (4), the cylinder and the cone have been unrolled (inside-up). A seam edge for these figures is along a specified line or element on its surface. A cylinder unfolds/unrolls as a rectangle with its length equal to its circumference (3.141 × dia.) and its height equal to its altitude. A cone develops as a portion of a circle (sector) with a radius equal to its slant height and an arc length equal to the circumference of its base.

The edges of a prism and pyramid correspond to the bend lines of the development. For a cylinder and cone, elements are established along their surface and bend lines are not required. Bend lines are normally specified by the drafter as required.

Objects that are composed of plane surfaces, such as prisms and pyramid shapes, along with single-curved surfaces, such as cylinders and cones, are developable. In other words, they can be laid flat and constructed of one single piece of material. Double-curved and warped surfaces on the other hand are considered to be undevelopable. Spheres, paraboloids, oblique helicoids, and cylindroids are examples of undevelopable surfaces. In most cases approximate methods can be used to adequately develop these types of surfaces.

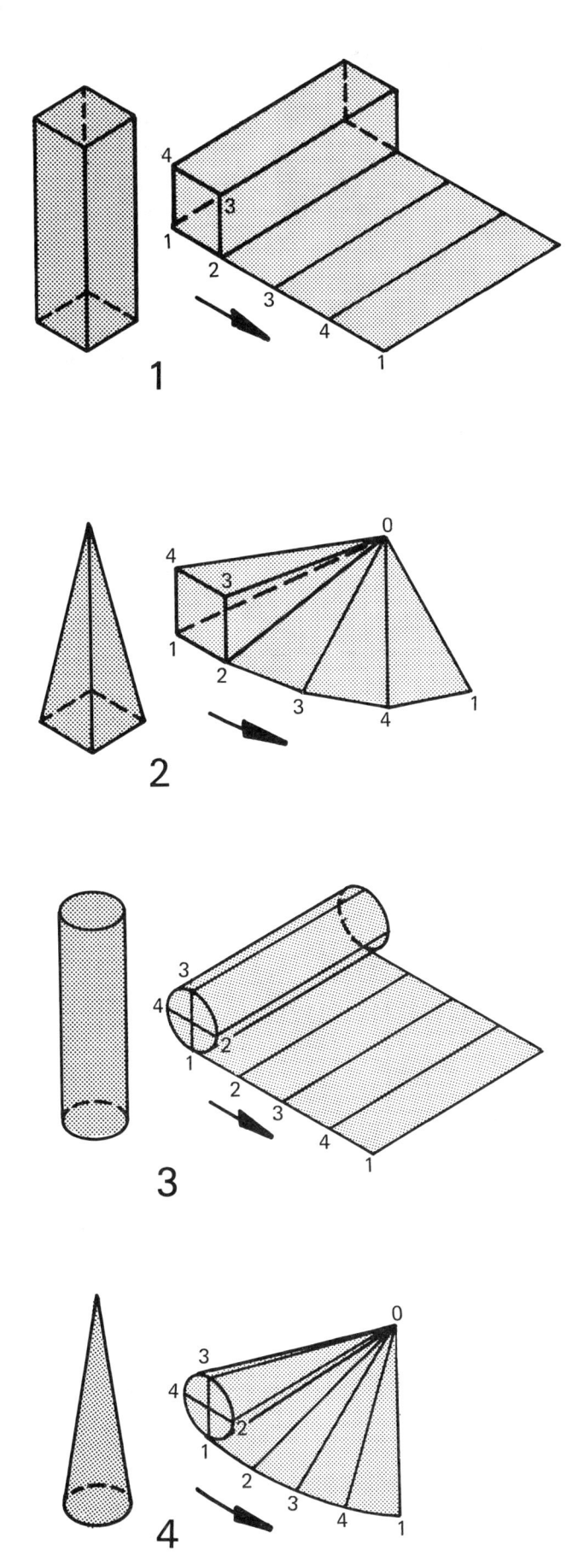

FIG. 7-3 Basic developments: 1. prism, 2. pyramid, 3. cylinder, 4. cone.

FIG. 7-4 Cylindrical satellite. *(Courtesy NASA.)*

FIG. 7-6 Conical design of rocket. *(Courtesy NASA.)*

7.3 TYPES OF DEVELOPMENTS

There are four basic types or classifications of developments. The division of developments is based on the shape of the surface and/or the method employed to construct its development.

1. **Parallel line:** Forms that are composed of parallel straight-line elements or edges, cylinders, prisms, Fig. 7-4.
2. **Radial line:** Forms whose edges or elements define triangular surface areas, pyramids, cones, Fig. 7-6.
3. **Triangulation:** Forms whose surfaces must be broken into triangular areas, transition pieces, Fig. 7-5.
4. **Approximate:** Forms whose surfaces cannot be truly developed, warped and double-curved surfaces, spheres, Fig. 7-7.

FIG. 7-5 Transition piece for air filtration system.

FIG. 7-7 Twelve-foot spherical explorer satellite. *(Courtesy NASA.)*

FIG. 7-8 Sheet metal development.

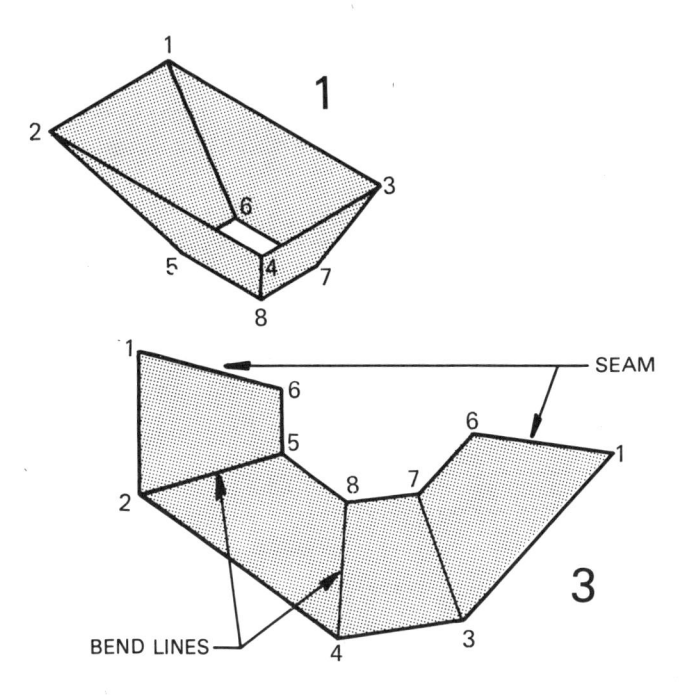

7.4 SHEET METAL DEVELOPMENTS

Many complex three-dimensional shapes are fabricated from flat sheet materials. The shape to be formed is subdivided into its simplest elements, which often individually have the shapes of prisms, cylinders, cones, pyramids, and spheres. All of these shapes can be formed from a two-dimensional sheet of material by first cutting to the proper pattern and then folding or rolling the material into its three-dimensional form. Sheet metal is a typical material used for developable products or parts. HVAC ducting, transition pieces, and aircraft and spacecraft bodies are made from sheet metal developments.

The fabricator, working from the design engineer's drawings and specifications, develops pattern drawings of each component to be produced in the fabrication shop. These patterns are usually made to the full size of the object and can only be made after the true lengths of all lines that will lie on the pattern have been determined. A pattern is a drawing composed entirely of true length lines. Therefore all patterns are true shape/size. Each development must be drawn accurately so that the final product is of the correct shape and within given tolerance limits. A bend allowance is usually added to the pattern drawing to accommodate the space taken by the bending process. An extra tab or lap must be added to the pattern so that the two adjoining edges which form the seam may be attached. The width of this tab/lap depends on the type of joining process. The length of the lap is normally established along the shortest edge so as to limit the amount and length of the seam.

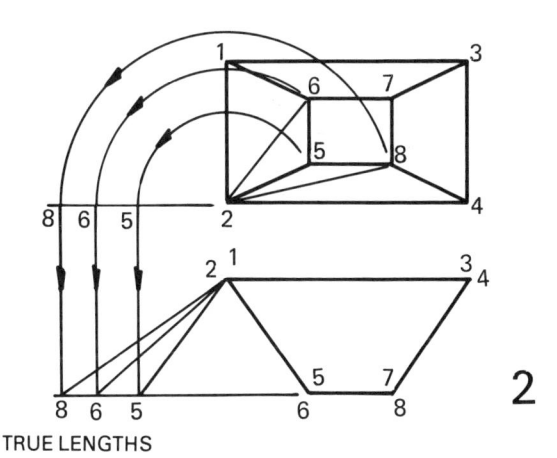

Throughout this text a bend allowance and a lap have been eliminated for the problems and example illustrations so as not to confuse the beginner. Each development will be a true development.

In Fig. 7-8, a transition piece made of sheet metal is shown as a pictorial illustration (1), orthographic projection (2), and development (3). This transition piece could be the end part of a hopper-like duct as shown in Fig. 7-9 of the air filtration system. Notice that the true length of each edge line has been established by revolution (2). The edge lines correspond to the bend lines in the development (3). Line 1-6 will be the first and the last line of the development and thus the seam when formed and joined.

FIG. 7-9 Transition piece and prism-shaped air duct.

FIG. 7-10 Sheet metal intersection and development models: two prisms, pyramid and a prism, cone and a horizontal prism, cone and a vertical cylinder, two cylinders, cone and a horizontal cylinder.

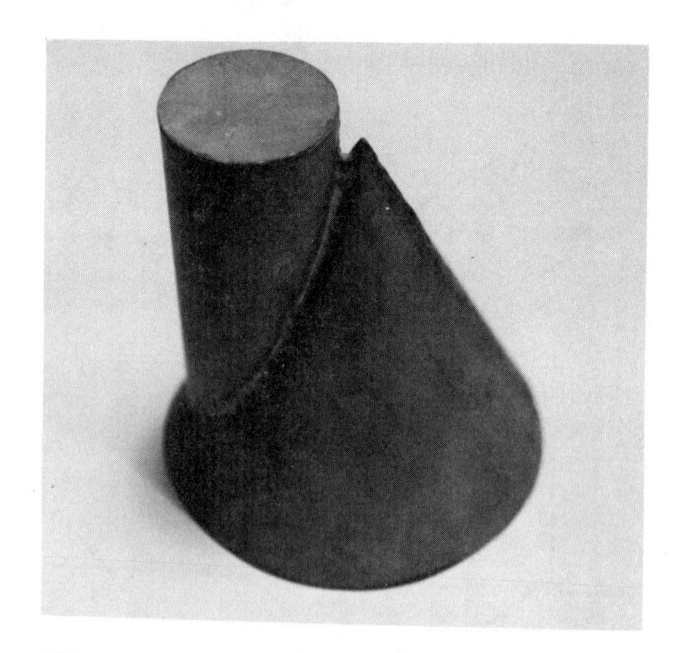

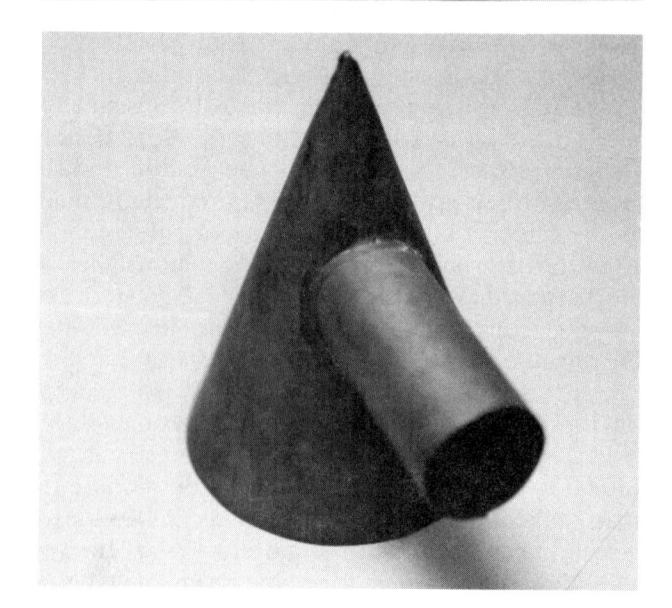

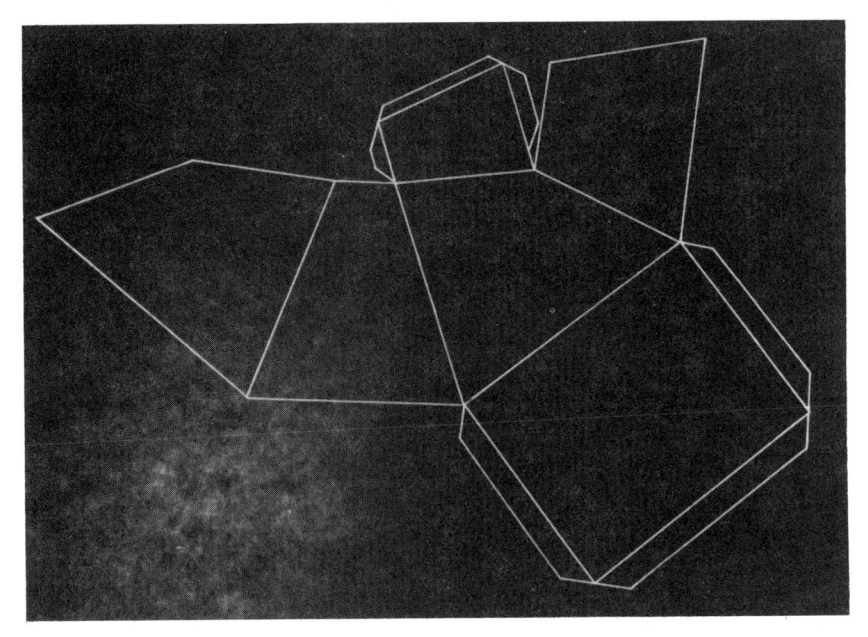

FIG. 7-11 Pattern drawing.

FIG. 7-12 Development model.

7.5 DEVELOPMENT OF MODELS

Though a pattern is normally drawn full size, a reduced scale model is also made by the designer during the refinement, analysis, and implementation stages of the design process. Small-scale, accurate models are constructed for design analysis and to explain design variations to the fabricator or purchaser. Models have a distinct advantage over pictorial drawings in that they can be viewed from any angle and are always seen in their true three-dimensional form.

Models of development problems should be made of a variety of configurations when completing projects from the text. The orthographic drawing is needed before a development and pattern can be made.

In Fig. 7-11, a pattern is drawn on thin cardboard. Lightweight cardboard, such as file folders, is an excellent material for making small models. The pattern outline and bend lines are easily transferred onto it, and it folds well, making sharp corners. Note that tabs must be added along seam edges so as to join the form by gluing or taping. Transfer the pattern onto the cardboard from a carefully executed projection by small pin pricks at controlling points (end points of edge/bend lines) or by the use of carbon paper. Draw the pattern on the cardboard and cut along the outline. The resulting cardboard pattern, Fig. 7-13, is then folded along bend lines and joined along the tabs. Fig. 7-12 and Fig. 7-14 show the finished form. Note that the top and bottom surfaces were included.

FIG. 7-13 Cardboard model pattern before gluing.

FIG. 7-14 Completed model.

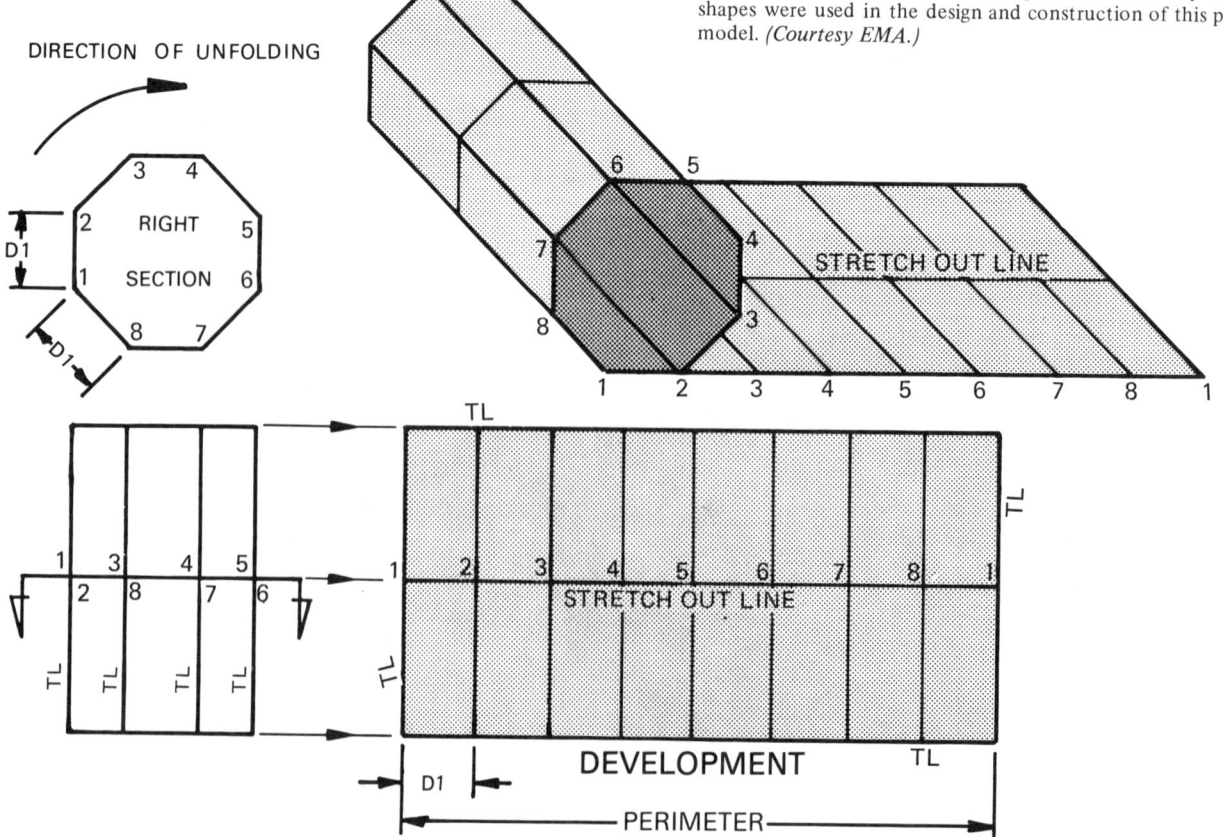

7.6 DEVELOPMENT OF A RIGHT PRISM

Parallel line developments, prisms, and cylinders are developed by completing the following steps. Fig. 7-15 is used to illustrate this process.

1. Draw the object as an orthographic projection. Fold lines between views are not necessary unless an auxiliary view is required.
2. Project any views necessary to show the true lengths of the object's edge lines or elements (for a cylinder).
3. Show the prism or cylinder as a right section. This provides the width dimensions of the faces of the prism or distances between elements for a cylinder, dimension D1.
4. Draw a **stretch-out line** perpendicular to the object's lateral edges (parallel to the edge view of the right section taken in the frontal view). The length of the stretch-out line is equal to the perimeter of the prism or circumference of the cylinder to be developed. The stretch-out line is used to establish the width of each face of a prism or the distance between elements along the lateral surface of a cylinder.
5. Unfold the lateral surface by measuring and marking off the widths of the faces or distances between elements taken from the right section view and laying them off along the stretch-out line, dimension D1. When transferring the face widths, the point to point measurements should be taken in a clockwise direction. This is called the "*direction of unfolding*." The length of the development/pattern is equal to the length of the stretch-out line.

6. Project the edge lengths from the frontal view and connect them to form the development's complete outline. Note that the development starts and finishes with the same edge line (line 1). In sheet metal developments the seam edge will be the shortest edge wherever possible.

FIG. 7-15 Development.

FIG. 7-16 Intersections and developments of a variety of geometric shapes were used in the design and construction of this power plant model. *(Courtesy EMA.)*

298

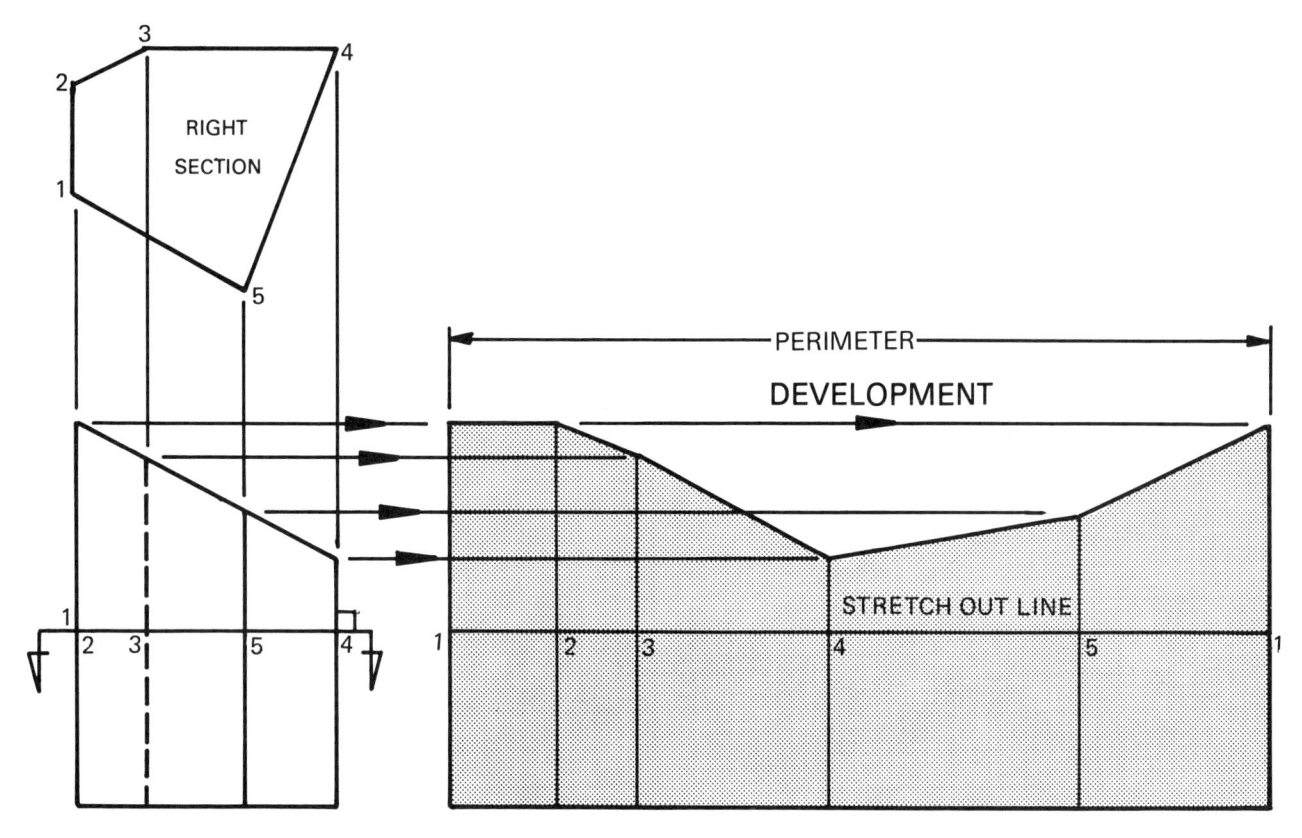

FIG. 7-17 Development of a right truncated prism.

7.7 DEVELOPMENT OF A TRUNCATED RIGHT PRISM

The first step in drawing a parallel-line development is to find the true lengths of each edge and the width of each face plane. A right section view shows the perimeter of the object. The length of the development is equal to the perimeter of the prism as measured in the right section view. A right section is always taken perpendicular to the true length edge lines of a prism or the axis line of a cylinder. The distance between each edge line/element is measured where they appear as points on the right section. The width of each lateral surface is equal the straight-line distance between points on the right section and is transferred directly to the stretch-out line. In Fig. 7-17, the distance between point 1 and point 2 in the right section view is transferred to the stretch-out line to establish the width of the first plane face. The prism is unfolded clockwise using the shortest edge as the seam if required. In Fig. 7-17, edge line 1 is used as the seam line. The stretch-out line is projected perpendicular to the edge view of the right section in the frontal view. The edge lengths are projected from the frontal view. The outline of the development is then completed by connecting the end points of the edge lines. Edge lines in both the frontal view and the development are true length. The development itself is made up completely of true length lines. Therefore, each lateral surface (plane face) is true shape/size as is the total development. The length of the development can be checked by measuring the perimeter of the prism (the distance around the right section view). The development length must equal the perimeter.

Sheet metal developments and patterns are used to create many types of configurations. In Fig. 7-18, each portion of the turbine outlet casing was rolled and formed from a pattern and then welded to complete the unit.

FIG. 7-18 The turbine outlet casing is a sheet metal development. *(Courtesy Magee-Bralla, Inc.)*

7.8 DEVELOPMENT OF A TRUNCATED AND CUTOUT RIGHT PRISM

The truncated triangular right prism in Fig. 7-19 has a cutout top. The frontal and horizontal right section views are given. A development of its lateral surface is required. Edge line 1 is to be used as the seam line.

1. Draw the stretch-out line perpendicular to the true length edge lines in the frontal view and parallel to the edge view of the right section.
2. Transfer the distance between the point view of edge line 1 and line 2, as measured from the right section, to the stretch-out line. Repeat this procedure using the distances between the point view of edge lines 2 to 3 and from 3 to 1.
3. The distances between each of the edge lines have been established along the stretch-out line. Project the length distances from the frontal view.
4. Connect points 1^1, 2^1, 3^1, and 1^1 to form the lower part of the outline. Edge lines 1-1^1, 2-2^1, and 3-3^1 are drawn as shown. Points 4 and 5 must be located before the upper border can be completed.
5. Measure the distance between edge line 1 and point 4 and between edge line 1 and point 5 in the right section view. Transfer these dimensions to the stretch-out line.
6. Project points 4 and 5 from the frontal view and extend projection lines from the corresponding points on the stretch-out line to locate points 4 and 5 on the development.
7. Connect points 1, 4, 2, 3, 5, and 1 to form the upper outline of the development.

The surface effect ship modeled in Fig. 7-20 has many parts that are made from sheet metal developments including the two stacks and the bridge.

FIG. 7-19 Development of a truncated and cutout right prism.

FIG. 7-20 Surface effect model. *(Courtesy Magee-Bralla, Inc.)*

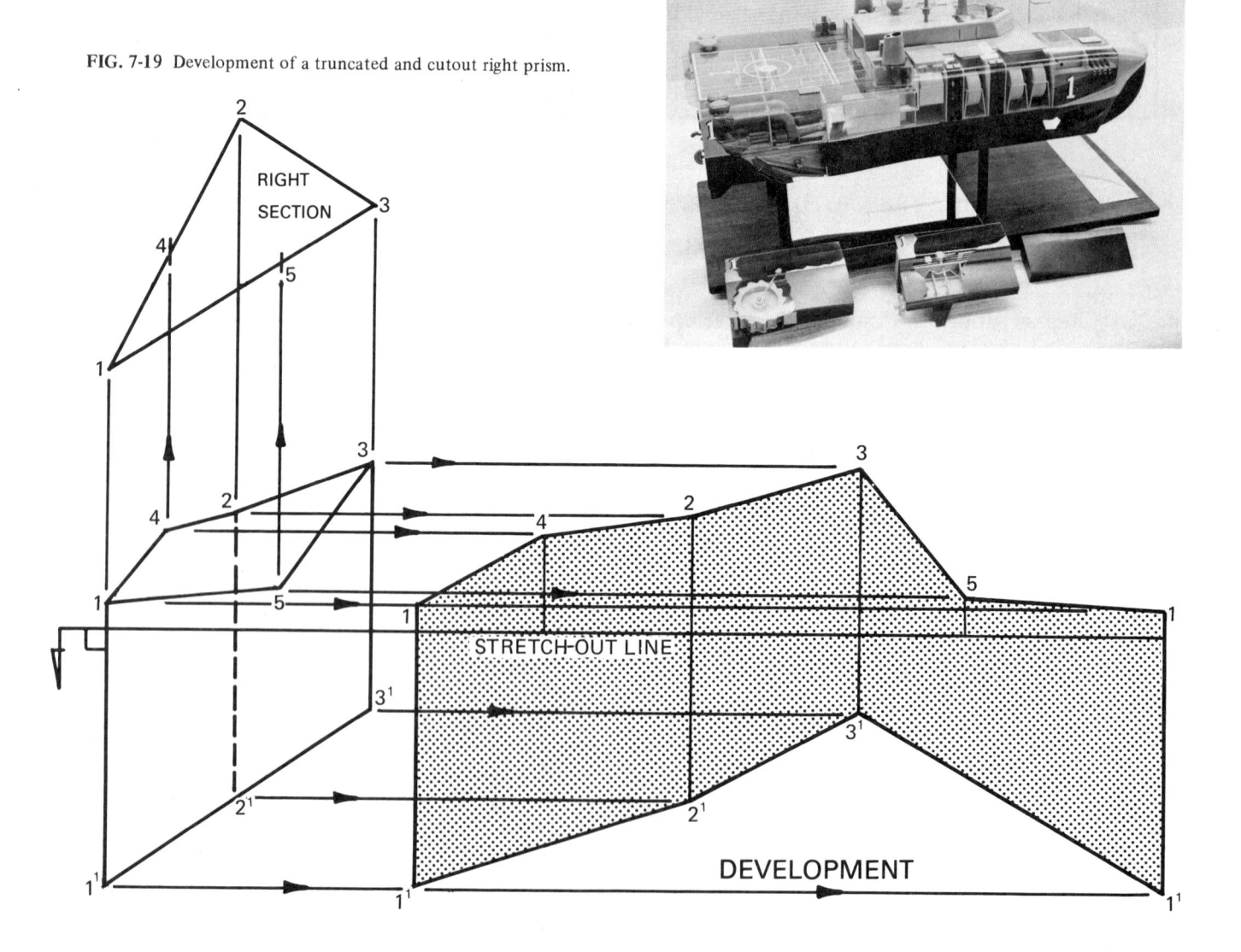

PROB. 7-1

PROB. 7-2

PROB. 7-3

PROB. 7-4

PROBS. 7-1 through 7-4. Develop and make a model of each given prism as assigned. Show all notation and completely label each problem.

7.9 DEVELOPMENT OF AN OBLIQUE PRISM

When a development is to be made of an oblique prism, a right section view and an auxiliary view showing the lateral edges as true length must first be projected. After the true length view and the right section view are drawn, the procedures for completing the development are the same as for a right prism. The right section view provides the distances between the edge lines, and therefore the widths of the prism's plane faces. A stretch-out line is drawn parallel to the edge view of the right section, perpendicular to the true length edge lines. Width dimensions are set off along the stretch-line. Edge line lengths are projected from the true length view to establish the lengths of the edge (bend) lines in the development. Perpendicular lines, representing the edge lines are drawn through the distances set off along the stretch-out line until they intersect the end points of the prism edge lines projected from the true length view.

In Fig. 7-21 the frontal and horizontal view of the sheet metal oblique prism are given.

1. Draw F/A parallel to the oblique edge lines and project auxiliary A. All four parallel edge lines appear as true length.
2. Draw A/B perpendicular to the true length edge lines and project auxiliary B. The prism appears as a right section. The distances between the point views of the edge lines are the true length widths of the prisms face, e.g. 1-2, 2-3, 3-4, and 4-1.
3. Draw the stretch-out line parallel to the edge view of the right section.
4. Transfer the edge widths from the right section view to the stretch-out line. Use edge line 1 as the seam line since it is the shortest edge.
5. Project the edge line's end points from the true length view. Draw the edge lines and connect the end points to complete the development outline.

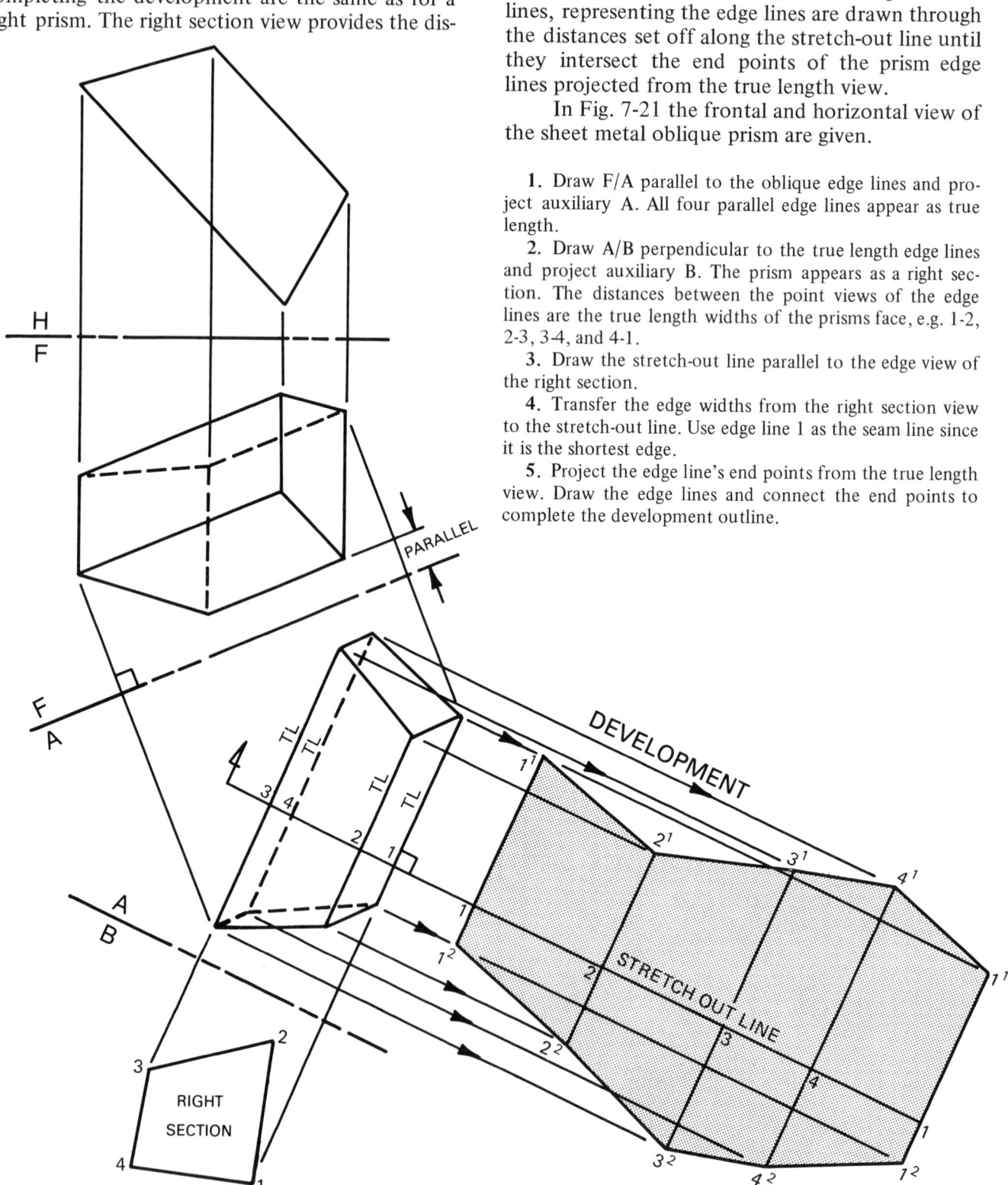

FIG. 7-21 Development of an oblique prism.

7.10 DEVELOPMENT OF A PRISM (TOP FACE AND LOWER BASE INCLUDED)

When one end face of a prism is perpendicular to its edge lines, a true shape end view takes the place of a right section. The stretch-out can be projected parallel to the edge view of an end surface if that surface is perpendicular to edge lines of the prism. The stretch-out line will form one complete edge of the development outline, Fig. 7-22.

When the lower base and the upper face are required, a view showing these surfaces as true shape must be provided. The true shape of an end surface is established by projecting an auxiliary view perpendicular to it, whether it be the base or top face. Each end surface is attached to an appropriate upper or lower border line on the development.

A development's stretch-out line can be established at any convenient location on or off the paper. When this procedure is used, distances above and below the stretch-out line are transferred from the true length view to establish edge line (bend line) lengths on the development. The face widths are, as before, taken from the right section (or true shape end view).

In Fig. 7-22, the development of the prism is required. The bottom surface and the top face are to be included as part of the development. Line 1 is to be used as the seam.

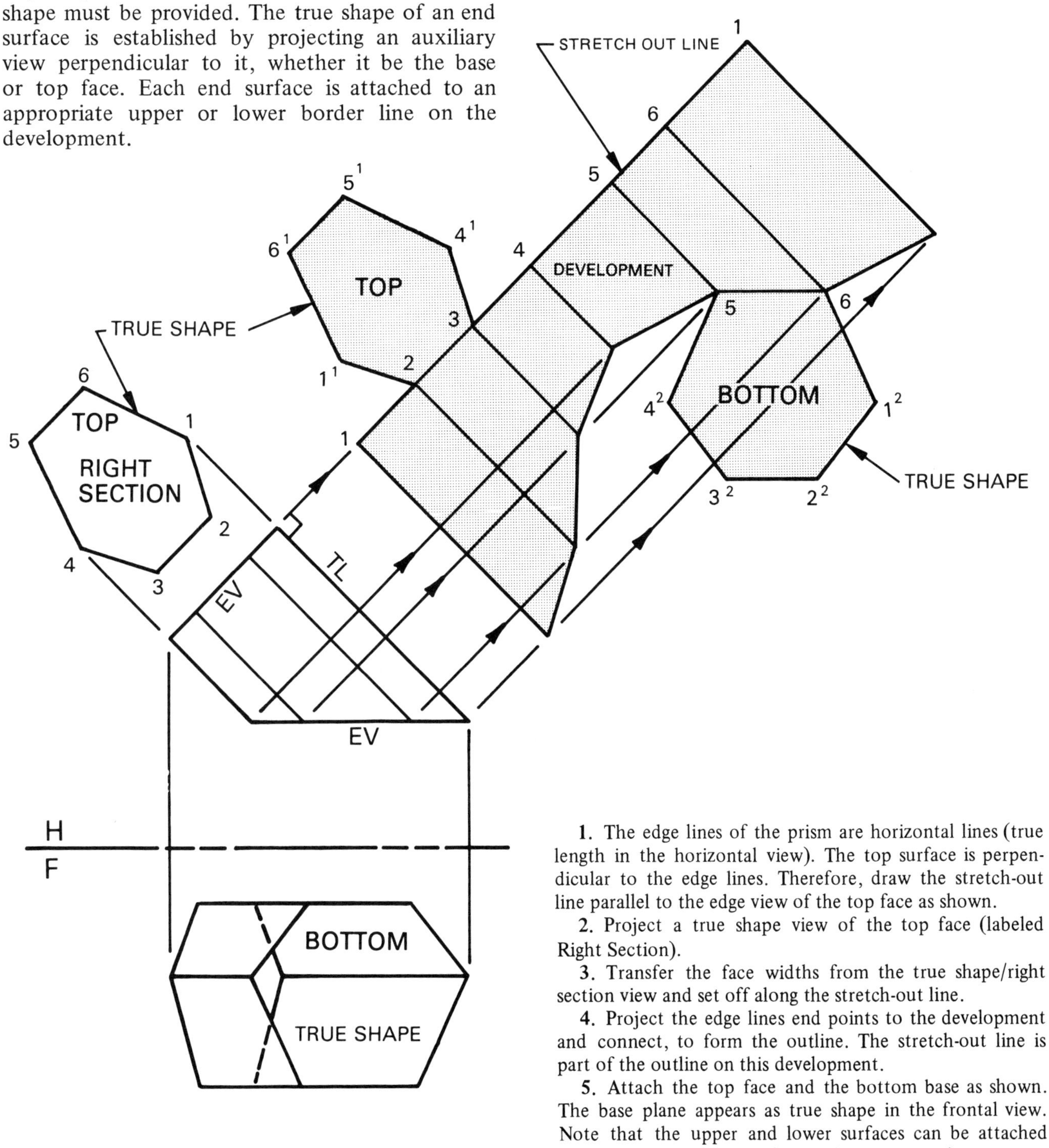

FIG. 7-22 Prism development (end surfaces included).

1. The edge lines of the prism are horizontal lines (true length in the horizontal view). The top surface is perpendicular to the edge lines. Therefore, draw the stretch-out line parallel to the edge view of the top face as shown.

2. Project a true shape view of the top face (labeled Right Section).

3. Transfer the face widths from the true shape/right section view and set off along the stretch-out line.

4. Project the edge lines end points to the development and connect, to form the outline. The stretch-out line is part of the outline on this development.

5. Attach the top face and the bottom base as shown. The base plane appears as true shape in the frontal view. Note that the upper and lower surfaces can be attached along any related line on the development's outline.

FIG. 7-23 Development of an intersected prism.

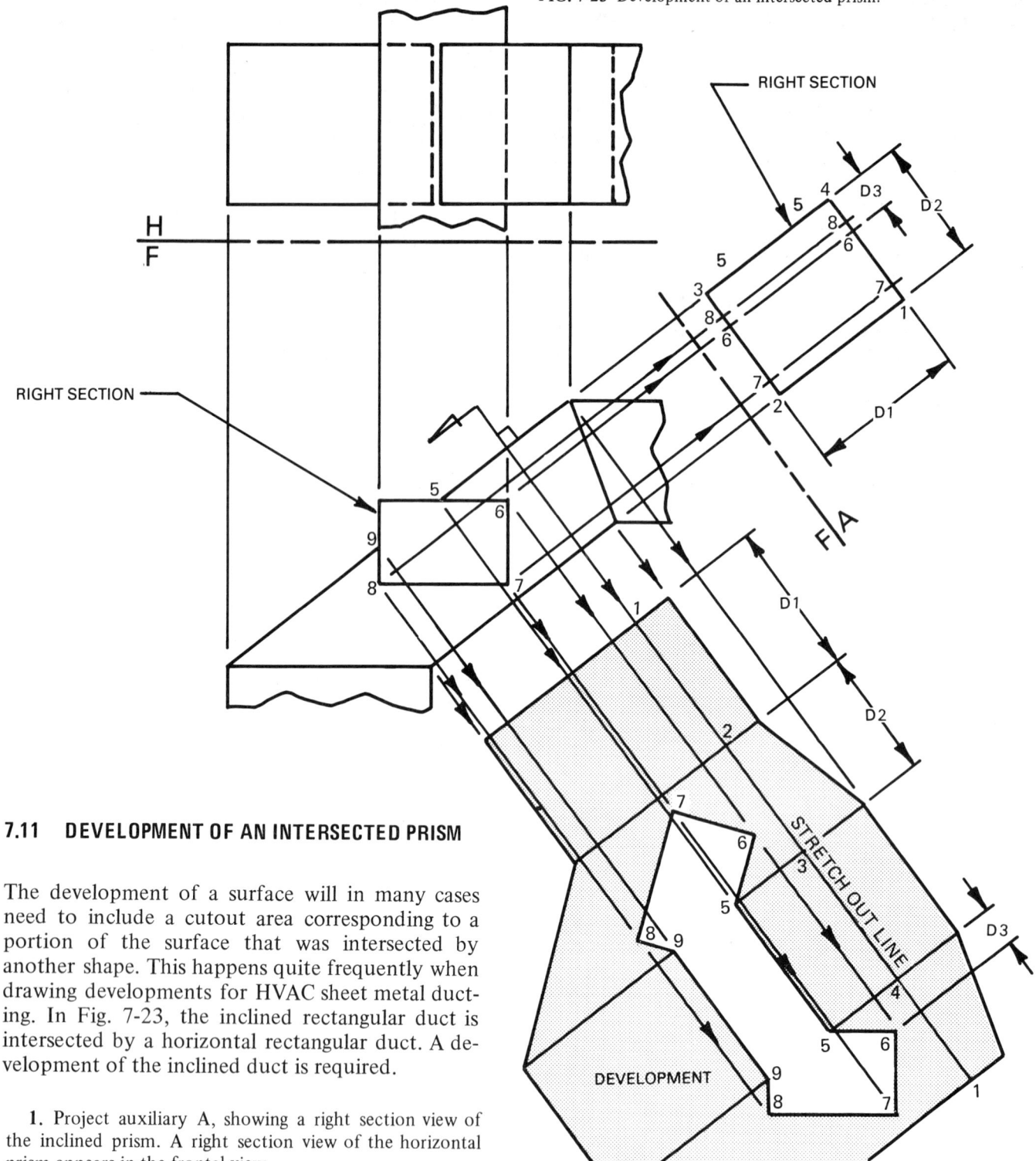

7.11 DEVELOPMENT OF AN INTERSECTED PRISM

The development of a surface will in many cases need to include a cutout area corresponding to a portion of the surface that was intersected by another shape. This happens quite frequently when drawing developments for HVAC sheet metal ducting. In Fig. 7-23, the inclined rectangular duct is intersected by a horizontal rectangular duct. A development of the inclined duct is required.

1. Project auxiliary A, showing a right section view of the inclined prism. A right section view of the horizontal prism appears in the frontal view.
2. Project the points at which the edge lines of the horizontal prism intersect the inclined prism in the frontal view to auxiliary A.
3. Draw the stretch-out line parallel to the edge view of the section as shown.
4. Set off the face widths along the stretch-out line starting with edge line 1, since it is the shortest line. Dimensions D1 and D2 locate the first two edge lines.
5. Project the end points of the edge lines to the development. Connect the end points of the edge lines to form the outline of the development.
6. Project all intersection points to the development. Point 5 and point 9 lie on edge lines 3 and 4. Use D3 to locate point 6 in both positions. Measurements for points 7, 8, and 9 are transferred from the right section view in similar fashion.
7. Connect the intersection points to form the cutout portion of the development.

304

PROB. 7-5

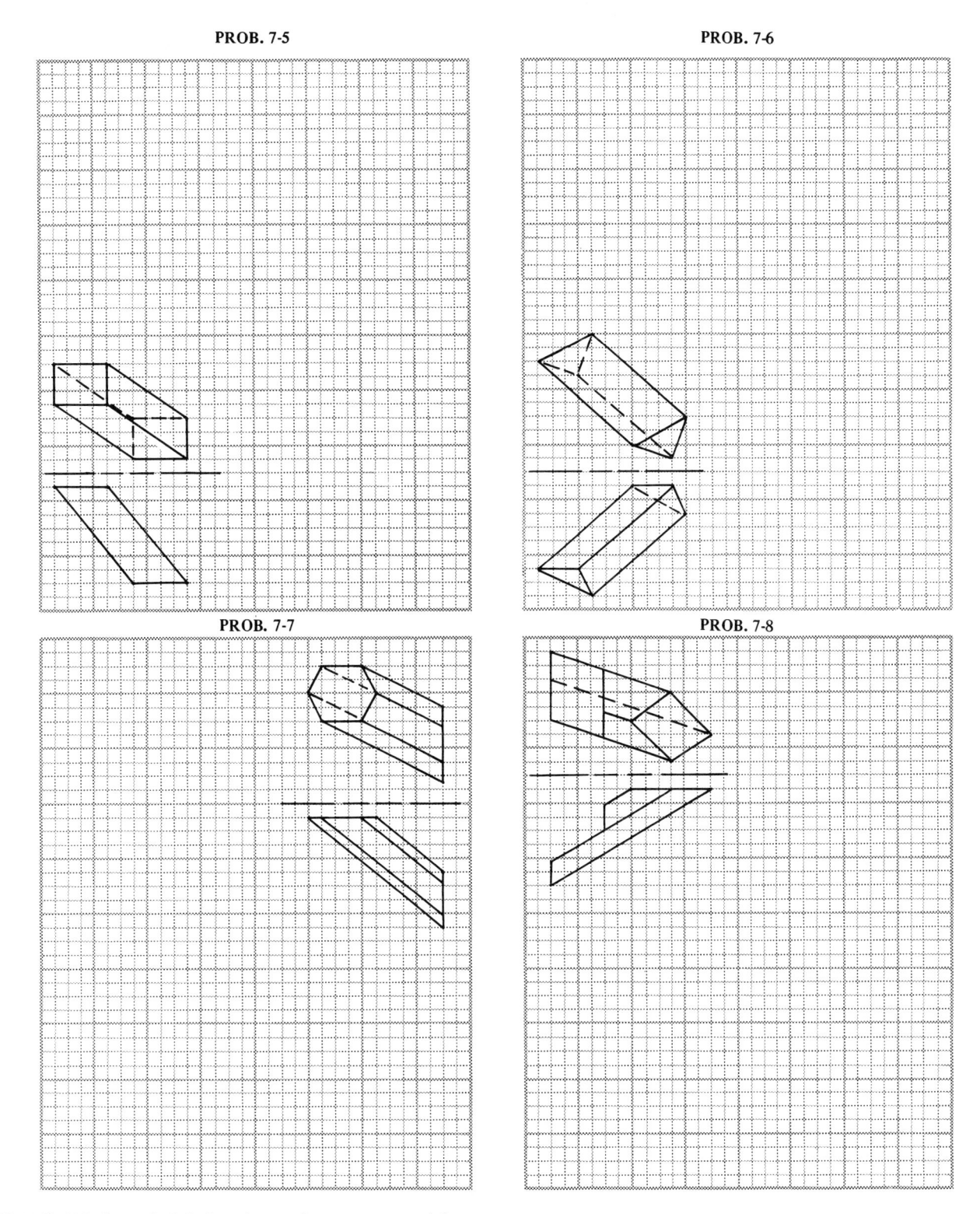

PROB. 7-6

PROB. 7-7

PROB. 7-8

PROBS. 7-5 through 7-8. Develop and construct models as assigned.

FIG. 7-24 Development of a pyramid.

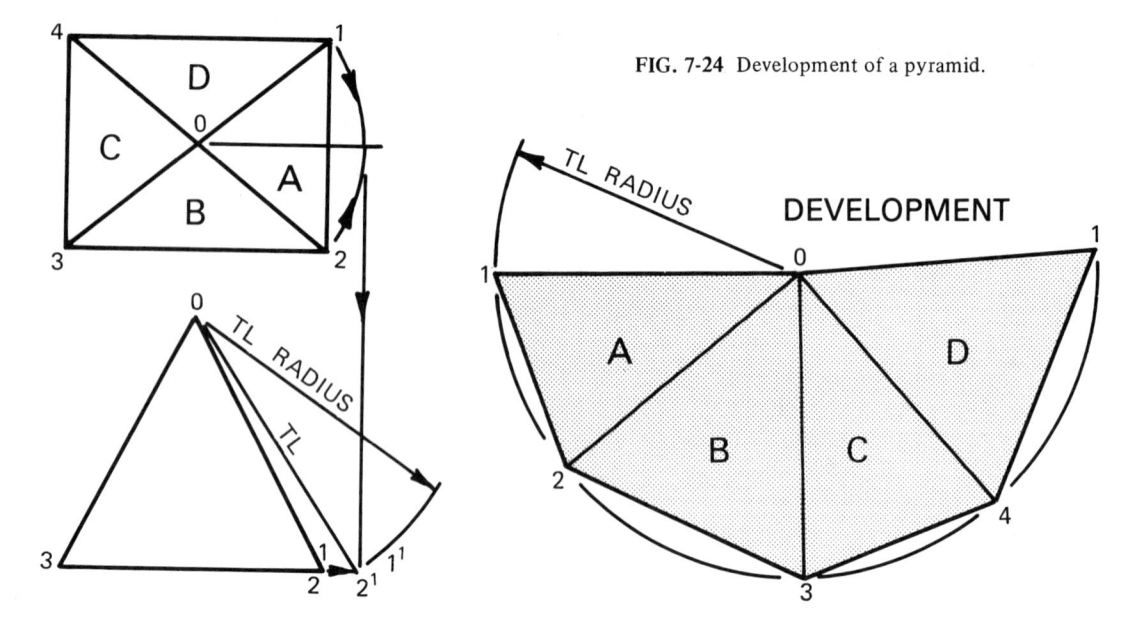

7.12 DEVELOPMENT OF A RIGHT PYRAMID

Developments of surfaces that are composed of triangular planes, such as pyramids, or that can be divided into small triangular areas, cones, are considered radial line developments. Each lateral edge of a pyramid, or element of a cone, radiates from the vertex point.

To develop a pyramid it is necessary to establish the true length of each of its lateral edges and base lines. The development of a pyramid consists of laying out the true shape of each lateral surface in successive order. The development pattern is normally unfolded so that it is inside up. One edge line is used as the seam. Each successive plane face is joined along a common edge line. If the pyramid is a right pyramid, each of its lateral edges will be of equal length. Therefore the true length of only one lateral edge is necessary.

In Fig. 7-24, the perimeter of the base is true length in the horizontal view. Revolve an edge line until parallel to the frontal plane to obtain its true length in the frontal view. Use this true length edge line as the *true length radius*. To start the development locate vertex point 0 at a convenient location. Swing an arc, using the TL radius an indefinite length, from point 0. Starting with point 1, lay off the true length distances transferred from the base edges in the horizontal view, lines 1-2, 2-3, 3-4, and 4-1. Connect each point with vertex point 0 and draw straight-line chords between the points to establish the base perimeter on the development.

In Fig. 7-25, the true length radius is determined by rotating one of the equal lateral edges. The perimeter is true length in the horizontal view. Swing the TL radius as shown to establish the lengths of the lateral edge lines on the development. Step off distances 1-2, 2-3, and 3-1 on the arc and connect as chord lines. Connect each point with vertex point 0 to establish the lateral edge (bend) lines.

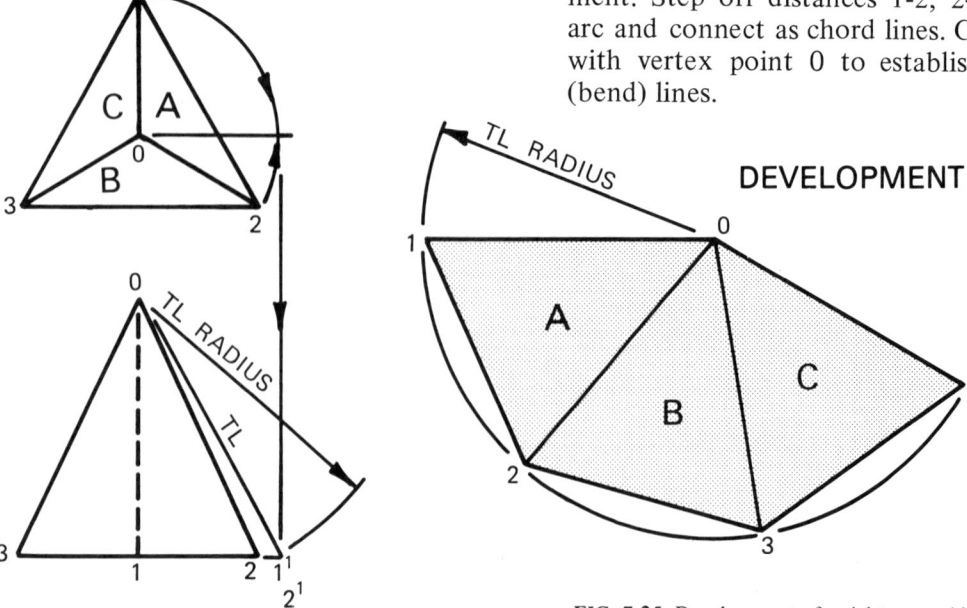

FIG. 7-25 Development of a right pyramid.

306

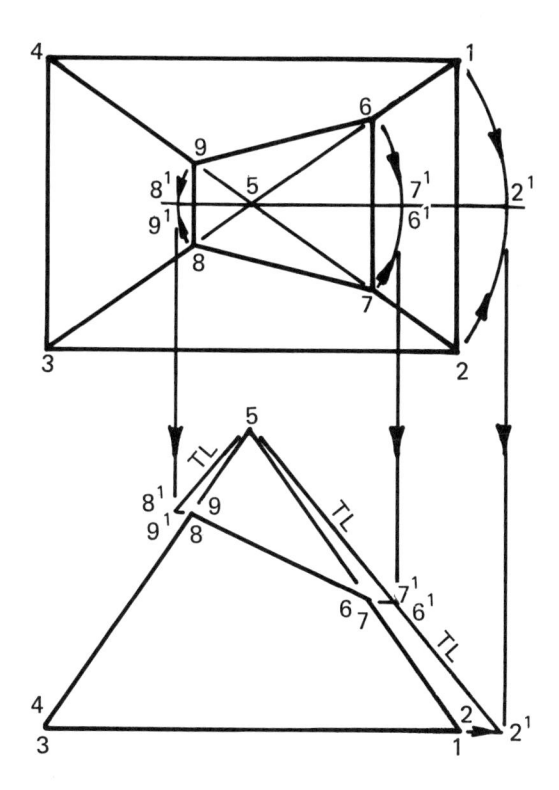

FIG. 7-26 Development of a truncated right pyramid.

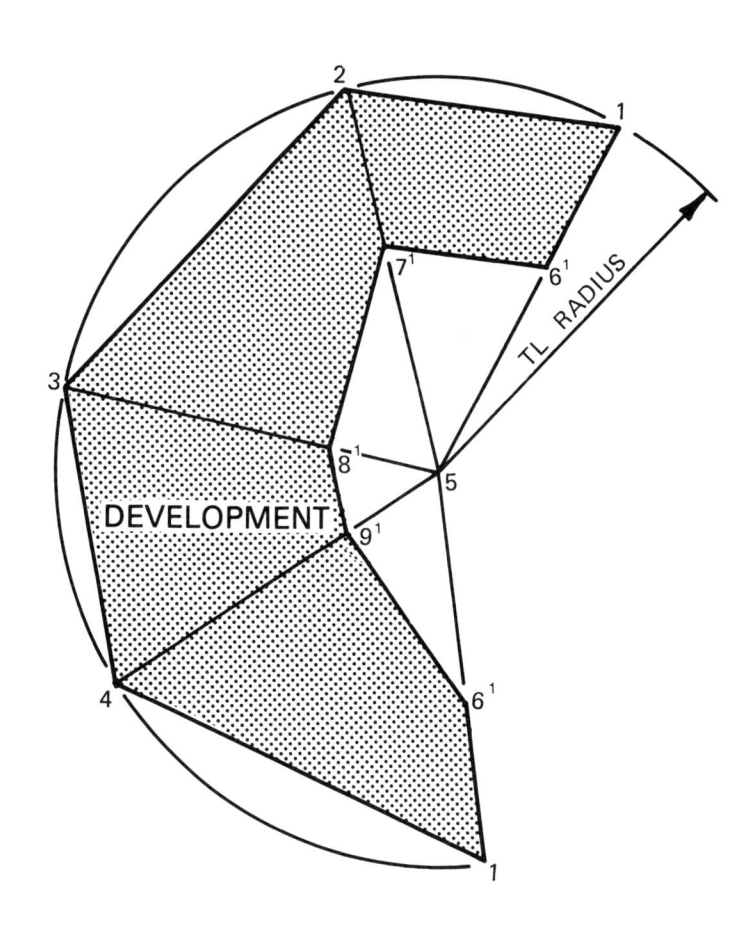

7.13 DEVELOPMENT OF A TRUNCATED RIGHT PYRAMID

To develop a truncated right pyramid, the true lengths of the lateral edges, from the base points to the vertex, must be determined first. The true lengths of the truncated edges are also required. The distance along the pyramid's edges, from the vertex to the truncated points, can be substituted for the truncated edge line measurements. This procedure is shown in Fig. 7-26, where the true length distances from vertex point 5 to points 6, 7, 8, and 9 have been determined using revolution.

Since a right pyramid has edges of equal length, only one edge line length needs to be established, line 5-2[1]. This true length can easily be found by revolution. The true length edge distance is used as the TL radius when constructing the development, 5-2[1].

To develop a pyramid, its triangular plane surfaces are unfolded in successive order with their matting edges joined. One edge line is predetermined as the seam edge, line 6[1]-1. The development begins and ends with the seam line. To start laying out the development, the vertex point is conveniently located. Using the true length edge length as the TL radius, line 5-2[1], an arc of indefinite length is swung. The true length base lines are stepped off along this arc. Distances 1-2, 2-3, 3-4 and 4-1 are taken from the horizontal view. Each point on the arc is connected to the vertex point. Note that the truncated edges must be determined before the bend lines can be drawn as visible lines. Lines 1-5, 2-5, 3-5, 4-5, and 1-5 are the extended edge lines of the development. The arc points are then connected in sequence as chord lengths representing the edges of the base, lines 1-2, 2-3, 3-4, and 4-1.

The true lengths of the truncated edges or the distances from the vertex point to the cut points, are determined by revolution. Points 6, 7, 8, and 9 are revolved in the horizontal view and appear as true length lines in the frontal view. Distance 5-6[1] is transferred to the development along line 5-1. Distances 5-7[1], 5-8[1], and 5-9[1] are also transferred to their respective edge lines on the development. Points 6[1], 7[1], 8[1], 9[1], and 6[1] are connected to complete the inside outline of the development.

307

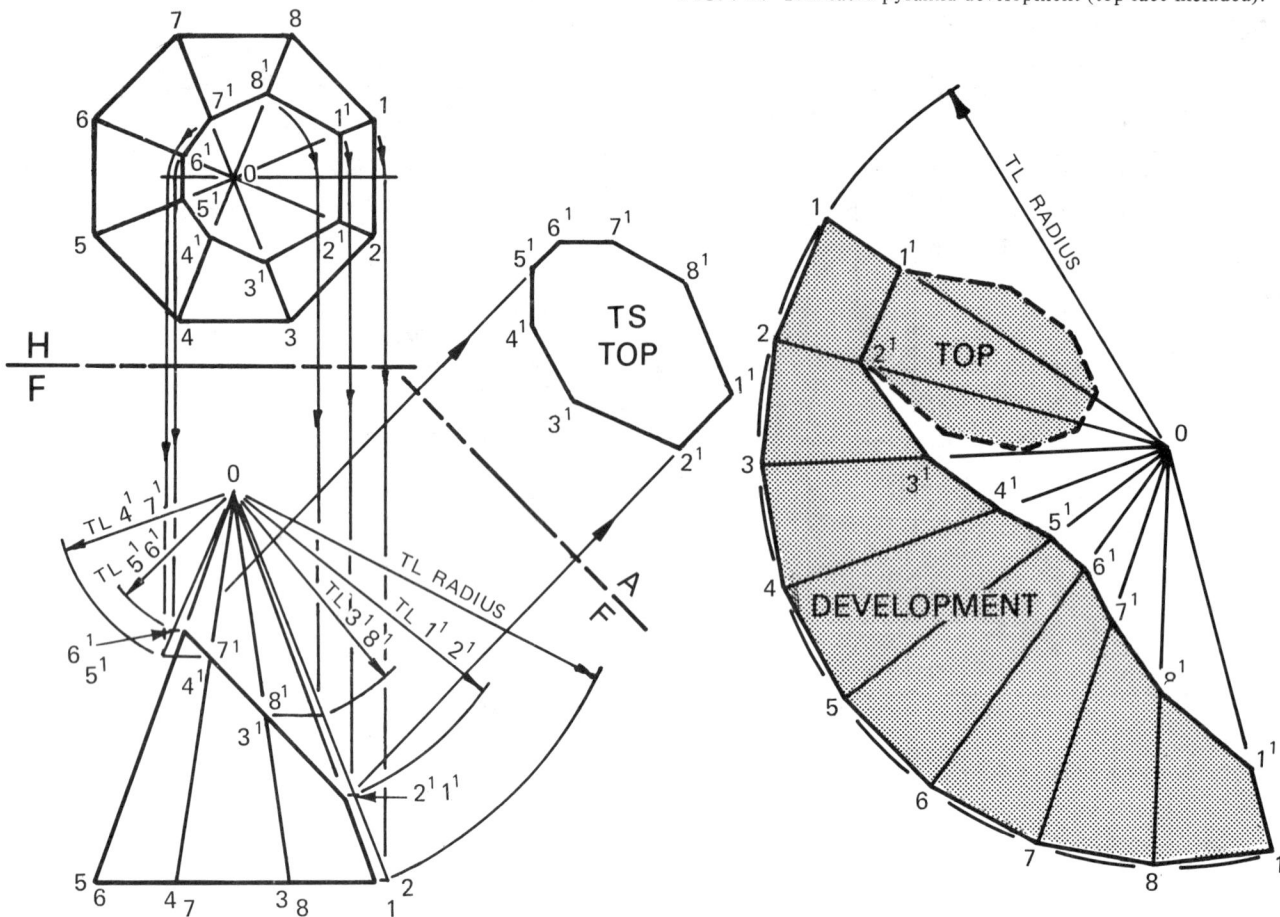

FIG. 7-27 Truncated pyramid development (top face included).

7.14 DEVELOPMENT OF A TRUNCATED RIGHT PYRAMID (TOP FACE INCLUDED)

A development of a pyramid is a radial-line development. Each of the edge lines of the pyramid radiates from the vertex point on the development. A right pyramid can be developed by simply swinging an arc equal in length to its edge lines and stepping off the base line distances as chordal distances.

In Fig. 7-27, the frontal and horizontal view of the truncated right pyramid is given. A development, including its upper face (truncated surface), is required.

1. Draw F/A parallel to the edge view of the truncated face and complete auxiliary A. The upper/top face appears as true shape.
2. Establish the true length of edge line 0-1 by revolution. The true length of line 0-1 is equal to all other edge lines and is used as the TL radius.
3. Solve for the true lengths of the distances from vertex point 0 to where each edge line has been cut. Points 1^1, 2^1, 3^1, 4^1, etc. represent the points at which the lateral edge lines have been cut. True length distances $0-1^1$, $0-2^1$, $0-3^1$, etc. are used to establish the upper outline of the development.

4. Locate vertex point 0 at a convenient location. Swing the TL radius (radius 0-1) an indefinite length.
5. Line $1-1^1$ is the shortest edge. Therefore it is used as the seam. Draw line 0-1 on the development and step off the base line distances along the arc. Distances 1-2, 2-3, 3-4, 4-5, 5-6, 6-7, 7-8, and 8-1 are laid off along the arc. Note that each base length is equal since the base plane is an octagon. Therefore distance 1-2 can be used for each chord length.
6. Connect the base points to vertex point 0. Draw these lines as construction lines only. The actual bend lines include the distance from the base points to the cut points only.
7. Connect the base points in sequence as straight-line chords to establish the lower outline of the development.
8. Transfer distance $0-1^1$ to line 0-1 on the development. Repeat this procedure to locate each cut point on the development. Connects points 1^1, 2^1, 3^1, 4^1, etc. to form the development's upper outline.
9. Attach the true shape of the top surface to the development along a common line. The top surface is true shape in auxiliary A.

FIG. 7-28 Development of an oblique pyramid.

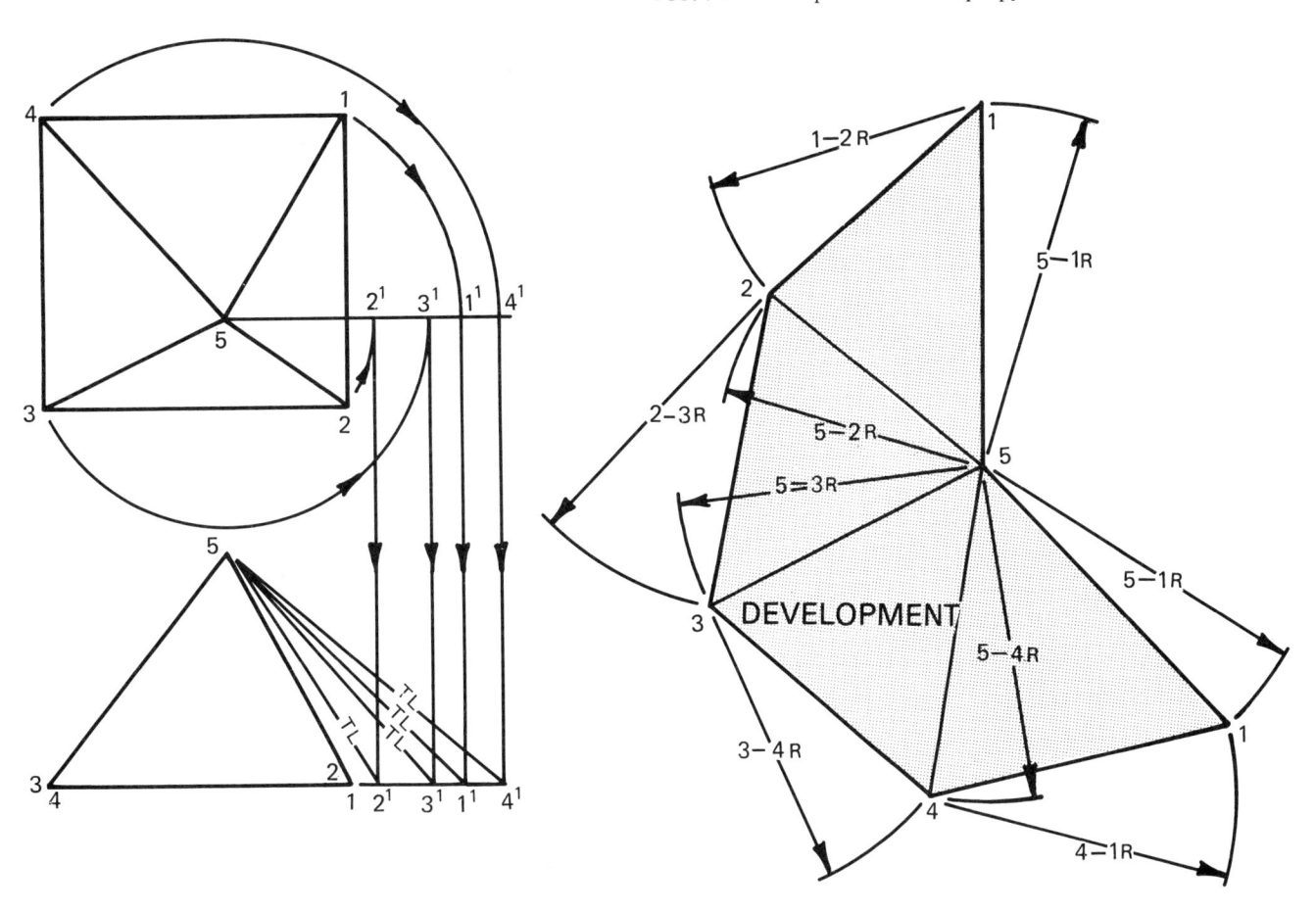

7.15 DEVELOPMENT OF AN OBLIQUE PYRAMID

The development of an oblique pyramid is similar to that of a right pyramid. The lateral edges of an oblique pyramid are unequal. Therefore a TL radius cannot be used to speed the development process. The true length of each lateral edge must be determined separately. Two methods are commonly used: the true length diagram and the revolution method. In this section the revolution method has been employed.

All lines on a development are true length. Therefore, besides the edge lines, the true lengths of the base lines are also required. The base plane normally appears as an edge in the frontal view and parallel to the horizontal plane. When this is the case, the true shape of the base plane shows in the horizontal view. A true shape view provides the true length of the base's perimeter.

The development is constructed by drawing each triangular lateral surface as true shape with common edges joined. In Fig. 7-28, the development of the oblique prism is required.

1. Revolve each lateral edge line about vertex point 5 in the horizontal view and show in the frontal view as true length measurements. Lines 5-1^1, 5-2^1, 5-3^1, and 5-4^1 are the true lengths of the lateral edge lines. Note that on the development each of these lines is used as a true length radius; the superscripts have been eliminated on the development.

2. The base lines are shown as true length in the horizontal view. Start the development by swinging radius 5-1R to establish line 5-1. 5-1R is taken from the frontal view: true length edge line 5-1^1. From point 5 swing arc 5-2R. From point 1, swing arc 1-2R (this is the true length of base line 1-2, taken from the horizontal view) until it intersects arc 5-2R at point 2. The lateral surface bounded by points 5-1-2 is true shape, with its inside up.

3. From vertex point 5 swing arc 5-3R. Using the true length of base line 2-3 as radius 2-3R swing an arc until it intersects arc 5-3R at point 3. Lateral surface 5-2-3 is true shape. Line 5-2 is a bend line.

4. Repeat this process to lay out the remaining two surfaces.

PROB. 7-9

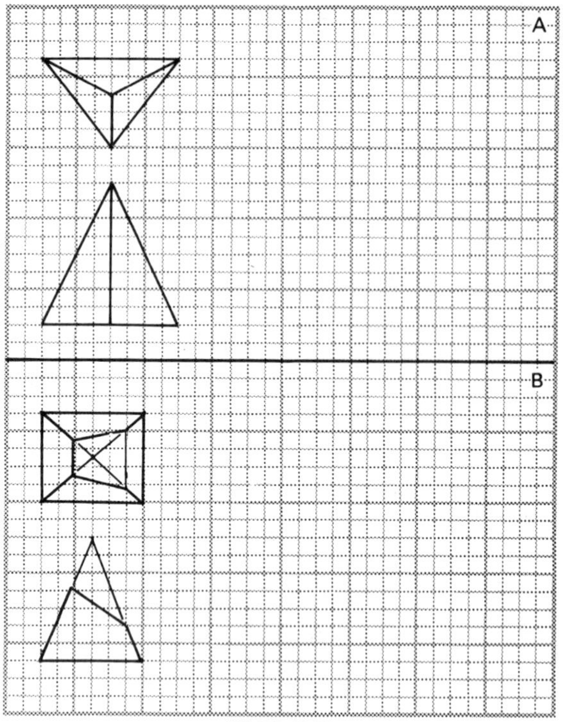

PROB. 7-10

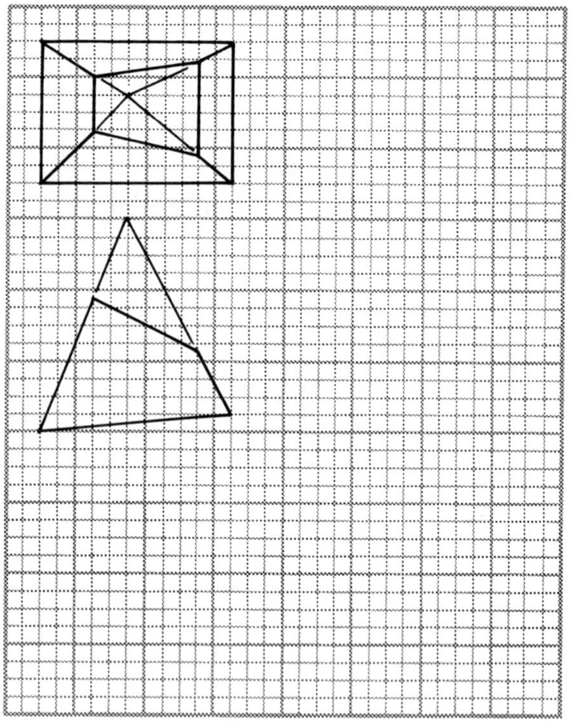

PROB. 7-11

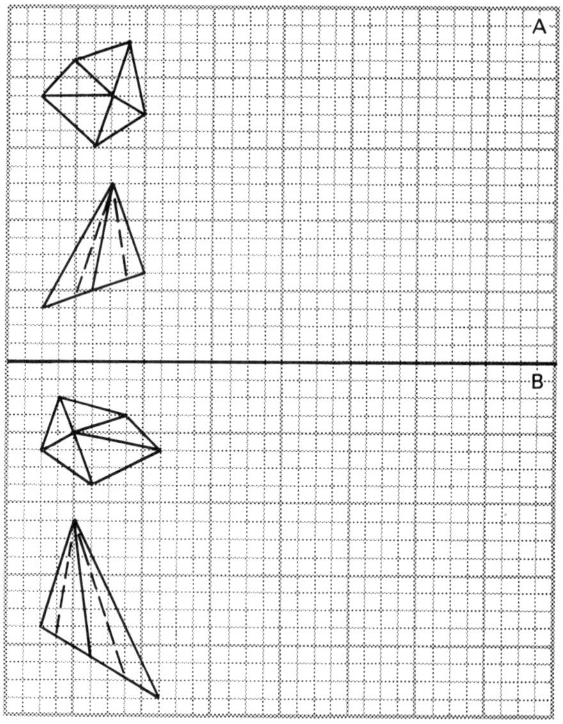

PROB. 7-12

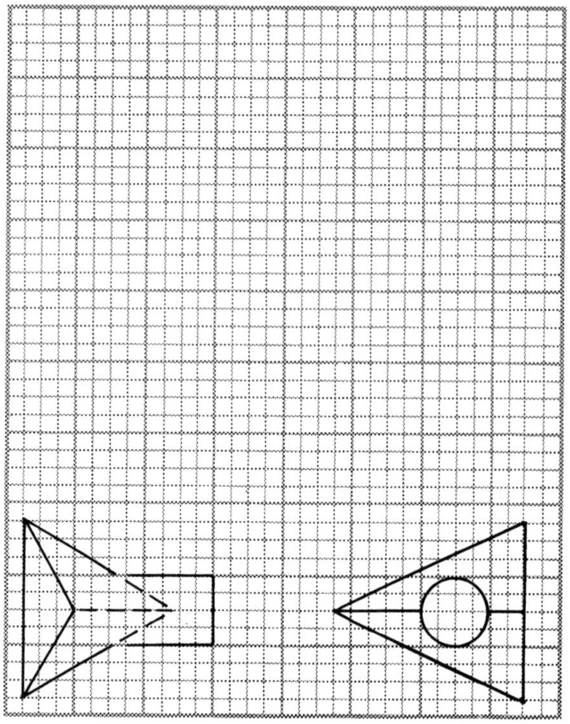

PROBS. 7-9 through 7-11. Develop each figure and construct a model as assigned.
PROB. 7-12. Complete the intersection and develop both pieces. Construct a model of the complete figure.

PROB. 7-13

A

B

PROB. 7-14

A

B

PROB. 7-15

A

B

PROB. 7-16

A

B

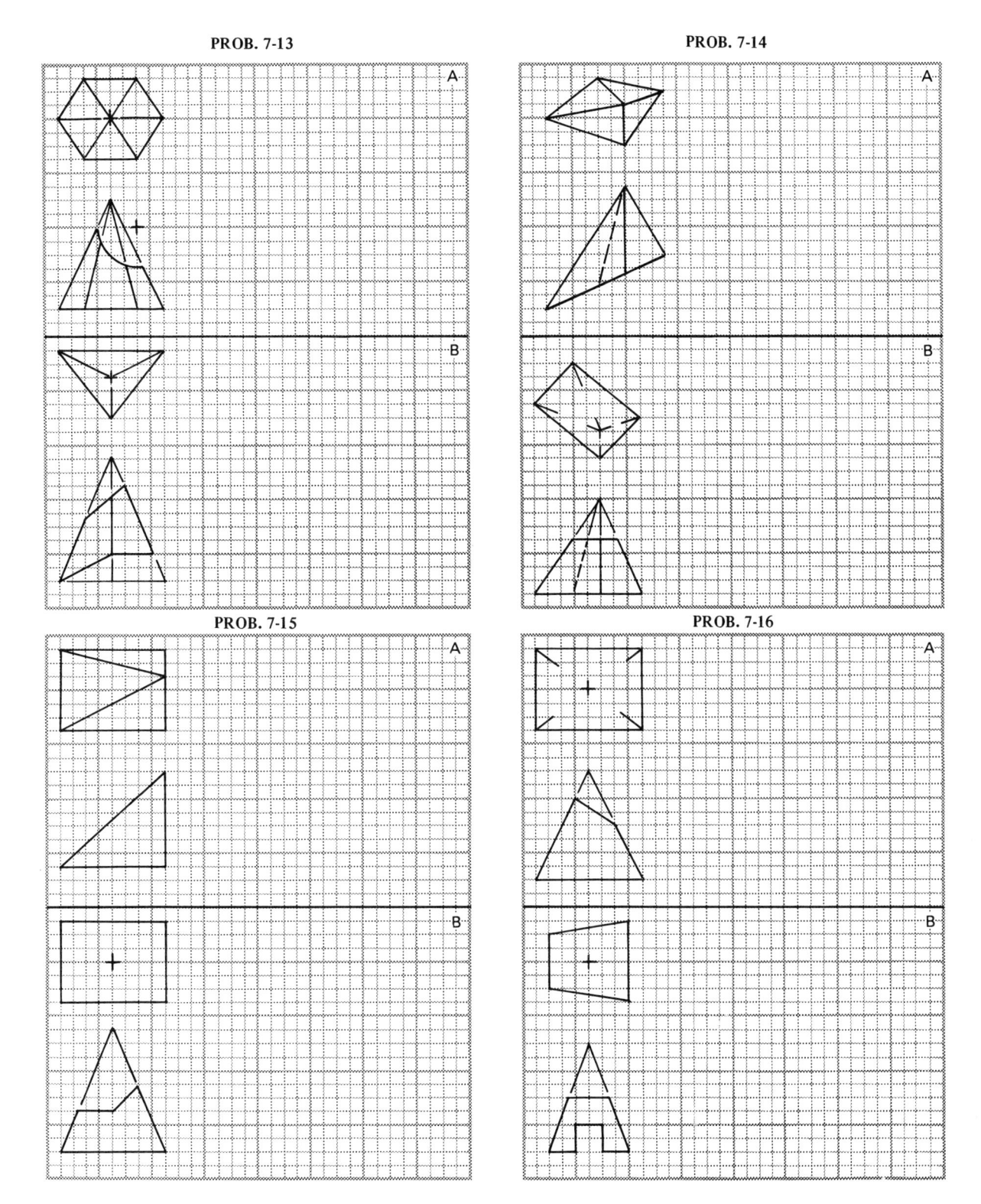

PROBS. 7-13 through 7-16. Complete the views and do a development and model as required.

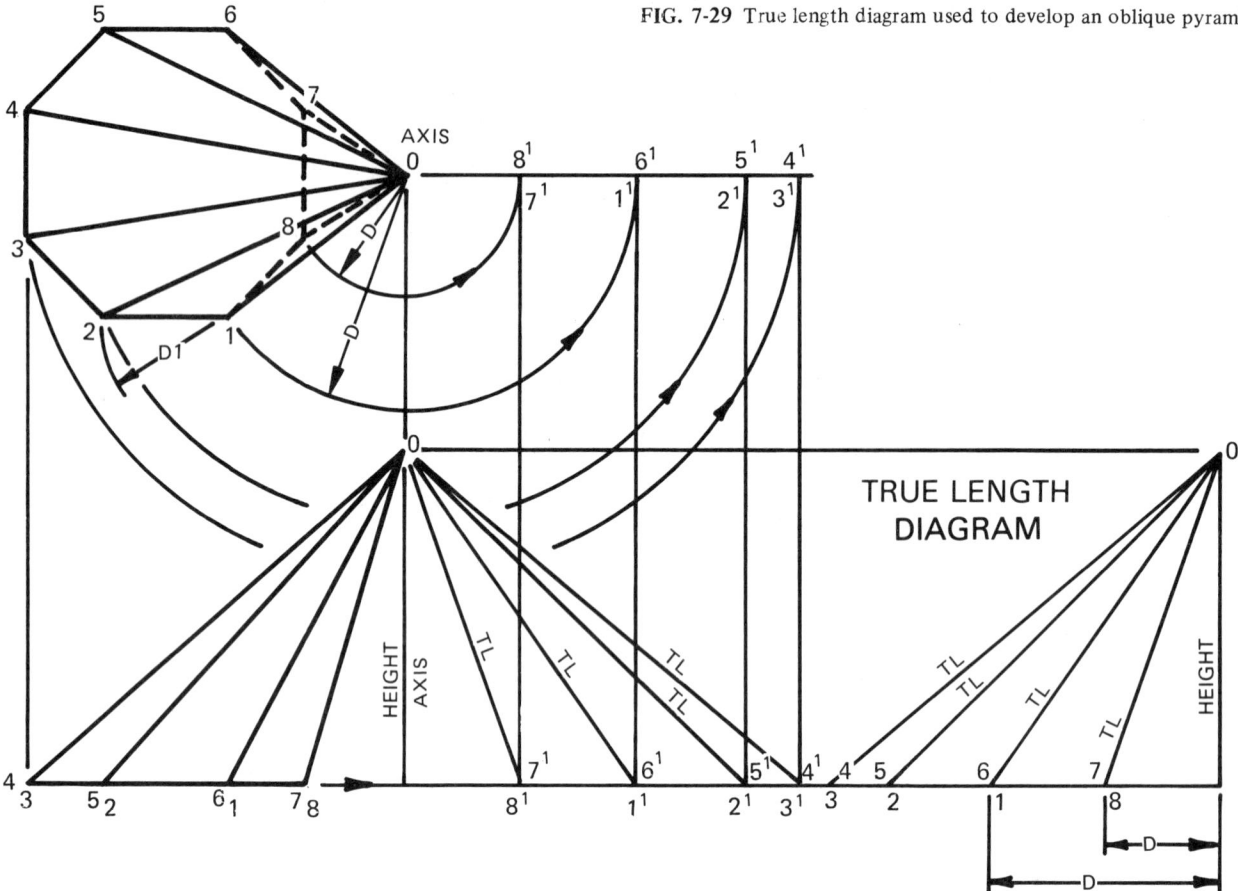

FIG. 7-29 True length diagram used to develop an oblique pyramid.

7.16 TRUE LENGTH DIAGRAMS

To develop a surface composed of numerous edges, a *true length diagram* can be used to eliminate some confusion. Since the true length of each edge is necessary, the revolution method may not be adequate. In Fig. 7-29, the revolution and the true length diagram methods are illustrated.

As can be seen, the revolution method takes up more room and requires that the given views be used to revolve the lines. A true length diagram can be constructed anywhere on or off the paper. One or more TL diagrams can be used as required for clarity if the edge lengths are very similar in length or too numerous.

The TL diagram method establishes the true length of each edge line by creating a right triangle. The height dimension is drawn representing a vertical line dropped from the vertex point to the base. The base dimension for each right triangle is transferred from the horizontal view as the straight-line distance from vertex point 0 to the base points, dimensions D. The hypotenuse equals the true length of a corresponding edge line.

The development is laid out using the true length edge lines and the true length base lines as shown. The base of the oblique pyramid is an octagon. Therefore D1 can be used for every base length when drawing the development.

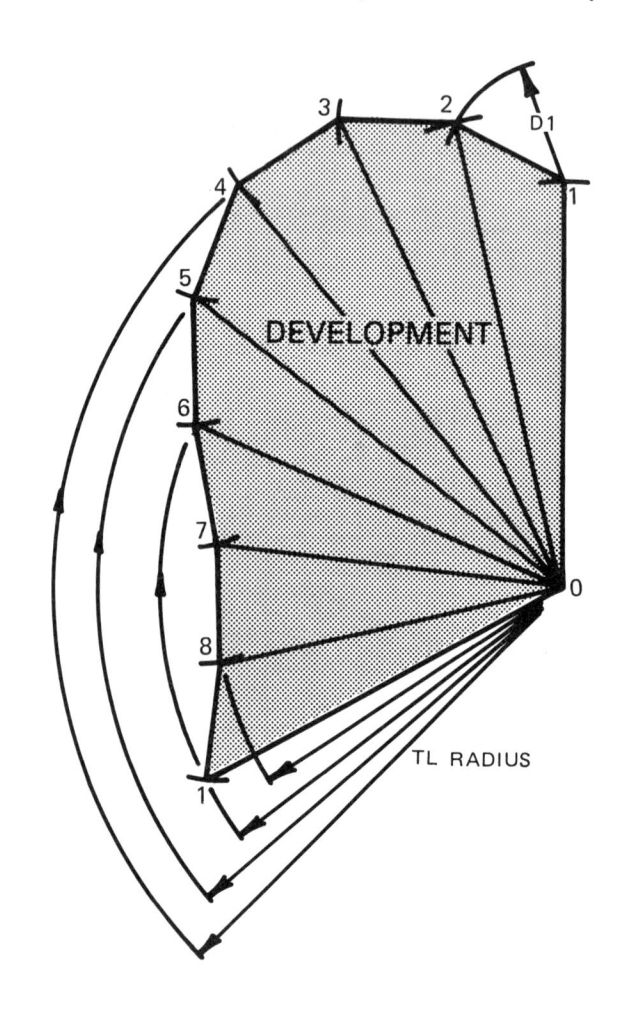

312

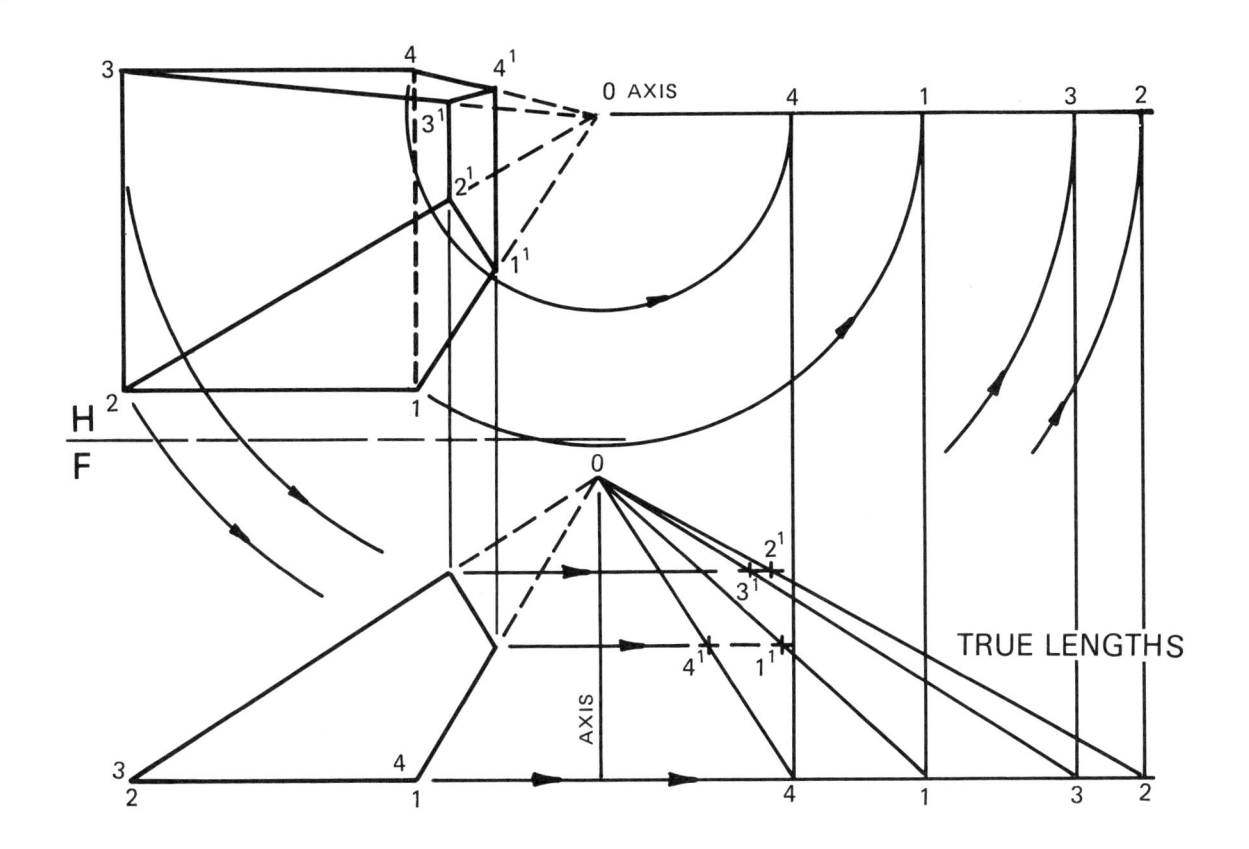

7.17 DEVELOPMENT OF A TRUNCATED OBLIQUE PYRAMID

A truncated oblique pyramid is easily developed if the vertex point can be established on the drawing. The true lengths of the edge lines from the vertex point to the base points must be determined first. In Fig. 7-30, the frontal and horizontal views of the oblique prism are given. A development is required.

1. Extend the lateral edge lines to establish the vertex point if it is not given.

2. Revolve each extended lateral edge line in the horizontal view and show as true length in the frontal view (or use a TL diagram).

3. Establish the true lengths of the bend lines (cut edges) by projecting each cut point in the frontal view, perpendicular to the axis line until it intersects its related true length line. Lines 1-1¹, 2-2¹, 3-3¹, and 4-4¹ are true length.

4. Start the development by drawing edge line, 0-1. Using base line 1-2 as the radius, swing arc 1-2 from point 1. Swing an arc from vertex point 0 using line 0-2 as the radius, until it intersects arc 1-2 at point 2. Triangular plane 0-1-2 is one panel/face of the development. Complete the remaining triangular faces of the development.

5. Transfer 0-1¹, 0-2¹, 0-3¹, 0-4¹, and 0-1¹ to their respective edge lines in the development. Connect points 1¹, 2¹, 3¹, 4¹, and 1¹ to form the upper outline of the development.

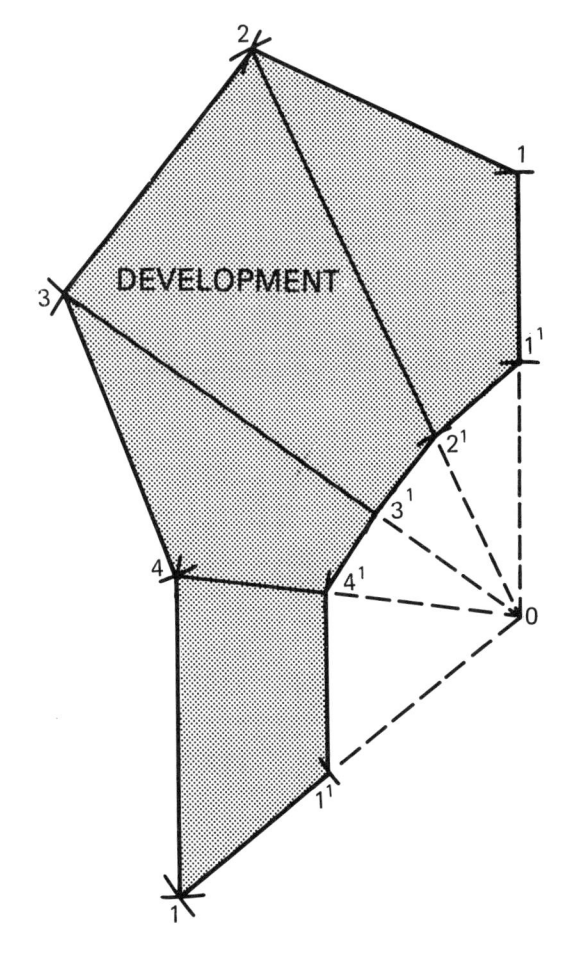

FIG. 7-30 Development of a truncated oblique pyramid.

PROB. 7-17

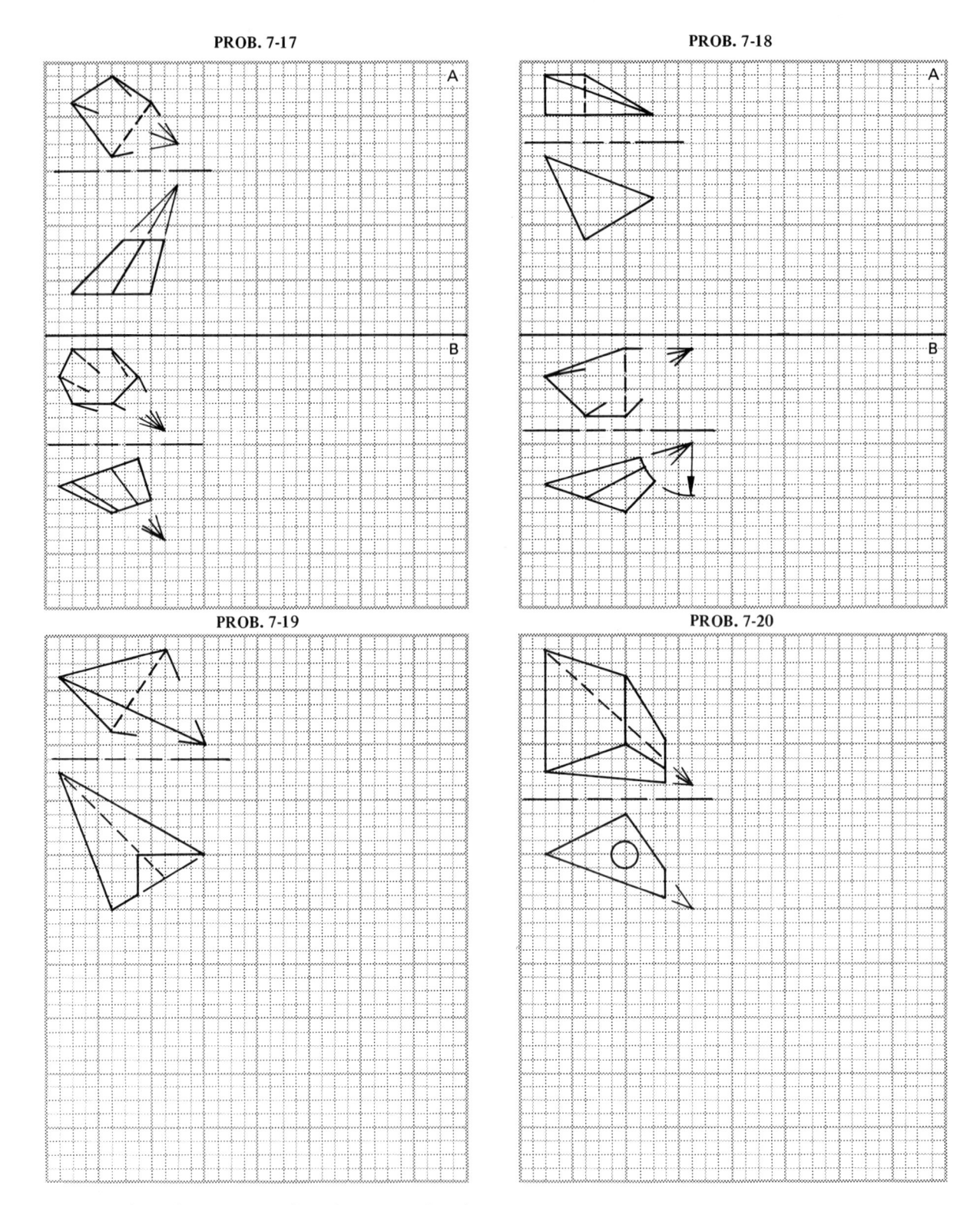

PROB. 7-18

PROB. 7-19

PROB. 7-20

PROBS. 7-17 through 7-20. Complete the views of each figure. Using a true length diagram, develop each figure and make a scale model as assigned.

FIG. 7-31 Water tower with spherical tank, conical base, and cylindrical stem.

7.18 CURVED SURFACES

In the preceding sections the development of geometric forms composed of straight lines and plane surfaces were introduced. The development of forms whose surfaces are curved is also an important part of engineering design work. *Curved surfaces fall into two basic categories: single-curved* and *double-curved*.

A *single-curved* surface is a ruled surface, since it can be generated by the movement of a straight line. Cylinders, cones, and convolutes are the three types of single-curved surfaces. In Fig. 7-31 the base of the water tower is a cone, as is its transition piece from the cylindrical stem to the spherical tank. The space probe in Fig. 7-32 incorporates both cones and cylinders into its design. Single-curved surfaces are the most common.

A *double-curved surface* is generated by the movement of a curved line. The sphere, spheroid, torus, paraboloid, and hyperboloid are examples of double-curved surfaces. Double-curved surfaces can only be approximately developed (single-curved surfaces can be accurately developed). The spherical water tank in Fig. 7-31 and the half sphere domed space station in Fig. 7-33 are examples of double-curved surfaces.

FIG. 7-32 Conical domes and a cylinderlike body make up this space probe. *(Courtesy NASA.)*

All curved surfaces are generated by the movement of a curved or straight line. The line which generates a surface is called a "*generatrix.*" Any one position of the generatrix is an *element* of the surface. The generatrix moves according to the "*directrix,*" which is a line or lines that define the direction and motion of the generatrix.

FIG. 7-33 Cylindrical space station with a half sphere dome top. *(Courtesy NASA.)*

FIG. 7-34 Single-curved surfaces are found throughout a petrochemical facility.

7.19 DEVELOPMENT OF SINGLE CURVED SURFACES

Cylinders, cones, and convolutes are the three types of single-curved surfaces. *A **single-curved surface** is generated by the movement of a straight line so that each of its two closest positions are in the same plane.* Any two consecutive positions (elements) will therefore be parallel (as in a cylinder) or will intersect (as in a cone or convolute). All single-curved surfaces are ruled surfaces and can be accurately developed. A variety of cones and cylinders can be seen in Fig. 7-34 and Fig. 7-35.

A *cylinder* is generated by a straight-line generatrix moving around a curved directrix. The directrix is normally a closed curve (ellipse, circle, etc.). All positions of the generatrix (elements) are parallel to one another. A cylinder develops as a parallel-line development.

A *cone* is generated by the movement of one end of a straight-line element (generatrix) around a curved directrix (normally closed). The other end of the generatrix is fixed at one point: the vertex/apex. The positions of the generatrix establish elements on the surface of the cone. All elements intersect at one point, the vertex. A development of a cone is a radial-line development since each of its elements radiate from the vertex point.

Convolutes are generated by a straight-line generatrix which moves in accordance and tangent to a double-curved line (directrix). Two consecutive, but never three, elements intersect. Aircraft wings and fuselages, piping and ducting transition pieces, and automobile bodies are a few examples of the use of convolutes in industry.

FIG. 7-35 Chemical tanks designed using cones and cylinders.

316

7.20 DEVELOPMENT OF A RIGHT CIRCULAR CYLINDER

Cylinders are ruled surfaces and will develop as parallel-line developments. The positions of the generatrix on the cylinder will establish elements on its surface. Each of a cylinder's elements is true length on the development. A cylinder is developed by unrolling its surface, normally inside-up. The length of the development is equal to the perimeter of its right section. A right circular cylinder has a development equal to its circumference: diameter × 3.141 (pi), Fig. 7-36. A right section (axis as a point) and a view showing the axis as true length are necessary to develop a cylinder. The edge view/right section determines the shape of the cylinder and provides a view where elements can be established on its surface. A true length view of the cylinder's axis shows all elements on its surface as true length. A development is made by rolling the lateral surface of the cylinder onto a plane. All elements will be true length with a development length equal to its perimeter (circumference, when a circular cylinder).

In Fig. 7-36, the right section of the cylinder is shown in the horizontal view. Elements are established along its surface by dividing the right section into a number of equal parts. The elements are located by evenly (radially) dividing the circumference of the circular section as shown; 12, 16 and 24 radial divisions are commonly used.

Each division is projected to the true length view to establish the elements on the lateral surface. The stretch-out line is drawn perpendicular to the TL view. The base perimeter may be used as the stretch-out line if it is perpendicular to the cylinder's axis as in the example. The length of the stretch-out line is equal to the perimeter and is divided into the same number of equally spaced parts as the right section and labeled accordingly. The true length of each element is projected to the development, from the true length view, to establish its outline. In Fig. 7-36, both bases are perpendicular to the axis. Therefore, all elements are the same length and the development unrolls as a rectangle, its height equaling the altitude of the cylinder and length equal to the circumference. Cylinders are a single surface, therefore the elements are drawn as thin construction lines in all views and on the development.

The space vehicle in Fig. 7-37 is composed of cylindrical shapes.

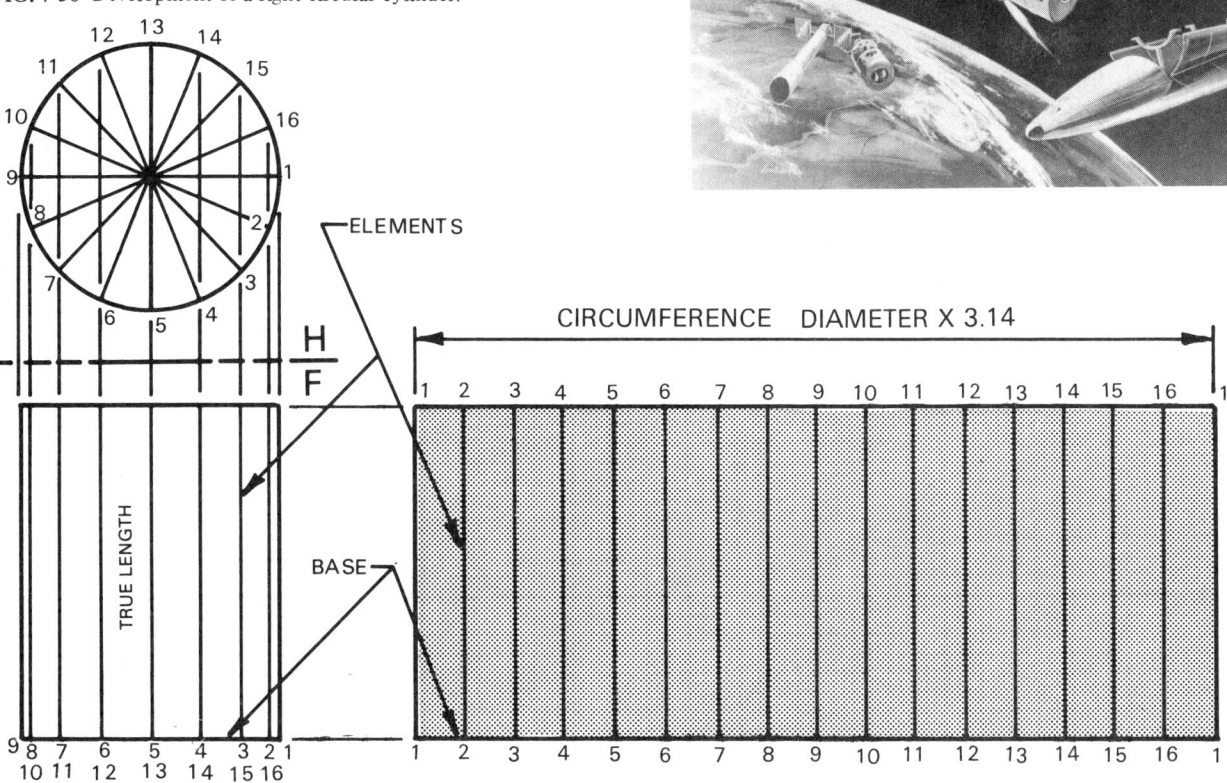

FIG. 7-37 Cylindrical space tug. *(Courtesy NASA.)*

FIG. 7-36 Development of a right circular cylinder.

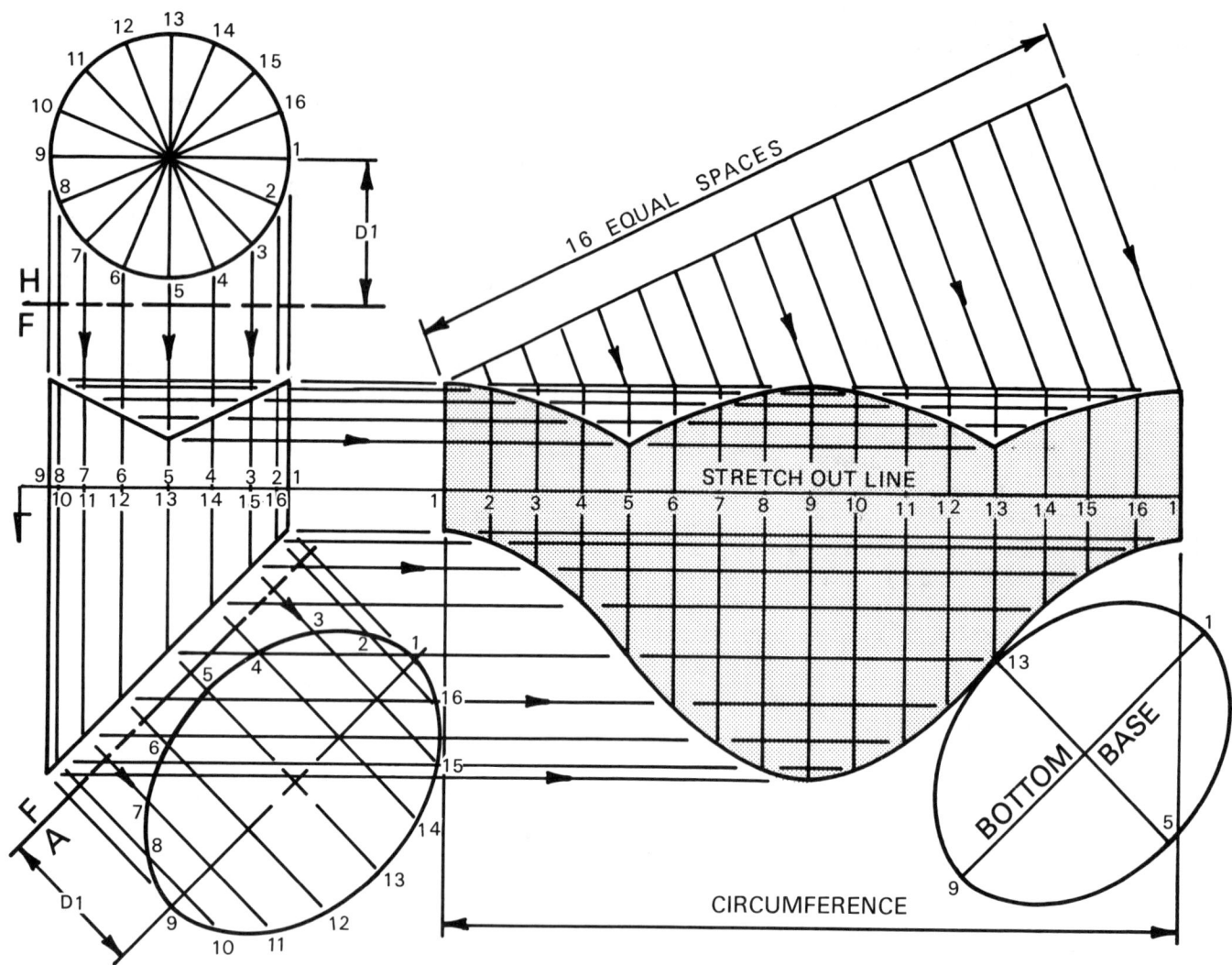

FIG. 7-38 Development of a truncated cylinder.

7.21 DEVELOPMENT OF A TRUNCATED RIGHT CIRCULAR CYLINDER

The development of a cylinder is accomplished by simply unrolling its lateral surface along a stretch-out line equal in length to the perimeter of its right section. In Section 7.20 it was explained how to establish elements along the lateral surface of the cylinder. The true lengths of these elements are projected to the development to establish the upper and lower outline. The distance between the elements is equal to the arc distance between the divisions established on the right section. The elements on a circular cylinder can be established by simply dividing the stretch-out line into an equal number of parts, corresponding to the divisions of the right section. When the cylinder's right section is anything but a circle, the chordal distance between divisions on the right section must be used instead. There will be a discrepancy between the real length and the length established by stepping off chord distances, since a chord is shorter than its related arc. This difference can be minimized by dividing the right section into small enough parts where the arc and chord distances are nearly equal.

In Fig. 7-38, the development of a truncated cylinder, including its lower base is required.

1. Divide the right section into 16 equal parts and project to the frontal view as true length elements on the cylinder's surface.
2. Draw F/A parallel to the edge view of the lower base and project auxiliary A.
3. Draw the stretch-out line perpendicular to the right section. Use the cylinder's calculated circumference as the length of the stretch-out line.
4. Divide the stretch-out line into 16 equal spaces corresponding to the number of elements established in right section and label accordingly. Note that the cylinder is to be developed inside-up with its shortest element as the seam line.
5. Project the true lengths of the elements from the true length view (frontal view) and draw the elements on the development. Draw a smooth curve through the end points of the elements.
6. Connect the base, tangent to the lower outline at element 13, since it is the most convenient and can be drawn at the same angle as in auxiliary A.

PROB. 7-21

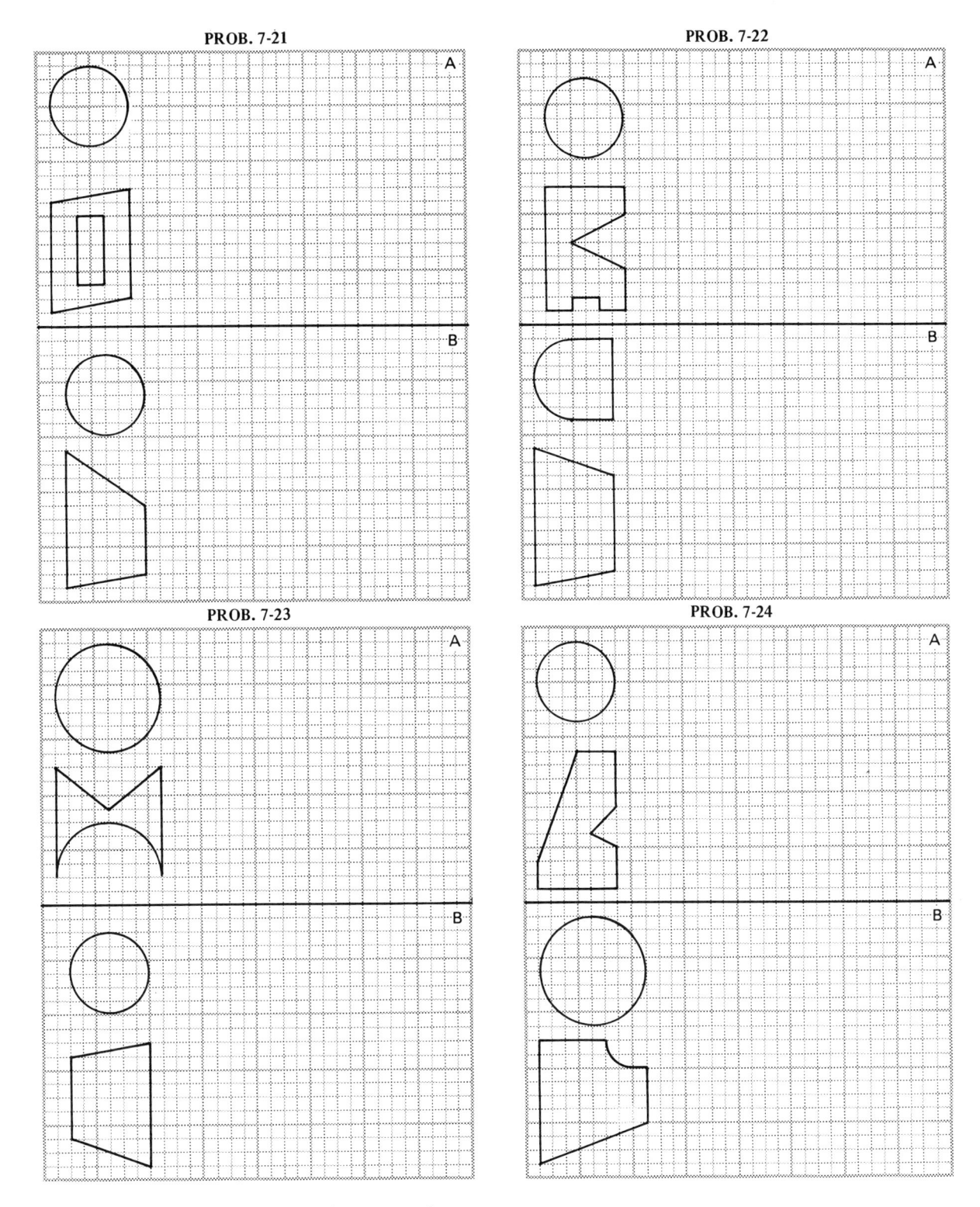

PROB. 7-22

PROB. 7-23

PROB. 7-24

PROBS. 7-21 through 7-24. Do a full development and construct a model of each cylinder.

319

7.22 DEVELOPMENT OF AN OBLIQUE CYLINDER

To develop a cylinder, a view showing its elements (and axis) as true length and a right section view (axis as a point) are necessary. When the given cylinder is oblique, both a true length and a right section view must be established before the development can be drawn. A right section of a cylinder must be taken perpendicular to the true length axis line. Thus, the true length view is projected first. The development is accomplished in the same manner as described in the preceding sections. When the cylinder to be developed has a right circular section the stretch-out line is equal to the circumference of the circle, Fig. 7-39. When the section is not a circle, chordial distances (D1) are used to lay off the elements along the stretch-out line. The length of the stretch-out line equals the accumulation of chord distances, D1 × eight for Fig. 7-39.

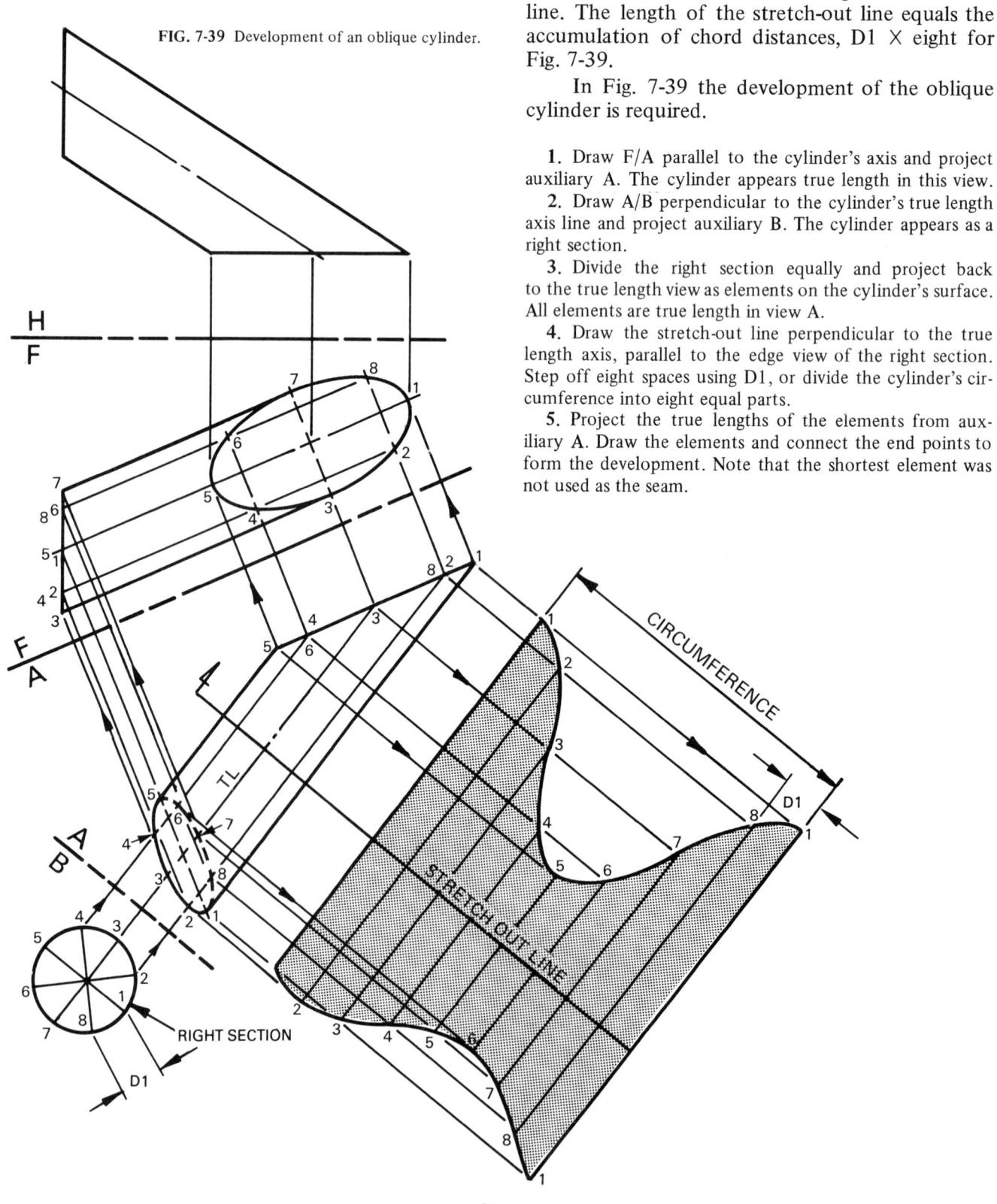

In Fig. 7-39 the development of the oblique cylinder is required.

1. Draw F/A parallel to the cylinder's axis and project auxiliary A. The cylinder appears true length in this view.
2. Draw A/B perpendicular to the cylinder's true length axis line and project auxiliary B. The cylinder appears as a right section.
3. Divide the right section equally and project back to the true length view as elements on the cylinder's surface. All elements are true length in view A.
4. Draw the stretch-out line perpendicular to the true length axis, parallel to the edge view of the right section. Step off eight spaces using D1, or divide the cylinder's circumference into eight equal parts.
5. Project the true lengths of the elements from auxiliary A. Draw the elements and connect the end points to form the development. Note that the shortest element was not used as the seam.

FIG. 7-39 Development of an oblique cylinder.

PROB. 7-25

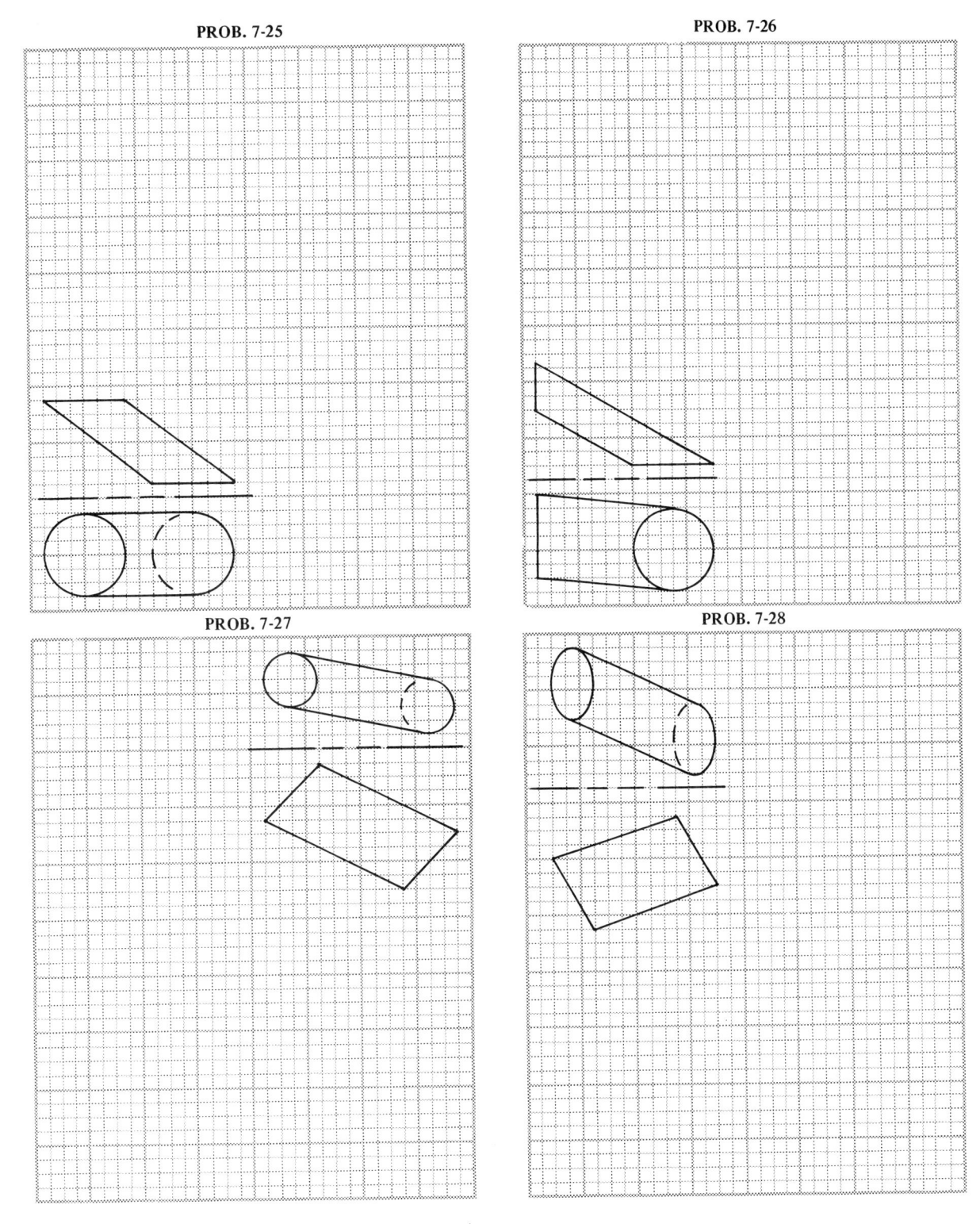

PROB. 7-26

PROB. 7-27

PROB. 7-28

PROBS. 7-25 through 7-28. Complete a model and a development of each oblique cylinder.

321

7.23 DEVELOPMENT OF INTERSECTING CYLINDERS

In Fig. 7-40, the line of intersection in the frontal view is established as in Chapter 6. The frontal and profile views are given. Determine the line of intersection and develop both cylinders.

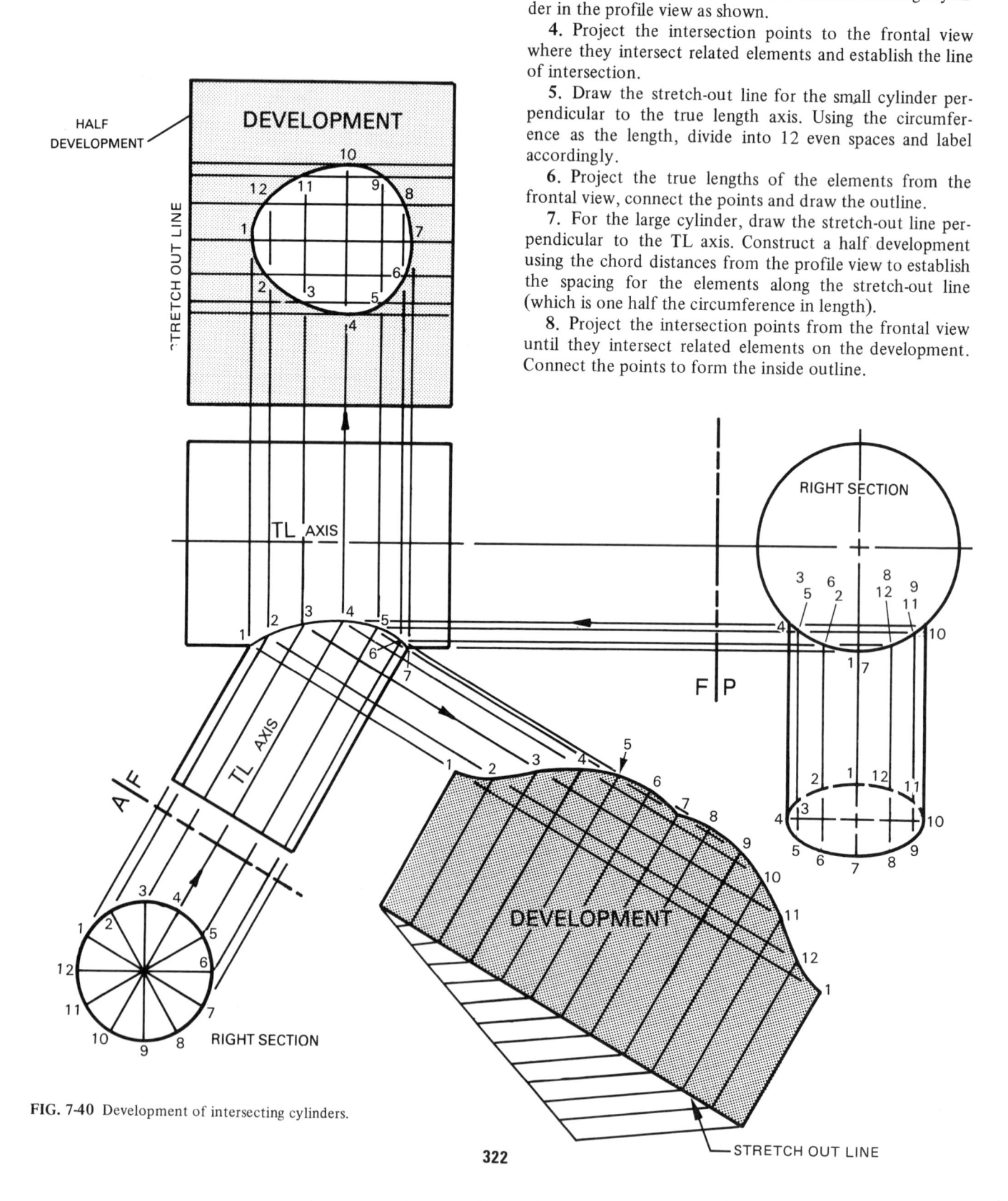

1. Draw F/A perpendicular to the small cylinder. Auxiliary A is a right section/end view.
2. Pass a series of evenly spaced cutting planes parallel to the axis of the small cylinder in the right section (this is the same as dividing the circumference equally). Each CP cuts an element on the surface of the cylinder in the frontal view. These elements are used to develop the cylinders as well as establish the line of intersection.
3. Locate the elements on the small cylinder in the frontal and profile views. The elements intersect the large cylinder in the profile view as shown.
4. Project the intersection points to the frontal view where they intersect related elements and establish the line of intersection.
5. Draw the stretch-out line for the small cylinder perpendicular to the true length axis. Using the circumference as the length, divide into 12 even spaces and label accordingly.
6. Project the true lengths of the elements from the frontal view, connect the points and draw the outline.
7. For the large cylinder, draw the stretch-out line perpendicular to the TL axis. Construct a half development using the chord distances from the profile view to establish the spacing for the elements along the stretch-out line (which is one half the circumference in length).
8. Project the intersection points from the frontal view until they intersect related elements on the development. Connect the points to form the inside outline.

FIG. 7-40 Development of intersecting cylinders.

FIG. 7-41 Elbow joint.

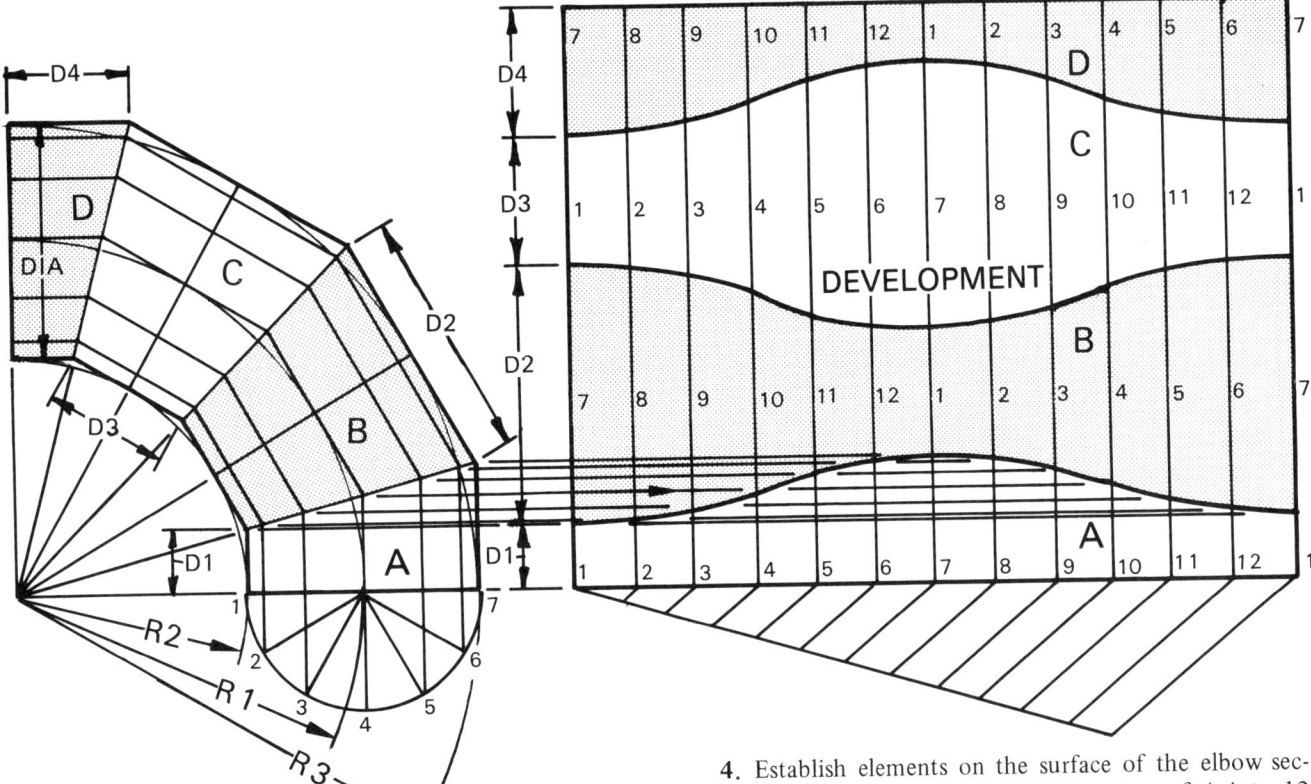

7.24 DEVELOPMENT OF AN ELBOW JOINT

In Fig. 7-41, the 90-degree four-section mitered elbow is developed using the same pattern for each of its four sections. The diameter of the cylindrical sections remains constant and the radius of the centerline (R1) is normally $1\frac{1}{2} \times$ the diameter of the cylindrical form. Each section consists of right circular cylinders cut to a specific angle by a plane. The angle of the bend or sweep angle can be any required angle and divided into any number of sections. To draw the elbow the sweep angle is divided into equal angles; each equal angular division of the bend is two less than two times the number of pieces of the elbow. A four-section elbow is divided by six, two for each interior section and one for each end section, Fig. 7-41. The following steps describe the process of drawing a mitered elbow and doing a development, Fig. 7-41.

1. Determine sweep angle and number of sections. Swing R1 based on design requirements. Swing R2 as the inside radius and R3 as the outside radius.
2. Divide the sweep angle into equal parts. Two for each interior section and one for each end piece, six in the example.
3. Draw perpendicular lines tangent to R2 and R3 as shown. Angular divisions alternate as tangent lines and section joints.
4. Establish elements on the surface of the elbow sections by dividing the revolved half section of A into 12 equal parts and locating the elements on all sections by projection.
5. Draw the stretch-out line perpendicular to section A and divide into 12 equal parts. Project the true lengths of section A to the development as shown.
6. Transfer the true lengths of remaining sections to the development and connect the points as smooth curves.

In Fig. 7-42 a 90° three-section mitered elbow is shown during the welding stage.

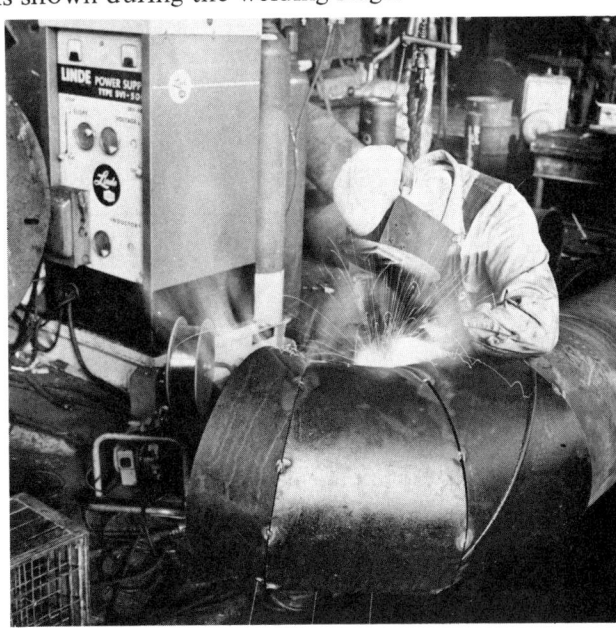

FIG. 7-42 Elbow joint being welded.

PROB. 7-29

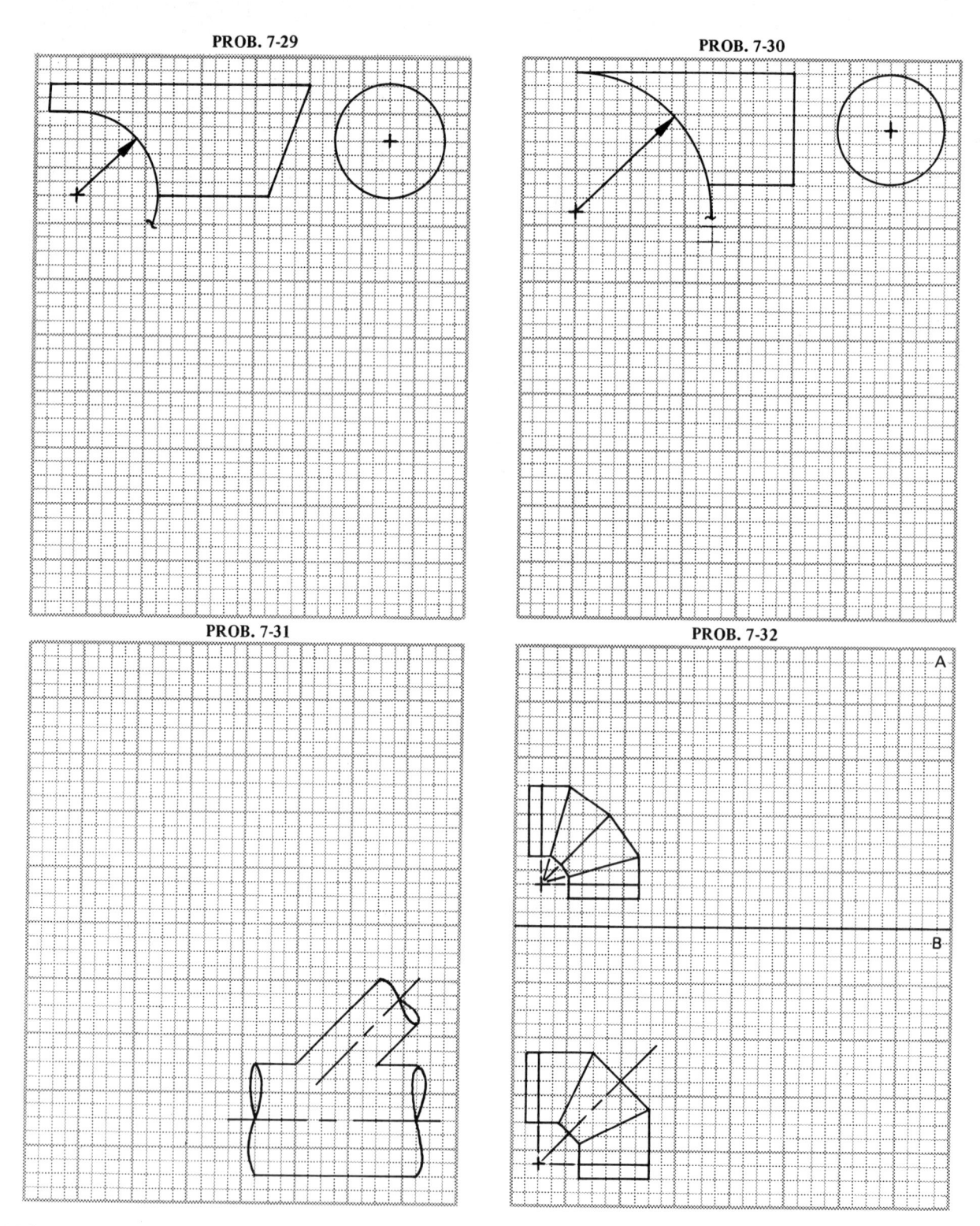

PROB. 7-30

PROB. 7-31

PROB. 7-32

PROBS. 7-29 through 7-30. Develop the intersected cylinders.

PROB. 7-31. Complete the intersection and develop both pipes.

PROB. 7-32. Develop the mitered elbows.

FIG. 7-43 Aircraft configurations. *(Courtesy NASA.)*

7.25 DEVELOPMENT OF CONES

Cones are used in the design of a variety of industrial products, airplane configurations (Fig. 7-43), storage tanks, ducting and piping transitions, and numerous structural, architectural, and mechanical designs. *A **cone** is a single-curved surface generated by the movement of a straight-line generatrix, fixed at one end and intersecting a curved directrix.* The fixed point is the vertex and the directrix is normally a closed curve (usually a circle or ellipse). Each position of the generatrix establishes an element on the surface of the cone. All elements of a cone intersect at the vertex point. Therefore the development of a cone is a radial-line development, all elements radiating from a single point/vertex. The generatrix of a cone is a straight line. Thus, a cone is a ruled surface and can be accurately developed.

There are three general types of cones: right circular, oblique, and open. A **right circular cone** is a *cone of revolution* generated by revolving the generatrix about an axis line with a circle as a directrix and an axis perpendicular to the base plane (directrix plane). An **oblique cone** has an axis that is not perpendicular to its base plane. Its directrix is a closed curve. An **open cone** has an open single-curved or double-curved line as a directrix.

In many designs a cone is cut below its vertex, "truncated." The exit cone of the motor shown in Fig. 7-44 is a truncated (in this case a frustrum) right circular cone.

FIG. 7-44 Exit cone of the solid propellant motor. *(Courtesy NASA.)*

FIG. 7-45 Development of a right circular cone.

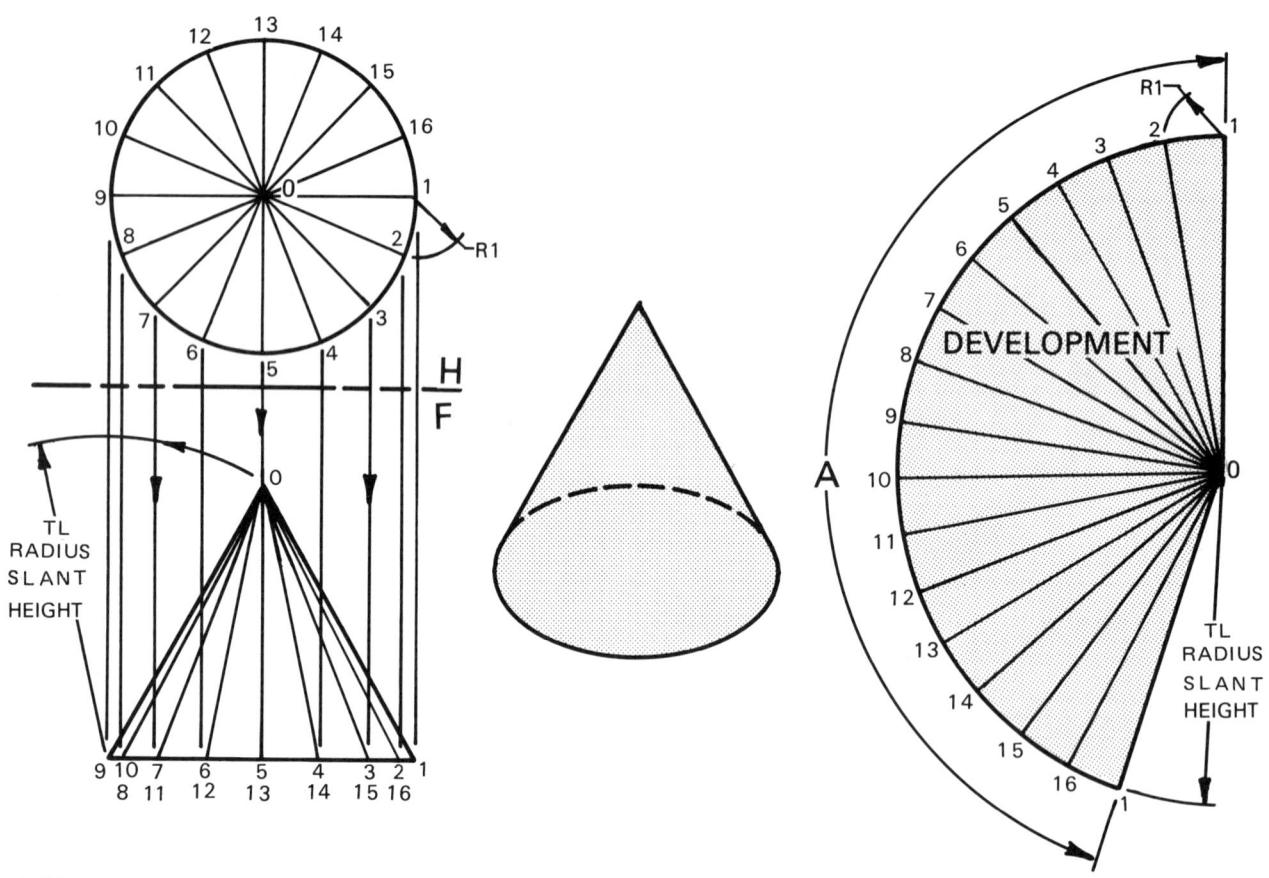

7.26 DEVELOPMENT OF A RIGHT CIRCULAR CONE

A cone can be developed by unrolling its lateral surface on a plane. Each element of a right circular cone will be of equal length and will intersect at the vertex point. Therefore the development is a radial line development, with each of the elements radiating from the vertex. A right circular cone develops as a sector of a circle, the radius of which is equal to the slant height of the cone with an arc length equal to the length of the cone's circumference.

The development of a right circular cone can be drawn using two methods. The graphical method involves dividing the cone's base circle into equal parts. In Fig. 7-45 the base circle is radially divided into 16 equal parts. An element on the cone's surface is drawn at each division. All elements are of the same length. The true length of an element equals the slant height of the cone. For the development, the slant height is used as the TL radius, which is swung an indefinite length. Distances between the base divisions, chord measurements, are stepped off along the development arc, R1. This method produces a development pattern with an arc length (A) slightly smaller than a true development since the chord distance between base divisions is smaller than the arc distance.

When an accurate development is required of a right circular cone, the arc angle (A) can be calculated. Angle A is the sector angle of the development. The sector angle (angle A) equals the radius of the cone's base divided by the slant height. Angle A = 360° radius(of base)/slant height. The development is drawn using the computed sector angle to establish the length of the development's arc.

The Apollo space capsule shown in Fig. 7-46 was designed with a conical configuration.

FIG. 7-46 Conical Apollo spacecraft model in wind tunnel. *(Courtesy NASA.)*

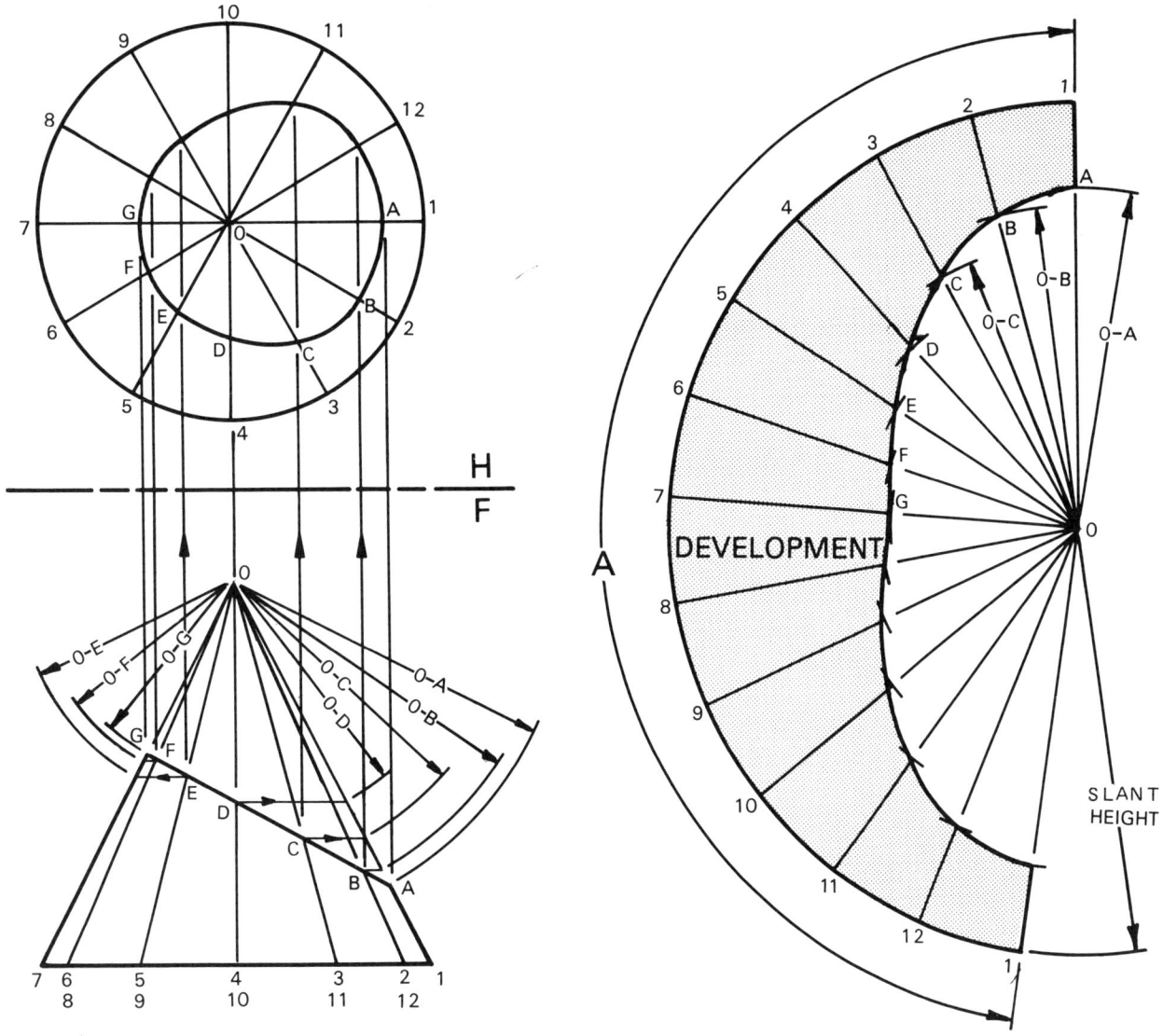

FIG. 7-47 Development of a truncated cone.

7.27 DEVELOPMENT OF A TRUNCATED RIGHT CIRCULAR CONE

The development of a truncated right circular cone can be established by drawing the development's sector as in Section 7.26. The upper outline of the development, corresponding to the truncated surface, can be determined using the same general method as that for a truncated right pyramid. A right circular cone will have equal elements. The length of the cut elements, the distance from the base division to where the element is cut, can be determined by establishing the distance from the vertex to the cut point. A cut cone element equals the extended element minus the vertex to cut point distance 1-A = 0-1 minus 0-A. The true length of each cut element, 0-A, 0-B, 0-C, etc, is transferred to the development as a true length radius from the vertex point along its related element.

In Fig. 7-47, a development of the truncated right circular cone is required.

1. Divide the cone's base into 12 evenly spaced parts to establish the surface elements and project to the F view.

2. Label the elements and the cut points along the elements as shown. Only the front half need be labeled, since the cone is symmetrical.

3. All extended elements are true length, therefore the cut points may be moved perpendicular to the axis until they intersect element 1 or 7 (which appear true length in the frontal view). This procedure is simply the revolution of each cut point in the horizontal view and its true length projection in the frontal view.

4. Using element 0-1 as the slant height, swing an arc from vertex 0 to start the development. Establish the sector of the development by the graphical method, stepping off the cone's base chord distances on the sectors arc, 1-2, 2-3, 3-4, etc.

5. Transfer the true lengths of the upper portions of the elements to their related elements on the development and connect the cut points to form the upper outline. 0-A is transferred to element 0-1. 0-B is transferred to elements 0-2 and 0-12. 0-C is transferred to elements 0-3 and 0-11.

PROB. 7-33

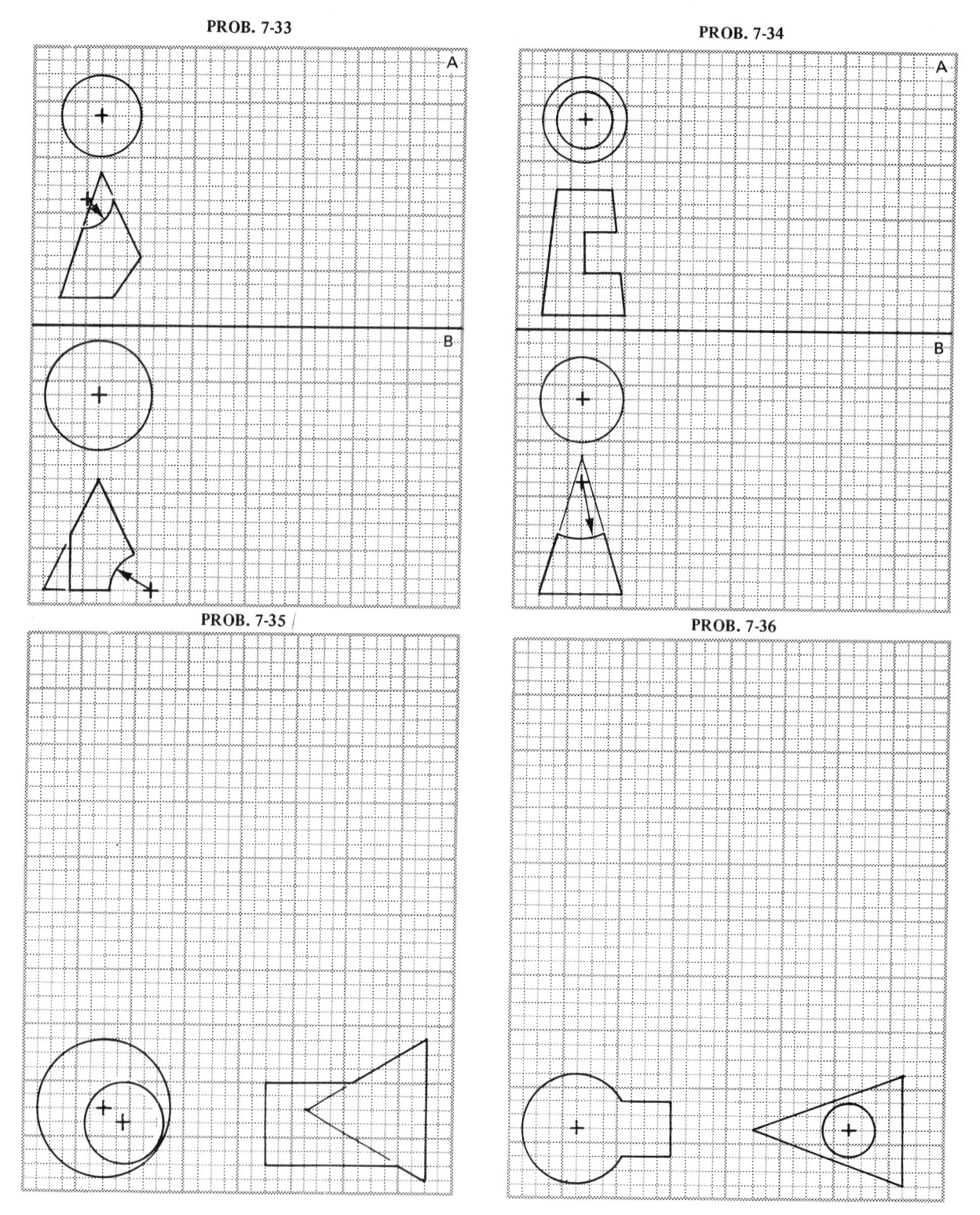

PROB. 7-34

PROB. 7-35

PROB. 7-36

PROBS. 7-33 through 7-36. Complete the given views, do a development and make a model as assigned.

FIG. 7-48 Development of an oblique cone.

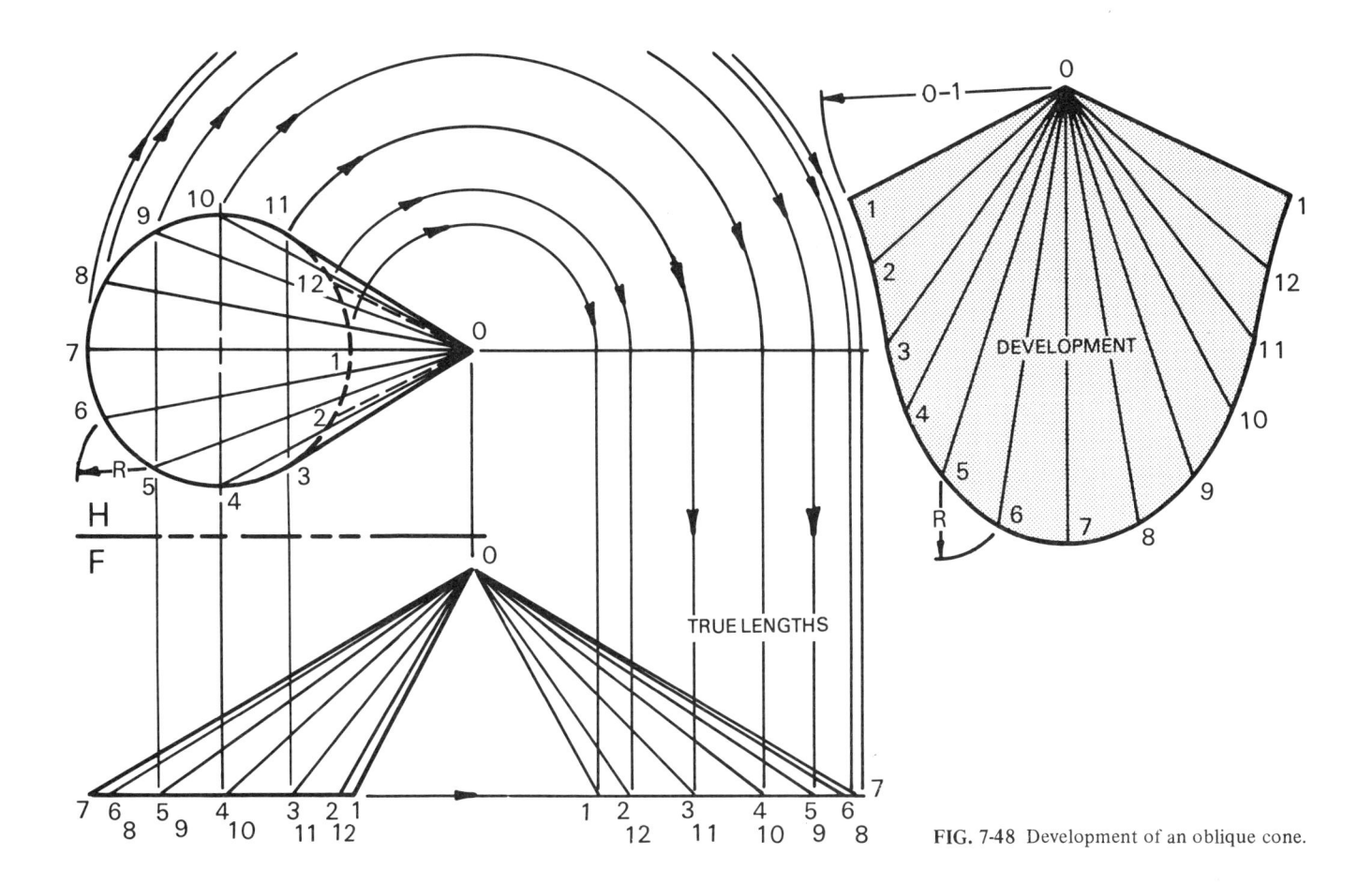

7.28 DEVELOPMENT OF AN OBLIQUE CONE

The development of an oblique cone is similar to the development of an oblique pyramid. Elements are established on the cone's surface by evenly dividing the base curve, as in Fig. 7-48. Since the cone is oblique, the elements are of different lengths. Therefore the development is not a sector of a circle. Two adjacent elements and their corresponding chordal distance define a series of triangular planes on the cone's surface. The development of the cone involves laying out each successive triangle with common edges joined. The true length of each element is determined before the development is started. Revolution or a true length diagram can be used to establish the true length of the elements.

In Fig. 7-48, a development of the oblique cone is required.

1. Divide the cone's base in the horizontal view into 12 equal parts. Draw elements from vertex 0 to each point on the base. Show the frontal and horizontal views of the elements.
2. Determine the true length of each element.
3. Use the shortest element as the seam and draw the development inside-up. Start the development by drawing element 0-1.
4. From point 1 on the development, swing an arc (1-2) equal to the chordal distance R. All chords equal R.
5. Using the true length of element 0-2 swing an arc from vertex 0 until it intersects arc 1-2 (R) and locates point 2. Triangular plane 1-0-2 is the first of 12 successive planes representing the unrolled surface of the cone.
6. Continue laying out the triangular planes as shown.
7. Connect the end points of the elements with a smooth curve to complete the outline of the development.

The nose section of the aircraft in Fig. 7-49 is an oblique cone.

FIG. 7-49 The nose section of this aircraft is an oblique cone. *(Courtesy NASA.)*

329

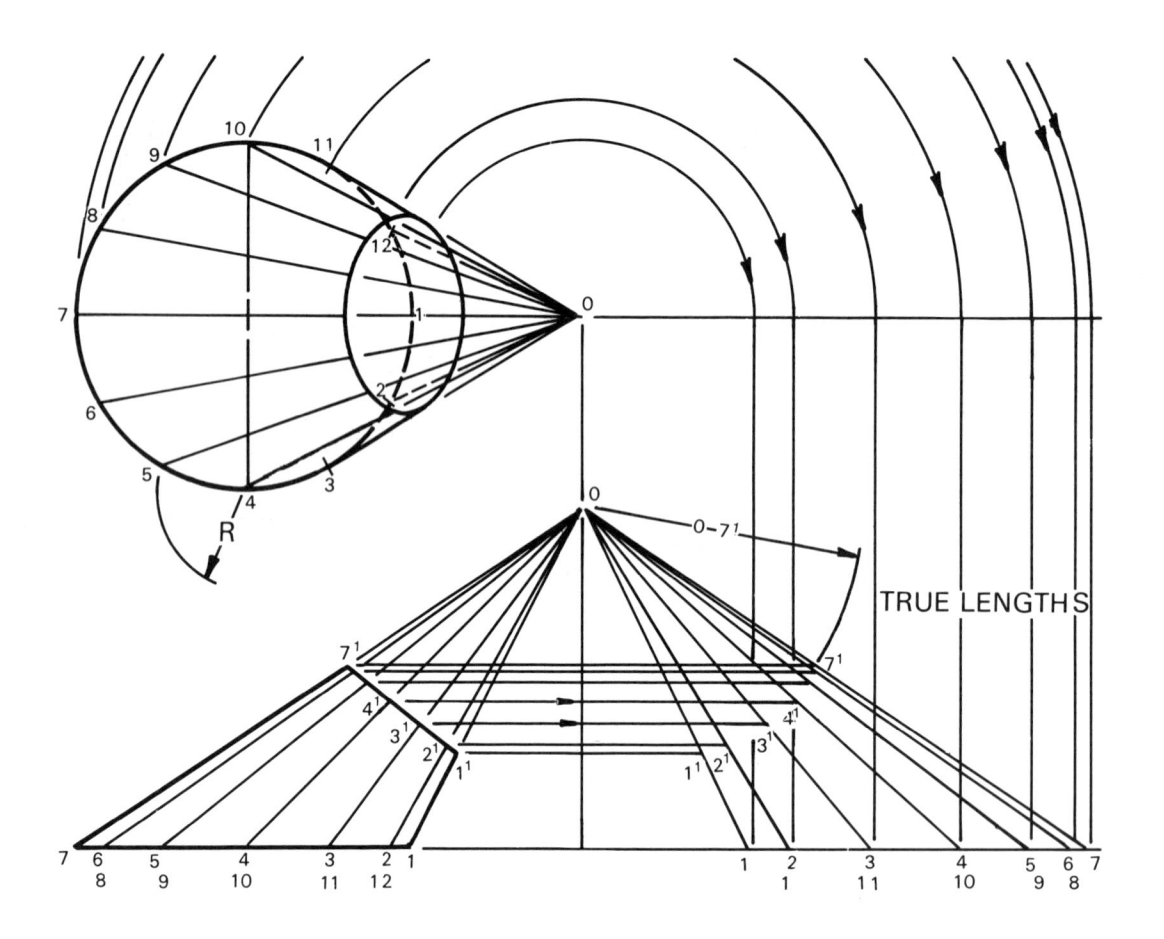

7.29 DEVELOPMENT OF A TRUNCATED OBLIQUE CONE

A truncated oblique cone is developed using the same method as for a truncated oblique pyramid. The surface of the cone is divided into triangular planes. Each triangle is bounded by two surface elements and a chord distance. The surface of the cone, without the truncated outline, is developed as in the previous section, 7.28. The true lengths of the upper portion of the elements, from the vertex point to the cut point, must be established before the truncated portion of the development outline can be determined. In Fig. 7-50, the base of the cone is evenly divided in order to establish surface elements. The true length of each total element and its upper (cut) portion is determined by revolution. The development is laid out using the true length elements and the equal base chord lengths (R). Line 0-1 is drawn first since it will be the seam edge. Triangle 0-1-2 is laid out, followed by each successive triangle until all points along the outer outline are located. The truncated outline is determined by locating the cut point on each element, e.g. arc 0-7^1 is transferred to element 0-7 on the development to locate point 7^1. Connect the points with a smooth curve.

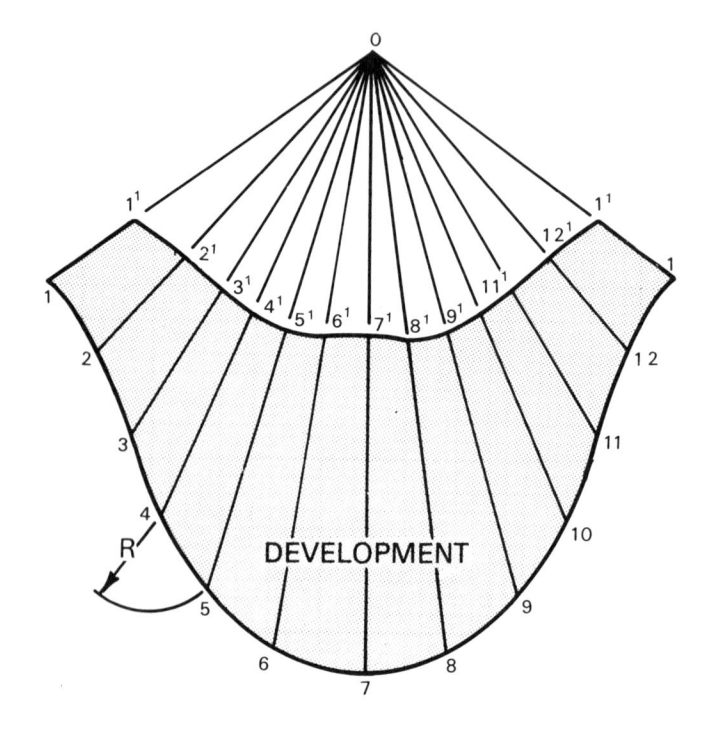

FIG. 7-50 Development of a truncated oblique cone.

7.30 DEVELOPMENT OF A CONICAL OFFSET

A conical offset is *transition piece* between two circular pipes of differing diameters on different axes. Since this type of transition piece is actually a frustrum of an oblique cone, it can be developed as described in Section 7.28. In order for the offset piece to be a frustrum it must be cut by parallel planes. In other words, the upper and lower base planes must be parallel. Therefore, the two given pipes are intersected by parallel planes as shown in Fig. 7-51. The resulting transition connection of two circular pipes, of different diameters and axes and intersected by parallel planes will be conical.

In Fig. 7-51, the frontal and horizontal views of the conical offset are given. Since the offset is symmetrical, only a half development need be drawn. The vertex is located by extending the edge lines of the offset until they intersect at vertex 0. Elements are established on the surface of the offset by evenly dividing the horizontal view of the offset, where it appears as a circle, and drawing elements from the vertex through each division. The elements are projected to the frontal view. A true length diagram is constructed in order to establish the true lengths and the frustrum portion of each element. Since the lower base of the offset is at an angle to the horizontal plane, the base end of the elements on the true length diagram will be at different elevations. The height dimensions can be projected from the frontal view as shown.

The true length chordal distance between divisions on the offset's base cannot be determined in the given views. The lower base is revolved in the frontal view until parallel to the horizontal plane. A true shape view of the offset base is projected as shown. The true chordal distances, as represented by R1 can now be used to lay out the base outline of the development.

Start the development by locating the vertex point and drawing the shortest element 0-1 as shown. The lower leg of each thin triangular plane (representing the surface to be developed) is equal to the base divisions, e.g., 1-2, 2-3, 3-4, etc. Layout the development using the true lengths from the TL diagram and the base divisions.

FIG. 7-51 Development of a conical offset.

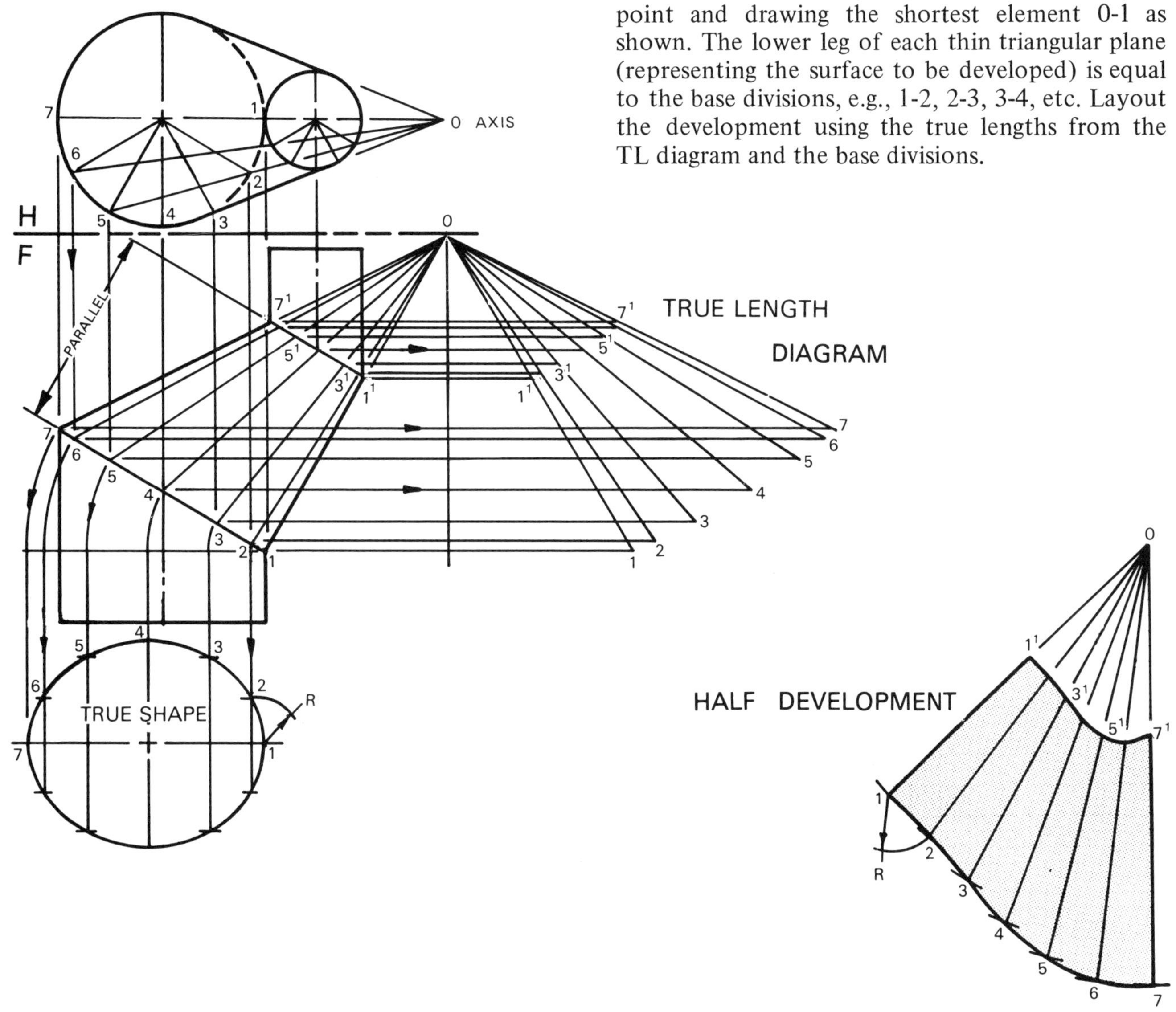

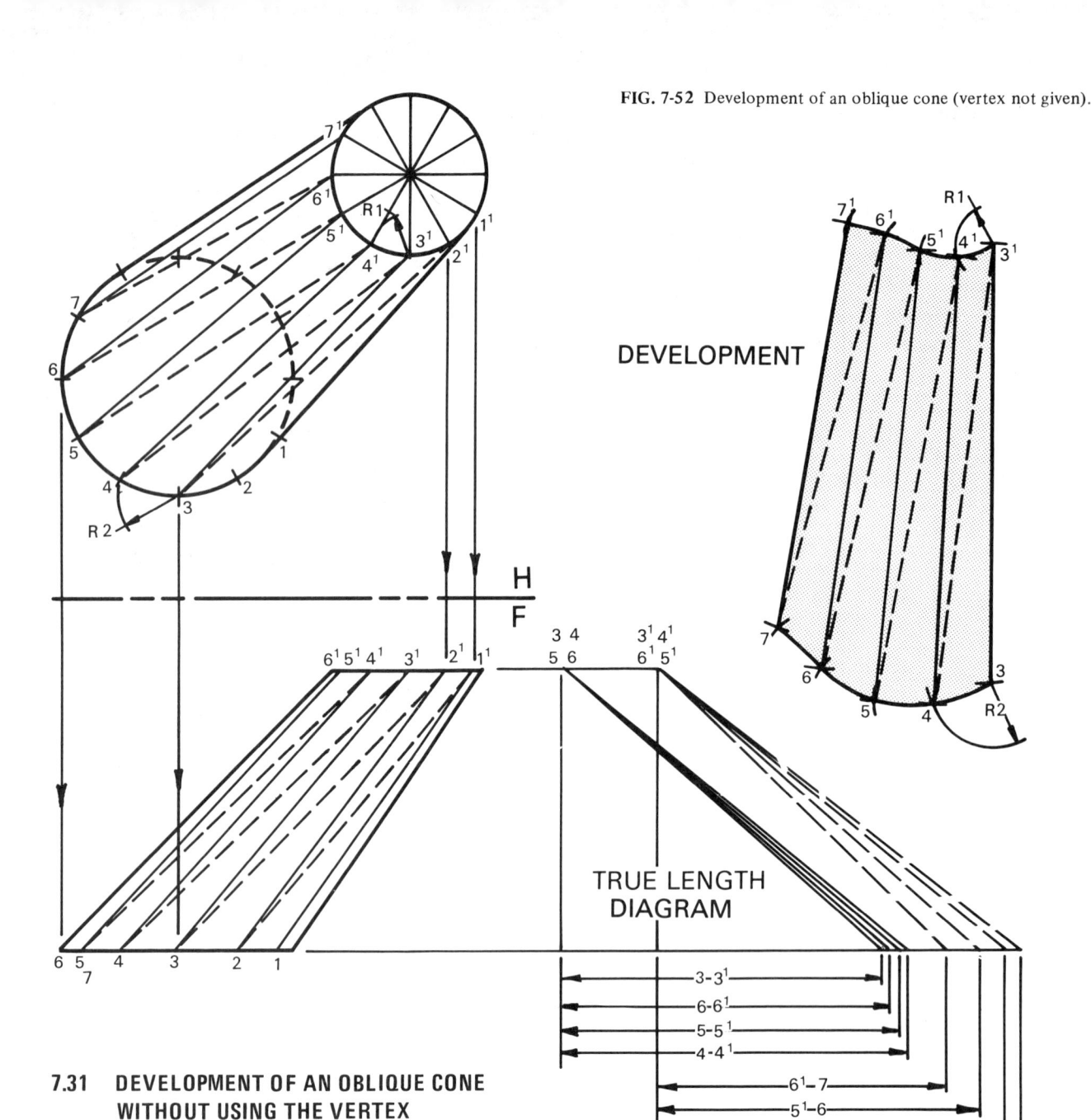

FIG. 7-52 Development of an oblique cone (vertex not given).

7.31 DEVELOPMENT OF AN OBLIQUE CONE WITHOUT USING THE VERTEX

In many cases, the vertex of a cone is unavailable to use for the construction of a development. The size and length of a truncated cone may be such that its vertex cannot lie on the paper. The development of such a truncated cone is accomplished by dividing its surface into a number of trapezoidal areas. The trapezoids are then divided diagonally. The resulting triangular planes are used to draw the development.

Establish the trapezoidal areas of the cone's surface by passing a series of vertical cutting planes through both bases. In Fig. 7-52, the circular bases of the oblique conical frustrum are divided into the same number of equal parts. Related divisions are connected to form elements along the cone's surface. The elements (solid lines) are projected to the frontal view as shown. Two adjacent elements create a trapezoidal area on the cone's surface. A diagonal (dashed lines) is drawn through each trapezoid. The trapezoidal elements and the diagonal elements are shown in both views. The true lengths of all elements are established by a true length diagram.

Chord distances R1 and R2 are used to establish the upper and lower development outlines. Start the development by laying out each successive triangular area using the true lengths of the elements and the chord distances. Only a partial development is shown in the example. Line 3-3^1 is drawn vertical. Triangle 3-3^1-4 is constructed using elements 3-3^1 and 3^1-4, and chord distance R2. The remaining triangles are laid out in consecutive order to complete the development.

PROB. 7-37

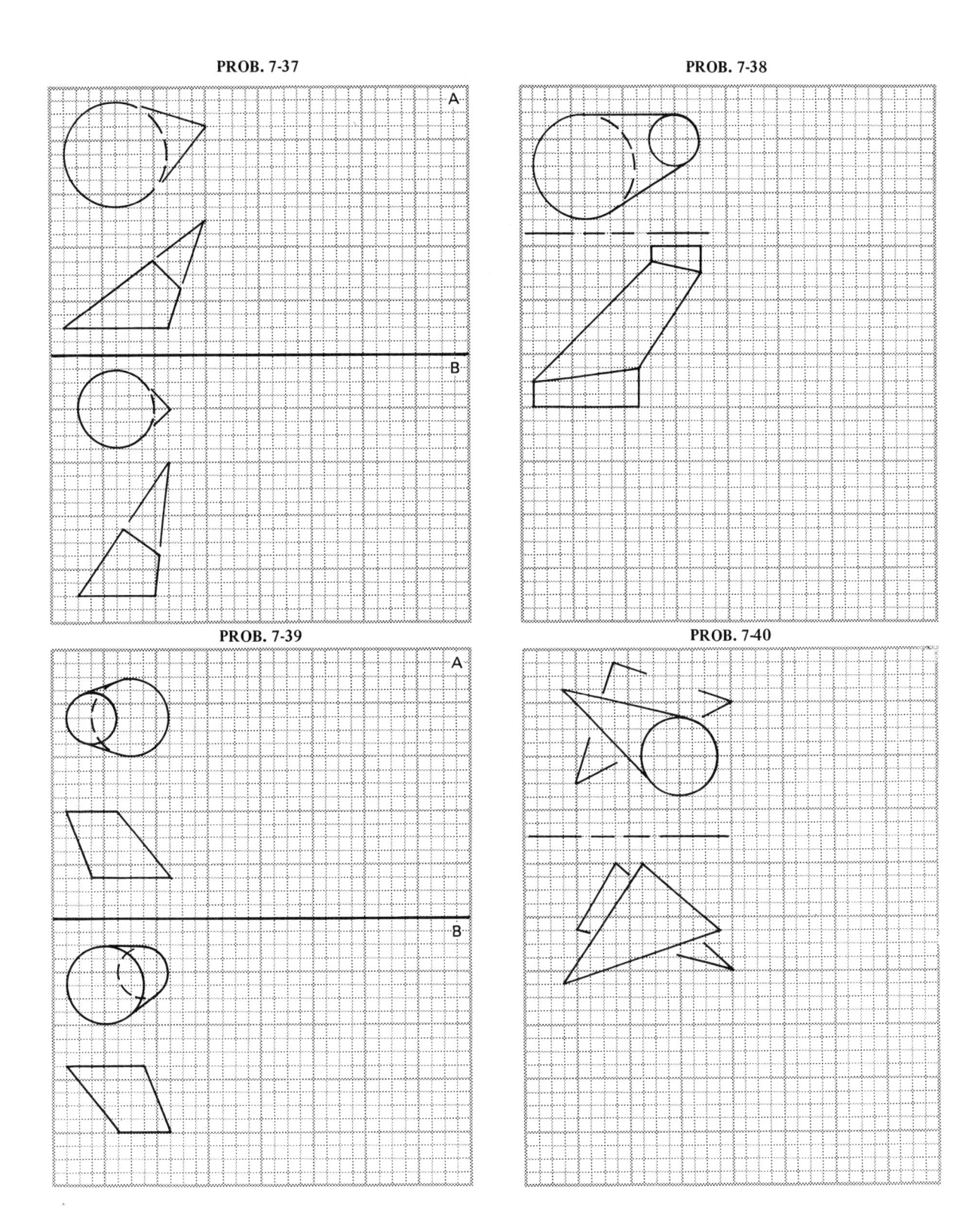

PROB. 7-38

PROB. 7-39

PROB. 7-40

PROBS. 7-37 through 7-39. Complete the given views where necessary. Develop the figure and construct a model as assigned.

PROB. 7-40. Complete the intersection of the plane and the cone and do a development of the cone showing the line of intersection.

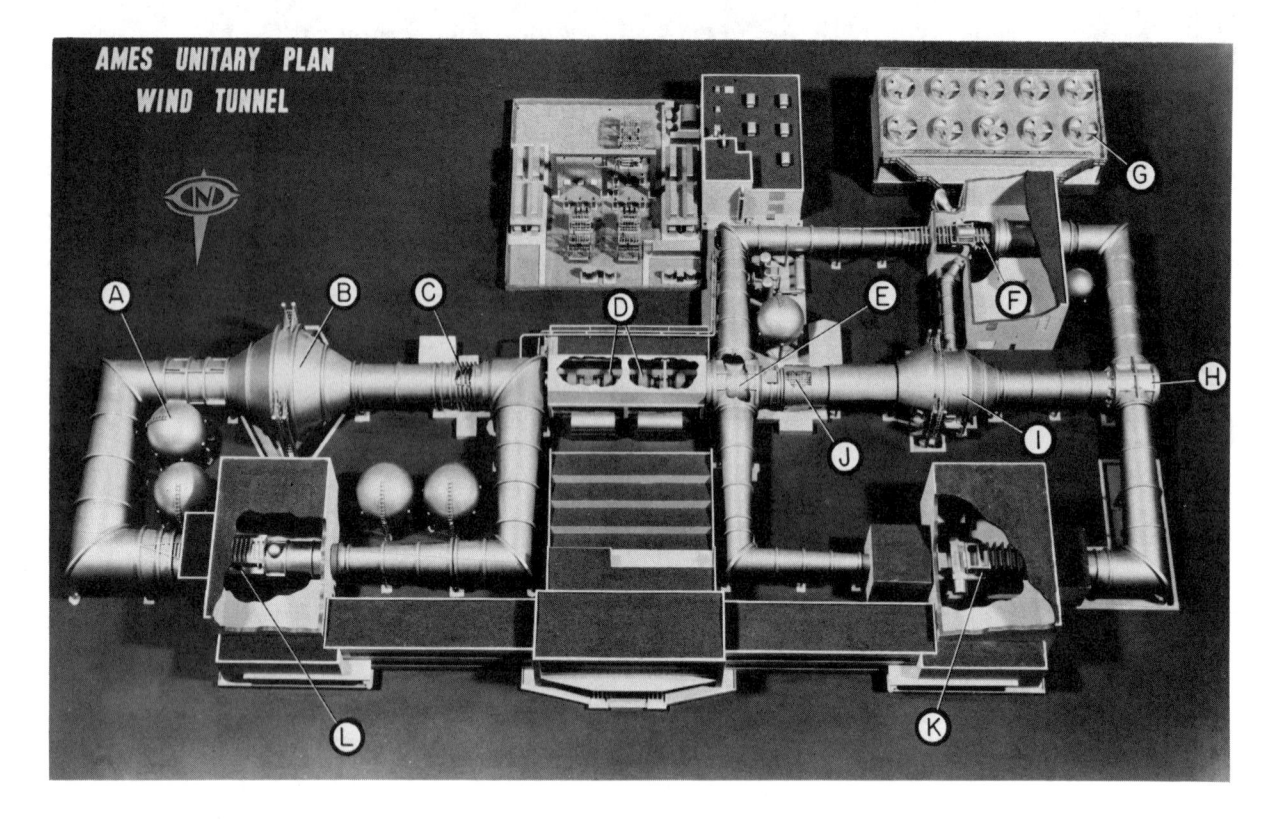

A. Dry Air Storage Spheres
B. Aftercooler
C. 3-Stage Axial Flow Fan
D. Drive Motors
E. Flow Diversion Valve
F. 8- by 7-Foot Supersonic Test Section
G. Cooling Tower
H. Flow Diversion Valve
I. Aftercooler
J. 11-Stage Axial Flow Compressor
K. 9- by 7-Foot Supersonic Test Section
L. 11- by 11-Foot Transonic Test Section

FIG. 7-53 NASA wind tunnel model. *(Courtesy NASA.)*

7.32 TRANSITION PIECES

*A general definition of a **transition piece** would include all shapes which connect two or more forms of different size.* This broad definition would thus include types of developments already covered under cones and pyramids. *A more precise definition of a **transition piece** would be a connecting piece which smoothly joins two or more dissimilar forms of differing shape and size.* Again, even this definition is inadequate since two circular (Fig. 7-53), elliptical, square, or two proportionate rectangular shapes must at times be joined, and these are also types of transition pieces. Since each of these types of connections have been covered in previous sections, transitions as presented here are connecting devices which join two or more dissimilar shapes. With the exception of transitions between forms composed of planes, square to rectangular (Fig. 7-54), etc., all transitions are only approximately developable.

*Transition pieces are developed by **triangulation**: dividing the surface of the piece into triangles.* Triangulation has already been used to develop a variety of shapes in preceding sections. Elements are drawn on the surface of the form to be developed and connected by diagonals if adjacent elements do not intersect. The development is laid out as a series of joined triangular areas.

In Fig. 7-53, the wind tunnel is composed of circular pipes whose transition pieces are circular cones. Each turn of the tunnel is made with a 90-degree mitered elbow. The actual test sections are rectangular, F, K, and L. The connection between a circular pipe section and a rectangular test section is a smooth transition piece.

FIG. 7-54 Cardboard model of a transition piece.

FIG. 7-55 Examples of transition pieces.

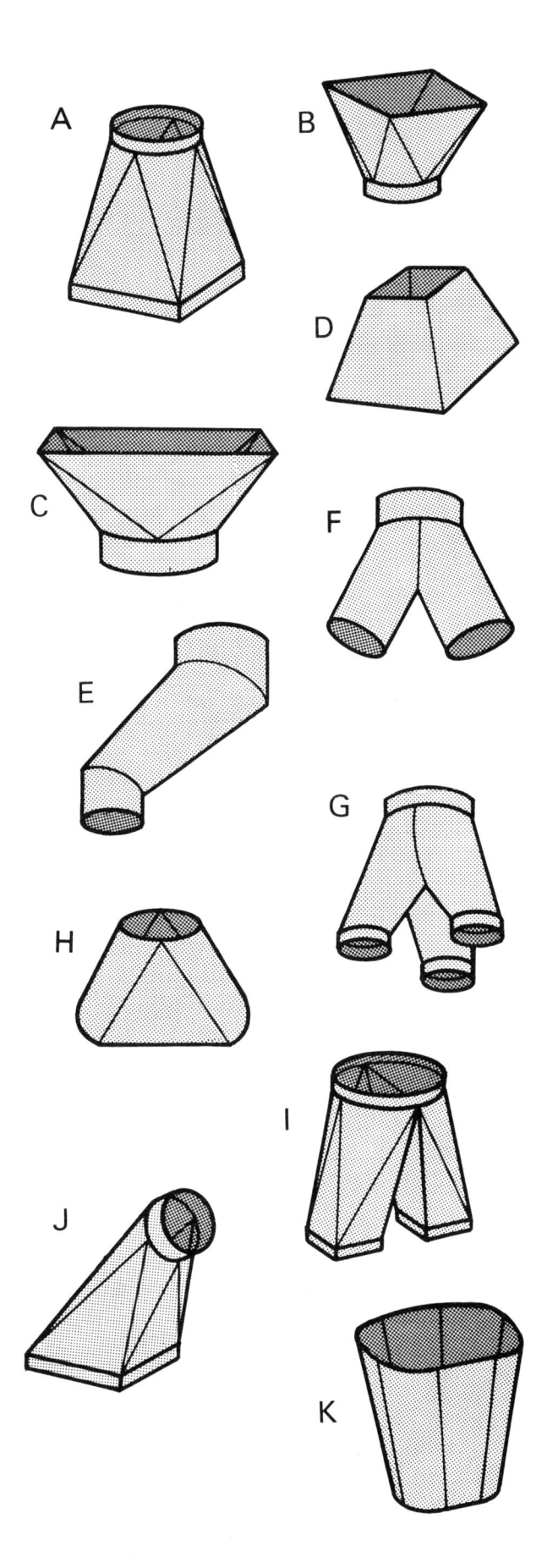

7.33 DEVELOPMENT OF TRANSITION PIECES

A *transition piece* joins two or more geometric forms of different shapes. Therefore, each opening of the transition piece will be a different configuration. In general, transition pieces are designed to be formed from sheet metal and welded along a common seam. Transition pieces are used to join a variety of materials and objects. Pipe shapes and HVAC ducting utilize transition pieces throughout their design. Hoppers, warped funnels, and vessel bottoms of all types have transition pieces integrated into their design. The conical, convolute, or warped surface configuration of an aircraft's forward section is a transition piece between the nose and the fuselage.

In Fig. 7-55, eleven possible variations of transition pieces are provided. The possibilities of shapes and sizes are limited only to the designer's imagination and the financial and production feasibilities.

A and B are symmetrical square to round transitions. This type is one of the more common variations. C is a rectangle to round, and is developed with the same general method as A and B.

D is a square to rectangle transition. It is composed of plane surfaces and can therefore be accurately developed. Note that his type is really a frustrum of a right pyramid. Its given surfaces are developed by triangulation if the vertex is unavailable.

The next three examples all involve the connecting of two or more circular or elliptical shapes. E is a conical offset, connecting two separate pipes of differing diameters and axes. F is a *Y* fitting, connecting two round pipes to one pipe of a larger diameter. G is three stream transition into a single large-diameter pipe.

H, I, J, and K are specialized variations of transition pieces: round to oblong (H), two square ducts to one round (I), square to round transition at an angle (J), and a hopper-type (K).

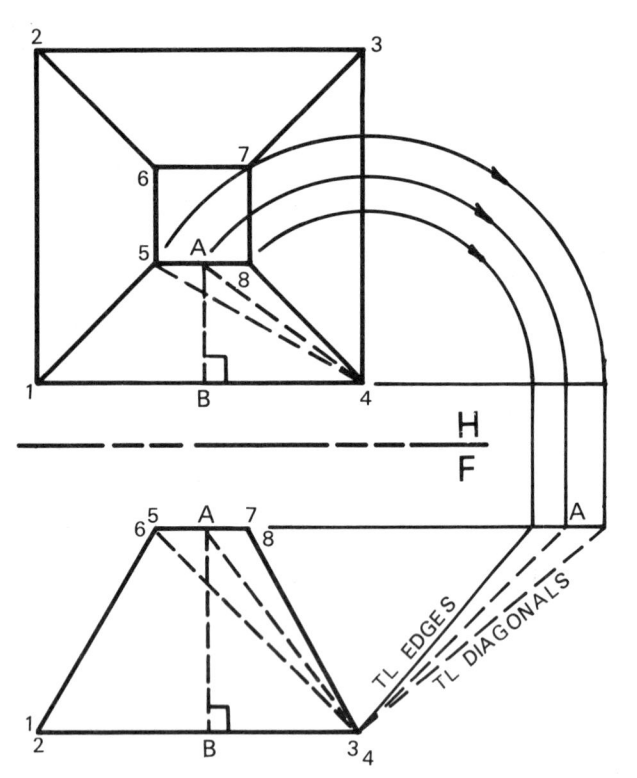

FIG. 7-56 Triangulation.

DEVELOPMENT

7.34 TRIANGULATION

In Fig. 7-56 the transition piece is developed using *triangulation*. The sheet metal hopper shown in Fig. 7-57 is an example of an industrial application of such a transition piece.

The square to square form developed in Fig. 7-56 does not fit into the narrow definition of a transition piece. It does not have dissimilarly shaped openings, and its edges can be extended to locate a vertex. Normally when such a piece is to be developed, methods are used which utilize the vertex, and the development is constructed as is a frustrum of a pyramid. This form is used here only to provide a simple illustration of the triangulation of a surface. Each surface of the object is identical. Therefore only one surface need be divided into a triangular area. A diagonal is drawn so as to divide one of the equal trapezoidal shapes into two triangular planes, 4-5. The true lengths of the hopper's edges and diagonals are established by revolution. The true lengths of the upper and lower openings appear in the horizontal view and can be transferred directly to the development.

To establish the shortest seam, divide the front surface in half. Line A-B will be used as the seam edge. This placement of the seam makes the joining method easier, quicker, and along the shortest line. This area must also be divided into a triangle. Draw a diagonal from point A to point 4 and establish its true length by revolution.

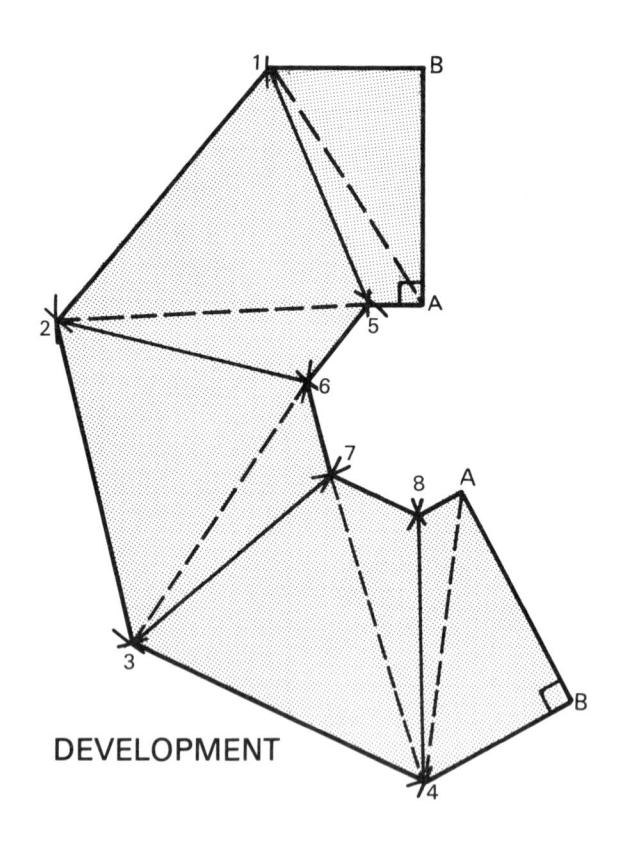

Start the development by drawing line A-B. Using the true lengths of the edges, diagonals and upper and lower opening edge lines as arc lengths, complete the development as shown. Swing arc A-4 and B-4 to locate point 4. Arc A-8 and arc 4-8 intersect at point 8.

FIG. 7-57 Air filtration transition piece.

FIG. 7-58 Triangulation used to develop a transition piece.

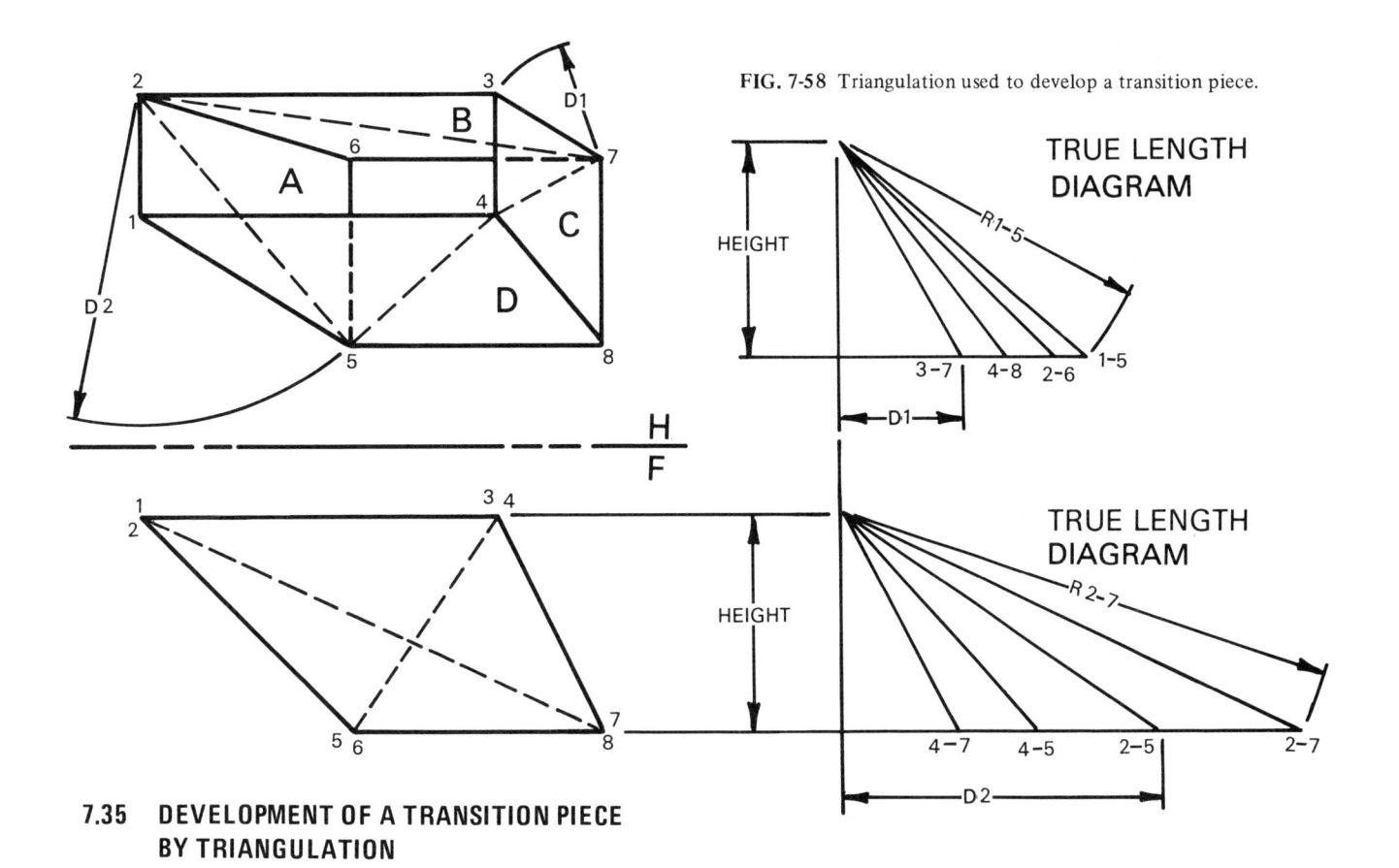

7.35 DEVELOPMENT OF A TRANSITION PIECE BY TRIANGULATION

The transition piece in Fig. 7-58 is developed by triangulation. Two true length diagrams are constructed. The upper diagram establishes the true lengths of the lateral edges. The lower diagram determines the true lengths of the diagonal elements. All height dimensions are identical and can be transferred or projected to the TL diagrams as shown. D1 and D2 illustrate the process of transferring the base dimensions for the right triangles, whose diagonals (hypotenuse) equal the true lengths of the required lines.

The upper and lower openings appear as true shape in the horizontal view, the true lengths of the base edges may be taken directly from this true shape projection.

The transition piece is composed of plane surfaces. The two openings are dissimilar and the piece is not symmetrical. Therefore all four lateral planes are different sizes. Divide the surfaces into triangular areas as shown and solve for their true lengths as described above.

Start the development by drawing line 1-5 as the seam line. Lay out the triangular areas using the true length elements, edges and opening edges as TL arcs, e.g., edges R1-5 and R2-7, base edges 1-2 and 5-6.

Draw line 1-5 and swing arc 1-2. From point 5 swing arcs 5-2 and 5-6. Arc 5-2 intersects arc 1-2 at point 2. Arc 2-6 intersects 5-6 at point 6. Complete the development by laying out each successive surface, A, B, C, and D.

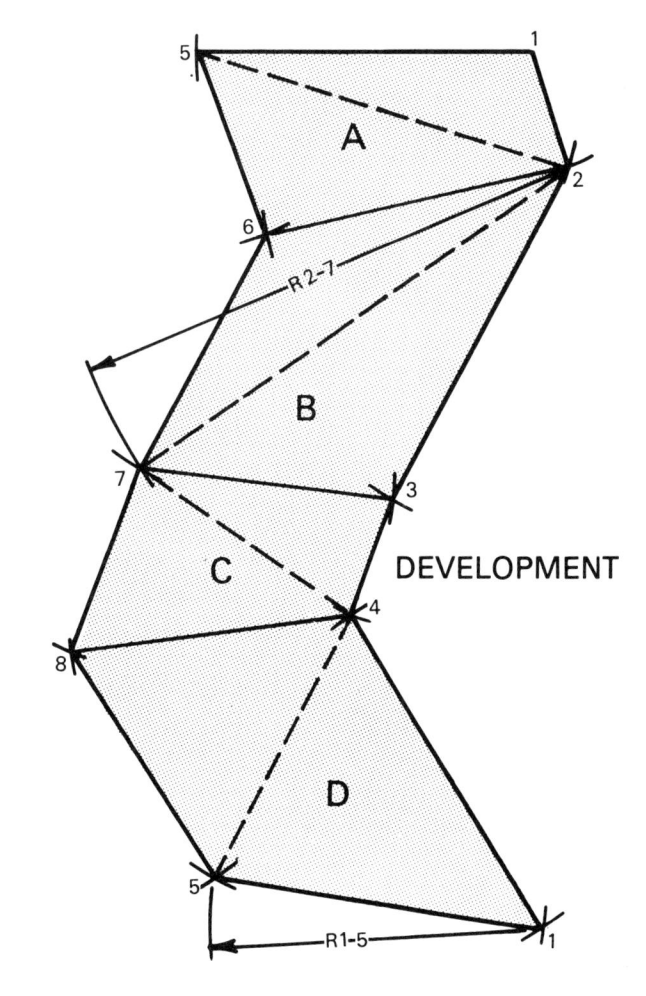

337

7.36 DEVELOPMENT OF A TRANSITION PIECE: CIRCULAR TO RECTANGULAR

A transition piece connecting a circular to rectangular geometric form can be developed by dividing its surface into triangular areas. In Fig. 7-59 the surface of the transition piece is composed of four isosceles triangles and four conical surfaces. The bases of the isosceles triangles form the lower base of the transition piece. The four conical surfaces are portions of an oblique cone. The first step in the development of a circular to rectangular transition is to divide the conical surfaces into triangular areas. In Fig. 7-59, the circumference of the circular base is divided into 12 equal parts. Points 1, 4, 7, and 10 already exist as divisions since they correspond to the vertex points of the isosceles triangular areas of the piece's surface. All other points divide the conical surfaces into three separate areas. Since the transition piece is symmetrical, each of the four conical surfaces and their triangular divisions are identical. Therefore, the true lengths of only one set of elements need be established.

A true length diagram is constructed as shown to establish the true lengths of the four elements. The true lengths of the lower rectangular base can be found in the horizontal view as can the chord distances between divisions on the upper circular base. The seam line is established by dividing the frontal triangular surface in half. Line 1-A is used as the seam line.

Start the development by drawing line 1-A. Using the true lengths of the elements, the chord distances and the lower base lengths, as arc lengths, lay out each successive triangular surface to complete the drawing. From point 1 swing arc 1-B. Swing arc A-B from point A. Arc 1-B intersects arc A-B and locates point B. Triangle 1-B-2 is laid out next. Each successive triangle is constructed so that the transition piece is unrolled clockwise and inside up.

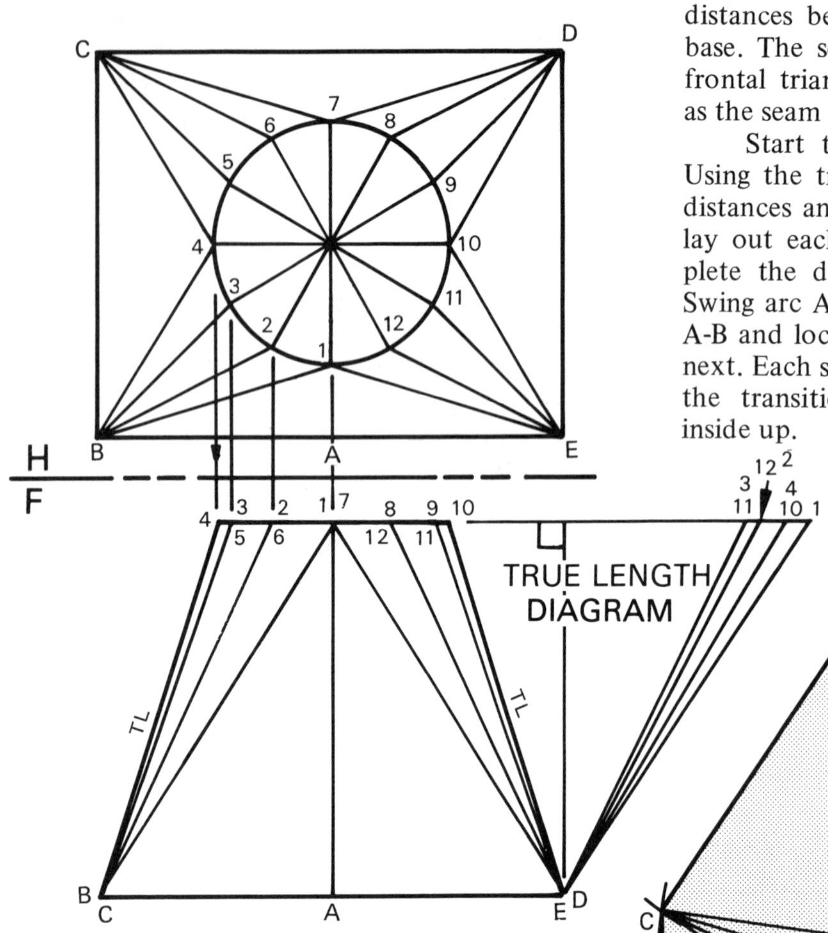

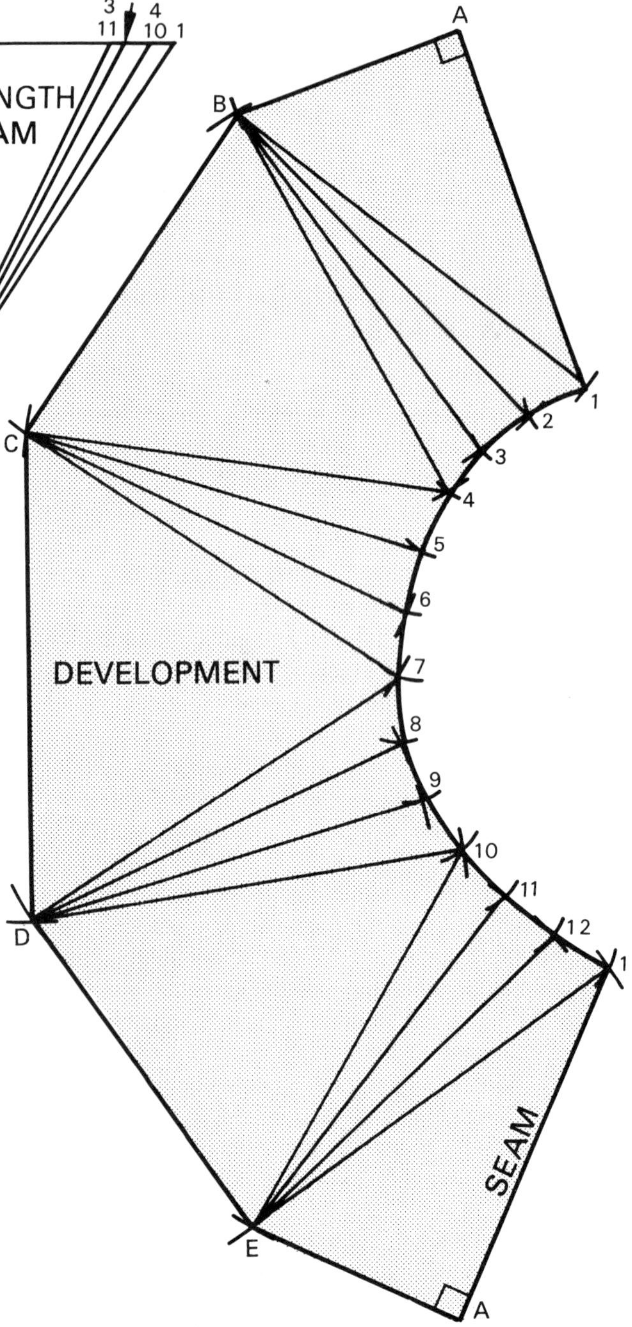

FIG. 7-59 Transition piece development: circular to rectangular.

FIG. 7-60 Transition piece development.

7.37 TRANSITION PIECE DEVELOPMENT

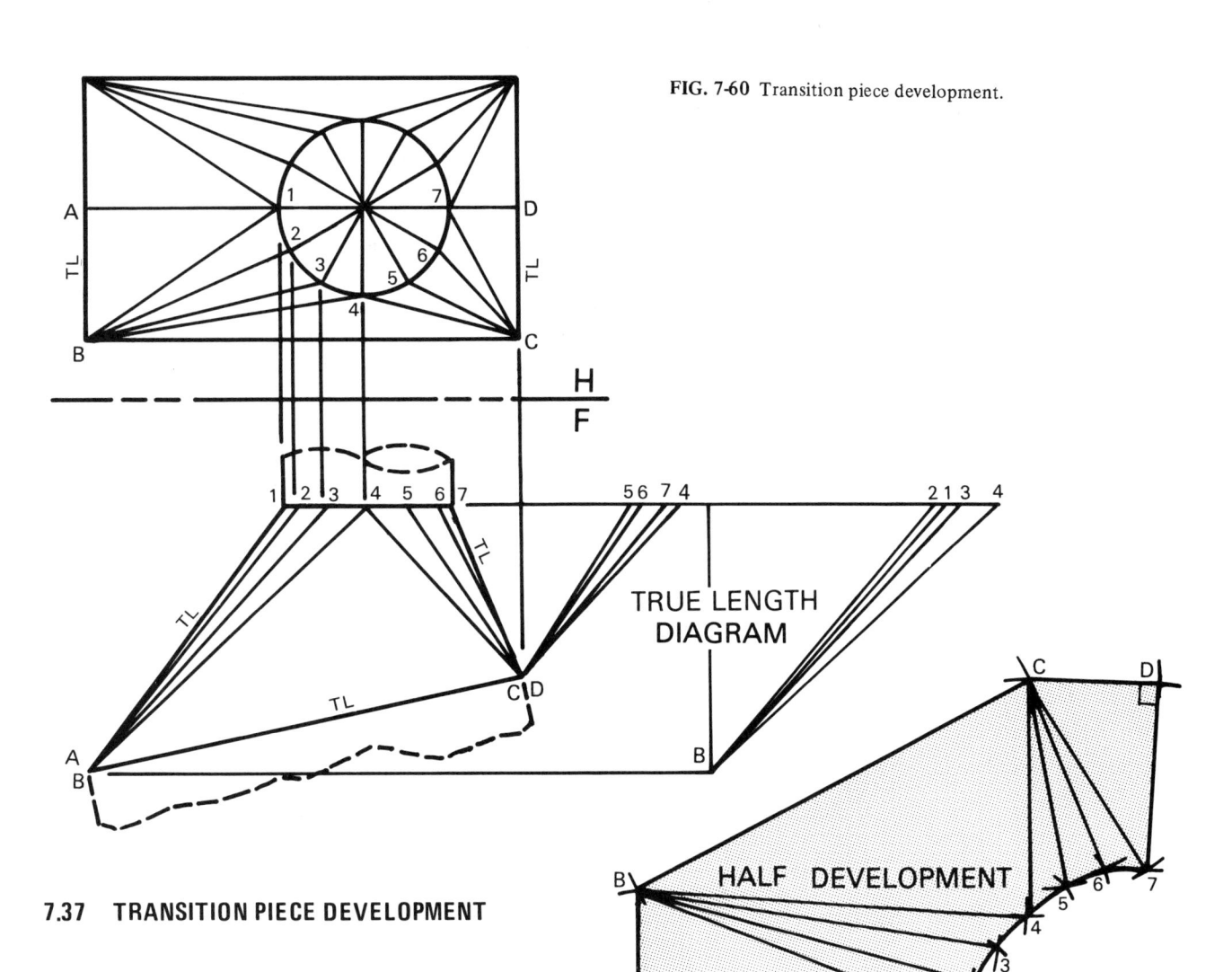

The rectangular to circular transition piece shown in Fig. 7-60 has an angled base edge. This figure is developed using the same general method as in the previous example. The transition piece is composed of four triangular lateral surfaces whose base lines form the lower base edge of the figure. The corners of the piece are portions of oblique cones. The development is constructed by dividing the surface into triangular areas which approximate the surface of the piece. Each triangle is then laid out in successive order with common elements joined. Note that this and the development in Fig. 7-59 are approximate developments since the given forms are basically warped surfaces.

The circumference of the upper base circle is divided into equal parts. Elements which define triangular areas on the conical surface are drawn through the division points and connected to one of the lower base corners. These elements correspond to bend lines when the piece is formed by rolling a flat piece of sheet metal that was cut to the outline of the pattern. Since the lower base is at an angle and the circular base is not centered left to right, as was Fig. 7-59, there will be a total of eight separate element lengths to establish before the development can be started. To avoid confusion, two true length diagrams are drawn as shown. Note that revolution could have been used to determine the true lengths of the elements.

The true lengths of the lower base edges can be seen in the horizontal view for the right (C-D) and left (A-B) edges. The front (edge line B-C) and rear edge appear true length in the frontal view. The true length chord distances of the upper base are transferred directly from the horizontal view to the development.

In Fig. 7-60 a half development is constructed, since the piece is symmetrical. The half development can be flipped over to complete the full pattern. Line D-7 is used as the seam edge line since it is the shortest line. The true lengths of the elements, base lines, and chord distances are used to lay out the development.

Start the development by drawing line A-1 horizontal to the paper. Line A-B is drawn perpendicular to line A-1 to form the first triangular area, A-1-B. Triangle 1-B-2 is laid out next. Complete the half development by laying out each successive triangular area as shown. To check the accuracy of the development, line D-7 must be perpendicular to line C-D at the end of the construction.

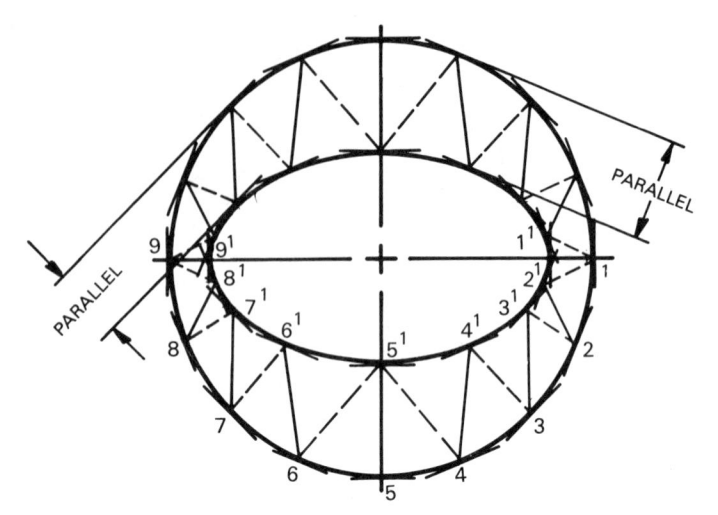

7.38 DEVELOPMENT OF A CONVOLUTE TRANSITION PIECE

A *convolute* is one of the three types of single-curved surfaces. Therefore its surface can be accurately developed. This type of transition piece is widely used in the aircraft industry as a smooth transition between the nose cone and the fuselage. A convolute is generated by a straight-line generatrix. Each position of the generatrix establishes an element on its surface. Two adjacent elements will intersect, but never three.

A convolute is formed by elements created by a tangent plane as it is moved along the two curved directrices. The convolute shown in Fig. 7-61 has a lower base (directrix) in the form of a circle and an upper base (directrix) in the shape of an ellipse. Elements are drawn on the surface of the convolute by dividing the lower circular base into equal parts, 1 through 9. Draw a line tangent to the lower base at each division point. Draw another line tangent to the upper base parallel to its related tangent on the lower base. Connect the points to form an element on the surface, e.g. a tangent line is drawn through point 6 on the lower base and another tangent line parallel to it on the upper base to establish point 6^1. Connect the points to form element $6\text{-}6^1$. This construction is repeated to determine a series of elements on the convolute. Two adjacent elements define a four sided (quadrilateral) area. Note that the closer and more numerous the elements the more accurate the development.

To develop the convolute, draw a diagonal through each quadrilateral to divide the surface into a triangular area. Solve for the true lengths of each element as shown.

Using the true lengths of the elements, and the true length chord distances on the upper and lower base perimeters, lay out each successive triangular area to complete the development.

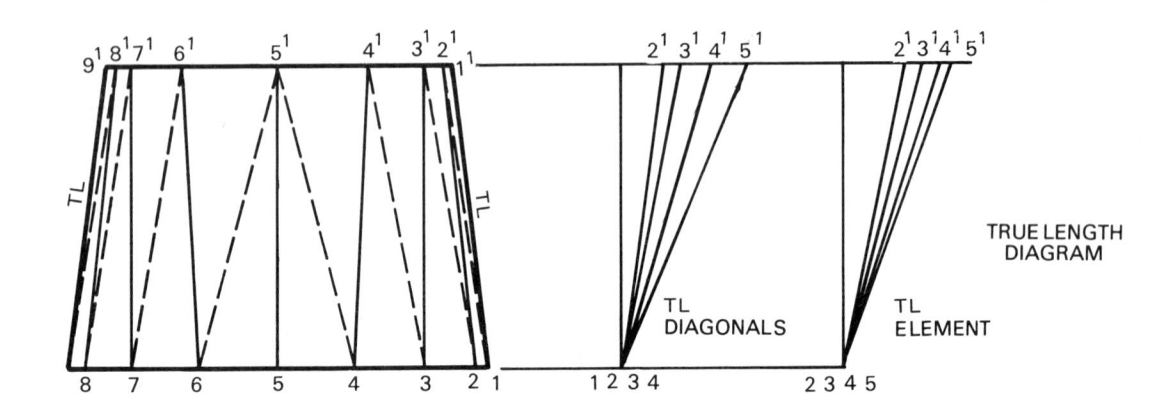

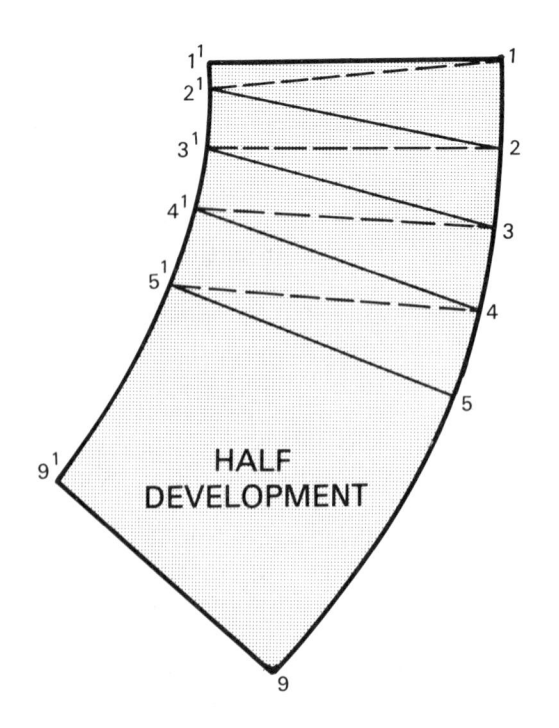

FIG. 7-61 Development of convolute transition.

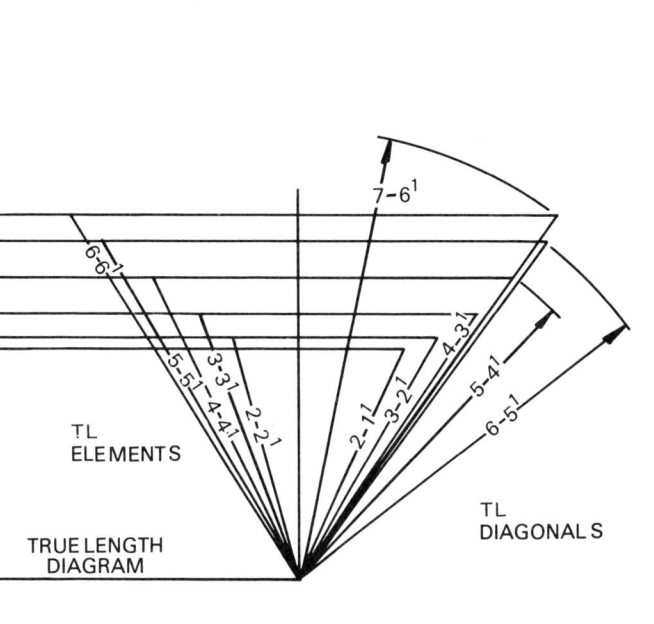

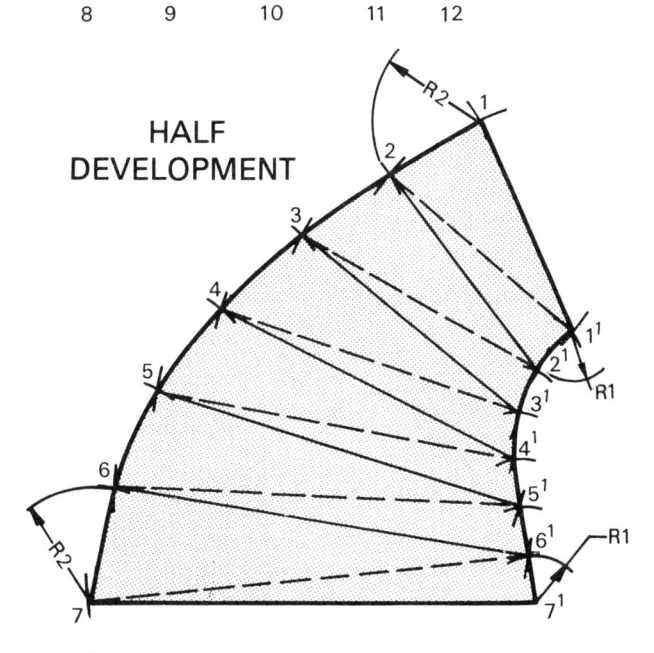

7.39 DEVELOPMENT OF A WARPED TRANSITION PIECE

A warped surface is considered to be theoretically nondevelopable. By dividing its surface into a series of thin triangles, a warped transition piece can be adequately developed by approximation.

The warped transition piece in Fig. 7-62 connects a large circular shape to a smaller circular shape at an angle. The lower base is divided into 12 equal parts. The upper base is divided into the same number of equal divisions and labeled in accordance with related points along the lower base. An auxiliary view showing the true shape of the upper base is projected in order to divide conveniently the circumference and to determine the true chord distances between divisions, R1. The chord distances on the circumference of the lower base appear as true length in the horizontal view, R2. Related points on the upper and lower bases are connected to establish elements. Diagonal elements are drawn between adjacent points on opposite bases, e.g. 7-6^1. The true length of each element and diagonal is determined by constructing a true length diagram.

Using the true lengths of the elements, diagonals, and the upper and lower chord distances as true length arcs, lay out the development as shown. Line 1-1^1 is used as the seam line since it is the shortest element. Draw line 1-1^1 to start the development. Swing arc R2 from point 1 until it intersects arc 1^1-2 at point 2. This will lay out triangle 1-1^1-2. Continue to construct each successive triangular area unfolding the transition piece clockwise, inside-up. A half development is all that is necessary since the piece is symmetrical.

FIG. 7-62 Development of a warped transition piece.

PROB. 7-41

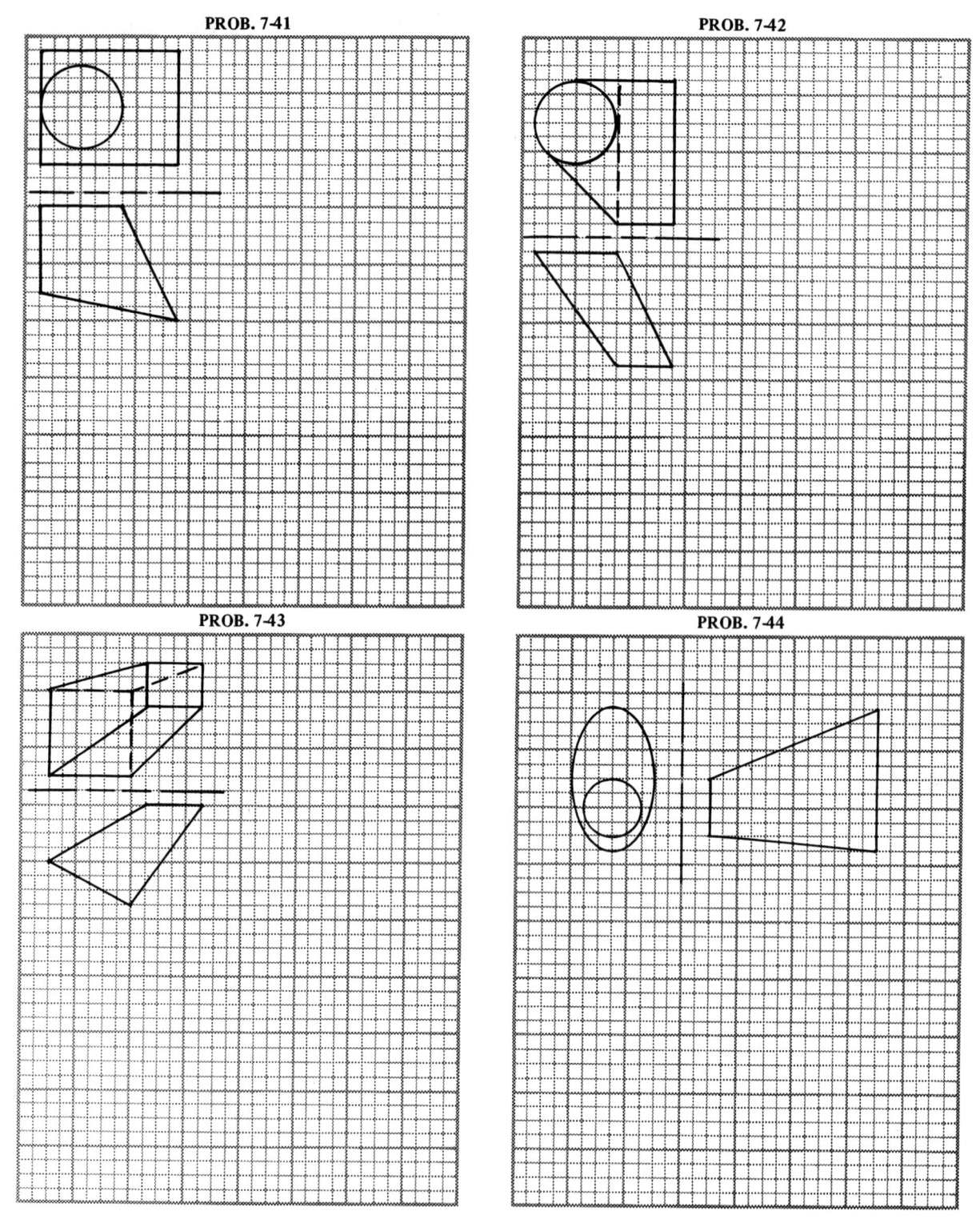

PROB. 7-42

PROB. 7-43

PROB. 7-44

PROBS. 7-41 through 7-44. Develop the given transition pieces. Make a model of each development as required. Draw a half development of each.

342

FIG. 7-63 Water tower designed in the shape of an oblate ellipsoid.

7.40 DOUBLE-CURVED SURFACES

Double-curved surfaces are divided into two basic types: surfaces of revolution and double-curved surfaces of the general type. **General types** of double-curved surfaces are composed of curved lines or contours drawn at predetermined spacings. Contour maps as covered in Chapter 8, topographic models, and fair surfaces of ships, airplanes, automobiles, and spacecraft as described in Chapter 7 are examples of the general type of double curved surfaces.

Double-curved surfaces of revolution *are generated by the movement of a curved line generatrix moving about a straight line axis (directrix).* A double-curved surface is composed solely of curved lines, therefore it is theoretically nondevelopable. Approximate developments are constructed from double-curved surfaces by enclosing them in portions of cones and cylinders. There are no straight lines on a double-curved surface. The intersection of a plane and a double-curved surface, perpendicular to its axis line, cuts a curved element on its surface. A plane passed parallel to its axis cuts a section showing the outline of the piece.

Double-curved surfaces of revolution include the following shapes: *sphere, annular torus* (Fig. 7-64), *spheroid/ellipsoid* (Fig. 7-63), *paraboloid,* and the *serpentine* (spring). A double-curved surface can be made by stretching flat sheet metal, which has been cut to a specific set of patterns, until it approximates the desired form. Surfaces of revolution can also be turned on a lathe if the finished piece is to be a solid. In general the sphere is the most common form of double-curved surface that will require developing.

FIG. 7-64 Segment of a torus-shaped space colony. *(Courtesy NASA.)*

FIG. 7-65 Refinery model.

7.41 SPHERES

Spheres are double-curved surfaces of revolution that are generated by a revolving curved line (circle) generatrix about a straight-line axis (directrix). A variety of photographs of industrial uses of spheres are provided throughout the text. In Fig. 7-65, the chemical reactor tower is topped by a dome in the shape of a half sphere.

Spheres can be developed by two methods. The *gore method* (meridian method) divides the surface of the sphere into a number of meridians. A *meridian* is established by passing a plane through the axis of the sphere. Two adjacent radial meridians define a section/panel. Meridians are evenly spaced (radially) so that each panel of the development is identical (Fig. 7-66). Therefore only one panel need be established, since it can be used as a pattern for the remaining sections. A panel is really a section of a cylinder that encloses the sphere between two adjacent meridians.

The *zone method* of developing a sphere passes a series of evenly spaced parallel planes perpendicular to the axis. Two adjacent cutting planes establish a horizontal section. Each section approximates the surface of the sphere. A horizontal section can be thought of as a frustrum of a cone whose vertex is at the intersection of the extended chords which define the frustrum's sides.

FIG. 7-66 Liquid nitrogen vessel constructed of welded gore sections.

FIG. 7-67 Sphere development (gore method).

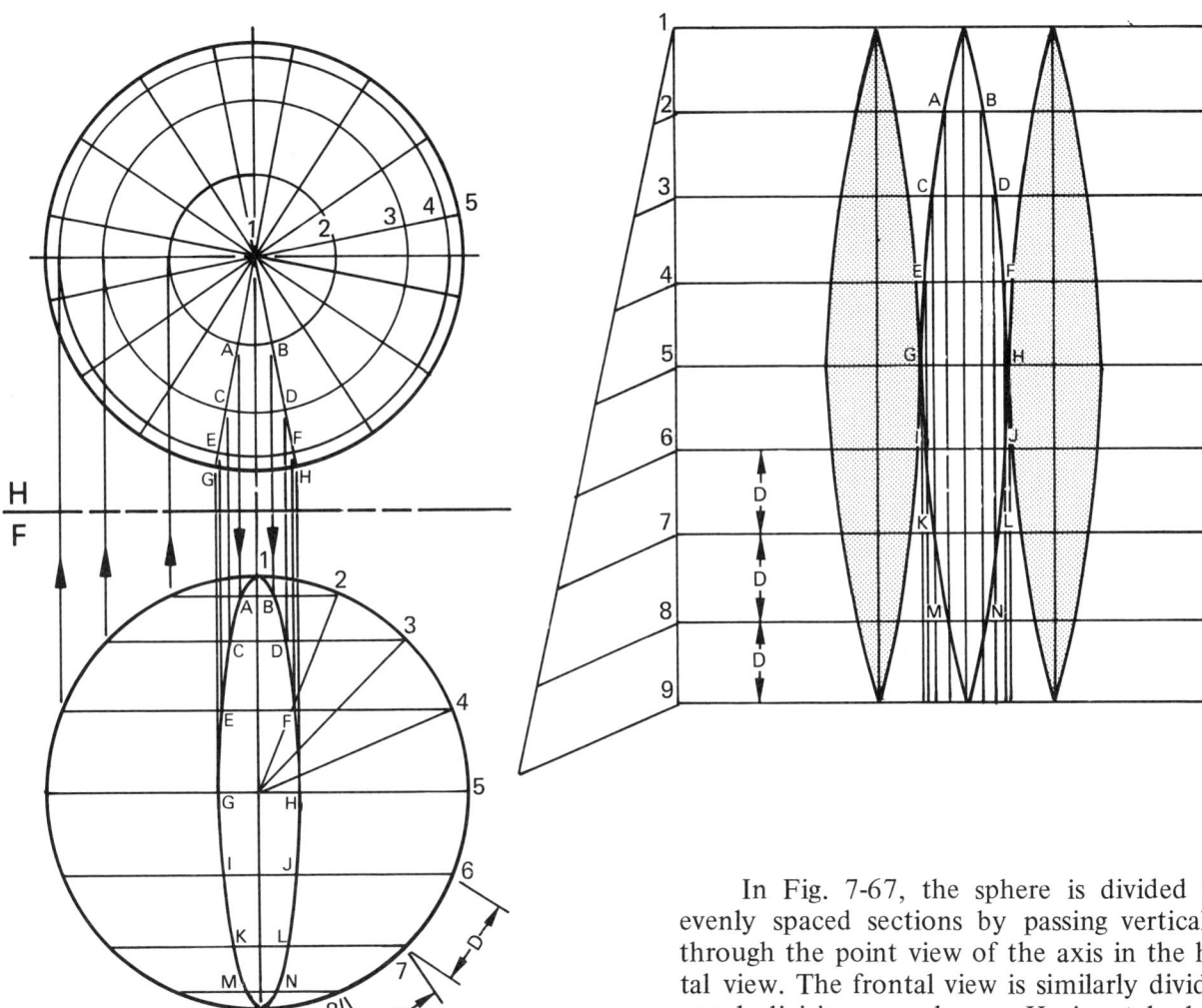

7.42 DEVELOPMENT OF A SPHERE (GORE METHOD)

A *sphere* is a double-curved surface which can only be approximately developed. The *gore* method of development divides the sphere into an equal number of sections (gores). Sections are established by passing equally spaced vertical planes through the axis. Each plane cuts a meridian on the sphere's surface. Two adjacent meridians form a section. Each section can be considered a section of a cylinder. The development of one section is all that is necessary, since it can be used as a pattern for the remaining sections. The greater the number of sections, the more accurate the development and spherically perfect the final piece. An increase in sections will also increase the number of sheet metal pieces to be cut and seams that need be joined.

In Fig. 7-67, the sphere is divided into 16 evenly spaced sections by passing vertical planes through the point view of the axis in the horizontal view. The frontal view is similarly divided into equal divisions as shown. Horizontal planes are passed through divisions in the frontal view, 1 through 9. Each horizontal plane appears as an edge in the frontal view and projects as a small circle element on the sphere in the horizontal view. The chord distance between horizontal planes, dimension D, is equal for all frontal divisions.

The vertical planes (meridian elements) and the circle elements intersect in the horizontal view, points A through N. Each intersection point is projected to the frontal view as shown to establish the gore (meridian) section.

The development is constructed by developing one section/panel. Start by drawing the stretch-out line equal to one half of the sphere's circumference. Divide the stretch-out line into eight equal spaces and label the lines 1 through 9 corresponding to the horizontal divisions. Each division should be equal to dimension D. Points A through N can be transferred to the development along related horizontal lines. The widest part of the section is at the equatorial line (5). Points G and H are transferred from the frontal view by measuring their distance from the axis line, which is the centerline of the section/panel.

FIG. 7-68 Development of a sphere (zone method).

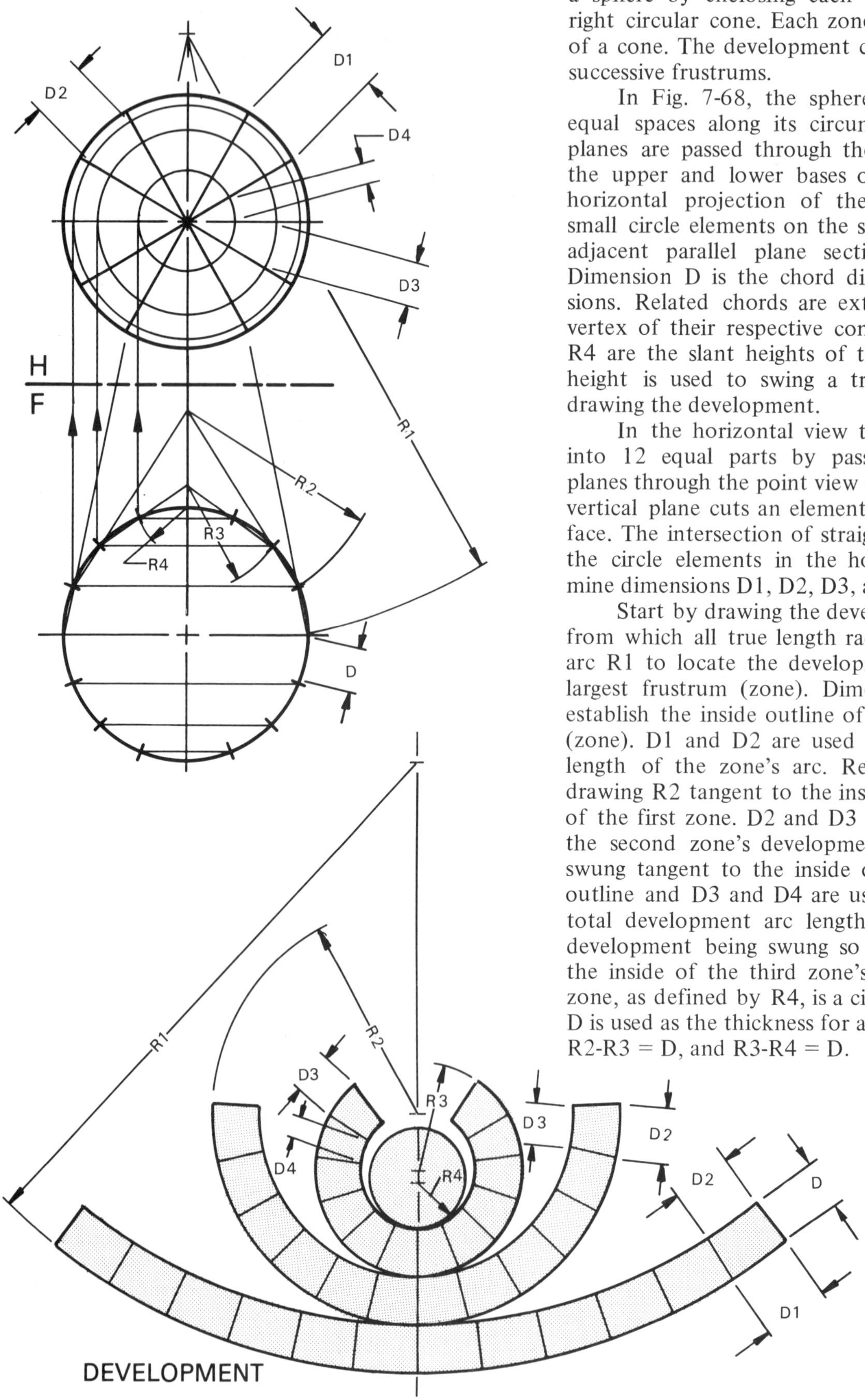

7.43 DEVELOPMENT OF A SPHERE (ZONE METHOD)

The *zone method* of developing a sphere involves dividing the surface of the sphere into horizontal zones. This procedure approximates the surface of a sphere by enclosing each horizontal zone in a right circular cone. Each zone is really a frustrum of a cone. The development consists of developing successive frustrums.

In Fig. 7-68, the sphere is divided into 16 equal spaces along its circumference. Horizontal planes are passed through the divisions to define the upper and lower bases of the frustrum. The horizontal projection of the plane sections are small circle elements on the sphere's surface. Two adjacent parallel plane sections define a zone. Dimension D is the chord distance between divisions. Related chords are extended to locate the vertex of their respective cones. R1, R2, R3, and R4 are the slant heights of the cones. Each slant height is used to swing a true length arc when drawing the development.

In the horizontal view the sphere is divided into 12 equal parts by passing vertical cutting planes through the point view of the axis line. Each vertical plane cuts an element on the sphere's surface. The intersection of straight-line elements and the circle elements in the horizontal view determine dimensions D1, D2, D3, and D4.

Start by drawing the development's centerline from which all true length radii are swung. Swing arc R1 to locate the development outline for the largest frustrum (zone). Dimension D is used to establish the inside outline of the largest frustrum (zone). D1 and D2 are used to establish the true length of the zone's arc. Repeat this procedure drawing R2 tangent to the inside development line of the first zone. D2 and D3 are used to establish the second zone's development arc length. R3 is swung tangent to the inside of the second zone's outline and D3 and D4 are used to determine the total development arc length. R4 completes the development being swung so as to be tangent to the inside of the third zone's outline. The fourth zone, as defined by R4, is a circle. Note dimension D is used as the thickness for all zones, R1-R2 = D, R2-R3 = D, and R3-R4 = D.

DEVELOPMENT

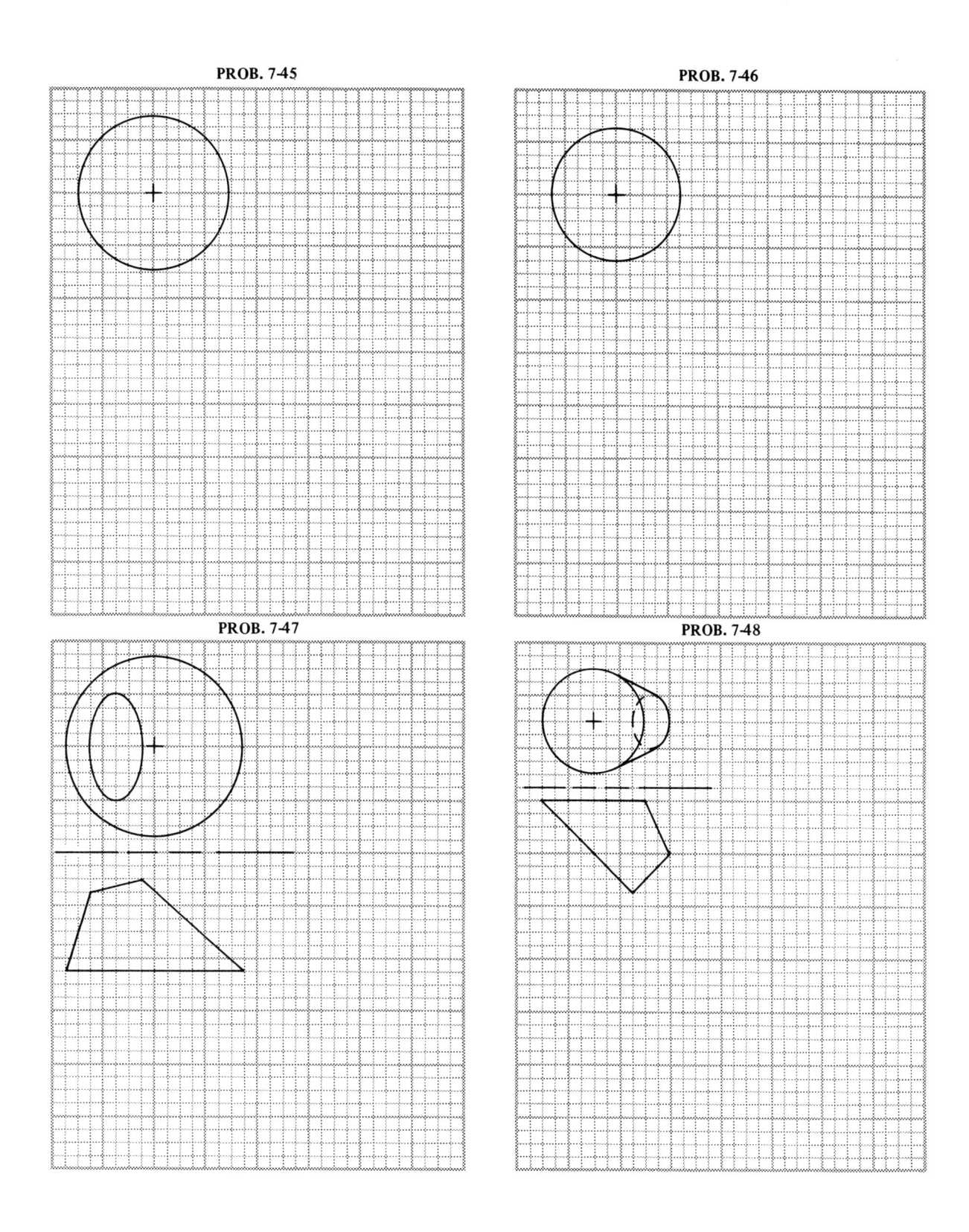

PROB. 7-45. Develop the sphere using the gore method. Show a minimum of three panels.
PROB. 7-46. Develop the sphere using the zone method.

PROBS. 7-47 and 7-48. Do a half development for 7-47 and a full development for 7-48.

347

PROB. 7-49

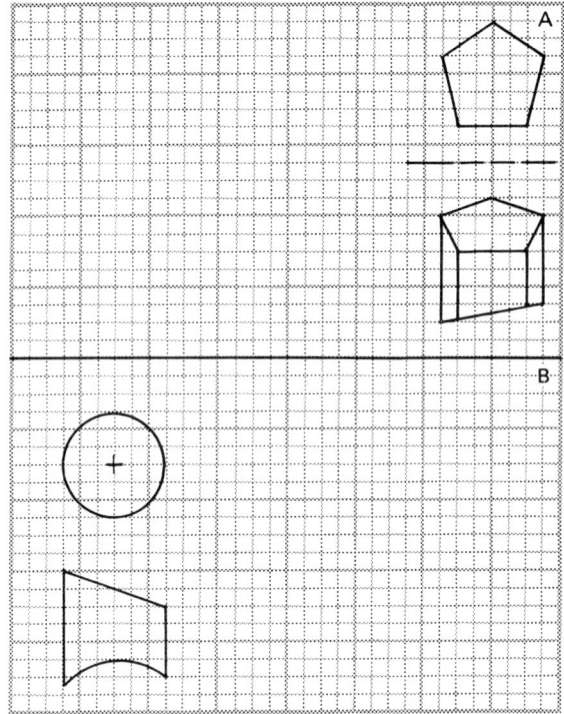

PROB. 7-50

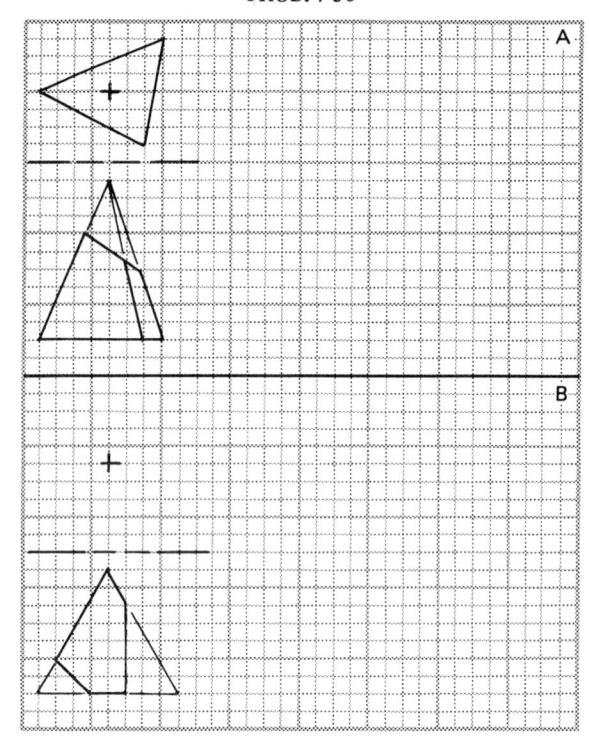

PROB. 7-51

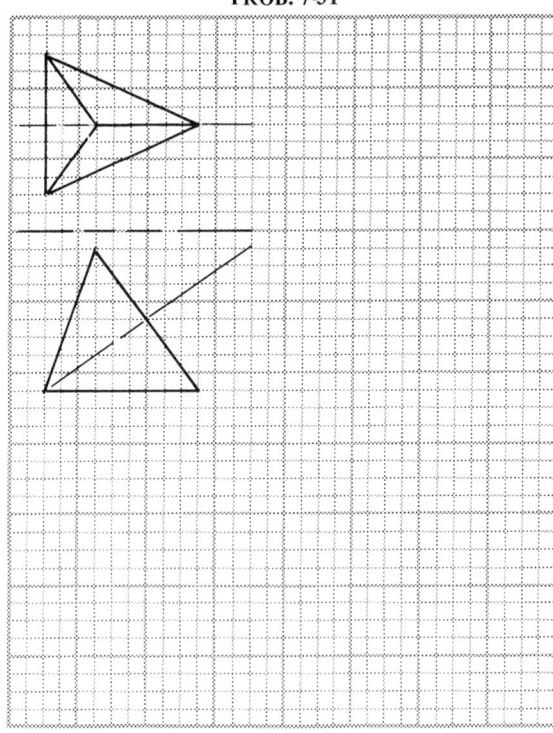

PROB. 7-49B. Develop the inside pattern for the cylindrical shape. Do not develop the ends. Use 15-degree segments.

PROB. 7-50A. Complete the figure and develop the inside pattern of the five surfaces of the truncated tetrahedron.

PROB. 7-50B. Complete the given views and develop the inside pattern of the cone.

PROB. 7-51. The given tetrahedron is intersected by a .75" (19.05mm) diameter cylinder whose centerline is shown in both views. Complete the views and solve for the intersection. Develop the inside pattern of the tetrahedron and cylinder, and construct a cardboard model. Submit the drawing and model.

QUIZ

1. What is a true length diagram and how is it used?
2. How is a stretch-out line used?
3. All surfaces in a developed view will appear _____.
4. Define *parallel line development*.
5. Sheet metal patterns are normally developed _____ up.
6. What is a transition piece and how is it used?
7. What type of figures must be approximately developed?
8. What is triangulation?
9. Define *radial line developments*.
10. It is necessary to have the _____ of each element and line in order to develop a figure.

TEST

PROB. 7-49A. Develop the inside pattern for the pentagonal prism. Do not develop the ends. Start the rollout with the shortest edge.

PROB. 7-52A through O

EXTRA ASSIGNMENTS

PROBS. 7-52A through 7-52O. Instructor can use these figures as alternative assignments. Problems are to be scaled from the text and drawn in an enlarged scale using SI or English units as required. In many cases the views of the figure must be completed.

PROB. 7-53A through L

EXTRA ASSIGNMENTS

PROBS. 7-53A through 7-53L. Same as 7-52.

350

PROB. 7-54A through H

A B C

D E F

G H I

EXTRA ASSIGNMENTS

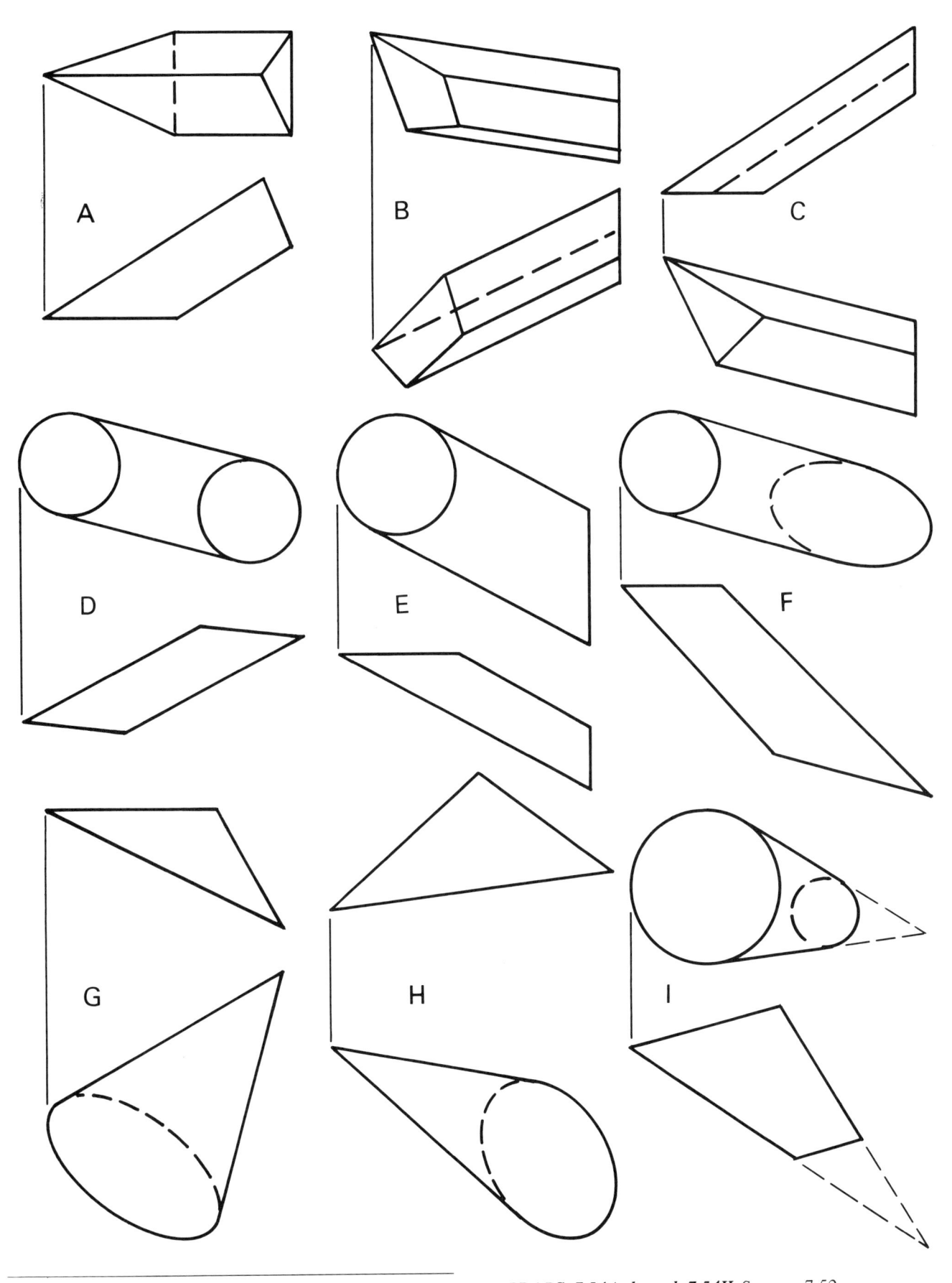

PROBS. 7-54A through 7-54H. Same as 7-52.

PROB. 7-55A through F

A

B

C

D

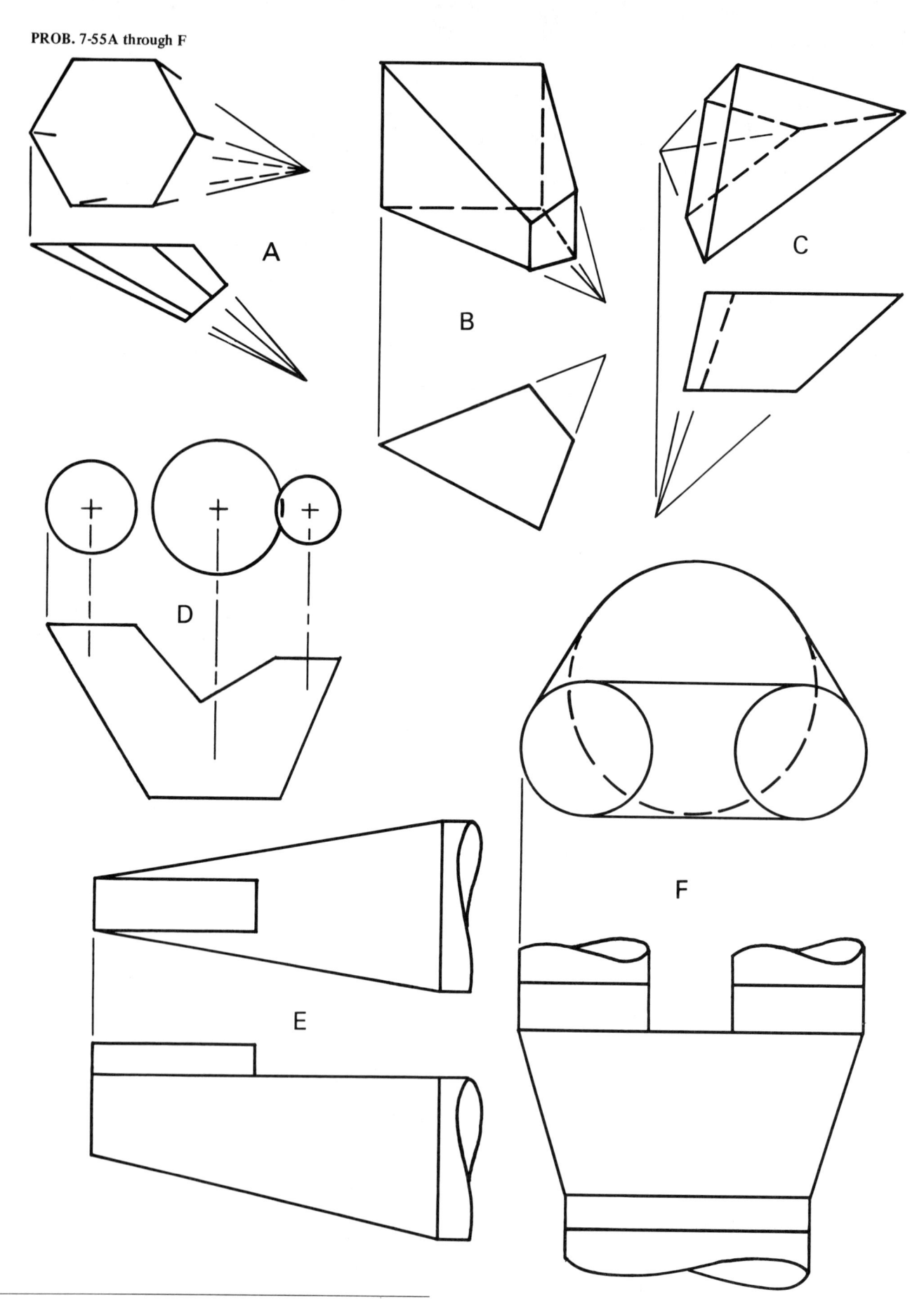

E

F

EXTRA ASSIGNMENTS

PROBS. 7-55A through 7-55F. Same as 7-52.

8

MINING AND GEOLOGY

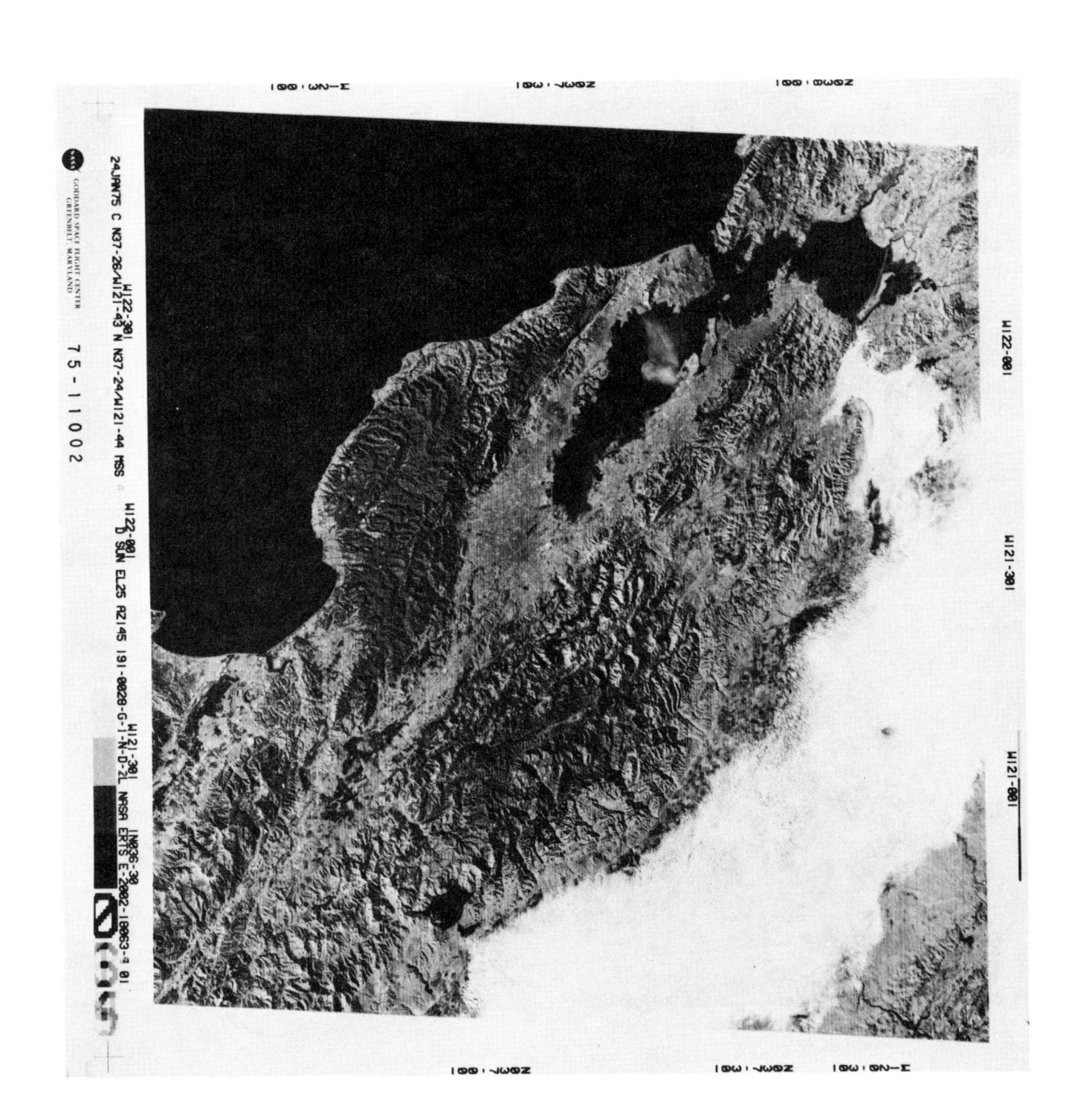

8.1 MINING AND TOPOGRAPHIC APPLICATIONS

The principals of orthographic projection are used continually in the real world of engineering construction and mining. Topographical and mining problems involving land contours, surface and subsurface earthworks, and their specific applications in construction technologies utilize a variety of descriptive geometry principals, practices, and procedures in their solutions. This chapter demonstrates this relationship in the use of contour maps and profile drawing to show the existing conditions of the earth's surface and in mining applications showing stratified subsurface configurations involving ore veins and strata. Surface construction (including roads, pipelines, railways, bridge piers, dams), building site construction (excavation, foundations, retaining walls), and water services (such as irrigation, drinking supply, reservoir placement, and water treatment facilities) all use techniques and drawing practices derived from orthographic projection and descriptive geometry. Subsurface construction and mining applications include location of valuable ore veins; tunnel construction; oil well drilling and deposit location; identification of slope/dip and extent of rock strata, coal beds, mineral deposits, faulted veins; and strike, dip, and thickness determination of strata. Subsurface construction and mining are also aided by applications of descriptive geometry to project the expected locations of intersecting mineral veins (which are normally bounded by parallel planes of rock strata) and the location of outcroppings (surface exposure of a stratum or ore vein).

FIG. 8-1 Cross country pipeline.
(Courtesy Grove Valve and Regulator.)

FIG. 8-2 Oil tanker loading in an Alaskan bay.

For both surface and subsurface construction the plan-profile drawing is essential. The plan-profile drawing is used extensively in highway, dam, and pipeline construction, Fig. 8-1. The construction of roadways and dams requires that existing earth contours be modified to the requirements of that construction. To guide this construction, drawings are prepared to show both the original earth conditions and the contours and levels of the completed construction. The application of descriptive geometry in which a topographic survey (horizontal/plan view) and vertical sections (front/profile, and other vertical section auxiliary views) are used together, permits the preparation of accurate scale drawings that show exactly what construction work is done. A plan-profile drawing shows the actual topographic configuration of a particular portion of the earth. The existing and after-construction configuration are established by contour lines which represent the intersection of horizontal cutting planes (at specified intervals) with the earth's surface.

The tanker in Fig. 8-2 is loading oil from the Alaskan pipeline. The design and construction of pipelines requires the application of plan-profile drawings derived from descriptive geometry procedures.

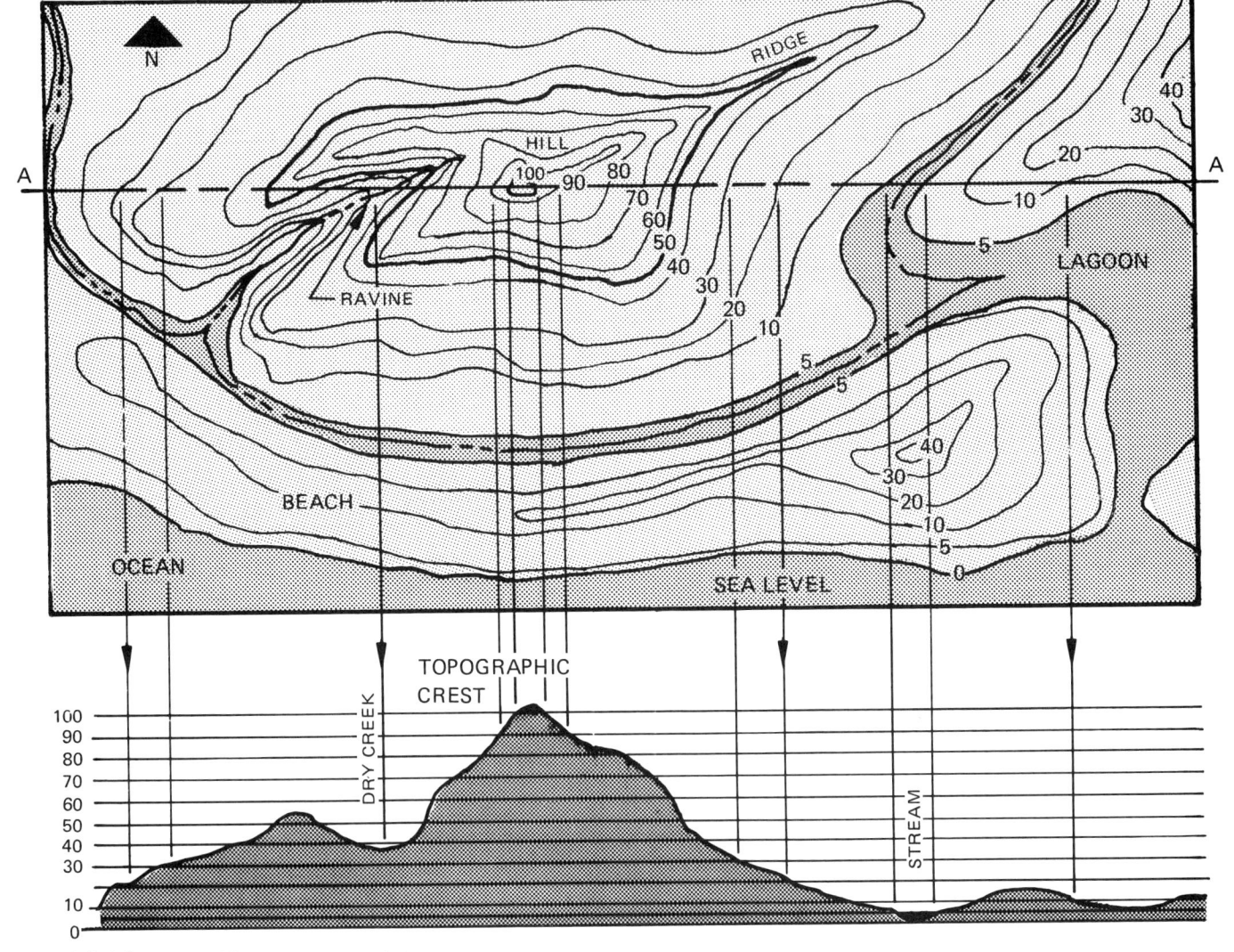

FIG. 8-3 Topographic contour map and plan-profile.

8.2 CONTOUR MAPS AND PLAN-PROFILES

The *plan-profile* drawing is based on a topographic map of a portion of the earth's surface. This map is drawn from a surveyor's topographical notes and calculations and or an aerial photo survey. Therefore a topographic map is a horizontal view/plan view of the earth's surface. Variations in the surface of the earth within the limits of the survey are represented by *contour lines*. *Contours are lines of intersection of a series of evenly spaced horizontal cutting planes and the irregularities of the ground surface.* Therefore a contour line represents a line on the earth's surface at a particular elevation/level. Normally contour lines are established with vertical divisions (intervals) of 10 ft or 100 ft, depending on the size of the area being drawn.

The frontal view of a topographical map will show the actual *profile* of the ground's surface and each horizontal cutting plane (representing the elevation view of the contour lines) as an edge view. Contour spacing (contour interval) can be seen here. Contours are established from a designated or standard starting elevation, usually sea level (0 ft).

The frontal view is called a *profile*, since it shows the "profile" of the ground's surface. In most cases, the profile view is actually a projection of a vertical cutting plane passed through the plan view at a specified location in order to determine the outline/profile of the ground along a particular line of interest.

In Fig. 8-3, the profile view is a vertical section taken along cutting plane A-A. The vertical and horizontal scales are the same (though they need not be if one wishes to exaggerate the profile configuration). The profile view is drawn by projecting points of intersection of contour lines and section line A-A in the plan view to the edge view of the contour lines in the frontal/profile view. In the plan (horizontal) view, changes in the slope of the terrain are represented by the irregular contour lines. Steep surfaces appear as contour lines close to one another and mildly sloping areas are shown by wide spacing. The high point of elevation is called the topographic crest. Crowded contour lines extending out from a hill or mountain represent ridges. Contour lines that extend in toward the higher elevation are ravines or canyons. Stream bed lines are drawn as shown.

FIG. 8-4 Plan-profiles and vertical sections.

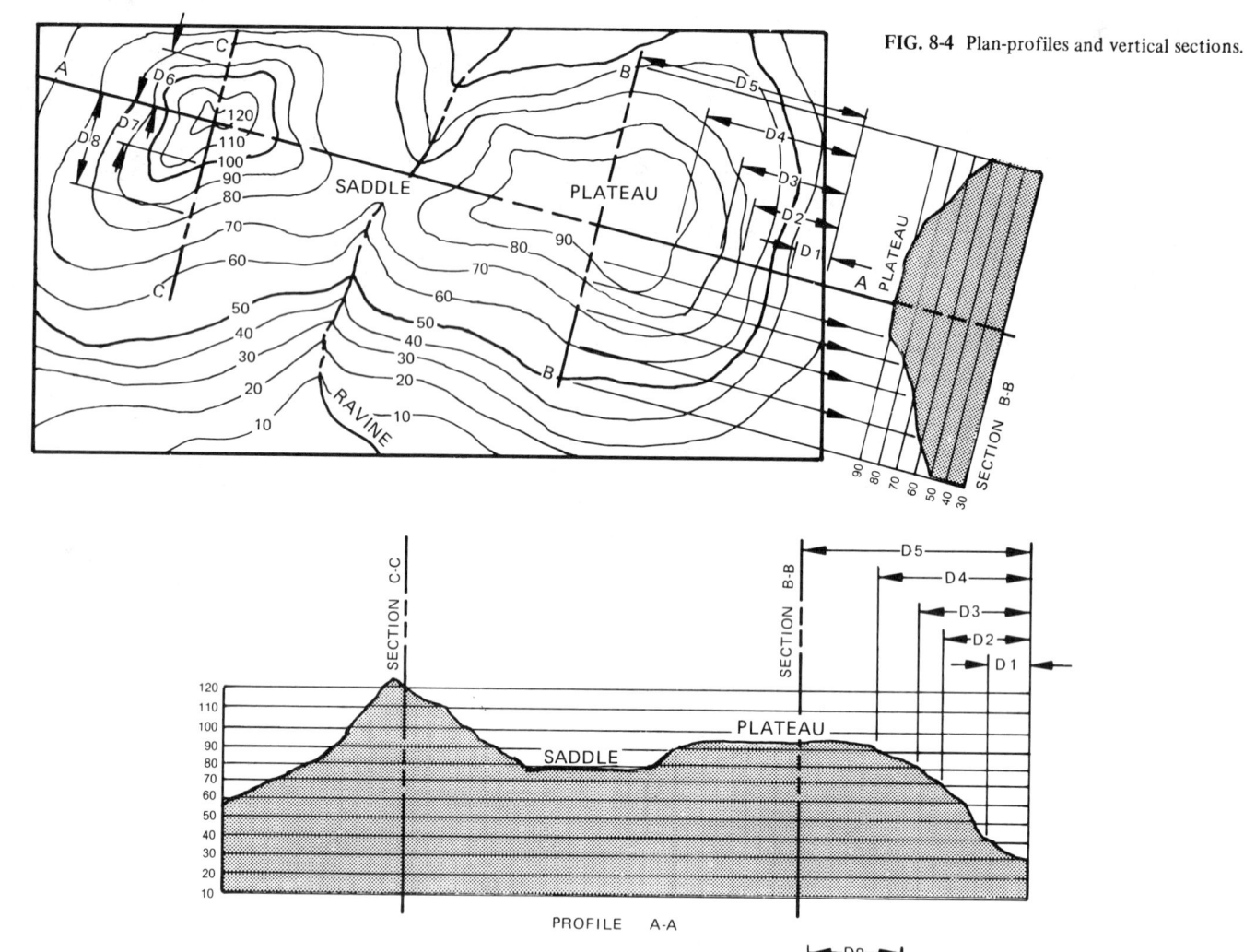

8.3 PLAN-PROFILES AND VERTICAL SECTIONS

*A **profile** is an elevation projection of a vertical section (cutting plane) of the earth's surface along a specified line.* The line can be a straight line, curved line, segments of straight lines at angles to one another, or a straight-curved line combination representing a pipeline, road, etc. The profile section is not usually a true frontal projection. The true length of the profile section includes its total distance (length) including curved lengths and angled straight distances.

A ***section*** is a vertical section (cutting plane) of the earth's surface normally taken perpendicular to the profile section as shown in Fig. 8-4. Sections represent the intersection of a vertical cutting plane and the earth's surface along a specific line, usually 90° to the profile section. In this example sections B-B and C-C are taken at 90 degrees to profile section A-A. A section is drawn as is a profile. The same contour intervals are shown (between usable elevations) as for the profile section.

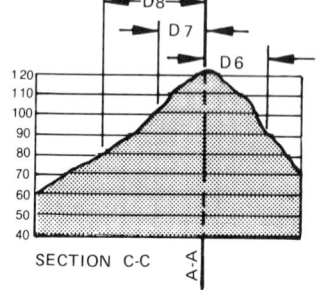

The profile is drawn by determining the true length of section line A-A in the plan view. Contours are established at 10-ft intervals and shown in the profile view as evenly spaced horizontal planes (edges). The intersection of each contour and profile section A-A in the plan view determines the outline of the profile view. Dimensions D1 through D5 illustrate this procedure.

Section B-B is taken perpendicular to A-A so that it passes through the center of the plateau. Section C-C is taken through the highest peak. D6, D7, and D8 are intersection points of section line C-C and the contours in the plan view, and are used to draw section C-C. Section B-B is projected directly from the plan view, therefore transfer dimensions are not required.

FIG. 8-5 Bearing, slope, and grade.

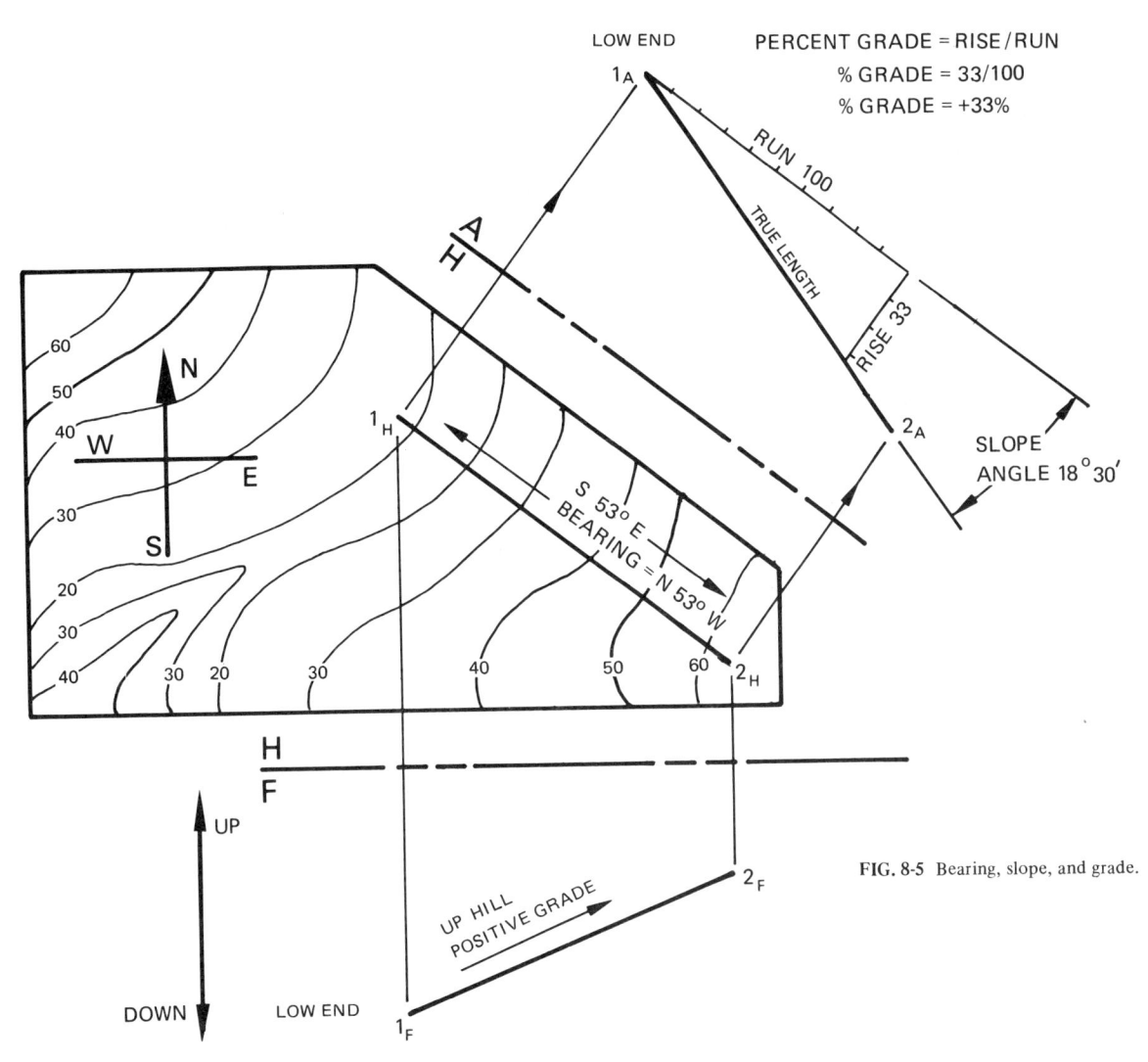

8.4 BEARING, SLOPE, AND GRADE OF A LINE

A line representing a road, pipeline, etc. can be verbally described so as to establish it at a particular location on a survey or contour map. The location of a line can be described in the plan view by providing its compass direction and a starting point. *The compass direction of a line is called its "bearing."* The bearing of a line is taken from the north or south point. The line's deviation from the north or south toward the east or west, as measured in degrees, completes its bearing description. In Fig. 8-5, line 1-2 has a bearing of N 53°W or S 53°E. This means that it deviates from north 53 degrees toward the west or from the south 53° toward the east. If the bearing is given as N 53° W it was taken from north, but if one assumes the direction of the line to be from point 1 towards point 2 it would be S 53°E. In some engineering fields, the bearing is measured from the high end of the given line toward the low end (in the plan view of course). In this case it could be taken from the south or north. This usage is not universally accepted since it can cause confusion. As an example, a road will follow the contour of the land in most cases. If the road is a straight line and is constructed over hills and through valleys, which is its low end? The concept of low end can be applied to a segment of a pipeline that must slant in one direction, but in general, should not be applied to topographical problems involving lines.

The angle a line makes with the horizontal plane is called its "slope." The slope can be described by its slope angle, grade, or a ratio of rise to run. The slope of a line is measured in an elevation view where it appears true length and the horizontal plane appears as an edge. The slope or "grade" of a line can be positive or negative depending on its intended direction. In Fig. 8-5, the slope/grade of the line is taken from point 1 towards point 2. Point 1 is below point 2 therefore it is a positive slope/grade. The true length of line 1-2 appears in auxiliary A. The angle it makes with the horizontal plane is its slope angle, +18°30'. The grade is given as a percentage of its rise/run. Using 100 units for the run, the rise is measured in the same units, 33/100. The percent grade equals 33/100 × 100 = +33%. Note that the percent grade equals the tangent of the slope angle times 100: tangent of 18°30' is .33 (.33 × 100 = 33%).

357

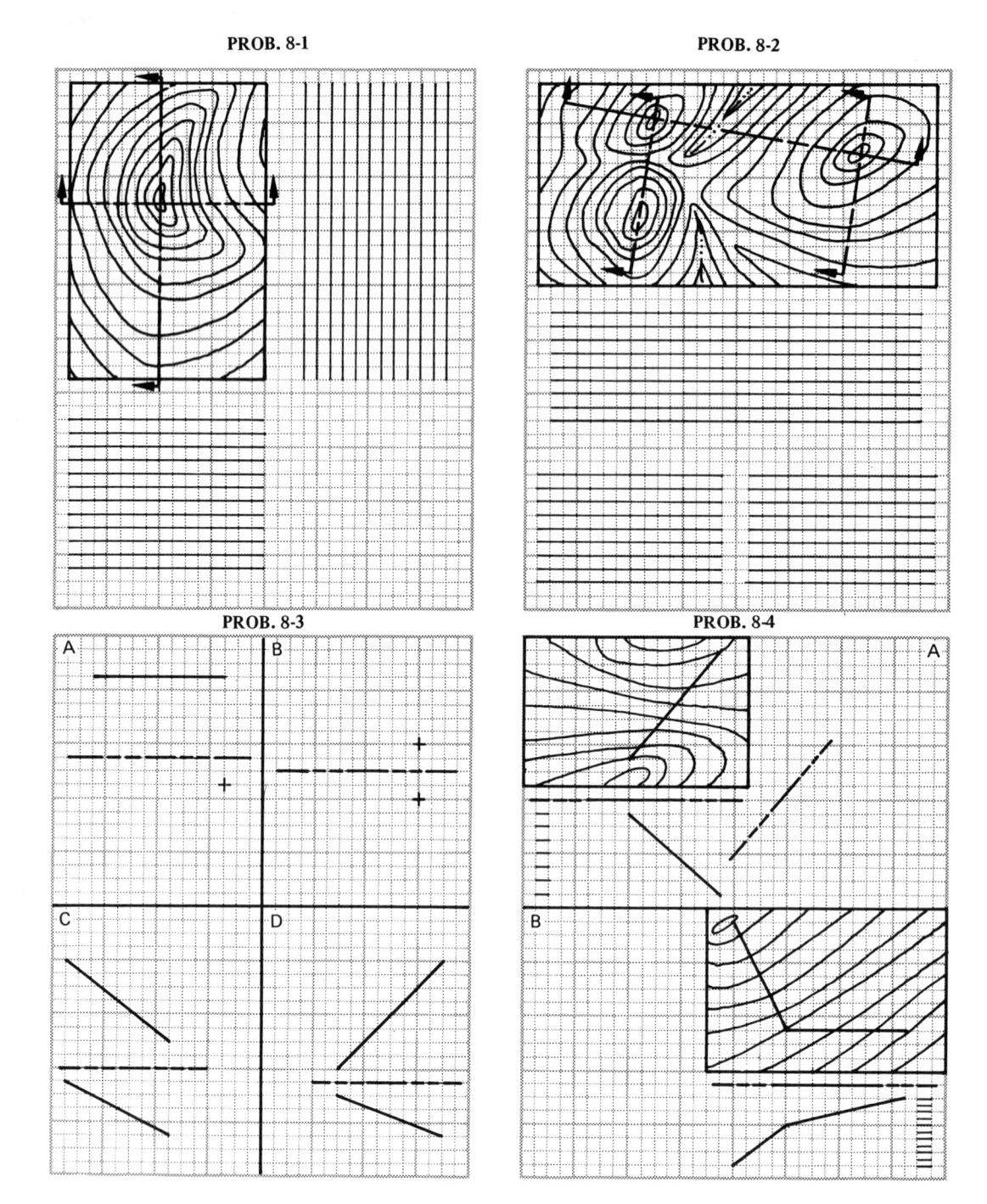

PROB. 8-1. Draw the profile and section of the contour map. The highest elevation is 1500 ft. Contours are plotted at 10-ft intervals. Vertical scale is 1″ = 40′. Completely label views.

PROB. 8-2. Draw a profile and the two required sections of the contour map. The peak at the right is at 5000 ft. Contours are at 10-ft intervals. Vertical scale is 1″ = 40′. There are three peaks on the map.

PROB. 8-3A. The given line is at a +33° slope from point 1 to point 2. Point 2 is shown in the front view. Calculate the bearing, azimuth, and grade.

PROB. 8-3B. Line 1-2 has a bearing of N 70°W, with a negative slope of 28 degrees and a length of 50 ft. Point 1 is given. Construct an auxiliary view to establish the true length, and slope. Scale is 1″ = 200′. The direction of the line is from point 1 towards point 2. Complete the H and F views.

PROB. 8-3C. The given line slopes down from point 1. Solve for the bearing, slope, and grade.

PROB. 8-3D. Line 1-2 has a positive grade. Calculate the azimuth, bearing, slope, and grade.

PROB. 8-4A. Line 1-2 slopes downward from 100-ft elevation. Solve for the bearing, slope, and grade. Contour lines are spaced at 10-ft intervals.

PROB. 8-5B. Find the grade, slope, and bearing of each line. The low point is at 2000-ft elevation. Contour lines are at 10-ft intervals.

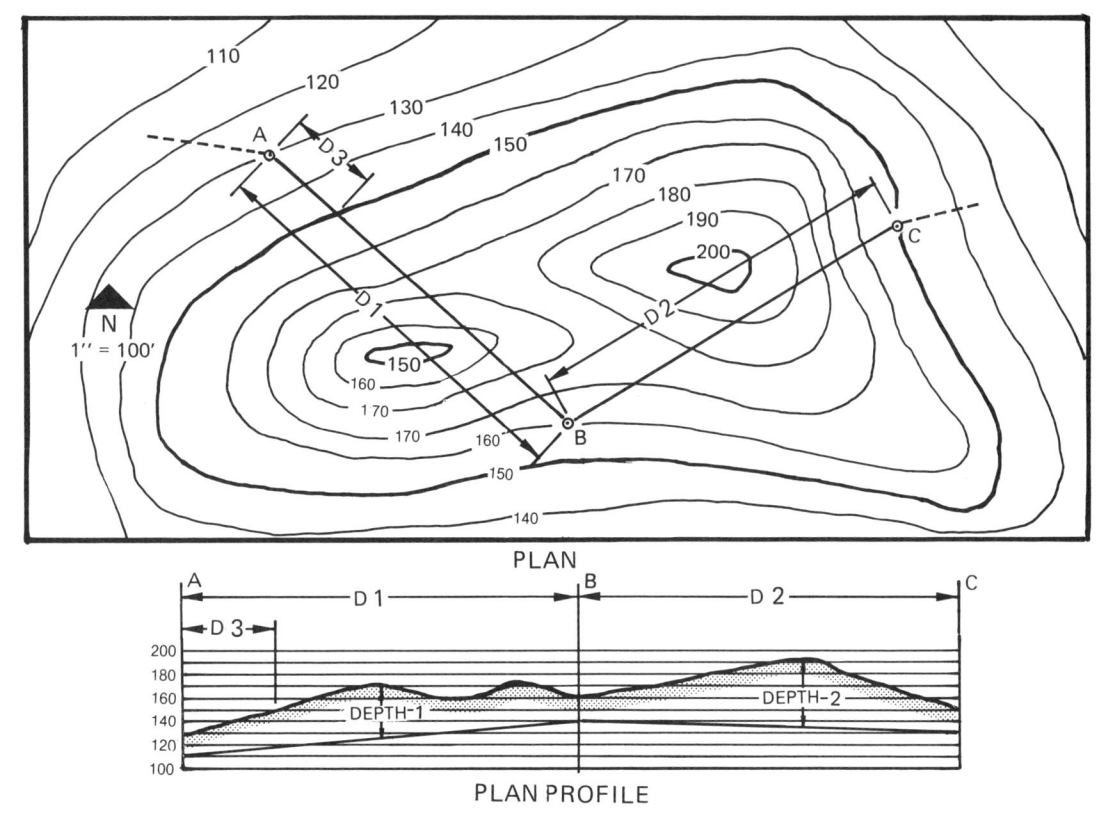

FIG. 8-6 Plan-profile of a pipeline.

8.5 PLAN-PROFILE OF A PIPELINE

A plan-profile drawing is used to describe the three-dimensional conditions along the length of various earthwork construction projects such as the construction of a highway or cross-country pipeline, Fig. 8-7.

In Fig. 8-6 a plan-profile of a pipeline is illustrated. The plan (horizontal) view is given. The pipeline is to be laid along the line indicated through station points A, B, and C. *Station points* are specific points established during the survey and shown on the topographic/contour map. The profile (vertical section) of the earth's surface is taken along lines A-B and B-C using a vertical scale equal to the horizontal scale. Note that it is normal practice to exaggerate the vertical scale in most cases, especially where the total area involved would make elevation differences extremely small and thus the profile outline indiscernible.

The depth of the pipeline at each station point is 20 ft. Draw the pipeline in the profile view and determine the grade and the maximum depth below grade in each section of the pipeline (A-B and B-C). The length of the total pipeline between point A and point C is used as the length of the profile section, dimensions D1 plus D2. The vertical scale, representing the edges of the horizontal planes (contour intervals), starting at elevation 100 is drawn using 10-ft intervals. The intersection points of the pipeline and the contours in the plan view are transferred to the profile section to determine the earth's profile/outline, D3. After the complete profile outline is established, the depth of the pipeline at the stations is measured at 20 ft. These points are connected to form the profile of the pipeline. Each pipe section is true length, therefore the grade can be measured in this view. Depth 1 and depth 2 are the maximum depth below grade for each section of the pipeline.

FIG. 8-7 Desert pipeline being laid.
(Courtesy Grove Valve and Regulator.)

PROB. 8-5

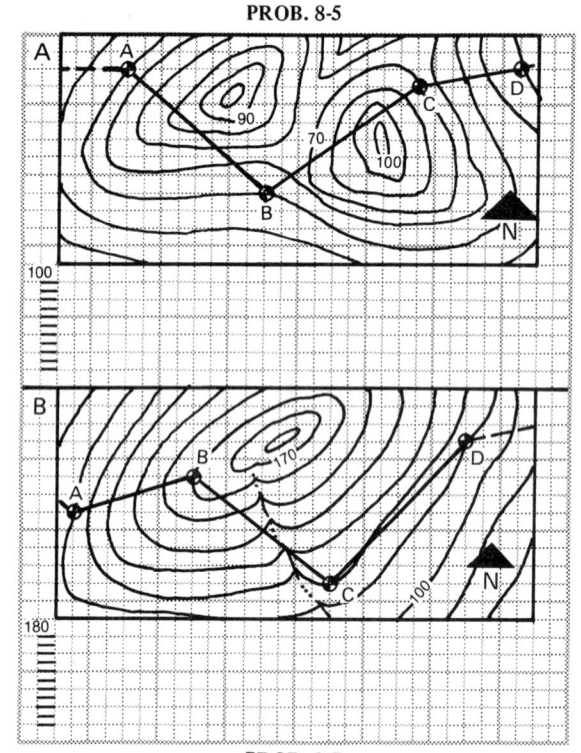

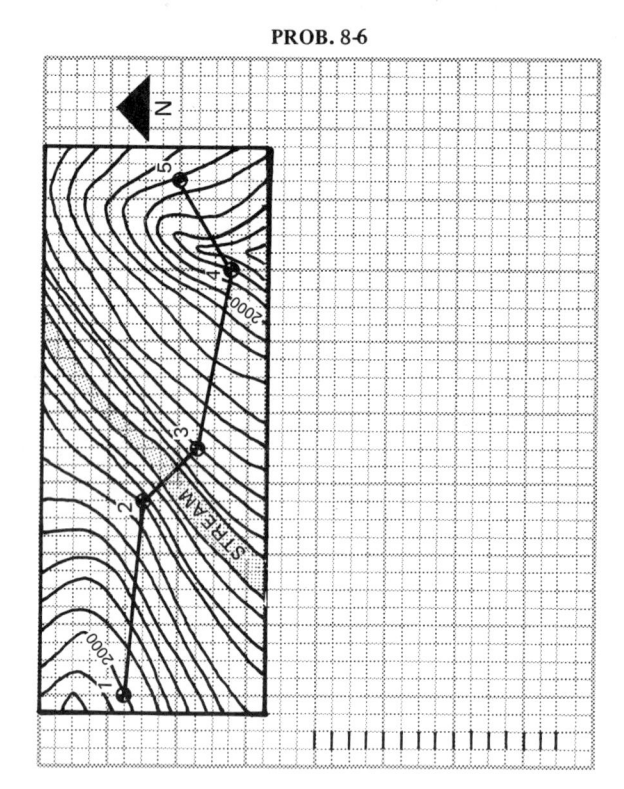

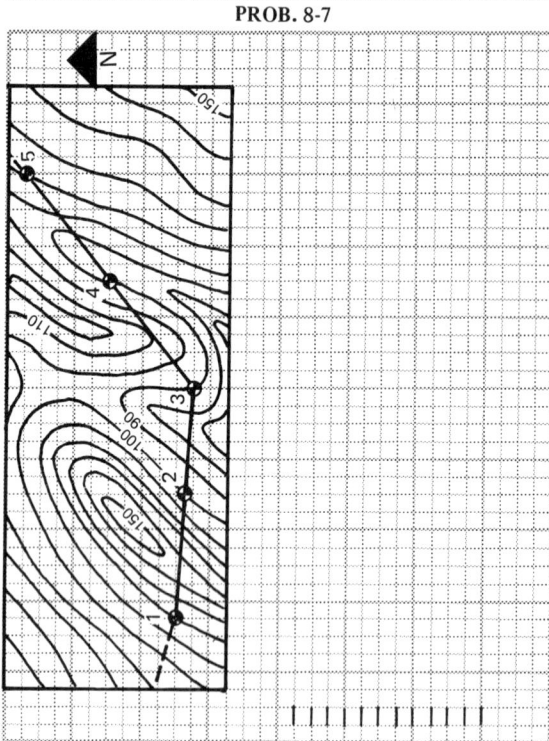

PROB. 8-6

PROB. 8-7

PROBS. 8-5A, 8-5B. For both problems a pipeline is to be laid along a line indicated through the station points A, B, C, and D. Draw the profile (vertical section) of the earth's surface from A to B, B to C, and C to D using a vertical scale of 1″ = 80′. The pipeline is to be 10 ft below the surface at each station point in problem A and 20 ft below in problem B. Draw the pipeline in the profile view and determine the bearing, grade of the pipeline, and maximum depth below grade in each section of pipe (between station points). Contour lines are spaced at 10-ft intervals.

PROB. 8-6. A pipeline is to be laid along the given line through the specified station points. The river is at 1000-ft elevation and the pipeline is to be 10 ft below grade at each station point. Draw a profile section through each of the points as in 8-5A and 8-5B; calculate the grade, bearing, and true length of each section along with the maximum depth of the pipeline for each section. How much clearance is there between the pipe and the river? The vertical scale is 1″ = 40′ and the contour lines are spaced at 10-ft intervals.

PROB. 8-7. Draw a profile of the power line stretched between poles located at the specified station points. The power line is 35 ft above the ground at each station point. Vertical scale is 1″ = 40′ and the contour lines are spaced at 10-ft intervals. What is the closest distance between the line and the ground for each section?

8.6 CUT AND FILL FOR A LEVEL ROAD

*The **cut and fill** for roadways, dams, and railways are described by showing the existing and the new contour lines within the area of construction.* The slopes of cut and fill vary according to soil and use requirements. The slope of a loose material is called its *angle of repose,* the angle at which the material will pile naturally without sliding. Each type of material will have a different angle of repose depending on its consistency, moisture content, and composition. Knowledge and understanding of the angle of repose are necessary in construction work involving cuts and fills, as they will help determine the *cut angle* and the *fill angle.*

FIG. 8-8 Cut and fill.

*A **cut** is a passageway created by the removal of ground material along a specified route which is completed at a required angle to the horizontal. Thus a cut will slice an angle from the surrounding ground surface. A **fill** is a section of a passageway that has been created by the raising of a portion of a specified route by the addition of loose material, rock, gravel.* The added material will form an angle with the horizontal equal to its angle of repose. Material removed during a cut can be used for the fill. Therefore, if possible, the cut and fill should be somewhat equal in volume.

The cut and fill contours are determined by taking a section of the passageway perpendicular to its centerline as in Fig. 8-8. The section shows the parallel contour intervals and the required angle of cut and fill drawn from the edge of the road. The intersection of the cut lines and the fill lines with the intervals is projected to the plan view to establish the contour lines of the cut and fill areas. The cut and fill contours are parallel to the road. The intersection points of the cut contours, and the fill contours with existing earth contours, are connected to establish the cut and fill areas (which are shaded differently) on the contour map.

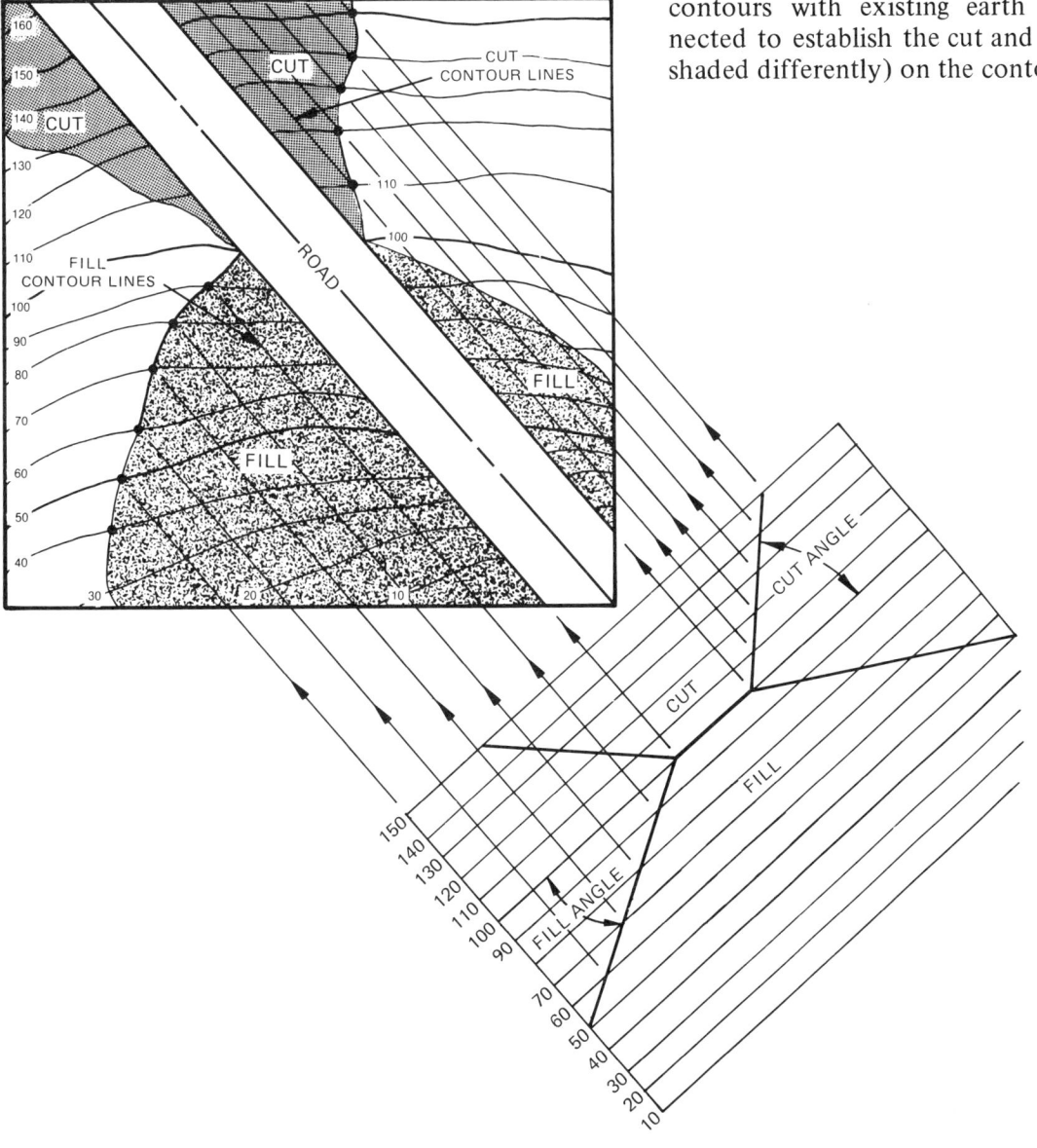

361

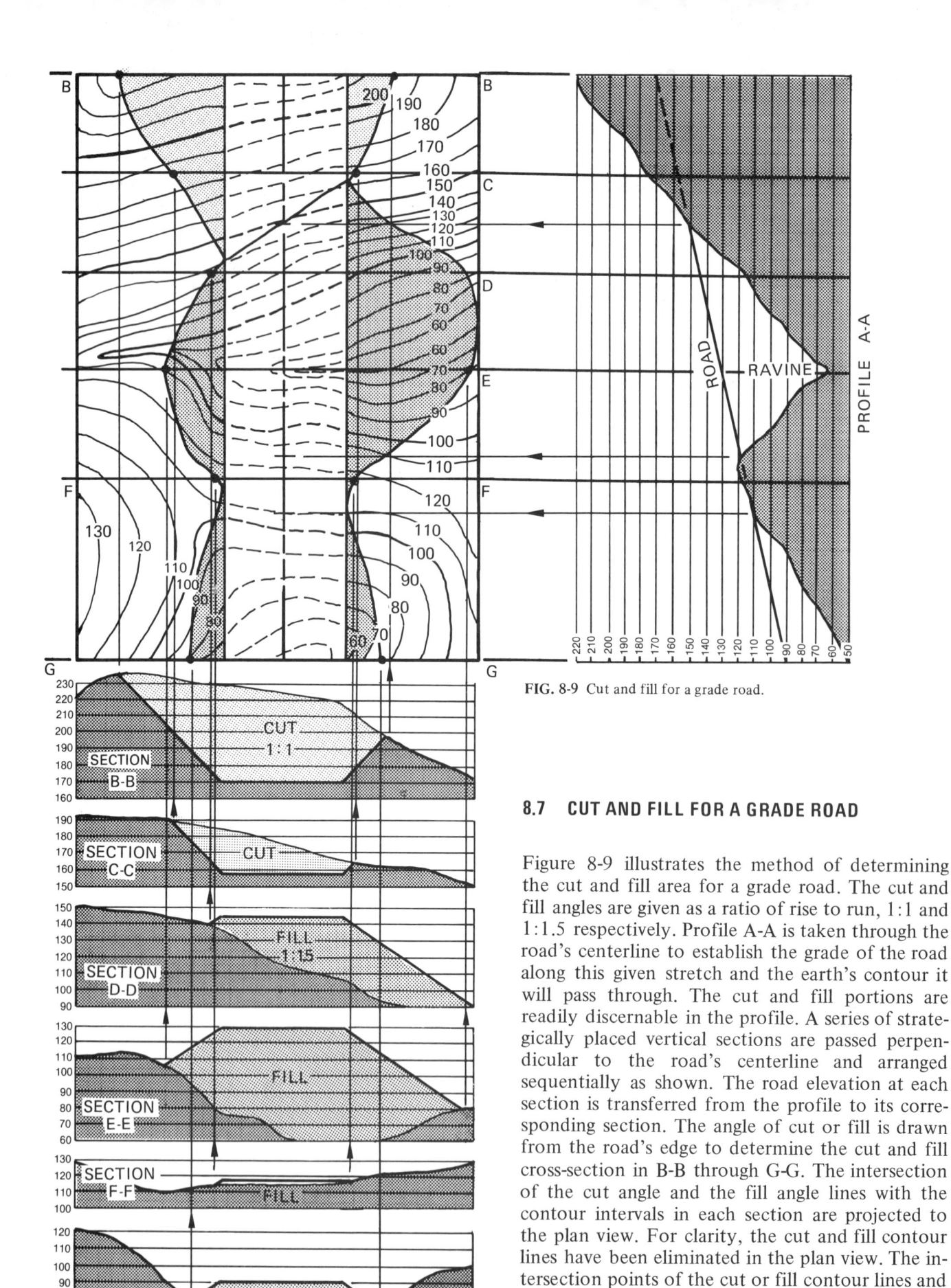

FIG. 8-9 Cut and fill for a grade road.

8.7 CUT AND FILL FOR A GRADE ROAD

Figure 8-9 illustrates the method of determining the cut and fill area for a grade road. The cut and fill angles are given as a ratio of rise to run, 1:1 and 1:1.5 respectively. Profile A-A is taken through the road's centerline to establish the grade of the road along this given stretch and the earth's contour it will pass through. The cut and fill portions are readily discernable in the profile. A series of strategically placed vertical sections are passed perpendicular to the road's centerline and arranged sequentially as shown. The road elevation at each section is transferred from the profile to its corresponding section. The angle of cut or fill is drawn from the road's edge to determine the cut and fill cross-section in B-B through G-G. The intersection of the cut angle and the fill angle lines with the contour intervals in each section are projected to the plan view. For clarity, the cut and fill contour lines have been eliminated in the plan view. The intersection points of the cut or fill contour lines and the existing natural terrain determine the cut and fill areas in the plan view. The fill area is the darkly shaded area.

FIG. 8-10 Cut and fill for a roadway and a dam.

8.8 CUT AND FILL FOR A ROADWAY AND A DAM

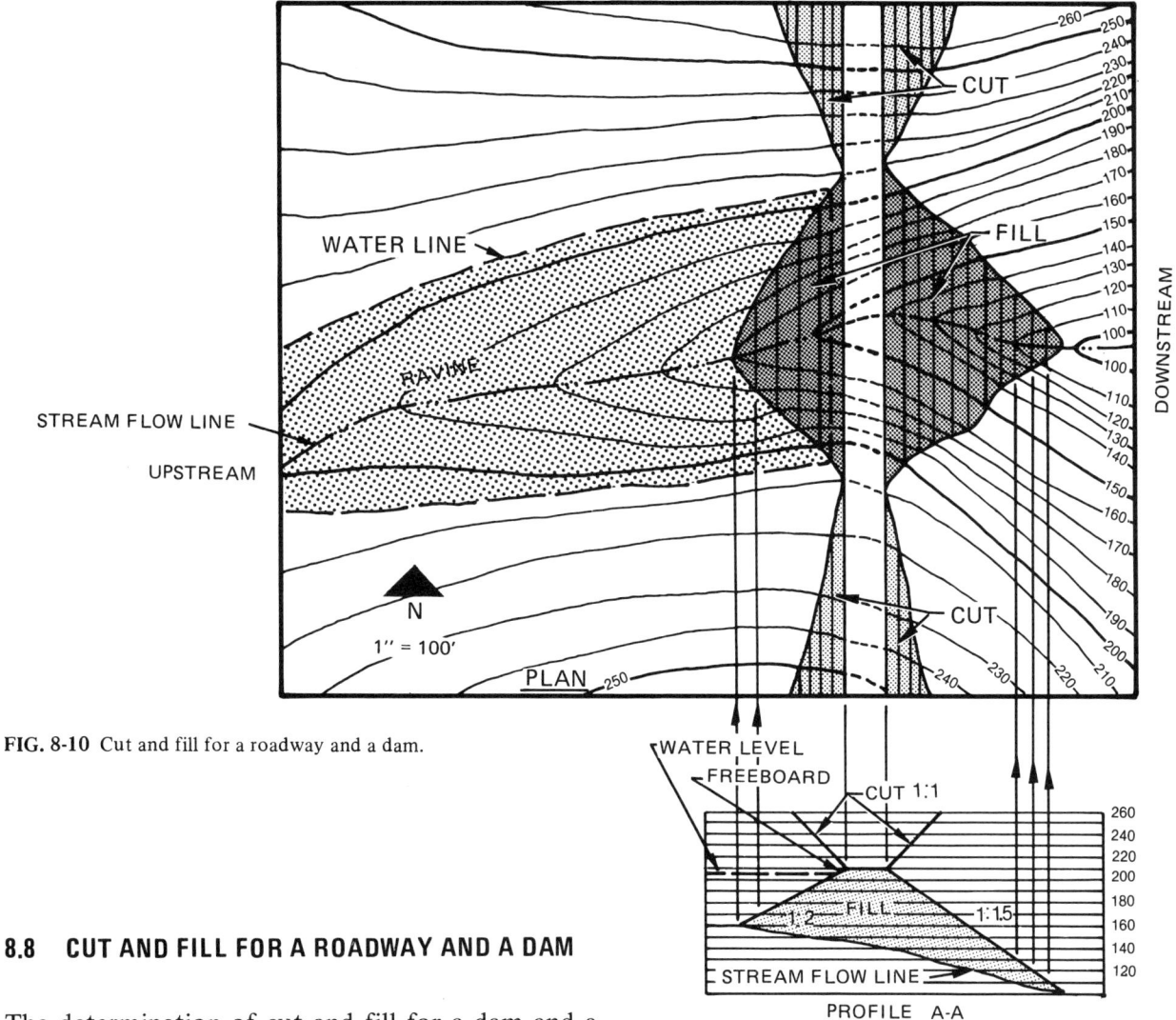

The determination of cut and fill for a dam and a road involve the same process. In Fig. 8-10 the fill portion of a roadway is to be used to create an earthen dam across a small ravine. The road has a constant elevation of 210 ft. The slope ratio of the cut areas of the road is 1:1. The slope ratio of the upstream side of the fill area is 1:2 and the downstream side is 1:1.5. The freeboard is 5 ft. The *freeboard* is the height above the highest water level to the top of the dam wall/structure; here the road will be the high point of the dam. The vertical profile section A-A is taken along the stream flow line. Complete the vertical section by drawing the contour intervals at the same scale as the plan (horizontal view). Establish the roadway and draw the slope lines. Then, on the plan, show all new contour lines and the limits of cut and fill using solid lines. Show the water line behind the dam with a broken line and shade the water area. Label the areas of cut and fill and shade differently. Note the maximum depth of fill on the downstream side and the maximum depth of water.

The profile view is drawn using the contour intervals between 100 ft and 260 ft, corresponding to the lowest and highest contours in the plan. The road is drawn at 210 ft and the cut and fill angles are drawn from its edges as shown. Show the stream flow line and the water level line in the profile. The water level will be 5 ft below the dam crest corresponding to the freeboard of 5 ft. The maximum depth of water is measured in the profile as a vertical dimension from the intersection point of the stream flow line and the 1:2 fill angle line and the water line, 45 ft. The maximum depth of fill is a vertical dimension measured from the edge of the road on the 1:1.5 fill side to the stream flow line, 70 ft.

Project the points of intersection of the cut and fill angle lines in the profile to the plan view to establish the cut and fill contours parallel to the road/dam. The cut and fill contours intersect the existing natural contours in the plan view and establish the cut and fill areas as shown. Since the dam/road is constructed through a ravine, there are cut areas on either side of the fill areas. Draw the water area between the 210-ft and the 200-ft contours in the plane view using a dashed line.

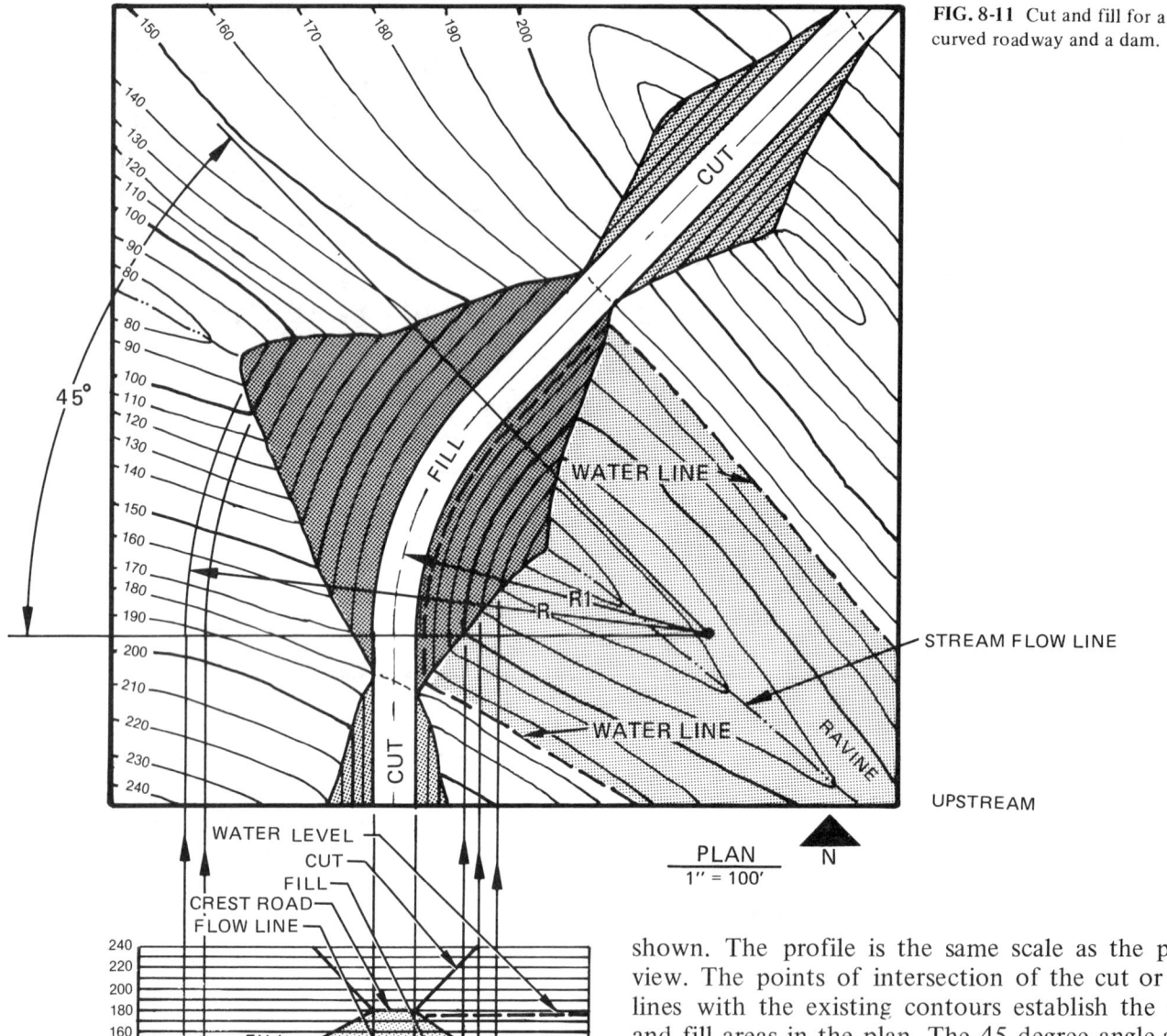

FIG. 8-11 Cut and fill for a curved roadway and a dam.

8.9 CUT AND FILL FOR A CURVED ROADWAY AND A DAM

In Fig. 8-11 the roadway crossing a small ravine is to be used to create an earthen dam. The road is to be 40' wide, level, and at an elevation of 180 ft (this corresponds to the crest of the dam). The curve has a radius of 300 ft, R1 (its center is given), and the direction change of the road is 45°. The slope ratio of the cut areas is 1:1, of the upstream fill is 1:1.5, and of the downstream fill is 1:2. The freeboard is 5 ft. Construct a vertical profile section taken along the stream flow line in the plan. On the vertical section draw the road at the given elevation and show the cut and fill angles and the stream flow line.

The cut and fill contours are determined in the profile view and projected to the plan as shown. The profile is the same scale as the plan view. The points of intersection of the cut or fill lines with the existing contours establish the cut and fill areas in the plan. The 45 degree angle and arcs swung from the center point of the curve (as is R1 and R) are used to draw the fill contour lines and to establish the cut area in the upper right corner of the plan. The cut contours are straight lines, parallel to the road. The fill contours are arcs parallel to the road.

Draw the water level in the plane between the 170-ft and 180-ft contours, and shade differently than the cut and fill areas. Note that the stream flow line and the original natural contours have been covered by the cut and fill contours.

The maximum depth of fill is measured as 82 ft and the maximum depth of the water is 63 ft. The area of the section can be calculated by finding the area of A, B, and C. V equals *vertical* and H equals *horizontal*.

$$A = 82V \times \frac{196H}{2} = 8,036 \text{ sq ft}$$
$$B = 80V \times 40H = 3,200 \text{ sq ft}$$
$$C = 78V \times \frac{104H}{2} = 4,056 \text{ sq ft}$$

$$\text{Total} = 15,292 \text{ sq ft}$$

PROB. 8-8

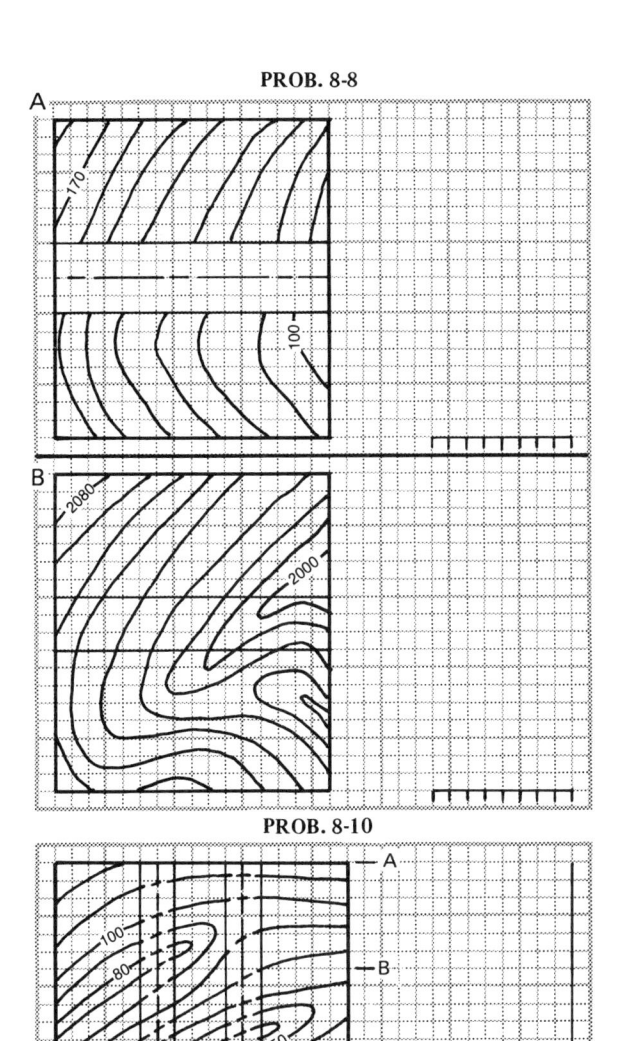

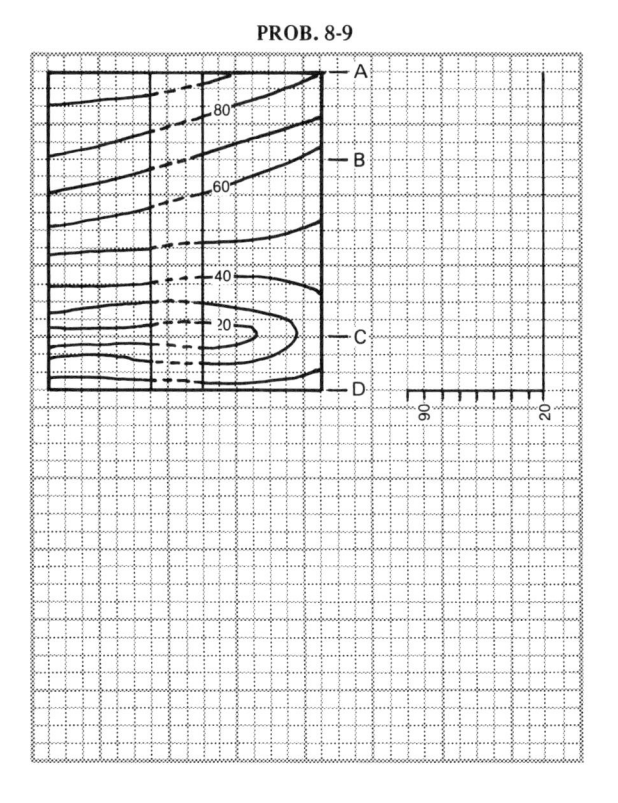

PROB. 8-9

PROB. 8-10

PROB. 8-8A. The given level road is at 130-ft elevation and has a cut angle of 45° and a fill angle of 30°. Draw the cut and fill areas on the contour map by means of a side profile section. Vertical scale is $1'' = 40$ ft and the contour lines are spaced at 10 ft.

PROB. 8-8B. The proposed roadway will be level and at an elevation of 2030 ft. The spacing of the contour lines is 10 ft. The cut will be 1:1 and fill 1:1.5. Draw the resulting cut and fill using a vertical scale of $1'' = 40$ ft. Plot the cut and fill areas in the plan view. Shade the cut and fill differently for clarity.

PROB. 8-9. This future highway has a grade of 12% with a cut of 1:1 and a fill of 1:1.5. Draw the side profile and required four sections. The road's low end elevation is 30 ft at section D. Contour line spacing is at 10-ft intervals and the vertical scale is $1'' = 40$ ft.

PROB. 8-10. The proposed divided highway has a grade of 10% with a low end elevation of 60 ft at section D. Solve for the cut and fill areas by drawing the profile section and sections A, B, C, and D. The cut will be 1:1 and fill 1:1.5. Shade the cut and fill differently. The vertical is $1'' = 40$ ft. Fill the area between the lanes to 5 ft below road grade. Contour line spacing is at 10-ft intervals.

PROB. 8-11

PROB. 8-12

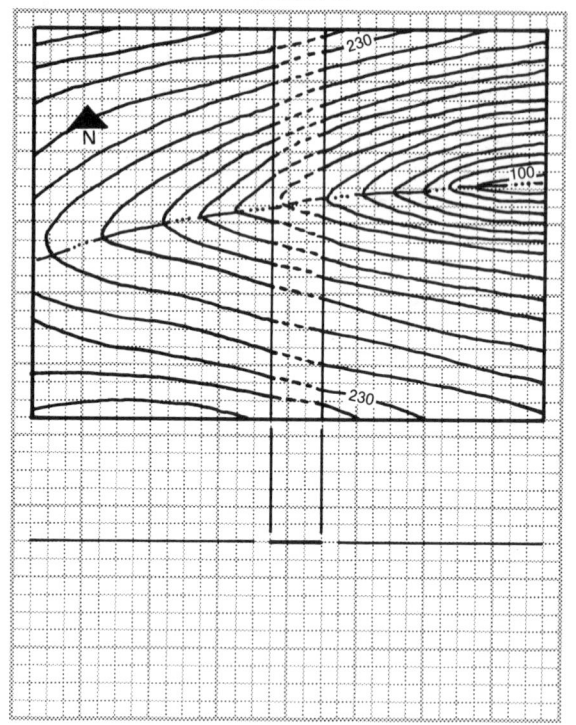

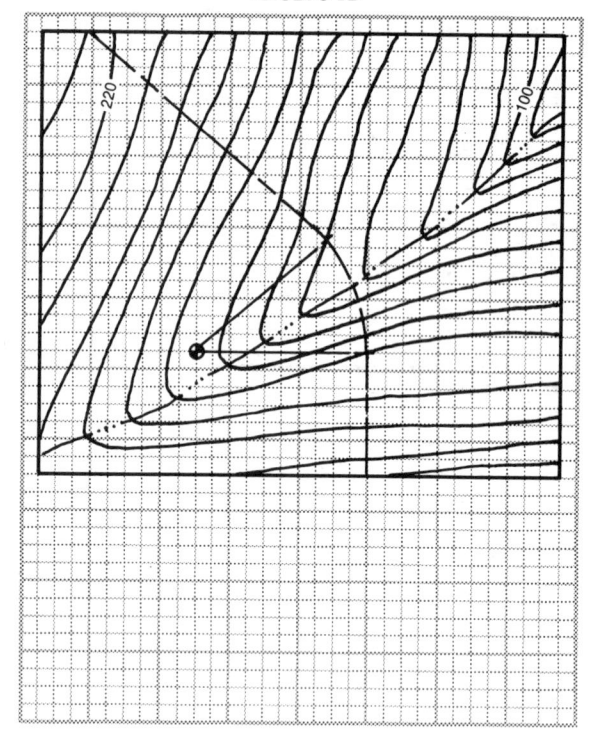

PROB. 8-11. Cut and fill for roadways and dams are described by showing the existing and new contour lines within the area of construction. The slopes of cut and fill vary according to soil and use requirements. In this problem the roadway has a constant elevation of 220 ft. The slope ratio of the cut areas of the roadway is 1:1. The slope ratio of the upstream side of the fill area is 1:2 and the downstream side is 1:1.5. The freeboard is 5 ft. Complete the vertical section (front view) by drawing the contour intervals at the same scale as the plan view (horizontal view) and by drawing the slope lines. Label the slope lines. On the plan view, show all new contour lines and the limits of cut and fill using solid lines. Show the water line behind the dam with a broken line and shade the water area. Label the areas of cut and fill. Give the maximum depth of water and maximum depth of fill.

PROB. 8-12. The roadway crossing a small ravine is to be used to create an earthen dam. The roadway is to be 45 ft wide, level, and at an elevation of 190 ft. The centerline of the roadway is given. The curve has a radius of 250 ft (its center is at the target) and the direction change of the road is 40°. The slope ratio of the cut areas is 1:1, of the upstream fill is 1:1.5, and of the downstream fill is 1:2. The freeboard is to be 15 ft. Solve for all new contour lines of both cut and fill, and the maximum height water line. The contour lines are to be solid lines and the water line is to be broken. Show the limits of cut, fill, and maximum water area and label. On the vertical section, show the flow line of the stream as well as the cut and fill slope lines. In the lower left corner of the sheet, note the maximum depth of fill, the maximum depth of water, and your estimate of the cross-sectional area of the fill at the flow line of the stream.

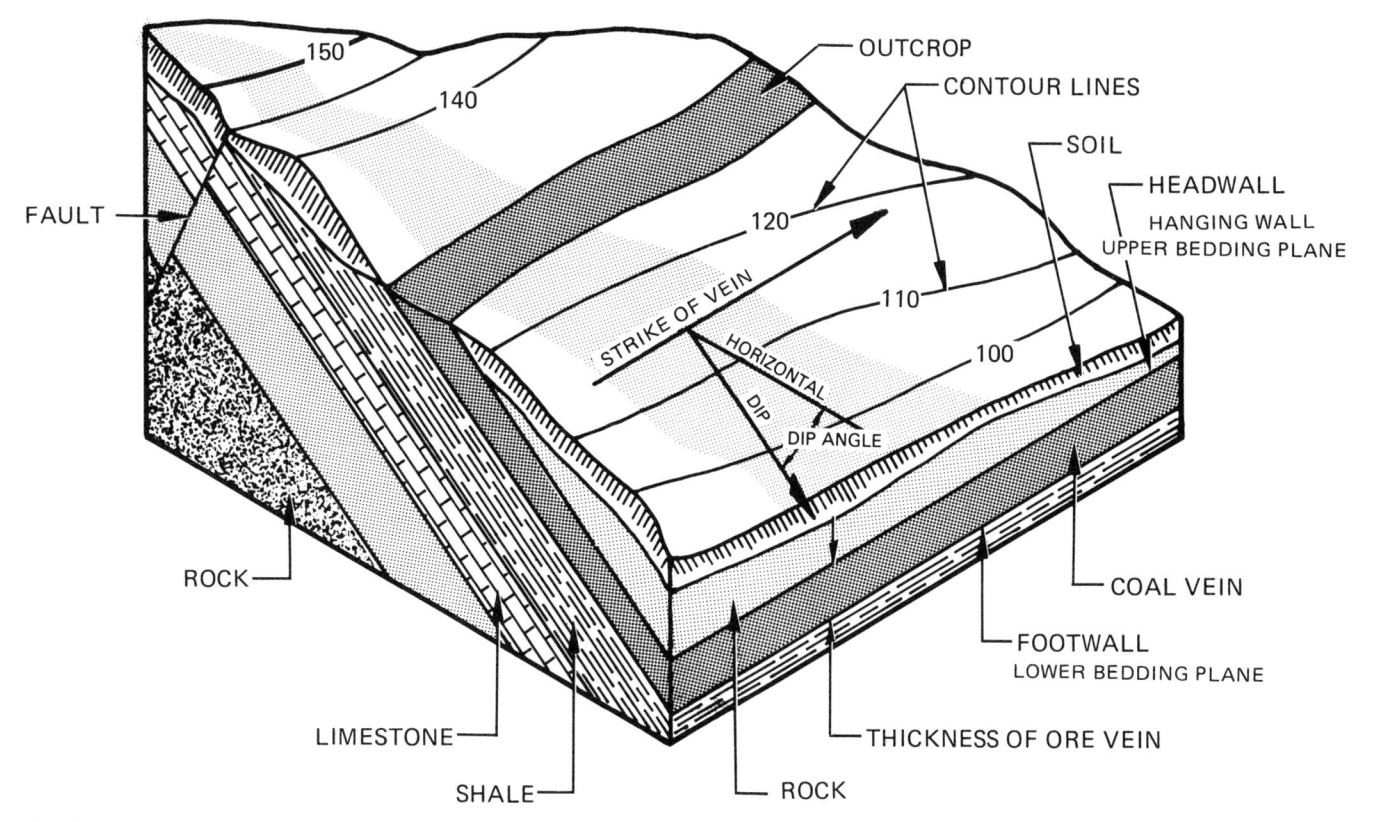

FIG. 8-12 Block diagram showing ore vein, outcrop, and strata.

8.10 MINING APPLICATIONS

A geologist or geological engineer uses concepts and procedures from descriptive geometry to solve various types of mining and geology problems. The earth's surface is covered with a thin layer of soil and vegetation. Beneath this layer lie a series of stratified layers (*strata*) of sedimentary rock. These layers were formed of mechanical (sandstone, limestone), chemical (salt, gypsum), or organic (coal) sediment. The process of *sedimentation* includes the transportation and the forming/solidification (cementing, bonding) of the sediments into solid layers/beds of rock. Sedimentary rock was formed in the earth's oceans, and subsequent upheavals and disturbances have faulted, sheared, folded, tilted, fractured, and distorted the original plane tabular formations. Therefore most strata are inclined and cover only limited areas, Fig. 8-12. The upper and lower surfaces of these strata are assumed to be more or less parallel with a uniform thickness, within limits. Layered rock formations/beds/strata/veins may contain valuable minerals, especially at the intersection of two strata. Veins of valuable minerals will sometimes fill openings and fissures between intersecting strata. A vein of this type can have any type of configuration. Other strata, for instance coal beds, are normally bounded by parallel surfaces and will at times intersect the surface of the earth along *outcroppings*, this type of stratum is also referred to as a "*vein*." The line along which a vein or stratum intersects the surface is called an *outcrop line*. Identification and location of outcrops plays an important part in the finding and mining of valuable ore deposits.

Contour maps are used to describe and establish the limits of a particular deposit of rock. The strike, dip/slope, and thickness of a stratum of rock describe its physical orientation. Since a stratum is a sheetlike mass of sedimentary rock which lies between two stratum of different compositions, its strike can be determined by measuring the bearing of a level line on either of its bounding surfaces (*upper bedding plane/headwall, lower bedding plane/footwall*). The surface of a bedding plane can be located by three or more points. Points on the upper or lower plane surfaces are found by drilling *boreholes*. The slope of a deposit/stratum is established by finding the angle it makes with the horizontal plane. The slope is referred to as the dip of a plane and includes the general direction of its tilt. Therefore, the dip of a vein/stratum includes its slope angle and dip direction. The dip direction is established by drawing a line perpendicular to the strike line in the direction of tilt. The strike and dip are measured as compass directions deviating from a north/south line towards the east or west. The strike is always given from the north towards the east or west, whereas the dip is given from the north or south depending on its orientation to the strike and low side of the plane.

8.11 MINING AND GEOLOGY TERMS

The following terms are used throughout the chapter as applied to mining and geology. The block diagram in Fig. 8-12 illustrates many of the terms. Figures on the following pages provide examples and applications of the terms and definitions presented here.

1. **Bedding plane:** The upper or lower surface of a stratum or vein. The thickness of a stratum is measured as the perpendicular distance between two bounding bedding planes. The location of a bedding plane of a stratum can be determined by locating three or more points on its surface. Bedding planes are considered to be plane surfaces and parallel to one another.
   Upper bedding plane: The top surface of a stratum/vein. Also known as the *hanging wall* or *headwall*.
   Lower bedding plane: The bottom surface of a stratum, usually called the *footwall*.
2. **Borehole:** Holes drilled from the surface are called boreholes. A borehole establishes the depth below the surface of a point on a bedding plane. Three points on the surface of a bedding plane define a plane which can be described by its *strike* and *dip*.
3. **Contour:** A line representing the surface of the earth at a specific elevation. Contour lines are drawn by connecting points established during the survey that are at a constant elevation. Contours are lines of intersection of horizontal planes with the natural terrain.
4. **Dip:** The dip of a plane is its general direction of tilt combined with its angle of slope. The dip is stated as a slope angle and its direction of slant toward the low side of the plane; e.g., 30° SW means the plane slopes 30 degrees toward the southwest.
5. **Dip angle:** The slope of a plane when referring to geological formations.
6. **Dip direction:** The general direction of tilt of a plane, measured as the compass direction of a line drawn perpendicular to the strike line toward the plane's low end/side.
7. **Fault:** A fracture and slippage of the earth's crust. A fault will sever all strata along its fault line. The plane formed by the fracture is called the fault plane. A stratum cut by a fault will be displaced in a direction parallel to the fracture.
8. **Footwall:** The *lower bedding plane* of a stratum.
9. **Headwall:** The *upper bedding plane* of a stratum. Also known as the *hanging wall*.
10. **Hanging wall:** Headwall/upper bedding plane of a stratum.
11. **Outcrop:** A portion of a rock stratum/formation that is at the surface is called the outcrop. An outcrop is an area bounded by the line of intersection of the upper and lower bedding planes and the surface of the ground. An outcrop can be seen at the surface if the soil covering has been removed by erosion. Outcrops are used to identify the location of valuable mineral deposits, which many times can be surfaced mined.
12. **Outcrop line:** The outcrop line defines the extent/area of an outcrop. If both surfaces of a vein/stratum intersect the ground as an unconsolidated deposit of bedrock, two outcrop lines can be plotted on a contour map.
13. **Stratum/strata:** A sheetlike layer of rock, normally bounded by two strata of differing compositions. The thickness of a stratum is consistent within a limited area. Strata are usually at an angle to the horizontal and may be faulted, bent, and otherwise distorted by the movement of the earth's crust throughout geological history. The upper plane surface is its upper bedding plane. Also called a headwall or hanging wall. The lower bounding surface is the lower bedding plane/footwall.
14. **Sedimentary rock:** Layered formations of mineral or rock strata formed by the depositing of sediment by water movement. Sedimentary rock can be of chemical, organic, or mechanical origination.
15. **Shaft:** An opening/passage to a mine. Normally vertical or slightly inclined to the surface. The shaft provides access to the tunnels and workings of a mine.
16. **Slope:** The angle that a plane makes with the horizontal. In mining, the slope angle of a plane is called its dip angle.
17. **Strike:** The compass direction of a level line on a plane. A strike is the bearing of a horizontal/level line (true length in the horizontal view) lying on the surface of a plane. The strike is measured from the north towards the east or west. The strike of a stratum establishes its orientation on a contour map.
18. **Thickness of vein:** The thickness of an ore vein or rock strata is the perpendicular distance between bedding planes.
19. **Tunnel:** In mining, a horizontal or slightly inclined excavation used to mine below the surface. A tunnel opens at one end to a shaft or to daylight on the side of a hill or mountain.
20. **Vein:** A bed, stratum, or deposit of useful or valuable mineral matter (gold, silver, uranium, coal, etc.). A vein can be bounded by parallel bedding planes in the form of a stratum or coal vein, or can take the shape of a fissure or crack it fills.
21. **Works/workings:** An excavation or group of excavations associated with mining, tunneling, or quarrying.

FIG. 8-13(1) Strike, dip, and thickness of ore vein.

8.12 STRIKE, DIP, AND THICKNESS OF AN ORE VEIN

The *strike, dip,* and *thickness* of an ore vein is determined by drilling four bore holes to establish three points on the upper surface (upper bedding plane/headwall) of the stratum and one on the lower surface (lower bedding plane/footwall). The upper and lower surfaces are assumed to be parallel. Points 1, 2, 3, and 4 are thus located on the plan/contour map in Fig. 8-13(2). Note that point 1 is at the surface, therefore it is on the outcrop line. To determine the strike, dip, and thickness of the plane, draw a frontal view showing plane 1-2-3 and point 4 using the elevations of the four points to set up the view. Label the low side of the plane and elevation 900 ft at point 1 on the outcrop line.

The strike of a plane is a bearing measurement of a horizontal (level) line on the plane. The strike is always measured from the north as a compass direction. In Fig. 8-13(1), plane 1-2-3 has been established as three points on the upper surface of an ore vein. Level (horizontal) line 2-5 is drawn parallel to H/F in the frontal view and projected to the horizontal view where it appears as true length. The strike is measured as N 78°E.

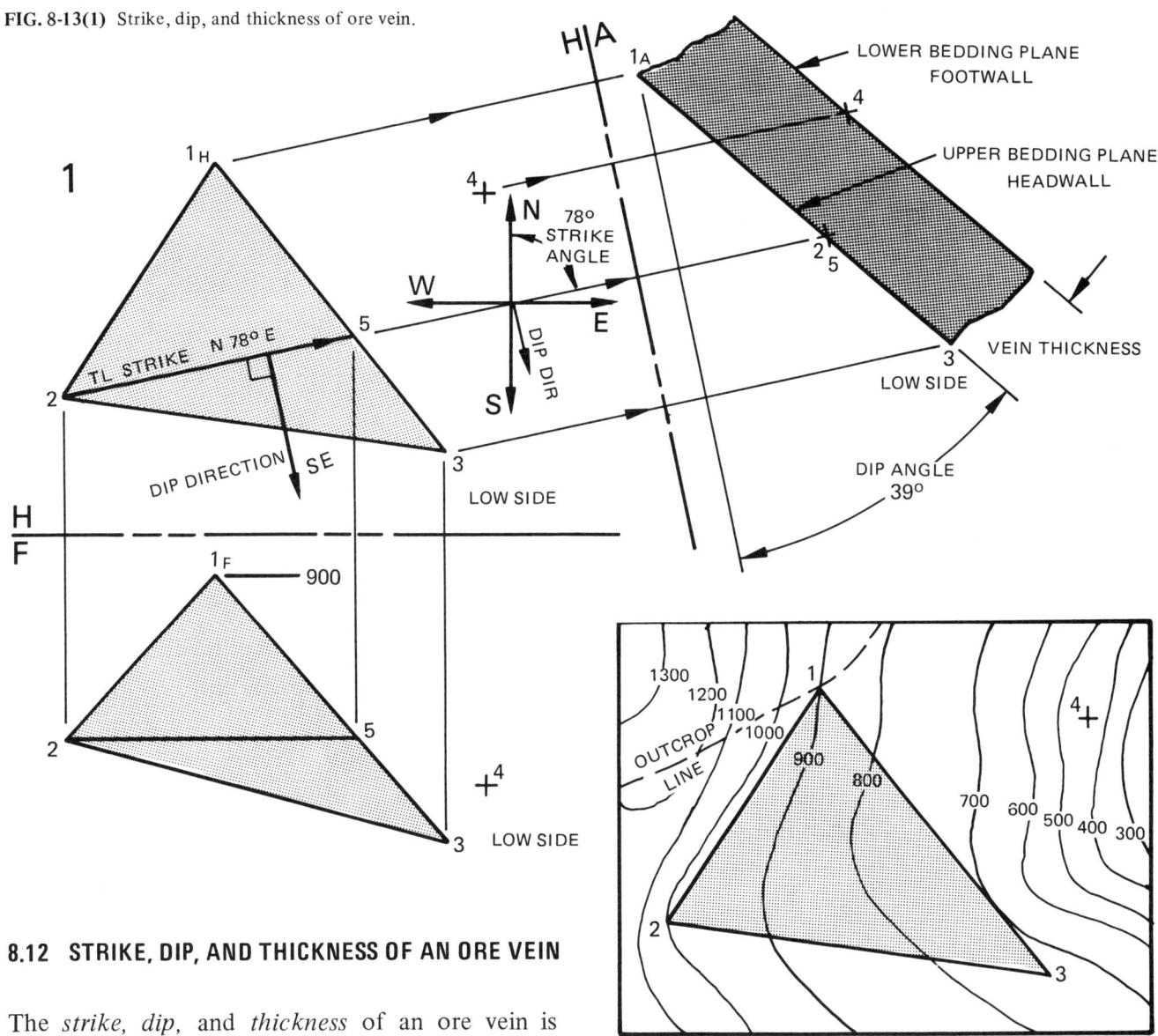

FIG. 8-13(2)

The dip of a plane is a geologist's term for its slope. The dip is measured as an angle the plane makes with the horizontal. Draw H/A perpendicular to horizontal line 2-5 and project auxiliary A. Plane 1-2-3 appears as an edge. Point 4 on the lower bedding plane determines a plane drawn parallel to and beneath plane 1-2-3 (upper bedding plane). The thickness is the perpendicular distance between the two planes, measured as shown.

The dip angle is 39° measured as a slope angle from the horizontal. The dip direction is the direction of slope of the plane. The dip direction is determined by drawing a line perpendicular to the strike line towards the low side of the plane in the plan view. The general compass direction of this line is the dip direction, SE in the example. The dip is normally stated as a combination of the dip angle and dip direction, 39°SE.

FIG. 8-14 Strike, dip, and outcrop of vein.

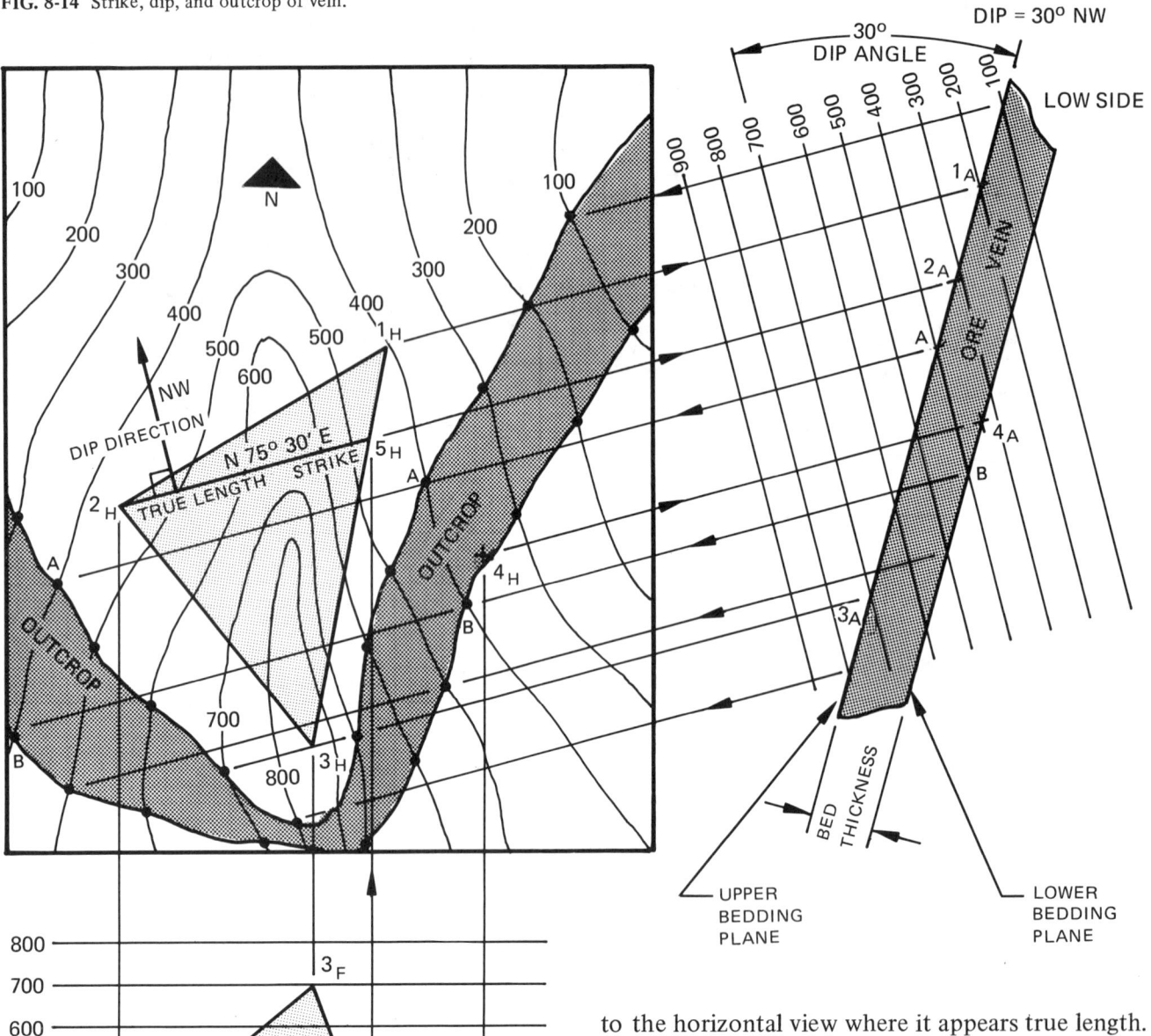

8.13 STRIKE, DIP, AND OUTCROP OF AN ORE VEIN

Test borings are often used to locate strata of valuable minerals. The data obtained can be used to plot the location of the ore vein on the ground surface, which may be hidden by top soil.

In Fig. 8-14 three points have established the upper bedding plane. A borehole at point 4 located a point on the lower bedding plane. The frontal view is drawn using the elevations of points 1, 2, and 3 to fix the location of the plane. Horizontal line 2-5 is drawn in the frontal view and projected to the horizontal view where it appears true length. The bearing of level line 2-5 equals the strike of the plane, N 75°30'E. A view taken perpendicular to line 2-5 shows the plane as an edge. Contour intervals appear as evenly spaced horizontal lines in this view. Point 4 is on the footwall. The lower bedding plane is drawn through point 4 and parallel to the edge view of plane 1-2-3. The upper and lower bedding planes are extended so as to show the intersection of the stratum/vein with each contour interval. The thickness of the stratum/bed is measured as a perpendicular distance between the upper and lower bedding planes.

The outcrop area is determined by projecting each point of intersection of a bedding plane and a contour line to the plan view as shown. The upper bedding plane intersects the 400-ft level contour at point A. Point A is projected to the plan view, where it intersects the 400-ft contour at two points along the outcrop line. Point B is established by the intersection of the lower bedding plane and the contour at 400-ft elevation.

FIG. 8-15 Outcrop and distances to ore vein.

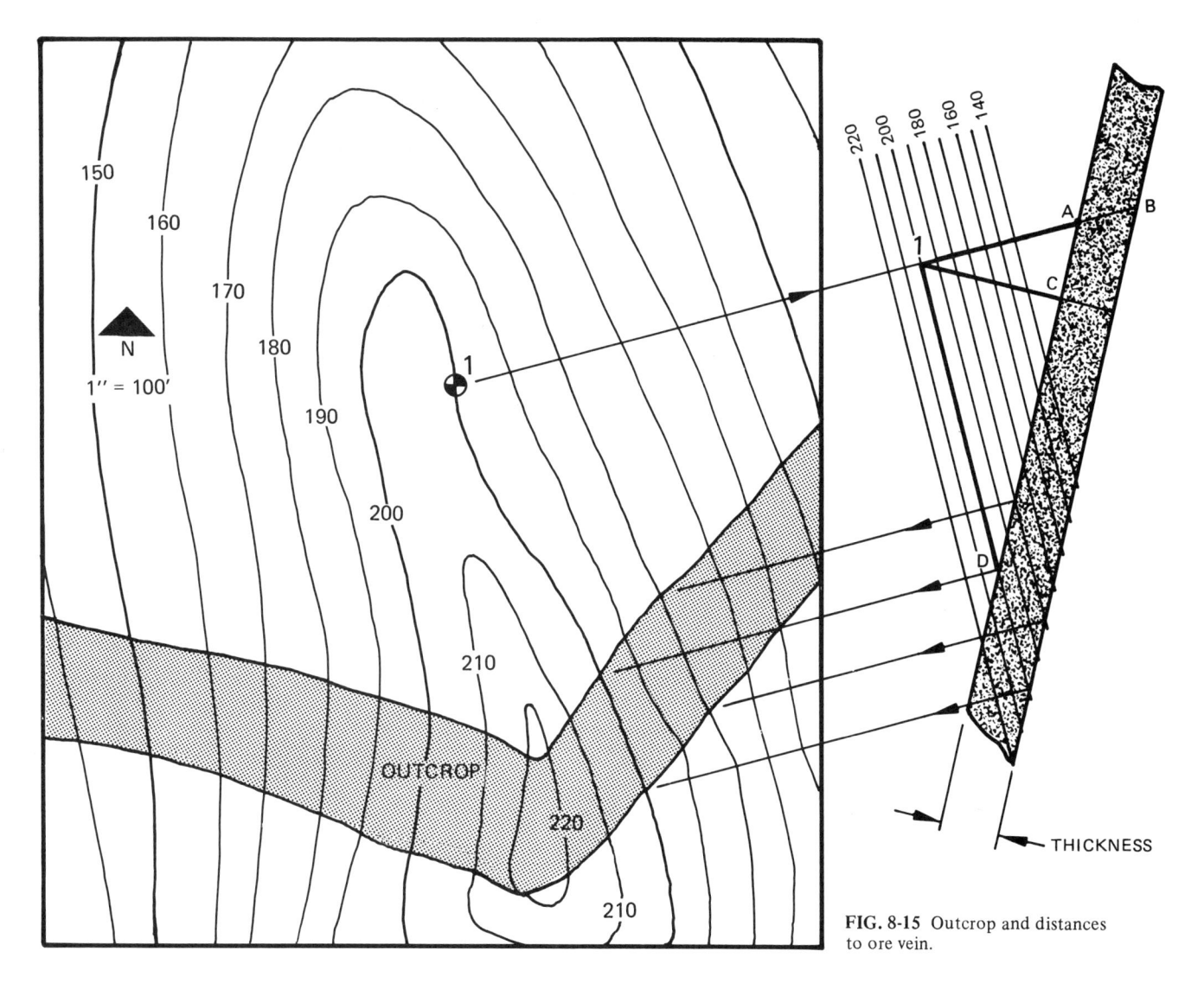

8.14 OUTCROP AND DISTANCES TO ORE VEIN

The strike and dip of a stratum along with a point on its bedding plane can be used to establish its outcrop in the plan view. The vertical, horizontal, and shortest distance from a point on the earth's surface to the upper bedding plane can be determined in a view showing the stratum as an edge.

In Fig. 8-15, a test boring has been made at point 1 to determine the position of a stratum of coal. The strike and dip are know to be N 76°E and 28°NW respectively. The top/headwall of the vein was found to be 100 ft below point 1. Continued drilling located the bottom/footwall of the vein at 140 ft below point 1.

Plot the outcrop of the coal vein and label it. Determine the shortest perpendicular distance and the shortest horizontal distance from point 1 to the headwall of the vein. Measure the thickness of the vein. Point 1 is given in the plan view.

The elevation view is constructed by drawing a line from point 1 in the plan with a bearing of N 76°E. The contour intervals are drawn perpendicular to this strike line. The vertical and horizontal scale are equal. Point 1 is located on the 200-ft contour line in the elevation view. The upper bedding plane is located at 100 ft below point 1, point A. The lower bedding plane is established 140 ft below point 1, point B. The upper bedding plane is drawn through A at 28°, with its low end determined in the plan as NW (perpendicular to the strike line). The lower bedding plane is drawn parallel to the upper bedding plane. The thickness can be measured as shown, 35 ft.

The shortest vertical distance is given as 100 ft. The shortest perpendicular distance is measured along a line drawn perpendicular to the upper bedding plane, line 1-C, 88 ft. The shortest horizontal distance is a line drawn from point 1 parallel to the contour lines, and in this case it falls on the 200-ft contour level, line 1-D, 188 ft.

The outcrop area is determined as in Fig. 8-14. Each intersection of the bedding plane and the horizontal contour lines in the elevation view is projected to the plan/contour map. These points lie on the contour and the outcrop line. Therefore they establish the intersection of the coal bed with the surface of the ground.

PROB. 8-13

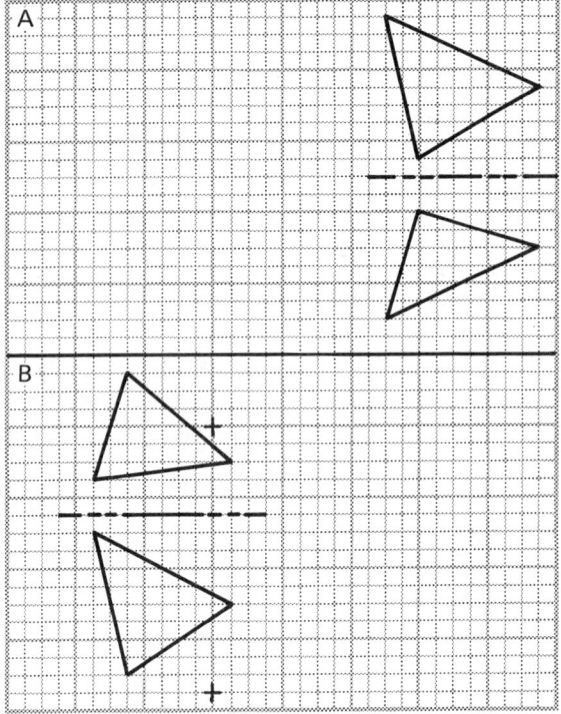

PROB. 8-14

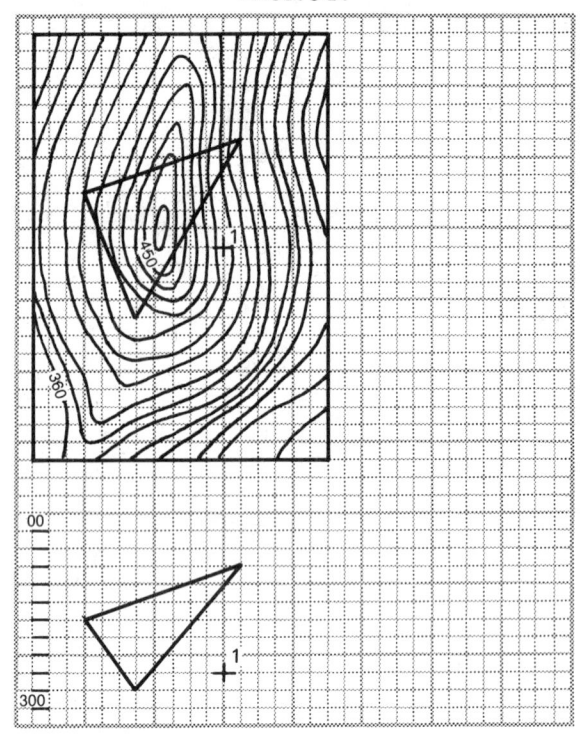

PROB. 8-15

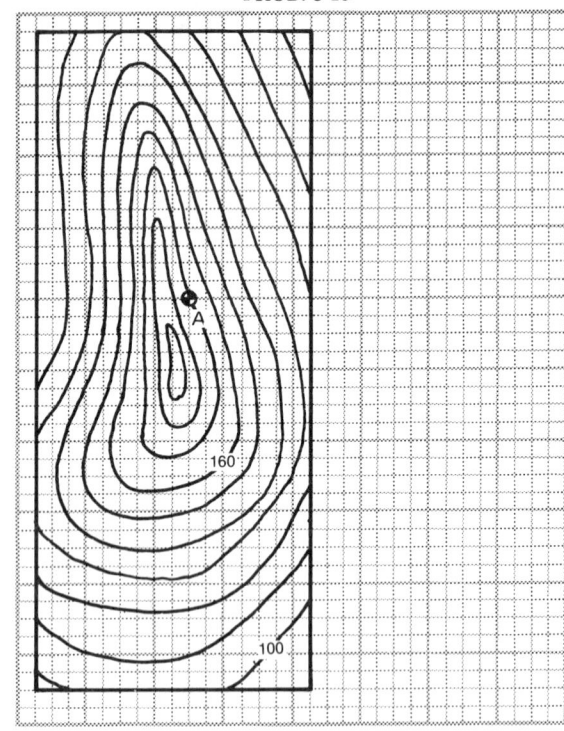

PROB. 8-13A. Solve for the strike and dip of the ore vein plane. Show all notation and label completely including low side, dip angle, dip direction, and directional arrows.

PROB. 8-13B. Do the same as for problem A. The given point represents the lower bedding plane. Show the thickness of the vein in the auxiliary view.

PROB. 8-14. The plane represents three points of the top bedding plane of an ore vein. The given point is on the lower bedding plane. Give the strike, thickness, and dip of the plane. Using an auxiliary view solve for the thickness and plot the outcrop area in the plan view. Shade the outcrop area for clarity. The scale is 1″ = 40 ft.

PROB. 8-15. Test borings are often used to locate strata of valuable minerals. The data obtained can be used to plot the location of the intersection of the ore vein and the ground surface, which may be hidden by topsoil. In this problem a test boring has been made at point A to determine the position of a stratum of coal. The strike and dip are known to be N 77°W and 23°NW respectively. The top of the vein was found to be 30 ft below point A. Continued drilling located the bottom of the vein at 50 ft below point A. Plot the outcrop of the coal vein and label it. Determine the shortest perpendicular distance from point A to the top of the coal vein, the shortest horizontal distance from point A to the top of the vein, and the thickness of the vein. Label the drawing completely. Scale is 1″ = 40 ft and the contour line spacing is at 10 ft. Show the outcrop area in the plan view and shade.

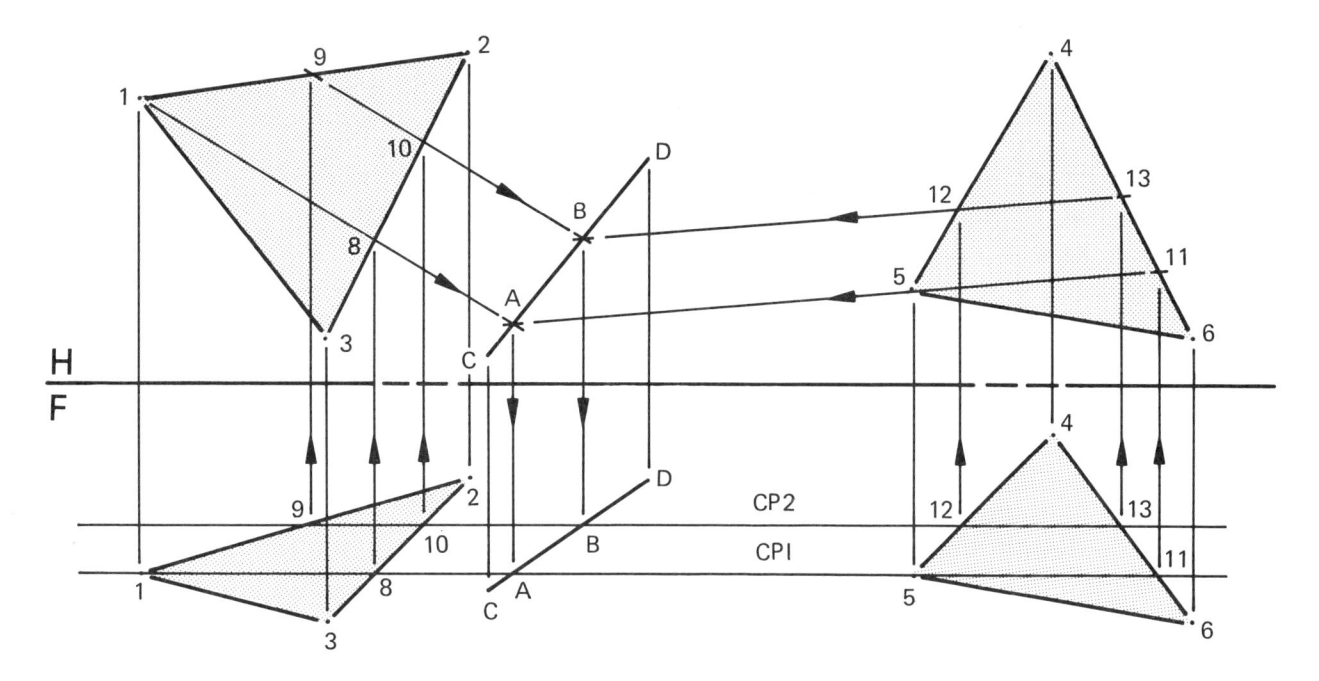

FIG. 8-16 Intersection of two unlimited planes (mineral strata).

8.15 INTERSECTION OF UNLIMITED PLANES (MINERAL STRATA)

The intersections of mineral strata are often important clues to the location of valuable ore deposits. If the position of two adjacent strata are unknown, their line of intersection can be determined by using the method for establishing the intersection of two infinite planes, Chapter 6, Section 6.6.

The intersection of two infinite planes is determined by passing a third plane through the given planes. This locates one point along the line of intersection. Two such cutting planes determine two points on the line of intersection. Connecting the two points establishes the line of intersection.

In Fig. 8-16, planes 1-2-3 and 4-5-6 represent two strata/bedding planes. Two horizontal cutting planes, CP1 and CP2 are passed through both planes in the frontal view. Intersection points of the cutting plane and the given planes are projected to the horizontal view.

In the horizontal view line 1-8 and line 5-11 are extended until they intersect at point A. Point B is located at the intersection of lines 9-10 and 12-13. Note that in the horizontal view the cutting planes cut parallel lines on the given planes. Each of the lines are true length/horizontal lines.

Points A and B are connected. Line A-B is the of intersection. Points A and B are located in the frontal view by projection. Line A-B is extended to show a longer line of intersection, line C-D.

Surveys and mapping are aided by the use of satellite and aircraft photographs, Fig. 8-17.

FIG. 8-17 Satellite photo used for mapping and geological surveys. *(Courtesy NASA.)*

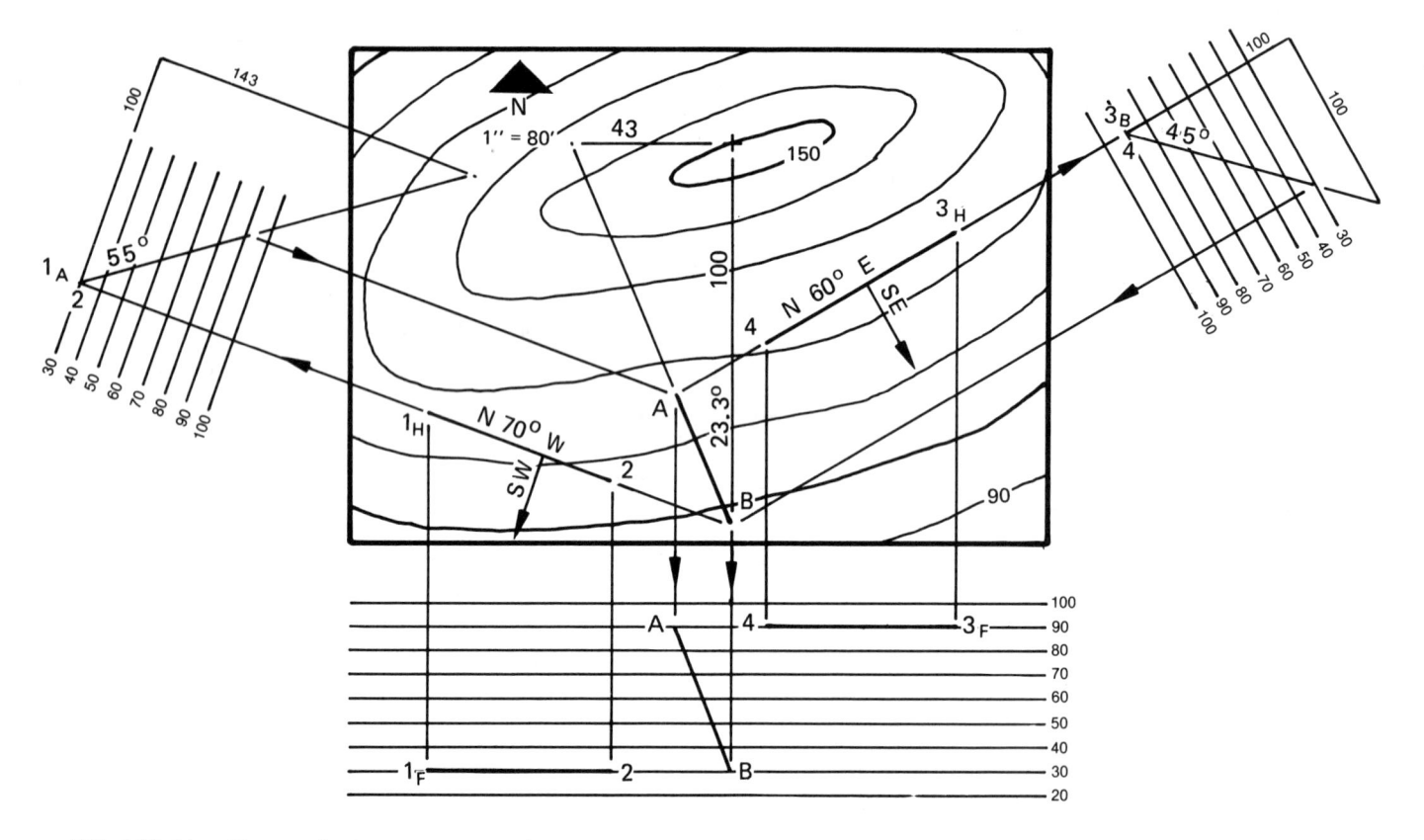

FIG. 8-18 Line of intersection between two ore veins.

8.16 LINE OF INTERSECTION BETWEEN TWO ORE VEINS

The intersection of two ore veins can be quickly established knowing only the strike, dip, and elevation of the strike line of each vein.

In Fig. 8-18, field data has located two ore veins that are represented by strike line 1-2 and 3-4. Both lines are given in the plan view. The dip of the vein at line 1-2 is 55°SW and at line 3-4 is 45°SE. Line 1-2 is at 30-ft elevation and line 3-4 is at 90-ft elevation. Determine the line of intersection (from elevation 30 ft to elevation 90 ft) and give its bearing.

Draw the frontal view by constructing the contour intervals as shown. Lines 1-2 and 3-4 are located in the frontal view by projection at their given elevations.

Lines 1-2 and 3-4 are true length in the horizontal view since they are strike lines (level). Project auxiliary views showing each as a point view. The auxiliary views are taken perpendicular to the strike line with a parallel line of sight. The contour intervals from elevation 30 ft through 100 ft are drawn perpendicular to the strike lines. The point view of line 1-2 is at elevation 30 ft and the point view of line 3-4 is at 90 ft.

The dip direction of each vein is drawn on the plan as a perpendicular to the strike line pointing towards the low side.

Draw a line from the point view of line 1-2 which makes an angle of 55° with the horizontal (the edge view of the contour lines). This line intersects each contour elevation and establishes the edge view of the vein. From the point view of line 3-4 draw a line at 45° to the horizontal. This line will also intersect all levels of the contour elevations and establish the edge view of its corresponding vein. Note that the grade can be used to construct the dip of the veins, 45° equals a grade of 1:1.

The line drawn from the point view of line 1-2 intersects contour elevation 90. This intersection point is projected to the plan view to where it intersects extended line 3-4 thus establishing point A. Line 3-4 is also at elevation 90 ft. Point A is one point along the line of intersection between the two veins. Point B is located by projecting the intersection point of the line drawn from the point view of line 3-4 where it crosses the 30-ft contour elevation line, to the plan view. It intersects extended line 1-2 at point B.

Both point A and B are located in the frontal view by projection. Line A-B represents the line of intersection between the two veins. Its bearing is measured in the plan view as shown.

FIG. 8-19 Intersection of two planes.

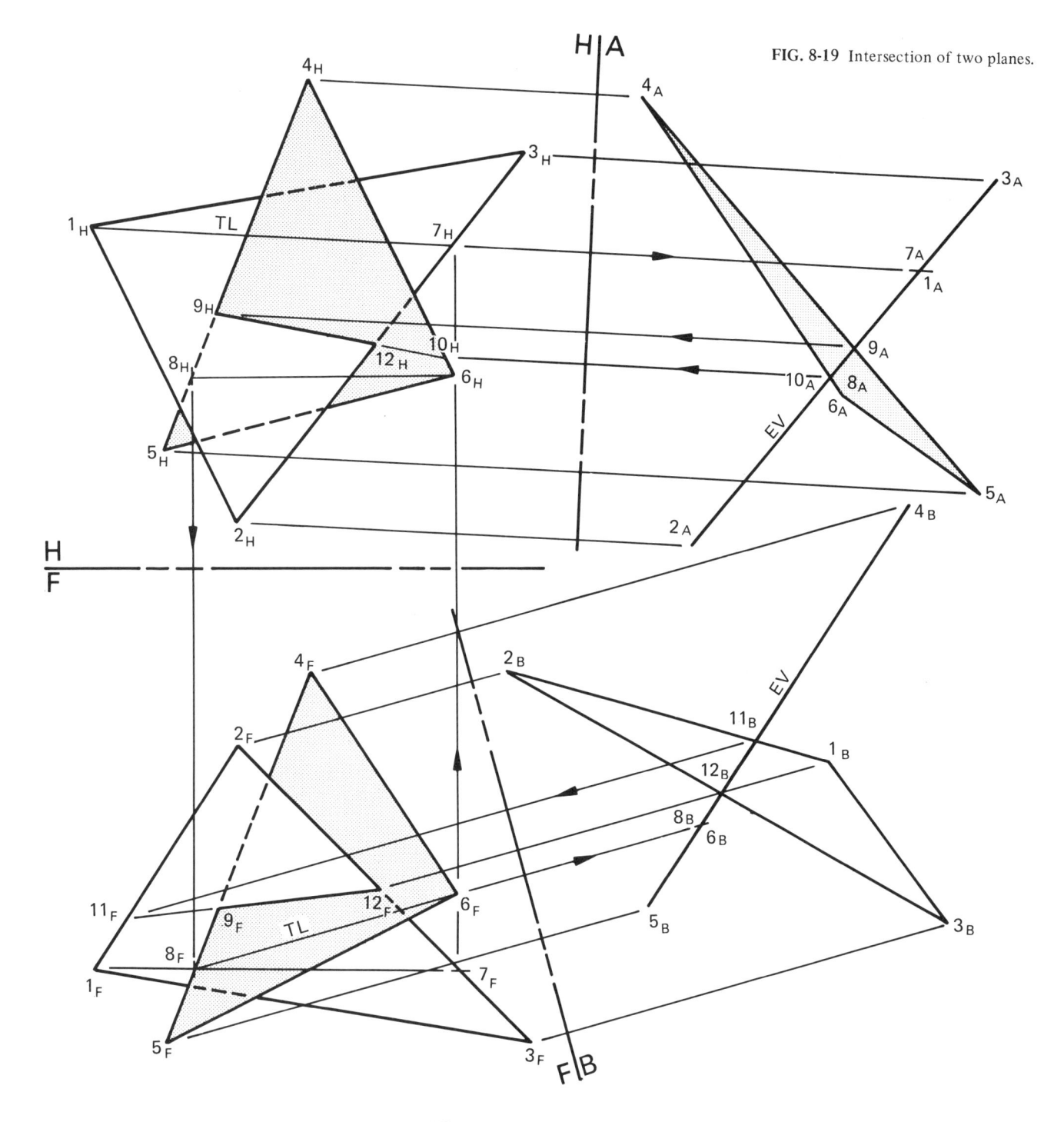

8.17 LINE OF INTERSECTION BETWEEN TWO OVERLAPPING ORE VEINS

The intersection of two or more overlapping strata or ore veins can be determined by methods provided in Chapter 6, Intersection of Planes. In Fig. 8-19, this procedure is illustrated using the edge view method. Planes 1-2-3 and 4-5-6 are given in the frontal and horizontal views. Determine the line of intersection using the edge view method.

1. Draw horizontal line 1-7 on plane 1-2-3 in the frontal view and project to the horizontal view as true length.

2. Draw H/A perpendicular to line 1-7 and project auxiliary A.

3. Plane 1-2-3 appears as an edge in auxiliary A. The edge view of plane 1-2-3 intersects the oblique view of plane 4-5-6 at points 9 and 10.

4. Project points 9 and 10 to the horizontal view. Line 9_H-10_H is the extended line of intersection. Line 9-10 intersects line 2-3 at point 12_H. Line 9_H-12 is the line of intersection.

5. Project points 9 and 12 to the frontal view and draw the line of intersection.

6. Determine visibility and complete the views.

An auxiliary view projected from the frontal view could have been used to establish the line of intersection, auxiliary B.

375

PROB. 8-16

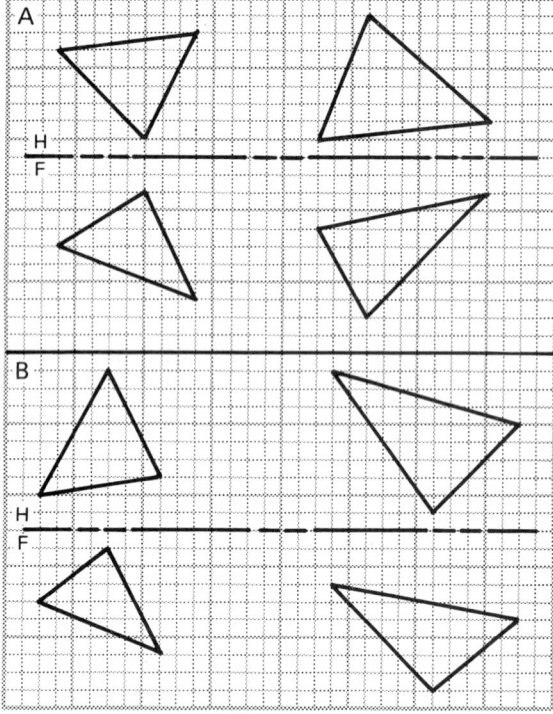

PROB. 8-17

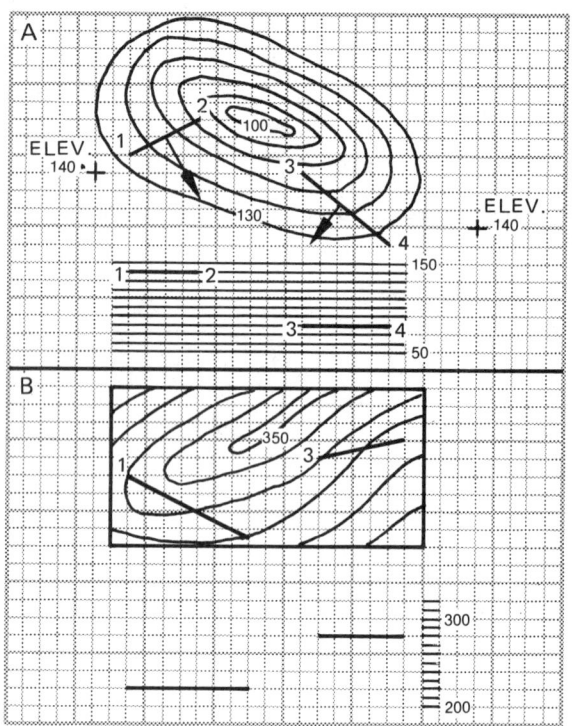

PROB. 8-18

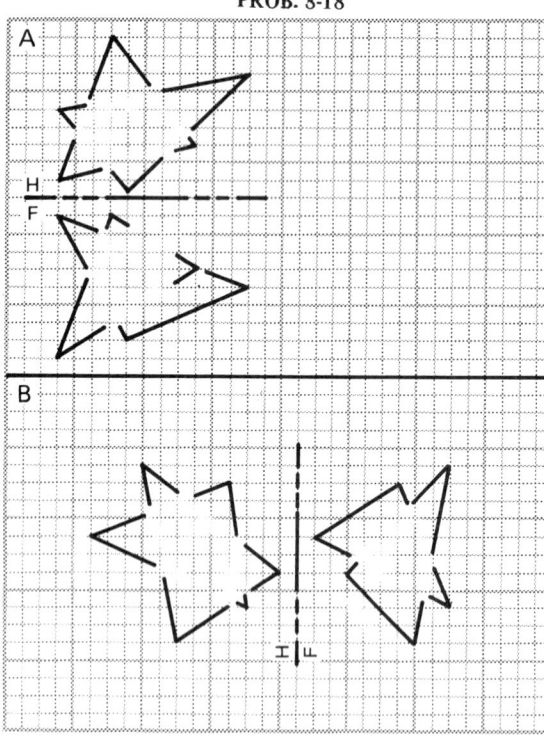

PROBS. 8-16A, 8-16B. The intersections of mineral strata are often important clues to the location of valuable ore. If the positions of two adjacent strata are known, their line of intersection is easily plotted. Using the cutting plane method, determine the line of intersection of the two infinite planes shown in each problem. Use two horizontal cutting planes.

PROB. 8-17A. Field data has located two ore veins that are represented by the strike lines 1-2 and 3-4. The dip of the vein established by line 1-2 is 50°SE and 40°SW using line 3-4. Determine the line of intersection between elevation 80 and 140 ft. Measure the bearing of the line of intersection.

PROB. 8-17B. The two given lines represent strike lines of ore veins. The dip of the vein at line 1-2 is 60°SW and at 3-4 is 48°SE. Determine the horizontal and frontal views of the line of intersection (from elevation 220 ft to elevation 280 ft) and give its bearing.

PROBS. 8-18A, 8-18B. Establish the line of intersection between the two overlapping planes. Use the edge view method for the problem A and the cutting plane method for problem B.

376

9

WARPED SURFACES

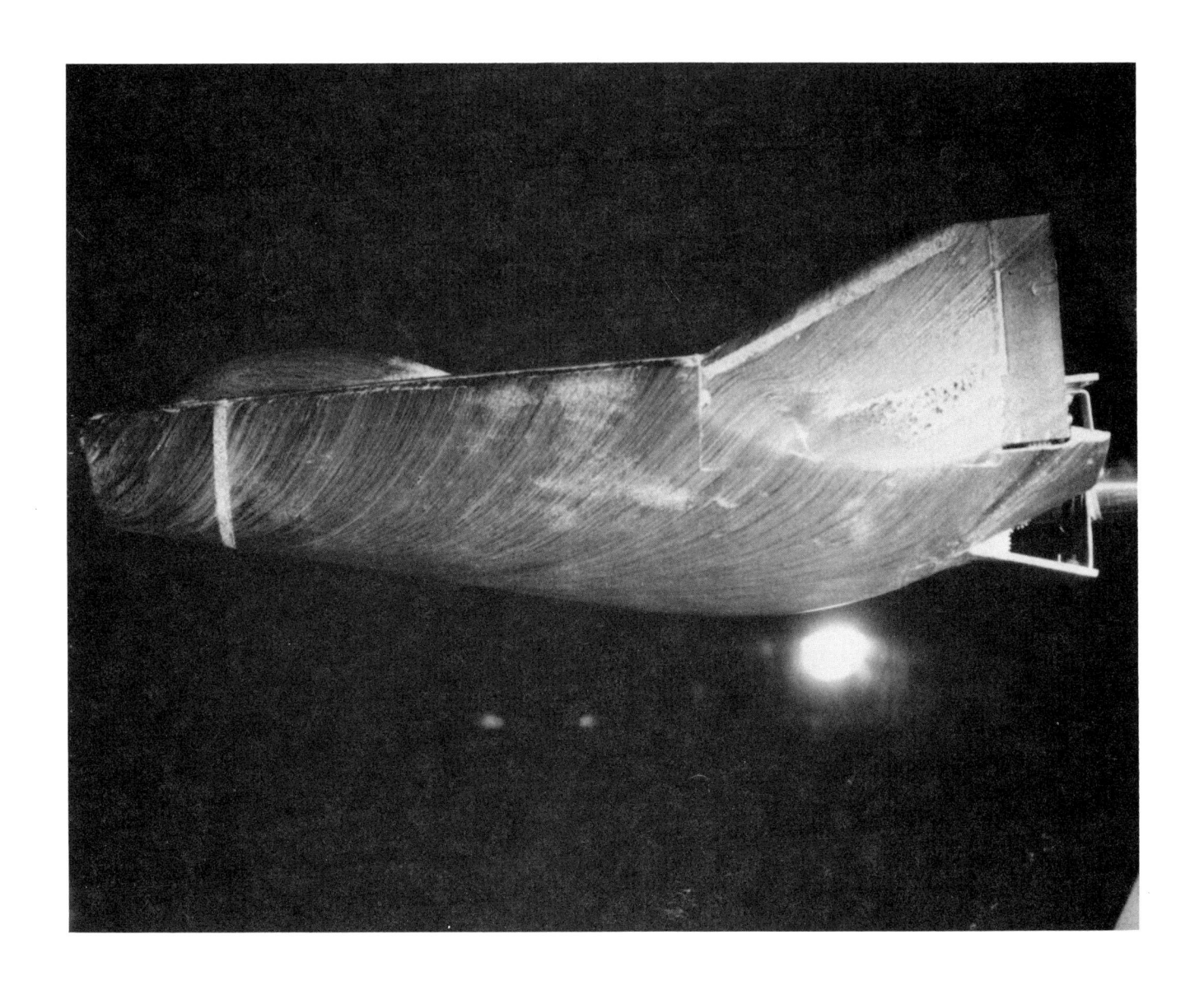

FIG. 9-1 The two cooling towers are hyperboloids of revolution.

9.1 WARPED SURFACES

A *warped surface* is a *ruled surface* whose adjacent elements are not parallel and do not intersect. **Warped surfaces** are generated by the movement of a straight line (generatrix) which does not lie in the same plane in any two consecutive positions. Solids bounded by warped surfaces include the cow's horn, conoid, cylindroid, warped cone, helicoid, and the hyperbolic paraboloid. An example of each is provided in this chapter. Note that this list is only a portion of the many variations of warped surfaces, most of which are based on these typical examples. The helicoid is covered in Chapter 10, The Helix.

A warped surface can only be approximately developed. Triangulation is used to construct a development of a warped surface as shown in Chapter 7, Section 7.39, Development of a Warped Transition Piece.

The intersection of lines or planes with warped surfaces can be solved using cutting planes and elements established on their surface. Methods provided in Chapter 7, Intersections, can be applied to the intersection of warped surfaces and lines, planes, or solids. Since a warped surface is a ruled surface, any point on its surface is located by drawing an element through the point. All elements of a warped surface will lie totally on the surface. Therefore a point is established on the surface by locating it on an element which lies on the surface.

Warped surfaces use three **directrices** to guide the movement of the generatrix (element). Normally one of the directrices is called a "*plane director.*" A generatrix will move parallel to the plane director and intersect the two remaining directrices. The two directrices may be curved or straight lines. Therefore two curved- or straight-line directrices and a plane director will determine the movement of the generatrix (straight-line element).

Warped surfaces can also be generated by three directrices. The three directrices may be any combination of straight-line or curved-line elements: hyperboloid of revolution, warped cone, cow's horn. Warped surfaces generated by two directrices and a plane director include the helicoid, cylindroid, conoid, and the hyperbolic paraboloid.

Warped surfaces are illustrated by showing their bounding edges, which include their directrices and the generated curved edges, along with a series of evenly spaced elements that lie on the surface.

In Fig. 9-1 the two cooling towers for the power plant were designed using hyperboloids of revolution as their surface configuration. In Fig. 9-2, the machinest is drilling reference marks on a semifinished propeller-blade model for a 6-inch airfoil machine. The blade is a portion of a warped surface.

FIG. 9-2 Propeller-blade model for airfoil machine. *(Courtesy NASA.)*

9.2 COW'S HORN AND WARPED CONE

A cow's horn is generated by the movement of a straight-line generatrix in contact with two curved-line directrices and one straight-line directrix. In Fig. 9-3 the cow's horn is generated by two parallel half circle directrices and a straight-line directrix which is perpendicular to the two curved directrices. Line A-B is the straight-line directrix which appears as a point view in the frontal view. A-B is located one half the distance between the center points, C and D, of the two semicircle directrices.

To draw the cow's horn, start by establishing a series of radially spaced cutting planes which originate at the point view of directrix A-B in the frontal view. The cutting planes intersect the circle directrices and establish elements between related intersection points. One CP intersects the directrix, whose center is at point C, at point 1 and the directrix, whose center is at point D, at point 1^1. These points are projected to the horizontal and profile views to establish one element on the surface. This process is repeated until a sufficient number of elements are established to draw a smooth-curved surface edge in the profile view.

FIG. 9-4 Warped cone.

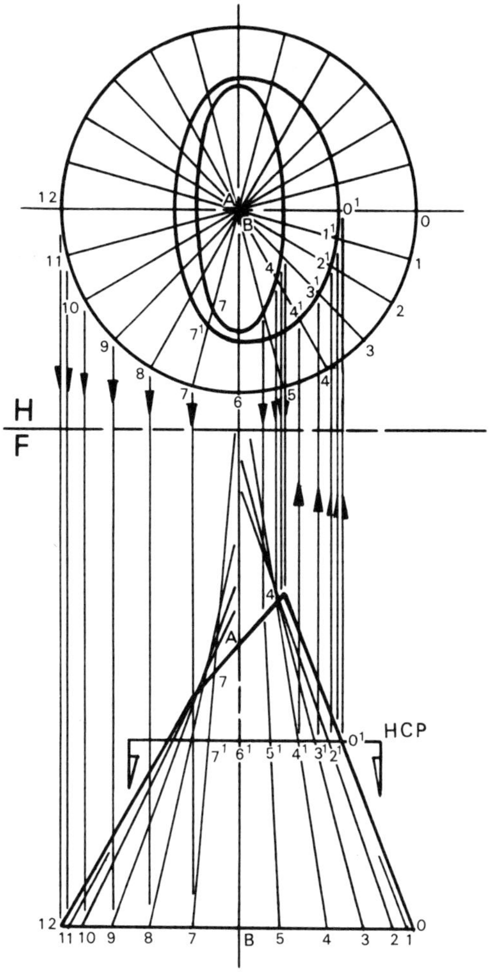

FIG. 9-3 Cow's horn.

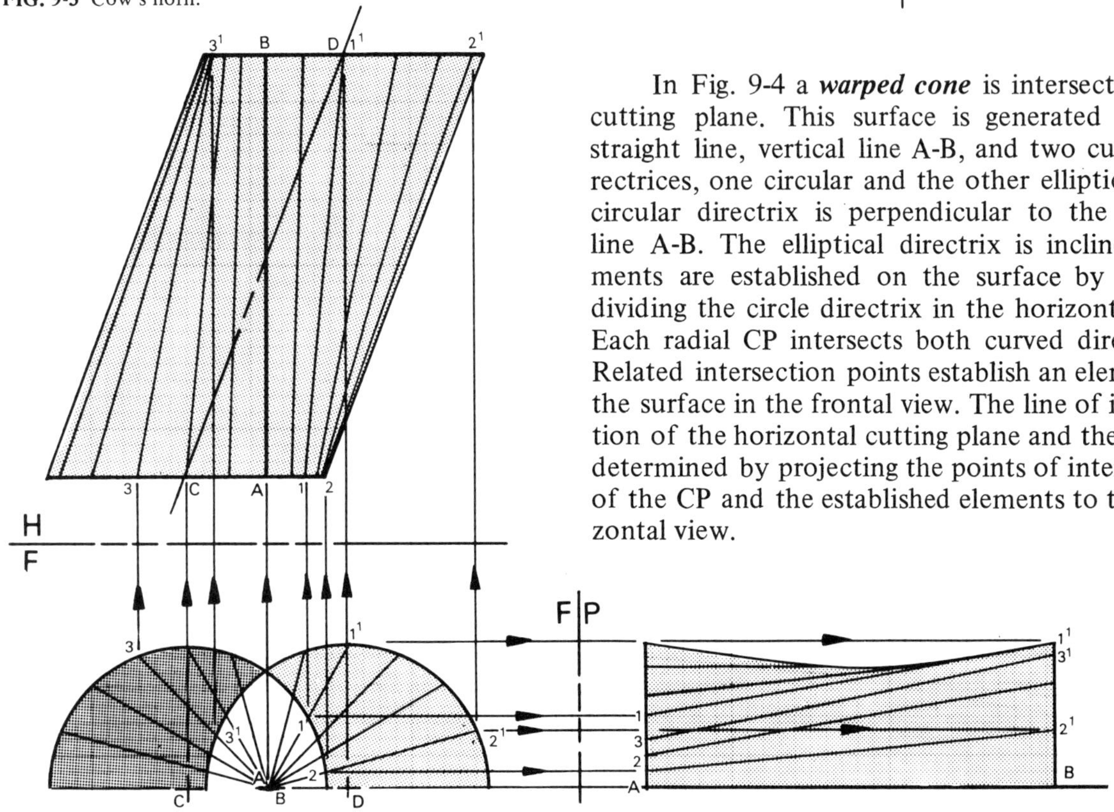

In Fig. 9-4 a *warped cone* is intersected by a cutting plane. This surface is generated by one straight line, vertical line A-B, and two curved directrices, one circular and the other elliptical. The circular directrix is perpendicular to the vertical line A-B. The elliptical directrix is inclined. Elements are established on the surface by radially dividing the circle directrix in the horizontal view. Each radial CP intersects both curved directrices. Related intersection points establish an element on the surface in the frontal view. The line of intersection of the horizontal cutting plane and the cone is determined by projecting the points of intersection of the CP and the established elements to the horizontal view.

9.3 THE CONOID

*A **conoid** is a warped surface generated by the movement of a straight-line generatrix which intersects a curved directrix and a straight-line directrix and moves parallel to a plane director.*

In Fig. 9-6, the given warped surface is a right conoid, since its straight-line directrix A-B is perpendicular to the plane director. In the frontal and profile views all elements on the surface are parallel to the plane director. In the horizontal view all elements converge at the point view of directrix line A-B. Each element intersects both the curved and straight-line directrices.

Section A and Section B represent vertical planes that intersect the conoid parallel to the curved and straight line directrices and perpendicular to the plane director. Each section cuts an elliptical line of intersection. Section A intersects element 3 in the horizontal view. This point of intersection is transferred to the profile view using

FIG. 9-5 Portions of air and spacecraft are sometimes composed of warped surfaces. *(Courtesy NASA.)*

D1. D2 is used to transfer the intersection of Section A and element 4. The same procedure is used to determine points along the line of intersection of Section B and the elements with which it intersects in the horizontal view. All points are located in the profile view and a smooth curve is drawn through them to complete the line of intersection.

The space shuttle concept illustrated in Fig. 9-5 is an example of warped surfaces used in the design of the body configuration of a spacecraft.

FIG. 9-6 Warped surface: the conoid.

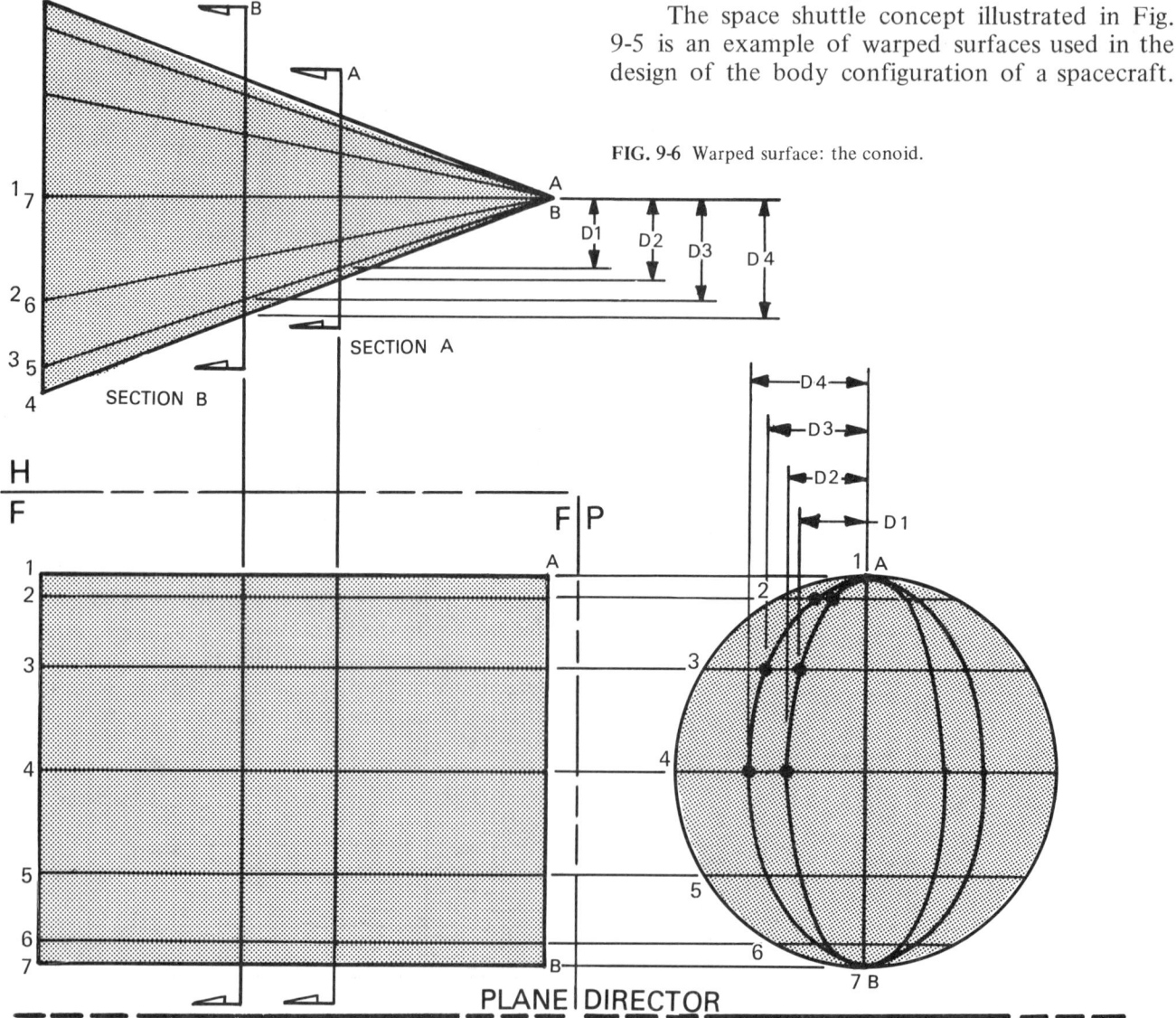

PROB. 9-1

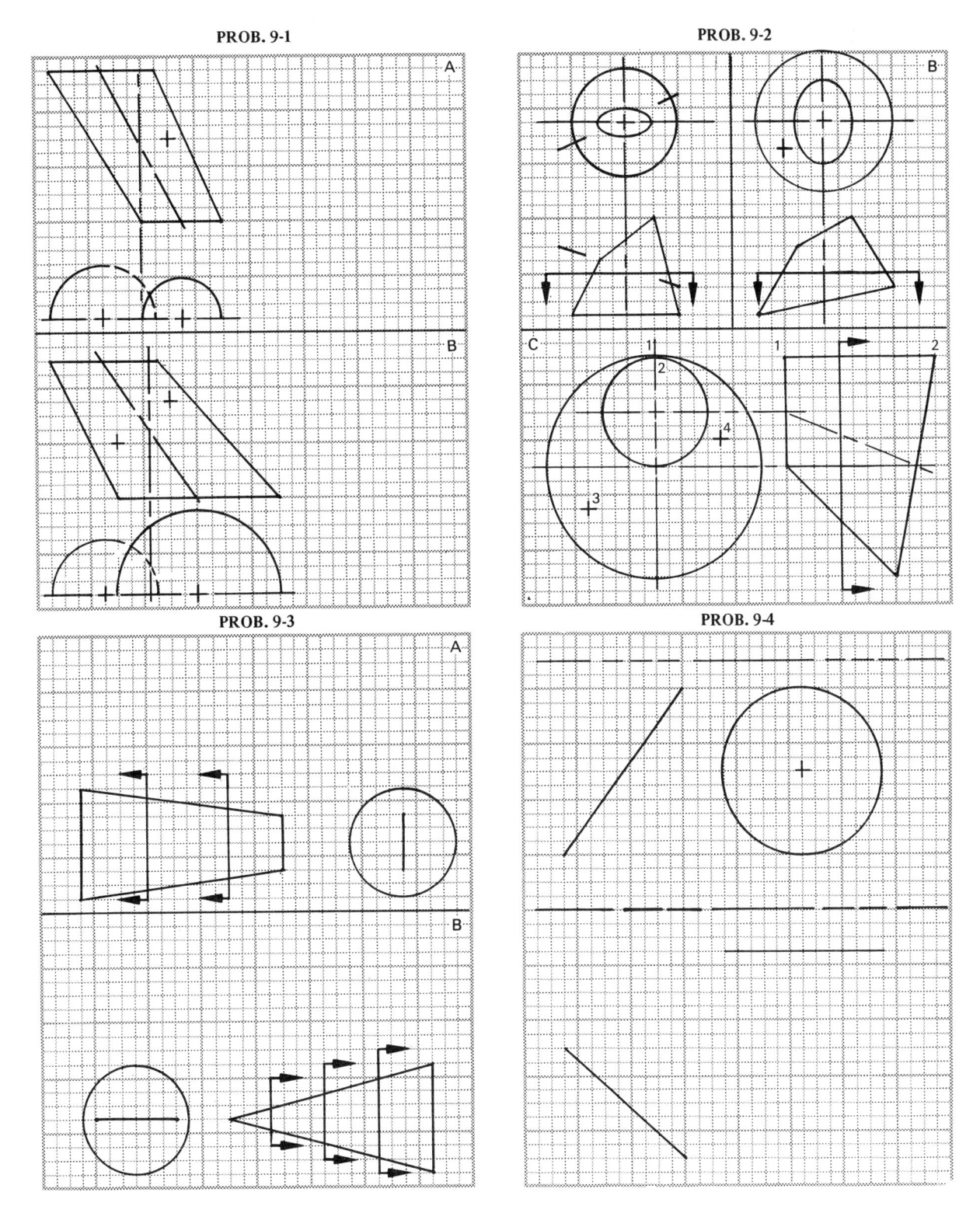

PROB. 9-2

PROB. 9-3

PROB. 9-4

PROB. 9-1A. Complete three views of the cow's horn using the two curves as its directrices. Show both visible and hidden elements. Use 12 straight-line evenly spaced elements and locate the given point on the figure in all views.

PROB. 9-1B. Use 24 equally spaced straight-line elements to draw the three views of the cow's horn. The two given points represent the intersection of a line and the figure, establish the line and points of intersection in all views. Do not show hidden elements.

PROB. 9-2A. Section the given warped cone using 24 evenly spaced straight-line elements. Establish the piercing points for the line and figure in both views.

PROB. 9-2B. Draw the required section in the top view and show the given point on the warped cone in the front view.

PROB. 9-2C. Use axis 1-2 as the directrix and complete the warped cone in both views. Show the required section and the intersection of line 3-4. Points 3 and 4 are on the cone's surface.

PROB. 9-3A, 9-3B. Using 24 elements, draw all three views of the conoids and sections.

PROB. 9-4. Draw the oblique conoid using 24 straight line elements evenly spaced on the given curve. The plane director is given.

FIG. 9-7 Warped surface: the cylindroid.

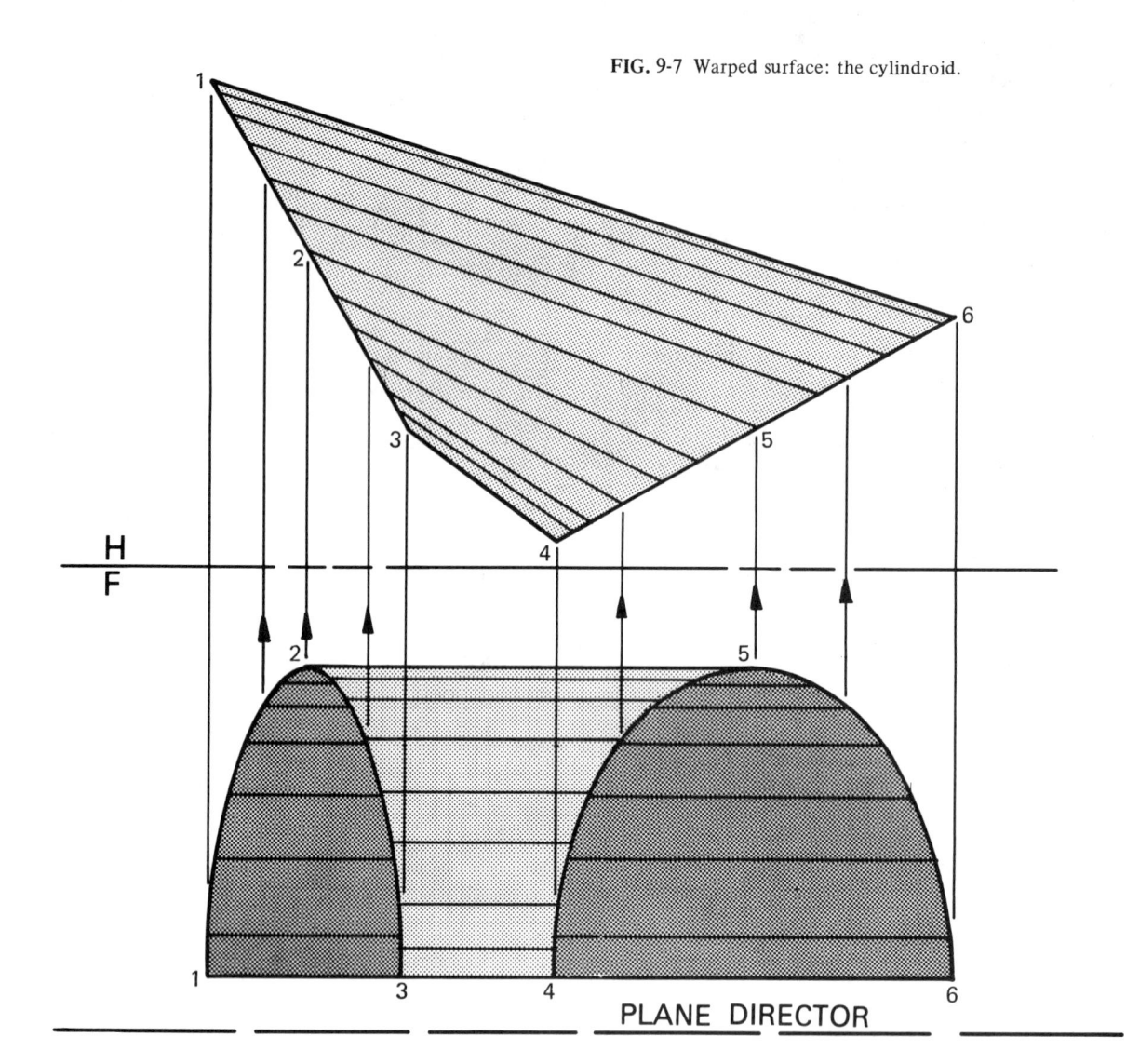

9.4 THE CYLINDROID

A cylindroid is generated by the movement of a straight-line generatrix which intersects two curved directrices and remains parallel to a plane director. The curved directrices can be any type of curved line but must not be parallel.

The elements of a cylindroid are parallel to the plane director, therefore the extreme point on the curved directrices are at the same elevation and equal in distance from the plane director. In Fig. 9-7, points 2 and 5 are the extreme points of the cylindroid. Both are the same distance from the plane director. Therefore element 2-5 is parallel to the plane director as are all other elements.

A cylindroid is represented by drawing the elements of its surface. The cylindroid in Fig. 9-7 is easily constructed by drawing a series of parallel elements in the frontal view. Each element is then projected to the horizontal view. The positions of the elements in the horizontal view illustrate the warp of the cylindroid.

Note that the curved directrices may be open or closed and that the warp of the figure may require showing both hidden and visible portions of the elements. In Fig. 9-7, all elements are visible in the horizontal view. In Fig. 9-3 of the cow's horn, the hidden elements were not indicated.

Both cylindroids and conoids are used in the aircraft and ship building industries. The model of the aircraft in Fig. 9-8 utilizes warped surfaces for portions of its fuselage.

FIG. 9-8 Aircraft model. *(Courtesy NASA.)*

382

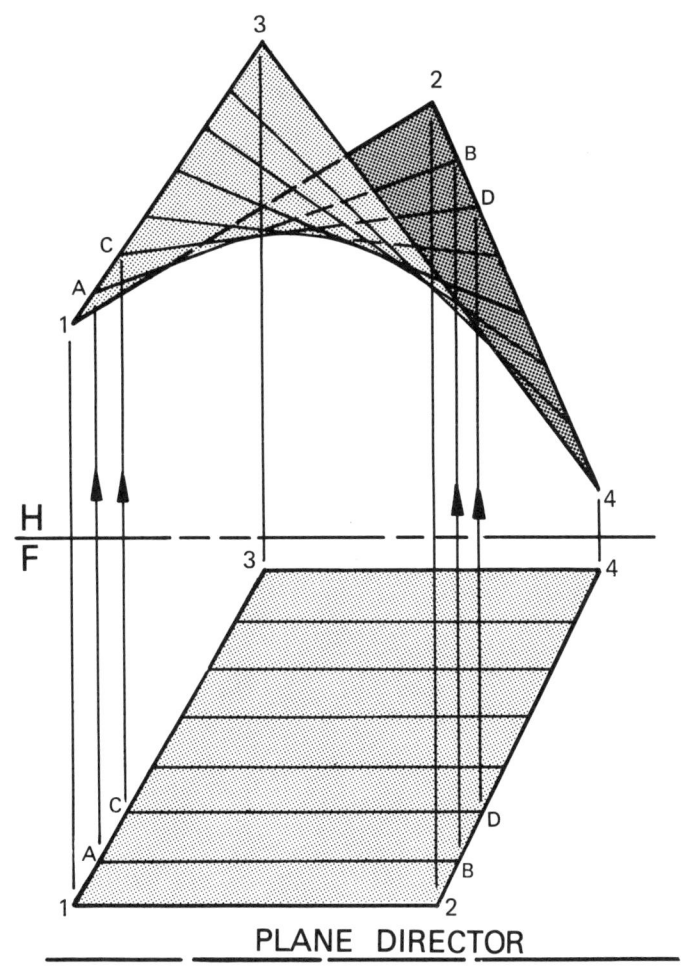

FIG. 9-9 Warped surface: hyperbolic paraboloids.

9.5 HYPERBOLIC PARABOLOIDS

A *hyperbolic paraboloid* is a warped surface, generated by a straight-line generatrix which intersects two straight-line directrices (skew lines) and remains parallel to a plane director. The straight-line directrices cannot be parallel or intersect. The plane director must not be parallel to either directrix. A hyperbolic paraboloid is in fact a *double-curved warped surface*, since it can be doubly ruled by using the first and last elements as the straight-line directrices and establishing a new plane director.

In Fig. 9-9 the views of the two directrices and the plane director are given. Lines 1-3 and 2-4 are the directrices. Directrix 1-3 is divided into equal parts. Draw the elements in the frontal view through the points with lines parallel to the plane director. Locate the elements in the horizontal view by projection. Element A-B and C-D are established in the frontal view and projected to the horizontal view as shown. The horizontal view shows the warp of the figure. Note that hidden elements are drawn as dashed lines. The more elements that are used, the smoother the outline of the warp.

Figure 9-9 can be double ruled by using line 1-2 and line 3-4 as the directrices and changing the plane director. The resulting shape would be identical to that generated by directrices 1-3 and 2-4.

Hyperbolic paraboloids are frequently used in the construction of dams, bridge piers, and other concrete structures. Architects also make use of this warped shape to design roofs for buildings. Concrete forms can be easily constructed using straight timbers along element lines for each generation of the surface. The compressor blade in Fig. 9-10 is a portion of a hyperbolic paraboloid.

The intersection of a plane with a hyperbolic paraboloid results in a straight line, a hyperbola, or a paraboloid.

FIG. 9-10 Airfoil mill, capable of sculpturing in metal the complex curve of a twisted blade of an axial flow compressor. *(Courtesy NASA.)*

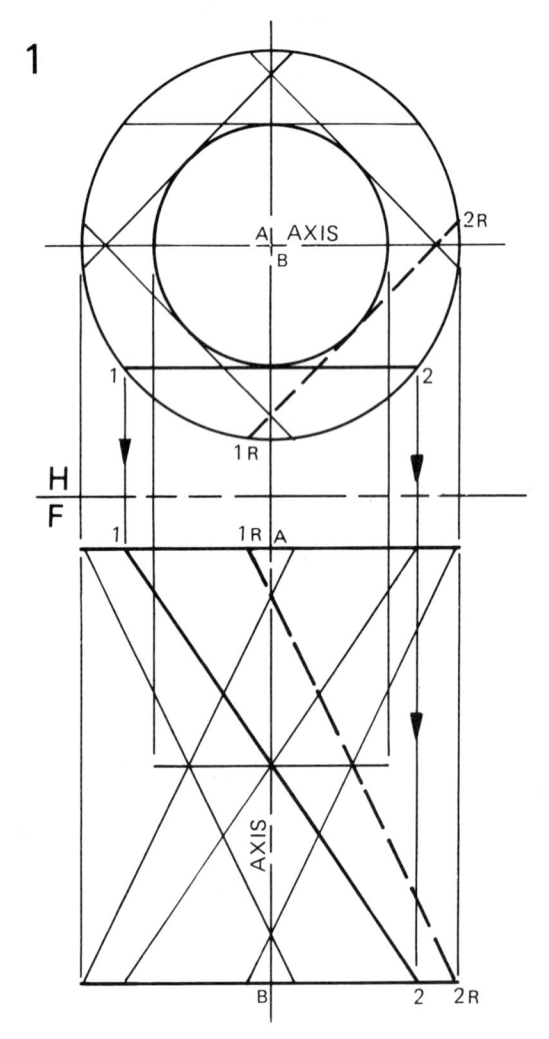

FIG. 9-11 Hyperboloid of revolution.

9.6 HYPERBOLOID OF REVOLUTION

*A **hyperboloid of revolution** is generated by a straight-line generatrix which revolves about a straight-line directrix, axis.* The generatrix does not intersect the axis. Three directrices are required: One is the axis line, and two are parallel circles, normally closed and of the same diameter. The cooling towers in Fig. 9-1 are examples of a hyperboloid of revolution whose circular directrices are of differing diameters.

The generatrix of a hyperboloid of revolution intersects the two circular directrices and revolves about the third, straight-line directrix (axis). The revolving element is tangent to a small inner circle called the *circle of the gorge.*

Hyperboloids can be double ruled, as could the hyperbolic paraboloid, by reversing the elements of the first generation. The hyperboloid is therefore a *double-curved warped surface.* The hyperbolic paraboloid and the hyperboloid can not be developed.

In Fig. 9-11(1), the three directrices are given along with the circle of the gorge. The two circle directrices are the same diameter. Axis A-B is the straight-line directrix about which the generatrix revolves.

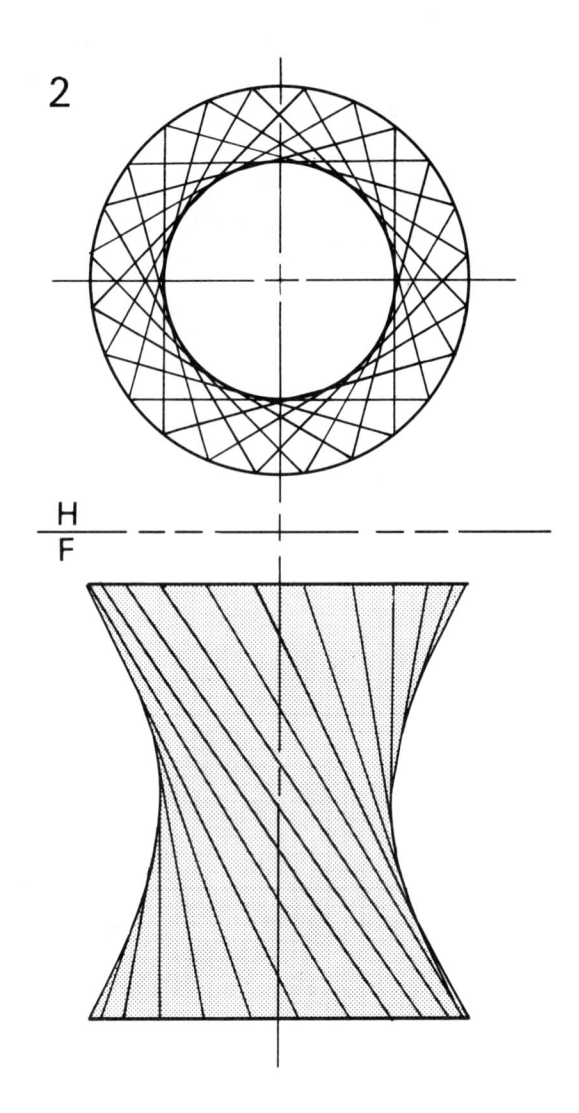

Element 1-2 is drawn as shown. Four positions of element 1-2 are shown in both views. An element intersects the two circle directrices and the circle of the gorge at one point each, along its length. Point 1 is on the upper circle, point 2 is on the lower circle. The element's midpoint is on the gorge circle. Note that position 1R-2R is not hidden in the frontal view. The dashed lines are used to illustrate its revolved position.

Figure 9-11(2) shows the completed warped surface. Enough positions of the revolving element must be provided to establish a smooth warped outline. Positions for the generatrix (element) are established by dividing the circular directrix into equal divisions as shown. The hidden elements are not shown in this view. Only a single generation of elements is shown. A double-ruled surface can be established by reversing the given elements.

The intersection of a plane, passed parallel to the axis line, will show a line of intersection in the shape of a hyperbola. A plane passed perpendicular to the axis results in a circle.

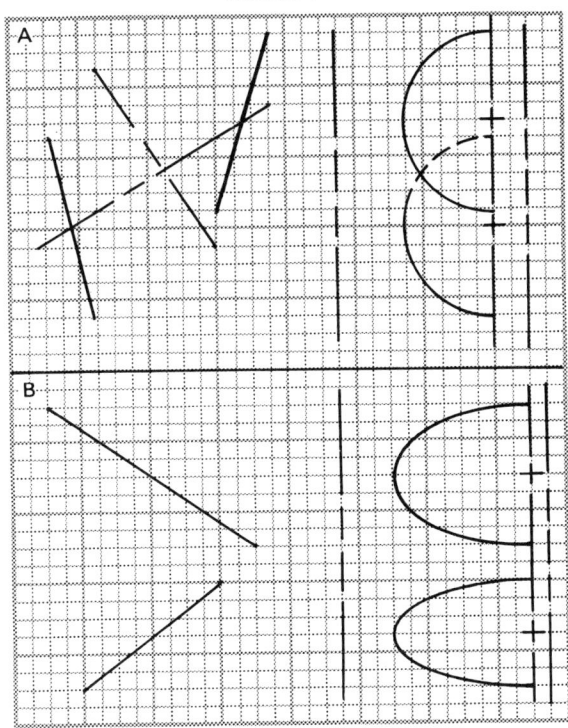

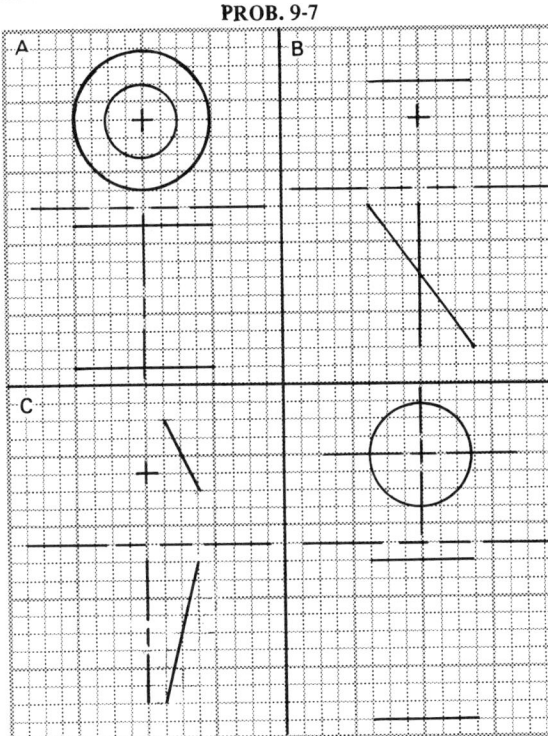

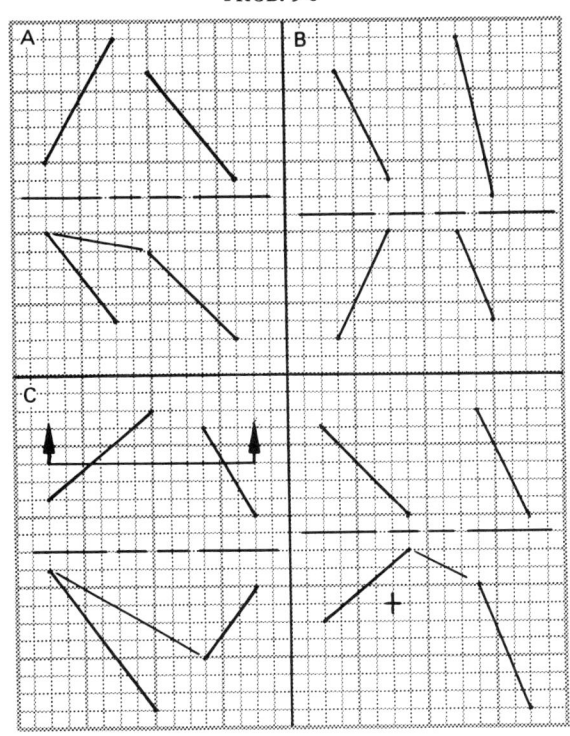

PROB. 9-5A. Construct a cylindroid using 24 equally spaced straight-line elements. Show the required section in the front view. The directrix is given.

PROB. 9-5B. Complete the two views of the cylindroid. Divide the given circular curve into 24 equally spaced straight-line elements. The directrix is given.

PROBS. 9-6A through 9-6D. For each of these problems draw a hyperbolic paraboloid. A, C, and D have a given element. The plane director is parallel to the given element for A, C, and D. B requires a vertical plane director parallel to the fold line. C requires the construction of a section in the front view, and D needs the given point shown in the top view. Use 15 equally spaced elements.

PROB. 9-7A. Complete the hyperboloid of revolution using the given O.D. and circle of the gorge. Use 12 equally spaced straight-line elements.

PROB. 9-7B. The vertical line is the axis and the other given line the generatrix; complete the hyperboloid of revolution using 24 equally spaced straight-line elements.

PROB. 9-7C. Draw a hyperboloid of revolution using the vertical line as the axis and the other given line as the generatrix. Use 24 equally spaced straight-line elements.

PROB. 9-7D. Draw a hyperboloid of revolution with a .375" (9.5mm) dia. circle of the gorge. O.D. and height are given. Use 12 equally spaced straight-line elements.

WORD PROBLEMS

PROB. 9-8. Draw a hyperboloid of revolution having equal diameter directrices of 3.5" and a circle of gorge 1" dia. The height is 4.625". Use 24 equally spaced straight-line elements.

PROB. 9-9. Construct a hyperboloid of revolution with a height of 100mm, O.D. of 70mm, and a circle of the gorge with a diameter of 34mm. Use 24 equally spaced straight-line elements.

385

10

THE HELIX

10.1 HELICES

FIG. 10-2 Left-handed conical helix.

A ***helix*** is a ***double-curved line*** *drawn by scribing the movement of a point as it revolves about an axis of a cylinder.* The resulting curve is traced on the cylinder by the revolution of a point crossing its right sections at a constant oblique angle. The point must travel about the cylinder at a uniform linear and angular rate. The linear distance (parallel to the axis) traveled in one complete turn is called the **lead**. This type of helix is called a **cylindrical helix**. If the point moves about a line that intersects the axis, it is a **conical helix**. The generating point's distance from the axis line changes at a uniform rate. A helix can be either right- or left-handed.

A helix of either type is drawn by radially dividing the end view (curve) into an equal number of parts. The lead is divided into the same number of parts. In Fig. 10-1, the right-handed cylindrical helix is drawn by dividing the circular end view into 16 equal divisions. The lead is also divided into 16 parts. Label the points on both views as shown. The end view divisions are projected to the front view as vertical elements on the surface of the construction cylinder. The lead divisions are drawn as horizontal elements. The intersection of related elements in the front view establishes a series of points on the surface of the cylinder. Each point represents a position of the generating point as it rotates about the axis.

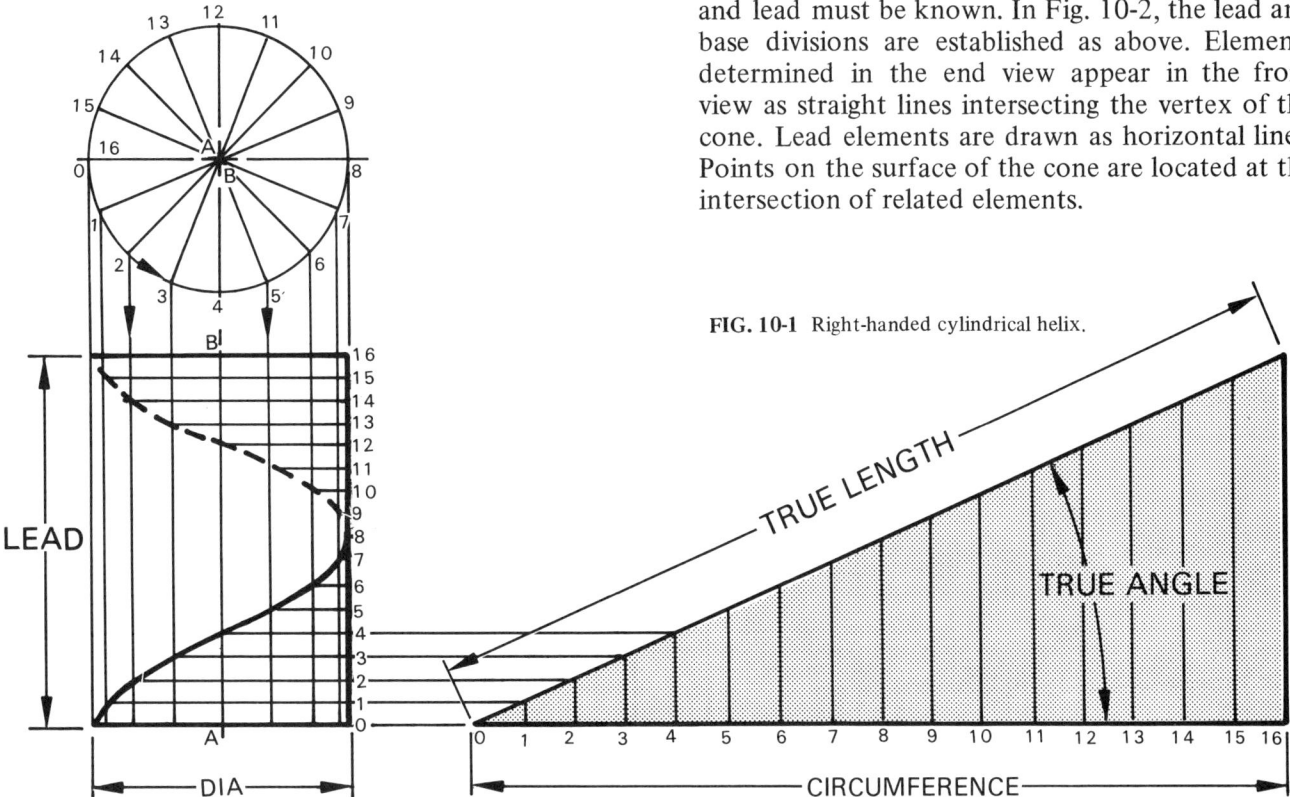

The cylindrical helix is developed by unrolling the cylinder's surface. The helix line is a straight line on the development. The angle the helix line makes with the base line is called the **helix angle** (true angle).

To construct a conical helix, its taper angle and lead must be known. In Fig. 10-2, the lead and base divisions are established as above. Elements determined in the end view appear in the front view as straight lines intersecting the vertex of the cone. Lead elements are drawn as horizontal lines. Points on the surface of the cone are located at the intersection of related elements.

FIG. 10-1 Right-handed cylindrical helix.

387

10.2 THE RIGHT HELICOID

A helix is used in the design of a variety of industrial products, structures and machine parts: screw threads, conveyor screws, spiral stairways. Some of these applications utilize the right helicoid in their design. A helicoid can only be approximately developed.

*A **right helicoid** is a **warped surface** generated by the movement of a straight-line generatrix perpendicular to a straight-line axis directrix and intersecting a helical directrix.* The right helicoid has a plane director. The generatrix (element) is at a right angle to the axis directrix making a constant angle with the plane director. The plane director is perpendicular to the axis.

The generatrix of a right helicoid intersects the axis at a right angle. The other end of the generatrix intersects the helix directrix. The generating line revolves about the axis at a uniform rate and along the axis (lead) at a uniform linear rate. In most cases the generatrix will not extend to the axis. In Fig. 10-3, the screw conveyor is right-handed right helicoid. Note that its generating element (generatrix) is perpendicular to, but not intersecting, the axis directrix, A-B. This type of right helicoid has two related helices as directrices. The elements intersect the inner cylinder along a helix traced on its surface and the outer cylinder along a helix traced on its surface. The lead of a helicoid is the distance the generatrix travels along (parallel to) the axis in one turn. The inner and outer helices have the same lead.

To draw a right heloid, the lead and the diameters of the inner and outer helices must be known. Whether the helicoid is right- or left-handed must also be stated. The inner helix has a steeper helix angle since it travels the same linear distance with a shorter helix line.

In Fig. 10-3, two complete turns of right-handed right helicoid are illustrated. The inner and outer helix diameters and the lead are given. The horizontal view is drawn showing the inner and outer circles. In the frontal view the axis line and the inner helix cylinder are drawn as shown. Elements are established on the surface of the helicoid in end view (horizontal view) by radially dividing the circumference into equal angles about the axis. Here 16 divisions are used. In the frontal view the lead (parallel to the axis) is divided into the same number of distances. Since the generatrix (generating element) must remain perpendicular to the axis, all elements are drawn as horizontal lines (perpendicular to axis A-B). The end points of the elements are projected from the horizontal view as shown. The end points are connected as a smooth curve. This establishes two coaxial helices in the frontal view. In the frontal view elements 4 and 12 appear as point views and elements 0 (16) and 8 are true length. All elements are true length in the horizontal/end view.

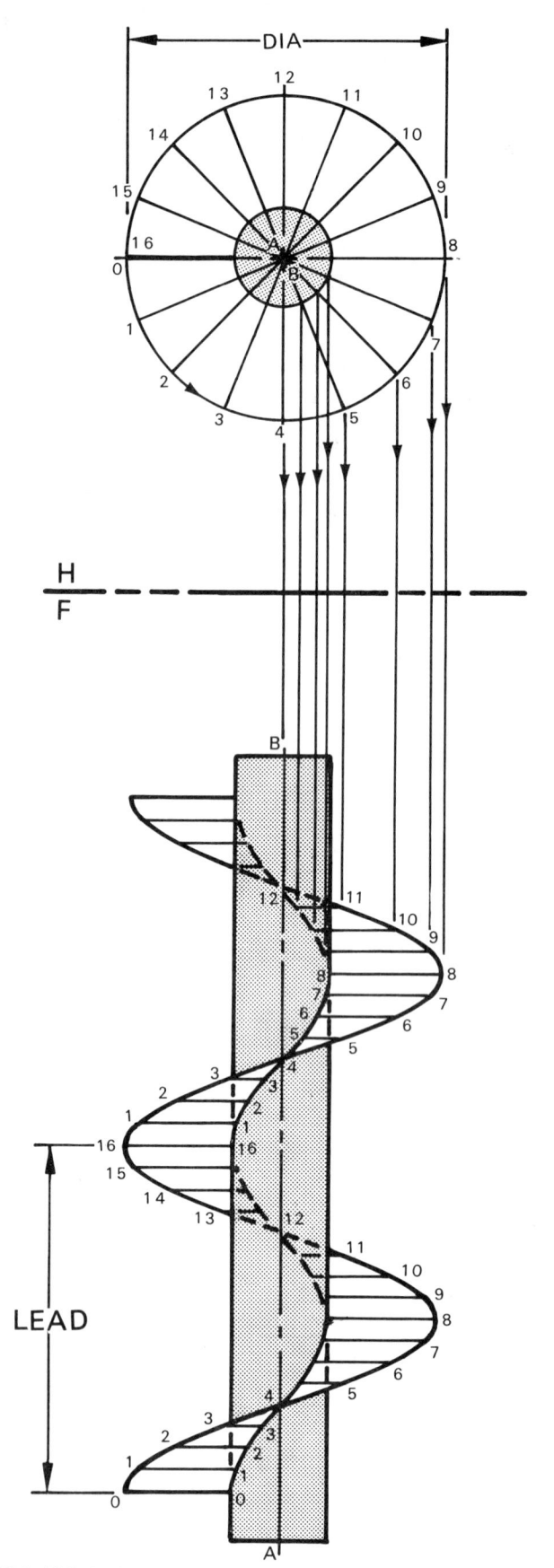

FIG. 10-3 Right helicoid.

FIG. 10-4 Oblique helicoid.

10.3 THE OBLIQUE HELICOID

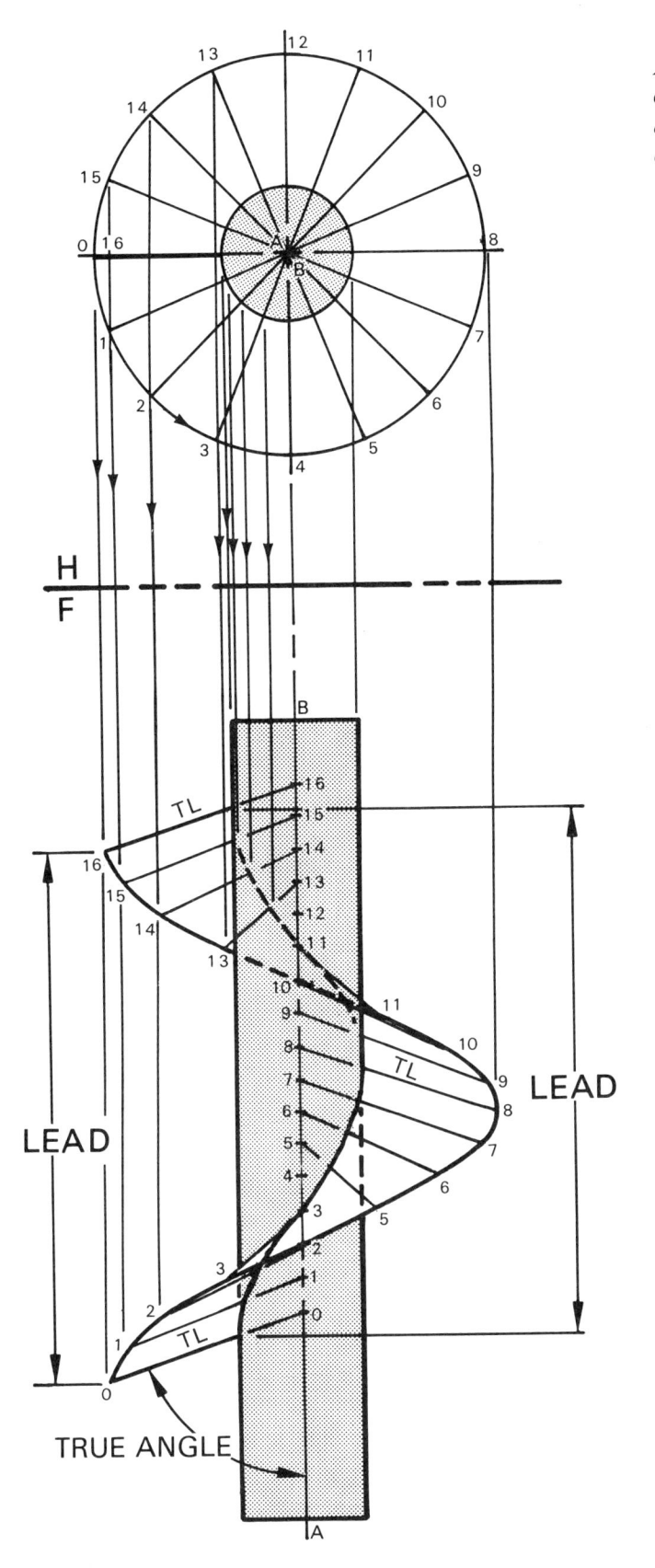

*An **oblique helicoid** is generated by the movement of a straight-line generatrix which makes a constant angle with an axis directrix and intersects a helical directrix.* The plane director for an oblique helicoid is a cone. To draw an oblique helicoid the lead, diameter of the helix directrix, and the generatrix angle must be given along with whether it is to be right- or left-handed. When the oblique helicoid is limited to the area between two helices, both the inner and outer helix (cylinder) diameters must be given. When this is the case the directrices of the helicoid are both helices. Note that the leads for both the inner and outer cylinder remain the same. The end points of the generatrix trace a helix on the inner and outer limiting cylinders. Since the oblique helicoid is a warped surface it can only be approximately developed.

The generating element travels along the lead at a uniform rate and about the axis at a uniform rate. All positions of the generatrix (elements) intersect the axis at a constant angle. In Fig. 10-4, the inner and outer limiting cylinders are given along with the generatrix angle (true angle) for the right-handed oblique helicoid.

Draw the inner and outer circles to establish the end view. The axis is used as the center point to divide the circle's circumference into 16 equal angles. An element is drawn from the point view of axis A-B through each division. Note that all elements are inclined, therefore they are not true length in the end (horizontal) view. The axis and the inner cylinder representing the shaft is drawn in the frontal view. The lead is divided into 16 equal parts. A true length element is drawn first. Element 0-0 is a frontal line, therefore it can be drawn at the required true angle in the frontal view. The outer helix is then constructed as in Sections 10.1 and 10.2.

Element 0-0 (and 8-8, 16-16) is true length. The intersection of element 0-0 and the inner cylinder in the frontal view determines the starting point for the inner helix lead divisions. The inner helix can easily be established by dividing the axis into 16 divisions corresponding to the lead, and connecting the divisions with related points on the outer helix. The inner element's end points are projected from the horizontal view as shown. Each end point is connected with a smooth curve to complete the helicoid. The visibility of the helicoid must be determined by inspection of the location of each element in the horizontal view.

PROB. 10-1

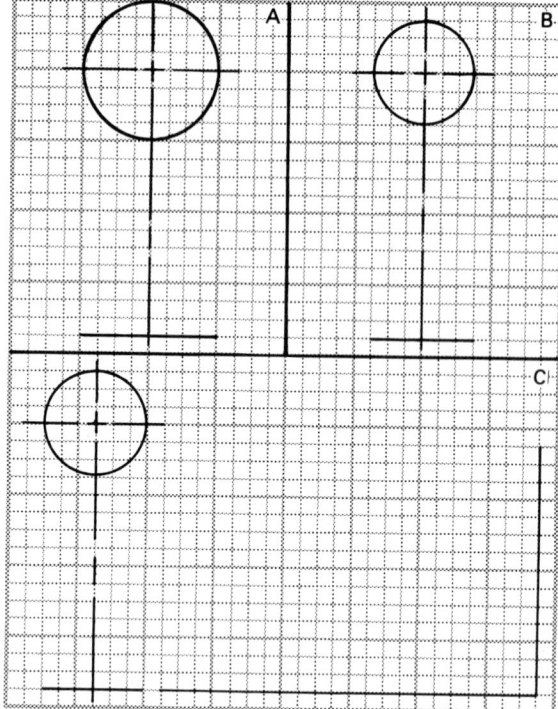

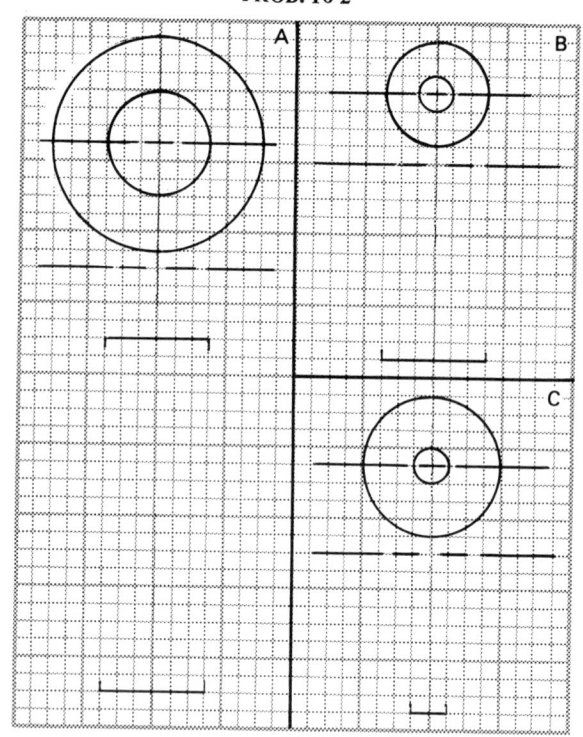

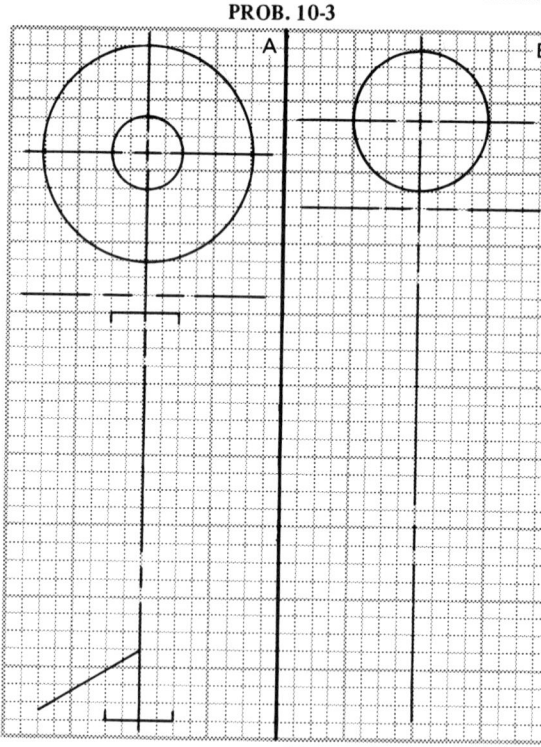

PROB. 10-2

PROB. 10-3

PROB. 10-1A. Draw a left-handed one-revolution cylindrical helix that gains 1/8″ (3.1mm) of height with each 15 degrees of revolution. Start from the lower front center.

PROB. 10-1B. Construct a right-handed conical helix with a lead of 1.75″ (44mm) and a 2.5″ (63.5mm) cone length. Start with the lower right side.

PROB. 10-1C. Draw a 2.75″ (69.8mm) helix with a lead of 1.375″ (34.9mm). Start with the lower front center. Develop the helix and note its true length.

PROB. 10-2A. Draw a left-hand, right helicoid with a straight line vertical axis. The lead is 2″ (50.8mm) and the helicoid will run the length of the given inner cylinder. Start from the lower right side.

PROB. 10-2B. Construct a right helicoid with a lead of 2.25″ (57.1mm). Draw both views showing only one complete turn. Start at the lower front center.

PROB. 10-2C. A left-handed spiral stairway is to be designed around the given center post. It will have a lead of 2″ (50.8mm). The stairway is a right helicoid with the outer diameter given. Start from the lower left side and draw one turn.

PROB. 10-3A. Draw two views of a right-hand oblique helicoid using the given diameters as the outer and inner limits. The lead is 4″ (101.6mm), with a 30-degree angle generating line which slopes downward. Draw one complete turn starting from the given true length element.

PROB. 10-3B. Using the given circle as the outside cylinder diameter, construct one turn of an oblique left-handed helicoid with a lead of 2.5″ (63.5mm) and an inside diameter of .5″ (6.3mm). Start at the top right side of the inner cylinder. Use a downward sloping 20-degree angle for the generating line.

11
TANGENCIES

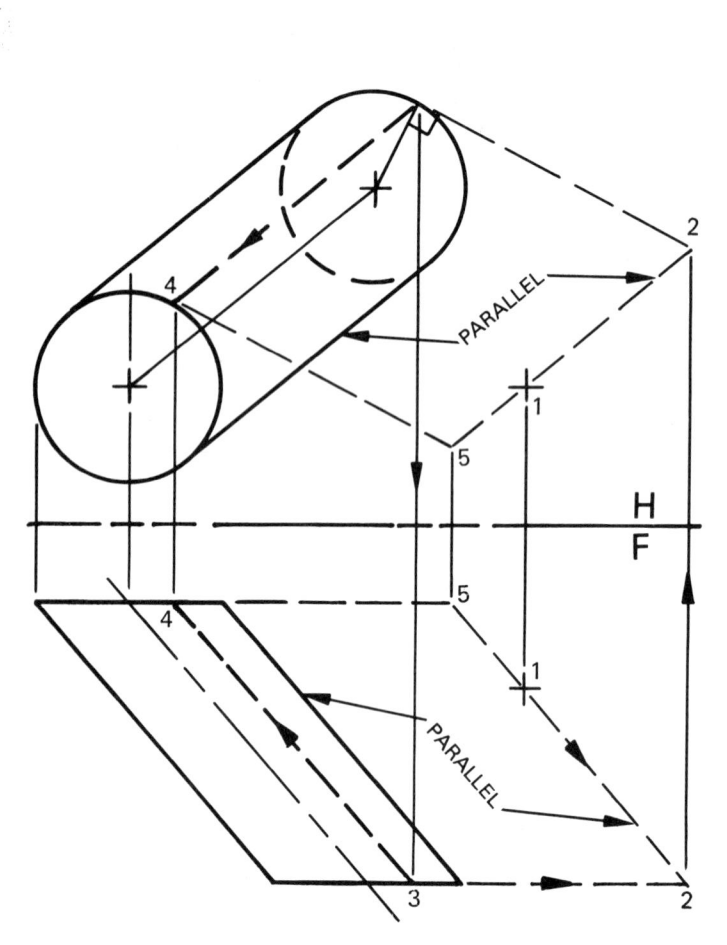

FIG. 11-1 Plane tangent to a cylinder through a given external point.

FIG. 11-2 Plane tangent to a cylinder through a given point on its surface.

11.1 TANGENCIES

In many cases, it is necessary to establish a plane tangent to a surface. Single-curved surfaces—cones, cylinders, convolutes—are ruled surfaces, therefore a plane can be constructed so as to contain one element of their surface. The ruled surface and the plane are tangent since they share a common line. Planes drawn tangent to a double-curved surface—sphere, torus, etc.—are constructed so as to contain one point of the curved surface. The plane and the surface share a common point, therefore they are tangent.

In general, two situations determine the placement of a plane tangent to a curved surface: if the plane is to be passed through an existing point on the surface or if the plane is to be passed through an existing external point.

In Fig. 11-1, a plane tangent to the given oblique cylinder and through point 1 is required. Draw a line through point 1 and parallel to the cylinder's axis in both views. Points 2 and 5 are established in the frontal view where the line intersects the extended upper and lower base planes. Lines 2-3 and 4-5 are drawn tangent to the cylinder's circular base in the horizontal view. Note that two positions are possible. Lines 2-3 and 4-5 could be drawn tangent to the front of the base circle. Line 3-4 is an element on the surface of the cylinder, therefore a common line. The frontal view of line 3-4 is determined by projection.

In Fig. 11-2, a plane is to be constructed tangent to the oblique cylinder and through point 1. Point 1 lies on the cylinder's surface and is given in the horizontal view. Draw element 2-3 through point 1, parallel to the cylinder's axis and project to the frontal view. Draw construction lines from the base's center points, through points 2 and 3 in the frontal view. Line 2-4 and line 3-5 will be perpendicular to these construction lines. Line 4-5 is parallel to the axis of the cylinder in both views. The horizontal view is completed by projection. Line 2-4 and line 3-5 are parallel to the cylinder's base planes in the horizontal view. Line 4-5 is used to limit the plane; it could have been any convenient distance from the cylinder. The plane and the cylinder are tangent since they both contain line 2-3.

11.2 PLANE TANGENT TO A CONE

A plane is tangent to a cone if they share a common line. This common line is an element on the cone's surface. Only one element of the cone will be on the plane. The plane intersects the base plane of the cone along a line tangent to the base curve.

In Fig. 11-3, points 1 and 2 and the cone are given. Points 1 and 2 lie on the surface of the cone and are given in the horizontal view. Draw a plane tangent to the cone and through point 1, and another tangent to the cone and through point 2.

A line is drawn from vertex V and through point 1 until it intersects the base circle at point 3. Line V-3 and point 1 are located in the frontal view by projection. Line V-3 is an element of the cone. Line 4-5 is drawn tangent to the base circle and perpendicular to line/element V-3 in the horizontal view. Line 4-5 can be any desired length. In the frontal view line 4-5 is parallel to the base plane. Plane V-7-8 containing point 2 and tangent to the cone is established by the same procedure.

In Fig. 11-4, point 1 and the oblique cone are given. Construct a plane containing point 1 and tangent to the cone.

Draw a line from vertex V through point 1 in both views. V-1$_F$ intersects the base plane in the frontal view at point 2 when extended. Point 2 is projected to the horizontal view as shown. Two tangent planes are possible. Lines are drawn from point 2 tangent to the cone's base circle in the horizontal view. Lines V-3 and V-4 are elements of the cone. Point 3 and point 4 are tangent points. Plane V-2-3 and plane V-2-4 contain point 1 and are tangent to the cone.

FIG. 11-3 Plane tangent to a cone through a given point on the cone.

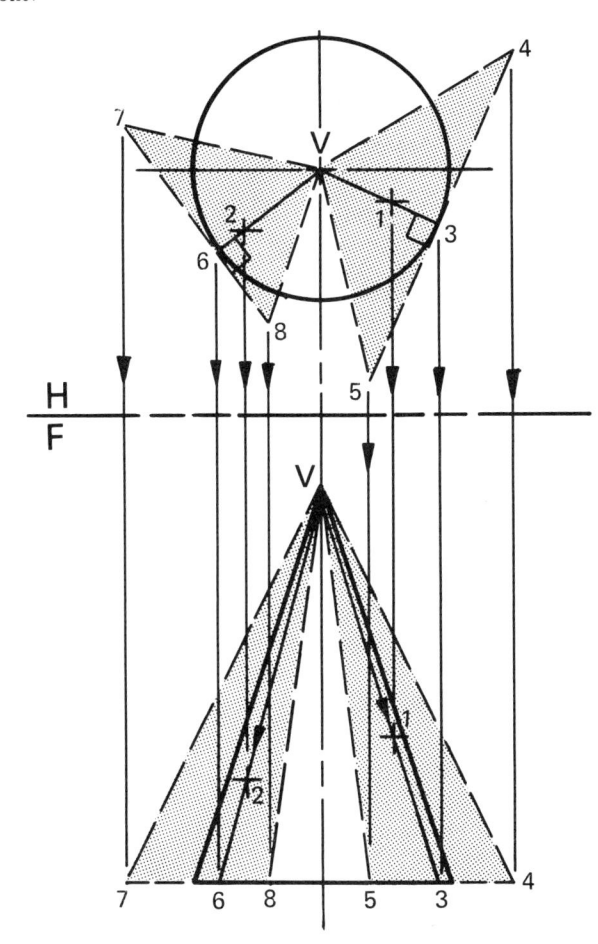

FIG. 11-4 Plane tangent to an oblique cone through a given external point.

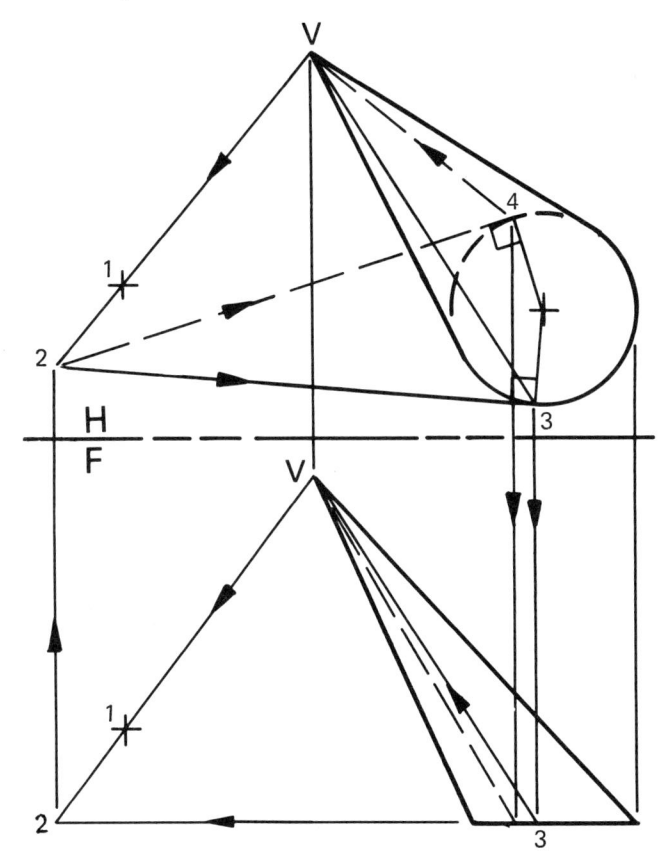

11.3 PLANE TANGENT TO A SPHERE

Planes tangent to a double-curved surface have only one common point. Therefore, a plane tangent to a sphere contains one point of the sphere. This being the case, a tangent plane is perpendicular to a radius line drawn from the sphere's center through the common point. Two intersecting lines determine a plane. If the lines are tangent to the sphere and intersect at one point on a sphere, they determine a plane tangent to the sphere.

In Fig. 11-5, point 1 and the sphere are given and a tangent plane is required. A line must be established containing point 1 and tangent to the sphere. Any plane containing this line (and not intersecting the sphere) is tangent to the sphere. Therefore, a line containing the common point and tangent to the sphere must be perpendicular to the radius line. A view showing either the line or the radius line as true length shows one or both of the lines as true length.

Draw a line (line 3-4) through point 1 and perpendicular to the radius line in the horizontal view. This line can be of any convenient length. Line 3-4 is a horizontal line (true length) and thus appears as parallel to H/F in the frontal view. Draw line 6-7 through point 1 and perpendicular to the

FIG. 11-6 Plane tangent to a sphere through a given external point.

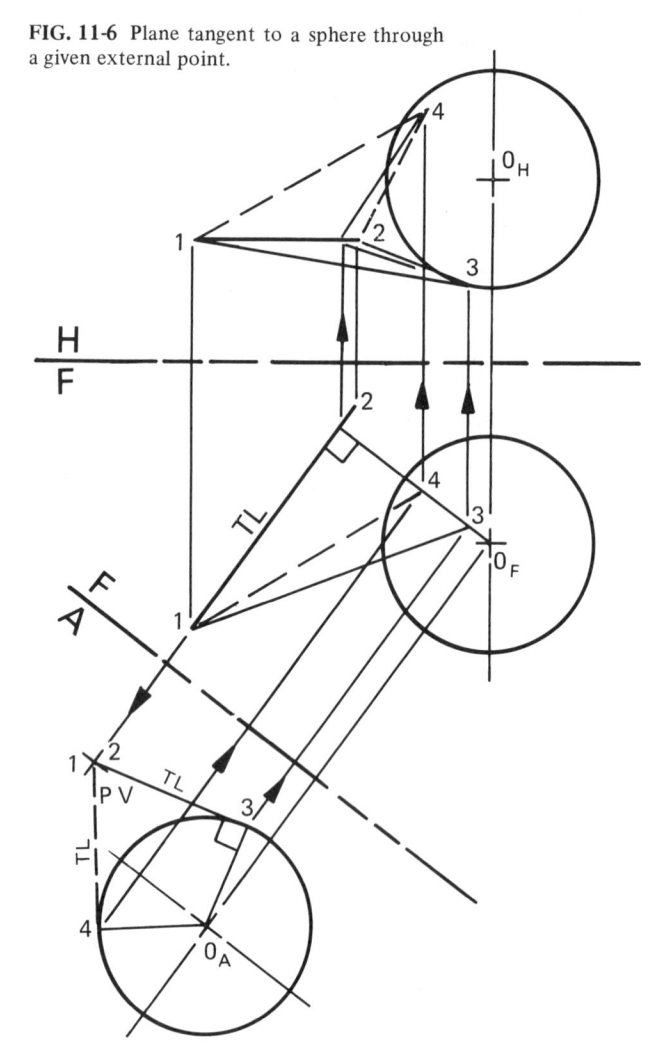

radius line in the frontal view. Line 6-7 is any convenient length. The horizontal view shows line 6-7 as parallel to H/F since it is a frontal line. Lines 3-4 and 6-7 determine a plane that is tangent to the sphere at point 1.

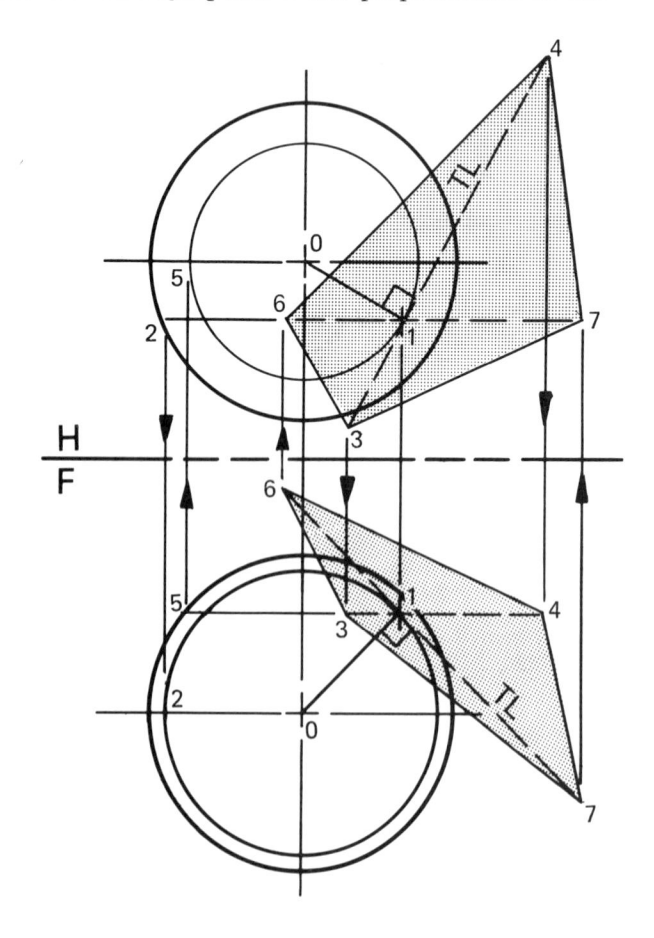

In Fig. 11-6, the sphere and point 1 are given. A plane is to be constructed containing point 1 and tangent to the sphere. Given line 1-2, this problem could have been the construction of a plane tangent to a sphere and containing a given line.

Draw line 1_H-2_H parallel to H/F at any desired length. Line 1_F-2_F is true length. The placement of line 1_F-2_F determines the line of sight for a point view projection. If line 1-2 is given, it already establishes this line of sight. Draw F/A perpendicular to line 1_F-2_F and project auxiliary A. Draw a line from the point view of line 1-2 tangent to the sphere; two positions are possible. These tangent lines are true length and therefore project as parallel to F/A in the frontal view. Points 3 and 4 are fixed by projection in the frontal view and located by transferring distances in the horizontal view. Plane 1-2-3 and plane 1-2-4 are tangent to the sphere.

FIG. 11-5 Plane tangent to a sphere through a given point on the sphere.

11.4 PLANE TANGENT TO A CONVOLUTE

In many cases it is necessary to attach a plane surface to a curved one. In Fig. 11-7 the space shuttle concept model has a plane wing section that is tangent to the curved fuselage.

A plane is tangent to a surface if they have a common point or one common line. In Fig. 11-8, a plane tangent to a convolute is required. A convolute is a single-curved surface and therefore a ruled surface. If a plane contains an element of a convolute, the plane and convolute are tangent. Remember, the plane can contain only one element (line) of the convolute to be tangent, otherwise they will intersect. Note that *a convolute can be generated by the movement of a plane tangent to the two curved directrices.*

FIG. 11-7 Space shuttle wind tunnel model.
(Courtesy NASA.)

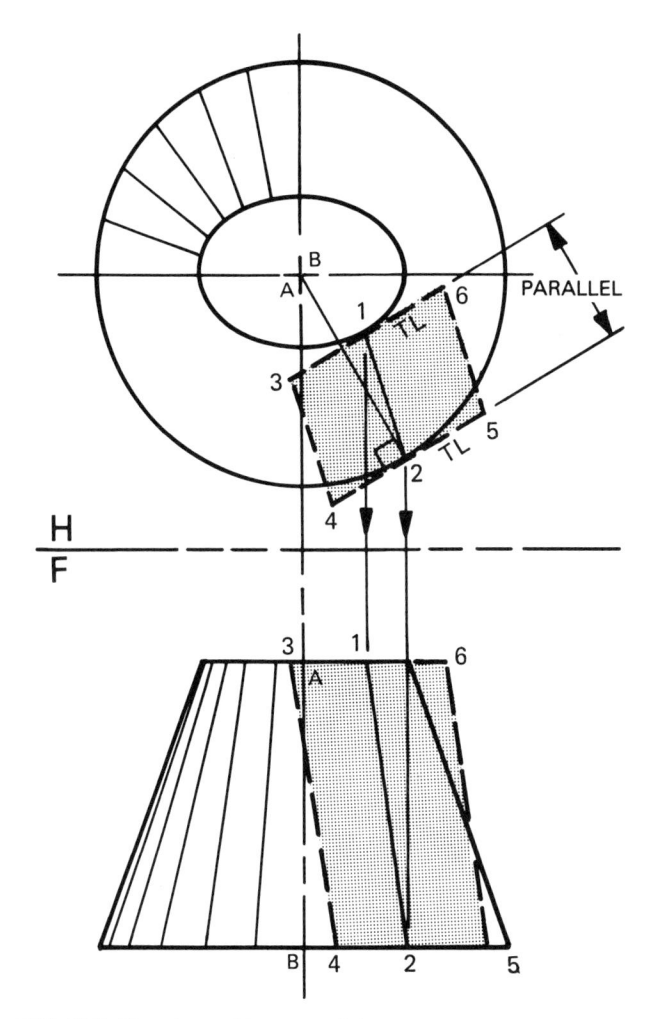

FIG. 11-8 Plane tangent to a convolute.

In Fig. 11-8, line 1-2 is an element of the convolute. The required plane is to contain this line. Draw a line from the point view of axis line A-B, in the horizontal view, to point 2. Draw line 4-5 through point 2 and perpendicular to line B-2 in the horizontal view. Line 3-6 is drawn parallel to line 4-5 and through point 1. Plane 3-4-5-6 and the convolute are tangent since they share a common line, line 1-2. The frontal view of the plane is completed by projection. Lines 3-6 and 4-5 are parallel to the convolute's bases. Lines 1-2, 3-4, and 5-6 are parallel in this example. Note that the plane's shape can be any desired form so long as line 1-2 is its "only" common line.

PROB. 11-1

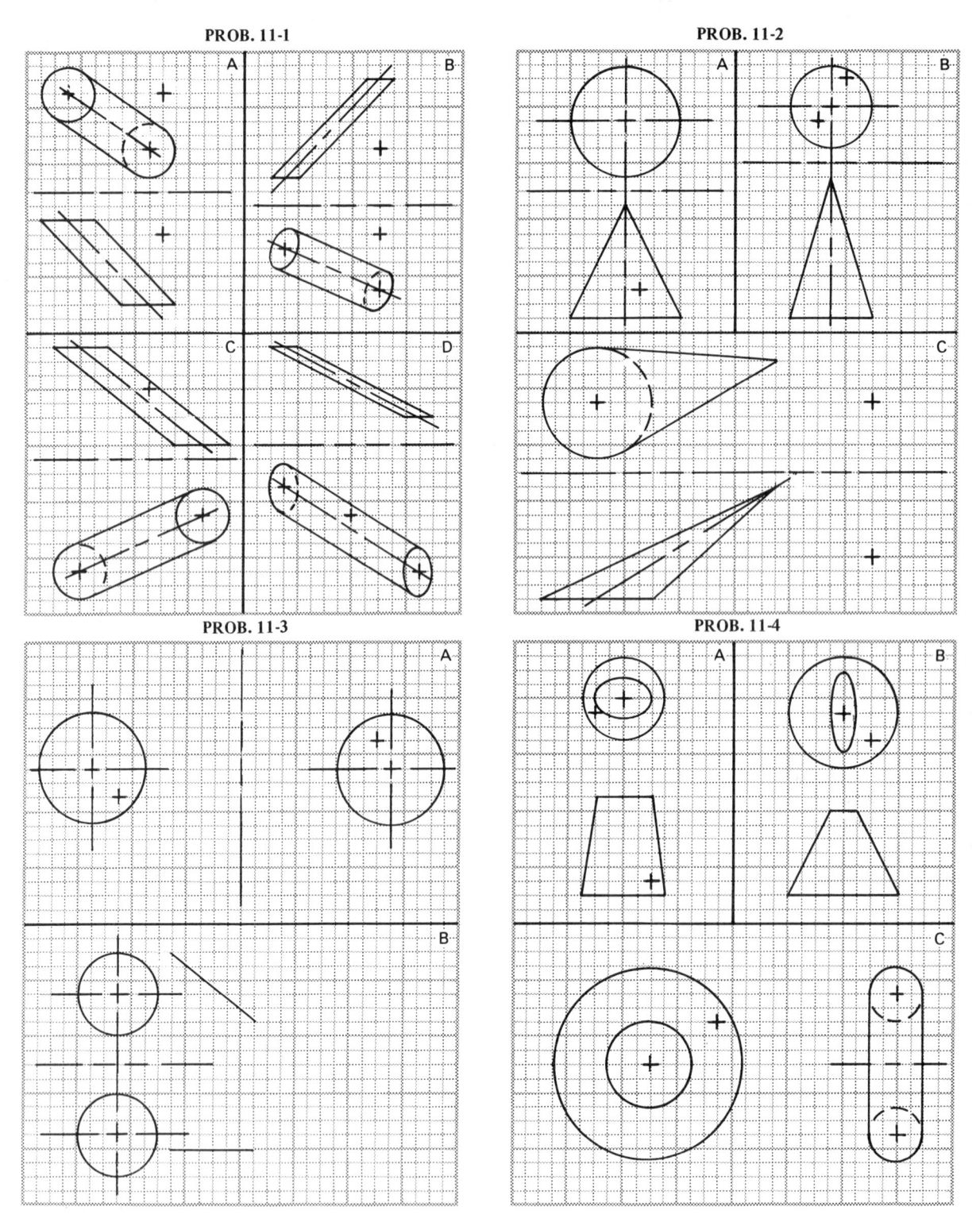

PROB. 11-2

PROB. 11-3

PROB. 11-4

PROBS. 11-1A, 11-1B. Draw a plane through the given point and tangent to the oblique cylinder.

PROB. 11-1C. Show the given point in its two possible positions in the front view. Pass a separate plane through each point and tangent to the cylinder.

PROB. 11-1D. Pass a plane tangent to the cylinder and through the given point which lies on the front of the figure.

PROB. 11-2A. Establish a plane tangent to the cone and through the given points. Show both possible answers.

PROB. 11-2B. Pass separate planes through each point and tangent to the cone.

PROB. 11-2C. Pass a plane through the point and tangent to the oblique cone. Show two possible solutions.

PROB. 11-3A. Pass separate planes tangent to the sphere and through the visible points on the sphere.

PROB. 11-3B. Draw a plane through the given line and tangent to the sphere. Show two solutions.

PROBS. 11-4A, 11-4B. Pass tangent planes through each point on the given convolute.

PROB. 11-4C. Construct a tangent plane through the point on the top of the torus and containing the given point. Rotate the plane so that it projects as an edge in the side view.

12

SHADES AND SHADOWS

FIG. 12-1 Shades and shadows.

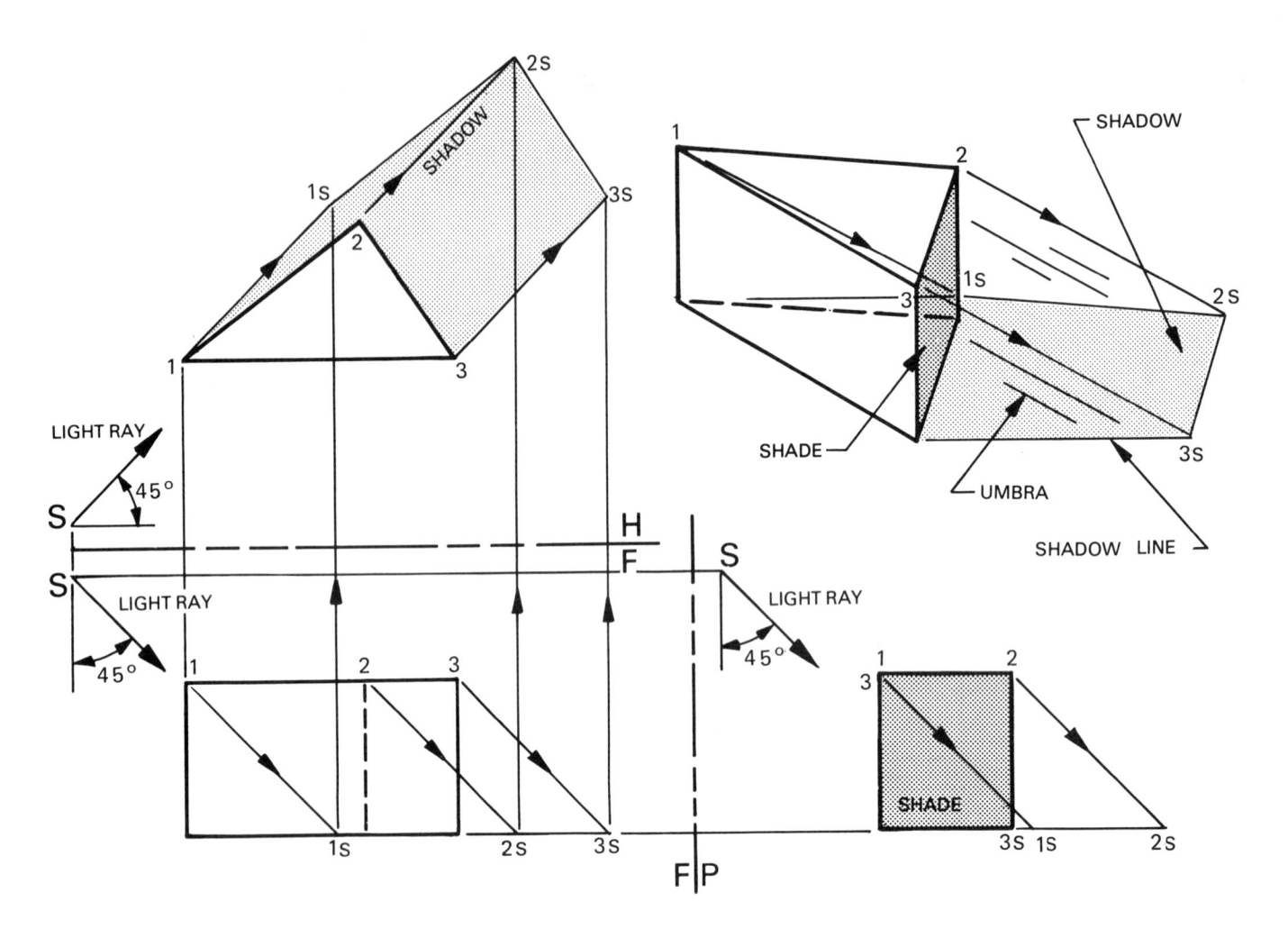

12.1 SHADES AND SHADOWS

The shaded area of, and the shadow cast by, an object can be established by applying procedures from Chapter 6, the intersection of a line and a plane. Shades and shadows are added to a drawing to emphasize its three-dimensional qualities. Normally engineering drawings will not contain the shade or shadow of the form represented. Architectural renderings, pictorial drawings, and technical illustrations for books, manuals, technical advertising, brochures, etc., all make use of shades and shadows in order to present the figure in a more recognizable form. By presenting the figure in a lifelike, natural way, communication of the intended appearance and configuration is heightened. In many situations a photograph cannot be made of the form, which may not even exist yet. Since people are accustomed to comprehending forms as a camera records/captures the image, the addition of shading and the shadow cast by the object adds realism, depth, and familiarity to the presentation.

*The **shade** of an object is that portion of its surface that has been excluded from the light source*, Fig. 12-1: in other words, the area of relative darkness on an object due to the interception of **light rays** by itself. *The **shadow** of an object is an area of relative darkness cast by the blocking of light rays as they are intercepted from the light source*. A shadow can be cast on the plane surface that the object rests upon, or another object.

The lines that bound a shadowed area are called *"**shadow lines**"* and the bounding lines of a shaded area are referred to as *"**shade lines**."* The actual volume of space (which lies between the shade and shadow areas) is the **umbra**.

The shade and shadow of an object are created by the exclusion of light rays within an area blocked by the object. Light rays are assumed to be *parallel* since the light source, normally the sun, is considered to be at infinity. Conventional practice establishes the light rays' direction as a diagonal within a cube from the upper left front corner to the lower right rear corner. Therefore, the light rays make a 45-degree angle with each projection plane as in Fig. 12-1. Note that the pictorial drawing of the object in this figure uses a different light ray direction to illustrate the terms defined above.

12.2 SHADES AND SHADOWS: CYLINDERS

Since a light source is assumed to be at infinity, all light rays are parallel. Light rays are intercepted by the bounding edges of an object. The points of interception are projected to the surface on which the object rests or to an intervening surface of another form placed between the resting surface and the object that intercepts the light rays.

In Fig. 12-2, the shadow of an object appears in the horizontal view and the shade area in the frontal view only. The light ray direction is drawn as a 45-degree line in both views as shown. The light rays are intercepted by the edges of the combined cylinder/prism. Each point of interception is projected to the horizontal plane on which the object sits. Points 3, 4, and 5 are corner points which are projected to the horizontal base plane (in the frontal view) and labeled as points of intersection, 3S, 4S, and 5S.

Since the upper portion of the object is circular, it is easier to locate the center points on the horizontal base plane than to plot a series of intersection points projected from the circular edges. Points 1S and 2S are points of intersection of the light rays and center points 1 and 2 (as if they existed separately from the cylinder itself). All points determined by the intersection of extended light ray lines and the horizontal base plane are then projected to the horizontal view.

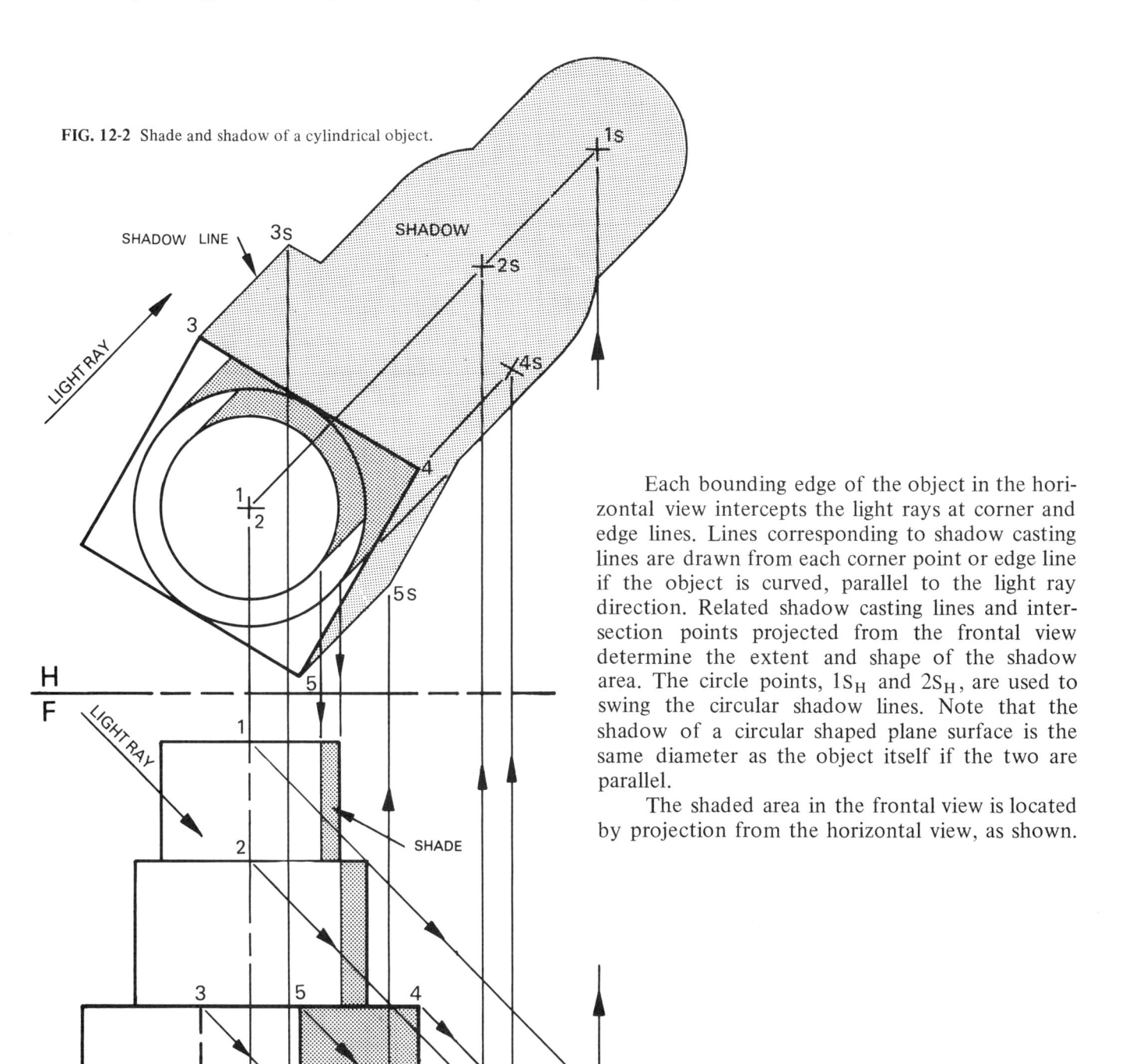

FIG. 12-2 Shade and shadow of a cylindrical object.

Each bounding edge of the object in the horizontal view intercepts the light rays at corner and edge lines. Lines corresponding to shadow casting lines are drawn from each corner point or edge line if the object is curved, parallel to the light ray direction. Related shadow casting lines and intersection points projected from the frontal view determine the extent and shape of the shadow area. The circle points, $1S_H$ and $2S_H$, are used to swing the circular shadow lines. Note that the shadow of a circular shaped plane surface is the same diameter as the object itself if the two are parallel.

The shaded area in the frontal view is located by projection from the horizontal view, as shown.

FIG. 12-3 Shade and shadow of a pyramid.

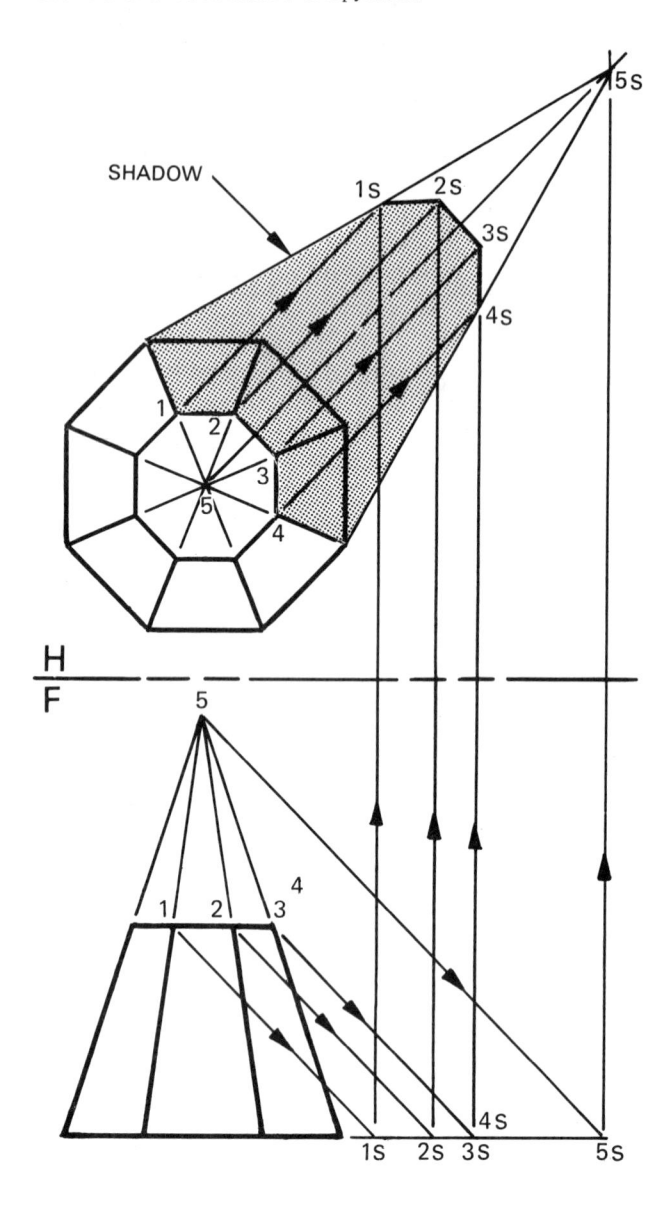

12.3 SHADES AND SHADOWS: CONES AND PYRAMIDS

The extreme point of the shadow line of a pyramid (or cone) is determined by locating the point of intersection of its vertex point, as it intercepts a light ray, and the horizontal base plane on which it rests.

In Fig. 12-3 the shade and shadow of the frustrum of the pyramid is illustrated. The light direction determines the angle of the parallel projectors drawn from the corner points in the frontal view. Each projector intersects the horizontal base plane and locates points 1S through 5S. The intersection points are then projected to the horizontal view to determine the extent of the shadow line.

In the horizontal view, parallel lines are drawn from each corner point of the object and vertex 5. The intersection of these parallel lines with related points projected from the frontal view establish the shadow area. Point $5S_H$ is located first. The limiting shadow lines are determined by drawing lines from the outer corners of the frustrum to point $5S_H$.

There will not be a visible shade area because of the frustrum's orientation.

In Fig. 12-4, the shadow area is determined by finding the point of intersection of the light ray intersected by vertex point 2 and the horizontal base plane on which the object rests, $2S_H$. The portion of the object's surface which is cast in shadow is determined by points of tangencies of the parallel light rays and the cone as shown. The center point, point 1, of the upper base of the frustum of a cone is projected to the horizontal base plane and then to the horizontal view so as to locate the circular portion of the shadow line. The shade of the figure is established by projection of the tangent points from the horizontal view.

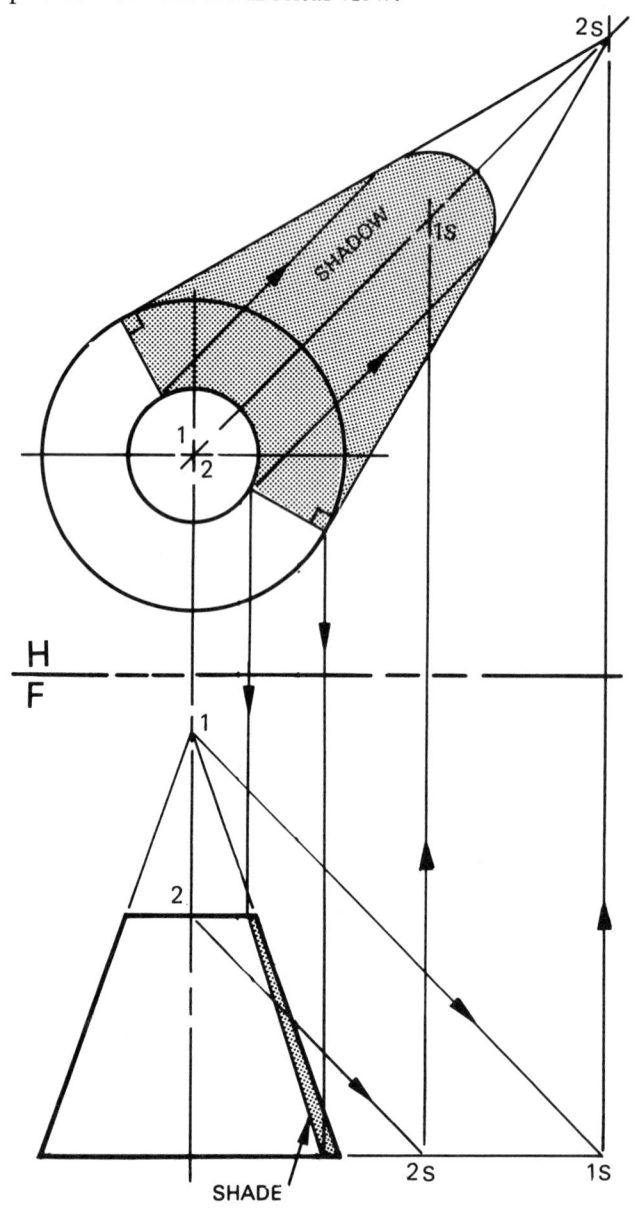

FIG. 12-4 Shade and shadow of a cone.

400

12.4 OBLIQUE PICTORIAL: SHADE AND SHADOW

The concepts applied to the projection of shades and shadows for objects shown in multiview projection can also be applied to pictorials. The first step in the construction of the shade and shadow for a pictorial is to establish the light ray direction. The light ray direction for an oblique pictorial is determined by drawing an oblique cube, Fig. 12-5. The conventional light ray direction is at 45 degrees to the frontal and horizontal views, and is drawn as a diagonal from the upper left front corner of the cube to the lower rear right corner. Line 1-2 is the pictorial light ray direction. Line 1-2 projects onto the horizontal plane as line 3-2 and onto the vertical plane as line 4-2. All intercepted light rays extended from points on the pictorial oblique object are parallel to line 1-2, e.g. lines 1-1S, 2-2S, 3-3S, etc., in Fig. 12-6. Vertical lines projected onto horizontal surfaces are parallel to line 3-2, e.g. lines 1^1-1S, 2S-3S in Fig. 12-6. Horizontal lines projected onto vertical surfaces are parallel to line 4-2, e.g. line 10-10S in Fig. 12-6.

The shade and shadow of Fig. 12-6 are established by determining the intersection of each intercepted corner point and edge line (curved or straight) and the horizontal plane. Point 1S is located at the intersection of a pictorial light ray projected from point 1 (parallel to 1-2 in Fig. 12-5) and the horizontal projection of the light ray from point 1^1 (parallel to 3-2 in Fig. 12-5).

A point on the oblique pictorial is projected to the horizontal plane by drawing a pictorial light ray from it to the intersection of its related horizontal light ray projection line. The horizontal light ray projector is drawn from a point on the base of the object established by dropping verticals from the objects points until they intersect an edge line of the base.

A pictorial light ray extended from a point on the object along with its horizontal projector and vertical line determine a triangular plane. Each plane is parallel and proportional to vertical plane 1-$1^1$1S. Therefore it is parallel and proportional to plane 1-2-3 of Fig. 12-5.

The curved portion of the shadow line is determined by assuming points along its circle edges and projecting them to the horizontal plane as above.

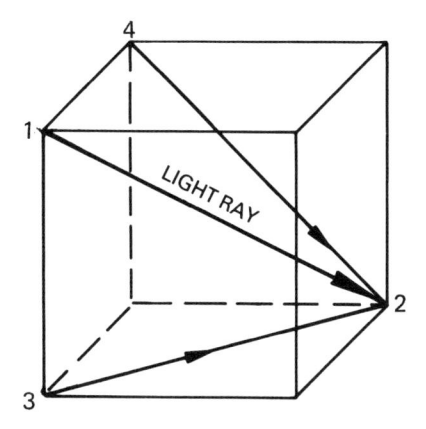

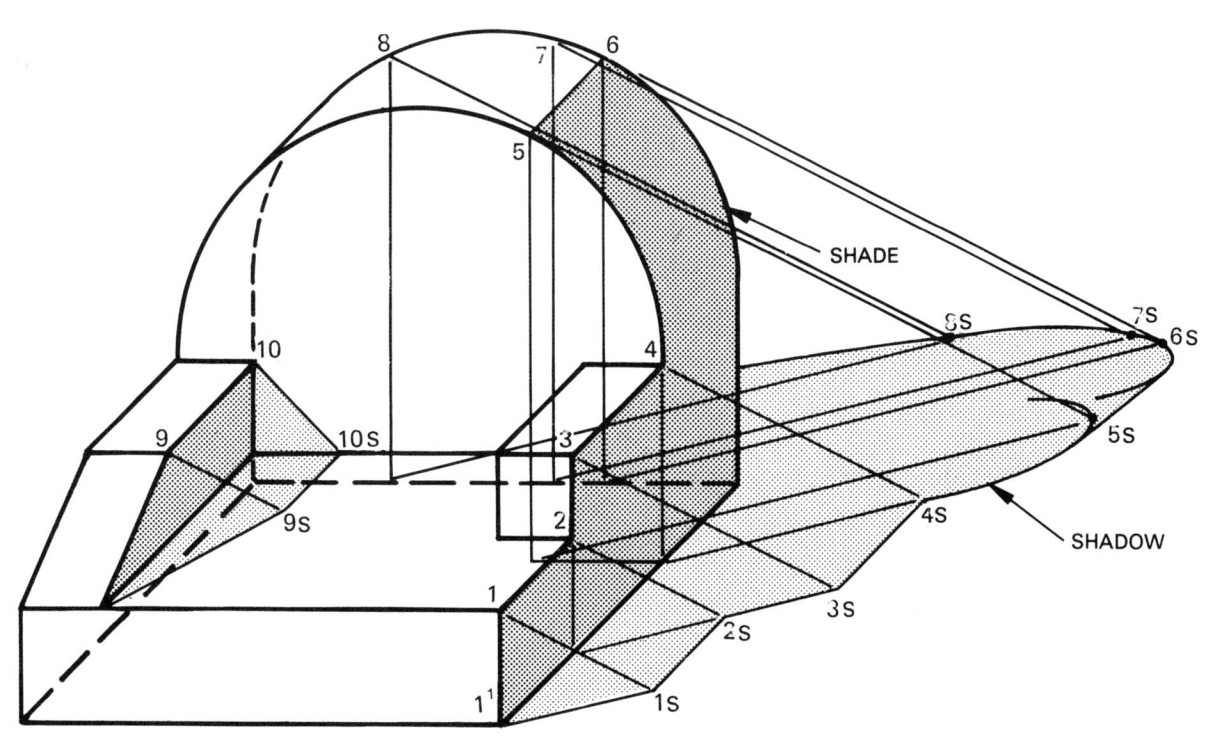

FIG. 12-5 Oblique box used to determine light ray direction.

FIG. 12-6 Shade and shadow of an oblique object.

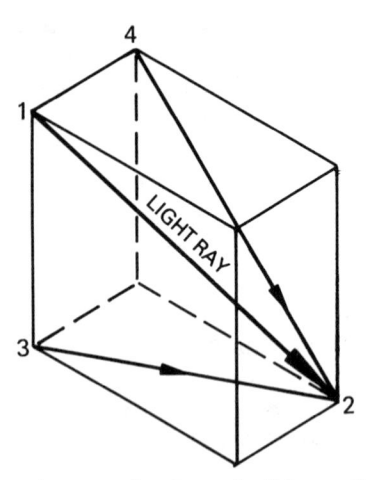

FIG. 12-7 Isometric box used to determine light ray direction.

12.5 ISOMETRIC PICTORIAL: SHADE AND SHADOW

The light ray direction for a pictorial isometric is established by constructing an isometric box as in Fig. 12-7. The depth of the cube is foreshortened by half in order to provide a more realistic shadow outline direction.

The pictorial light ray direction corresponds to line 1-2. Its horizontal projection is line 3-2 and its vertical projection, line 4-2. All intercepted light rays extended from points on the pictorial isometric object are parallel to line 1-2, e.g. lines 2-2S, 4-4S, 5-5S, 6-6S, 7-7S, etc., Fig. 12-8. Vertical lines projected onto horizontal surfaces are parallel to line 3-2, e.g. line 0^1-0S, Fig. 12-8. Horizontal lines projected onto vertical surfaces are parallel to line 4-2, e.g. line 3-3^1, Fig. 12-8.

In Fig. 12-8, intercepted pictorial light rays are drawn from each point on the object (parallel to 1-2 in Fig. 12-7). Verticals are drawn from each point until they intersect a horizontal edge of the base. A horizontal projection line from each related point on a horizontal edge line is drawn parallel to 3-2 in Fig. 12-7. The intersection of pictorial light ray projectors and their related horizontal projection lines determine points along the shadows outline. The three related lines, vertical, horizontal projector, and pictorial light ray projector, form a vertical plane. All planes are parallel and proportional to plane 0-0^1-0S. Therefore, they are parallel and proportional to plane 1-2-3 of Fig. 12-7.

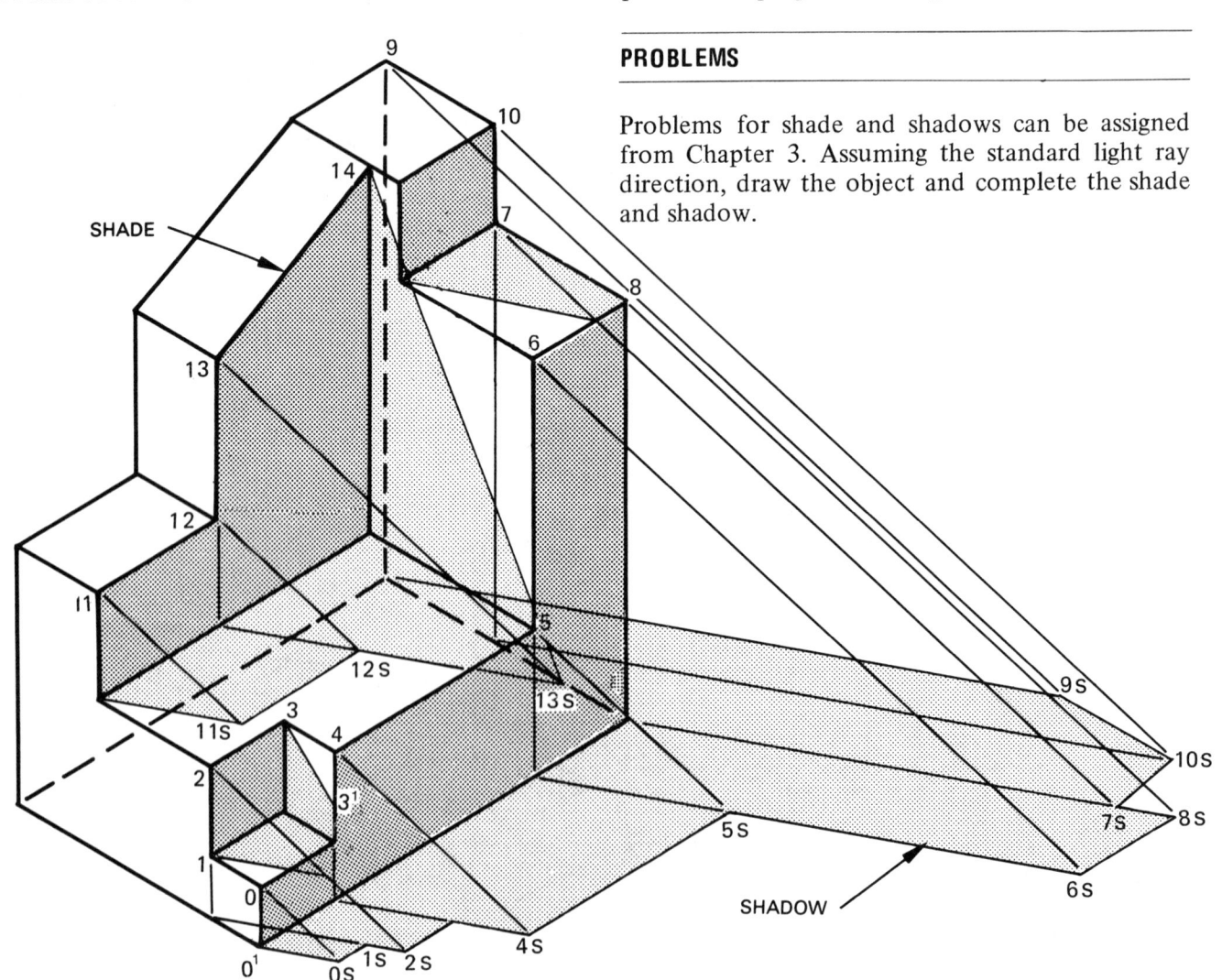

PROBLEMS

Problems for shade and shadows can be assigned from Chapter 3. Assuming the standard light ray direction, draw the object and complete the shade and shadow.

FIG. 12-8 Shade and shadow: isometric object.

13

VECTORS AND GRAPHICAL STATICS

WALTER N. BROWN

Instructor, Santa Rosa Junior College

Santa Rosa, California.

FIG. 13-1 Construction backhoe/loader. *(Courtesy J. I. Case.)*

FIG. 13-2 Rotating ship's crane.

FIG. 13-3 Construction crane and building trusses.
(Courtesy NASA.)

PART I: VECTOR REPRESENTATIONS OF FORCES FOR GRAPHICAL SOLUTIONS TO STATICS PROBLEMS

The design of structures such as roof trusses, highway bridges, transmission towers, dock loading cranes, earth movers, etc., requires a complete analysis of all of the forces acting on the complete structure as well as the forces acting in each member and connection of the structure. Graphical representation of these forces, known as vectors, provides a powerful tool for the analysis of these force systems. The structures in Fig. 13-1 through 13-5 can be designed using graphical analysis. A graphical solution that has been completed using reasonable drafting skill will provide answers equal in usefulness to the answers of a straight mathematical solution. Graphical and mathematical solutions must be thought of as complementing each other. One solution method can always be (and should be) used as an accuracy check of the other.

FIG. 13-5 Construction crane models. *(Courtesy Engineering Model Associates, Inc.)*

FIG. 13-4 Electrical transmission towers.

A strong positive feature of a graphical solution is its ability to be visually checked for correctness of procedures and projections. *Statics* is the study of force systems that are in a state of equilibrium. The principles and practices of descriptive geometry are the tools necessary to make graphical solutions to statics problems.

13.1 DEFINITIONS OF TERMS

Vector geometry introduces new terms and concepts that must be understood before the techniques of graphical vector analysis can be applied. The following definitions are used in this chapter. Additional definitions will be introduced in the text at appropriate times.

Force: *An influence (as a push or a pull) that attempts to cause motion or a change of motion.* There are four characteristics of a force.

1. *Magnitude:* The amount of the force measured in standard units, such as pounds, newtons, etc.
2. *Point of application:* The place where a force is applied to an object.
3. *Line of action:* The line that positions the force as it acts at the point of application.
4. *Sense:* The tendency of motion along the line of action caused by the force, either toward the point of application (called *compression*) or away from the point of application (called *tension*).

The term *"direction of force"* is often used in nongraphical applications but is not exact enough for use in the graphical analysis of structures. Direction is a combination of the *point of application, line of action,* and *sense* of a force.

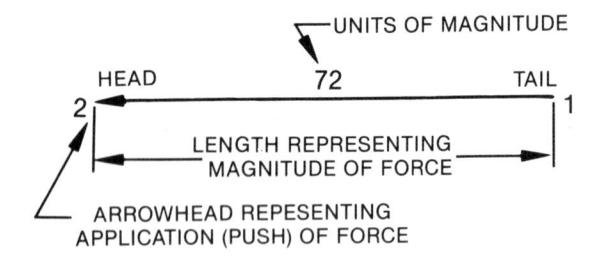

FIG. 13-6 The vector 1-2.

Vector: A graphical representation of a force. It is a straight line with an arrowhead at one end, as shown in Fig. 13-6. Vectors are drawn to scale with length units representing force units (as 1mm = 10 newtons). This scale length is the magnitude of the vector. The arrowhead indicates the application of the force along its length and is determined from the sense of the force. Labeling gives the *tail to head* relationship of a vector. The vector 1-2 shown in Fig. 13-6 acts from tail end 1 to head end 2. A vector 2-1 would act from 2 to 1 and would represent a force opposite to that shown in Fig. 13-6.

Resultant force: A single force acting on an object that will produce the same effect as several forces acting on that object. The resultant force is the summation of the several individual forces.

Equilibrium: A state of balance between opposing forces or actions. The summation of all forces is equal to zero when these forces are "in equilibrium."

Equilibrant force: A force (or forces) that will counteract all of the other forces acting on an object so as to create equilibrium. An equilibrant force is equal in magnitude but opposite in action to the resultant force of several individual forces.

Component forces: Two or more forces that will produce the same effect as a single force. The summation of the two or more individual forces is equal to the given force. Component forces are often positioned parallel to the Cartesian coordinates and are then called the X component, Y component, and Z component.

Bow's notation: A systematic method for notation of a *structural diagram* and its *force diagram*. The spaces between structural members are numbered and these numbers are used to label vectors. This permits consecutive labeling of vectors. The advantages of this method of notation will become apparent in problem solutions given in the text.

Pin connection: A connection of structural members that does not restrain the movement of the members around the connection, as with a loose bolt. This is the connection used with axially loaded members. A *pin connection* is also called a *pin connected joint* or *pin joint.*

13.2 DEFINITIONS OF FORCE SYSTEMS

The analysis of structures and their force systems utilizes both two-dimensional and three-dimensional space. The following terms are used in this chapter to define these space relationships.

Collinear forces act along a common line of action. These forces have parallel lines of action, although their actions may be opposed. A two-force system in equilibrium is always collinear, with the two forces being equal in magnitude and opposite in action.

Coplanar forces all lie in one plane (two-dimensional space). The number of forces is not limited as long as they lie in a single plane. A three-force system in equilibrium is always coplanar. The magnitudes and actions of these forces are the subject of Part II of this chapter.

Noncoplanar forces lie in more than one plane (three-dimensional space). The number of forces is not limited. A noncoplanar force system in equilibrium has a minimum of four forces. Noncoplanar force systems are the subject of Part III of this chapter.

Concurrent forces have lines of action that pass through a common point. Concurrent force systems may be coplanar or noncoplanar. Nonparallel three-force systems in equilibrium are always concurrent coplanar force systems.

Nonconcurrent forces have lines of action that do *not* pass through a common point. Forces with lines of action that are parallel, skewed, or have multiple intersections are nonconcurrent. These force systems may be coplanar or noncoplanar.

13.3 DEFINITIONS OF DIAGRAMS

The graphical solution of engineering problems utilizes specific related diagrams that perform analytical functions similar to simultaneous equations in mathematical solutions. The following diagrams are used in these graphical solutions. Examples of these diagrams are shown in Fig. 13-22 on page 415.

Structural diagram: A view[1], drawn to scale, of an object that accurately shows the position of all structural parts of the object, the connections of these parts, and the external forces acting on the object. Often a *structural diagram* is simplified by showing only the centerlines of each member. *Space diagram* is another name used for this simplified view.

Free body diagram: A view of an isolated connection or part of an object. A free body diagram is used in the analysis of all forces acting at a single connection or on a single part of the object.

Force diagram: A view of the vectors, drawn to scale, of the forces acting on an object and of the force acting in each part of the object. The length unit used represents a unit of force. *Vector diagram* is another name used for this view.

String diagram: This diagram is defined and used in the section on coplanar nonconcurrent force systems. *Funicular diagram* is another name used for this view.

13.4 NECESSARY CONDITIONS FOR EXPECTED RESULTS

Graphic problem solutions, like mathematical solutions, are only as good as the skills used by the problem solver. The accuracy of the solution is a direct result of the accuracy used in the drafting of the diagrams and of following the sequence of steps that lead to the solution. Careful layout and sharp linework and lettering will produce a solution that is visually easy to check for procedure and accuracy by the one who draws it as well as by anyone who reviews or checks it. These solutions will have all of the useful accuracy of any mathematical solution.

The solutions for the structural design requirements of any structure or machine become a part of its design documentation and a reference part of the contract documents that are used to bring the structure or machine into reality. Therefore these solutions must stand up to the rigors of accepted engineering practice, design and code review by governing agencies, contract law, and possible later modifications to the design. This can easily span several years and involve many individuals who will be reviewing and using these solutions. It is not possible to overemphasize the value of good design documentation—documentation that includes these solutions. It is because of this need that, in critical situations, mathematical methods are valuable for checking graphical solutions and graphical methods do the same for mathematical solutions. Attention to the following items will aid in the consistent production of useful solutions and design drawings.

---

[1] In two-dimensional systems, only one view (the true size view) is used. In three-dimensional systems, a minimum of two adjacent views are necessary to show the space relationships and true lengths.

Proper equipment: Graphical problem solutions require precision drafting of exact line lengths and angles. This is only possible with quality equipment properly used. Lines must be drawn exactly parallel to other lines and line lengths must be measured with precision. A solid T square and adjustable triangle used on a true-edged drawing board are minimum requirements. Drafting on quality vellum is recommended as it allows changes with minimum damage to the paper. Quality vellum will produce excellent reproductions by any method if dark and sharp linework is used.

Systematic procedure: There is a systematic sequence of steps for the solution of each force problem, just the same as there is for any other descriptive geometry problem. The solutions of force problems are applications of the skills and methods learned in descriptive geometry. The recommended procedures for each problem solution are given in the examples.

Big is better: The selection of an appropriate scale for a drawing is a compromise between the degree of accuracy required, the smallest unit that must be measured, and the space available for drawing the view. While calculators routinely give readings to eight places, professional design engineering practice rarely requires readings beyond three significant figures. This accuracy is readily obtainable with views that can be arranged on standard sizes of paper. The largest scale that available space permits should be selected for ease and accuracy of measuring lengths. Measurements of less than one hundredth of an inch (one half of the smallest division on the "50" engineers scale) are not desirable or realistic.

Summarization of data: As previously discussed, problem solutions as design documentation must be able to stand the test of time and review. It becomes critical that in addition to being accurate, the solution must also be explicit in presenting the given design parameters and the results of the analysis. This is effectively done by a summarization of each force acting in the structural system. The summarization lists both the given forces and the determined forces. More discussion of summarization is given in the examples.

13.5 COPLANAR VECTOR SUMMATION

The summation of two or more vectors is their resultant vector. Coplanar vectors appear true length in a single view and their true length projections are used to determine the resultant vector.

Three structural members connected by a loose pin at point A are shown in Fig. 13-7. The force acting in each member and the angles between members are given. Also an X axis and Y axis have been superimposed through the pin at point A. Two graphic methods—parallelogram and polygon—are illustrated, as well as a mathematical method.

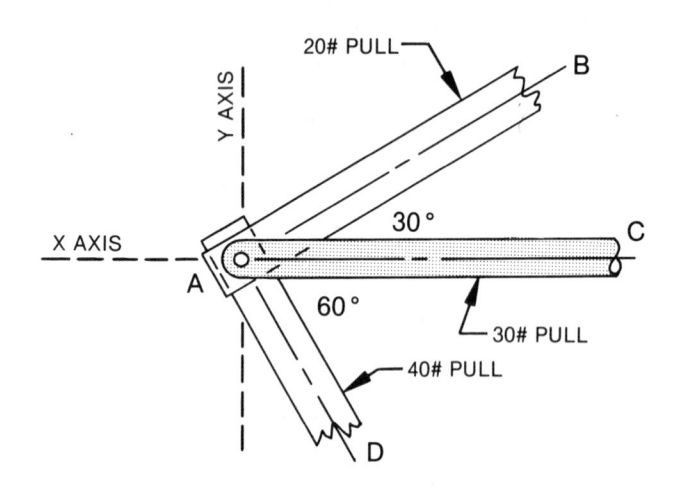

FIG. 13-7 Structural diagram.

The *parallelogram method* is shown in Fig. 13-8. Vectors **AB**, **AC** and **AD** are drawn exactly parallel to members **AB**, **AC**, and **AD** with their lengths representing their respective magnitudes and with arrowheads indicating their respective actions on the pin at point A. In this method, only two vectors can be summed into a resultant vector in each step. The first step is to determine the resultant of the vectors **AB** and **AC**. The parallelogram ABCE is drawn to find point E and the vector **AE** is the resultant of vectors **AB** and **AC**. Next, the parallelogram AEFD is drawn to find point F and vector **AF** is the resultant of resultant vector **AE** and vector **AD**. Therefore, vector **AF** is the resultant vector of vectors **AB**, **AC**, and **AD**. Its magnitude and angle with the positive X axis can now be measured. It is good practice to draw given vectors and resultant vectors so they look different. This permits easy visual identification when reading the completed diagram. Here, the broken line vector is the resultant vector.

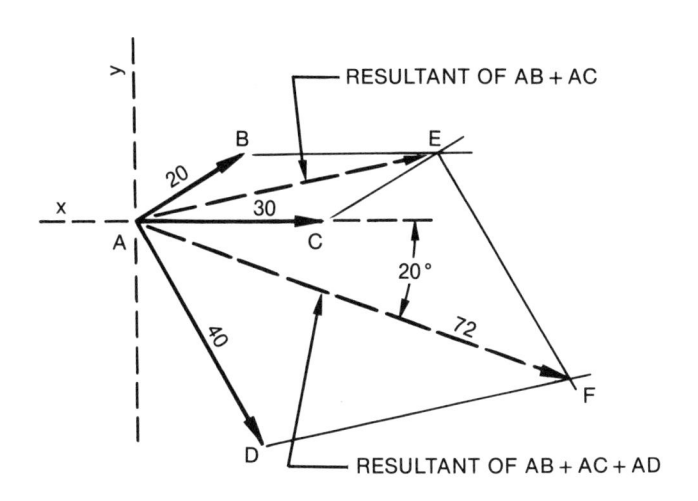

FIG. 13-8 Parallelogram method.

The *polygon method* is shown in Fig. 13-9 and is the method used in drawing *force diagrams*. To make a systematic identification of the vectors possible in the sequence of their summation, the use of **Bow's Notation** is introduced. A simplified sketch of the centerlines of members AB, AC, and AD is drawn and labeled. Then the spaces between pairs of members are numbered. These numbers give *automatic* labeling so that vector **1-2** acts in member AB, vector **2-3** acts in member AC, and vector **3-4** acts in member AD. The great advantage to this is that the label of the tail of one vector is the same as the label of the head of the succeeding vector in the *vector polygon*. The use of Bow's Notation gives a logical and consistent sequence to vector summation, as well as simplifying vector labeling. Additional advantages will become apparent as different problems are examined.

The *vector polygon* is constructed by drawing vector **1-2** exactly parallel (either it's parallel or it isn't!) to member AB. Next, vector **2-3** (parallel to member AC) is drawn starting at point 2 of vector **1-2**. Finally, vector **3-4** (parallel to member AD) is drawn starting at point 3 of vector **2-3**. The resultant of vectors **1-2**, **2-3**, and **3-4** is the vector **1-4** which can now be drawn (to look different) and measured for magnitude and angle.

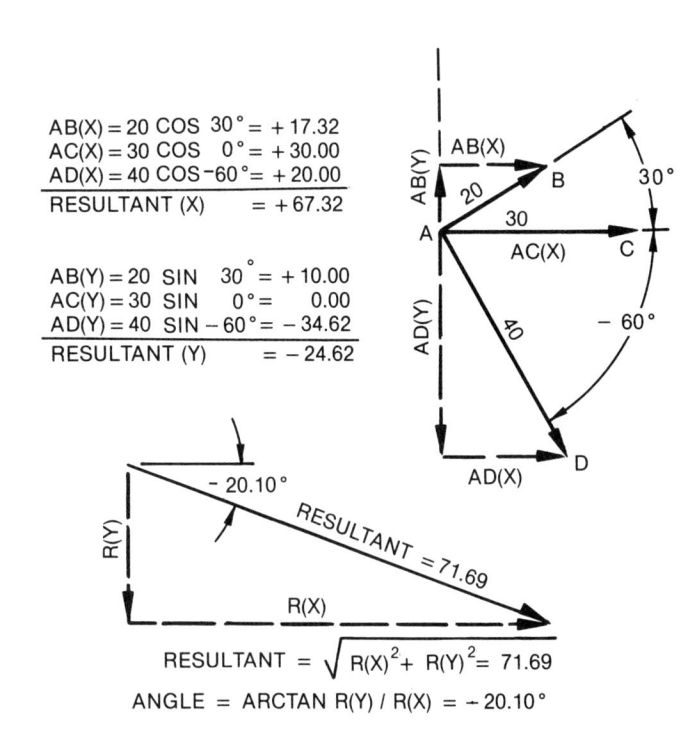

$$AB(X) = 20 \cos 30° = +17.32$$
$$AC(X) = 30 \cos 0° = +30.00$$
$$AD(X) = 40 \cos -60° = +20.00$$
$$\text{RESULTANT (X)} = +67.32$$

$$AB(Y) = 20 \sin 30° = +10.00$$
$$AC(Y) = 30 \sin 0° = 0.00$$
$$AD(Y) = 40 \sin -60° = -34.62$$
$$\text{RESULTANT (Y)} = -24.62$$

$$\text{RESULTANT} = \sqrt{R(X)^2 + R(Y)^2} = 71.69$$
$$\text{ANGLE} = \arctan R(Y)/R(X) = -20.10°$$

FIG. 13-10 Mathematics method.

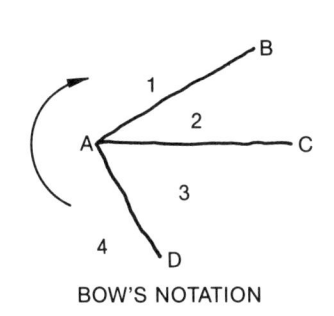

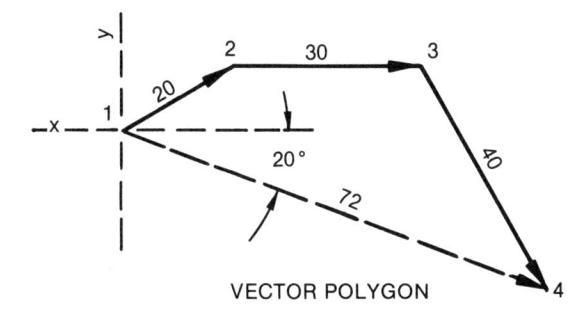

FIG. 13-9 Polygon method.

A *mathematics method* is illustrated in Fig. 13-10 using the summation of forces parallel to the horizontal X axis and to the vertical Y axis. The X axis component and the Y axis component of each force acting in members AB, AC, and AD is individually calculated and summed. These sums are the X and Y components of the resultant force. While this method gives two (or as many as the calculator will permit) decimal places of accuracy, three-significant-figure accuracy (obtainable in the graphic solutions) is all that is required for most professional engineering practice.

PROBLEMS 13-1 through 13-6

Each problem is to be solved on one sheet of 8 1/2" X 11" drafting quality paper. Regular tracing vellum is recommended and is regularly used by practicing engineers. Before drawing a final problem solution, make a quick semi-freehand study of the views that will be used in the solution. This rough drawing will provide a check on the proposed method of solution, will aid in selecting the largest scale the sheet size will permit, and will help in view placement so as to make a neat, logically organized and easily read drawing. Use this "study of views" method for all problems in this chapter—it is a time saver for producing quality work.

Problems 13-1 through 13-3 are of a coplanar system of force vectors that appear true length in Fig. 13-11.

PROB. 13-1. Using the parallelogram method, determine the magnitude and angle (measured from the positive X axis) of the resultant vector **OR** that is the combinational result of the given vectors **OA, OB,** and **OC.** The given vector magnitudes are: **OA** = 120 force units, **OB** = 60 force units, and **OC** = 70 force units. The angles are given in Fig. 13-11. (The term "force units" is used here to emphasize that any decimal scalar unit can be used in the solution of these problems—SI, U.S. Customary, or any other. The only requirement is that a unit of force can be assigned to a unit of length.)

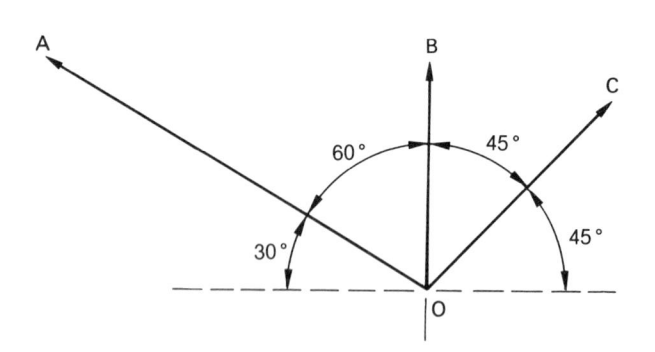

FIG. 13-11 Problems 13-1, 13-2, and 13-3.

PROB. 13-2. Solve problem 13-1 using the polygon method.

PROB. 13-3. Solve problem 13-1 mathematically.

PROBS. 13-4 through 13-6 are of a noncoplanar system of force vectors that are represented in the horizontal and frontal views of Fig. 13-12. The background grid represents 10 force units per square. The solution of these problems uses the same techniques used in problems 13-1–13-3 with the following additions: (a) Each view is summed separately to determine the resultant vector projection in that view and (b) The order of summing must be the same in each view—points must project between views as in all descriptive geometry problems.

PROB. 13-4. Using the parallelogram method, project the horizontal and frontal projections of the resultant vector **OR** that is the combinational result of vectors **OA, OB,** and **OC.** Using the rotation method, determine the magnitude of the resultant vector **OR** and measure its angle relative to the horizontal and frontal planes. Upward and forward give positive angles.

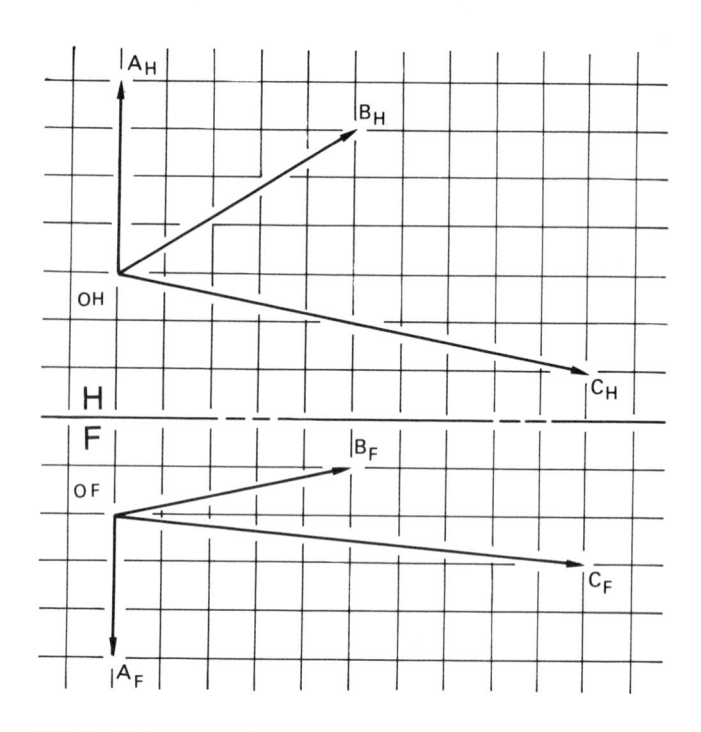

FIG. 13-12 Problems 13-4, 13-5, and 13-6.

PROB. 13-5. Solve problem 13-4 using the polygon method and using an auxiliary view to determine the magnitude of the resultant vector **OR.**

PROB. 13-6. Solve problem 13-4 mathematically.

FIG. 13-13 Concurrent force systems in action.

PART II: COPLANAR FORCE SYSTEMS

Many kinds of load lifting equipment such as mobile construction cranes and towers, earthmoving loaders and backhoes, forklifts, and tow trucks are primarily applications of coplanar force systems or combinations of coplanar force systems. Highway and building beams and trusses also will usually function as coplanar systems. While these objects are physically three dimensional, the main loads acting on and in the objects, as a result of the work being done—called *working loads*—normally will form coplanar force systems. Secondary considerations of materials, stability, and buckling necessitate the three-dimensional shape.

The forces acting as a result of design working loads are the subject of this chapter. The equipment and structures themselves will be considered to be weightless (not true) and their materials will be considered infinitely strong (also not true). Consideration of these factors is a refinement of the principles presented here. To include them now would only obscure basic concepts that need to be understood first.

Section 13.6 is a discussion of two-force systems that may also be called one-dimensional systems. The two forces necessarily act along a common line of action. Sections 13.7, 13.8, and 13.9 discuss three-force systems that are necessarily two-dimensional coplanar systems. The force systems in Section 13.7 and 13.8 are also concurrent systems, another necessary condition when three lines of action are nonparallel. The special condition of three parallel forces, a system that is coplanar and nonconcurrent, is treated in Section 13.9.

The *string diagram* and how it is applied in the solution of systems of four (or more) nonconcurrent coplanar forces is introduced in Section 13.10. This method works equally well for parallel and nonparallel forces. The *Maxwell Diagram* and its use in solving for the force acting in each member of a building or bridge truss is introduced in Section 13.11.

13.6 TWO FORCES ACTING ON AN OBJECT

Two methods of supporting a 10-pound block are shown in Fig. 13-14: a column under the block and a chain suspending the block. These two systems are graphically shown in drawings (a) and (c) of Fig. 13-15. This section examines how the column and the chain transmit the weight of a 10-pound block to each of the supporting scales. The column and the chain will both be considered to be weightless so that the weight of the block and the readings of the two scales can all be considered equal. A first examination of force systems often uses the assumption of weightless members. However, the weight of the members is included in later refined calculations.

In Fig. 13-15, drawing (a), the forces acting on the column are shown collinear with the centerline of the column. If this is not true and the column is not otherwise supported, the system collapses. The push of the block at point A must be counteracted by an equal and opposite push by the column (Newton's Third Law!), as shown in drawing (b). A similar equal and opposite push is transmitted from grain to grain along the length of the column to the end at point B. Here the column now exerts a push on the scales and the scales pushes back with an equal force, as indicated by the dial reading in Fig. 13-14.

In Fig. 13-15, drawing (c), the forces acting on the chain are shown collinear with the taut chain. This collinearity is easily demonstrated with a simple string model. The downward pull of the hook on the block at point D must be counteracted by an equal and opposite upward pull by the chain (Newton again) as shown in drawing (d). A similar equal and opposite pull is transmitted from link to link of the chain until the top link exerts a downward pull equal to the weight of the block.

These two examples demonstrate important principles that apply to two-force structural members in compression or in tension.

COMPRESSION IN A COLUMN

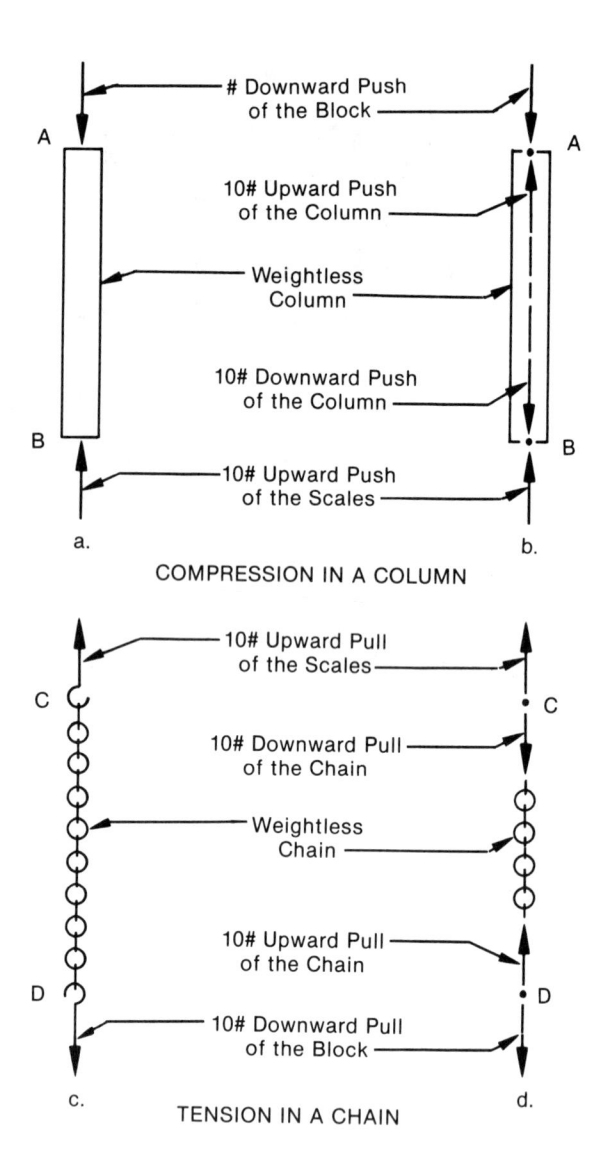

TENSION IN A CHAIN

FIG. 13-15 Two-force systems.

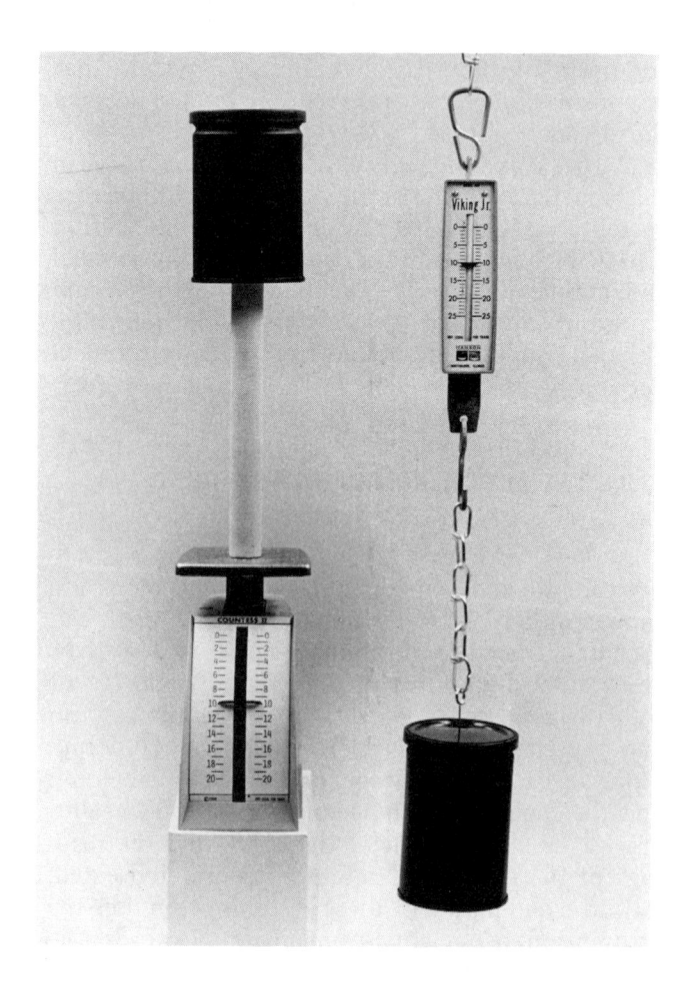

FIG. 13-14 Compression and tension forces.

1. When two forces act on a member in equilibrium, the lines of action of the two forces are collinear, equal in magnitude, and opposite in action. The system is one dimensional.
2. A member in compression exerts a pushing force at each end of the member.
3. A member in tension exerts a pulling force at each end of the member.
4. A member that receives a compression force at one end will deliver the same compression force to another member at its other end.
5. A member that receives a tension force at one end will deliver that same tension force to another member at its other end.

13.7 THREE FORCES AT A PIN CONNECTION

A three-force member support bracket is shown in Fig. 13-16. This bracket has pin connectors at each end of each member. It supports the gravity load shown. The problem requires determining the nature of the force acting in each member of the bracket when the system is in equilibrium.

FIG. 13-16 Three-forces acting at a pin connection.

The **structural diagram**, as shown in Fig. 13-17, must accurately reproduce the shape of the support bracket. The *structural diagram* is the source of the line of action of the force (and of the vector) acting in each member. Therefore, if the *force diagram* is to produce useful results, graphical accuracy in this diagram is mandatory.

The first step in solving any problem is to examine the problem statement and to determine exactly what needs to be done to produce the required answers. In this problem, it is to determine the force acting in each member of the bracket. In the *structural diagram*, letter names are given to each *connection* (also called a *joint*) of the bracket so that the members may be identified as AB, AC, and AD. If the bracket is made of adequate materials so as to support the specified load, it will be in equilibrium and all of its members and connections will also be in equilibrium. The condition of equilibrium is the key to the solution of this problem (and to most other structural problems). Is there any one place where all forces act and must be in equilibrium? Point A is the connection that is common to all three members—AB, AC, and AD. Therefore, solving for equilibrium at joint A should (and will) provide the necessary answers. The technique used is to consider joint A as an isolated part of this structure (a free body) that is in equilibrium. The solution then finds the magnitude and sense of the force in each member common to point A. These forces hold point A in equilibrium.

KNOWN AND UNKNOWN CHARACTERISTICS OF FORCES IN Fig. 13.17			
CHARACTERISTICS OF EACH FORCE	FORCE IN STRUCTURAL MEMBERS		
	MEMBER AB	MEMBER AC	MEMBER AD
POINT OF APPLICATION	Point A	Point A	Point A
LINE OF ACTION	Parallel to Member AB	Parallel to Member AC	Parallel to Member AD
MAGNITUDE	UNKNOWN	UNKNOWN	85 pounds
SENSE	UNKNOWN	UNKNOWN	Tension (-)

FIG. 13-18 Force parameters.

A concise summary of the given problem parameters is shown in Fig. 13-18. A tabulation like this is helpful to plan strategies for the solution to the problem and to assure that significant facts are not overlooked. This tabulation is made as soon as the acting forces are identified. It is a worksheet which may not be included with the design documentation but it will be a significant help in efficient problem solving.

Examination of any selected joint starts with the use of a *free body diagram*. This may be drawn freehand (it is only used visually) but should closely approximate the angles of the members (and of the lines of action of their forces). Figure 13-19(a) shows the initial layout and labeling of the *free body diagram*. It is a freehand version of the

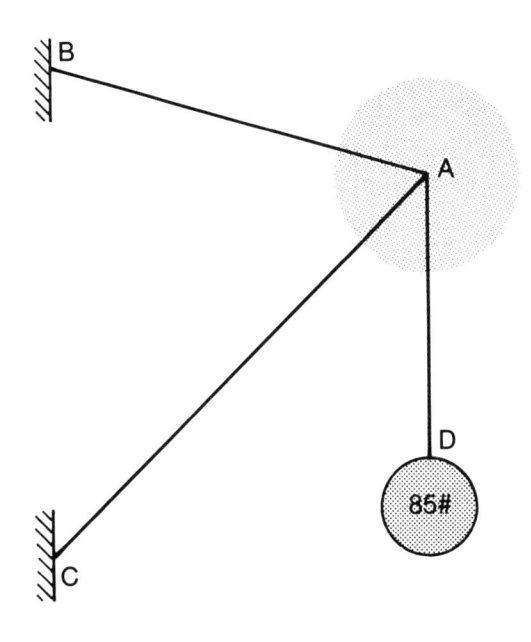

FIG. 13-17 Structural diagram.

413

bracket within the shaded area of Fig. 13-17. An arrowhead is added to indicate how the known force acts on point A. The gravity force attached at point D of member AD causes a tension force to act in member AD and in turn, to apply a tension force pulling away from point A along the line of member AD (Principle #5 of Section 13.6). Next, Bow's Notation is added to give logical sequence and notation to the vectors of the *force diagram*. Numbering of spaces should always result in vector 1-2 representing the known force.

a. INITIAL LAYOUT   b. BOW'S NOTATION

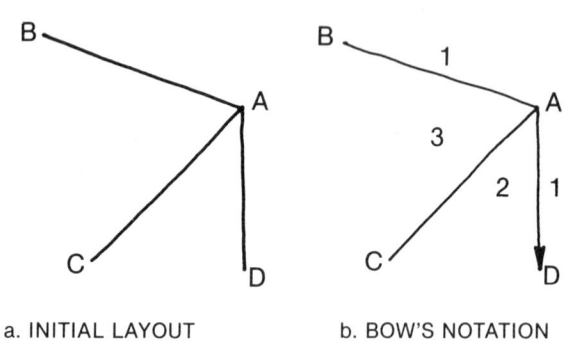

FIG. 13-19 Free body diagram.

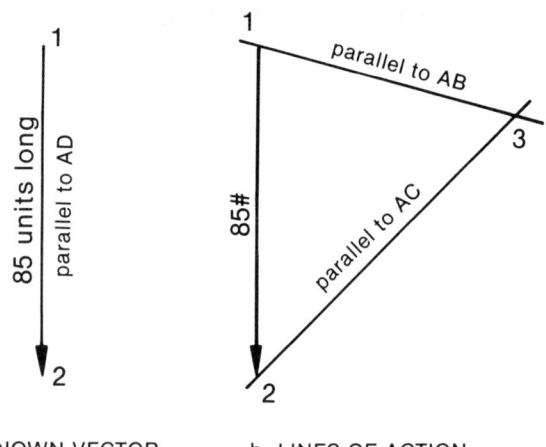

a. KNOWN VECTOR   b. LINES OF ACTION

FIG. 13-20 Force diagram.

The *force diagram* is started in Fig. 13-20(a) by drawing the known vector 1-2 exactly parallel to its member AD. This is its correct line of action. Its length is accurately scaled to represent its known magnitude.

Vector names are taken from Bow's Notation in the *free body diagram* in a continuous order of rotation. In this solution, the rotation around point A is clockwise. The labeling places #1 at the vector tail and #2 at the vector head so that reference to vector 1-2 shows that the push of the force if from 1 to 2. Bow's Notation will cause point 2 also to be the tail of vector 2-3 and point 1 to also be the head of vector 3-1.

The *lines of action* of vectors 2-3 and 3-1 can now be drawn, even though point 3 is as yet unknown. Actually, finding point 3 is the first reason for drawing this force diagram. Vector 2-3 has the line of action of member AC and is drawn through point 2 of vector 1-2 exactly parallel to member AC. Similarly, the line of action of vector 3-1, parallel to member AB is drawn through point 1 of vector 1-2. As shown in Fig. 13-20(b), the intersection of these two lines of action is the desired point 3.

Vector problem solutions always require the completion of an unique triangle from some given information. In this example, two angles are known, as well as their common side. (Actually all angles are known because all lines of action are known.) This is normally the case in structural problems with rigid members. Other problems could have applications of two sides and their included angle known or all three sides known. Vector geometry is as simple as this, once the required triangle is identified.

With point 3 located, the vectors 2-3 and 3-1 can be completed by drawing appropriate arrowheads as shown in Fig. 13-21. The condition of equilibrium requires that the summation of all of the vectors in the force diagram be zero. This means that all vectors are connected head to tail, which is also the result of correct use of Bow's Notation. The magnitude of each vector can now be measured and should be recorded on the diagram. The sense of these forces is unknown at this moment but is determined by completing the *free body diagram*. The arrowhead of a vector indicates the push of the force along the line of action of its corresponding member as it acts on joint A. This push indication, transferred to the member lines of the *free body diagram*, give the sense of the force in relation to point A. If the implied motion is toward point A, the member is in *compression*. If the implied motion is away from point A, the

FORCE DIAGRAM   FREE BODY DIAGRAM

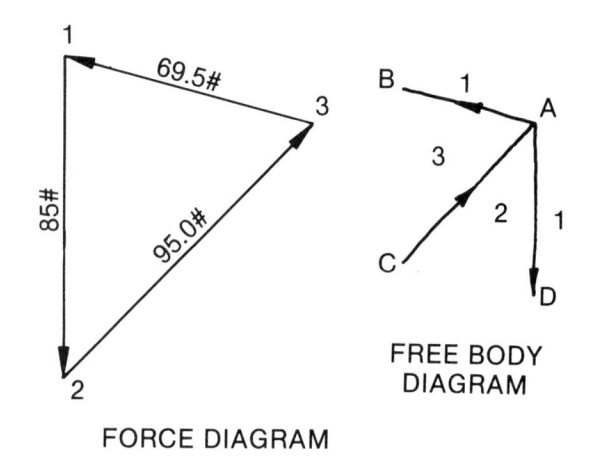

FIG. 13-21 Completed diagrams.

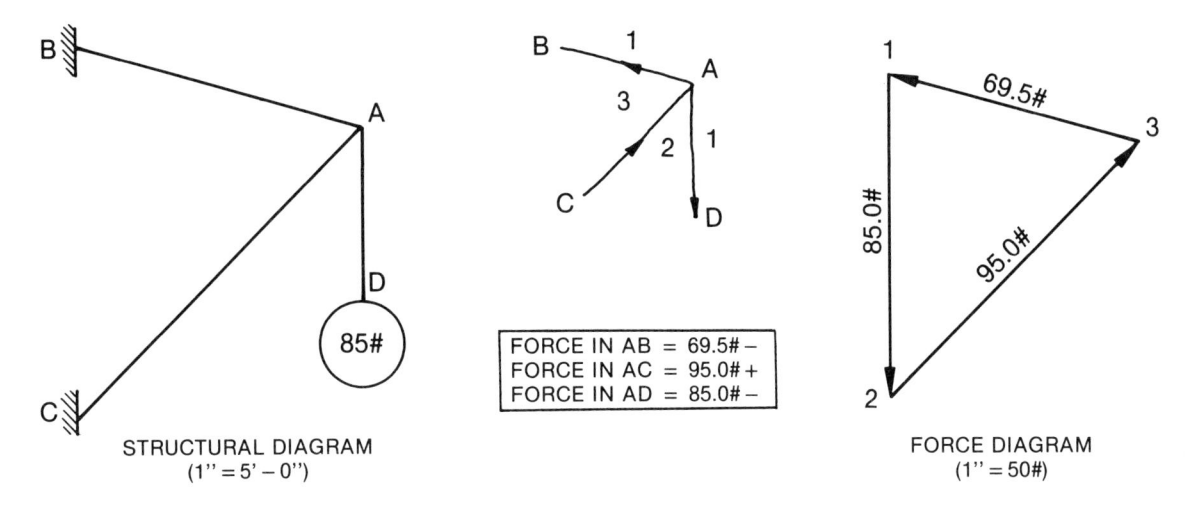

FIG. 13-22 Vector solution of a coplanar concurrent structural problem.

FORCE IN AB = 69.5# −
FORCE IN AC = 95.0# +
FORCE IN AD = 85.0# −

member is in *tension*. To improve readability and make known and found senses look different, the arrowheads on lines AB and AC are placed in the middle of their length. (Note that these arrowheads imply motion of the member. Arrowheads of vectors in a force diagram are always at the ends of the vectors.)

The *free body diagram* and the *force diagram* now have the answers of the problem statement. However, it takes considerable searching by another reader or reviewer to find them (also for the preparer at some later time). This brings up a cardinal rule of all engineering calculations: *A solution is unfinished (and unacceptable) until it contains a clearly identified summary of all results asked for in the problem statement!* The problem asks for the force acting in each member of the bracket, not for the magnitude of vector 1-2, vector 2-3 and vector 3-1 and how they act on joint A. The *free body diagram* and the *force diagram* are necessary for the solution of the problem, but the user of the answers wants to know how many pounds of compression or tension act in each member. The summary should answer *this* question and might read: Force in member AB = 69.5 pounds tension. This is somewhat wordy and can better be abbreviated to: Force in AB = 69.5#−. The '−' sign indicates tension (not a negative number). Compression is indicated with a '+' sign and must be given (it is shorthand for compression, not of a positive number). These are standard conventions and are used throughout this chapter.

The complete solution of this problem is shown in Fig. 13-22 and should be carefully examined for subtle but significant features that contribute to a solution that is easy to understand, easy to check visually, and easy to use. Some items to note are:

1. Natural left to right sequence of diagrams.
2. Structural diagram drawn to scale and the scale noted.
3. Scale of force diagram noted.
4. Free body diagram, as a link between members in the structural diagram and vectors in the force diagram, is positioned between these diagrams.
5. Values are recorded where they are found: magnitude in force diagram and sense in free body diagram. This is valuable for visual checking.
6. Summary of values is boxed so that it "jumps out" from the other drawings.
7. In addition to the above, precision drafting and accurate reading of values are assumed.
8. Does the solution "look right"?

FIG. 13-23 Three nonparallel forces acting on a beam.

13.8 THREE FORCES ACTING ON A BEAM

A stiff member capable of resisting bending is often used in simple lifting applications. This stiff member, better known as a beam, may be combined with two-force members to make a useful mechanism. Figure 13-23 shows such an application with a beam supporting a gravity load at one end and being supported at its opposite end and at an intermediate point. To design this system, it is necessary to determine the forces acting in the two-force members and to determine the force acting at each of the three connections of the beam. It is assumed that the system will be in equilibrium.

Fig. 13-24 is a simplified *structural diagram* of this system. A little thought about the problem should suggest that the real problem is to determine what happens at points A, B, and C of the beam. The force acting in member AE is already known and the force acting in member BD will be known when the conditions at point B are determined. However, the three separate connections to beam ABC raise some questions that do not seem answerable by the methods used with the last problem. In it, the common pin connection was very conveniently at the point of concurrency, but in this problem there are three connection points. Figure 13-25 is a concise tabulation of the knowns and unknowns of the forces acting on beam ABC.

As in the last problem, the magnitude and sense of two of the forces are unknown. In addition, the line of action of the force at point C is only partially known—only that it passes through point C. To resolve this, consider the necessary conditions of a three-force system: A three-force system is coplanar and concurrent. But concurrent where? Not at points A, B, or C! However, concurrency does exist and is easily found. Members AE and BD are both two-force members, and forces carried by them must have lines of action through

KNOWN AND UNKNOWN CHARACTERISTICS OF FORCES IN Fig. 13.24			
CHARACTERISTICS OF EACH FORCE	AT POINT A	AT POINT B	AT POINT C
POINT OF APPLICATION	Point A	Point B	Point C
LINE OF ACTION	Parallel to Member AE	Parallel to Member BD	Through Point C
MAGNITUDE	85 pounds	UNKNOWN	UNKNOWN
SENSE	Downward	UNKNOWN	UNKNOWN

FIG. 13-25 Force parameters.

their respective connections. Extending these lines of action until they intersect, as shown in Fig. 13-26, establishes the point of concurrency at point 0. Drawing line OC gives the desired line of action at point C.

A part of the required answer to the problem is the angle of the force acting at point C. It is recommended that the tangent method be used to measure this (and all other) angles. One side of the triangle should always be drawn 100 units long, as shown in Fig. 13-26. Then the reading of the other side of the triangle, divided by 100, is the tangent of the angle adjacent the 100 side. The division by 100 is easily made by moving the decimal two places to the left. Then only one value need be entered into the calculator (much safer than entering two values and dividing). Angles of forces acting *on* members are measured from the positive X axis with clockwise measurements as minus degrees and counterclockwise measurements as plus degrees. This is consistent with mathematical practice.

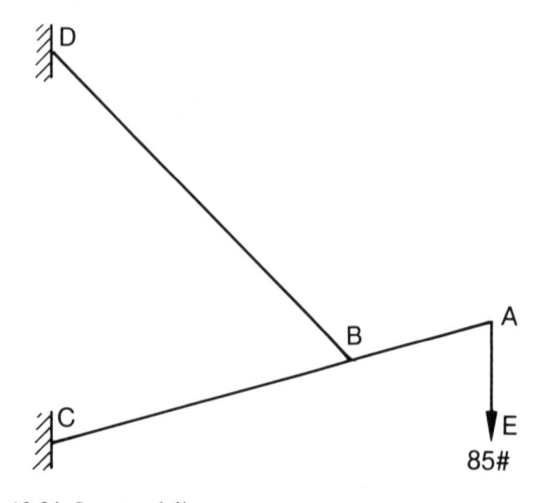

FIG. 13-24 Structural diagram.

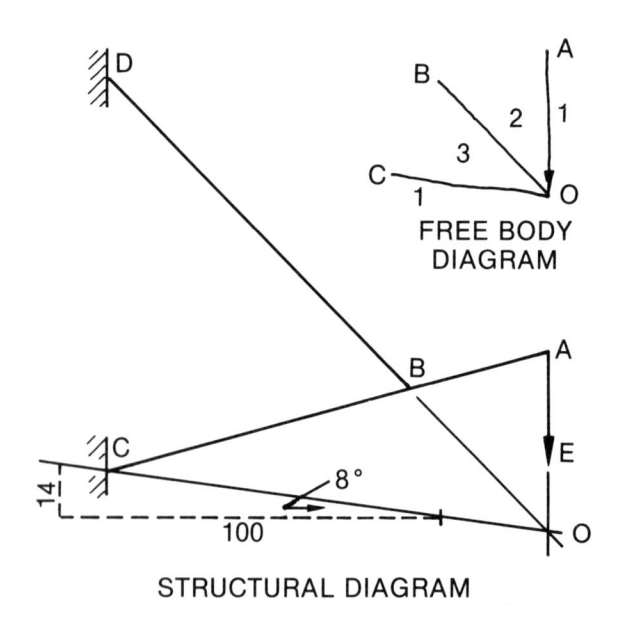

FIG. 13-26 Point of concurrency.

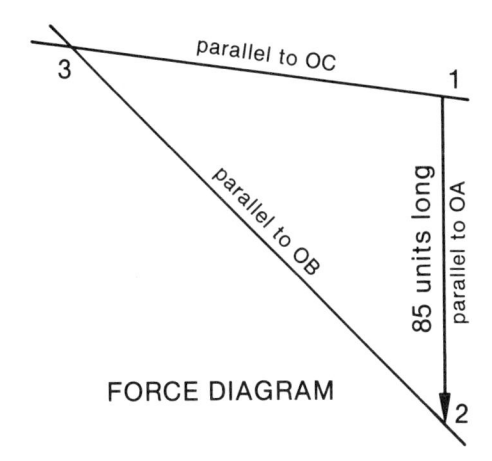

FIG. 13-27 Lines of action.

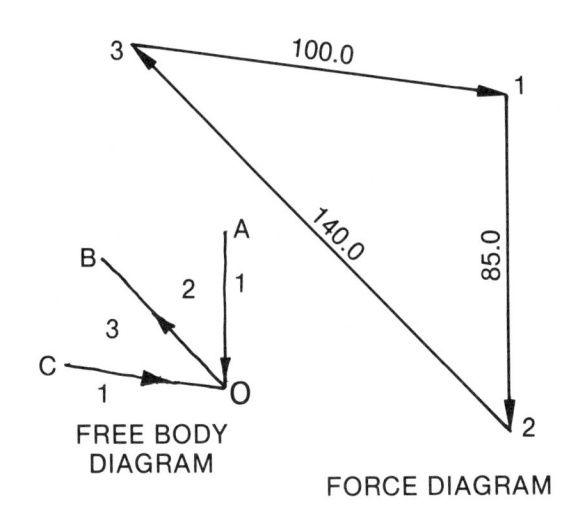

FIG. 13-28 Completed diagrams.

The *free body diagram* is similar in form to the one used in the last problem but uses some slightly different concepts. It considers the point O (the point of concurrency) as the free body and then projects the forces as they act *on the beam*. The free body diagram with Bow's Notation is shown in Fig. 13-26. Note that Bow's Notation will make the known force **vector 1-2**. The rotation around point O is counterclockwise in this example. A clockwise rotation would give identical values, but the force diagram would have vector **1-2** on the left side. This rotation selection is more of a sheet composition consideration than anything else.

The *force diagram* is started with drawing vector **1-2** and the lines of action of vectors **2-3** and **3-1** to position point 3, as shown in Fig. 13-27. The force diagram is then completed and the magnitudes of the forces recorded. This is the same as was done in the last problem. The free body diagram is completed with the push arrowheads of the found forces drawn in the middle of the OB and OC lines, as shown in Fig. 13-28. Remember that, in this problem, these indicate how the forces at points B and C act *on* the beam.

Figure 13-29 shows the three diagrams of this solution and the necessary tabulation of results. Using the results of this solution, a design engineer will be able to select materials and choose the member sizes that will produce a safe and economical mechanism.

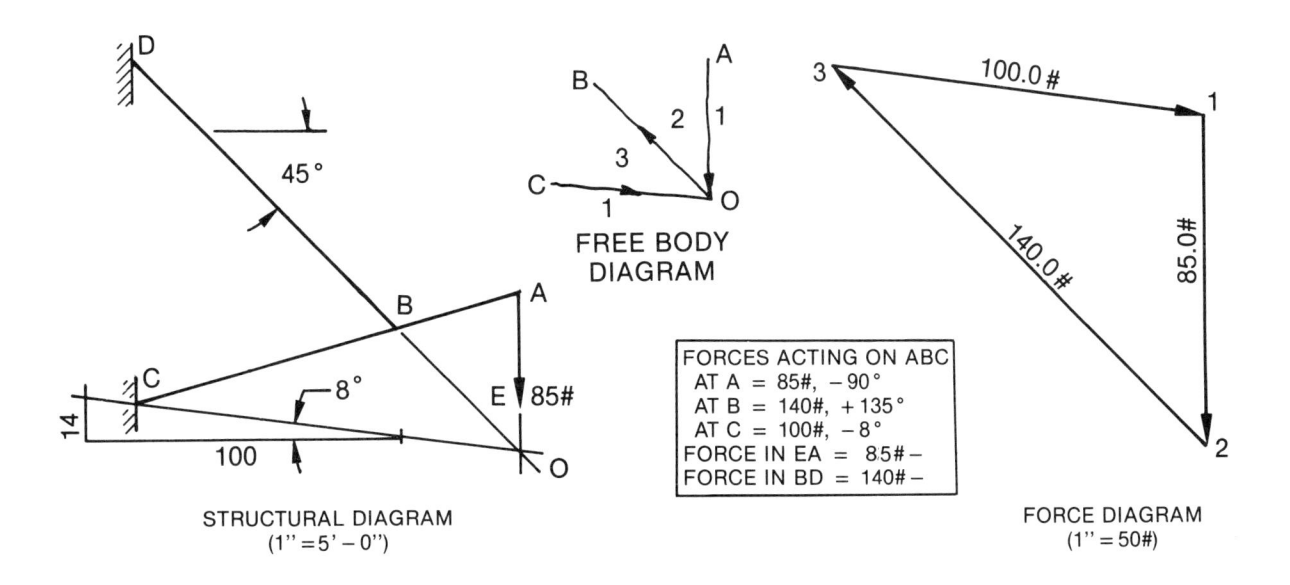

FIG. 13-29 Vector solution of a three-force beam problem.

PROBLEMS 13-7 through 13-12

Each problem is to be solved on one sheet of 8 1/2" × 11" paper. Make a quick semi-freehand drawing of the structural diagram, free body diagram, and force diagram to aid in making appropriate scale selections and a pleasing sheet arrangement.

PROB. 13-7. The bracket ABCF supports a load of 400 kilograms at point F. Determine the magnitude and sense of the force acting in member AB, member AC, and member AF.

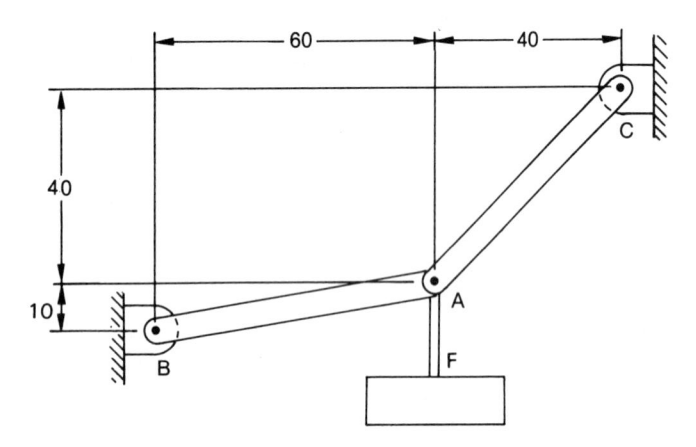

FIG. 13-30 Problem 13-7.

PROB. 13-8. A tow truck lifts a load of 1000# on its hook at point H. Determine the magnitude and sense of the force acting in members AC and AD and in the cable BAH. The pulley at A is frictionless so that the force acting in the cable is constant throughout its length.

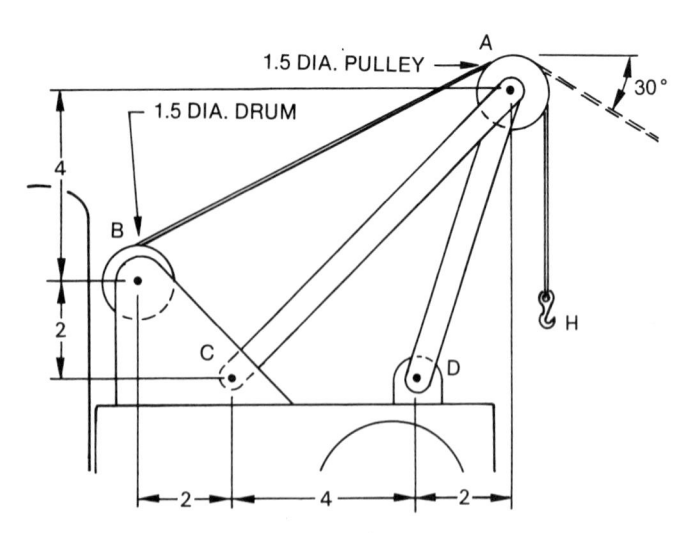

FIG. 13-31 Problems 13-8 and 13-9.

PROB. 13-9. Solve problem 13-8 with the hook extending to the right so that cable AH is 30 degrees below horizontal (as shown by the dashed lines).

PROB. 13-10. A boom ABC operated by a piston BD is lifting a load of 600 kilograms with the hook at point E. The angle of lift is shown. Determine the magnitude and sense of the necessary force in piston BD to keep the system in equilibrium. Determine the magnitude and direction (measured from the positive X axis) of the force acting on the boom at point C.

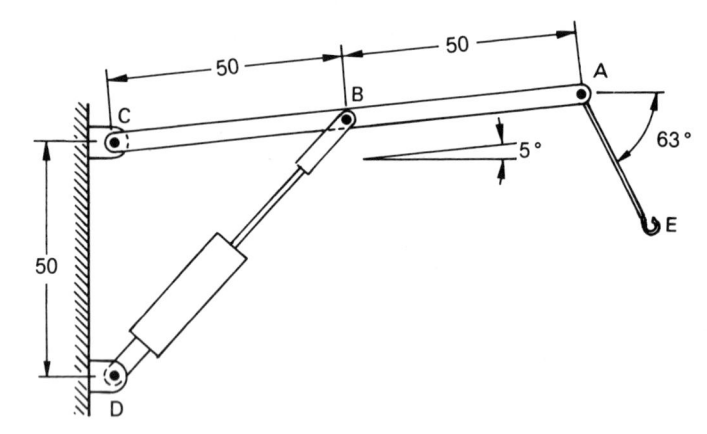

FIG. 13-32 Problem 13-10.

PROB. 13-11. A rotating ship's crane is used to unload cargo. It is lifting an 8-kip load in the sling at point F. Determine the magnitude and direction of the forces acting on the boom ABC at points A, B, and C. The cable AF passes over a pulley (frictionless) at A and extends downward to the left, parallel to the boom, to an electric winch mounted inside of the boom.

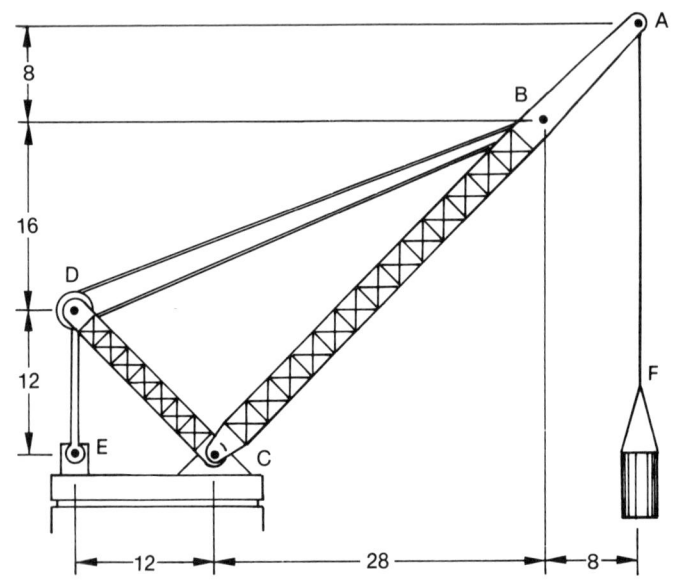

FIG. 13-33 Problems 13-11 and 13-12.

PROB. 13-12. Using the answers from problem 13-11, determine the magnitude and sense of the force in member CD and member DE and the reaction at point C that results from the forces acting in member CD and the boom ABC.

FIG. 13-34 Three parallel forces acting on a beam.

13.9 THREE PARALLEL FORCES ACTING ON A BEAM

Beams that carry gravity loadings frequently have forces acting along parallel lines of action. Figure 13-34 shows this condition: The known load is a gravity load and acts vertically, and the support at one end of the beam is also known to act vertically. The problem statement asks for the magnitude

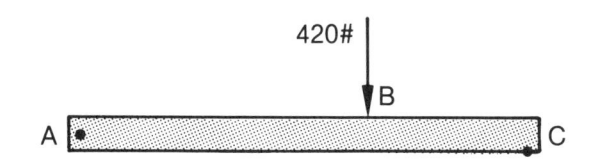

FIG. 13-35 Structural diagram.

and angle of the force acting at each support. As with all static (at rest) problems, the system is in equilibrium.

The simplified *structural diagram* for this system is shown in Fig. 13-35 with a pin connected support at point A, a known vertical line of action support at point C, and a known vertical load acting at point B. The known and unknown characteristics of the forces are summarized in Fig. 13-36. The line of action at point A is given as "through point A" because a fixed pin connection is capable of accepting a force acting in any direction. In this case, however, the force at A acts with a vertical line of action. Why? The forces at points B and C, having vertical lines of action, cannot exert any horizontal component forces; therefore, the force at point A cannot have any horizontal component either and necessarily acts vertically.

KNOWN AND UNKNOWN CHARACTERISTICS OF FORCES IN Fig. 13.35			
CHARACTERISTICS OF EACH FORCE	FORCES ON MEMBER ABC		
	AT POINT A	AT POINT B	AT POINT C
POINT OF APPLICATION	Point A	Point B	Point C
LINE OF ACTION	THROUGH Point A	VERTICAL	Through Point C
MAGNITUDE	UNKNOWN	420 pounds	UNKNOWN
SENSE	UNKNOWN	Downward	UNKNOWN

FIG. 13-36 Force parameters.

If the forces are parallel, they do not intersect (some would say that parallel lines intersect at infinity) and there is no point of concurrency. However, a special kind of *force diagram* can be used to solve this problem. First Bow's Notation is added to the structural diagram as shown in Fig. 13-37, and it now also serves as a *free body diagram*. The force diagram is started by duplicating the *position* of points of application and lines of action of the space diagram. It is suggested that the beam be lightly drawn so as to keep all points of application relative to each other. Then the known vector **1-2** is drawn on its line of action. The next construction will divide the magnitude of vector **1-2** into the magnitude of the two supports and will also assign each magnitude to its respective support. This construction is shown in Fig. 13-38. Two parallel horizontal lines are drawn through points 1 and 2 so that these lines intersect lines of action 2-3 and 3-1 to form a rectangle. Then a corner to corner diagonal is drawn so that both "1" labels are on

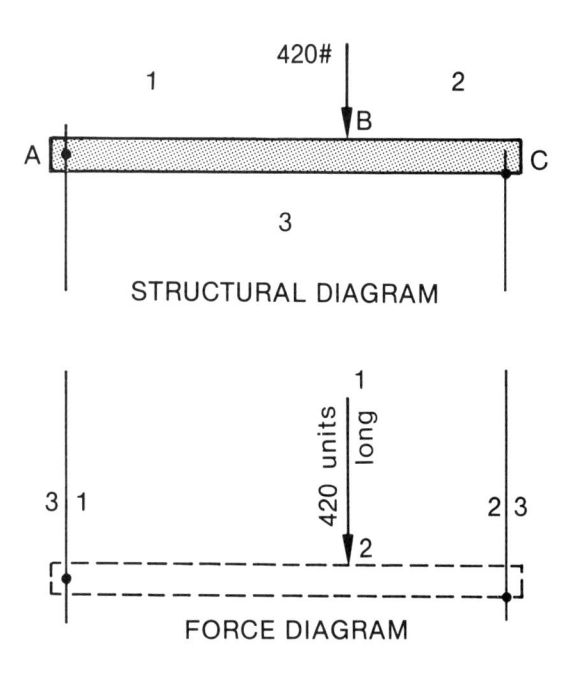

FIG. 13-37 Lines of action.

one side of it and both "2" labels are on the other side. The diagonal line intersects vector 1-2 at the temporary point T3. This forms the equilibrium force diagram of vectors 1-2, 2-T3, and T3-1 but it is difficult to read because vectors 2-T3 and T3-1 are superimposed over vector 1-2. To make the diagram more visual, the vector 2-T3 is moved to line of action 2-3 and vector T3-1 to line of action 3-1, by drawing a horizontal line through point T3. Arrowheads and labels complete the force diagram.

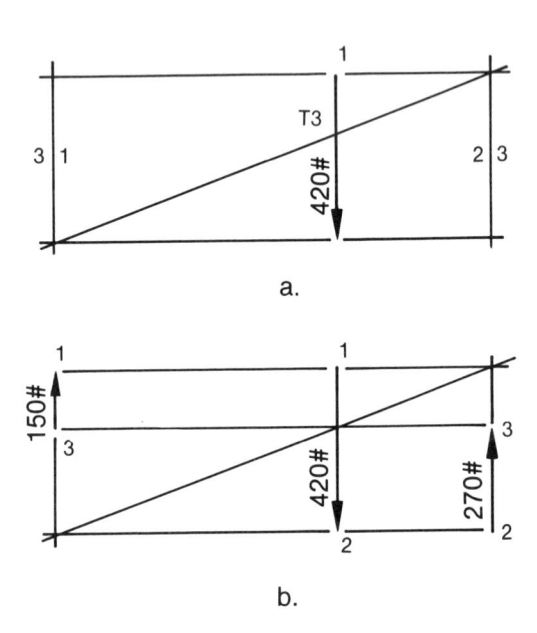

The structural diagram and force diagram of the complete solution, along with the necessary tabulation of results, are shown in Fig. 13-39. This problem could also be solved by the method given in the next section.

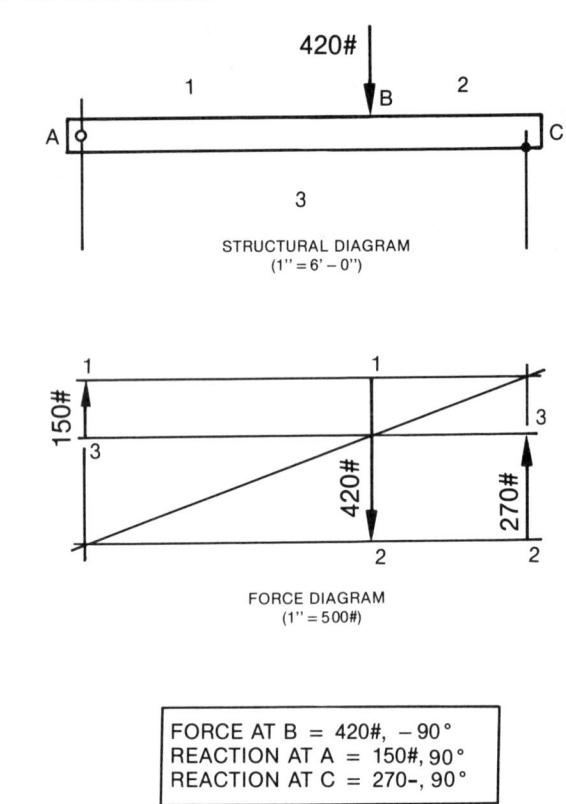

FIG. 13-38 Completing force diagram.

FIG. 13-39 Three parallel forces.

FIG. 13-40 Four forces acting on a beam.

FIG. 13-41 Multiforce beam supporting a loader bucket. *(Courtesy J. I. Case.)*.

420

13.10 FOUR OR MORE FORCES ACTING ON A BEAM

Beams are used in building construction and in mechanical equipment to carry combinations of loads that may act along many lines of action. The example in this section gives a solution for the problem when the force system is coplanar. The bent beam in Fig. 13-40 is supported with a fixed pin joint at the left end and a roller support at the right end. This kind of support is commonly used with beams because it allows for thermal expansion and contraction in the beam. The fixed pin support can accept a force acting along any line of action whereas the roller support can only act along a line of action that is normal (perpendicular) to the supporting surface. Mechanical equipment uses beams with similar supports except that the beam may pivot at the fixed pin support and be supported by a two-force member at the movable support, with the force acting along the line of action of the two-force member. The loader/backhoe in Fig. 13-41 has two examples of such a beam, at the loader and at the backhoe.

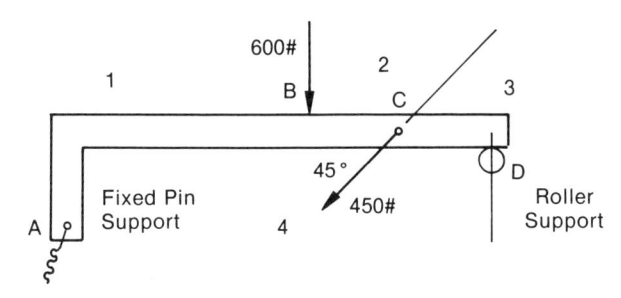

FIG. 13-42 Structural diagram.

The simplified *structural diagram* for the bent beam of Fig. 13-40 is shown in Fig. 13-42. Several important features of the lines of action in this structural diagram need to be examined. The fixed pin support at A has an *unknown* line of action and this is indicated by the wavy line. The box sitting on the beam exerts its gravity force vertically through its center of mass onto the beam. This is represented by the force arrow (and line of action) at B. The block and tackle attached at the hole in the beam pulls downward and to the left. The force arrow at C shows this relationship. (More about its line of action later.) The roller support at D can only act normal to the supporting surface (bottom face of the beam) and therefore has a vertical line of action.

Bow's Notation is placed on the structural diagram in nonconcurrent beam problems. The structural diagram then also serves as a modified *free body diagram*. As in the concurrent force problems, Bow's Notation gives order to the sequence of the vector summation. This sequence needs to be examined. The construction of a *force diagram* dictates that the known force vectors be graphically summed first, before the reaction vectors (the term "reaction" is often used for the forces acting at supports of beams) can be determined. This means that the spaces adjacent to the forces must be numbered first. They must also occur sequentially. That is why the line of action at C has been extended above the beam. The last restriction on Bow Notation is that only *one* space is permitted between the lines of action of the two reactions. A summary of the known and unknown characteristics of the four forces acting on the beam are tabulated in Fig. 13-43. Of the four forces, two act vertically, one diagonally, and the other has an unknown line of action—a nonconcurrent system of forces.

KNOWN AND UNKNOWN CHARACTERISTICS OF FORCES IN Fig. 13.40				
CHARACTERISTICS OF EACH FORCE	FORCES ACTING ON THE BEAM			
	AT POINT A	AT POINT B	AT POINT C	AT POINT D
POINT OF APPLICATION	Point A	Point B	Point C	Point D
LINE OF ACTION	Through Point A	Vertical	45° Diagonal	Vertical
MAGNITUDE	UNKNOWN	600 pounds	450 pounds	UNKNOWN
SENSE	UNKNOWN	Downward	Downward	UNKNOWN

FIG. 13-43 Force parameters.

One attack on this problem could be to first combine the known applied forces into one resultant applied force and then use a three-force solution method. Determining the magnitude and angle of the resultant force is graphically easy. However, determining the point of application of this resultant force (so as to position its line of action) requires the use of an additional diagram that will be introduced now. The method given will avoid making a determination of a resultant applied force (unnecessary information) and go directly for the determination of the two reactions.

The additional diagram that is needed to solve this problem is the *string diagram* (also often called the *funicular diagram*). The string diagram is an accurate representation of the lines of action of all of the forces in the system and is constructed using an expanded force diagram. The steps of this construction are:

1. The *string diagram* is started in Fig. 13-44(a) by locating the points of application in their exact space relationship. It is recommended that the shape of the beam be lightly drawn (to scale) and then the points of application located and labeled. This will provide for easy visual checking against the structural diagram. The line of action of each force is drawn through its respective point of application and

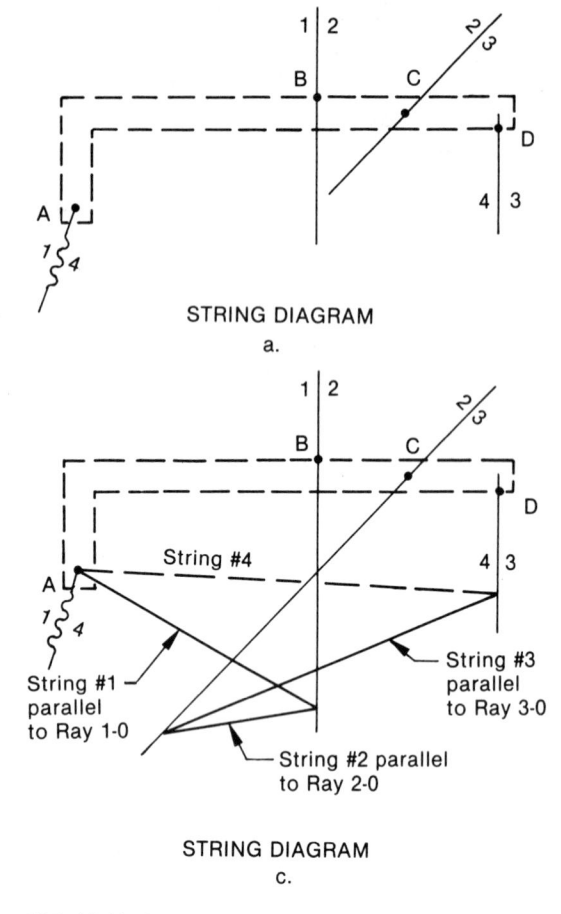

FIG. 13-44 Construction of string and force diagrams.

the lines of action are labeled according to Bow's Notation. Because the line of action at the point of application A is unknown, it is indicated with a wavy line. The *strings* of the string diagram will be drawn between these lines of action using information from the force diagram.

2. The *force diagram* is started in Fig. 13-44(b) by graphically summing the vectors **1-2** and **2-3** of the known forces acting at points B and C. The line of action of the reaction at point C is known and its corresponding vector *line* is drawn through point 3. All that is known about point 4 is that it lies *somewhere* on this line. The determination of point 4 is the reason for using the string diagram.

3. The *force diagram* is expanded to include *rays*, which are used to give information about the *strings* of the string diagram. These rays are lines extending from the ends of the vectors to a *pole point*, labeled 0 in Fig. 13-44(b). The positioning of pole point 0 is arbitrary and does not effect the results of the solution, but it does effect the shape (and the ease of drawing) of the *string diagram*.

4. Once the rays from point 0 to the ends of all of the known vectors are established in the *force diagram*, they are used to give direction to corresponding strings in the *string diagram*. Note that ray 2-0 is common to vector 1-2 and vector 2-3. Also ray 3-0 is common to vector 2-3 and vector 3-4 even though point 4 is not yet located. It follows that ray 1-0 is common to vector 1-2 and the unknown vector 4-1 and that a ray 4-0 will be common to vector 3-4 and vector 4-1. The determination of ray 4-0 will position point 4 on the vector line 3-4 and permit completing the force diagram. Ray 4-0 is determined after finding its string in the string diagram.

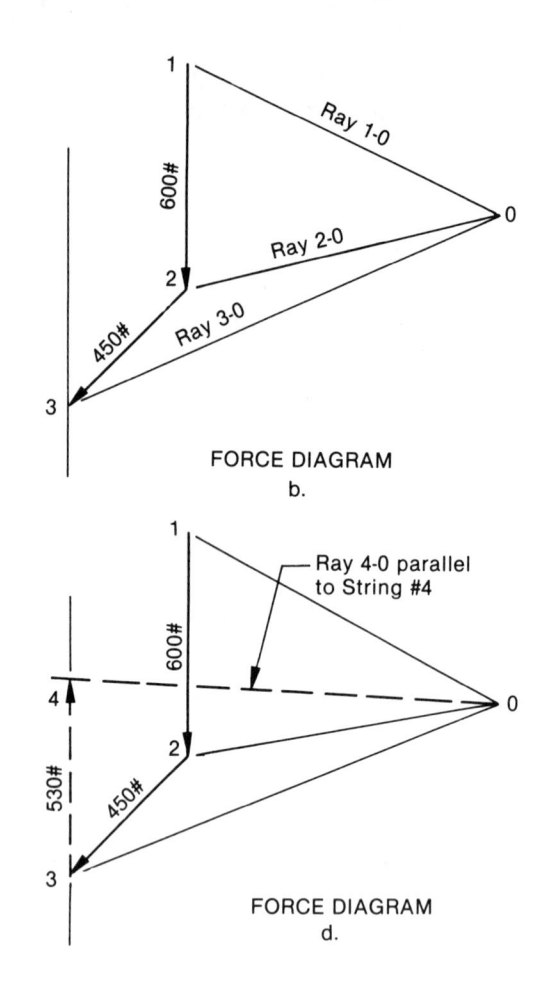

5. Returning to the *string diagram* in Fig. 13-44(c), the strings are drawn sequentially between their lines of action *and* parallel to their corresponding rays. How is it possible to draw from the line of action of the force at point A when the line of action 4-1 is as yet unknown? Well, it isn't quite totally unknown for it *must pass through* point A and therefore, point A is *on* line of action 4-1 regardless of its angular position. Thus it is *always* necessary to start the strings of the string diagram at the point of application of the fixed reaction. Starting at point A, string #1 is drawn (parallel to ray 1-0) to line of action 1-2. Next, string #2 is drawn to line of action 2-3 and string #3 to line of action 3-4. Note that the number of each string is reflected in the names of the two lines of action it intersects. Because this force system is in equilibrium, the string diagram always closes. This

means that the strings form a closed polygon. String #4 is drawn using this principle. It is recommended that the closing string be drawn to look different (helps in visual checking).

6. The necessary information is now available to complete the *force diagram.* The ray 4-0, Fig. 13-44(d), is drawn parallel to string #4 and intersects vector line 3-4, giving position to point 4 and allowing completion and measuring of vector 3-4. All that remains to be done is to close the equilibrium force diagram with vector 4-1 and to measure the magnitude and angle of vector 4-1.

The completed solution is shown in Fig. 13-45 with the required tabulation of all forces acting on the beam. Note that the angles of the forces reflect how they act *on the beam,* not how the beam acts on the imposing loads and the reactions. Drawing the found vectors, found string, and found ray with broken lines (to look different) makes the solution easier to read and mentally (and visually) to reconstruct once it has become "cold" with the passage of time. Problems with three or more known loads (five or more forces including the two reactions) are solved using the same principles that are illustrated here, remembering that the vectors of all known forces must occur sequentially in Bow's Notation and in the force diagram.

This example was designed to illustrate important concepts for the solution of coplanar nonconcurrent force systems using the *string diagram*.

1. The relative position of all of the points of application of the acting forces must be maintained in the string diagram and all known lines of action must accurately represent the action line of each force.
2. Because the point of application of a fixed reaction will be the only known point on this reaction's line of action, and because strings are drawn between lines of action, the first string to be drawn must always start at this point of application.
3. Because the strings are drawn between two adjacent lines of action, as shown by Bow's Notation on the structural diagram, they must extend to these same lines of action in the string diagram even if it requires crossing over other lines of action. (In other words, don't just draw to the first line of action intersected. If lines of action cross, the first one probably won't be the right one. This is where complete labeling pays off!)
4. A thorough knowledge of the steps and procedures of a problem solution, coupled with a neatly composed, accurately drawn and completely labeled analysis allows the design engineer to make the final check of any solution: ***"Does it look right?"***

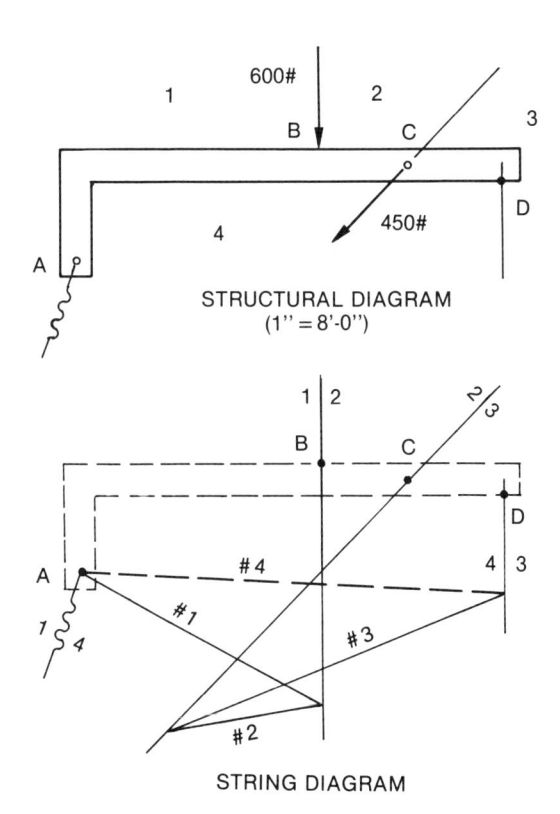

FIG. 13-45 Four forces acting on a bent beam.

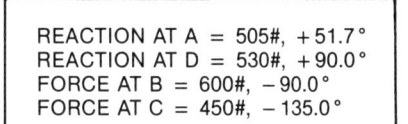

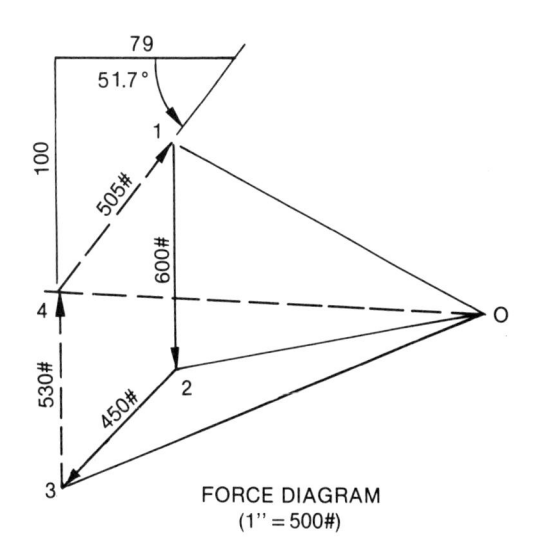

13.11 TRUSSES AND THE MAXWELL DIAGRAM

A truss is a structure constructed of many relatively short and lightweight two-force members. The members are arranged to create connected triangular shapes which give rigidity to the truss (a triangle cannot be distorted when its three sides maintain their respective lengths). A truss can support the same loads as a beam with a great saving of weight and material. Additionally, trusses allow a large load carrying structure to be built using small individual pieces—an advantage when transportation and construction space limitations are present.

The *Maxwell Diagram* is a method used to determine the magnitude and sense of the forces acting in the members of a truss. The complete analysis of a truss requires that the magnitude and direction of each reaction be determined first and then the magnitude and sense of the force acting in each member of the truss can be determined. The analysis to find reactions of trusses is exactly the same as that used in Section 13.10 to find the reactions of beams.

After the reactions are determined, the next step in the analysis of a truss is to determine the stress acting in each member of the truss. The method to be used here requires that each connection of the truss be pin connected, meaning that the centerlines of all members act through a common point and the members could be connected by a single pin (bolt) through each connection. This prevents rotational forces from acting in the connection and causes each member to carry only axial loads (tension or compression).

The analysis of a truss requires that each connection of the truss be examined independently to determine the magnitude and sense of the forces acting at that connection. As forces in individual members are determined, they become known forces and can be used to determine the forces acting in the adjacent connection. The order of examination of each connection is determined by the number of unknown forces acting at any connection—*only two* unknown forces can be resolved at each connection as it is investigated.

A truss fabricated from wood members is shown in Fig. 13-46. The loads it carries are shown in Fig. 13-47. The "k" after the number of each load is an abbreviation for kilopound (kip) and is equal to a force of 1000 pounds. The truss in turn is supported on masonry walls. The problem is to determine the magnitude and sense of the force that acts in each of the members of the truss.

The first step in the solution of the problem is to reduce the problem to its simplest form. This

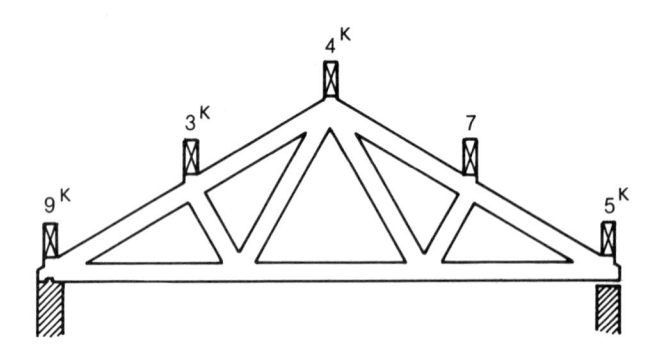

FIG. 13-47 Structural diagram.

can best be accomplished by the use of a *structural diagram* where all structural members are represented by their centerlines and all imposing forces are represented by the force arrows as shown in Fig. 13-48. Before proceeding further, the imposing forces are examined to determine if any of these forces do *not* contribute to the forces acting in the members of the truss. In this example, the 9k force acting vertically over the support at R1 and the 5k force acting vertically over the other support at R2 act directly on the supports and do not contribute

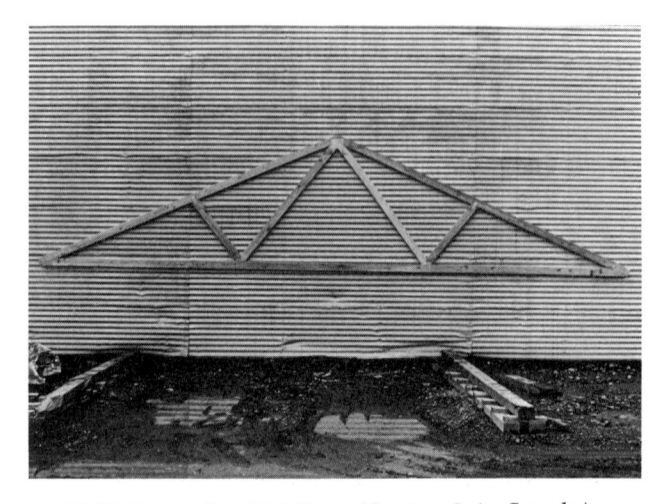

FIG. 13-46 Twenty-foot Fink Truss. *(Courtesy Boise Cascade.)*

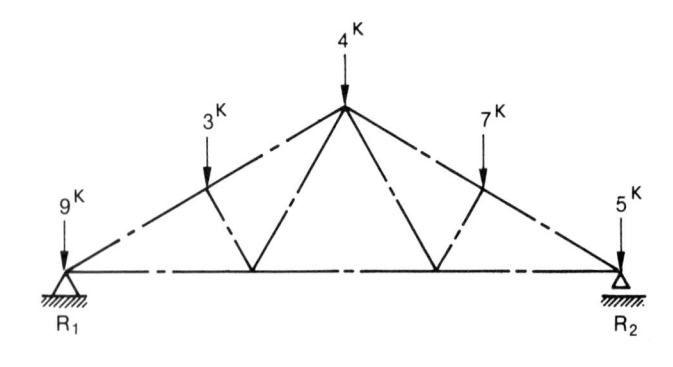

FIG. 13-48 Simplified diagram.

to forces acting in the truss. Therefore, they can be eliminated from this analysis. The structural diagram shown in Fig. 13-49 shows only those forces that contribute to the stresses in the truss members Also, the magnitude and direction of the reactions have been determined using the analysis of Section 13.10 and are shown here as reaction forces.

In Fig. 13-49, Bow's Notation has been added by labeling the external spaces between the imposing loads and reactions with letters and the internal spaces of the truss with numbers. The labeling of

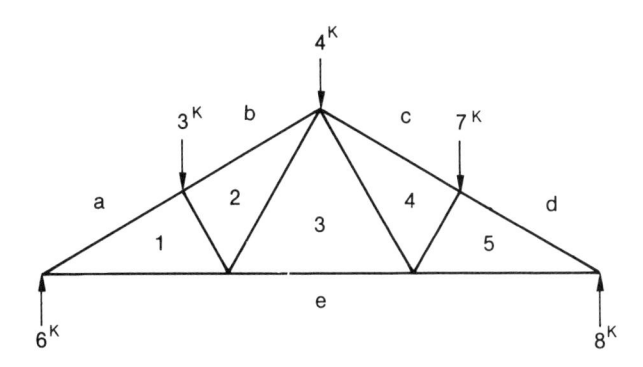

FIG. 13-49 Structural diagram with bow's notation.

external spaces is clockwise and starts between the left reaction force and a loading force, as was done in the solution for the reactions. The labeling of internal spaces is from left to right, taking each space in order. This is the only labeling normally done on trusses and each member is named for the two spaces it is between. (The connections are not labeled as in other vector solutions.) This system of identification results in all external truss members (chord members) being identified by a letter and a number, all internal truss members (web members) being identified by two numbers, and all imposing loads and reactions being identified by two letters—a systematic identification that immediately gives information about each member. The vector for each member will be identified in the same manner.

The forces acting at each joint (connection) can now be found for those joints that do not have more than two unknown forces. The only joints having this condition are at the two ends of the truss. Selecting the joint at the left end, a free body diagram, Fig. 13-50(a), is started using the notation from the structural diagram, and a Force Diagram, Fig. 13-50(b), is also started. The known force (vector **EA**) is drawn parallel to its known line of action and at a selected force scale. It is labeled using the clockwise sequence of the structural diagram so that its tail is at E and its head is at A. Then the line of action of vector **A-1** is drawn through point A of vector EA (parallel to member

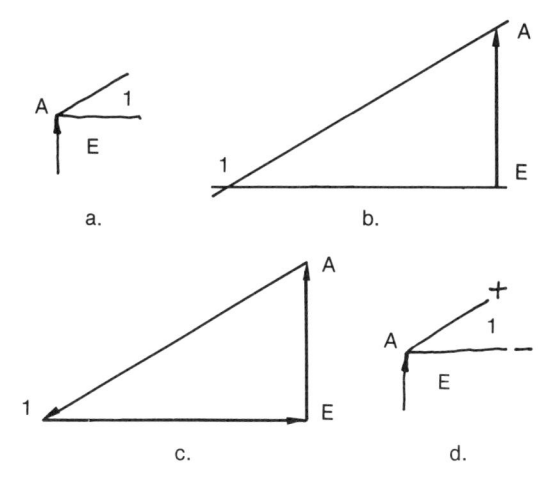

FIG. 13-50 Forces at joint EA1.

A-1) and the line of action of vector 1-E is drawn through point E of vector EA (parallel to member 1-E). The intersection of these two lines of action locates the point 1, and this force diagram can be completed, Fig. 13-50(c), by drawing the vectors so that the joint is in equilibrium (all vectors in the diagram are head to tail and the diagram is closed). Knowing the direction of "push" of the forces as they act on this joint, the free body diagram can be completed and the sense of each force can be noted, Fig. 13-50(d).

The solution of this joint now makes the forces in members A-1 and 1-E known and they can be used in the solution of the next joints. Again, a solution requires that not more than two unknown forces act through the joint. Therefore, joint E-1-2-3 is ruled out (members 1-2, 2-3, and 3-E are all unknown) and joint 1-A-B-2 must be examined next.

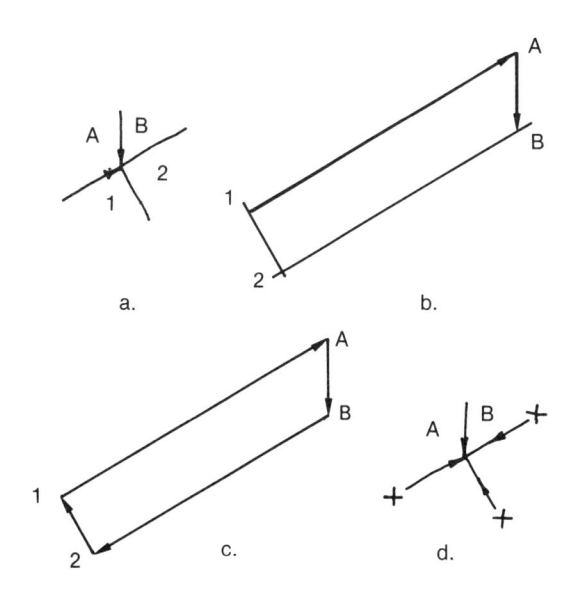

FIG. 13-51 Forces at joint 1AB2.

425

A new free body diagram of this joint is started showing the lines of action for the four forces acting on this joint and the direction given for the *two* known forces, Fig. 13-51(a). Note that the direction for the "push" of the force in member 1-A is now opposite the direction used in the free body diagram of joint EA1 (compression members "push" at each end of the member and in exactly opposite direction—tension members "pull" at each end as was shown in Section 13.6). Then the force diagram is started by first drawing the two known forces (1-A and AB in this order) and by drawing the lines of action of B-2 and 2-1, Fig. 13-51(b). The completion of the force diagram, Fig. 13-51(c), and the free body diagram, Fig. 13-51(d), determines the forces in members B-2 and 2-1. Joints E-1-2-3 and 3-2-B-C-4 are solved using these same procedures as shown in Fig. 13-52 and Fig. 13-53 respectively.

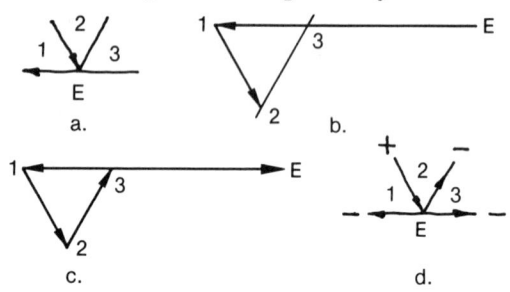

FIG. 13-52 Forces at joint E123.

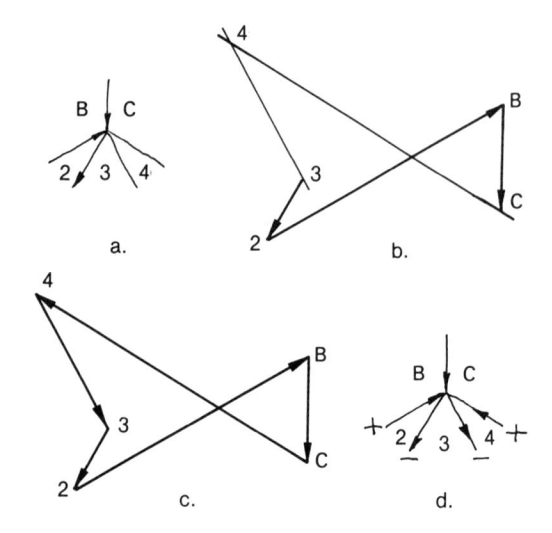

FIG. 13-53 Forces at joint 32BC4.

Before continuing further with these solutions, a comparison of the force diagrams for joints E-A-1, 1-A-B-2, E-1-2-3, and 3-2-B-C-4 shows that the force vector for each truss member in the left half of the truss has appeared twice, identical in line of action and length (magnitude) but acting in opposite directions. If these force diagrams were stacked one over the other, some drawing could be eliminated and in addition a more concise solution for all of the joints could be made. This is the basis for drawing a *Maxwell Diagram*. A series of individual force diagrams are drawn for each joint so that each is over the previous diagram. In this way, each force diagram for a joint can use vectors determined at a preceding joint. The imposing loads and the reactions must form an equilibrium force diagram before starting the individual joint force diagrams. Figure 13-54 shows the resulting stack of equilibrium force diagrams for all joints of the truss. Note that when the force diagrams are superimposed in this manner, the vectors of all forces in all members in a truss would have an arrowhead at each end. To avoid this, the arrowheads are not drawn on vectors for truss members in the Maxwell Diagram but the arrowheads for all impos-

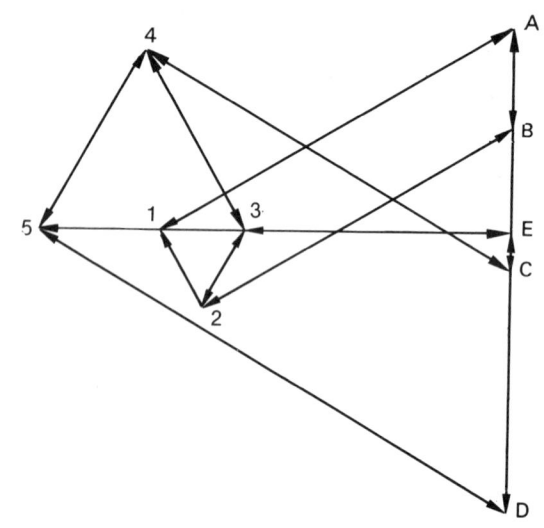

FIG. 13-54 Stacked force diagrams.

ing loads and reactions are drawn as shown in Fig. 13-55. However, the existence of arrowheads on each vector acting through a joint under investigation is assumed and is necessary to determine the

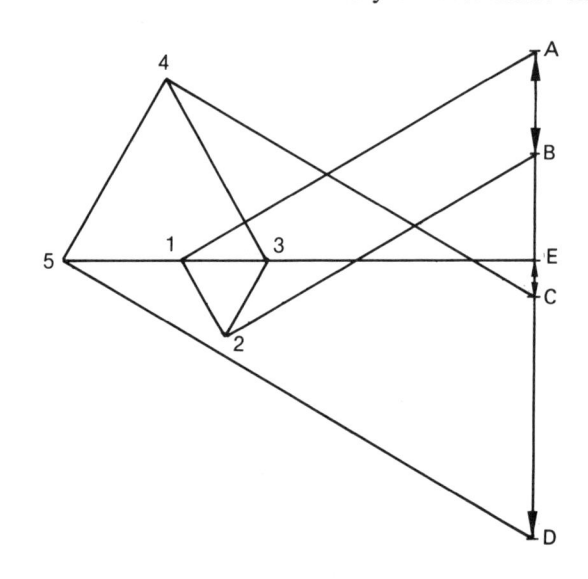

FIG. 13-55 The Maxwell diagram.

sense of the force in that member. Also, a free body diagram is not usually drawn for each joint, although the structural diagram can be marked to serve this purpose.

Figure 13-56 shows a complete Maxwell Diagram solution to the original problem (after the reactions have been determined). The final step in completing the problem is the tabulation of the values determined. List the external truss members (chords) first in alphabetical order, followed by the internal (web) members in numerical order. Give the sense of each force after its magnitude.

The solution for a similar truss having two vertical imposing loads and the third imposing load at an angle is shown in Fig. 13-57. The solution for the reactions R1 and R2 is the first step of this problem.

Graphical solutions of external forces acting on beams and trusses and solutions of internal forces acting in trusses can provide adequate results for the design and construction of these equilibrium structural systems if reasonable care is used in their projections. The principal requirements are that the design conditions be accurately known, that each vector be drawn exactly parallel to its line of action, that all vectors be carefully scaled (both drawing and reading), and that all force diagrams close. Any significant error of closure of a force diagram when lines of action are known (as in the Maxwell Diagram) casts doubt upon the accuracy of the results.

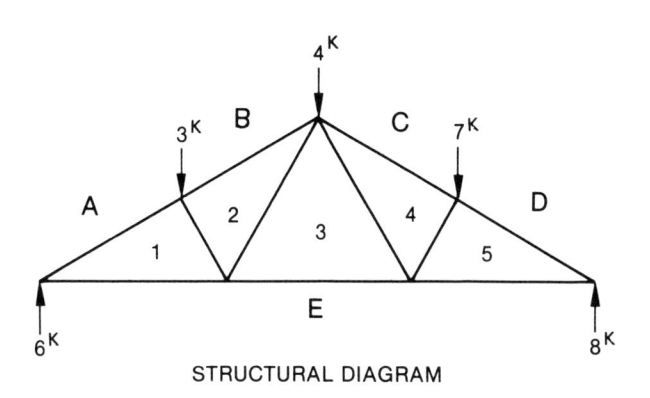

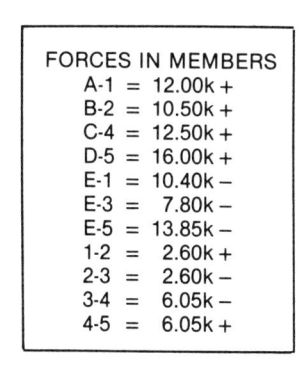

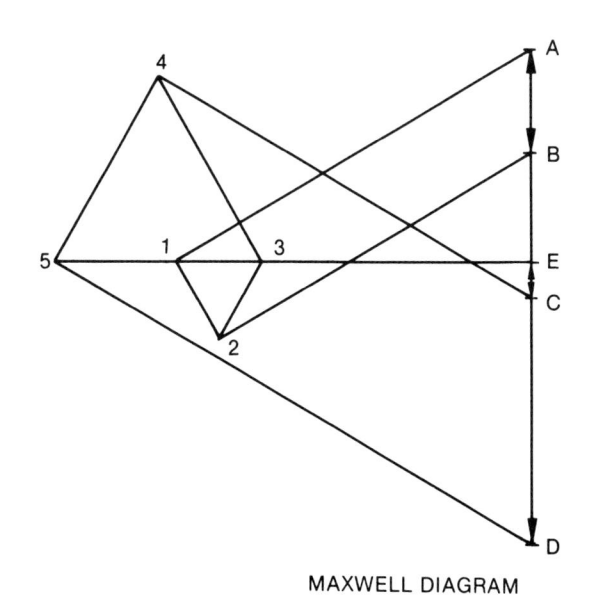

```
FORCES IN MEMBERS
  A-1 = 12.00k +
  B-2 = 10.50k +
  C-4 = 12.50k +
  D-5 = 16.00k +
  E-1 = 10.40k −
  E-3 =  7.80k −
  E-5 = 13.85k −
  1-2 =  2.60k +
  2-3 =  2.60k −
  3-4 =  6.05k −
  4-5 =  6.05k +
```

FIG. 13-56 Forces acting in a fink truss.

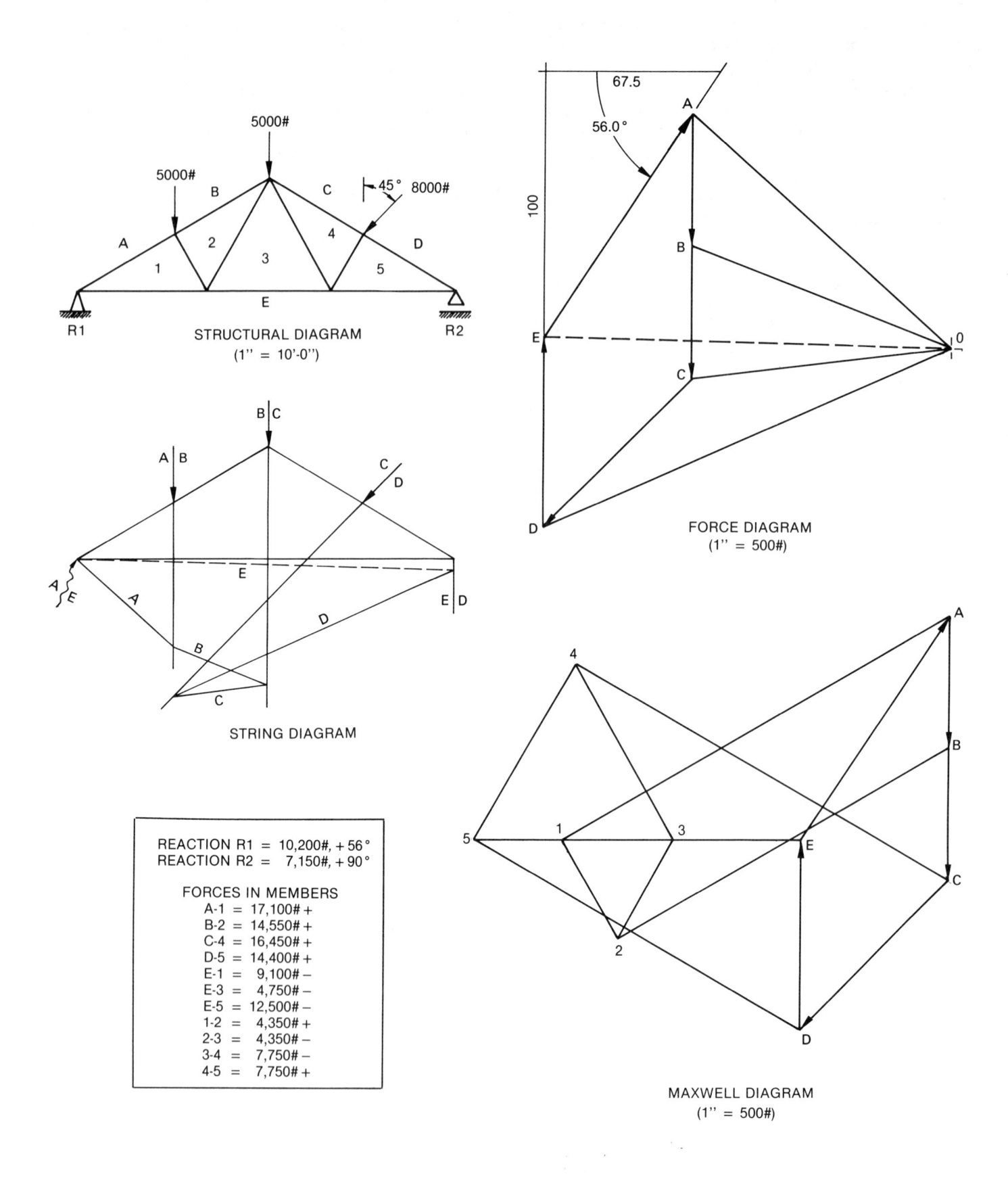

FIG. 13-57 Solution of reactions and stress in members of a truss.

PROBLEMS 13-13 through 13-18

Each problem is to be solved on one sheet of 8 1/2" × 11" paper. Make a quick semi-freehand study to aid in sheet layout and scale selection. In all of these problems, the reaction R1 is fixed and can accept a load in any direction. The reaction R2 is movable and can only act vertically. Measure all reaction directions from the positive X axis (up/vertical is +90 degrees). All truss joints are pin connected.

PROB. 13-13. Determine the reactions of the bent beam supporting the 900# weight.

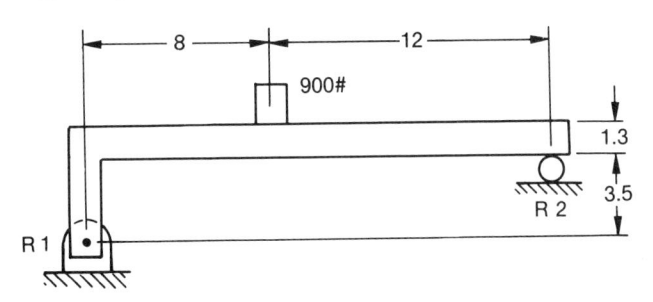

FIG. 13-58 Problem 13-13.

PROB. 13-14A. Determine the reactions of the Fink roof truss when the loads are: X = 5 kips, Y = 6 kips and Z = 8 kips. All loads are vertical.
PROB. 13-14B. Determine the magnitude and sense of the force in each member of the Fink roof truss using the reactions found in part A.

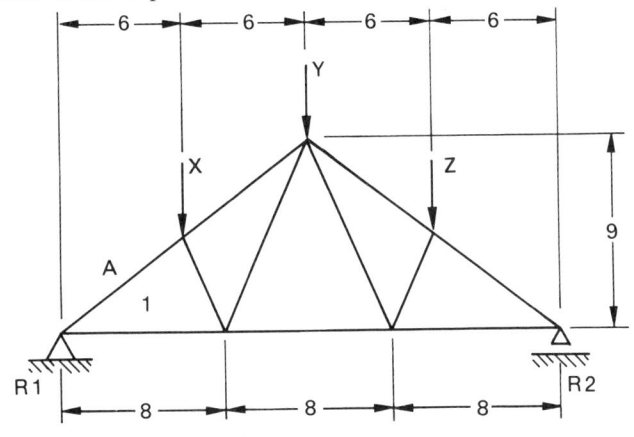

FIG. 13-59 Problem 13-14.

PROB. 13-15A. Determine the reactions of the Howe roof truss when the loads are: X = 6 kips, Y = 4 kips and Z = 8 kips. Loads X and Y are vertical and load Z is perpendicular to the top chord of the truss.
PROB. 13-15B. Determine the magnitude and sense of the force in each member of the Howe roof truss using the reactions found in part A.

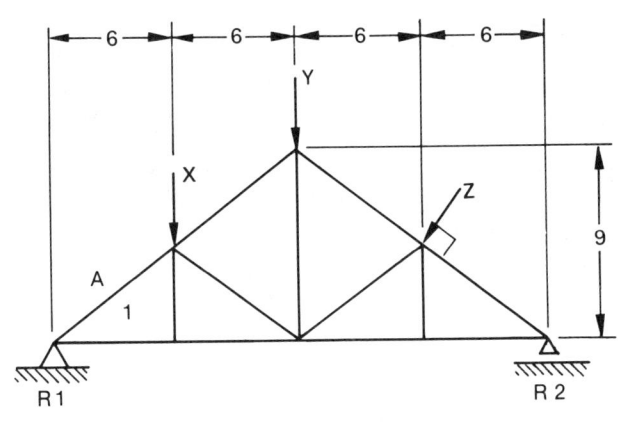

FIG. 13-60 Problem 13-15.

PROB. 13-16A. Determine the reactions of the Howe bridge truss when the loads are: X = 10 kips, Y = 15 kips and Z = 18 kips. The angle of each load is the result of traffic applying brakes.
PROB. 13-16B. Determine the magnitude and sense of the force in each member of the Howe bridge truss using the reactions found in part A.

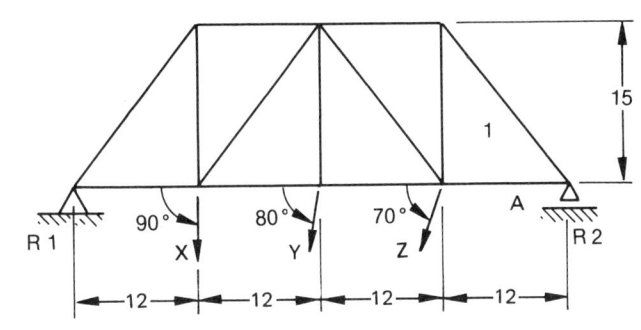

FIG. 13-61 Problem 13-16.

PROB. 13-17A. Determine the reactions of the modified Fink roof truss when the loads are: X = 7.5 kips, Y = 7.5 kips and Z = 5 kips. Loads X and Z are vertical and load Y is perpendicular to the top chord of the truss.
PROB. 13-17B. Determine the magnitude and sense of the force in each member of the modified Fink roof truss using the reactions found in part A.

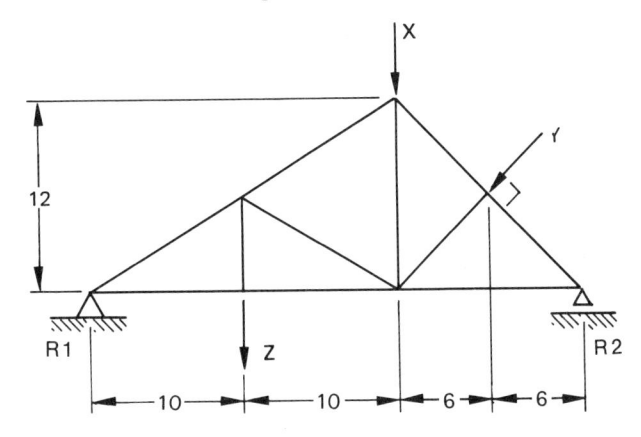

FIG. 13-62 Problem 13-17.

429

FIG. 13-63 Ship model with pivoting cargo loading cranes. *(Courtesy Magee-Bralla, Inc.)*

PART III: NONCOPLANAR FORCE SYSTEMS

Force systems that act in a single defined plane (two dimensions) have been explored. The principles used there will now be extended to three-dimensional systems; or more specifically, noncoplanar concurrent systems. These systems all are in the general form of a tripod supporting a fourth member that exerts a known force at the point of concurrency. Applications of noncoplanar concurrent force systems occur in stationary support structures that are designed to carry loads applied from various three-dimensional directions. Examples are found in electrical transmission towers, ship and dockside cranes, tow trucks, portable shop floor cranes, and automobile jack stands, to name only a few. The observant student will identify many more.

As in the discussion of two-dimensional force systems, only the forces acting as a result of working loads will be investigated. The equipment and structural members themselves will again be considered weightless and infinitely strong. It is only after the forces resulting from working loads are determined that the selection of the individual members can be made. Then a second more complete analysis is carried out that *does* include the weights contributed by the structure itself. This last analysis is easily accomplished (using these same techniques) by determining the center of gravity and mass of each part and including this known force in the system of forces acting on each member.

The principal difference between the analysis of two-dimensional and three-dimensional systems is the introduction of depth into the views. In coplanar (two-dimensional) analysis, a structural view is selected that projects all members true length, resulting in all vectors projecting in true length. In noncoplanar (three-dimensional) analysis, two structural views are required to establish the space relationships of the members. From two given views, it may be necessary to project auxiliary views to find necessary true lengths before starting the force analysis. Section 13.12, The Special Case, gives the solution method when the two given structural views are sufficient for the problem solution (Fig. 13-65). Section 13.13, The General Case, expands on the special case to provide a solution method when auxiliary structural views are necessary (Fig. 13.66).

FIG. 13-64 Heavy duty construction cranes.

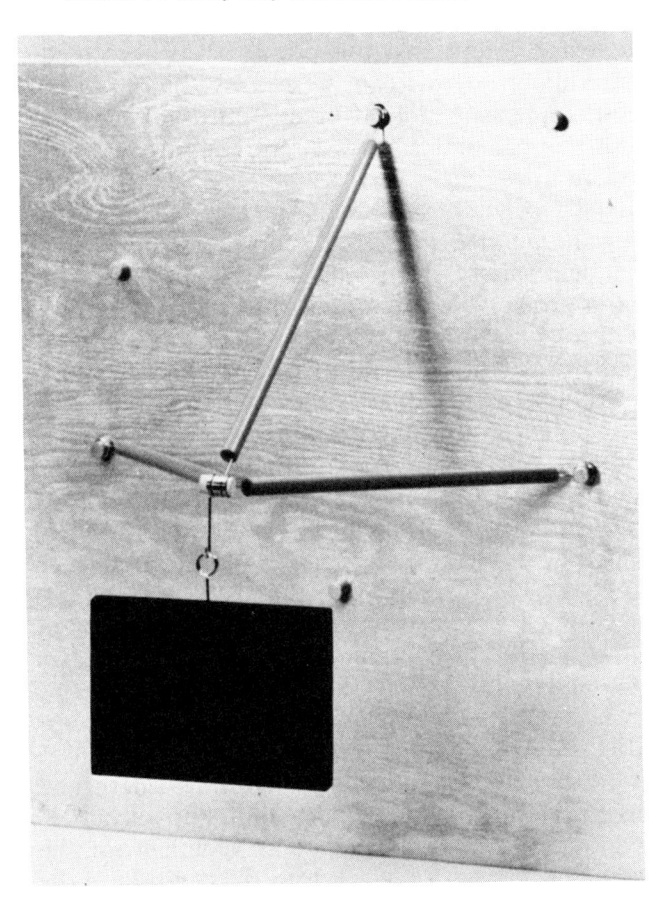

FIG. 13-65 Special case—noncoplanar concurrent force system.

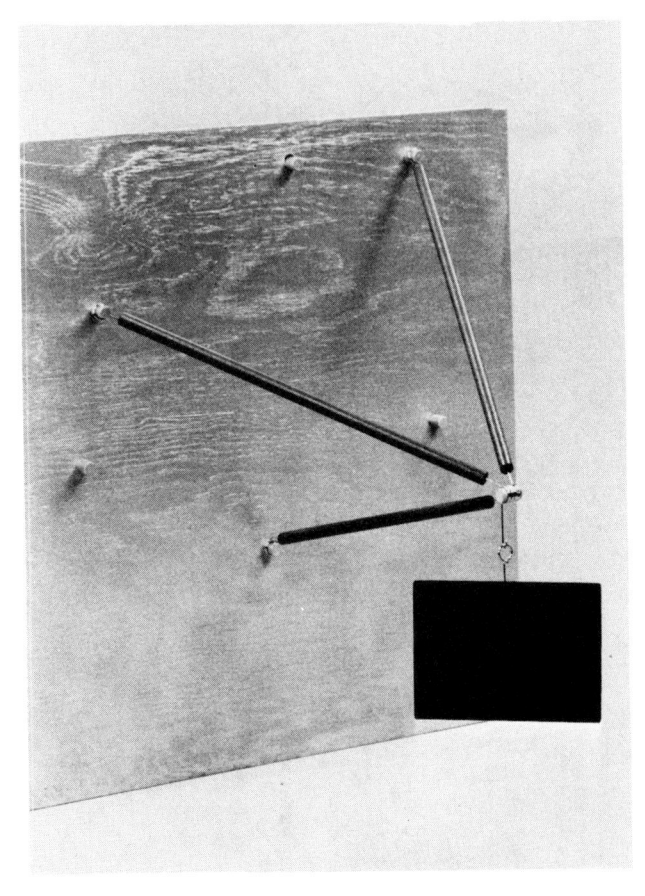

FIG. 13-66 General case—Noncoplanar concurrent force system.

13.12 THE SPECIAL CASE

The *special case* solution is possible when the given structural diagram views will provide the necessary lines of action to draw the force diagram views. It is important to keep in mind that (1) the force diagram is *three dimensional,* the same as the structural diagram (the line of action of each force is parallel to the member in which it acts!); and (2) the space relationships of the structural diagram views are reflected in the force diagram views.

The previous coplanar problem solutions have all been based on one basic principle: the ability to draw a unique triangle when two angles and the included side are known. This same principle is applied to each of the views in the solution of noncoplanar problems and must satisfy these two general conditions:

1. Not more than three lines of action are projected in each view.
2. One measurable force vector can be drawn along one of the projected lines of action.

These two conditions will be expanded upon in the next section.

FIG. 13-67 Electrical transmission line support brackets.

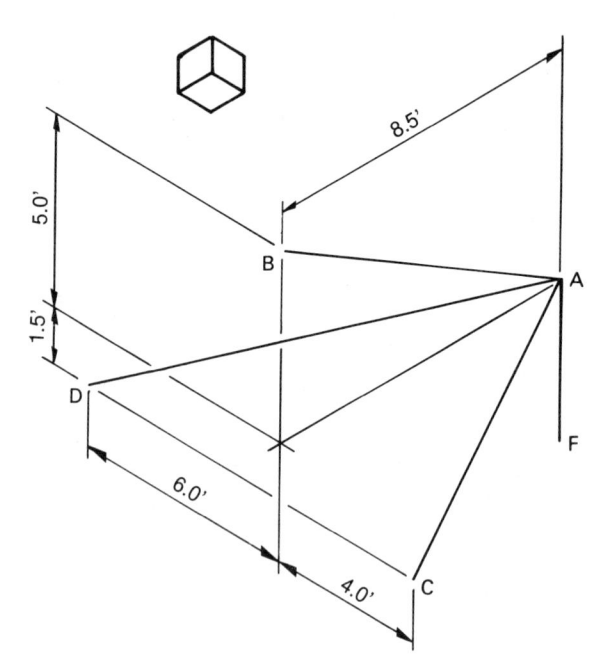

FIG. 13-68 Special case bracket.

A transmission line tower supports an electrical cable with an insulator as is shown in Fig. 13-67. The problem is to determine the force acting in each of the three members that provide support for the insulator. The space relationships for a similar problem are given in Fig. 13-68. To eliminate coincidences that might be interpreted by the reader as necessary conditions, this problem has been made nonsymmetrical (as can be seen by the 6.0-ft and 4.0-ft dimensions). The force applied at point F is 1000 pounds and is a gravity force acting vertically downward.

To solve this problem in the most straightforward manner, structural views should be selected that will satisfy the conditions for lines of action and a projectable force vector. The first condition of three lines of action for four members (AB, AC, AD, and AF) may seem to present a quandary but is easily satisfied either if one member is projected into point projection (its line of action is a point and effectively disappears in that view—it will also be true length in the adjacent view which may be an advantage) or if two members are projected so that they coincide (one lying over the other—their projected lengths are not required to be equal), so that their force vectors will share the same line of action in that projection. If a frontal view is drawn so that the plane ACD is in edge view by putting the line DC in point projection, this condition is satisfied. At the same time member AF, which carries the given load, is projected true length and satisfies the second condition of a known projectable length for the known force vector. Note, the member carrying the given load must *not* have the same line of action as another member.

432

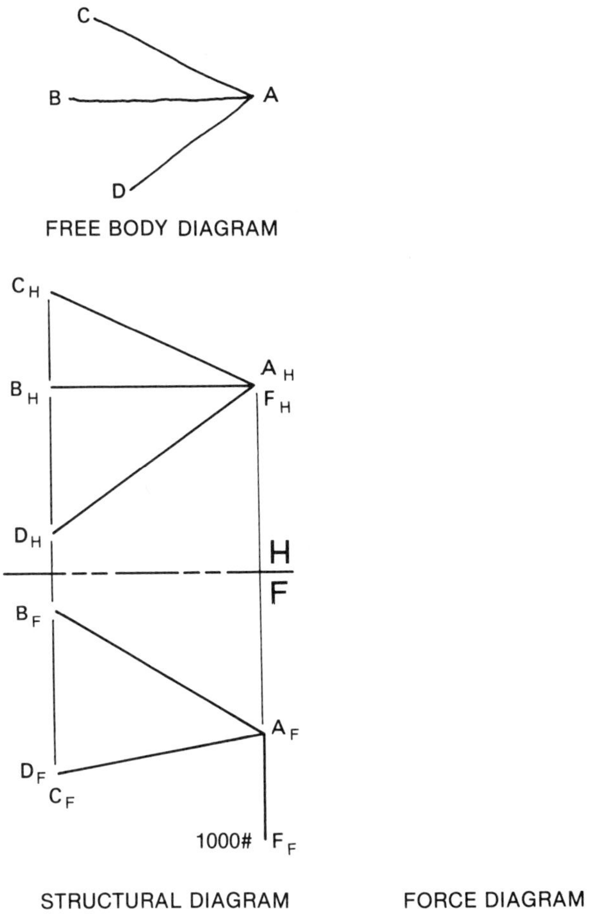

FREE BODY DIAGRAM

STRUCTURAL DIAGRAM   **FORCE DIAGRAM**

FIG. 13-69 Structural diagram.

This frontal projection and the horizontal projection are drawn in the **structural diagram** of Fig. 13-69. The horizontal view also meets the "three lines of action" requirement and will get a "projectable vector length" from the frontal view. These two projections are the "special" views that provide all of the information necessary to solve the problem (without projecting additional space views) and this is called "*The Special Case.*"

Before starting the three-dimensional force diagram, it is necessary to decide the order of summation of the force vectors and to insure that this order will be followed in each view of the force diagram. A free body diagram is used for this purpose, as well as for determining the sense of each of the forces. The *free body diagram* has been started in Fig. 13-69. The free body diagram (only one view) is sketched parallel to one of the space views, normally the space view having one member in point projection. This permits putting this point projection member into any of three positions so as to assign the most desirable vector sequence. This sequence requires that each view of the force diagram form a triangular shape. This is turn means that the vectors of members that coincide in a space view must occur sequentially—one immediately after the other. In this problem, this prohibits placing line AF between lines AC and AD. Either of the other two choices are acceptable, the only difference is in how the force diagram will appear. Member AF is placed between members AB and AC in Fig. 13-70 and Bow's Notation is added to complete the free body diagram sufficiently for use in drawing the force diagram. Bow's Notation assigns the known force in member AF to the vector **1-2** which necessarily is drawn first in the force diagram.

The force diagram has frontal and horizontal projections that parallel the structural diagram projections. These will be drawn and completed by using lines of action from the structural diagram and by drawing the known vector to scale length. The *force diagram* is started in Fig. 13-70 by drawing the frontal view of vector **1-2** true length (1000# at a scale of 1″ = 500#—the illustration is about one half real problem size). The horizontal view of vector **1-2** is drawn in point projection.

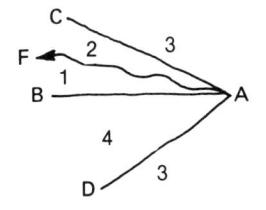

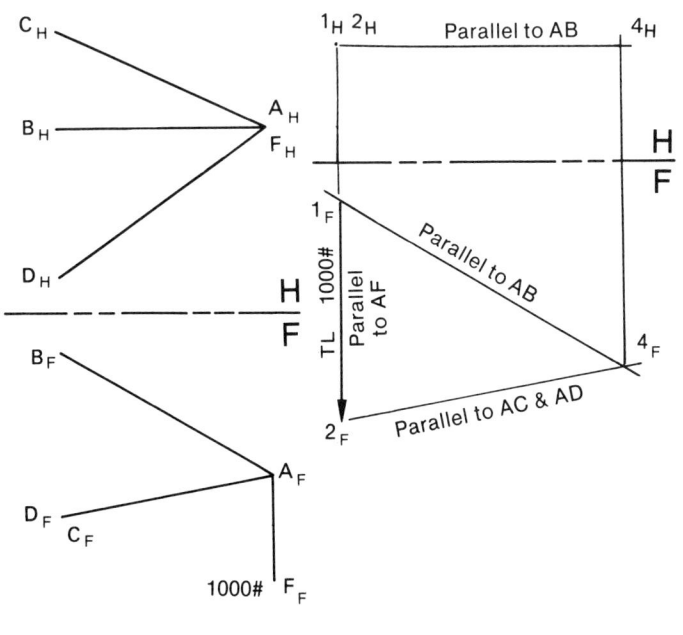

FREE BODY DIAGRAM

STRUCTURAL DIAGRAM   **FORCE DIAGRAM**

FIG. 13-70 Free body diagram.

433

FORCE IN AB = 1695#
FORCE IN AC =
FORCE IN AD =
FORCE IN AF = 1000# –

FREE BODY DIAGRAM

STRUCTURAL DIAGRAM   FORCE DIAGRAM

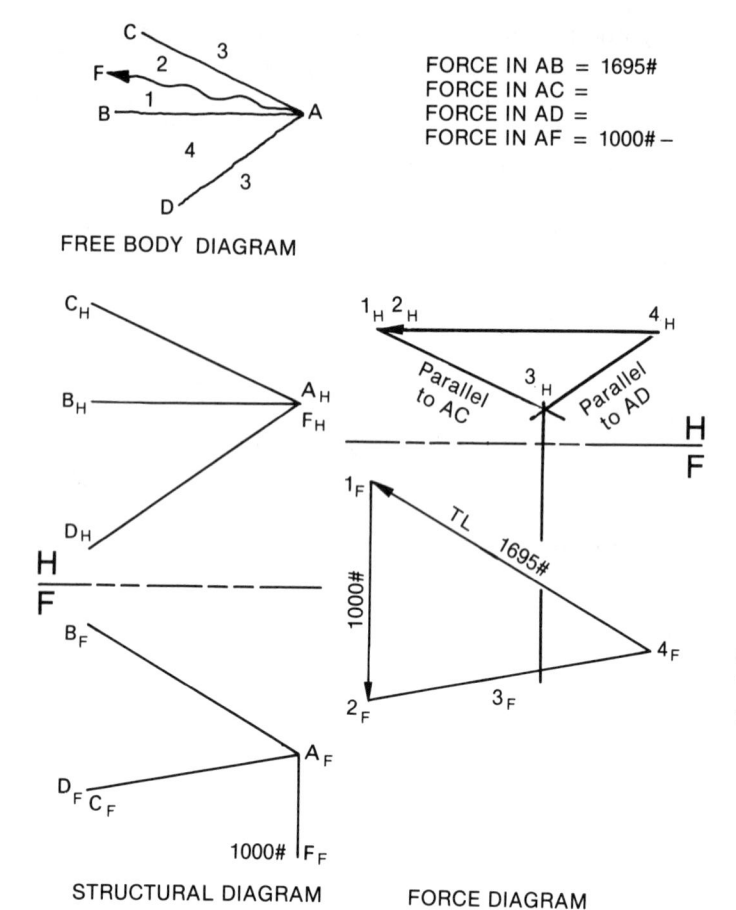

FIG. 13-71 Force diagram.

Then, in the frontal view, the line of action of vector 2-3 is drawn, parallel to member AC through point 2_F. Vector 3-4 is collinear with vector 2-3 (in the frontal view only) because members AC and AD are combined in the frontal structural view. All that is known now is that both points 3_F and 4_F lie on this line—somewhere. The line of action of vector 4-1 is drawn through point 1_F and the intersection of these two lines of action is point 4_F. This must be point 4_F (and not 3_F) because 4 is common to both lines of action. Point 4_H is located by projecting from point 4_F across the fold line onto the line of action of vector 4-1 in the horizontal view. Vector 4-1 is completed, and being true length in the frontal view, it is measured.

Locating point 4_H now gives a projected length of vector 4-1 in the horizontal view, making it possible to find point 3_H by projecting the line of action of vector 2-3 in the horizontal view through point 2_H and of vector 3-4 through point 4_H as shown in Fig. 13-71. Their intersection is point 3_H and it can be projected into the frontal

view to locate point 3_F on the line 2_F-4_F. The horizontal and frontal views of the force diagram are now completed by adding arrowheads to the vectors. The arrowheads must agree (project) in both views!

The free body diagram is now completed by transferring arrowheads from the horizontal force diagram view onto the corresponding member lines as shown in Fig. 13-72. The sense of each force can now be determined by its action on point A. The magnitude of vector 4-1 is directly measurable in the frontal view where it is true length. Vectors 2-3 and 3-4 are foreshortened in both views. Their true lengths are determined in an auxiliary view, by rotation, or by true length diagram. It is convenient in this problem to use the auxiliary view method. It is only necessary to project those vectors being measured—to project the other vectors adds no useful information.

The problem is completed with the customary summary of results that answer the original question, "What is the force (magnitude and sense) acting in each member at point A?"

FREE BODY DIAGRAM

FORCE IN AB = 1695# –
FORCE IN AC = 990# +
FORCE IN AD = 715# +
FORCE IN AF = 1000# –

STRUCTURAL DIAGRAM   FORCE DIAGRAM

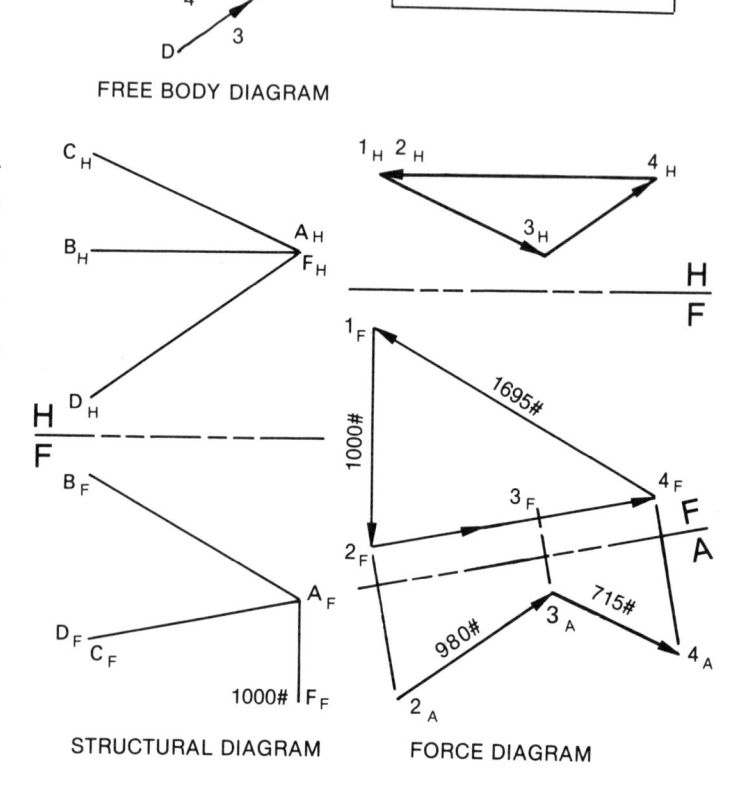

FIG. 13-72 The special case.

434

Another Special Case problem is presented in Fig. 13-73, where a 16-ft tall pole is supported by three nonsymmetrically positioned guy wires. It is known that the force acting in guy wire AB is 140#−. What are the forces in the other two guy wires and in the pole? The guy wires and the pole are weightless.

The complete solution to this problem is given in Fig. 13-74 and the reader is encouraged to examine it carefully. This solution is shown at the minimum recommended drawing size. Note that the known force in member AB is projectable in both views of the force diagram and that one of the unknown force members appears in point projection in the structural diagram.

FIG. 13-73 Isometric view of pole problem.

FORCE IN AB = 140#−
FORCE IN AC = 72#−
FORCE IN AD = 110#−
FORCE IN AE = 268#+

STRUCTURAL DIAGRAM
(1" = 8'-0")

FORCE DIAGRAM
(1" = 50#)

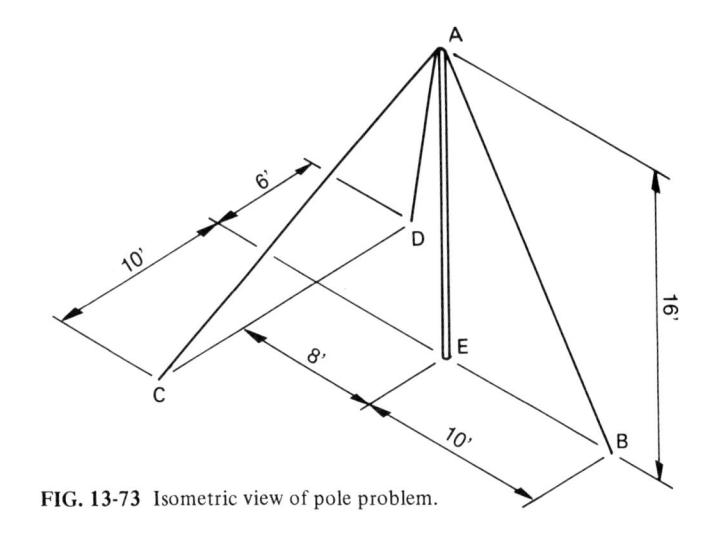

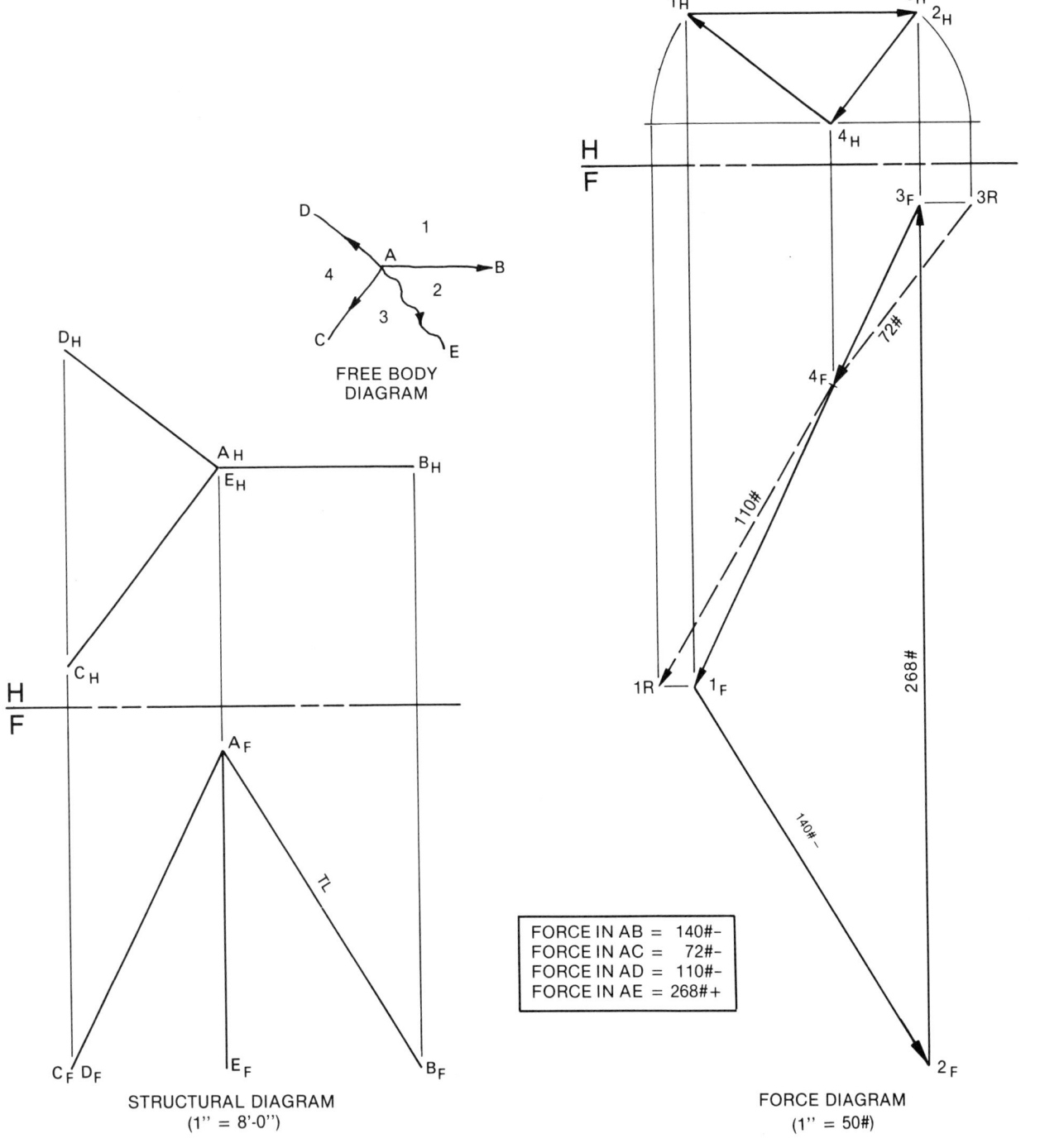

FIG. 13-74 A pole supported by guy wires.

13.13 THE GENERAL CASE

The *general case* solution is necessary whenever the given structural views project more than three lines of action from the members in one or both views. Such views cannot be used directly for drawing a force diagram view. The first steps of the general case solution are to project an auxiliary view (or views) until two *adjacent structural diagram* views meet the two conditions stated in Section 13.12:

1. Not more than three lines of action are projected in each (adjacent) view.
2. One measurable force vector can be drawn along one of the projected lines of action.

When two adjacent structural diagram views meet these requirements, the force diagram is drawn in the same way that it is drawn in the *special case*, except that the vector projections in the force diagram are drawn parallel to the member projections of *these* two adjacent views.

A bracket with nonsymmetrically positioned members is shown in Fig. 13-75. This bracket is supporting a vertical gravity load of 1200 pounds at point F. What is the force in each member when the bracket is in equilibrium? Frontal and horizontal views are shown in Fig. 13-76. The frontal view projects a separate line of action for each of the four members and cannot be used (condition #1). The horizontal view does satisfy condition #1 and can be used if an adjacent first auxiliary view can be projected that will satisfy both necessary conditions. Because the known force member AF is in point projection in the horizontal view, it will project true length in *any* adjacent view. Therefore, an auxiliary view that projects any two other members with a common line of action will satisfy the conditions. The auxiliary view shown is drawn by projecting an edge view of the plane defined by members AB and AC. (Horizontal line CX, lying on plane ABC is projected into point projection.) The horizontal view and this first auxiliary view now satisfy both condition #1 and condition #2 and provide the lines of action and a measurable vector length projection necessary to complete the force diagram.

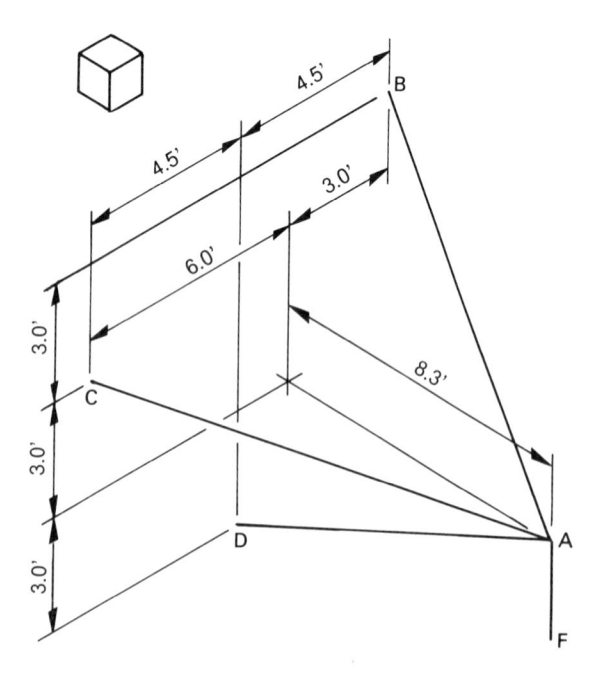

FIG. 13-75 A general case bracket.

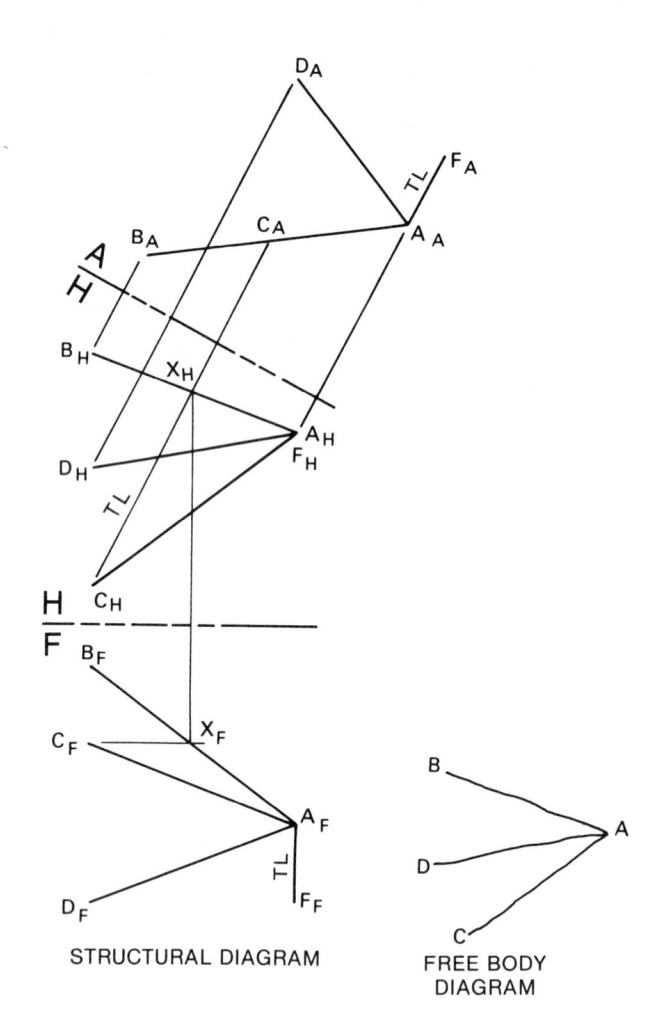

FIG. 13-76 Structural diagram.

STRUCTURAL DIAGRAM

FREE BODY DIAGRAM

436

in Section 13-12. How is it determined that this is point 3 and not point 4? Point 4 is then found as shown in Fig. 13-78 and the two views of the force diagram are completed by adding arrowheads to show equilibrium.

The vectors **2-3**, **3-4**, and **4-1** all are foreshortened in both views of the force diagram. Therefore, true length projections of the vectors must be completed to determine their magnitudes. In the special case problem, it was convenient to project an auxiliary view. Here space limitations made rotation a desirable method and is shown in Fig. 13-79. Also, the free body diagram is completed to determine the sense of each force. Finally, the results of the solution are summarized and boxed. Note that Fig. 13-76 through 13-78 are shown at about half size. The complete solution in Fig. 13-79 is shown full size.

FIG. 13-78 Force diagram.

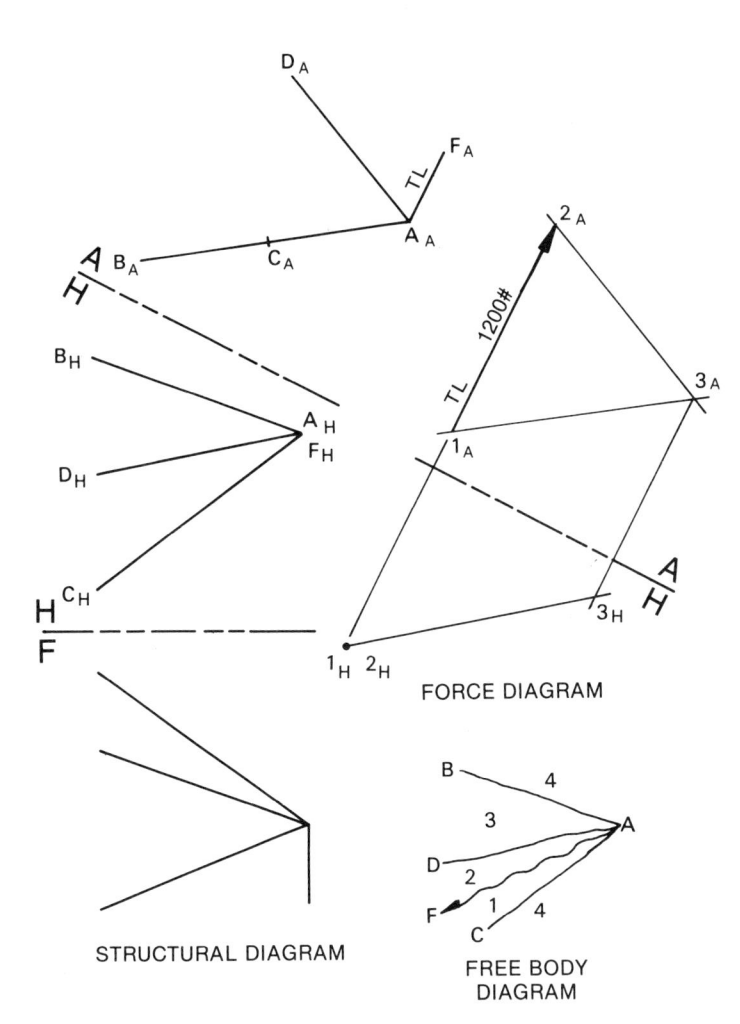

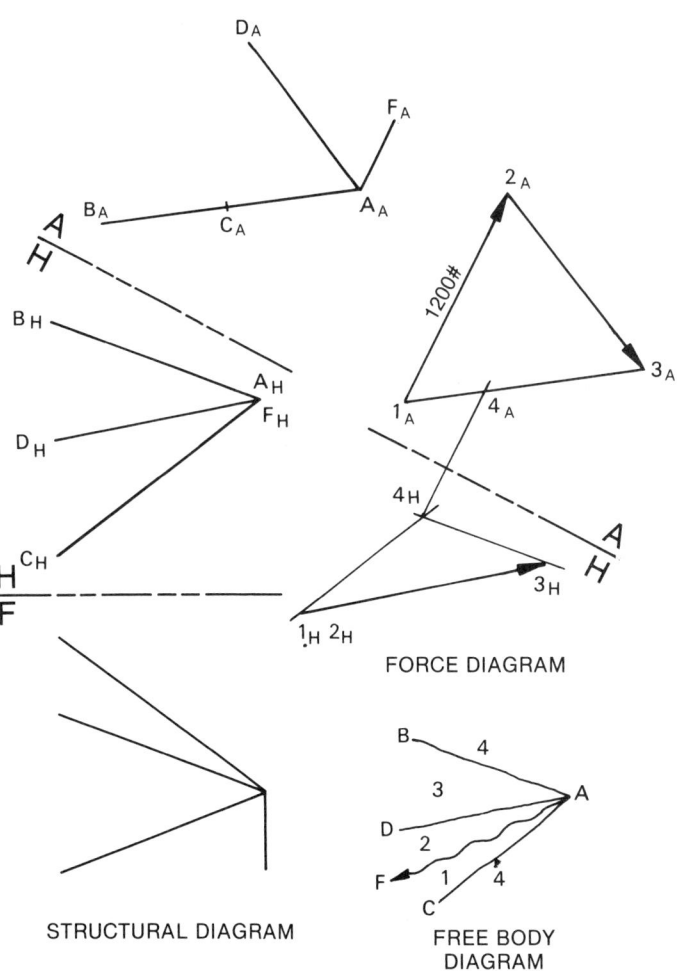

FIG. 13-77 Free body diagram.

Before starting the force diagram, the free body diagram is drawn to establish vector names and vector sequence for the force diagram. The *free body diagram* shown in Fig. 13-77 is drawn so that the vectors **3-4** and **4-1** (acting in members AB and AC that are combined in the auxiliary view) occur sequentially—the necessary condition for constructing triangular shaped force diagram views. Now the *force diagram* can be drawn, parallel in *all* respects, to the horizontal and auxiliary views. The frontal view is *not* used for any lines of action. (That is why it is shown incomplete in Fig. 13-77 and Fig. 13-78.) Now vector **1-2** is drawn and the position of point 3 is found using the method given

437

FIG. 13-79 The general case.

FORCE IN AB = 870# −
FORCE IN AC = 855# −
FORCE IN AD = 1440# +
FORCE IN AF = 1200# −

STRUCTURAL DIAGRAM
(1" = 4'-0")

FORCE DIAGRAM
(1" = 500#)

FREE BODY DIAGRAM

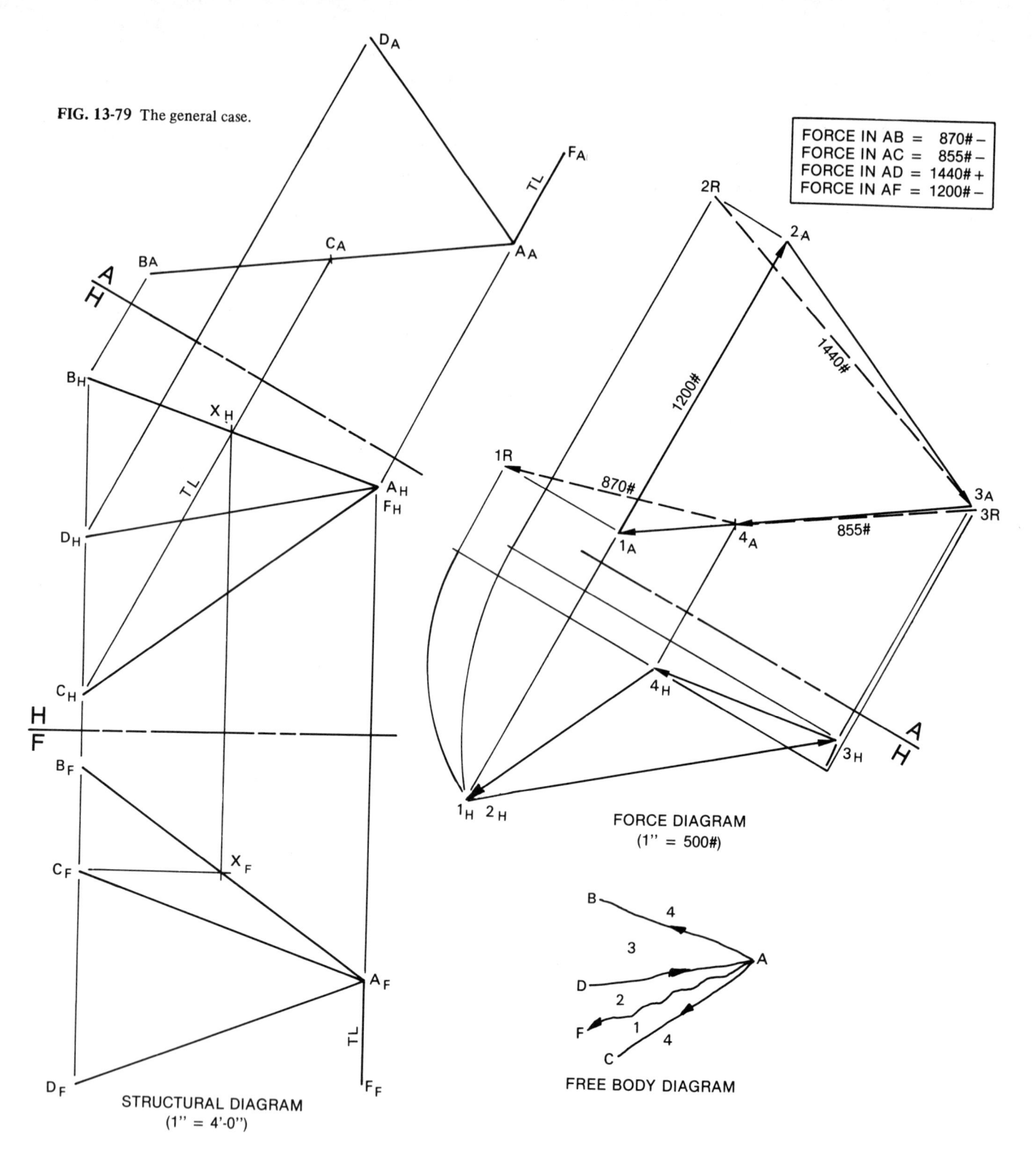

The horizontal and frontal views of Fig. 13-80 show the space relationships of a tripod structure that supports a tension force of 1250 pounds at point F. The line of action of the force is collinear with member AF. To satisfy the two conditions for drawing a force diagram, it is necessary to first project member AF into true length in auxiliary view A. Then auxiliary views B and C are projected so that not more than three lines of action are represented in either view and these views are used to draw the force diagram. Why was it necessary to draw the true length projection of member AF in auxiliary view A first?

This example illustrates the "worst possible" general case where the given views show no member in true length and more than three lines of action are represented in each given view. To obtain usable results in a problem with this number of successive projections, precision drafting is a must. Projection errors are magnified in each successive view! Drawing views as large as possible along with making careful projections and measurements will give values with three significant figure accuracy—sufficient for most applications in professional engineering practice.

FORCE IN AB = 295#+
FORCE IN AC = 1205#−
FORCE IN AD = 1435#+
FORCE IN AF = 1250#−

FREE BODY DIAGRAM

STRUCTURAL DIAGRAM
(1" = 6'-0")

FORCE DIAGRAM
(1" = 500#)

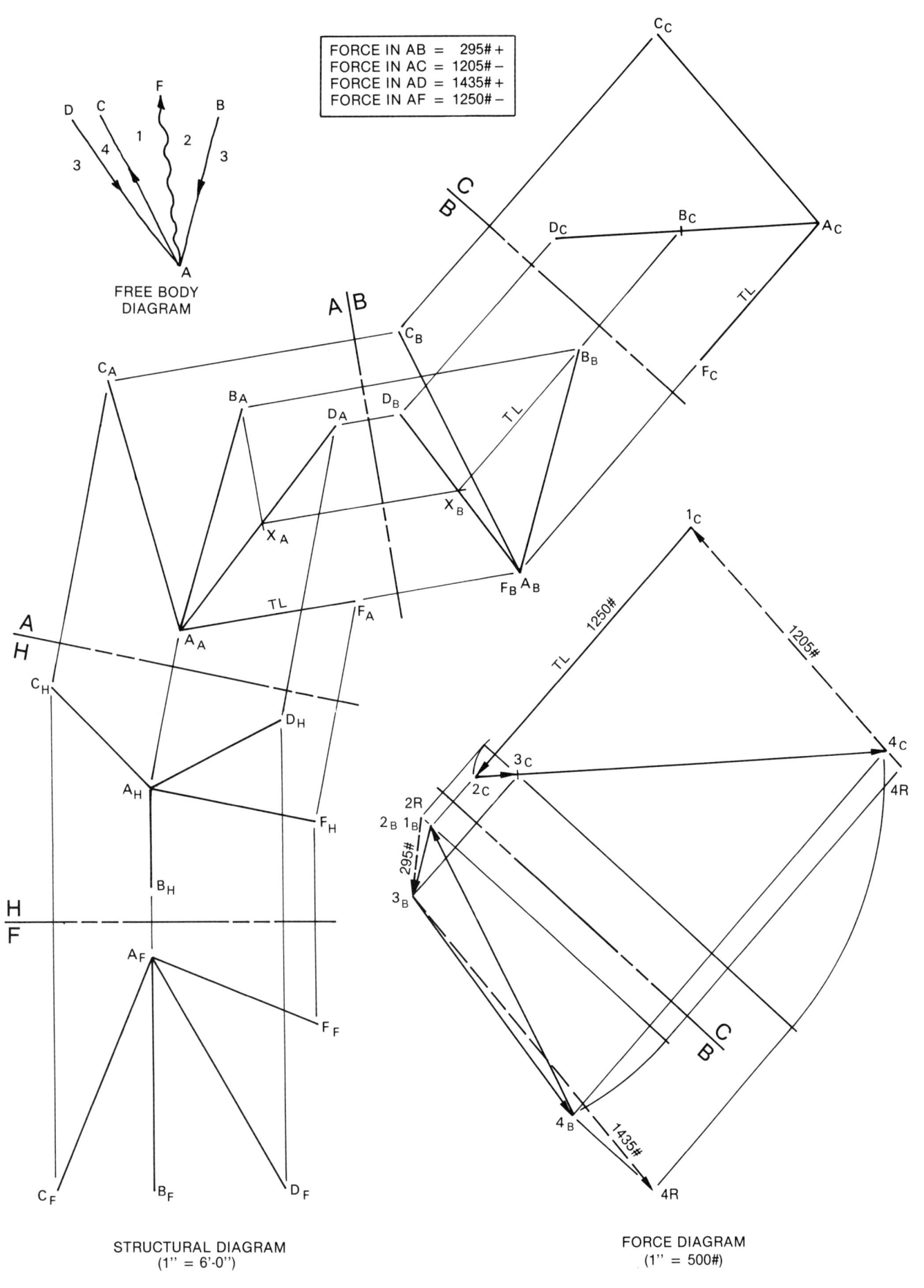

FIG. 13-80 The general case—three auxiliary views required.

439

PROBLEMS 13-18 through 13-22

Each problem is to be solved on an 8 1/2" × 11" sheet of paper. A quick semi-freehand study of the necessary views is necessary to make a workable sheet layout. All joints of these structural systems are pin connected.

PROB. 13-18. Determine the magnitude and sense of the force acting in each member of the spaceframe. Member AF supports a gravity load of 125# at point F.

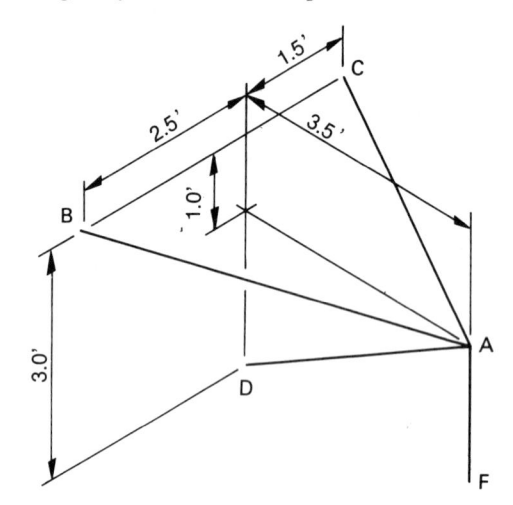

FIG. 13-81 Problem 13-18.

PROB. 13-19. Determine the magnitude and sense of the force acting in each member of the spaceframe. Members AB and AC lie in a horizontal plane. Member AF is vertical and is in compression. The magnitude of the force in member AF is 1100#.

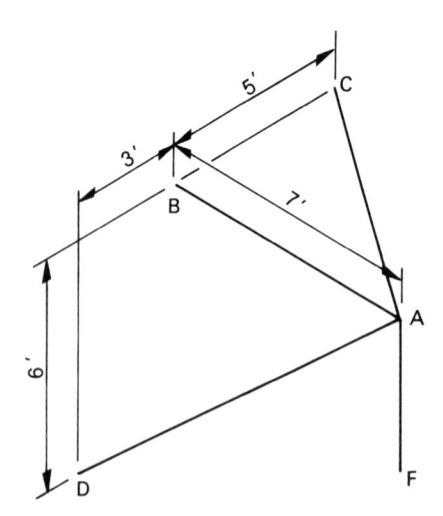

FIG. 13-82 Problem 13-19.

PROB. 13-20. A pole AB is supported by guy wires. The pole will buckle if the compressive stress in it exceeds 15,000#. Determine the maximum allowable force in each guy wire.

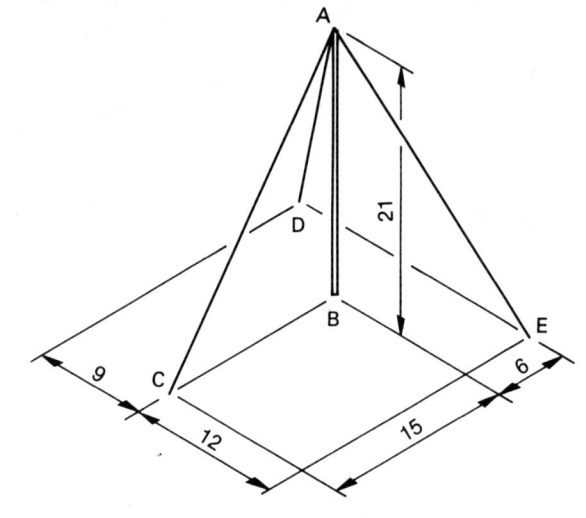

FIG. 13-83 Problem 13-20.

PROB. 13-21. Determine the magnitude and sense of the force acting in each member of the spaceframe. Member AB is horizontal. Member AF supports a gravity load of 130#.

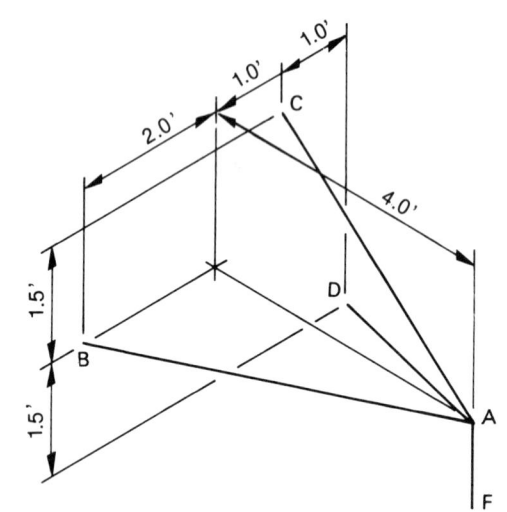

FIG. 13-84 Problem 13-21.

PROB. 13-22. Determine the magnitude and sense of the force acting in each member of the spaceframe. Member AE is parallel to the plane described by points B, C, and D and makes a 60-degree angle with the horizontal plane. Member AC is horizontal. A tension force of 12000# acts in member AE.

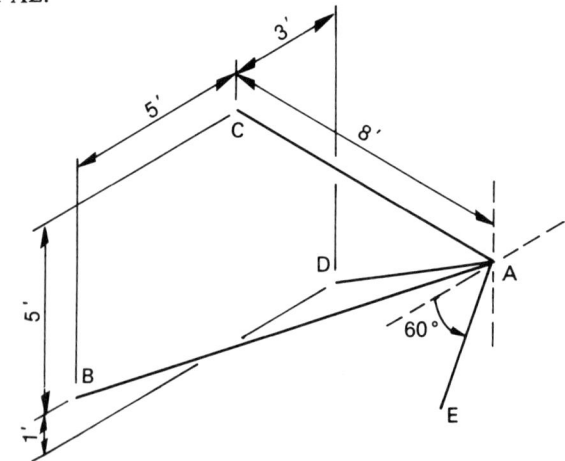

FIG. 13-85 Problem 13-22.

440

INDEX

A

Abbreviations, 6, 14
Across corners, 25
Across flats, 25
Acute angle, 20
Adjacent views, 51
Angle:
  between line and plane, 111, 174-77
  between lines, 109-10
  between planes, 179-81
Angle of repose, 361-64
Angles, 20
Angle symbol, 20
Annular torus, 245, 343
Approximate development, 294
Arc, 20
Architect scale, 13
Auxiliary view, 6, 44-48
  lines, 66
  planes, 143
  points, 61
Axis of rotation, 122, 197
Azimuth, 114

B

Base angle, of cone, 131-33
Beam, 416-23
  four or more forces, 421-23
  three forces acting, 416-17
  three parallel forces, 419-20

Bearing, 113-14, 118
Bedding plane, 367-71
Bend lines, 293-95
Bevel, 117
Bisectors:
  angles, 23
  lines, 23
Bore holes, 367-71
Bow's notation, 406, 409, 414, 425
Break line, 5
Buttock lines, 286-87

C

CADD, 38, 64
Chord, 20
Circle, 238
Circle of the gorge, 384
Circle on planes, 154-55
Circles, 20, 26-30
  rectify, 30
Circular line of intersection, 244-45
Circular planes, 152, 154-55
Clearance, 196
Collinear forces, 406, 412
Common line, 212
Compass, 17
Complementary angle, 20
Complementary angle method, 176
Compression, 406, 412
Concentric circle, 20
Concurrent forces, 407

Cone, 131-33, 235-41, 261
   development, 293, 316-18, 325-30
   locus, 131-33
   oblique, 235
Cone of revolution, 131-33, 235, 325
Conical, 261
Conical offset, development, 331
Conic section, 238
Conid, 380
Coning, 124
Contour, 366
  interval, 355-56
  lines, 355-57
Contract documents, 407
Convolutes, 316, 340
Coplanar forces, 407
Coplanar force systems, 411
Cow's horn, 379
Curved lines, 16-19
Curved surfaces, development, 315
Cut, 361-64
Cut and fill, 361-64
Cut and fill, dam, 363-65
Cut angle, 361-64
Cutting plane line, 5
Cutting plane method, 202, 207, 215, 249, 275
Cylinder, 131, 255, 228-33, 316-23
   development, 293
   oblique, 229
Cylinder of revolution, 228
Cylindroid, 382

D

Design documentation, 407
Development, 292-351
   approximate, 294
   cone, 315-18, 325-30, 332
   conical offset, 331
   curved surfaces, 315
   cylinder, 315-23
   double-curved surfaces, 343-46
   elbow joint, 323
   models, 297
   parallel line, 294
   prism, 298-304
   pyramid, 306-14
   radial line, 294
   sheet metal, 295
   sphere, 342-46
   transition piece, 331-41
   triangulation, 294, 336-41
Development element, 5
Development models, 297
Dihedral angle, 179-81
Dihedral angle by revolution, 191
Dimension line, 5

Dimensions, 6, 39
Dip, 158, 367-74
  angle, 158, 368-74
  direction, 158, 368-74
Directrix, 325, 343, 378
Dividers, 17
Double-curved surfaces, 34, 286, 315, 343, 383
  development, 315
  sphere, 343-46
Double-curved warped surfaces, 384
Double revolution, 187-88, 198
  of solid, 198
Drafting leads, 7

E

Eccentric circle, 20
Edge view, 5-6
  of plane, 149-55, 158
  by revolution, 184
Edge view of plane, 149-55, 158
Elbow joint, 323
Ellipse, 31, 152, 238
  angle, 152
Ellipsoid, 343
Elliptical cone, 235, 237
Engineering practice, 407
Engineering scale, 413
Equilibrium, 406
Erasing, 8
Extended line of intersection, 213, 216
Extended point of intersection, 203
Extension line, 5

F

Fairing, 286
Fair surfaces, 286-87
Fault, 386
Fill, 361-64
  angle, 361-64
First angle projection, 41
Fold line, 5, 39-48
Footwall, 367-71
Force, kind:
  component, 406
  equilibrant, 406
  resultant, 406
Force, properties:
  line of action, 406
  magnitude, 406
  point of application, 406
  sense, 406
    compression, 406
    tension, 406

Force Diagram, 407, 414, 417, 419, 422
Force systems:
  collinear, 406
  concurrent, 407
  coplanar, 407
  nonconcurrent, 407
  noncoplanar, 407
Frame line, 286-87
Freeboard, 363-64
Free body diagram, 407, 413, 416-17, 419, 421
French curve, 18
Frontal auxiliary view, 45-46
Frontal line, 67, 71
  on planes, 146
Frontal plane, 139
Frontal view, 6, 39-48
Funicular diagram, 407, 421

G

General case, coplanar:
  concurrent force system, 436-39
Generatrix, 325, 378-84, 387-89
Geometric construction, 16-32
Glass box method, 38-48
Gores, 345
Grade, 116-19
Great circle, of sphere, 242

H

Hanging wall, 367-71
Headwall, 367-71
Helicoid, 388, 399
  oblique, 389
  right, 388
Helix, 387-89
  conical, 387
  cylindrical, 387
Hidden line, 5
Hoppers, 335
Horizontal auxiliary view, 45, 47
Horizontal line, 19, 67, 72
  on planes, 146
Horizontal view, 6, 39-48
Hyperbola, 238
Hyperbolic paraboloid, 383
Hyperboloid of revolution, 131, 384

I

Image plane, 39-48
Inclined line, 67, 69
Inclined plane, 139, 141

Individual line method:
  piercing points, 203
  plane intersections, 215-16
Instrument drawings, 7
Intersecting lines, 82, 86
Intersection line, 5
Intersection of ore veins, 374-75
Intersection point, 6
Intersections, 212-85
  cone and cylinder, 264-69
  cone and prism, 270
  cones, 262
  cutting plane method, 215
  cylinder and prism, 258-59
  cylinder and sphere, 273
  cylinders, 255
  edge view method, 214
  infinite planes, 217
  line and cone, 236
  line and plane, 219
  line and sphere, 243
  ore veins, 374-75
  pictorial, 275-87
  plane and cone, 237-40
  plane and cylinder, 231-33
  plane and prism, 223-24
  plane and pyramid, 226-27
  plane and sphere, 244
  plane and torus, 245
  planes, 213, 373
  prism and pyramid, 251-54
  prism and sphere, 272
  prisms, 247
Intersections, pictorial, 275-87
  line and plane, 275-76
  perspective, 284
  plane and solid, 280-81
  planes, 276-79
Isometric projection, 37
Isosceles triangle, 238

L

Labeling, 4, 14
Lead, 387-99
Lead selection, 7
Lettering, 9-12
Lettering forms, 9-12
Level lines, 67-68
Light rays, 398
Line:
  dividing, 21-22
    adjacent method, 21
    parallel method, 21
    proportional division, 22
    vertical method, 22
Line extension, 204

Line key, 5
Line of action, of force, 406
Line of intersection, 212
Line of sight, 39, 43, 45-48
Line on plane, 145-49
Lines, 19, 64-134
   foreshortened, 67
   frontal, 67, 71
   horizontal, 67, 72
   inclined, 67, 72
   intersecting, 82, 86
   level, 67-68
   nonintersecting, 87
   nonparallel, 90
   oblique, 67, 74
   parallel, 82, 89-91
   perpendicular, 82, 93-96
   on planes, 145-49
   point view, 78
   principal, 67
   profile, 67, 69
   skew, 182
   true length, 67, 75-76
   vertical, 67-68
   visibility, 83-85
Line types, 8
Line weights, 4, 7
Linework, 7
Lofting, 286-87
Lower bedding plane, 37-71

M

Magnitude, of force, 406
Major diameter, 152-53, 155
   ellipse, 152-55
Maxwell Diagram, 416, 424-30
Mechanical engineer's scale, 13
Metric scale, 13
Mining, 354, 367-76
'—' (minus) sign, 415
Multiview projection, 37, 38

N

Noncoplanar forces, 407
Noncoplanar force systems, 430
   general case, 438
   special case, 434
Noncurrent forces, 407
Nonintersecting lines, 87
Nonparallel lines, 90
Normal planes, 139-41
Notation, 4-5

O

Object line, 5
Oblique cone, 325, 329-30
Oblique cutting plane, 269
Oblique lines, 67, 74-76
   true length, of, 75-76
Oblique planes, 141
Oblique projection, 37
Obtuse angle, 20
Open cone, 325
Orthographic projection, 37-55, 65
Outcrop, 367-71
Outcrop line, 367-71

P

Parabola, 238
Paraboloid, 343
Parallelism, 161-63
Parallel line developments, 294
Parallel lines, 5, 19, 82, 89-91
   on planes, 147
Path of rotation, 122-29
Patterns, 292, 295, 297
Pencil technique, 9
Percent grade, 116
Perimeter, 298
Perpendicularity:
   of lines, 82, 93-96
   of planes, 165-72
Perpendicular lines, 5, 19
Perspective, 37
   intersection, 284
Pictorial intersecting, 275-87
Piercing point, 5, 200-8
Pin connection, 405
Plane director, 379-84
Planes, 137-208
   auxiliary views, 142
   edge view, 149-55
   frontal, 139
   horizontal, 139
   inclined, 141
   oblique, 141
   profile, 139
   revolution, 183-98
   slope, 158
   strike, 157
   true shape, 151-55
   true size, 151-55
   vertical, 140
   visibility, 160
Plane tangencies, 392-95
Plan-profile, topographic, 355-56, 359, 362
Point of application, of force, 406

Point of convergence, 263, 269
Point on line, 80-81
Point on plane, 144
Points, 6, 58-63
   auxiliary view of, 61
   rectangular coordinates of, 60
   views of, 59
Point view, 6, 122-23
Point view of line, 67
Pole point, 422
Polygons, 24-25
Polyhedra, 33
Polyhedron, 219
Primary auxiliary view:
   lines, 66
   points, 61
Primary auxiliary views, 45-50
Principal lines, 67
Principal projection planes, 137, 139
Principal views, 40-42
Prism, 219-24
   development, 293
Prisms, 33
Problem page set-up, 14
Problem specifications, 14
Profile, topographic, 355-56, 361-62
Profile auxiliary view, 48
Profile lines, 67, 73
   on plane, 146
Profile planes, 139
Profile view, 6, 39-48
Projection, 37-55
   glass box method, 38-48
   natural method, 38
   normal method, 38
Projection lines, 39
Projection plane, 39-43, 64, 137
Pyramids, 33
   development, 293

R

Radial line development, 294
Ray, 422
Reaction, 421
Rectification, 30
Reference line, 5
Regular polygons, 24-25
Related views, 51
Restricted revolution, 196
Resultant force, 406
Revolution, 121-33, 183-98
   lines, 124-33
   locus, 131-33
   planes, 183-98
   points, 122-33

Revolved point, 6
Rise, 116-17
Ruled surfaces, 34
Run, 116-17

S

Scales, 15
Seam, 293, 295
Secondary auxiliary view, 44, 50
   lines, 66
   points, 62
Sedimentation, 367-68
Sense, of force, 406
Serpentine, 343
Shade line, 398-99
Shade point, 6
Shades, 398-402
   cones, 400
   cylinders, 399
   isometric, 402
   oblique, 401
   pictorial, 401-2
   pyramids, 400
Shading, 14, 398-402
Shadow line, 398-99
Shadow point, 6
Shadows, 398-402
Shaft, 368
Sheet metal developments, 295-96
Shortest distance,
   between lines, 99-104, 106
   line method, 99, 101
   plane method, 100, 102
   point on plane, 172-73
Sight line, 83
Single-curved surfaces, 34
   development, 315
Skew lines, 82
Slant height, 131-33
Slope, 115-19
   direction, 158
   ratio, 117
   by revolution, 127
Slope angle of cone, 131
Slope by revolution, 127
Small circle of sphere, 242
Space diagram, 407
   concurrent force systems, 432-35
Spheres, 242, 272-73
   development, 344-46
   gore method, 345
   zone method, 346
Spheroid, 343
Statics, 405
Steepest connection between lines, 105
Strata, 367-68

Stretch-out line, 298
Strike, 157-58, 368-75
  angle, 369
String diagram, 407, 411, 421
Strings, 422
Structural diagram, 407, 413, 416, 419
Successive auxiliary views, 44
  lines, 66
  points, 62
Summarization, of data, 408
Summary of results, 415, 417, 420, 423, 434, 437
Supplementary angles, 20
Surfaces:
  curved, 34
  double curved, 34, 384
  ruled, 34, 378
  single curved, 34
  warped, 35, 378-84
Symbol key, 5

T

Tangencies, 392-95
  plane and cone, 393
  plane and convolute, 395
  plane and cylinder, 392
  plane and sphere, 394
Tangency, 27
Tangent arcs, 28
Tension, 406, 412
Thickness of vein, 368-74
Third angle projection, 41-48
Three-dimensional force systems, 430
Three-force system, concurrent, 413-15
Topographic, 354-57, 359
Torus, 245, 343
Transition piece, 294, 331-44
  development, 331-41
Transverse section, 286-87
Triangles, 24, 29
  circumscribing circle of, 29
  inscribed circle of, 29
Triangulation, 294, 334-41
True length diagram, 77, 312
True length line, 5-6
  of oblique lines, 75-76
  on planes, 146
  by revolution, 122, 124-26
True shape, 5-6, 151-55
  of plane:
    by double revolution, 188
    by revolution, 185-88
True size, 5-6, 151-55

Trusses, 424-30
Tunnel, 368
Two-force system, 411

U

Umbra, 398

V

Vector:
  definition, 406
  diagram, 407
  polygon, 409
  uses, 405
Vector summation:
  mathematics method, 409
  parallelogram method, 408
  polygon method, 409
Vein, 367-71
  thickness, 367-71
Vertex, 235
Vertex angle, 131-33
Vertex line, 263, 269
Vertical lines, 19, 67-68
Vertical plane, 139-40
View identification, 6
View subscript, 6
Visibility:
  inspection method, 84
  of lines, 83-85
  of planes, 83-85
  sight line, 83
  visibility test, 83-85
Visibility of lines, 83-85
Visibility of planes, 160

W

Warped cone, 379
Warped surfaces, 35, 378-84, 388
Warped transition piece, 341
Water line, 286
Working loads, 411

Z

Zones, sphere development, 346